41st European Physical Society Conference on Plasma Physics

(EPS 2014)

IOP Science: Plasma Physics and Controlled Fusion
Volume 57, Number 1 (2015)

Berlin, Germany
23-27 June 2014

Editors:

R. O. Dendy
M. Koepke

ISBN: 978-1-5108-0713-6
ISSN: 0741-3335

Printed from e-media with permission by:

Curran Associates, Inc.
57 Morehouse Lane
Red Hook, NY 12571

Some format issues inherent in the e-media version may also appear in this print version.

Printed by Curran Associates, Inc. (2015)

For permission requests, please contact the Institute of Physics
at the address below.

Institute of Physics
Dirac House, Temple Back
Bristol BS1 6BE UK

Phone: 44 1 17 929 7481
Fax: 44 1 17 920 0979

techtracking@iop.org

Additional copies of this publication are available from:

Curran Associates, Inc.
57 Morehouse Lane
Red Hook, NY 12571 USA
Phone: 845-758-0400
Fax: 845-758-2634
Email: curran@proceedings.com
Web: www.proceedings.com

Table of contents

Volume 57

Invited papers from the 41st European Physical Society Conference on Plasma Physics (Berlin, Germany, 23–27 June 2014)

Preface

010301

41st European Physical Society Conference on Plasma Physics FREE ARTICLE
Svetlana Ratynskaia and Sylvie Jacquemot

Invited Papers

014001

Challenges in understanding and predictive model development of plasma-assisted combustion Igor V Adamovich and Walter R Lempert....1

014002

Plasma-based creation of short light pulses: analysis and simulation of amplification and focusing C Riconda, S Weber, L Lancia, J-R Marquès, G Mourou and J Fuchs....9

014003

Studying planetary matter using intense x-ray pulses K Appel, M Nakatsutsumi, A Pelka, G Priebe, I Thorpe and Th Tschentscher....18

014004

Physics in the magnetic configuration space of W7-X J Geiger, C D Beidler, Y Feng, H Maaßberg, N B Marushchenko and Y Turkin....23

014005

Electric propulsion for small satellites Michael Keidar, Taisen Zhuang, Alexey Shashurin, George Teel, Dereck Chiu, Joseph Lukas, Samudra Haque and Lubos Brieda....34

014006

Energetic ion behaviour in MAST M Cecconello, O M Jones, W U Boeglin, R V Perez, D S Darrow, I Klimek, S E Sharapov, M Fitzgerald, K G McClements, D L Keeling, S Y Allan, C A Michael, R J Akers, N J Conway, R Scannell, M Turnyanskiy, G Ericsson and the MAST Team....44

014007

Enhanced efficiency of plasma acceleration in the laser-induced cavity pressure acceleration scheme J Badziak, M Rosiński, S Jabłoński, T Pisarczyk, T Chodukowski, P Parys, P Rączka, E Krousky, J Ullschmied, R Liska and M Kucharik....55

014008

Symbolic regression via genetic programming for data driven derivation of confinement scaling laws without any assumption on their mathematical form A Murari, E Peluso, M Gelfusa, I Lupelli, M Lungaroni and P Gaudio....66

014009

Excited-state kinetics and radiation transport in low-temperature plasmas G Colonna, G D'Ammando, L D Pietanza and M Capitelli....71

014010

Industrial plasmas in academia Ch Hollenstein, AA Howling, Ph Guittienne and I Furno....86

014011

Alfven wave acceleration of particles in the aurora F Mottez....95

014012

The influence of finite ion temperature on plasma blob dynamics P Manz, G Birkenmeier, D Carralero, G Fuchert, H W Müller, S H Müller, B D Scott, U Stroth, T T Ribeiro and E Wolfrum....102

014013

Kinetics of general electromagnetic fluctuations in unmagnetized plasmas: aperiodic thermal noise R Schlickeiser and P H Yoon....108

014014

Flow damping in stellarators close to quasisymmetry Iván Calvo, Felix I Parra, José Luis Velasco and J Arturo Alonso....115

014015

The physics and detection of nanodust in the solar system N Meyer-Vernet, I Mann, G Le Chat, P Schippers, S Belheouane, K Issautier, A Lecacheux, M Maksimovic, F Pantellini and A Zaslavsky....122

014016

Influence of the external electrical circuit on the regimes of a nanosecond repetitively pulsed discharge in air at atmospheric pressure Fabien Tholin and Anne Bourdon....132

014017

Fast sawtooth reconnection at realistic Lundquist numbers S Günter, Q Yu, K Lackner, A Bhattacharjee and Y-M Huang....144

014018

Quantification of the impact of large and small-scale instabilities on the fast-ion confinement in ASDEX Upgrade B Geiger, M Weiland, A Mlynek, M Reich, A Bock, M Dunne, R Dux, E Fable, R Fischer, M Garcia-Munoz, J Hobirk, C Hopf, S Nielsen, T Odstrcil, C Rapson, D Rittich, F Ryter, M Salewski, P A Schneider, G Tardini and M Willensdorfer....150

014019

Charging of multiple grains in subsonic and supersonic plasma flows D Block and W J Miloch....165

014020

Non-linear MHD modeling of edge localized mode cycles and mitigation by resonant magnetic perturbations François Orain, M Bécoulet, J Morales, G T A Huijsmans, G Dif-Pradalier, M Hoelzl, X Garbet, S Pamela, E Nardon, C Passeron, G Latu, A Fil and P Cahyna....173

014021

Doppler tomography in fusion plasmas and astrophysics M Salewski, B Geiger, W W Heidbrink, A S Jacobsen, S B Korsholm, F Leipold, J Madsen, D Moseev, S K Nielsen, J Rasmussen, L Stagner, D Steeghs, M Stejner, G Tardini and M Weiland....183

014022

Shock ignition: a brief overview and progress in the design of robust targets S Atzeni, A Marocchino and A Schiavi....193

014023

Implosion dynamics in direct-drive experiments D T Michel, R S Craxton, A K Davis, R Epstein, V Yu Glebov, V N Goncharov, S X Hu, I V Igumenshchev, D D Meyerhofer, P B Radha, T C Sangster, W Seka, C Stoeckl and D H Froula....203

014024

Energetic particles and multi-scale dynamics in fusion plasmas F Zonca, L Chen, S Briguglio, G Fogaccia, A V Milovanov, Z Qiu, G Vlad and X Wang....210

014025

H-mode transitions and limit cycle oscillations from mean field transport equations Gary M Staebler and R J Groebner....220

014026

Thomson scattering on non-equilibrium low density plasmas: principles, practice and challenges Emile Carbone and Sander Nijdam....228

014027

Magnetic perturbations as a viable tool for edge turbulence modification N Vianello, C Rea, M Agostini, R Cavazzana, G Ciaccio, G De Masi, E Martines, A Mazzi, B Momo, G Spizzo, P Scarin, M Spolaore, P Zanca, M Zuin, L Carraro, P Innocente, L Marrelli, M E Puiatti and D Terranova....239

014028

Dynamics of low–intermediate–high-confinement transitions in the HL-2A tokamak Y Xu, J Cheng, J Q Dong, Y B Dong, M Jiang, W L Zhong, L W Yan, Z B Shi, Z H Huang, Y G Li, L Nie, M Xu, K J Zhao, X Q Ji, D L Yu, Y Liu, Z Y Cui, W Chen, Q W Yang, X T Ding, X R Duan and Yong Liu....248

014029

Pedestal and edge localized mode characteristics with different first wall materials and nitrogen seeding in ASDEX Upgrade P A Schneider, L Barrera Orte, A Burckhart, M G Dunne, C Fuchs, A Gude, B Kurzan, W Suttrop, E Wolfrum and the ASDEX Upgrade Team....255

014030

Advanced scheme for high-yield laser driven nuclear reactions D Margarone, A Picciotto, A Velyhan, J Krasa, M Kucharik, A Mangione, A Szydlowsky, A Malinowska, G Bertuccio, Y Shi, M Crivellari, J Ullschmied, P Bellutti and G Korn....267

014031

Theoretical description of heavy impurity transport and its application to the modelling of tungsten in JET and ASDEX upgrade F J Casson, C Angioni, E A Belli, R Bilato, P

Mantica, T Odstrcil, T Pütterich, M Valisa, L Garzotti, C Giroud, J Hobirk, C F Maggi, J Mlynar and M L Reinke....274

014032

Electromagnetic stabilization of tokamak microturbulence in a high-β regime J Citrin, J Garcia, T Görler, F Jenko, P Mantica, D Told, C Bourdelle, D R Hatch, G M D Hogeweij, T Johnson, M J Pueschel and M Schneider....284

014033

Sawtooth control in JET with ITER relevant low field side resonance ion cyclotron resonance heating and ITER-like wall J P Graves, M Lennholm, I T Chapman, E Lerche, M Reich, B Alper, V Bobkov, R Dumont, J M Faustin, P Jacquet, F Jaulmes, T Johnson, D L Keeling, Yueqiang Liu, T Nicolas, S Tholerus, T Blackman, I S Carvalho, R Coelho, D Van Eester, R Felton, M Goniche, V Kiptily, I Monakhov, M F F Nave, C Perez von Thun, R Sabot, C Sozzi and M Tsalas....295

014034

Explosive reconnection of the double tearing mode in relativistic plasmas with application to the Crab nebula J Pétri, M Takamoto, H Baty and S Zenitani....309

014035

Impact of magnetic perturbation coils on the edge radial electric field and turbulence in ASDEX Upgrade G D Conway, S Fietz, H W Müller, T Lunt, P Simon, W Suttrop, M Maraschek, T Happel and E Viezzer....316

014036

Topology bifurcation of a magnetic flux surface in toroidal plasmas K Ida, T Kobayashi, S Inagaki, Y Suzuki, S Sakakibara, K Itoh, H Tsuchiya, C Suzuki, M Yoshinuma, Y Narushima, M Yokoyama and S-I Itoh....325

014037

Characterising dust in JET with the new ITER-like wall OPEN ACCESS J C Flanagan, M Sertoli, M Bacharis, G F Matthews, P C de Vries, A Widdowson, I H Coffey, G Arnoux, B Sieglin, S Brezinsek, J W Coenen, S Marsen, T Craciunescu, A Murari, D Harting, A Cackett and E Hodille....334

014038

The H-mode density limit in the full tungsten ASDEX Upgrade tokamak M Bernert, T Eich, A Kallenbach, D Carralero, A Huber, P T Lang, S Potzel, F Reimold, J Schweinzer, E Viezzer and H Zohm....345

Plasma Phys. Control. Fusion **57** (2015) 010301 (3pp) doi:10.1088/0741-3335/57/1/010301

Preface

41st European Physical Society Conference on Plasma Physics

Svetlana Ratynskaia
Chairperson of the 2014 EPS Plasma Physics Conference Program Committee
Space & Plasma Physics, School of Electrical Engineering, Royal Institute of Technology (KTH), Stockholm, Sweden

I have been honoured to chair the program committee (PC) of the 41st European Physical Society (EPS) Conference on Plasma Physics and to act as a guest editor for this special issue of *Plasma Physics and Controlled Fusion* featuring peer-reviewed papers by the plenary and invited speakers of the conference.

The 41st EPS Plasma Physics conference, hosted by the Max Planck Institute for Plasma Physics, took place in the Berlin Congress Center, located at the famous Alexanderplatz in the centre of Berlin. The dedicated work of the local organizing committee, chaired by Professor Thomas Klinger, resulted in five flawless conference days. The event attracted 649 delegates from 40 countries. Over the last months of program preparation, I interacted closely with the conference secretary, Beate Kemnitz, and Matthias Hirsch who was responsible for the website, to whom I would like to express my gratitude for our efficient collaboration.

The PC paid close attention to the nominations made via the online open forum, which provided a valuable input of 350 suggestions from the plasma physics community. The PC also received and accommodated recommendations from the American and Japanese Physical Societies.

Following the EPS plasma physics division board request, a particular goal of the EPS 2014 program was to increase the participation in the typically under-represented low-temperature and dusty plasmas (LTDP) and basic and astrophysical plasmas (BAP) sessions. This was facilitated by joined sessions, which nurtured productive interactions between these plasma communities. Judging from the high attendance numbers, both the joined and special 'kinetics' sessions have received much interest from the participants. The PC has also recognized that a significant part of the delegates belong to the magnetic confinement fusion (MCF) community and that this has to be reflected in the number of talks allocated to the MCF session. To that end, the schedule has been arranged to accommodate parallel MCF talks. Composing a program echoing the interdisciplinary character of the conference was undoubtedly made possible thanks to the enthusiasm of the four sub-chairs, Paola Mantica (MCF), Alessandra Benuzzi-Mounaix (beam-plasma and inertial fusion; BPIF), Robert Bingham (BAP) and Giorgio Dilecce (LTDP), to collaborate and to search for a compromise in pursuing the common goal.

The submission and on-line proceedings system has been operated as smoothly as ever under the command of Boudewijn van Milligen, whose persistent 'behind-the-scenes' labor and attention have been vital to the work of the PC.

It is my sincere hope that the readers will find the present volume, which gathers contributions reflecting various aspects of contemporary plasma physics, stimulating and worth their attention.

Sylvie Jacquemot
Chairperson of the EPS Plasma Physics Division
Laboratoire d'Utilisation des Lasers Intenses, Ecole Polytechnique, Palaiseau, France

Outstanding scientists honored at the 41st European physical society conference on plasma physics

The 2014 **Hannes Alfvén Prize** was awarded by the Board of the Plasma Physics Division of the EPS to **Patrick Mora** (Centre de Physique Théorique, Ecole Polytechnique, France) *'for decisive results in the field of laser-produced plasma physics, in particular for illuminating descriptions of laser light absorption in plasmas, electron heat transport in steep temperature gradients and plasma expansion dynamics into vacuum'*.

The first remarkable result of Patrick Mora is a theoretical model, now included in all the major textbooks, which couples laser light absorption and energy transport to plasma

Figure 1. From left to right: E Highcock, C Hollenstein and P Mora.

hydrodynamics. With his colleague Jean-François Luciani, he then proposed a nonlinear and nonlocal theory of electron heat transport which enables interpretation of experimental results showing a strong departure from the classical Spitzer and Härm formalism, thus inspiring a variety of models now currently used for hydrodynamic computations of laser or x-ray created plasmas, especially in the context of inertial confinement fusion. Later, he drew, with Tom Antonsen, USA, a model for the propagation of a short laser pulse in under-dense plasmas, exhibiting its self-focusing character and its tendency to develop Raman-like instabilities. The numerical code he elaborated, named WAKE, is proved to be more efficient than usual Particle-in-Cell codes in numerous experimental situations and is now widely used. In particular it enabled the first simulation of complete cavitation of electron density in what was later named the 'bubble regime', and is shown to play a crucial role in relativistic quasi-monoenergetic electron beam acceleration. Finally his recent work on plasma expansion into vacuum is possibly his most important work as it gives a complete understanding of the flow dynamics, elucidating for instance the structure of the ion front and predicting the maximum ion velocity. This model was used to sustain a lot of recent experimental results on ion acceleration obtained in European laboratories.

The **2014 EPS Plasma Physics Innovation Prize** was awarded to **Christoph Hollenstein** (Centre de Recherche en Physique des Plasmas, EPFL, Switzerland) *'for instrumental contributions to the field of Plasma Processes in Industry and for his strong impact in spin-off activities of fusion R&D'*.

Among the various topics tackled by Christoph Hollenstein, one can first mention the development of a plasma process for thin film deposition using the plasma torch and, more significantly, all the work done on radio-frequency (RF) plasma reactors for large area deposition of amorphous or microcrystalline silicon for solar cell or flat display applications. As an additional outcome, he also contributed to the fundamental understanding of powder formation in reactive plasmas used for deposition of many important industrial coatings. Moreover, he studied the plasma chemistry of silicon-oxide deposition as barrier coating for packaging applications. More recently, a method was investigated to modify the functional and decorative properties—for instance the color—of a substrate under reactive plasma conditions, with potential applications in the watch industry. Electrical discharge machining was investigated as well, with a systematic study on the eroding spark physics performed thanks to a set of plasma diagnostics, including spectroscopic and imaging tools, which demonstrated that the spark is a cold and dense, weakly non-ideal, plasma, thus improving the understanding of the fundamental mechanisms underlying the industrial process. Christoph Hollenstein and his colleagues have also uncovered the physics behind the formation of plasmoids in RF plasma reactors, which exhibit a funnel-like behavior, and they proposed strategies to prevent damage induced by them. An additional experimental approach is currently used to explore arcing in spacecraft and spacecraft equipment, with special attention paid to arc ignition, propagation. Such studies lead to the development of safer electrical equipment. All these great developments and inventions were, and are, made in close collaboration with European industrial partners.

The **2014 EPS Plasma Physics Division PhD Research Award** has been judged by an external committee, comprising Francesco Pegoraro (IT), Miklos Porkolab (US), Karl-Heinz Spatschek (DE) and Elisabeth Rachlew (SE), who examined all the candidatures in a process managed by Elisabeth Wolfrum representing the EPS Plasma Physics Division. Based on their conclusions, this year's award goes to **Edmund Highcock** (Oxford University, UK) for his thesis entitled 'The zero turbulence manifold of fusion plasmas'.

EPS/PPCF/IUPAP Student Poster Prizes were also attributed during the conference to:

- **Livia Casali** (IPP Garching, Germany) for her poster 'Transport analysis of high radiation and high density plasmas at ASDEX Upgrade',
- **Brendan Kettle** (QUB, UK) for his poster 'XUV absorption of Warm Dense Aluminum',
- **Himank Anand** (EPFL, Switzerland) for his poster 'First experimental results of plasma shape control using real time equilibrium reconstruction in TCV', and
- **Alexandra Vallet** (CELIA, France) for her poster 'Semi-analytic modeling of shock ignition'.

In addition, the **10th Itoh Project Prize in Plasma Turbulence** was awarded by a selection committee, composed of S-I Itoh (Japan–chair), A Fasoli (Switzerland), I Furno (Switzerland), K Ida (Japan), K Itoh (Japan), V Naulin (Denmark) and U Stroth (Germany), to **Zhouji Huang** (EPFL, Switzerland) for his work 'Investigation of plasma turbulence and geodesic acoustic modes using tangential phase-contrast imaging in the TCV tokamak'. Two students also received high commendations from the judges: B Schmid (U. Stuttgart, Germany) for 'Zonal flow formation in the stellarator TJ-K' and M Madi (CEA/IRFM, France & American University of Beirut, Lebanon) for 'Effect of density fluctuations in the scrape-off layer on the lower hybrid power spectrum'.

IOP Publishing

Plasma Physics and Controlled Fusion

Plasma Phys. Control. Fusion 57 (2015) 014001 (8pp)

doi:10.1088/0741-3335/57/1/014001

Challenges in understanding and predictive model development of plasma-assisted combustion

Igor V Adamovich and Walter R Lempert

Nonequilibrium Thermodynamics Laboratory, Department of Mechanical and Aerospace Engineering, The Ohio State University, Columbus, OH 43210, USA

E-mail: adamovich.1@osu.edu

Received 20 June 2014, revised 22 July 2014
Accepted for publication 28 July 2014
Published 28 November 2014

Abstract

The key challenges to quantitative insight into fuel–air plasma kinetics, as well as plasma-assisted ignition and flameholding, are identified and assessed based on the results of recent experimental and kinetic modeling studies. Experimental and modeling approaches to address these critical issues are discussed. The results have major implications for the fundamental understanding of pulsed electric discharge dynamics, molecular energy transfer in reacting flows, plasma chemical reactions, and development of low-temperature plasma-assisted combustion technologies.

Keywords: plasma-assisted combustion, optical diagnostics, kinetic modeling

(Some figures may appear in colour only in the online journal)

1. Introduction

Over the last 10–15 years, there has been significant progress in demonstrating the utility of non-equilibrium plasmas for augmentation of combustion phenomena, such as reduction of ignition delay time and ignition temperature, as well as increase in flame stability and flammability limits [1, 2]. High peak voltage, nanosecond pulse duration discharges are of particular interest for plasma-assisted combustion, since they can generate diffuse non-equilibrium plasmas at high pressures ($\sim$1 bar) and high pulse repetition rates (up to $\sim$100 kHz), and are characterized by high peak reduced electric fields, E/N, of several hundred Townsends ($1\,\mathrm{Td} = 10^{-17}\,\mathrm{V\,cm^2}$). At these high E/N values, a significant fraction of discharge input energy goes into the population of excited states of molecules (vibrational and electronic), as well as molecular dissociation and ionization by electron impact. Collisional quenching of the excited states (including reactive quenching), and reactions of radical species generated in the discharge considerably expand the variety of chemical reactions in low-temperature fuel–air mixtures, resulting in fuel oxidation and ignition. From a fundamental kinetics perspective, however, the dominant energy transfer and chemical reaction processes in these plasmas are still not fully understood, especially when fairly complex hydrocarbon fuels are involved. Without such understanding, predictive kinetic modeling and analysis of plasma-assisted combustion phenomena remains problematic. The main objective of the present work is to identify and assess the most significant challenges to quantitative insight into fuel–air plasma kinetics, as well as plasma-assisted ignition and flameholding, based on the results of recent experimental and modeling studies.

2. Electric field and electron density: measurements and prediction

It is well known that energy partition among different energy modes in non-equilibrium, low-temperature, high-pressure plasmas (rotational, vibrational, electronic, dissociation, and ionization) is controlled primarily by the electron energy distribution function. Over a wide range of conditions, this partition can be characterized in terms of the reduced electric field, E/N. Electron density, n_e, is the other parameter that controls energy coupling to the plasma, such that specific power loading is proportional to the product of E/N and n_e. Thus, knowledge of these two key parameters

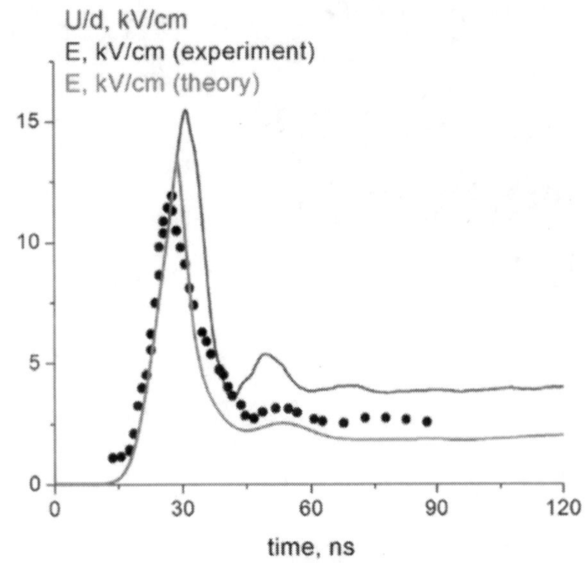

Figure 1. Experimental [6] and predicted electric field and applied voltage/electrode gap in a ns pulse, plane-to-plane discharge in N_2 at 0.25 bar, electrode gap of 1.2 mm.

is critical for insights into the kinetics of energy transfer and chemical reactions in these plasmas. Estimating E/N and n_e in the plasma from discharge voltages and current waveforms, especially in transient discharges that develop as ionization waves, may well result in significant uncertainties in predicting other plasma parameters, such as vibrational and electronic excitation, e.g. $N_2(X^1\Sigma, v = 0) + e \rightarrow N_2(X^1\Sigma, v > 0, A^3\Sigma, B^3\Pi, C^3\Pi) + e$, as well as molecular dissociation by electron impact, e.g. $O_2(X^1\Sigma) + e \rightarrow O(^3P, ^1D) + O(^3P) + e$. Thus, measurements of electric field, electron density, and electron temperature in transient plasmas sustained by nanosecond pulse duration discharges, and their comparison with modeling predictions, are critical for fuel–air plasma characterization and for the validation of kinetic models. Truly predictive models should incorporate, at the very least, realistic electrode geometry, the Poisson equation for the electric field, and equations for number densities of charge species (see [3, 4]), such that electric field and electron density distributions in the plasma could be predicted based on the experimental voltage waveform, and compared with the experimental results.

Four-wave mixing, and coherent anti-Stokes Raman scattering (CARS) and Thomson scattering are two powerful laser diagnostics that have been used recently for time-resolved electric field measurements [5–7], as well as time-resolved electron density and electron temperature measurements [8–10] in high-pressure, high specific energy loading molecular plasmas. Electric field measurements have been performed using both H_2 and N_2 as the active species, which are among the dominant species in plasma-assisted combustion environments. Sub-nanosecond time resolution electric field diagnostics are under development [7]. Figure 1 compares the experimental [6] and predicted electric field in a nanosecond pulse discharge in nitrogen at 0.25 bar, using the kinetic

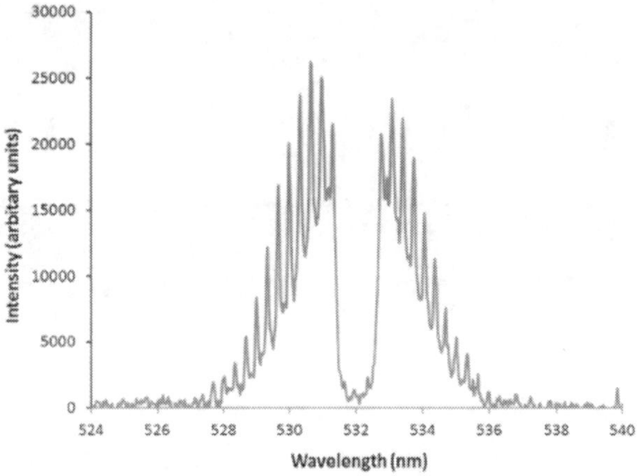

Figure 2. Thomson/Raman scattering spectrum in a 10% O_2–He mixture at 100 Torr, 100 ns after a ns pulse discharge current rise. Distance between spherical electrodes 1 cm, $n_e = 6 \times 10^{13}$ cm^{-3}, $T_e = 1.7$ eV [10].

model developed in [11]. Rapid electric field reduction during breakdown and the effect of cathode voltage fall on the field in the plasma after breakdown (such that the field is significantly lower compared to the ratio of voltage over distance) are readily apparent. Thomson scattering electron density measurements in molecular plasmas are quite challenging because of strong Rayleigh and Raman scattering [8]; however, Rayleigh scattering can be filtered out using a mask in a triple-grating spectrometer, and the Raman spectrum can be subtracted if the signal-to-noise and spectral resolution are sufficiently high [8]. Figure 2 plots a typical Thomson/Raman scattering spectrum in a 10% O_2–He mixture at 100 Torr, 100 ns after the beginning of the discharge pulse. The gap in the center of the spectrum, caused by the blocking of Rayleigh scattering, and the Raman lines of O_2 are apparent. The inferred electron density and electron temperature for these conditions are $n_e = 6 \times 10^{13}$ cm^{-3} and $T_e = 1.7$ eV.

3. Temperature and vibrational level populations: measurements and prediction

The parameter that has the most significant effect on chemical processes in fuel–air plasmas, in particular on the rates of chain branching fuel oxidation reactions, is gas temperature. Therefore time-resolved and spatially resolved temperature measurements in these plasmas, as well as comparison with kinetic modeling predictions, are extremely important for understanding the mechanism of plasma-assisted ignition. Since in molecular plasmas a significant fraction of discharge energy is loaded into the vibrational mode of nitrogen, with subsequent vibrational relaxation, time-resolved and spatially resolved measurements of N_2 vibrational temperature and vibrational level populations are also critical for predicting the rate of energy thermalization in the plasma. Recently, these measurements have become available due to progress in the development of picosecond CARS [12] and spontaneous Raman scattering laser diagnostics [13]. As expected, O_2 vibrational excitation is much less pronounced compared to that of N_2 [13].

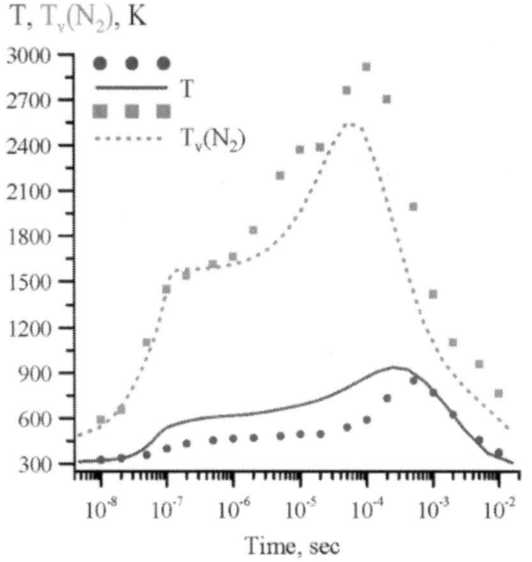

Figure 3. Experimental and predicted temperature and N_2 vibrational temperature during and after a ns pulse discharge in air between two spherical electrodes 1 cm apart at 100 Torr [12].

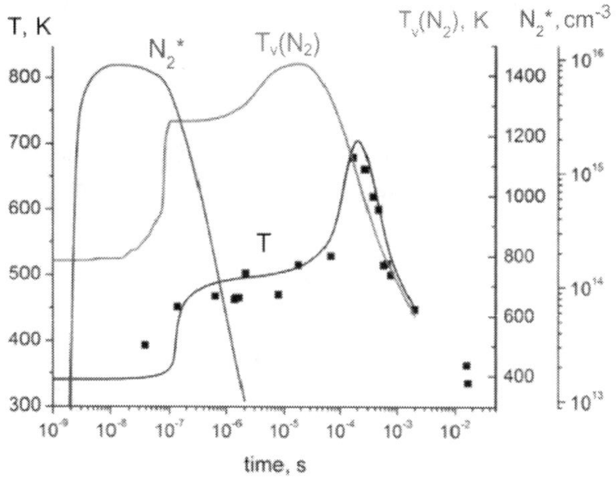

Figure 4. Experimental and predicted temperature during and after a ns discharge pulse in an H_2–air mixture ($\phi = 0.14$) between two spherical electrodes 0.9 cm apart at 40 Torr, plotted together with predicted number density of electronically excited N_2 molecules and $T\ v(N_2)$.

Figure 3 plots experimental picosecond CARS and predicted, time-resolved rotational/translational temperature and 'first level' N_2 vibrational temperature, $T_v = \omega_e(1 - 2 \times e)/\ln(N_{v=0}/N_{v=1})$, during and after ~100 ns duration discharge pulse in air, at 100 Torr [12]. A 'rapid' temperature rise, on a short time scale, $t \sim 10^{-7}$–10^{-6} s, up to $T \sim 500$ K (somewhat difficult to see in figure 3), is primarily due to an energy release during quenching of excited electronic states of nitrogen, $N_2(A^3\Sigma, B^3\Pi, C^3\Pi) + O_2 \rightarrow N_2\ (X^1\Sigma) + O + O$. This has been predicted by kinetic modeling [14] and has also been studied by optical emission spectroscopy in nanosecond pulse discharges at atmospheric pressure [15]. From figure 3, it can be seen that the discharge produces significant vibrational non-equilibrium, $T_{v\max} \sim 2900$ K, $T_{\max} \sim 850$ K. In these experiments, $N_2(v = 0-9)$ vibrational level populations have also been measured, indicating a strongly non-Boltzmann, 'bimodal' vibrational distribution [12]. Kinetic modeling calculations [12] demonstrated that a significant transient rise of N_2 vibrational temperature after the discharge pulse (from $T_{v\max} \sim 1500$ to ~ 2900 K at $t \sim 10^{-7}$–10^{-4} s, see figure 3) is due to a 'downward' vibration—vibration energy exchange, $N_2(v = 2) + N_2(v = 0) \rightarrow N_2(v = 1) + N_2(v = 1)$. T_v reduction at $t \sim 10^{-4}$–10^{-3} s, which coincides with a significant 'slow' temperature rise in the same time scale, from $T \sim 500$ K to ~ 850 K, is due to vibration–translation relaxation of N_2 by O atoms, $N_2(v) + O \rightarrow N_2(v-1) + O$. Obviously, both 'rapid' and 'slow' temperature increases, caused by the quenching of excited electronic states and the vibrational relaxation of nitrogen, respectively, have a significant effect on the energy balance and thus are critically important for the rates of fuel oxidation in nanosecond pulse fuel–air plasmas.

The effect of 'rapid' heating in nanosecond pulse fuel–air plasmas is illustrated in greater detail in figure 4, which compares experimental (ps pure rotational CARS) and predicted (using the kinetic model developed in [11]) temperatures during and after a nanosecond pulse discharge in a H_2–air mixture at 40 Torr. It can be seen that the time scales for the 'rapid' temperature rise and the decay of electronically excited N_2 molecules are the same (the 'slow' rise is due to N_2 vibrational relaxation by O atoms). 'Rapid' heating becomes considerably more pronounced at higher pressures [15], due to both higher specific energy loading in a localized discharge filament and faster quenching of excited electronic states of N_2. This raises the following questions: can plasma chemical reactions among excited electronic species and radicals generated in the low-temperature plasma compete with 'conventional' fuel–air chemical reactions, the rates of which rapidly increase with temperature? At what pressures does 'rapid' heating become the dominant effect in transient fuel–air plasmas, compared to low-temperature excited species and radical species chemistry? Answering these questions would have a considerable impact on the development of plasma-assisted combustion applications. This requires additional systematic studies in well-characterized fuel–air plasmas, over a wide range of pressures, as well as comparison with kinetic modeling calculations to evaluate the contributions of different mechanisms.

4. Effect of excited species and radicals on plasma chemistry

Reactive quenching of excited electronic states and vibrational relaxation in air and fuel–air mixtures may have a significant effect on plasma chemistry, generating radicals such as O, H, and OH, e.g. $O(^1D) + H_2 = H + OH$, and dissociating oxygen and fuel species, e.g. $N_2(A^3\Sigma, B^3\Pi, C^3\Pi) + H_2 \rightarrow N_2(X^1\Sigma) + H + H$, $N_2(A^3\Sigma, B^3\Pi, C^3\Pi) + O_2 \rightarrow N_2(X^1\Sigma) + O + O$. Plasma-generated radicals can initiate fuel oxidation chain reactions, such as $H + O_2 \rightarrow OH + O$, $O + H_2 \rightarrow OH + H$, and generate nitric oxide, $N_2^* + O \rightarrow NO + N$ (here the asterisk stands for N_2 excited electronic or vibrational states),

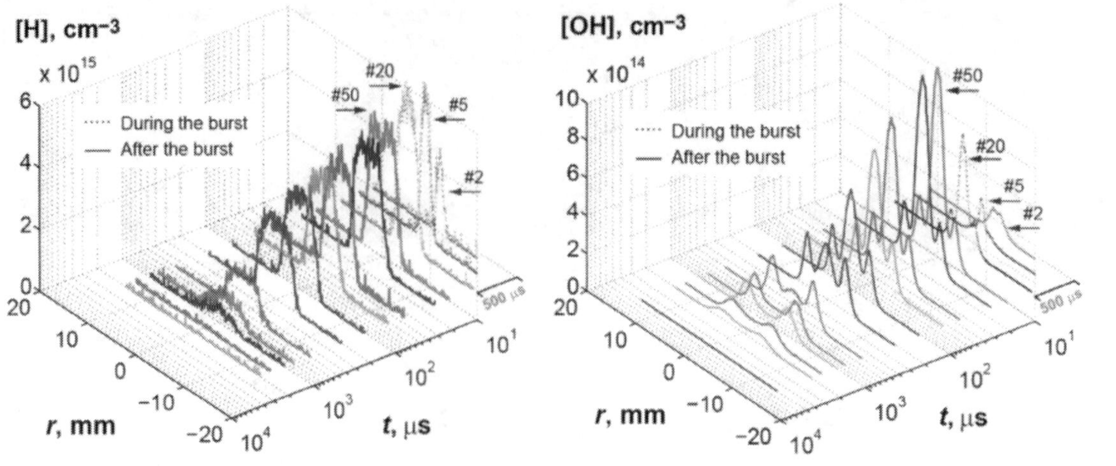

Figure 5. Radial distributions of absolute H (left) and OH (right) across a point-to-point discharge filament during and after a ns discharge burst (50 pulses) in Ar: O_2: H_2 = 80 : 20 : 2 mixtures at $P = 40$ Torr and $T_0 = 300$ K [19]. Pulse numbers are indicated in the figure.

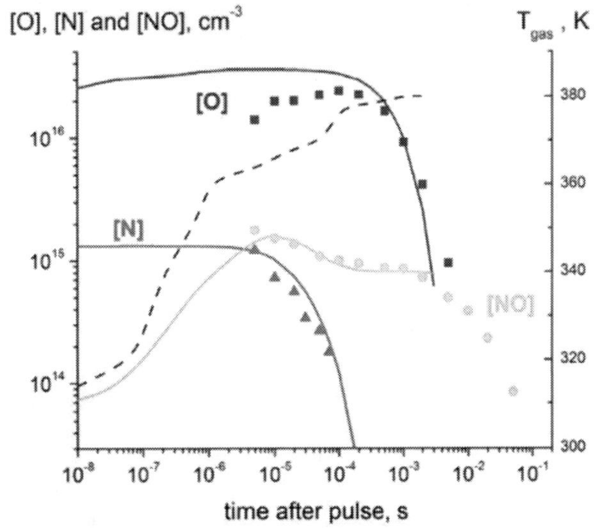

Figure 6. Experimental and predicted absolute O, N, and NO number densities after a ns pulse discharge in air between two spherical electrodes 1 cm apart at 100 Torr (low pulse energy, near room temperature conditions).

which may strongly affect radical concentrations. For this reason, absolute measurements of electronically excited (by cavity ring-down spectroscopy and calibrated optical emission spectroscopy) [16] and vibrationally excited N_2 molecules (by CARS [12] or spontaneous Raman scattering [13]) as well as those of N, O, NO, H, and OH radicals (by single-photon and two-photon absorption laser induced fluorescence (LIF) [16–18]) in transient fuel–air plasmas yield critical insights into the kinetics of chemical reactions of plasma-generated radicals. Spatially resolved LIF, two-photon absorption laser induced fluorescence (TALIF), and Rayleigh scattering measurements in transient plasmas with strong gradients [19] are especially desirable (see figure 5). Figure 6 compares experimental (LIF and TALIF) [18] and predicted [11] time-resolved absolute number densities of O, N, and NO after a ns pulse discharge in air at 100 Torr, at low pulse energy, and near room temperature conditions. In these experiments, $N_2(v = 0\text{–}4)$ populations have also been measured, by ps CARS. The results of these measurements demonstrated that (i) vibrationally excited $N_2(X^1\Sigma,v)$ molecules do not have a detectable effect on NO formation in ns pulse low-temperature air plasmas, and (ii) NO under these conditions is formed primarily by reactive quenching of a number of excited electronic states of N_2, including multiple singlet and triplet states.

The effects of vibrationally excited nitrogen molecules, $N_2(X^1\Sigma,v)$, on plasma chemical reactions have been discussed in the literature for quite some time. In fact, one of the objectives of simultaneous ps CARS, LIF, and TALIF measurements [18] was to quantify the effect of the $N_2(X^1\Sigma, v) + O \rightarrow NO + N$ reaction on the rate of NO formation in low-temperature transient air plasmas. However, the contribution of this reaction, compared to the reactive quenching of excited electronic states of N_2 appears to be small. This result does not imply that $N_2(X^1\Sigma, v)$ molecules cannot affect air and fuel–air plasma chemical reactions in general. Recent time-resolved measurements of OH in the nanosecond pulse discharge afterglow in preheated hydrocarbon–air mixtures [20], which demonstrated an anomalously long OH lifetime, led the authors to suggest that vibrationally excited nitrogen may delay OH recombination due to a near-resonance energy transfer to the HO_2 radical, $N_2(X^1\Sigma, v = 1) + HO_2 \rightarrow N_2(X^1\Sigma, v = 0) + HO_2(\nu_2+\nu_3) \rightarrow N_2+H+O_2$ [21]. However, gas temperature and N_2 vibrational temperature in these experiments have not been measured, and a more direct effect of heating the flow by the discharge on OH lifetime cannot be ruled out. Direct vibrational temperature measurements are critical due to a significant uncertainty in the state-specific rates of N_2 vibrational relaxation in reacting fuel–air mixtures. It may well be possible that reactions of vibrationally excited N_2 molecules in fuel–air plasmas may affect low-temperature fuel oxidation chemistry, and additional experimental studies (such as simultaneous N_2 vibrational CARS and OH LIF measurements) are needed to verify this hypothesis.

A closely related issue is the effect of long-lived metastable species generated in molecular plasmas, such as $O_2(a^1\Delta)$, on fuel-oxidizer plasma chemistry, compared to radicals participating in rapid reactions of fuel oxidation, such as O, H, and OH. Recent experiments show that metastable $O_2(a^1\Delta)$ molecules, which have a long lifetime and may accumulate in O_2-containing plasmas up to mole fractions of several per cent, have a significant effect on H_2–O_2 ignition delay length [22] and on C_2H_4–O_2–Ar flame speed [23]. In [23, 24], the effect of $O_2(a^1\Delta)$ was isolated from that of O atoms and ozone (which serves as a source of O and H radicals in the flame preheat zone). However, the predictive capability of kinetic models is limited by the lack of data on collisional quenching of $O_2(a^1\Delta)$ by radicals and hydrocarbon species [24, 25]. On the other hand, the effect of O atoms on CH_4 and C_2H_4 oxidation in low-temperature plasmas [26–28], as well as on the extinction limit of low-pressure CH_4 diffusion flames [28, 29], appears to be relatively well understood. In particular, at temperatures far below the ignition threshold, radical species generated in the plasma predominantly recombine and do not result in ignition, due to the near absence of chain branching processes.

Finally, the effect of low-temperature plasma-generated radicals on low octane number fuels exhibiting cool flame chemistry versus high octane number fuels for which cool flames are not observed, is not understood and requires further experimental and modeling studies. At temperatures below the self-ignition temperature, plasma-generated O atoms (as well as other radical or excited species) enhance alkyl radical (R) production (e.g. $RH + O \rightarrow R + OH$). The subsequent fate of the alkyl radical determines whether cool flame chemistry will occur or not [30, 31]. At high temperatures, the alkyl radical thermally decomposes to form smaller hydrocarbons, thus delaying significant radical branching and heat release until later in the reaction sequence and, more importantly, higher temperatures. At low temperatures, the alkyl radical may form the alkylperoxy radical by reacting with molecular oxygen ($R + O_2 \rightarrow RO_2$). The stabilization of RO_2, followed by internal isomerization to form the hydroperoxyalkyl radical (QOOH), and the addition of a second O_2 molecule to form O_2QOOH, results in significant radical branching during the decomposition and further reactions of O_2QOOH. Enhancing this low-temperature cool flame sequence, and enabling it over a broader range of conditions, is desirable for shortening the ignition delays of large hydrocarbon fuels. The molecular structure of the fuel, which affects the octane number, influences the rates of R and QOOH formation, and further study of how these processes are altered by the plasma chemistry is required [32]. The role of $O_2(a^1\Delta)$ is of particular uncertainty in this low-temperature reaction sequence.

5. Development of predictive plasma-assisted ignition/combustion chemistry mechanism

Quantitative predictions of ignition by low-temperature plasmas requires knowledge of the rates of the molecular energy transfer processes, such as electron impact excitation and dissociation of air and fuel species, vibrational relaxation, rates and products of reactive quenching of excited electronic states, reactions among charged species, as well as a mechanism of fuel oxidation via 'conventional' chemical reactions at low temperatures. Kinetic mechanisms of 'conventional' combustion [33, 34] have been developed and validated for high temperature conditions and may well be inapplicable at low temperatures typical for many plasma-assisted combustion environments. An illustration of this is found in a recent study [17] which compares absolute time-resolved OH and temperature measurements by LIF in lean H_2–air and hydrocarbon-air mixtures excited by a burst of nanosecond discharge pulses with kinetic modeling calculations, using conventional hydrocarbon–air combustion mechanisms. Although modeling predictions for H_2–air, CH_4–air, and C_2H_4–air agree with the data fairly well [17], the agreement between the model and the data in C_3H_8–air is quite poor. This demonstrates the need for the development of an accurate, predictive plasma-assisted combustion chemistry mechanism. The predictive capability of a plasma chemistry model involving such a large number of energy transfer processes and reactions requires extensive, preferably built-in, kinetic sensitivity analysis to identify the dominant processes. Identifying the reduced reaction mechanism is extremely critical, especially for hydrocarbons, since the kinetic modeling of coupled ns pulse discharge dynamics, energy transfer in the plasma, and plasma-assisted combustion requires the incorporation of a wide range of time scales, $\sim 10^{-12}$–10^{-2} s, and is very computationally intensive even in 1D geometry [35].

The experimental validation of such a mechanism would require comparison with the results of plasma-assisted ignition experiments with well-characterized conditions, using metrics that would be sufficiently sensitive to the kinetics used. The metrics used in conventional combustion kinetics include laminar flame speed, as well as ignition delay time (e.g. measured in shock tubes or a rapid compression machine). In low-temperature plasma-assisted ignition, these metrics are clearly insufficient since they also need to be correlated with measurements of parameters controlling both plasma chemistry and conventional chemistry reactions, such as the number densities of key radical species and temperature in the plasma. Since species number density measurements using laser diagnostics require significant signal accumulation, this requires the use of well-reproduced, repetitively pulsed plasma ignition cycles [36]. Time-resolved temperature, N_2 vibrational temperature (in case of significant vibrational non-equilibrium), key radical species (such as OH, O, or H), and at least one key dominant species (such as fuel or O_2) measurements during plasma-induced ignition (as illustrated in figure 7), appear to provide sufficient amounts of data for kinetic mechanism validation. Systematic validation would require these measurements to be made over a range or equivalence ratios, fuels, pressures, and energy coupled to the plasma (e.g. number of discharge pulses). Note that single-shot ignition delay time or ignition temperature measurements, such as those obtained in a shock tube or a rapid compression machine, do not lend themselves to time-resolved temperature and species number density measurements.

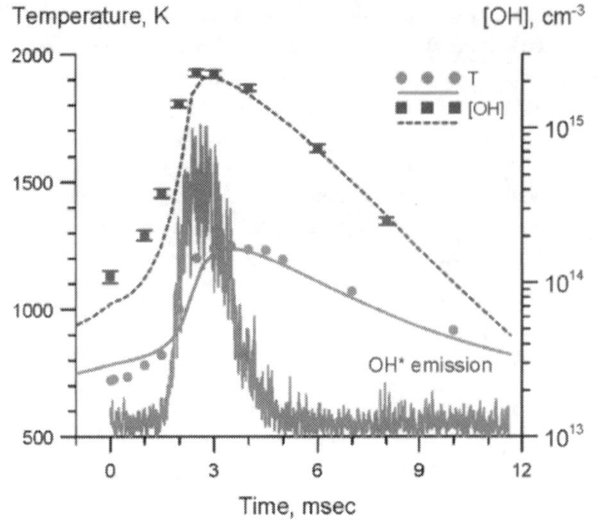

Figure 7. Experimental and predicted temperature and OH during ignition after a ns discharge burst in plane-to-plane geometry. H_2–air, $T_0 = 500$ K, $P = 80$ Torr, $\phi = 0.4$, 115 pulses [36]. Temperature and T_v (N_2) are measured by ps CARS at $P = 92$ Torr, 120 pulses [17].

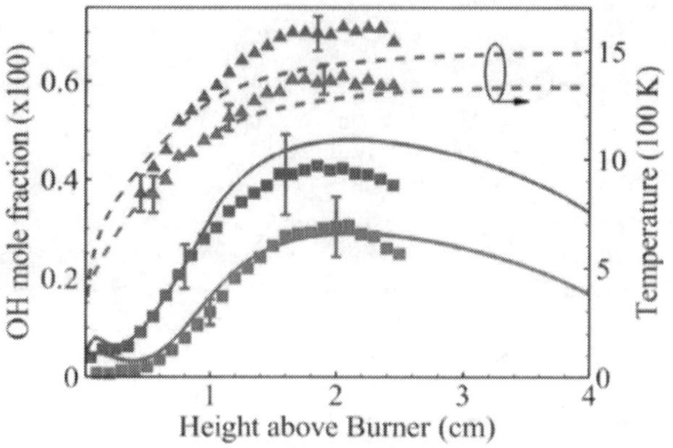

Figure 8. Experimental [37] and predicted [38] temperature and OH distributions across a burner stabilized H_2–O_2–N_2 flat flame ($\phi = 0.5$, $P = 25$ Torr), with (blue) and without (red) ns discharge burst (200 pulses) in plane-to-plane geometry. Squares and solid lines: OH; triangles and dashed lines: temperature.

An approach that is complementary to time-resolved measurements during repetitively pulsed ignition cycles could be spatially resolved measurements of temperature and species number densities across a premixed or diffusion laminar flame front, at conditions where radical species are generated by the plasma. These experiments would not require the achievement of well-reproduced repetitive ignition, which requires operation at very low discharge burst repetition rate, ~1 Hz, to remove combustion products from the test cell [36]. An example of these measurements (by OH LIF) [37], along with modeling calculations, is illustrated in figure 8 [38]. Note that the effect of the plasma on flame speed and reaction zone location appears to be relatively modest [23, 24, 37], and as such is unlikely to be a good metric for kinetic mechanism validation. On the other hand, measurement of radical species and temperature distributions in the preheat zone of the flame, especially in fuel-lean conditions when the effect of plasma is most pronounced [37], may well be used for this purpose.

6. Effect of plasmas on non-premixed turbulent flames and reacting compressible flows

The effect of radical generation in nanosecond pulse discharge plasmas on blow-off velocity and lean flammability limits in premixed and jet diffusion flames has been studied fairly extensively [39, 40], demonstrating stable combustion and flameholding at significantly higher velocities and lower equivalence ratios compared to baseline. Recent experiments also provided insight into the temporal response of a lean, premixed, weakly turbulent flame to nanosecond pulse discharge excitation [41], suggesting that the plasma may help to control combustion instabilities, such as coupling between heat release and acoustic oscillations. However, the effect of plasmas on non-premixed turbulent flames at high Reynolds numbers is still to be explored. In these flames, the presence of plasma may help prevent local extinction and/or accelerate re-ignition by producing a pool of radicals at the locations where the local temperature and reaction rates become too low to sustain combustion, due to flow field turbulence and its interaction with finite-rate chemistry. Additional work in this field, using quantitative laser diagnostics to measure flow field velocity and scalars in turbulent flames in the presence of non-equilibrium plasma is critical. In particular, high-frame rate (~10 kHz) turbulent flame imaging of velocity, temperature and species concentrations (e.g. particle imaging velocimetry [42], OH planar laser induced fluorescence (PLIF) [42], Rayleigh scattering [43], Raman scattering [44], CH PLIF [45], and CH2O PLIF [46]), is necessary in order to begin to understand the coupling between plasma kinetics and turbulent combustion dynamics.

One of the most critical issues of plasma-assisted ignition, combustion, and flameholding in non-premixed high-speed flows is whether the plasma needs to be located in the airflow, in the fuel injection flow, or in a partially premixed region near the fuel injector. Since mole fractions of hydrocarbon fuels in the mixture are typically fairly low (a few percent even at stoichiometric conditions), it appears that plasma excitation of airflow and generation of sufficient amounts of excited electronic species and radicals, N_2^* and O atoms, would be the obvious approach. However, the excitation of the airflow upstream of the fuel injection port may result in a significant loss of excited species and radicals, both due to quenching and recombination, and due to mixing with unexcited air, thus diluting the radical pool. A similar problem exists with plasma excitation of a partially premixed fuel–air region in the vicinity of a fuel injector [47, 48]. Since a rapid mixing of fuel and airflow is critical for stable combustion, the radicals generated in the discharge may also be significantly diluted in the flow, thus reducing the impact of plasma excitation. A recently suggested approach [49] is to sustain the discharge partially in the fuel injection line and terminate the current path in the combustor, such that the discharge filament would penetrate into the flow with the fuel jet. The advantages of this approach include (i) radical generation in high-speed fuel injection flow and their rapid transport into the airflow; (ii) discharge following a high-speed fuel jet and thus penetrating deep into the airflow; and (iii) fuel–air mixing enhancement. Simple analysis of high electron density plasmas in a high-speed flow [49], such that the local ionization/recombination balance is not affected by convection, demonstrates that a steady-state discharge filament may exist only if the current vector is parallel to the flow velocity vector in the filament, $\vec{j} \parallel \vec{u}$. Recently, it has also been demonstrated that fuel–air mixing can be enhanced significantly by unsteady oscillating discharge filaments [50]. Additional studies, both experimental and coupled electric discharge/reacting compressible flow modeling, are necessary to obtain quantitative insight into discharge behavior in high-speed compressible flows, and its effect on ignition and flameholding.

Finally, one of the least explored problems in non-premixed, high-speed compressible flows is the effect of plasma-assisted ignition on combustion instability development. Evidence of such instability, caused by strong coupling between pressure in the combustor and discharge parameters (voltage and current), and resulting in high-amplitude oscillations of pressure and flame front, was detected in recent supersonic combustion experiments [48]. A systematic experimental study of this coupling is necessary to obtain insight into the instability development mechanism, and possibly to control and suppress it.

7. Summary

An overview of recent experimental and kinetic modeling studies of low-temperature plasma-assisted combustion exhibits the following key challenges:

(i) Measurements and modeling predictions of electric field, electron density, gas temperature, and N_2 vibrational level populations in high-pressure pulsed fuel–air plasmas are necessary for predictive insight into discharge energy partition.
(ii) Measurements and modeling predictions of excited electronics states of N_2^* and key radicals (O, H, OH, and NO) are critical for quantifying their effect on pulsed fuel–air plasma chemistry, including fuels exhibiting cool flame chemistry.
(iii) Development of a predictive plasma-assisted combustion chemistry mechanism requires a set of temperature, N_2 vibrational temperature, and radical concentration data during a repetitively pulsed plasma-enhanced ignition process, or across a pulsed plasma-enhanced flame. The results would help quantify the effect of 'rapid' heating, and the possible reactions of vibrationally excited molecules, compared to reactions of plasma-generated radicals.
(iv) Kinetic sensitivity analysis is necessary to identify the reduced reaction mechanism, which can be incorporated into a practical discharge dynamics/energy transfer/plasma chemistry kinetic model incorporating a wide range of time scales and a realistic geometry.
(v) The effect of plasmas on non-premixed turbulent flames needs to be studied using high frame rate imaging of temperature and radical species concentration fields.
(vi) Electric discharge dynamics and combustion instability development in reacting compressible flows need to be studied experimentally, and by using coupled electric discharge/reacting compressible flow modeling.

Acknowledgments

This work has been supported by the US Air Force Office of Scientific Research MURI 'Fundamental Aspects of Plasma Assisted Combustion', Technical Monitor Dr Chiping Li. The authors would also like to thank R Yetter, J Sutton, S Leonov, N Popov, and N Aleksandrov for extensive technical discussions.

References

[1] Starikovskaia S M 2006 *J. Phys. D: Appl. Phys.* **39** R265
[2] Starikovskiy A and Aleksandrov N 2013 *Prog. Energy Combust. Sci.* **39** 61
[3] Tholin F, Lacoste D A and Bourdon A 2014 *Combust. Flame* **161** 1235
[4] Takana H and Nishiyama H 2014 *Plasma Sources Sci. Technol.* **23** 034001
[5] Ito T, Kobayashi K, Czarnetzki U and Hamaguchi S 2010 *J. Phys. D: Appl. Phys.* **43** 062001
[6] Boehm P, Luggenhoelscher D and Czarnetzki U 2013 Laser-spectroscopic electric field measurements in a ns-pulsed microplasma in nitrogen *66th Annual Gaseous Electronics Conf. (Princeton, NJ, October 2013)*
[7] Goldberg B, O'Byrne S and Lempert W R 2012 AIAA Paper 2012-0240 *50th AIAA Aerospace Sciences Meeting (Nashville, TN, January 2012)*
[8] van Gessel A F H, Carbone E A D, Bruggeman P J and van der Mullen J J A M 2012 *Plasma Sources Sci. Technol.* **21** 015003
[9] van Gessel B, Brandenburg R and Bruggeman P 2013 *Appl. Phys. Lett.* **103** 064103
[10] Roettgen A, Adamovich I V and Lempert W R 2014 AIAA Paper 2014–2113 *45th AIAA Plasmadynamics and Lasers Conf. (Atlanta, GA, June 2014)*
[11] Shkurenkov I, Burnette D, Lempert W R and Adamovich I V 2014 Kinetics of excited states and radicals in a nanosecond pulse discharge and afterglow in nitrogen and air *Plasma Sources Sci. Technol.* **23** 065003
[12] Montello A, Yin Z, Burnette D, Adamovich I V and Lempert W R 2013 *J. Phys. D: Appl. Phys.* **46** 464002
[13] Lo A, Cléon G, Vervisch P and Cessou A 2012 *Appl. Phys. B: Laser Opt.* **107** 229
[14] Popov N A 2001 *Plasma Phys. Rep.* **27** 886
[15] Rusterholtz D L, Lacoste D A, Stancu G D, Pai D Z and Laux C O 2013 *J. Phys. D: Appl. Phys.* **46** 464010
[16] Stancu G D, Kaddouri F, Lacoste D A and Laux C O 2010 *J. Phys. D: Appl. Phys.* **43** 124002
[17] Yin Z, Montello A, Carter C D, Lempert W R and Adamovich I V 2013 *Combust. Flame* **160** 1594
[18] Burnette D, Montello A, Adamovich I V and Lempert W R 2014 Nitric oxide kinetics in the afterglow of a diffuse plasma filament *Plasma Sources Sci. Technol.* **23** 045007
[19] Yin Z, Eckert Z, Adamovich I V and Lempert W R 2014 Time-resolved radical species and temperature distributions in an Ar–O_2–H_2 mixture excited by a nanosecond pulse discharge *Proc. Combust. Inst.* at press

[20] Wu L, Lane J, Cernansky N P, Miller D L, Fridman A A and Starikovskiy A Yu 2011 *Proc. Combust. Inst.* **33** 3219
[21] Starikovskiy A 2012 AIAA Paper 2012-0244 *50th AIAA Aerospace Sciences Meeting (Nashville, TN, January 2012)*
[22] Smirnov V V, Stelmakh O M, Fabelinsky V I, Kozlov D N, Starik A M and Titova N S 2008 *J. Phys. D: Appl. Phys.* **41** 192001
[23] Ombrello T, Won S H, Ju Y and Williams S 2010 *Combust. Flame* **157** 1906
[24] Ombrello T, Won S H, Ju Y and Williams S 2010 *Combust. Flame* **157** 1916
[25] Popov N A 2011 *Plasma Sources Sci. Technol.* **20** 045002
[26] Uddi M, Jiang N, Mintusov E, Adamovich I V and Lempert W R 2009 *Proc. Combust. Inst.* **32** 929
[27] Zuzeek Y, Choi I, Uddi M, Adamovich I V and Lempert W R 2010 *J. Phys. D: Appl. Phys.* **43** 124001
[28] Sun W, Uddi M, Won S H, Ombrello T, Carter C and Ju Y 2012 *Combust. Flame* **159** 221
[29] Sun W, Uddi M, Ombrello T, Won S H, Carter C and Ju Y 2012 *Proc. Combust. Inst.* **33** 3211
[30] Westbrook C K 2013 *Annu. Rev. Phys. Chem.* **64** 201
[31] Zador J, Taatjes C A and Fernandes R X 2011 *Prog. Energy Combust. Sci.* **37** 271
[32] Tsolas N, Togai K and Yetter R A 2015 Kinetics of plasma assisted hydrocarbon oxidation in flow reactors *53rd AIAA Aerospace Sciences Meeting (SciTech 2015) (Kissimmee, FL, 5–9 January 2015)*
[33] Wang H, You X, Joshi A V, Davis S G, Laskin A, Egolfopoulos F and Law C K 2007 High-Temperature Combustion Reaction Model of H_2/CO/C_1-C_4 Compounds *USC Mech Version II* (http://ignis.usc.edu/USC_Mech_II.htm May 2007)
[34] Konnov A 2000 *Detailed Reaction Mechanism for Small Hydrocarbons Combustion, Release 0.5.* http://homepages.vub.ac.be/~akonnov/
[35] Nagaraja S, Yang V and Adamovich I 2013 *J. Phys. D: Appl. Phys.* **46** 155205
[36] Yin Z, Adamovich I V and Lempert W R 2013 *Proc. Combust. Inst.* **34** 3249
[37] Li T, Adamovich I V and Sutton J A 2014 AIAA Paper 2014-0664 *52nd AIAA Aerospace Sciences Meeting (SciTech 2014) (National Harbor, MD, 13–17 January 2014)*
[38] Nagaraja S, Li T, Sutton J, Adamovich I and Yang V 2014 Nanosecond plasma enhanced $H_2/O_2/N_2$ premixed flat flames *Proc. Combust. Inst.* at press
[39] Pilla G, Galley D, Lacoste D A, Lacas F, Veynante D and Laux C O 2006 *IEEE Trans. Plasma Sci.* **34** 2471
[40] Bak M S, Im S-K, Mungal M G and Cappelli M A 2013 *Combust. Flame* **160** 2396
[41] Lacoste D A, Xu D A, Moeck J P and Laux C O 2013 *Proc. Combust. Inst.* **34** 3259
[42] Böhm B, Heeger C, Gordon R L and Dreizler A 2011 New perspectives on turbulent combustion: multi-parameter high-speed planar laser diagnostics *Flow Turbul. Combust.* **86** 313–41
[43] Patton R A, Gabet K N, Jiang N, Lempert W R and Sutton J A 2012 *Appl. Phys.* B **108** 377
[44] Gabet K N, Jiang N, Lempert W R and Sutton J A 2010 *Appl. Phys.* B **101** 1
[45] Gabet K N, Patton R A, Jiang N, Lempert W R and Sutton J A 2012 *Appl. Phys.* B **106** 569
[46] Jiang N, Patton R A, Lempert W R and Sutton J A 2011 *Proc. Combust. Inst.* **33** 767
[47] Do H, Cappelli M A and Mungal M G 2010 *Combust. Flame* **157** 1783
[48] Do H, Im S-K, Cappelli M and Mungal M G 2010 *Combust. Flame* **157** 2298
[49] Savelkin K V, Yarantsev D A, Adamovich I V and Leonov S B 2014 Ignition and flameholding in a supersonic combustor by an electrical discharge combined with a fuel injector *Combust. Flame* accepted
[50] Leonov S B and Yarantsev D A 2007 *Plasma Sources Sci. Technol.* **16** 132

IOP Publishing

Plasma Physics and Controlled Fusion

Plasma Phys. Control. Fusion **57** (2015) 014002 (9pp)

doi:10.1088/0741-3335/57/1/014002

Plasma-based creation of short light pulses: analysis and simulation of amplification and focusing

C Riconda[1], S Weber[2], L Lancia[3,4], J-R Marquès[5], G Mourou[6] and J Fuchs[5]

[1] LULI, Sorbonne Universités-UPMC-Ecole Polytechnique-CNRS-CEA, 75005 Paris, France
[2] Institute of Physics of the ASCR, ELI-Beamlines, 18221 Prague, Czech Republic
[3] SAPIENZA, University of Rome, Dip. SBAI, 00161 Rome, Italy
[4] INFN Ð Sez. Roma Ð SAPIENZA, University of Rome, 00185 Rome, Italy
[5] LULI, Ecole Polytechnique-CNRS-CEA-UPMC, 91128 Palaiseau, France
[6] IZEST, Ecole Polytechnique-CEA, 91128 Palaiseau, France

E-mail: caterina.riconda@upmc.fr

Received 18 June 2014, revised 23 July 2014
Accepted for publication 28 August 2014
Published 28 November 2014

Abstract

Plasmas can serve as damage-less optics for amplifying and focusing light pulses to very high intensity. This provides a way to overcome the limitations of solid-state optical materials as a damage threshold in the classical sense is absent. The amplification process relies on parametric processes in plasmas exploiting the coupling of transverse electromagnetic waves to a longitudinal plasma wave. The plasma response can either be an electron plasma wave (stimulated Raman scattering), an ion-acoustic wave (stimulated Brillouin scattering) or a more complicated non-resonant feature in the case of very short pulses.

Keywords: plasma-based amplification, PIC simulations , parametric instabilities

(Some figures may appear in colour only in the online journal)

1. Introduction

Standard solid-state optical materials have a damage threshold of the order of one joule per centimetre square. This implies very large surfaces for the creation of ultra-intense light pulses. Large gratings are difficult to produce and imply high costs. It has been known for quite some time that gratings can also be produced in plasmas by the beating of two electromagnetic waves [1]. Plasmas do not have a damage threshold in the classical sense as the material as lattice and chemical bonds are already completely broken down. In contrast to optical materials, however, plasmas are not stationary but evolve. Nevertheless they can be exploited for short-pulse purpose as the characteristic time scale of plasma evolution would be on nanosecond scale.

The feasibility of plasma-amplification using ion-acoustic waves has been shown in simulation [2–7] and experiment [8–10]. The mechanism is based on the collision of two laser pulses in a plasma: a pump pulse and a seed pulse. The pump pulse is several picoseconds long and provides the energy, the seed pulse is very short, a few tens to a few hundreds of femtoseconds, and provides the characteristic time scale for the final amplified seed. Due to the coupling with a longitudinal wave in the plasma (in the present case an ion-acoustic wave), a transfer of energy from pump to seed takes place and thereby increasing the intensity of the seed. The interaction of the three waves is a controlled parametric instability based on Brillouin backscattering (SBS—stimulated Brillouin scattering).

Standard SBS operates on characteristic time scales which are given as inverse of the ion-acoustic wave (IAW) frequency: $\omega_{\mathrm{IAW}} = k_{\mathrm{cs}}c_{\mathrm{s}} \approx 2k_{\mathrm{o}}\sqrt{ZT_{\mathrm{e}}/m_{\mathrm{i}}}$. Depending on the plasma temperature the times are of the order of 1–10 ps. However, a so-called strong-coupling (sc) version of SBS exists which has characteristic time scales of the order of tens of femtoseconds, sc-SBS [11]. Attaining this particular interaction regime requires a fulfilment of the following threshold condition:

$$\left(\frac{v_{\mathrm{osc}}}{v_{\mathrm{e}}}\right) > 4k_{\mathrm{o}}c_{\mathrm{s}}\omega_{\mathrm{o}}/\omega_{\mathrm{pi}}^{2}, \qquad (1)$$

with $v_{osc} \sim \sqrt{I_p}$ the oscillating velocity of the electron in the laser field, $\omega_{pi} \sim \sqrt{n_i}$ the ion plasma frequency and $v_e \sim \sqrt{T_e}$ the electron thermal velocity. By adjusting density n, temperature T_e and intensity of the pump pulse I_p one can operate in the sc-regime. In more practical units the condition is given as:

$$I_p\lambda_o^2 > 10^{-2}T_e^{3/2}\sqrt{\frac{Z}{A}}\left(\frac{n_c}{n_e}\right)\sqrt{1-\frac{n_e}{n_c}}. \quad (2)$$

Here, I_p is in units of 10^{14} W cm^{-2}, λ_o in micrometre and T_e in keV. Fulfilling the threshold condition induces a transition from an eigenmode regime of IAW, characterized by ω_{IAW}, to a driven quasi-mode regime, characterized by the frequency:

$$\omega_{sc} = \frac{1+\imath\sqrt{3}}{2}\left(k_o^2 v_{osc}^2\omega_{pi}^2/\omega_o\right)^{1/3}, \quad (3)$$

with $\imath$ the imaginary unit. The characteristic growth rate is give by $\gamma_{sc} = \mathcal{Im}(\omega_{sc})$, corresponding to a characteristic time scale as $\tau_{sc} = 1/\gamma_{sc}$. One has to always consider that the desired instability, sc-SBS, evolves in competition with other unwanted plasma instabilities such as filamentation, Raman backscattering, wavebreaking etc, and therefore it has to be optimized [5]. In the case of very short seed pulses with duration τ_s and such that $\tau_s \ll \tau_{sc}$, the amplification process takes place in a mixed-mode regime based on non-resonant modes (MMNR-regime) [3].

It is interesting to point out that one of the first experiments observing accidental amplification took place in 1966 using a Raman configuration [12].

2. The final stage of amplification: focusing with plasma mirror or plasma lens

The plasma amplification scenario cannot be separated from the subsequent focusing of the amplified seed pulse. At the final relativistic intensities obtained, of the order of 10^{18} W cm^{-2}, plasmas are advantageous for focusing. Two fundamental possibilities present themselves: plasma mirror (PM) and plasma lens (PL).

The use of plasma optics (PM, PL) has several advantages over conventional reflective optics: they are

- much cheaper,
- less sensitive to misalignment,
- not sensitive to target debris,
- increasing the distance between target and focusing optics
- improving the temporal contrast at the same time (in the case of PM) [13],
- very compact devices (about 10% in volume compared to solid-based counterparts).

Plasma amplification and plasma focusing are examples of the growing field of plasma optics [14] which will help to pave the way towards Zetta–Exawatt science and technology [15, 16].

The purpose of plasma amplification based on ion-collective modes and MMNR modes is not to achieve extremely high intensities. The idea is rather to get to relativistic intensities, $a_o = 1$–3, and then obtain the final intensity by focusing the pulse to an almost diffraction limited focal spot. This provides an intensity increase which is given by the ratios of the full width at half maximum (FWHM) before and after focusing. In order to obtain extremely high intensity one would perform the amplification process with ever larger transverse sizes of the laser pulses. A pump and seed beam of 1 mm^2 focused to 1 μm^2 gives an increase in intensity by a factor of 10^6 [6]. This requires of course that filamentation will not destroy the pump beam during the amplification process. Analytical analysis and PIC-simulations have shown that filamentation can be avoided by tailoring plasma density, laser intensity and interaction time [5, 17].

2.1. Plasma mirror analysis

Plasma mirrors are well established for the purpose of pulse cleaning and contrast enhancement [19]. An ellipsoidal plasma mirror operating at near critical density can also be used to focus a laser pulse in a very efficient way [20–22].

PMs can easily handle fluences of up to 1 kJ cm^{-2} [23] and relativistic fluences have been demonstrated [24–26]. A recent successful experimental set-up is reproduced in figure 1. Ellipsoidal plasma mirrors will also be an essential ingredient for focusing the forthcoming 10 PW laser beams [27, 28] to an almost diffraction limited focal spot in order to obtain intensities of the order of 10^{24} W cm^{-2}.

2.2. Plasma lens analysis

Relativistic self-focusing of a laser beam can be used in a controlled way to focus a pulse to a diffraction limited focal spot [29–35]. The characteristic required propagation length is given as [36, 37]:

$$L_{focus} = D_L\left(\frac{n_c a_o}{n_e}\right)^{1/2}, \quad (4)$$

with D_L the intensity FWHM of the laser pulse, $a_o = v_{osc}/c = eE_{max}/cm_e\omega_o = 0.85 I_o^{1/2}\lambda_o$ the relativistic amplitude and n_e/n_c the density in units of the critical density. Here I_o is in units of 10^{18} W cm^{-2} and λ_o in micrometres.

Present simulations of plasma amplification in the mixed-mode regime show the possibility of using laser pulses with FWHM of the order of 100 μm and final intensity of the order of 10^{18} W cm^{-2}, i.e. $a_o \approx 1$. If a density $n_e/n_c \approx 0.9$ were used, a plasma lens of 100 μm would be necessary to focus the pulse to a diffraction limited focal spot. Although these values seem difficult to achieve, it would allow compression of the amplified laser pulse by a further factor of 10^4 (corresponding to the ratio of the initial to final focal spot) to obtain intensities of the order of 10^{22} W cm^{-2}. A first step could be to use smaller seed pulses with FWHM of about $\sim$10 μm to reach 10^{20} W cm^{-2}.

However, there is a caveat. The laser pulse is at the same time subject to filamentation which operates on a very short time scale in high-density plasmas. The typical growth rate for filamentation is given by $\gamma_{fil}/\omega_o \approx 10^{-5} I_{14}\lambda_o^2(n_e/n_c)$ [38] with I_{14} the laser intensity in units of

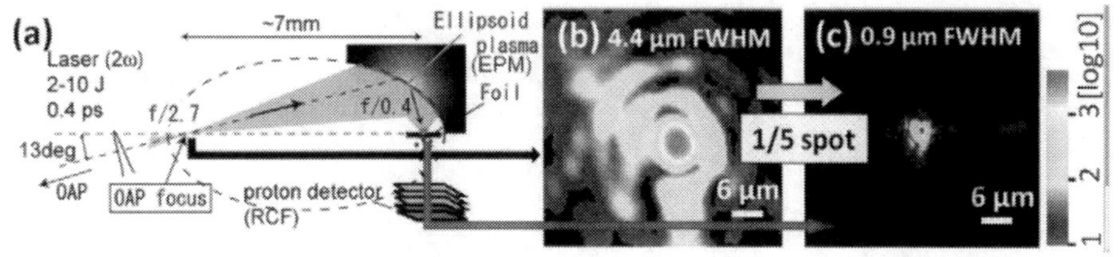

Figure 1. (*a*) Typical set-up and functioning of a focusing plasma mirror device (reprinted with permission from [21]), combining a standard off-axis parabola (OAP) with an ellipsoidal plasma mirror. (*b*) Focal-spot produced by the OAP. (*c*) Tighter focusing produced by the subsequent PM.

$10^{14}\,\mathrm{W\,cm^{-2}}$ and λ_o in micrometre. For the aforementioned characteristic parameters of the amplified seed pulse and plasma parameters, filamentation will occur over a distance of $L_{\mathrm{fil}} = c(1 - n_e/n_c)^{1/2}/\gamma_{\mathrm{fil}} \approx 0.6\,\mu\mathrm{m}$. To avoid filamentation considerably lower plasma densities need to be used which in turn will increase the distance for self-focusing. A trade-off implies setting the distances equal for self-focusing L_{focus} and filamentation L_{fil} to obtain the limiting density which in practical units has the value:

$$\frac{n_{\mathrm{limit}}/n_c}{1 - n_{\mathrm{limit}}/n_c} = \frac{3}{(D_L/\lambda_o)^2} \frac{1}{(I_{18}\lambda_o^2[\mu\mathrm{m}])^{2.5}}. \qquad (5)$$

For $\lambda_o = 1\,\mu\mathrm{m}$, $I = 10^{18}\,\mathrm{W\,cm^{-2}}$ and $D_L = 100\,\mu\mathrm{m}$ this gives a density of the order $n_{\mathrm{limit}}/n_c \approx 3 \times 10^{-4}$, which in turn results in a plasma lens of the order of $10^4\,\mu\mathrm{m}$ length. However, the exact nature of very short pulse filamentation is still unclear. Recent simulations and related experiments did not show filamentation even under more severe conditions [36]. Moreover, the use of a periodic plasma-vacuum structure might help to control the transverse modulation of the laser beam and thereby shorten the overall length of the plasma lens [39]. For this to work the plasma layers must have a width which is smaller than the typical length over which filamentation grows.

3. Simulation configuration

The simulations presented in this paper are summarized in table 1. The purpose of the present paper is to investigate two issues. (i) In the case of very short seed pulses with duration $\tau_s \ll \tau_{sc}$ the question is, which regime is more efficient, the mixed mode non-resonant regime (MMNR) or the stimulated Raman scattering (SRS) amplification regime? (ii) In the case of a seed duration shorter, but comparable to the growth rate of the strong-coupling SBS instability, where a large energy transfer from the pump to the seed was found [5], can one preserve the shape and transverse size of the seed in order to efficiently couple the amplified pulse to a plasma mirror or to a plasma lens?

The two-dimensional kinetic simulations were performed with the electromagnetic, fully relativistic PIC-code emi2d [40]. The code has Cartesian geometry and is two-dimensional in real space and three-dimensional in velocity space. The computational box is $\sim 800\,\mu\mathrm{m}$ in the propagation direction with open boundary conditions and $\sim 100\,\mu\mathrm{m}$ in the transverse direction employing periodic boundary conditions. The discretization parameters are: $\Delta t = 0.27\,\omega_o^{-1}$ for the time

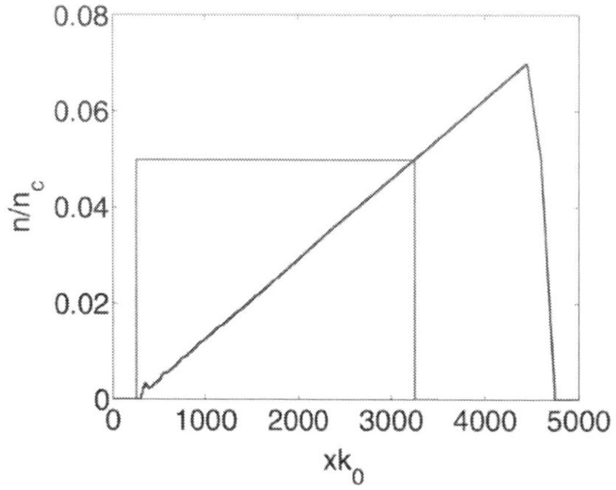

Figure 2. One-dimensional cuts of the plasma density profiles (blue: ramp, red: flat) used for the amplification process (see table 1). The profiles are homogeneous in transverse direction. The pump beam enters from the left and the seed beam from the right.

step and $\Delta x = 0.294\,\omega_o^{-1}$ and $\Delta y = 0.3\,\omega_o^{-1}$ for the parallel and transverse directions, respectively. The temperature is $T_e = 500\,\mathrm{eV}$ for the electrons and $T_i = 50\,\mathrm{eV}$ for the ions. The ion mass is $m_i = 3600\,m_e$, consistent with deuterium gas, and the charge is set to $Z = 1$. The pump pulse duration is 3.5 ps.

Two different density profiles were used (see figure 2). In the case of extremely short pulses and in order to compare with SRS simulations, a homogenous plasma at $n_e/n_c = 0.05$ and size $\sim 500\,\mu\mathrm{m}$ was considered, while in the other cases a triangular profile with maximum density $n_e/n_c = 0.07$ was used in order to minimize unwanted losses from the pump due to spontaneous Raman backscattering. For a pump at $I_p\lambda_o^2 = 10^{16}\,\mathrm{W}\mu\mathrm{m}^2\,\mathrm{cm}^{-2}$ as used in most of the simulations, and in the case of the homogenous plasma, we find that $\tau_{sc} = 249\,\omega_o^{-1} \sim 132\,\mathrm{fs}$, while $\gamma_{\mathrm{SRS}}/\omega_o \sim 4.27 \times 10^{-3}\left(\frac{n_e}{n_c}\right)^{1/4}(I_{14}\lambda_o^2)^{1/2}$ (in the limit $n_e/n_c \ll 1$) corresponds to a typical growth time for SRS of $\tau_{\mathrm{SRS}} = 50\,\omega_o^{-1} = 26\,\mathrm{fs}$.

It should be noted that in a PIC-simulation $\delta n/n_{\mathrm{pic}} \gg \delta n/n_{\mathrm{thermal}}$, i.e. the noise is strongly over-estimated. As a consequence the detrimental instabilities intervening have an unrealistic large growth rate, whereas the amplification process should be modelled realistically as the instability grows in a controlled way due to the seed pulse and is not excited from random noise. Unless stated otherwise in the text, the intensity of the seed pulse has been set to $I_s = 10^{17}\,\mathrm{W\,cm^{-2}}$. In the

Table 1. Summary of the simulations. FWHM in/out refers to the initial and final transverse size of the B_z-field of the seed pulse. 'Profile' refers to the plasma profile used for amplification, see figure 2. I_p is the pump intensity and τ_s the seed pulse duration. The cases e8 (MMNR), e9 and e10 are simulations with a seed frequency of $\omega_1 = \omega_o$, $\omega_1 = 0.8\,\omega_o$ and $\omega_1 = 1.2\,\omega_o$, respectively (see discussion in the text). Note: no amplification takes place in the case e10.

case	I_p[W cm^{-2}]	τ_s[fs]	FWHM in [μm]	FWHM out [μm]	'profile'
e7	10^{16}	80	32	22	ramp
f7	10^{16}	80	64	40	ramp
r4	10^{16}	30	32	16	ramp
r5	5×10^{15}	30	100	50–100	ramp
e8	10^{16}	13	32	~10	flat
e9	10^{16}	13	32	~10	flat
e10	10^{16}	13	32	32	flat

standard configuration, the pump and seed have exactly the same frequency unless stated otherwise.

In all the simulations the pump and seed beam collide head-on.

4. Comparing Raman with Brillouin in the mixed regime for very short pulses

In a previous publication [3] it was pointed out how sc-SBS and SRS behave differently with respect to the need of satisfying the resonant condition. In the case of sc-SBS one has the advantage that (a) the frequency spread of the seed pulse for a short pulse is sufficient to assure a coupling between the three waves involved and (b) there is a natural spread of the resonant frequencies in the sc-regime that relaxes the resonant condition. For this reason no downshift of the seed frequency with respect to the pump beam frequency is necessary as there is always sufficient overlap of the frequency distributions such that significant amplification occurs. This is not the case for most of SRS amplification regimes, where the mismatch condition needs to be fulfilled. However, if the seed pulse becomes very short and intense, i.e. $\tau_s/\tau_{sc} \ll 1$ [5], and the conditions of sc-SBS amplification are satisfied (i.e. no downshift between pump and seed), a large amplification of the seed pulse is found, which also shows a significant frequency spread and downshift from its initial value [3]. A comparison with immobile ions simulations shows that the amplification has its main contribution from electron motion, although ion motion is not negligible, and a pure ion amplification is developing behind the seed pulse [41, 42]. This regime is best characterized as a mixed-mode non-resonant regime (MMNR) [3]. To differentiate this regime from the SRS amplification regime a set of simulations was performed by considering 3 different values of the mismatch for the seed. In the first case the mismatch satisfies the SRS amplification condition, in the second case there is no mismatch and thirdly the seed is upshifted by ~20%, so that even for a short pulse sc-SBS cannot be excited (see cases e9, e8, e10 in table 1). For the case e9 the downshift follows from $\omega_1 = \omega_o(1 - \omega_{ep}/\omega_o)$ with $\omega_{ep}/\omega_o = \sqrt{n_e/n_c} \approx 0.22$ for $n_e/n_c = 0.05$. Here ω_{ep} is the electron plasma frequency. Exact resonance is not required due to the natural spread in frequency of the short pulse. As mentioned above, the plasma profile is homogeneous for this set of simulations. The results of the amplification as a function of time for the three cases as well as the initial spatial distribution of the B_z field for the seed are shown in figure 3. Figure 3(*b*) shows the maximum value of the seed intensity as it propagates inside the plasma which extends from $xk_o = 250$ to $xk_o = 3250$, and the seed propagates from right to left. The blue curve is the SRS-amplification case ($\omega/\omega_o = 0.8$, case e9), the red curve corresponds to the MMNR case ($\omega/\omega_o = 1$, case e8), and the black curve is the upshifted seed case ($\omega/\omega_o = 1.2$, case e10). As can be seen, the latter shows no amplification at all (nor self-focusing), but the maximum amplitude decreases and the results show some backscattering of the seed and no coupling with the pump. The SRS case instead starts to grow immediately but saturates around $I\lambda_o^2 = 7\times10^{17}$ Wμm^2 cm^{-2} and then oscillates around this value until it leaves the plasma. Some defocusing and refocusing is apparent at this stage. Finally in the case of the MMNR-regime the amplification process starts much more slowly than the SRS case, but grows to the maximum value of ~1.5×10^{18} Wμm^2 cm^{-2} comparable to amplification observed with other density profiles [3, 5]. Indeed, it is found that for the very short seed case where the pump/seed energy transfer is not optimized, the shape of the density profile is not crucial, provided that the plasma is short enough so that spontaneous losses are not dominant. The seed intensity for the MMNR case and for the SRS-amplification case at four different times as the seed propagates inside the plasma are shown in figures 3(*c*) and (*d*). SRS amplification starts immediately and is efficient from the beginning, however it quickly saturates in contrast to MMNR that keeps growing. As mentioned above, some focusing and defocusing of the pulse is clearly visible in figure 3(*d*), as well as development of wings.

The corresponding spatial distribution of the B_z field for the seed after it has exited the plasma is reproduced in figure 4. As one can see, the shape is very different in the MMNR regime and in the SRS case. In particular for the SRS case the seed has developed wings, and the phase front is not flat anymore. The situation is much better in the case of the MMNR regime, although the seed has not grown uniformly, so that it's transverse size is reduced to about half of the initial value. However, the phase front is still flat.

The fact that in the SRS case amplification starts earlier, and that the seed downshift allows the resonant condition for SRS to be satisfied, is confirmed by analysis of the phase space at the time when the seed has reached an intermediate position in space. This is illustrated in figure 5(*b*) which reproduces

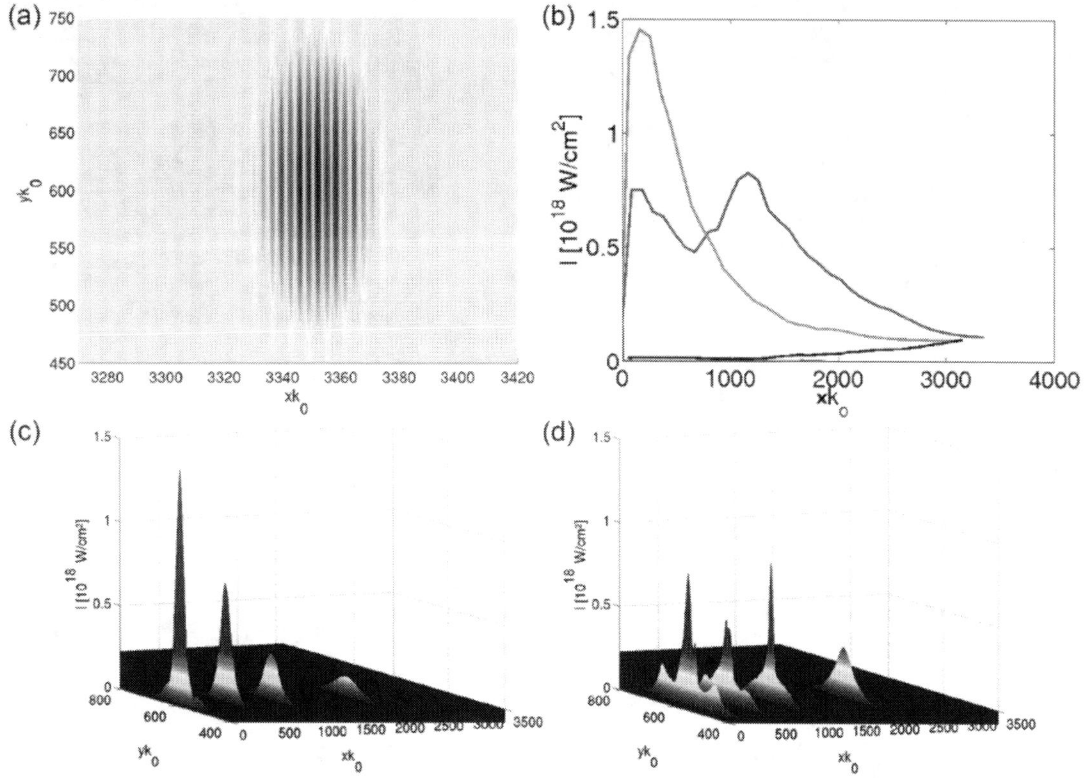

Figure 3. (*a*) Initial spatial distribution of the B_z-field for the Raman case. (*b*) Intensity growth curves: red: e8, blue: e9. The black curve corresponds to an upshifted seed pulse with $\omega_1 = 1.2\,\omega_0$, case e10. Plotted is the maximum intensity of the seed pulse as a function of its position in the plasma amplifier (see figure 2). The seed pulses are entering the simulation box from the right. (*c*) and (*d*) three-dimensional rendering of the seed pulse growth for the cases e8 and e9, respectively.

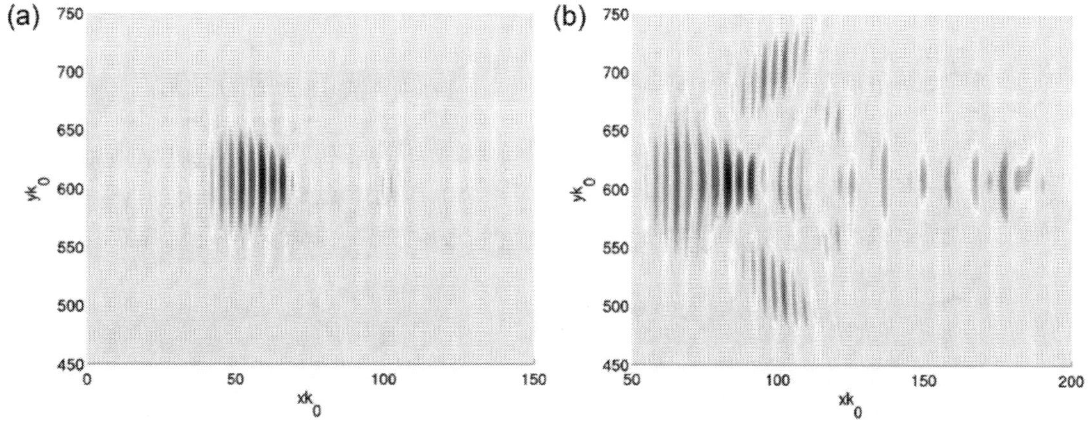

Figure 4. Spatial distribution of the B_z-field after amplification for the cases e8 (*a*) and e9 (*b*).

the normalized electron momentum parallel to the propagation of the laser (longitudinal). One can clearly see the effect of the seed/pump coupling resulting in large electron oscillations (and some wave breaking) between $xk_o = 1800$ (where the seed is at the time of the snapshot) and $xk_o = 2000$. In the front part, at smaller value of xk_o one can see some electron activity related to spontaneous thermal SRS from the pump alone. For comparison the same image for the MMNR case is shown in figure 5(*a*): although the electron activity is present as well in the seed location, it is weaker, while the spontaneous thermal SRS signal is the same. One finds the opposite behaviour in the ion phase space at the same time, shown in figure 6 around the seed location. No signature is visible in the SRS case(*b*), while for the MMNR(*a*) there is some ion activity. The same features are found at a later time, after the seed has crossed the whole plasma (figures 7 and 8): the electron activity is much stronger for the SRS case, while the ion activity is only present in the MMNR case.

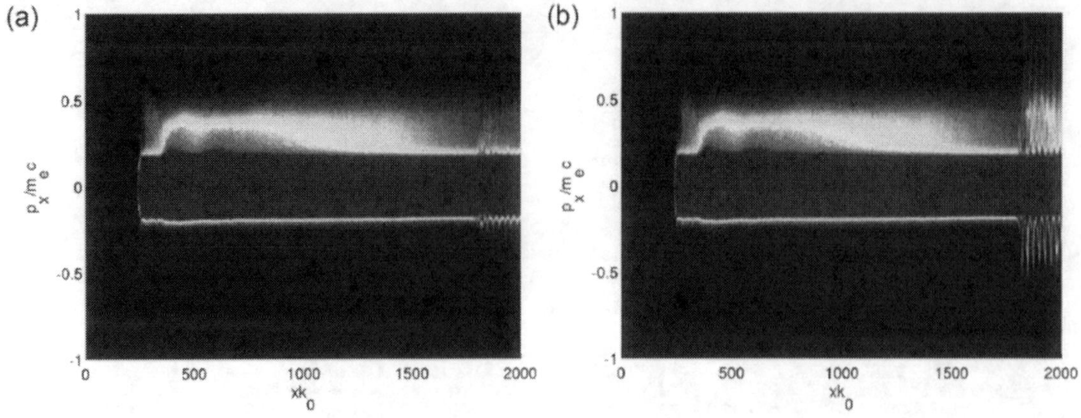

Figure 5. Electron phase-space activity at time $t\omega_o = 5600$ corresponding to the seed at $xk_o = 1800$ (see figure 4): (*a*) case e8 and (*b*) case e9. The central part of the phase space, corresponding to the thermal electrons, has been saturated in order to visually enhance the electron activity for large momentum values.

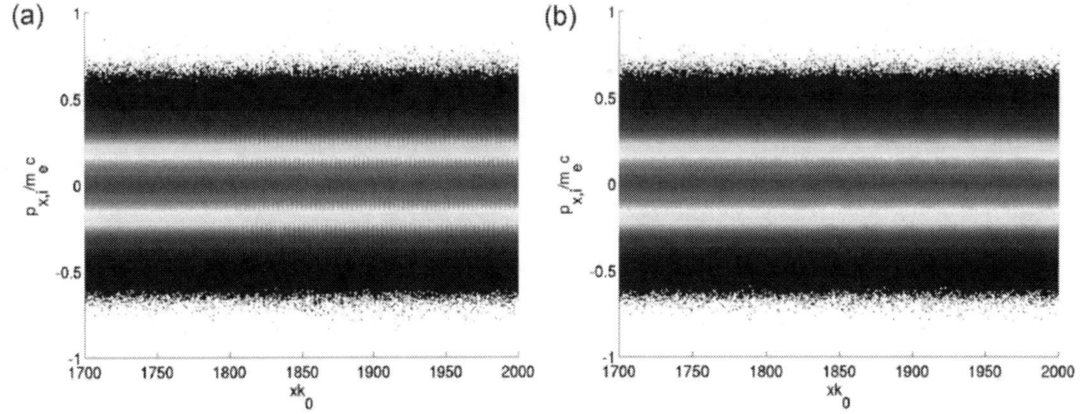

Figure 6. Ion phase-space activity at time $t\omega_o = 5600$ corresponding to the seed at $xk_0 = 1800$ (see figure 4): (*a*) case e8 and (*b*) case e9.

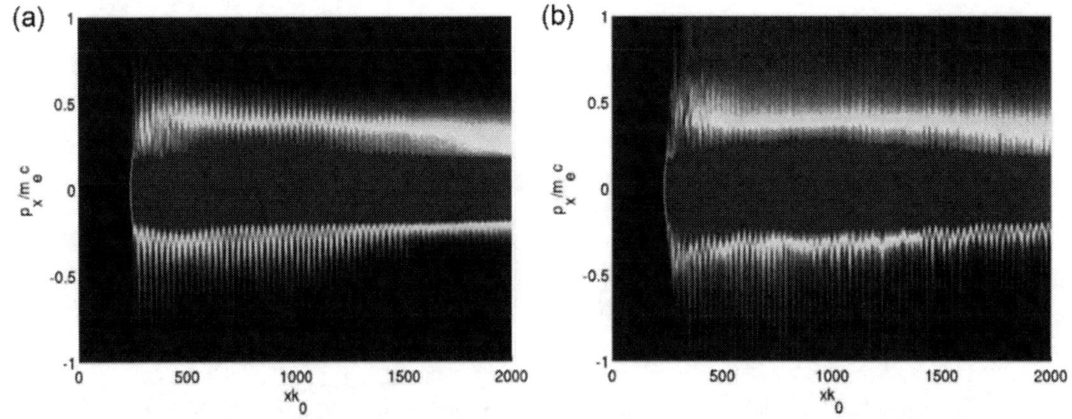

Figure 7. Electron phase-space activity at time $t\omega_o = 7200$ corresponding to the seed going out of the plasma (see figure 4): (*a*) case e8 and (*b*) case e9.

5. Analysis of wave front and spatial structure of amplified pulses

As discussed above it is important that the transverse size of the pulse is preserved, and that the wave front is plane. In order to satisfy these conditions, the pump has to be well behaved, and filamentation and thermal backward pump losses need to be minimized, so that the amplification happens smoothly over the whole pulse. In figures 9–10 the B_z component of the laser pulse resulting from amplification of two 80 fs seeds

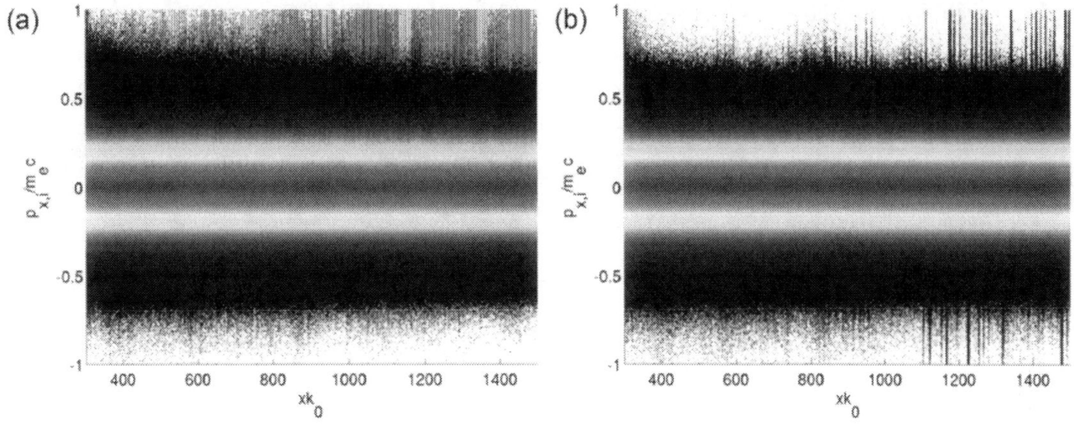

Figure 8. Ion phase-space activity at time $t\omega_0 = 7200$ corresponding to the seed going out of the plasma (see figure 4): (*a*) case e8 and (*b*) case e9.

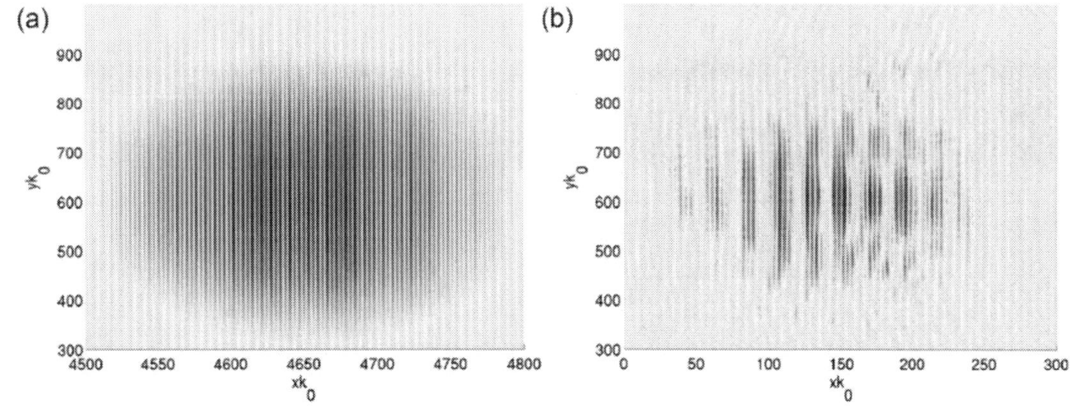

Figure 9. Spatial distribution of the B_z-field of the seed pulse before (*a*) and after (*b*) amplification for case e7 (see table 1). xk_0 is the direction of propagation and yk_0 is the perpendicular direction.

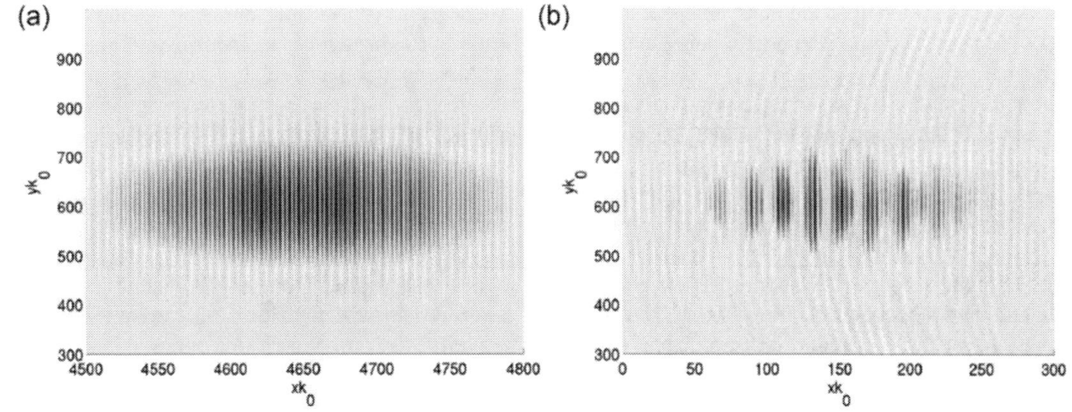

Figure 10. Spatial distribution of the B_z-field of the seed before (*a*) and after (*b*) amplification for case f7.

differing only in transverse size are reproduced, having FWHM of 32 and 64 μm, respectively. (Note: the transverse size of the B_z field is larger by a factor of $\sqrt{2}$ than the transverse side of the pulse intensity). As it can be seen for these cases, the wavefront and the transverse shape are preserved, although there is a reduction in the size of two thirds with respect to the initial values. If a shorter pulse is considered, as reproduced in figure 11 for the case of a 30 fs seed, the situation is somewhat different, since the central part of the seed grows to larger intensity, but the wings are not amplified as much. As a result there is an effective reduction of the transverse size from 32 μm to 16 μm. Slowing down the

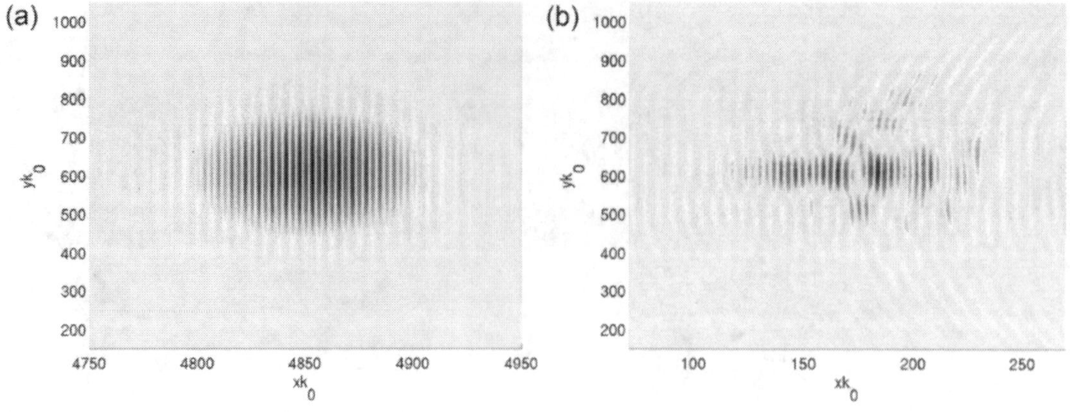

Figure 11. Spatial distribution of the B_z-field of the seed before (*a*) and after (*b*) amplification for case r4.

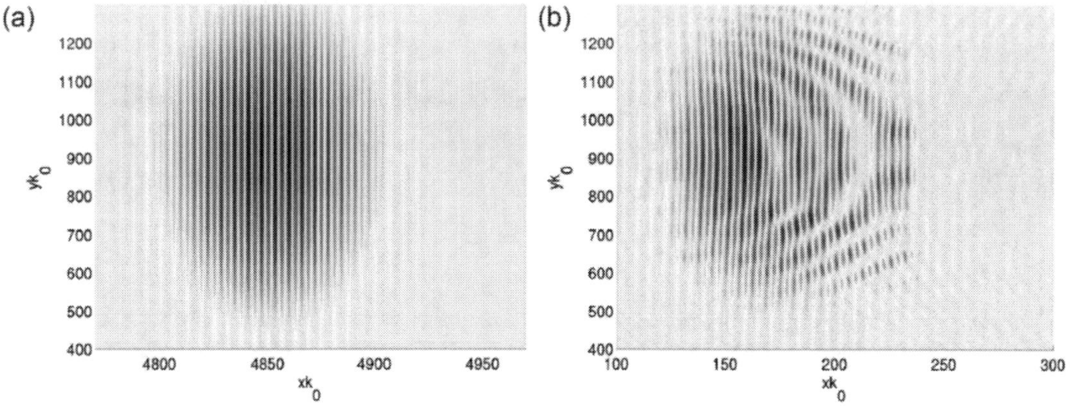

Figure 12. Spatial distribution of the B_z-field of the seed before (*a*) and after (*b*) amplification for case r5.

amplification process by considering a slightly less intense pump (5×10^{15} W cm^{-2}) results in a larger amplified seed. Figure 12 shows the amplification of a seed having an initial FWHM of 100 μm. Not only is the transverse size preserved, but the wavefront is plane and suitable for being focused by a plasma mirror.

Even if in many cases the amplification happens evenly in the transverse direction so that the size of the spot is roughly preserved, as the seed gets amplified and becomes relativistic, one cannot separate amplification from some self-focusing. A complex interplay of relativistic self-focusing induced from the mass increase of the electrons and the action of the ponderomotive force transferring energy and amplifying the seed sets up and requires further investigation. Ideally, once the relative contribution of amplification and self-focusing is clarified, one could envisage a set-up such that the amplification and the focusing stages both happen in a single pass. This will be a subject of further investigation.

6. Conclusion

Comparing simulations in the MMNR regime and in a strongly driven but otherwise standard SRS case shows that nonlinearities and kinetic effects acting rapidly in the SRS regime spoil the amplification process. SRS appears more suitable for amplification with low growth rates on very long plasmas [17] or in the superradiant regime [43]. At any rate, whatever the process, plasma amplification cannot be separated from subsequent plasma focusing. Combining the two in a single experiment remains one of the future experimental challenges. Experimental investigations concerning the physics of very short pulses of the order of a few tens of femtoseconds and large transverse beam size of several hundred micrometres are under way. The scheme of plasma amplification and subsequent focusing using plasma depends on the availability of moderately compressed high-energy lasers as a starting point. ELI-Beamlines [27] with its L4 laser system will find itself in a unique position for future experiments involving plasma optics.

Acknowledgments

Usage of the computing center CCRT of the *Commissariat à l'énergie atomique et aux énergies alternatives* is acknowledged. CR acknowledges support by grant ANR-11-IDEX-0004-02 Plas@Par. Support of SW by ELI-Beamlines (CZ.1.05/1.1.00/483/02.0061) and OPVK (CZ.1.07/2.3.00/20.0279) projects is acknowledged.

References

[1] Milroy R *et al* 1979 *Phys. Fluids* **22** 1922
[2] Andreev A *et al* 2006 *Phys. Plasmas* **13** 053110
[3] Riconda C *et al* 2013 *Phys. Plasmas* **20** 083115
[4] Riconda C *et al* 2013 *Proc. SPIE* **8779** 877921
[5] Weber S *et al* 2013 *Phys. Rev. Lett.* **111** 055004
[6] Frank A *et al* 2014 *Eur. Phys. J. Spec. Top.* **223** 1153
[7] Humphrey K *et al* 2013 *Phys. Plasmas* **20** 102114
[8] Lancia L *et al* 2009 *Proc. SPIE* **7359** 73590N
[9] Lancia L *et al* 2010 *Phys. Rev. Lett.* **104** 025001
[10] Lancia L *et al* 2014 *Phys. Rev. Lett.* submitted
[11] Forslund D *et al* 1975 *Phys. Fluids* **18** 1002
[12] Maier M *et al* 1966 *Phys. Rev. Lett.* **17** 1275
[13] Doumy G *et al* 2004 *Phys. Rev.* E **69** 026402
[14] Fuchs J *et al* 2014 *Eur. Phys. J. Spec. Top.* **223** 1169
[15] Mourou G *et al* 2014 *Eur. Phys. J. Spec. Top.* **223** 979
[16] Quinn M *et al* 2014 *Eur. Phys. J. Spec. Top.* **223** 985
[17] Trines R *et al* 2010 *Nature Phys.* **7** 87
[18] Trines R. *et al* 2011 *Phys. Rev. Lett.* **107** 105002
[19] Thaury C *et al* 2007 *Nature Phys.* **3** 424
[20] Kon A *et al* 2010 *J. Phys.: Conf. Ser.* **244** 032008
[21] Nakatsutsumi M *et al* 2010 *Opt. Lett.* **35** 2314
[22] Wittmann T *et al* 2009 *Rev. Sci. Instrum.* **77** 083109
[23] Stavroudis O *et al* 1992 *J. Opt. Soc. Am.* A **9** 2083
[24] Vincenti H *et al* 2014 *Nature Commun.* **5** 3403
[25] Gonoskov A *et al* 2011 *Phys. Rev.* E **84** 046403
[26] Streeter M *et al* 2011 *New J. Phys.* **13** 023041
[27] Extreme Light Infrastructure: www.eli-beams.eu
[28] APOLLON project: http://cilexsaclay.fr
[29] Fang F *et al* 2006 *Advanced Accelerator Concepts: 12th Workshop (Lake Geneva, Wisconsin, USA, 10–15 July 2006)* vol 877, p 721
[30] Chen X *et al* 1993 *Phys. Rev. Lett.* **70** 2082
[31] Chessa P *et al* 1998 *Phys. Plasmas* **5** 3451
[32] Hubbard R *et al* 2002 *Phys. Plasmas* **9** 1431
[33] Kurki-Suonio T *et al* 1989 *Phys. Rev.* A **40** 3230
[34] Tzeng K *et al* 1996 *Phys. Rev. Lett.* **76** 3332
[35] Wang Y *et al* 2012 *Phys. Plasmas* **19** 103109
[36] Bin J 2014 arXiv:1402.4301
[37] Wang H *et al* 2011 *Phys. Rev. Lett.* **107** 265002
[38] Max C *et al* 1974 *Phys. Rev. Lett.* **33** 209
[39] Shorokov O *et al* 2003 *Phys. Rev. Lett.* **91** 265002
[40] Code developed by Heron A and Adam J-C at CPHT, Ecole Polytechnique
[41] Lehmann G *et al* 2012 *Phys. Plasmas* **19** 093120
[42] Lehmann G *et al* 2013 *Phys. Plasmas* **20** 073112
[43] Ersfeld B *et al* 2005 *Phys. Rev. Lett.* **95** 165002

IOP Publishing Plasma Physics and Controlled Fusion

Plasma Phys. Control. Fusion **57** (2015) 014003 (5pp) doi:10.1088/0741-3335/57/1/014003

Studying planetary matter using intense x-ray pulses

K Appel[1], M Nakatsutsumi[1], A Pelka[2], G Priebe[1], I Thorpe[1] and Th Tschentscher[1]

[1] European XFEL, Hamburg, Germany
[2] Helmholtz-Zentrum Dresden Rossendorf, Dresden, Germany

E-mail: karen.appel@xfel.eu

Received 27 June 2014, revised 5 August 2014
Accepted for publication 22 August 2014
Published 28 November 2014

Abstract

Free-electron laser facilities enable new applications in the field of high-pressure research including planetary materials. The European x-ray Free Electron Laser (European XFEL) in Hamburg, Germany will start user operation in 2017 and will provide photon energies of up to 25 keV. The high-energy density science instrument (HED) is one of the six baseline instruments at the European XFEL. It is dedicated to the study of dense material at strong excitation in a temperature range from eV to keV and pressures >100 GPa which is equivalent to an energy density >100 J mm^{-3}. It will enable studying structural and electronic properties of excited states with hard x-rays. The instrument is currently in its technical design phase and first user experiments are foreseen for summer 2017. In this contribution, we present the x-ray instrumentation and foreseen x-ray techniques at HED and concentrate on prototype hard-condensed matter experiments in the field of planetary research as proposed during recent user consortium meetings for this instrument. These include quasi-isentropic (ramped) compression and shock compression experiments.

Keywords: European XFEL, shock and ramp compression, high-energy optical laser

(Some figures may appear in colour only in the online journal)

1. Introduction

Free-electron laser facilities such as FLASH [1], FERMI [2], LCLS [3], SACLA [4] allow the study of samples with high brightness, pulsed x-rays enabling new applications probing rapid temporal responses during, for example, shock-induced material changes. In addition, the FEL beam could be used as a pump or the FEL beam could be combined with high-energy optical laser (OL) systems for pump–probe experiments.

HED [5] at the European XFEL in Hamburg [6] will be a new, unique platform for experiments combining hard x-ray FEL radiation with the capability to put matter under extreme conditions of pressure, temperature or electric and magnetic field. Scientific applications will include the study of matter occurring inside exo-planets, new extreme-pressure phases and solid-density plasmas and structural phase transitions of complex solids in high magnetic fields. Besides the applications in planetary physics, HED will enable studying relativistic laser–matter processes and warm dense matter.

2. Studying matter at extreme conditions

The study of structure and composition of planetary materials with experimental methods poses enormous technical and experimental challenges. Diamond anvil cell (DAC) techniques have been used for decades in order to study for example elements at conditions present in the interior of the Earth. Until recently, they were limited to maximum pressures of 400 GPa which covers conditions down to the Earth's core. The introduction of double stage DACs considerably extended the upper pressure limit to 800 GPa [7]. Combined with pulsed-laser heating [8], new regimes of extreme conditions inside exo-plantes can be experimentally realized and probed with a suitable x-ray source. Another promising development are dynamic DACs which allow realizing pressure change rates of up to 500 GPa s^{-1} for the pressure range of conventional DACs [9]. Dynamic compression of matter—either by using dynamic DACs or lasers—will allow to access states in the P–T phase diagram not accessible to static high-pressure methods.

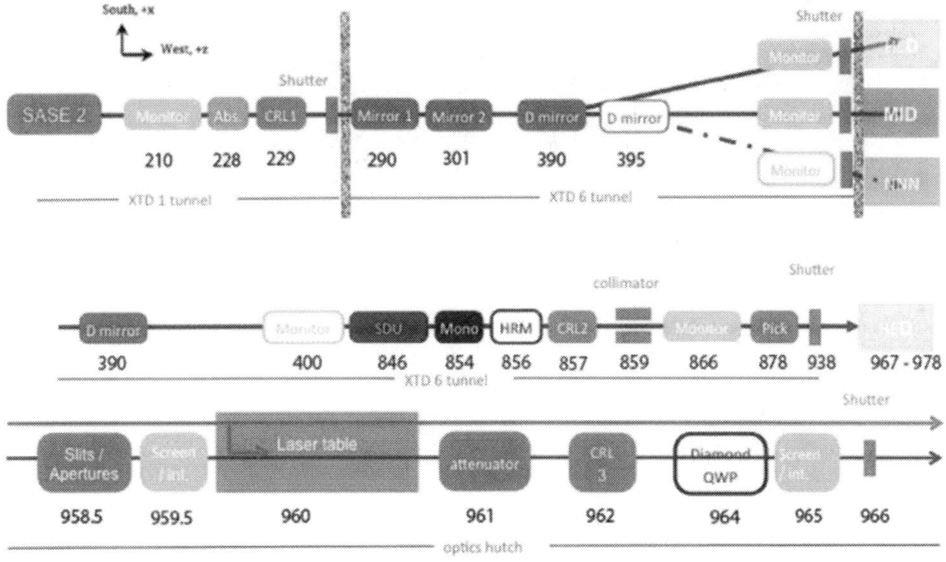

Figure 1. Overview of the x-ray beam transport of the HED instrument. The numbers are distances in metres from the source (not to scale).

Using high-energy OL pulses and depending on the temporal pulse shape one can compress matter using shocks or ramped pulses to very high internal pressures. With energies of 100 J, pressures of up to 1 TPa are feasible for ramp compression experiments with moderate laser beam sizes (100 μm) [10]. This will enable to realize conditions of the interior of Saturn, Neptune, Uranus, Earth, Venus [11]. At the HED instrument, DAC and laser-induced ramp and shock experiments will be applicable and samples can be probed with the ultra-fast x-ray FEL beam at a high-repetition rate.

3. High-energy density science instrument overview

The HED instrument will be located in a side branch of the x-ray beam transport at the SASE2 undulator of the European XFEL. Photon energies in the range of 3–25 keV can be covered in self-amplified spontaneous emission mode (SASE). The accelerator delivers electron bunch trains with up to 2700 bunches in 600 μs with a charge of 0.02–1 nC and operates at 10 Hz. This leads to a maximum repetition rate of 4.5 MHz for the photon pulses. The expected x-ray FEL radiation properties have been simulated in 2011 [12] and show that a pulse duration of 107–0.68 fs can be expected for 5–25 keV. For the maximum bunch charge of 1 nC, 2×10^{12}–4.5×10^{12} photons can be expected in one FEL pulse.

The HED x-ray beam transport at the SASE2 undulator is about 1 km long and is composed of three major sections where x-ray optical components and monitors are installed (figure 1). The first section starts at the separation of electron and photon beams and ends with the SASE2 beamline front end (XTD1: 150–290 m). This section hosts general components, to run and tune the FEL beam, as well as the first focusing unit (CRL1), shared by HED and the two other SASE2 instruments, Materials Imaging and Dynamics (MID) and an optional third instrument (NNN). In the second section (XTD6: 290–930 m), the beam transport is split into three in order to steer the three beams to the three instruments in the experiment hall (XHEXP). This allows to share the SASE2 beam among three dedicated instruments. Besides general components, HED specific optics like the split and delay unit and the monochromators are placed here. The third section of the HED beam transport starts in XHEXP and ends with the beam stop in the experiment hutch (HED-EXP). This section includes the optics hutch of HED.

Three OL systems will be available at the HED instrument for driving samples into extreme states. A 100 TW class ultrahigh-intensity OL does reach focused intensities far beyond 10^{18} W cm^{-2} allowing the study of relativistic laser–matter interactions. A pump–probe OL provides high-repetition rates thereby matching the x-ray delivery pattern with up to 100 mJ pulse energies. For the studies of planetary matter, a 100 J class high-energy OL is available. It has a pulse duration of a few to 20 ns with temporal shaping capability and thus enables shock and shock-less compression experiments. The expected repetition rate is >1 Hz.

It is planned to have a 1 MJ capacitor bank plus split coils available to allow for magnetic scattering experiments in pulsed magnetic fields of up to 50 T. Details on the experimental set-up are still under discussion as well as the position of the capacitor which might be placed at some distance from the experimental area.

Using the radiation from the x-ray free electron laser (FEL), it will be possible to study samples by x-ray diffraction, inelastic x-ray scattering, x-ray spectroscopy and imaging techniques. The main energy range is 5–25 keV, energies down to 3 keV can be reached within certain limitations. Due to the high excitation energies of up to 25 keV, diffraction patterns will have excellent quality and a wide range of elements ($Z > 18$) will be accessible by x-ray absorption spectroscopy. In addition, the high brilliance and coherence of the FEL radiation will enable spatially resolved studies.

Table 1. Specification requirements for the DiPOLE high-energy OL system for the HED instrument.

Laser energy	100 J at $\lambda = 1030$ nm (minimum), up to 100 J at $\lambda = 515$ nm (second harmonic), with an energy stability of <1%
Beam size	75×75 mm^2 after the final amplifier
Wavelength λ	Standard operation mode is $\lambda = 515$ nm (second harmonic) on sample, using LBO crystals
Repetition	>1 Hz (minimum), 10 Hz (goal)
Temporal structure	2–20 nm duration at square shape. Temporal pulse-shaping capability for ramped compression experiments (t^3 shape) but probably with reduced energy

4. X-ray beam transport

Three different mirrors will be installed at the HED beamline [13]. The first pair of mirrors are horizontal offset mirrors (mirrors 1 and 2; figure 1) which geometrically displace the photon beam from the electron beam. 100 m downstream of the second offset mirror, a distribution mirror (mirror 3; figure 1) guides the beam to the HED instrument. The focusing concept of the HED instrument is based on the use of compound refractive lenses (CRL1-3). The fundamental element of the CRL system is the beryllium (Be) lens [14]. The lenses are chromatic due to the large changes of the Be refractive index with energy. Multiple lenses will be stacked to achieve the desired focal length at the energy of interest. A relatively large variation of focus sizes is required for HED at the sample locations, classified into large (200–50 μm), medium (20–10 μm), and tight focusing (2–1 μm). As XFEL is a coherent source, the achievable minimum spot sizes are not limited by demagnification of the source but by diffraction. To reach the envisaged spot sizes at the sample position, three different lens positions were proposed in the HED CDR [5]. The current concept provides coincidence of x-ray focus and sample position for every third keV in the energy range of 5–25 keV.

The bandwidth of the natural SASE2 radiation is 10^{-3} ($\Delta E/E$). For applications in the energy range of 5–25 keV, a channel cut Si_{111} monochromator is foreseen, which provides an energy resolution of 10^{-4} [15]. This is essential for x-ray absorption studies and can be crucial for x-ray diffraction experiments. Some experiments, like inelastic x-ray scattering, demand for a still better energy resolution in the range of $<10^{-6}$. To enable such experiments, a high-resolution monochromator will be installed. A delay unit is planned for the HED instrument for the energy range 5–20 keV. This unit will enable time-resolved x-ray pump—x-ray probe experiments as well as sequential diffractive imaging on a fs to ps time scale [16]. Other important devices to realize experimental requirements are slits, attenuators and a pulse picker. The x-ray slits, located inside the HED-OPT hutch, will serve to define the beam size and to block the beam halo propagating with the main pulse. The attenuator is used to attenuate the x-ray beam in a controlled fashion as required for the experiment, for example, if the full intensity would be harmful to samples or instrumentation, or would at least create thermal distortions. The requirement for attenuation can be as high as eight orders of magnitude. A pulse picker is planned at SASE2 to allow operation at reduced pulse modes. At present, selection of pulses at rates of up to 10 Hz is envisaged. In the future, higher rates may serve the down-selection of pulses inside an x-ray pulse train.

5. High-energy long pulse laser

A 100 J class long pulse laser is planned for the HED instrument. It will be contributed by the Central Laser Facility (CLF) of STFC Rutherford Appleton Laboratory in Didcot, UK in the framework of the Helmholtz International Beamline for Extreme Fields User Consortium (HIBEF UC). This laser will be a diode-pumped solid-state laser (DPSSL) system. The DPSSL scheme uses advanced gain media and cooling schemes to overcome the limitations of flashlamp-pumped Nd : glass technology and is expected to provide reliable operation of a high-energy laser at repetition rates of up to 10 Hz (table 1). A DiPOLE (Diode Pumped Optical Laser for Experiments) prototype operates with 10 J pulse energy at 10 Hz repetition rate.

The required spot size on sample varies from 20 to 500 μm with uniform top-hat intensity distribution. Such spatial profiles can be achieved by using an appropriate phase plate. For focusing, $f/4$ or larger optics will be employed. As seen in figure 3, the high-energy OL beam can be split into two beams with an additional delay line for one arm. This capability is required in particular for two-side irradiation experiments. The arrangements of mirrors and focusing lenses inside the experiment chamber are summarized in figure 2. A high flexibility of incidence angles ($-20°$–$200°$) with respect to the x-ray beam will be provided. One specific requirement is the collinear geometry when using a DAC due to its small opening angle. Absolute 0° incidence is provided using a mirror and lens with a clearance hole at the centre.

For optical diagnostics of the laser-induced compression, a velocity interferometer system for any reflector (VISAR) will be provided at the instrument. The VISAR system measures the surface motion [17] and allows the determination of the shock breakout timing and shock velocity. It requires 10–50 ns duration and temporally coherent probe laser pulses. The required laser energy is 1–10 mJ.

6. X-ray techniques relevant to study planetary matter

For DAC experiments, XRD and x-ray spectroscopy will be used to study the phases in the cell. Due to the small sample volume in DACs, the x-ray beam should have a diameter of

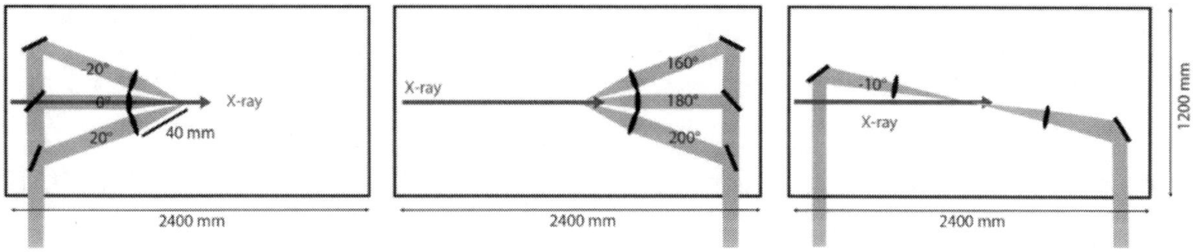

Figure 2. Some possible arrangements of mirrors and focusing lenses for the high-energy OL inside the experiment chamber. Simultaneous use of two split laser beams with adjustable delay is also provided (right graph).

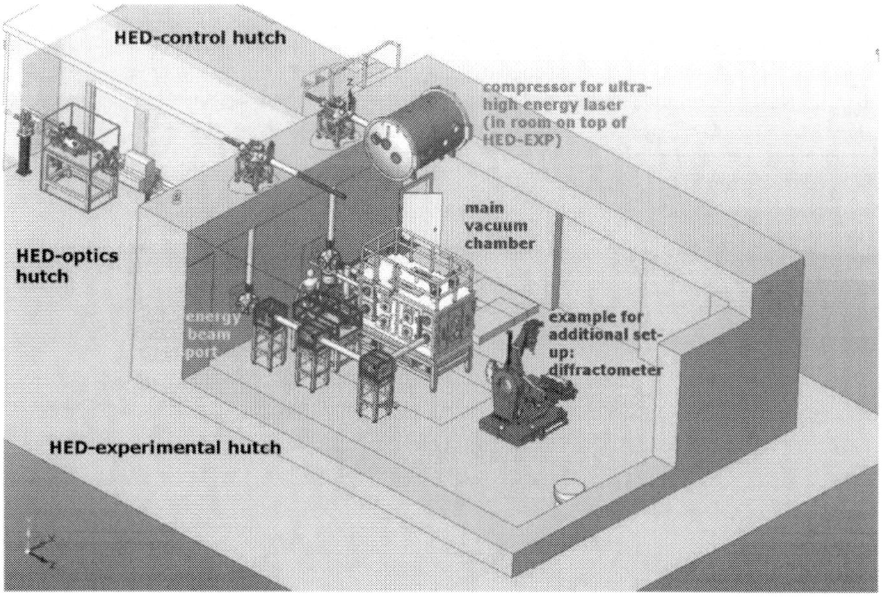

Figure 3. Birds eye view of HED-optics hutch (left) and HED experimental room (right) with the main vacuum chamber and the optical laser beam paths. The main vacuum chamber in 2.4 m long and 1.2 m wide. The diffractometer placed downstream can be replaced with other dedicated set-ups like for example the DAC set-up.

around 1 μm in order to be smaller than the sample. Excitation energies should be as high as 25 keV, in order to minimize absorption effects of the incoming beam and the outgoing signal in the diamond anvils. Detectors for XRD should be placed as close as possible to the DAC in order to cover a large two-theta angle and scope with the limited opening angle of the DACs. This will enable to record as many reflections as possible which is required to identify phases. X-ray spectroscopy will be performed in the energy-dispersive mode. To provide a high-energy resolution, smallest pixel sizes are required (25–50 μm) for the detector. In case that dynamic or pulsed-laser heating experiments are to be performed in the DAC, high read out repetition rates are a prerequisite for the detectors.

Laser-induced compression experiments will use all x-ray methods as listed above. XRD and x-ray spectroscopy set-ups will in principle have the same demands for detectors as for DAC experiments with the only difference that beam size and opening angle are now much larger. The set-up for small-angle x-ray scattering (SAXS) and phase contrast imaging (PCI) will be very similar to each other, with a detector of high dynamic range ($>10^4$) placed in the far field, several metre downstream of the interaction point. Especially for PCI, the size of detectors should be sufficiently large to image the complete sample.

Due to the expected high number of photons at the European XFEL, detectors will have to deal with instantaneous many photon deposition. This is different to conditions at modern synchrotron facilities, where detectors are generally operated in single photon counting modus. An equal data quality can be provided with charge integrating detectors with automatic gain switching capability. Several detectors with this technology are being developed at the moment: the Adaptive Gain Integrating Pixel Detector (AGIPD) is a 2D detector, which can be integrated in tiles of 128 × 512 pixels (pixel size 200 μm). It can be operated at the repetition rate of the European XFEL pulse trains and can store up to 350 frames for each pulse train. The Gain Optimizing microsTrip sysTem witH Analog ReaDout (GOTTHARD) detector is a 1D detector with a smaller pixel size (pixel size 50 μm) which can also run at 4.5 MHz. Both detectors are planned for HED for XRD and x-ray spectroscopy, respectively. Additional foreseen CCD detectors for HED run at a repetition rate which is matched to the high-energy OL i.e. 10 Hz maximum. The choice of detectors for HED is not yet finalized.

7. Experimental area

The inner size of the experimental hutch of HED is 11.4 m × 9 m with a height of 4 m. It hosts the main vacuum chamber (figure 3), which includes the interaction point between XFEL and OLs. Alternative, exchangeable set-ups either dedicated for DAC experiments or magnetic scattering are placed downstream. The delay lines for the lasers are positioned in the HED-EXP as well as the VISAR system.

8. Outlook

The HED instrument passed a technical design review in march this year and has now entered the detailed design phase. A technical design report of the instrument has been published [18]. The installation for the high-energy OL is planned to start within the second half of 2016. Commissioning is planned for mid-2017. The HED instrument is expected to be ready for user operation in late 2017.

References

[1] Ayvazyan *et al* 2006 *Eur. Phys. J.* D **37** 297
[2] www.elettra.trieste.it/FERMI/index.php?n=Main.CDRdocument
[3] SLAC National Accelerator Laboratory, Linac Coherent Light Source, LCLS (https://slacportal.slac.stanford.edu/sites/lclspublic/Pages/Default.aspx)
[4] Ishikawa T *et al* 2012 *Nature Photon.* **6** 540
[5] Nakatsutsumi M and Tschentscher Th 2013 XFEL.EU TR-2013-003 (www.xfel.eu/documents/technical_documents/)
[6] Altarelli *et al* 2007 *XFEL.EU Technical Report* 1–630
[7] Dubrovinsky L *et al* 2012 *Nature Commun.* **3** 1163
[8] Goncharov A F *et al* 2012 *J. Appl. Phys.* **111** 112609
[9] Evans W J *et al* 2007 *Rev. Sci. Instrum.* **78** 073904
[10] Fratanduono D E *et al* 2011 *J. Appl. Phys.* **110** 073110
[11] Swift D *et al* 2012 *Astrophys. J.* **744** 59
[12] Schneidmiller E A and Yurkov M V 2011 XFEL.EU TR-2011-006 (www.xfel.eu/documents/technical_documents/)
[13] Sinn H, Gaudin J, Samoylova L, Trapp A and Galasso G 2011 XFEL.EU TR-2011-001 (www.xfel.eu/documents/technical_documents/)
[14] Lengeler B *et al* 1999 *Appl. Phys. Lett.* **26** 3924
[15] Sinn H *et al* 2012 XFEL.EU TR-2012-006 (www.xfel.eu/documents/technical_documents/)
[16] Roling S *et al* 2012 *Proc. SPIE* **8504** 850407
[17] Celliers P M *et al* 2004 *Rev. Sci. Instrum.* **75** 4916
[18] Nakatsutsumi M, Appel K, Priebe G, Thorpe I, Pelka A, Muller B and Tschentscher Th 2014 TR-2014-004 (www.xfel.eu/documents/technical_documents/)

Plasma Phys. Control. Fusion 57 (2015) 014004 (11pp)
doi:10.1088/0741-3335/57/1/014004

Physics in the magnetic configuration space of W7-X

J Geiger, C D Beidler, Y Feng, H Maaßberg, N B Marushchenko and Y Turkin

Max-Planck Institut für Plasmaphysik, D-17491 Greifswald, Germany

E-mail: Joachim.Geiger@ipp.mpg.de

Received 21 June 2014, revised 20 August 2014
Accepted for publication 29 August 2014
Published 28 November 2014

Abstract

The neoclassical confinement and the bootstrap current are analysed in the configuration space of W7-X by self-consistent neoclassical transport simulations. Since the establishment of quasi-stationary operation is the most important goal for W7-X, the analysis concentrates on high-performance discharge scenarios in magnetic configurations which are adjusted so that bootstrap current vanishes, or, alternatively, on scenarios where the bootstrap current can be balanced by strong ECCD. Both scenarios lead to restrictions either in the configuration space or in plasma parameters and ECRH heating scenarios. Furthermore, the flexibility of the magnetic configuration space of W7-X is briefly described with emphasis on other physics topics of interest, for example, ballooning unstable configurations as well as configurations with a magnetic hill which might lead to interchange instability.

Keywords: stellarator, magnetic configuration, neoclassical transport, bootstrap current, optimization

(Some figures may appear in colour only in the online journal)

1. Introduction

The tokamak perspective for a reactor with stationary operation appears to be somewhat endangered, since a huge amount of recirculating power is necessary for ECCD (electron cyclotron current drive) and/or n-NBCD (negative neutral beam current drive). Furthermore, the appearance of disruptions and the (Greenwald) density limit restrict advanced tokamak discharge scenarios. All these problems, however, do not exist in stellarators (where the plasma has much less free energy than in tokamaks). On the other hand, an optimization of advanced stellarator configurations is mandatory: the unfavourable neoclassical transport at low collisionalities must be significantly reduced and the bootstrap current must be minimized, at least for the proper operation of an island divertor utilizing low-order rationals of the rotational transform, ι (forming the magnetic islands within the divertor). In principle, current drive schemes can also be applied for the current control, but this goal can be obtained by the optimization process of the magnetic configurations.

The Wendelstein 7-X stellarator (W7-X) has been optimized with respect to good MHD stability, improved neoclassical confinement, good confinement of fast particles, and low bootstrap current; for a brief overview of the optimization process, see [1, 2] and the references therein. W7-X belongs to the HELIAS line (HELIcal axis Advanced Stellarator) [3] where the 3D shaping of the magnetic axis allows for additional freedom for different optimization targets. (Also the name 'high-mirror advanced stellarator' was introduced for this type of magnetic configuration with good neoclassical confinement and low bootstrap current [4]). This paper concentrates on the scenario development with respect to good neoclassical confinement and low bootstrap current for quasi-stationary discharges (pulse duration up to 30 min) with high performance. Establishing such discharges is the main mission of W7-X to demonstrate the reactor perspective of highly optimized stellarators with an island divertor [5].

It must be mentioned that open issues remain which might prevent any realistic reactor perspective of W7-X-type stellarators: (i) the problem of impurity accumulation, and (ii) a sufficient density control (even density profile control). For high-performance discharges, the radial electric field, E_r, evaluated from the ambipolarity condition is typically negative leading to a strong inward convection of impurities

($\propto Z_I E_r/T_i$ where Z_I is the impurity ion charge state and T_i the ion temperature) in traditional neoclassical theory. Second, since a link exists between radial energy and particle fluxes, a corresponding link exists between power and particle sources (at least in neoclassical theory, here used in the transport simulations) which requires central refuelling by pellets. Furthermore, a neoclassical particle transport barrier might appear outside of the power deposition zone [6] for 'high-mirror advanced stellarators'. This topic has not been included in the W7-X optimization. W7-X must find solutions for both of these crucial issues.

The outcome of the W7-X optimization process focussed on one point in the magnetic configuration space (with high mirror and an edge $\iota_a = 1$; $N_p = 5$ field periods resulting in five islands in the divertor) produced by a set of non-planar coils of five different types with identical coil currents; this 'original' configuration is included in the configurational sensitivity analysis of section 5. The desire for experimental flexibility, i.e. the realization of both low- and high-mirror configurations as well as $\iota_a = 5/4$ and $\iota_a = 5/6$ divertor scenarios, resulted in a new coil design (with about half the original mirror size for identical currents in the modular coils) with additional planar coils. This coil system and the flexibility of the magnetic configuration space of W7-X is briefly described in section 2 as such a documentation is lacking to date. An equivalent configuration rather close to the 'original' one can nevertheless be realized with this 'new' W7-X coil system.

In this paper, quasi-stationary discharge scenarios are calculated by means of an iteration loop (up to sufficient convergence): equilibrium (by the VMEC code [7]), neoclassical transport coefficient database (by the DKES code [8]), and neoclassical transport simulation (by the NTSS code [9]); section 3. Here, 'anomalous' transport contributions are omitted for the bulk part of the plasma with sufficiently high temperatures. The reason is simply that no reliable theory-based 'anomalous' transport model yet exists for stellarators, and that the assumption of neoclassical transport in the bulk part of the plasma is supported by experimental power balance analysis; see e.g. [10–12]. Furthermore, the energy confinement time, τ_E, calculated with the assumption of neoclassical transport is in accordance [9, 13] with the ISS04 scaling [14]. Ignoring (feasible) turbulent contributions to the radial transport (it is very unlikely that such contributions affect the parallel transport), leads in all likelihood to upper limits for the energy confinement time, τ_E, and the volume averaged plasma pressure, $\langle\beta\rangle$. Decreased temperatures, e.g. due to additional turbulent transport, affect the bootstrap current densities (i) through influence of the equilibrium on the magnetic configuration (reduced $\beta(r)$), (ii) by the collisionality and radial electric field (see [15] for the mono-energetic bootstrap current coefficient), and (iii) by the energy convolution based on Maxwellians. All these dependencies are rather complex, but the total bootstrap current generally decreases with reduced temperatures.

Two discharge scenarios are analysed in section 4: first, 'conservative' scenarios without electron cyclotron current drive (ECCD) where the magnetic configuration of W7-X is adjusted to minimize the bootstrap current, $I_b \simeq 0$, and, second, scenarios with strong ECCD with $I_b + I_{ECCD} \simeq 0$. High-performance discharges, i.e. at high density with ECRH (electron cyclotron resonance heating) in O2-mode [16], require these 'conservative' scenarios since the ECCD efficiency is much too low to balance an appreciable bootstrap current. Strong ECCD is obtained only for X2-mode at intermediate or low densities, and the density, the ECCD power and the power deposition must be adjusted to achieve current balance. Since in W7-X only ECRH with up to 10 MW at 140 GHz (corresponding to $B \simeq 2.5$ T at second harmonic) is available for long pulses, neutral beam injection (NBI) scenarios are disregarded in this context (the NBI pulse duration is less than 10 s).

Finally, a configurational sensitivity analysis for the high-mirror configuration at $\iota_a = 5/5$ is performed (in section 5) showing the impact of (small) modifications of the currents in the W7-X coil system on the neoclassical energy confinement as well as on the bootstrap current.

2. Magnetic configuration space

The large variety of W7-X configurations is obtained by the very flexible coil current system. The classification of the different magnetic configurations, however, is based on the Fourier representation B_{mn} (e.g. in Boozer co-ordinates) of the magnetic field strength, B, on flux-surfaces,

$$B(r,\theta,\zeta)/B_0 = \sum_{m,n=0} b_{mn}(r)\cos(m\theta - nN_p\zeta)$$

where r is the flux-surface label identified with the minor radius, θ and ζ the poloidal and toroidal angle-like co-ordinates, respectively, $B_{mn} = B_0 \cdot b_{mn}$, and B_0 is a reference magnetic field (e.g. B_{00} at the outermost flux-surface).

2.1. The coil system

W7-X has a major radius of $R \simeq 5.5$ m, an aspect ratio of $A = R/r_a \simeq 10$ (r_a is the plasma radius), and $N_p = 5$ field periods. In each half field period, the coil system is composed of 5 modular and 2 planar (tilted) superconducting coils; see figure 1. The additional control coil (also called sweep coil) is also shown. This coil will only be used for fine-tuning of the island structure in the divertor and has nearly no impact on the magnetic configurations analysed here (and is therefore omitted in this context). The identical coils are arranged to insure stellarator symmetry in the other half of the field period. Finally, one trim coil per field period exists which is located at the outside of the cryostat vessel. These coils might be used for control of small error fields in the divertor (e.g. for suppressing a 1/1-island forming an envelope around the five 5/5-islands). Two poloidal cross sections of the flux surfaces are up-down symmetric: the 'bean-shaped' cross section between coils of type 1 and the 'triangular' one between the coils 5; see figure 2.

All coils of a half field period have independent power supplies resulting in a great variety of magnetic configurations. The W7-X 'standard' configuration is technically defined: the current ratios $i_n = I_n/I_1$ in all modular coils, i_1 to i_5, are identical, and there is no current in the planar coils,

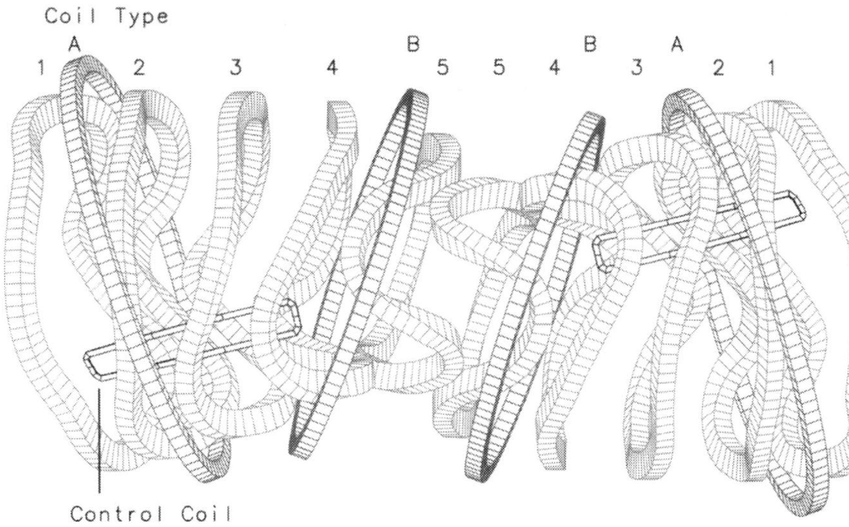

Figure 1. One field period of the W7-X coil system with modular (1 to 5) and planar (A and B) coil types.

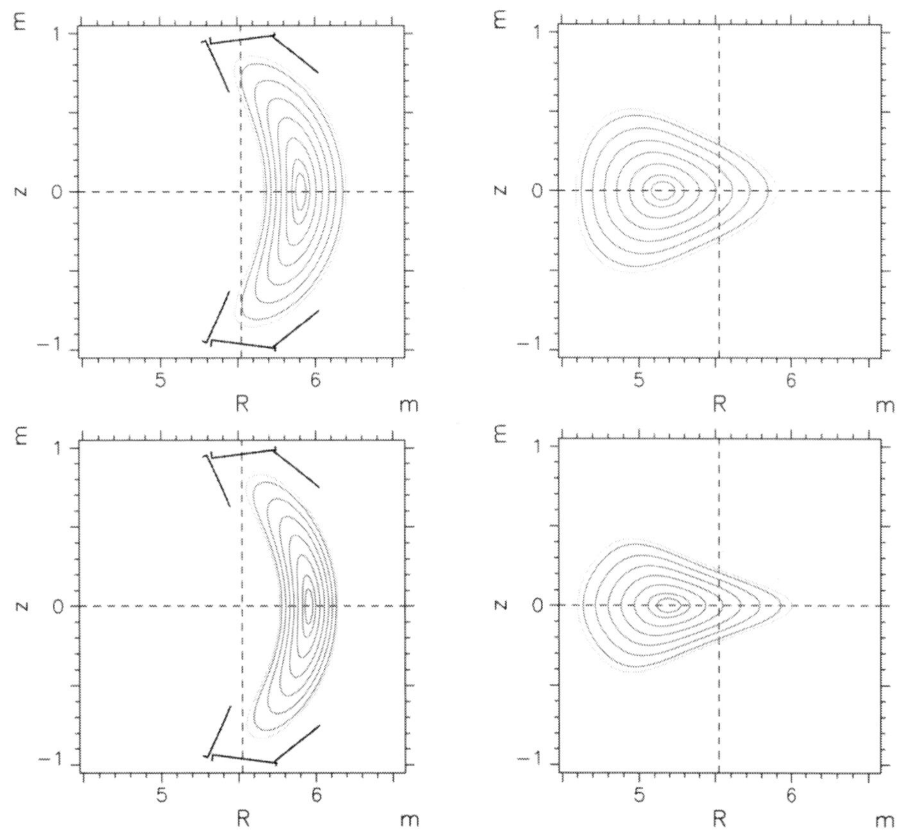

Figure 2. The cross section in the 'bean-shaped' (left) and in the 'triangular' plane (right) for the 'high-mirror' configuration at low-ι ($\iota_a = 5/6$, upper row) and at high-ι ($\iota_a = 5/4$, lower row). The current distribution in the modular coils is identical to the 'high-mirror' configuration with $\iota_a = 5/5$, but the toroidal mirror term b_{01} is reduced (increased) by about 10% (lower for low-ι and higher for high-ι, respectively).

$i_A = i_B = 0$ (see figure 3). For this configuration, the edge rotational transform is $\iota_a = 1$ with 5 independent islands in the divertor. The toroidal mirror term, b_{01}, is increased if the currents in the coils close to the 'bean-shaped' plane, where the local toroidal curvature (in real space) is largest, are increased compared with that in the other coils, i.e. $1 > i_2 > i_3 > i_4 > i_5$ (this might be different for special cases with shaping of the toroidal mirror form; see section 5). The (standard) 'high-mirror' configuration with $\iota_a = 1$ (see figure 3) is defined by $i_2 = 0.972$, $i_3 = 0.926$, $i_4 = 0.880$, $i_5 = 0.852$, and $i_A = i_B = 0$ being rather close to the original one found in the optimization process (see [1]). An opposite ordering of

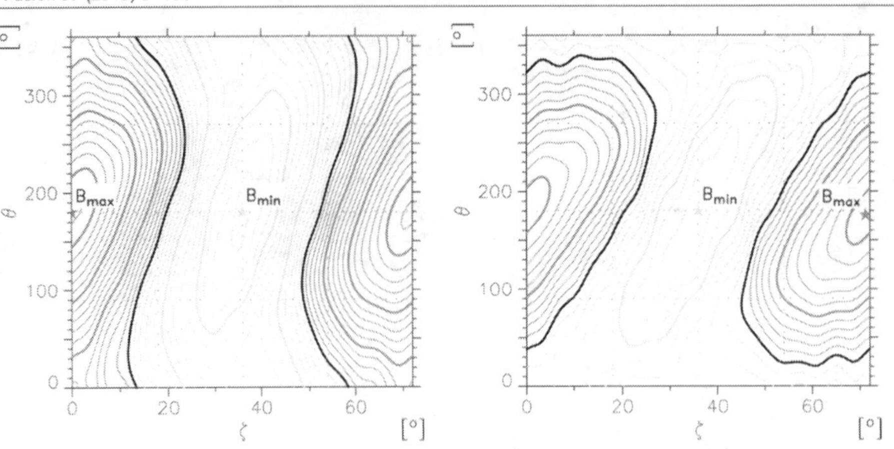

Figure 3. Isolines of B at half the plasma radius for the 'high-mirror' ($b_{01} \simeq 0.1$; on the left) and for the 'standard' configuration ($b_{01} = 0.047$; on the right) at $\iota_a = 1$. The 'bean-shaped' plane is at $\zeta = 0^\circ$ and the 'triangular' plane at $\zeta = 36^\circ$. $B = B_{00}$ is given by the black line, $B > B_{00}$ by the red and $B < B_{00}$ by the green lines; the increment $\Delta B/B_{00}$ is 0.01.

the modular coil currents, $1 < i_2 < i_3 < i_4 < i_5$, leads to a reduced b_{01}. For $i_A = i_B > 0$, the rotational transform, ι, is reduced (the field of the planar coils has no poloidal contribution) and it is increased for $i_A = i_B < 0$. For a fixed B_0, the magnetic field from the planar coils adds to the modular field and the currents in the modular coils must be reduced (and vice versa for $i_A = i_B < 0$). $i_A = -i_B > 0$ results in an inward shift of the plasma column, and $i_A = -i_B < 0$ to an outward shift, but affects also the toroidal mirror term, b_{01}.

One important ingredient of the optimization was the reduction of the average toroidal curvature given by the b_{10} Fourier mode in comparison to $\epsilon_t = r/R$. All W7-X configurations are characterized by a significantly reduced $\kappa = -b_{10}/\epsilon_t$ (roughly $0.3 < \kappa < 0.55$) corresponding to a very low averaged toroidal curvature. Since the magnetic field generated by the planar coils does not contribute to the κ-reduction, κ is lowest for high ι and highest for low ι in current-free configurations, respectively; see figure 2.

The vacuum magnetic configuration is completely determined by the normalized coil currents, i_n. Thus, I_1 is adjusted to the desired power deposition of the ECRH (at 140 GHz in X2- and O2-mode) close to the 'bean-shaped' plane; i.e. the average field in the central region, i.e. $B_{00}(0)$, is significantly lower than 2.5 T for high-mirror vacuum configurations.

2.2. The main Fourier modes of W7-X configurations

The stellarator optimization for W7-X with the constraint of a rather narrow b_{mn} Fourier spectrum 'converged' to one configuration with a rather high toroidal mirror term, $b_{01} \simeq 0.1$, and with the ratio of averaged toroidal curvature to helical curvature, $b_{10}/b_{11} \approx 0.5$ at $\iota_a = 1$; see [1, 17]. The realization of this optimization point in the configuration space of W7-X, however, is slightly different. Whereas b_{10} is nearly identical, the $|b_{11}|$ Fourier mode with dominant impact on ι was slightly increased (by a modification of the modular coils) to avoid the appearance of the rational $\iota = 5/6$ in the plasma core at finite β, resulting in a ratio of $b_{10}/b_{11} = 0.468$ (at half the plasma radius). As a consequence, the mono-energetic radial transport coefficients at low collisionality are increased whereas the bootstrap current coefficients are reduced; see figure 4 in [17] for the dependence of these coefficients on b_{10}/b_{11} and b_{01} for a b_{mn} model configuration with only these three modes. These configurations are also compared here in the sensitivity analysis of section 5.

The calculation of all Fourier modes in magnetic co-ordinates from the coil currents is straightforward: the B_{mn} as well as the Fourier representation of the flux-surfaces is performed by the VMEC code [7] and, in a second step, Boozer co-ordinates are determined by the JMC code [18]. The inverse problem of estimating the coil currents from the main b_{mn}-Fourier modes and ι_a, however, is impossible [19]. So far, a database of precalculated magnetic configurations is used to obtain a good guess, and the final coil current distribution is estimated with an additional fine-tuning (the existing function parametrization technique for VMEC equilibria [20] will be implemented in near future for this purpose). In fact, the back-transform from a Fourier representation with only the main b_{mn} modes is not unique. For example, the toroidal mirror term, b_{01}, can also be slightly adjusted by an inward- or outward-shift, $i_A = -i_B \neq 0$; see section 5.

The main b_{mn} modes of the W7-X configurations are the toroidal mirror term, $b_{01} \geqslant 0$, the averaged toroidal and helical curvatures, $b_{10} < 0$ and $b_{11} < 0$, respectively. In general, all other b_{mn} terms are much smaller. For high-mirror configurations, the impact of $\iota_a = 5/4$, 5/5 and 5/6 on all three b_{mn} modes is rather strong: $-b_{10}$ and $-b_{11}$ decrease with ι_a whereas b_{01} increases, and both $\kappa = -b_{10}R/r$ and b_{10}/b_{11} decrease with ι_a. For fixed ι_a, b_{01} has only a small impact on b_{10} and b_{11}. Furthermore, the extent of the minimum of B in the vicinity of the 'triangular' plane can be modified by the current distribution in the modular coils, which introduces higher harmonics b_{0n}: this extent becomes broader for $b_{02} > 0$ and narrower for $b_{02} < 0$. The impact of this mirror form on the three main b_{mn} modes, however, is fairly small.

The impact of deeply trapped particles on the neoclassical confinement properties is rather strong. For the high-mirror configurations, the confinement condition for these particles leads to the concept of quasi-isodynamic stellarators [21] with significant reduction in the radial component of their ∇B-drift [22]. In a simplified approach, this drift is roughly given by the poloidal variation of B in the 'triangular' plane at $\zeta = 36°$; see figure 3, left. In these scenarios, the minimum of B is quite pronounced and is located on the inner part of the flux-surfaces (i.e. at $\theta = \pi$). The simplified ansatz with $\mathrm{d}B(\theta, \zeta = 36°)/\mathrm{d}\theta = 0$ only for $m = 1$ and $n = -1, 0, 1$ leads to the same optimization condition,

$$b_{1-1} + b_{11} = b_{10}.$$

Consequently, the ratio b_{1-1}/b_{11} must be between -0.58 and -0.50 for the configurations close to the optimized configuration. Such strong b_{1-1} terms are not present in the W7-X configurations (and are neither supported by the outermost flux-surface in the optimization nor in the realized modular coil system). In W7-X, b_{1-1} is about one order of magnitude smaller than required for supressing the radial drift of deeply trapped particles (in the lowest order for only $m = 1$ modes). This 'missing' b_{1-1} term also has an important impact on the fast-particle confinement [23], one of the W7-X optimization criteria.

Also the 'standard' configuration of figure 3 (on the right) has a rather pronounced minimum of B at the inner side of the 'triangular' plane. An equivalent approach to minimize the variation of B in the helical direction with $\theta - N\zeta = 0$ with only the b_{11}, b_{10} and b_{01} terms taken into account leads to the condition,

$$b_{01} + b_{10} = 0.$$

This condition, however, can only be fulfilled at a specific radius, r, since b_{10} scales linearly with r whereas b_{01} is almost independent of r (at least in the axis expansion of the vacuum configuration). Although the b_{01} term can be adjusted in W7-X, a good approximation to a σ-optimized configuration [24, 25] cannot be realized for all r. Nevertheless, the neoclassical confinement at low collisionalities, characterized by the 'effective helical ripple', ϵ_{eff}, in the $1/\nu$-regime (for the definition, see [15] and the references therein), can be improved in comparison to the W7-X configurations with $b_{01} \simeq 0$. (A similar improvement is also obtained for the LHD 'inward-shifted' configuration with $R = 3.6$ m [15] with the dominant b_{21} mode locally coupling with the b_{11} and b_{31} modes which have different radial dependencies.)

W7-X vacuum configurations without a toroidal mirror, $b_{01} \simeq 0$, can be viewed as 'classical' stellarator configurations with strong optimization due to small averaged toroidal curvature, κ. This becomes obvious from the estimation of the 'effective helical ripple' where

$$\epsilon_{\text{eff}} \approx -b_{11}\,\kappa^{4/3}$$

is found with all b_{mn} modes included for the calculation of ϵ_{eff}. With finite β, however, the b_{01} mirror term becomes negative leading to a fairly pronounced minimum of B at the outer side of the 'bean-shaped' plane (i.e. at $\theta = 0$) where the (local) toroidal curvature dominates. A significant fraction of trapped particles in the vicinity of this minimum can exist dominating the radial transport and resulting in a strong degradation of the neoclassical confinement. Such configurations with an inverted mirror term, $b_{01} < 0$, are of additional interest for investigating the roles of ballooning instability and trapped electron modes [26].

In a quite different region of the W7-X configuration space at low-ι_a and very high $b_{01} > 0.2$, a vacuum magnetic hill appears [27]. These configurations are predicted to be Mercier unstable similarly to the inward-shifted LHD configurations [14, 28]. This analysis of the impact on confinement can also be performed in W7-X and compared to the LHD findings. High interest in such experimental investigations exists to assess the importance of a magnetic well which suppresses interchange activity at low β.

The impact of pressure, i.e. the β profile, is significant only for the b_{00} and the b_{01} Fourier modes which are reduced with rising β. In general, ι is only weakly affected (depending on the β-profile as well as on the configuration), as both the bootstrap current and the ECCD affect the rotational transform more strongly. The impact on b_{10} and b_{11} is rather small (both terms become slightly less negative). Both $\kappa = -b_{10}/\epsilon_t$ and b_{10}/b_{11} weakly decrease with $\langle\beta\rangle$. The reduction of the toroidal mirror term for high b_{01} leads to a significant improvement of neoclassical confinement. For W7-X, ϵ_{eff} decreases with at least moderate $\langle\beta\rangle$ which is a quite different behaviour to that of LHD configurations [29] with strong Shafranov shift [30].

Finally, the radial dependence of b_{10}/b_{11} has an important consequence for the bootstrap current density profile (see section 4.1). In agreement with the axis-expansion of the Fourier modes, $b_{mn} \propto r^m$, b_{10}/b_{11} is nearly constant for the inner radii, but increases at the outer radii. Following the analysis of figure 4 in [17], the bootstrap current coefficients become negative at small r and positive at outer r for an adjusted b_{01}-mirror term. In addition to the thermodynamic forces for neoclassical transport (i.e. density and temperature gradients as well as the radial electric field), this feature can lead to negative bootstrap current densities at the inner radii and to positive ones at the outer radii. For such a scenario, the shear, ι', may be increased.

3. Self-consistent transport simulations

The neoclassical mono-energetic transport coefficients are calculated with the DKES code [8] and depend only on the b_{mn}-spectrum for each radius and $\iota(r)$. In particular, the bootstrap current coefficient is strongly affected by ι, e.g. in the ECCD scenarios (roughly scaling with $1/\iota$ at low collisionalities; see section 4.2). The full thermal neoclassical transport matrix is obtained by energy convolution of the three mono-energetic transport coefficients stored in a database for each configuration as a function of radius, collisionality and radial electric field. In the energy convolution, parallel momentum corrections [31] are included to both the bootstrap current densities and the parallel conductivity (collision operator used by DKES does not conserve parallel momentum). These neoclassical coefficients are used in the transport simulations by the NTSS code [9] to calculate the radial particle and energy fluxes, the bootstrap current densities as well as the transient current diffusion (both on the skin-time of a few seconds and on the L/R-time which is typically one order of magnitude larger). The radial electric field entering both in the transport coefficients and in the thermodynamic forces is calculated from the ambipolarity condition of the radial particle fluxes by solving a diffusion equation (in this way, a unique E_r is obtained; see [9] for more details). Only at the outermost radii, with low temperature, is an additional 'anomalous' transport assumed. With the energy fluxes for all species, the energy balance is solved and the temperature profiles are calculated.

The ECRH power deposition and the ECCD are calculated with the TRAVIS ray-tracing code [32] which is coupled to NTSS. In TRAVIS, the ECCD can be calculated in the collisional and collisionless limits (with parallel momentum conservation), or finite collisionality effects [33] may be included. NBI deposition as well as NBCD are included in a simplified model assuming slowing-down on flux-surfaces. In the transport simulations, $Z_{\text{eff}} = 1.5$ (with C^{6+}) independent of r is assumed, however, the impurity flux is taken into account for evaluating E_r. From Z_{eff}, a rough estimate of radiative losses is obtained and included in the energy balance.

So far, the particle balance is disregarded for calculating the density profile; a fixed $n_e(r)$ is simply assumed. In the simulations presented here, the shape of the density profile is identical, and only its central value, $n_e(0)$, is varied. The reason for this approach is not related to the particle fluxes (they are used for estimating E_r), but to the particle sources. Particle sources at outer radii are gas puffing (which can be calculated by the EIRENE code [34]) and recycling (mainly from the island divertor targets). In order to calculate the recycling sources, a coupling with the edge transport modelling, i.e. the EMC3/EIRENE code [35, 36], is necessary, but really not at all trivial (e.g. the common boundary conditions at the plasma edge). For central power deposition, also central particle sources are needed [6], which might lead to density control problems at outer radii. Without these central particle sources, the density profiles are predicted to become very hollow as has been observed in LHD [12, 37]. For such a scenario, pellet injection might be mandatory for controlling the full density profile. (In the near future, a stellarator version of the HPI2 pellet ablation code [38] will be used for the transport simulations). With these restrictions, the transport simulations are not completely self-consistent.

The simulated β and current density profiles (bootstrap, ECCD and, in principle, NBCD) are the input for the VMEC equilibrium code [7], the results of which are expressed in a Boozer coordinate representation using the JMC code [18]. The new b_{mn}-spectrum and ι-profile are used by DKES for the next iteration step until sufficient convergence is reached (e.g. total current of a few kA is tolerable for island divertor operation).

In section 4, only two discharge scenarios are described. In a 'conservative' scheme, the total bootstrap current, I_b, is minimized by a proper adjustment of only the toroidal mirror term, b_{01}, since the ECCD efficiency is very low at high densities (ECRH in O2-mode), required in quasi-stationary discharges with high performance, i.e. high $\langle\beta\rangle$; see section 4.1. In the second scenario at lower b_{01} (section 4.2), e.g. in the 'standard' configuration, the bootstrap current is balanced by strong counter-ECCD significantly affecting the ι-profile. This scenario leads to constraints on the density and the ECRH power: n_e must be much lower (ECRH in X2-mode) and the ECCD power has to be increased. Here, the 'free' parameters in the iteration for vanishing total current are the density, the ECRH power and (partly) the ECCD deposition (i.e. all coil current ratios, i_n, are fixed). Here, more steps are needed in the iteration since the bootstrap current coefficients depend sensitively on the ι-profile as well as the balance of ECCD and I_b on the density and heating power.

4. Current-free discharge scenarios

For quasi-stationary operation of W7-X with a proper functioning of the island divertor, total current control is needed, i.e. the bootstrap current must vanish, $I_b \simeq 0$, or it must be balanced by counter-ECCD, $I_b + I_{\text{ECCD}} \simeq 0$. For the first scenario, the magnetic configuration must be adjusted, i.e. mainly the b_{01} term, which is one criterion of the W7-X optimization. The second scenario allows for much more freedom with respect to the magnetic configurations of interest.

4.1. Conservative scenarios without ECCD

Starting point for these 'conservative' scenarios is the selection of the central density for the final phase, the ECRH heating power and the island configuration, i.e. ι_a. High performance at high $\langle\beta\rangle$ requires densities above the X2 cut-off ($\simeq 1.2 \times 10^{20}\,\text{m}^{-3}$). Here, an O2-scenario with $n_e(0) = 1.5 \times 10^{20}\,\text{m}^{-3}$ at 5 MW ECRH power with peaked deposition close to the magnetic axis is chosen as final target of the iteration process. In each iteration step, the equilibrium calculation with the β-profile and the current density profile from the previous iteration step included are used for updating the DKES-database and for the new transport simulation. The iteration is assumed to be converged if nearly identical $I_b \simeq 0$, τ_E and $\langle\beta\rangle$ (and profiles) with respect to the previous step are obtained. For $\iota_a = 5/5$, the toroidal mirror term for $I_b \simeq 0$ is $b_{01} = 0.108$ in vacuum and slightly reduced for the converged configuration at $\langle\beta\rangle = 3.06\%$ ($\beta(0) \simeq 7\%$); see table 1. The finite $\beta(r)$ reduces ϵ_{eff} in the main confinement region (although ϵ_{eff} increases further towards the edge). The total bootstrap current is sufficiently small ($\simeq$3.4 kA) as the ion ($\simeq$13 kA) and electron ($\simeq -10$ kA) contributions nearly cancel each other; see figure 4 (lower left plot). The bootstrap current density, j_b, is negative at the inner radii and positive at the outer radii resulting in a small reduction of the inner ι values. The energy confinement time, τ_E, significantly exceeds the one from the ISS04 scaling [14] (the normalized $\tau_E^n = \tau_E/\tau_E^{\text{ISS04}}$ is also given in table 1 to eliminate the volume dependence).

In this 'conservative' scheme, all coil currents and launching angles of the different EC beams are identical in the early phase of the discharge. For the very early phase, a value of $n_e(0) = 0.3 \times 10^{20}\,\text{m}^{-3}$ was chosen also with 5 MW ECRH in X2-mode. For this case, only the vacuum configuration was used (no iteration). For the $\langle\beta\rangle \simeq 0.7\%$ found in this simulation, the small change in the b_{mn} spectrum can be ignored. The ECRH position for this low-β case is

Table 1. Conservative scenarios without ECCD for $\iota_a = 5/5$, $5/6$ and $5/4$. On-axis ECRH deposition in O2-mode is modelled only for the high density cases; the coil currents in each ι_a scenario are kept fixed for the different $n_e(0)$. The configuration parameters, $\epsilon_{\rm eff}$, κ and b_{10}/b_{11}, are given at half the plasma radius. The transport simulation results are obtained for 5 MW ECRH for both X2- and O2-mode.

ι_a	$n_e(0)$ [10^{20} m^{-3}]	I_b [kA]	τ_E [s]	τ_E^n	$\langle\beta\rangle$ [%]	Vol. [m^3]	$B_{00}(0)$ [T]	$b_{01}(0)$	$\epsilon_{\rm eff}$ [%]	κ	b_{10}/b_{11}
$\frac{5}{5}$	0.30	6.0	0.147	1.32	0.75	29.4	2.38	0.108	2.64	0.450	0.474
	0.75	5.0	0.351	1.98	1.84	29.3	2.32	0.100	2.41	0.447	0.479
	1.50	3.4	0.605	2.33	3.06	29.7	2.27	0.094	2.16	0.440	0.480
$\frac{5}{6}$	0.30	5.0	0.126	1.36	0.91	28.5	2.12	0.244	6.25	0.577	0.608
	0.75	−1.6	0.260	1.77	1.92	28.5	2.06	0.236	5.71	0.586	0.602
	1.50	2.1	0.463	2.10	3.15	28.4	2.08	0.230	5.10	0.586	0.614
$\frac{5}{4}$	0.30	6.0	0.125	1.32	0.70	23.8	2.51	0.040	0.89	0.353	0.388
	0.75	8.6	0.370	2.42	2.11	23.8	2.46	0.037	0.87	0.347	0.385
	1.50	0.2	0.596	2.70	3.32	23.9	2.39	0.027	0.70	0.331	0.375

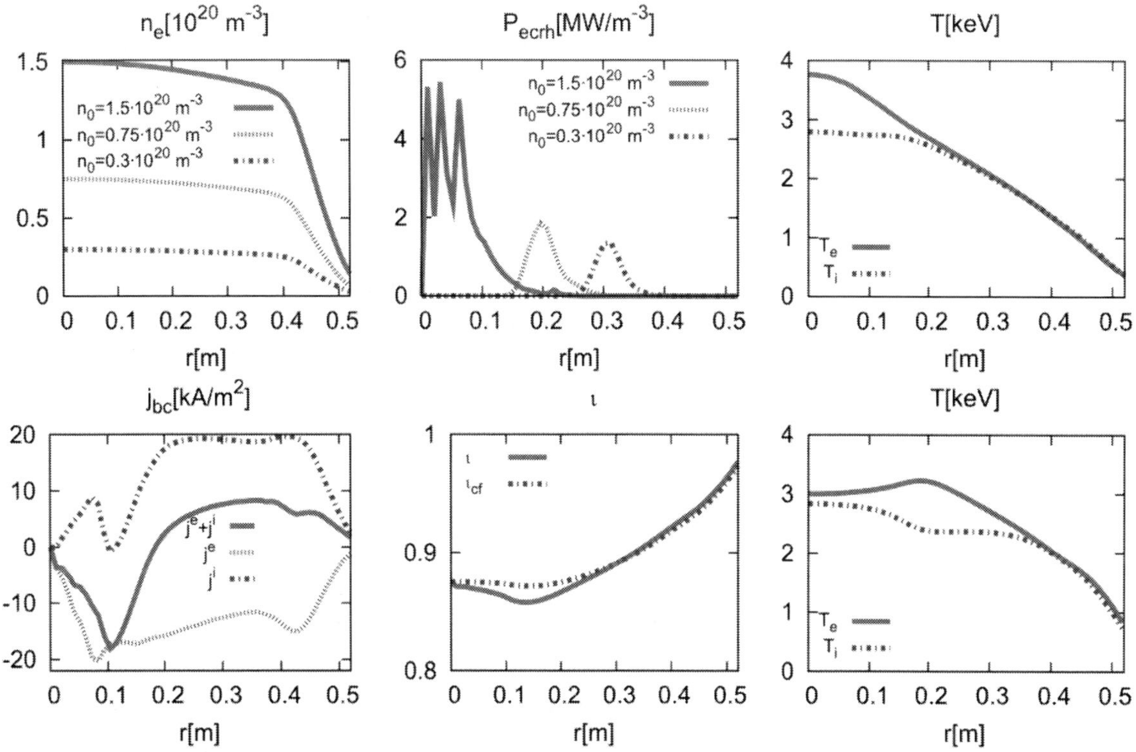

Figure 4. Upper line: density profiles (left), the ECRH power deposition profiles (for all scenarios; centre), the T_e and T_i profiles for the O2 case (right). Lower line: the bootstrap current density profiles for the O2 case (left), the ι-profile (where ι_{cf} is the current-free one), and the T_e and T_i profiles for the X2 case with $n_e(0) = 0.75 \times 10^{20}$ m^{-3} (right).

strongly off-axis on the low-field side to allow for central deposition in the high-β phase in O2-mode. For this low-density case, τ_E exceeds $\tau_E^{\rm ISS04}$ only slightly due to the strong off-axis ECRH deposition. Also for this low-density case, the bootstrap current is found to be sufficiently small.

Next, the density is increased to $n_e(0) = 0.75 \times 10^{20}$ m^{-3} in preparation for the transition from X2- to O2-mode by switching the polarization of the EC beams (simulation again with 5 MW power). This transition is analysed in detail in [16] for the 'standard' configuration. With increased β at this density, the ECRH deposition is shifted inwards compared to the vacuum scenario. Here, an iteration of the β-profile is performed; for the temperature profiles, see figure 4 (lower right plot). Again, the impact of $j_b(r)$ on the ι-profile is small. For all these cases with increased $n_e(0)$ and the transition from X2- to O2-mode by (fast) changing of the beam polarization, the bootstrap current remains small enough and no ECCD is needed to control the island configuration in the divertor.

For both $\iota_a = 5/4$ and $\iota_a = 5/6$, the same scenarios as for $\iota_a = 5/5$ are analysed. The high-ι configurations have both a smaller $\kappa = -b_{10}R/r$ and b_{10}/b_{11} ratio, and a smaller b_{01} mirror term should be sufficient for $I_b \simeq 0$. For the $n_e(0) = 1.5 \times 10^{20}$ m^{-3}, O2-mode scenario, a vacuum value of only $b_{01} = 0.04$ (and further reduced with finite β) is needed

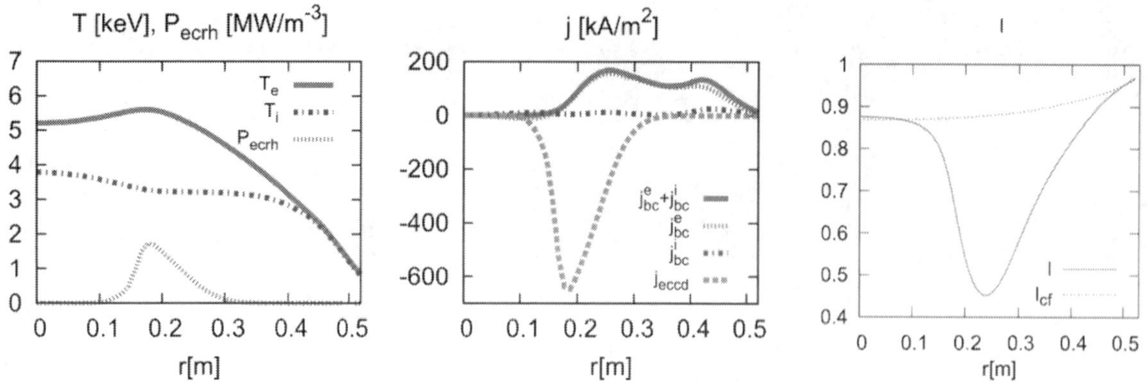

Figure 5. T_e and T_i profiles (power deposition is indicated), left, the current density profiles, j_{ECCD} and j_b (centre), and the ι-profile (right).

for $I_b \simeq 0$. For this rather low toroidal mirror term, ϵ_{eff} is much lower in the main confinement region compared to the other ι_a scenarios allowing for much higher $T_e(0)$ although the plasma volume is reduced. The electron and ion contributions to I_b cancel each other globally, but not locally. In the central region, $j_b < 0$ (the electron contribution dominates) resulting in $\iota < 1$. The electron contribution to I_b dominates at low density whereas the ion contribution predominates at the intermediate density; the impact on $\iota(r)$ is small in both cases.

Finally, results for the low-ι scenario with $\iota_a = 5/6$ are described. For the high-performance target with $n_e(0) = 1.5 \times 10^{20}\,m^{-3}$ with O2-mode, a very large $b_{01} \simeq 0.25$ term in vacuum is required for $I_b \simeq 0$. The low-ι configurations have both the largest κ and b_{10}/b_{11} in comparison with the higher-ι_a ones. For this configuration, $\epsilon_{eff} \geqslant 5\%$ in the main confinement region is rather high. The rather large $\langle\beta\rangle$ is attributed to the significantly reduced B_{00} for $B \simeq 2.5$ T at the ECRH deposition in the 'bean-shaped' plane. Also for these scenarios, the simulations show only a moderate impact of j_b on the ι profile.

4.2. Scenarios with strong counter-ECCD

Here, the 'standard' configuration with identical modular currents, I_n, and $I_A = I_B = 0$ for $\iota_a = 5/5$ (see section 2.1) is selected for scenarios with ECCD compensation of the bootstrap current, $I_b + I_{ECCD} \simeq 0$. ECCD in O2-mode is not efficient enough by far, but the ECCD efficiency is also not sufficient at higher density ($n_e(0) \simeq 10^{20}\,m^{-3}$) and for moderate power levels in X2-mode. Consequently, $n_e(0)$, the heating power, as well as the power deposition region are 'free' parameters in the iteration process, whereas $j(r) = j_b + j_{ECCD}$ (with a dominant impact of the EC driven current density, j_{ECCD}) and $\beta(r)$ determine the plasma response of the magnetic configuration. In these iterations, counter-ECCD significantly reduces ι at the ECCD position resulting in a strong increase of the bootstrap current coefficients (roughly scaling as $1/\iota$ at low collisionalities). Consequently, more iteration steps are needed for scenarios with strong ECCD compared to those without ECCD (section 4.1).

Within the iteration the central density has to be reduced, $n_e(0) = 0.67 \times 10^{20}\,m^{-3}$ (for the profile shape, see figure 4; upper left plot), and the ECRH power increased to 7 MW with a deposition at around one third of the plasma radius. Only for this rather strong off-axis deposition could the ι-reduction be limited to a minimum value of 0.45 (the change in $\iota(r)$ due to the local current scales with $\Delta\iota \propto 1/r^2$). The converged profiles are given in figure 5. The total bootstrap current is $I_b = 82.1$ kA (with the main contribution from electrons), and $I_{ECCD} = -81.0$ kA is found; the rather small difference is negligible for the island divertor operation. Even for this rather low n_e, $\langle\beta\rangle = 2.2\%$ (with $\beta(0) \simeq 4\%$) is fairly high due to the good neoclassical confinement with $\epsilon_{eff} \leqslant 0.6\%$ in the main confinement region. The energy confinement time is $\tau_E = 0.34$ s (and normalized $\tau_E^n = 2.7$). For higher density or lower ECRH power, no current balance $I_b + I_{ECCD} \simeq 0$ could be obtained in this fairly strong off-axis deposition scenario ($I_{ECCD} \propto P_{ECCD}/n_e$ behaves quite differently to the I_b dependence on power and density; see [16]).

The shape of $\iota(r)$ is characterized by negative shear $\iota' < 0$ inside the ECCD position and positive shear $\iota' > 0$ outside. (In the tokamak community, the sign convention is the opposite: $\iota' < 0$ is designated as 'positive' shear.) In W7-AS, strong mode activity was found mainly in the $\iota' < 0$ region for co-ECCD, but with coupling to other low-order ι-rationals with $\iota' > 0$ [39]. In equivalent discharges with strong counter-ECCD, i.e. with only strong positive shear, however, no mode activity could be observed. For the counter-ECCD scenario of figure 5 with $\iota = 1/2$ within the ι-profile, island formation might be expected at this low-order rational.

The situation becomes even more sensitive because of the ι-dependence of the bootstrap current and of the neoclassical confinement for an equivalent simulation with on-axis counter-ECCD. The ECCD current densities close to the axis become formally very large leading to strongly negative ι. Consequently, the role of the $\iota \simeq 0$ region on the confinement becomes essential, but there are two further issues. The first one is related to ergodization of the magnetic field, i.e. the assumption of nested flux-surfaces (in the VMEC code) is violated without rotational transform (equivalent to the start-up of tokamak discharges). Then, neither particle and energy confinement nor localized current densities can be expected. The second issue is a neoclassical one: particle and energy confinement are determined by the

Table 2. Impact of small configuration modifications on plasma performance compared to the (standard) 'high-mirror' configuration (reference configuration **A**); the definition of the different configurations is described in the text. The configuration parameters, ϵ_{eff}, κ and b_{10}/b_{11}, are given at half the plasma radius. The transport simulation results are obtained for 5 MW ECRH for both on- and off-axis deposition.

	on-axis ECRH				off-axis ECRH							
config.	I_b [kA]	τ_E [s]	τ_E^n	$\langle\beta\rangle$ [%]	I_b [kA]	τ_E [s]	τ_E^n	$\langle\beta\rangle$ [%]	Vol. [m^3]	ϵ_{eff} [%]	κ	b_{10}/b_{11}
A	18.4	0.386	2.23	2.10	10.6	0.376	2.12	1.95	29.4	2.04	0.441	0.473
B	28.9	0.420	2.39	2.17	19.7	0.395	2.21	1.95	29.4	1.40	0.429	0.461
C	10.5	0.330	2.11	2.04	2.1	0.310	1.95	1.83	27.1	2.74	0.452	0.484
D	17.1	0.352	2.29	2.12	9.5	0.342	2.18	1.96	26.6	2.04	0.440	0.471
E	19.2	0.352	2.33	2.16	11.2	0.342	2.22	2.00	26.2	1.97	0.445	0.480
F	8.0	0.358	1.99	1.87	3.1	0.349	1.91	1.74	29.5	2.45	0.438	0.475
G	25.9	0.390	2.35	2.22	15.5	0.378	2.24	2.05	29.2	1.72	0.444	0.472
H	40.7	0.398	2.30	2.16	32.0	0.388	2.19	2.01	29.4	1.98	0.441	0.473
I	44.4	0.427	2.81	2.53	29.0	0.416	2.67	2.35	27.4	1.15	0.457	0.510

plateau transport, which scales with $1/\iota$ and extends to very low collisionalities [15] (confirmed in the DKES calculations); also the bootstrap current coefficients scale with $1/\iota$ at low ν^*. These neoclassical scalings reflect the strong deviation of particle orbits from flux-surfaces (even if they exist). Because of these issues at very low (but finite) ι, the assumption of good confinement is questionable. Counter-ECCD discharges in W7-AS showed completely flat density and temperature profiles in the central region consistent with $\iota \simeq 0$ in the current balance simulations [39].

The transport simulations for W7-X counter-ECCD reflect these findings. Very large transport coefficients are assumed within the central region with e.g. $\iota < 0.05$, and only the neoclassical ones outside (i.e. with DKES data for $\iota \geqslant 0.05$). With the large neoclassical transport coefficients at low ι, the temperature gradients (and also E_r) are strongly decreased resulting in strongly reduced thermodynamic forces for driving the bootstrap current density in this region. Such a transport and current balance simulation is performed (also for the W7-X 'standard' configuration) with $n_e(0) = 0.5 \times 10^{20}\,\mathrm{m^{-3}}$ at 5 MW heating power in X2-mode with optimum launch angles for maximum central ECCD. In the central region $r < 0.21$ m with $\iota < 0.05$, the ECCD current density profile is assumed to be completely flat (and $j_b = 0$) allowing for $\iota \simeq 0$. A total ECCD current of $I_{ECCD} = -72.4$ kA is obtained, balancing the bootstrap current of roughly $I_b = 67.7$ kA (with dominant contribution by electrons); $\langle\beta\rangle = 1.6\%$, $\tau_E = 0.325$ s ($\tau_E^n = 2.4$), and $\epsilon_{eff} > 0.6\%$ in the main confinement region is found for the converged iteration.

The clear disadvantage of these strong counter-ECCD scenarios is the significantly reduced performance since they are restricted to fairly low densities; no access to high densities with ECCD in O2-mode exists. Furthermore, these scenarios are very sensitive to the current density as well as the density and temperature profiles. Consequently, such scenarios should be investigated in W7-X discharges, but the 'conservative' schemes of section 4.1 without strong ECCD will be more attractive for quasi-stationary operation with edge ι_a-control for the island divertor.

5. Configurational sensitivity analysis

The dependence of both the neoclassical confinement and the bootstrap current on the magnetic configuration is very complex. In general, the target function as well as the specific weights for both topics within the optimization process might force a compromise with respect to high τ_E, high $\langle\beta\rangle$ and low I_b. As is shown in section 4, the weight for a minimized I_b must be significantly increased compared to the 'original' W7-X optimization under the constraint of a proper functioning of the island divertor. In this section, the configurational sensitivity of τ_E, $\langle\beta\rangle$ and I_b is analysed with the (standard) 'high-mirror' configuration being the reference case **A** in table 2; the coil current ratios, i_n, are given in section 2.1. For all configurations, a fixed parabolic β-profile with $\langle\beta\rangle = 2\%$ is assumed, and an iteration for an equilibrium consistent with the transport simulations is omitted.

For the transport simulations, $n_e(0) = 0.75 \times 10^{20}\,\mathrm{m^{-3}}$ with the profile shape of figure 4 (upper left plot) with 5 MW ECRH (without ECCD) and both on-axis and off-axis deposition are chosen. Except for the case of b_{01} variation in configurations **B** and **C**, the toroidal mirror term is $b_{01} \simeq 0.1$ in vacuum and slightly lower at finite β. With central power deposition at this rather low density, a pronounced 'electron-root' feature (up to about 30% of the plasma radius) is obtained for all the 'high-mirror' configurations in table 2. This neoclassical 'electron-root' with strongly positive E_r and significantly improved electron heat confinement has been experimentally observed in numerous stellarators consistent with the theoretical predictions [10]. This 'electron-root' feature leads to an additional contribution of between 25% and 70% to the bootstrap current, I_b. For comparison, a second set of transport simulations is performed for off-axis power deposition (at $r \sim 0.18$ m) without this 'electron-root' feature.

For the (standard) 'high-mirror' configuration (case **A**) with on-axis ECRH deposition, I_b is significantly larger than for the X2-scenario of section 4.1 with a vacuum $b_{01} = 0.11$ (the same $n_e(r)$, but with strong off-axis power deposition). The equivalent transport simulation without the pronounced 'electron-root' feature leads to smaller I_b by a factor of about 2. The impact of the toroidal mirror term is demonstrated by the scenarios **B** with $b_{01} = 0.08$ and **C** with $b_{01} = 0.12$ (in vacuum): I_b is reduced with increasing b_{01}, but at the cost of degraded neoclassical confinement (τ_E^n and $\langle\beta\rangle$ are much less affected than ϵ_{eff} due to the strongly nonlinear temperature dependence of the neoclassical energy fluxes). For an increased b_{01}, the average B_0 must be reduced due to the ECRH resonance condition close to the 'bean-shaped' plane (at the maximum of B) largely compensating the impact on $\langle\beta\rangle$ (compare also section 4.1 for the high- and low-ι scenarios).

For the inward- (case **D**) and outward-shifted (case **E**) configurations with $i_A = -i_B \neq 0$ (see section 2.1), the $b_{01} = 0.1$ term is adjusted by varying the modular coil currents close to the 'bean-shaped' plane. With the same b_{01} term, the inward and outward shift has only a small impact on I_b and the neoclassical confinement both for on-axis and off-axis power deposition. Consequently, these shifts might also be used for a slight modification of the toroidal mirror term without changing the modular i_n.

The shape of the B-variation at the magnetic axis has a rather strong impact. In configuration **F**, the toroidal range of the B minimum is narrower, and broader for **G** with respect to the reference configuration **A**. Some freedom exists in the current distribution in the modular coils which can be used for an optimization process based only on this coil system. For both cases **F** and **G**, the narrower (broader) minimum B range is obtained by increasing (decreasing) the current i_3, whilst keeping all other current ratios nearly the same. In the b_{mn} spectrum, a second-harmonic term, $b_{02} = -0.021$ (**F**) and $b_{02} = 0.023$ (**G**), is responsible for the width of the minimum B range. With the additional b_{02} term, the average magnetic field for $r \simeq 0$ is changed ($(b_{max} - b_{min})/2$ is the toroidal ripple in this case). The neoclassical confinement characterized by ϵ_{eff} (for the $1/\nu$-transport regime) is significantly affected whereas the configurational parameters κ and b_{10}/b_{11} are nearly unchanged. In the case **F** with the narrow minimum range, I_b is significantly reduced, but the energy confinement is degraded, and vice versa for the broad minimum range (**G**). The fairly strong difference in $\langle\beta\rangle$ is partly attributed to the average field (higher for **F** and lower for **G**; this is different to the scenarios **B** and **C**).

Next, the effect of higher b_{mn} modes is analysed. In the case **H**, only the b_{01}, b_{10} and b_{11} Fourier modes are taken into account for DKES and the transport calculations. The higher b_{mn} modes have only a rather small effect on the neoclassical confinement reflected by τ_E and $\langle\beta\rangle$, but have a strong impact on I_b. This result is in agreement with the simplified analysis of [17] (figure 6) based on mono-energetic transport coefficients only for one radius.

Finally, the magnetic configuration (case **I**) obtained from the 'original' modular coil system as an outcome of the W7-X optimization process (with identical currents in the modular coils leading to $b_{01} = 0.10$ in vacuum; see section 1) is compared with case **A**, i.e. the (standard) 'high-mirror' W7-X configuration. The ratio b_{10}/b_{11} is increased with respect to case **A** resulting in a reduced radial transport, but increased bootstrap current coefficients; compare figure 4 in [17]. Again, this 'old' analysis is confirmed by the transport simulations for both power depositions: τ_E and $\langle\beta\rangle$ are increased, but I_b is significantly larger. In this sense, the weight for minimized bootstrap current in the optimization process is increased strongly for the existing W7-X coil system.

The configuration modifications **B** to **G** in the near vicinity of the (standard) 'high-mirror' configuration **A** at $\iota_a = 5/5$ demonstrate that the minimization of I_b is in conflict with neoclassical confinement improvement reflected by increased τ_E^n and $\langle\beta\rangle$. Transport simulations with 'artificial' quasi-isodynamic corrections in the b_{mn} spectrum, i.e. where the b_{1-1}, b_{20} and both $b_{21} = b_{2-1}$ terms in the case **A** spectrum are replaced by the constraints given in [22], leads to significantly improved confinement but with large $I_b > 0$. Consequently, quasi-isodynamicity mainly for deeply trapped particles is not sufficient to obtain high τ_E^n and $\langle\beta\rangle$ at low I_b (the QIPC configuration analysed in [15] is very different to W7-X configurations). A variation of the b_{10}/b_{11} ratio seems to be more efficient, but this ratio is nearly fixed by the W7-X coil system for a specific ι_a required for the island divertor.

A main part of this necessary compromise between neoclassical confinement optimization and minimization of the bootstrap current is also reflected in the ι_a-dependence; compare section 4.1. Here, the configuration parameters for the low-ι_a and high-ι_a cases with $b_{01} \simeq 0.1$ are briefly given. For $\iota_a = 5/6$ with $\epsilon_{eff} = 1.2\%$, $\kappa = 0.49$ and $b_{10}/b_{11} = 0.51$ (at half the plasma radius), a rather large $I_b > 0$ is obtained, whereas for $\iota_a = 5/4$ with $\epsilon_{eff} = 3.1\%$, $\kappa = 0.38$ and $b_{10}/b_{11} = 0.42$, even a fairly negative I_b is found. Consequently, configurations with very different mirror terms b_{01} are required for $I_b \simeq 0$ in the different ι_a scenarios.

6. Summary and conclusions

Different quasi-stationary discharge scenarios have been analysed. For W7-X configurations with small or intermediate toroidal mirror terms, b_{01}, the predicted bootstrap current, I_b, is rather large and can only be balanced for high-power ECRH with ECCD in X2-mode at low or intermediate densities. An exception is the high $\iota_a = 5/4$ configuration with a vacuum $b_{01} \simeq 0.04$, which allows for high density operation (in O2-mode) without ECCD. For high-performance discharges with high $\langle\beta\rangle \geqslant 3\%$, the density must exceed the X2 cut-off limit, but the ECCD efficiency in O2-mode is rather low and thus insufficient to balance I_b. Consequently, 'conservative' scenarios where the bootstrap current is controlled by the magnetic configuration are most promising. Except for the high-ι cases, only high-mirror configurations with $b_{01} \simeq 0.11$ for $\iota_a = 5/5$ and with $b_{01} \simeq 0.25$ for $\iota_a = 5/6$ are suited for scenarios with sufficiently small I_b.

Nevertheless, transport simulations with strong ECCD and $I_b + I_{ECCD} \simeq 0$ have been performed in which $I_{ECCD} < 0$ reduces significantly the $\iota(r)$ at the radial ECCD position (increasing the bootstrap current coefficients in this region). These scenarios are not qualified for quasi-stationary discharges with high-performance (i.e. high density). Furthermore, low-order rationals appear in the

ι-profile both with positive and negative shear, ι', which might lead to MHD activity.

For all the W7-X configurations under investigation, the minimization of I_b is in conflict with the neoclassical confinement improvement. Within any optimization process, qualified target functions (here, e.g., I_b, τ_E and $\langle\beta\rangle$ obtained from neoclassical transport simulations) together with the corresponding weights need to be defined (previous optimizations have been based on simple figures of merit, e.g. ϵ_{eff} and the collisionless limit of the mono-energetic bootstrap current coefficient [40]). Additional constraints, like the proper operation of the island divertor in W7-X (this was not included in the 'original' optimization process), affect the weights of the target functions: the minimization of I_b in the magnetic configuration space becomes the highest priority and a (moderate) degradation of the neoclassical confinement must be accepted. Furthermore, the role of ϵ_{eff} (characterising only the $1/\nu$ transport) is overestimated in comparison to τ_E and $\langle\beta\rangle$ (with smaller configurational dependence), where also the full nonlinear dependence on the simulated plasma parameters is taken into account. The ansatz of purely neoclassical transport modelling in most of the plasma volume leads to an upper limit for the confinement. Assuming strong turbulence would reduce the achievable τ_E and $\langle\beta\rangle$.

In the next step of the discharge scenario development, the particle balance for calculating a self-consistent density profile must be included in the iteration loop of equilibrium (VMEC code), neoclassical transport coefficients (DKES code), and self-consistent transport calculations (NTSS code). This extension, however, is not trivial: the particle sources mainly by recycling and pellet injection must be self-consistently calculated. Furthermore, a coupling to the edge plasma simulations (EMC3/EIRENE code) is essential.

Acknowledgments

The authors acknowledge very helpful discussions with P Helander, C Nührenberg and J Nührenberg. This project has received funding from the European Union's Horizon 2020 research and innovation programme under grant agreement number 633053. The views and opinions expressed herein do not necessarily reflect those of the European Commission.

References

[1] Grieger G *et al* 1992 *Phys. Fluids* B **4** 2081
[2] Lotz W, Nührenberg J and Schwab C 1991 *13th Int. Conf. on Plasma Physics Controlled Nuclear Fusion Research (Washington, DC, 1990)* vol 2 (Vienna: IAEA) p 603 (www-naweb.iaea.org/napc/physics/FEC/STIPUB844_VOL2.pdf)
[3] Nührenberg J and Zille R 1986 *Phys. Lett.* A **114** 129
[4] Beidler C D and Maaßberg H 1996 *Theory of Fusion Plasmas (Varenna 1996)* (Bologna: Editrice Compositori) p 375
[5] Renner H *et al* 2000 *Nucl. Fusion* **40** 1083
[6] Maaßberg H, Beidler C D and Simmet E E 1999 *Plasma Phys. Control. Fusion* **41** 1135
[7] Hirshman S P, van Rij W I and Merkel P 1986 *Comput. Phys. Commun.* **43** 143
[8] van Rij W I and Hirshman S P 1989 *Phys. Fluids* B **1** 563
[9] Turkin Y *et al* 2011 *Phys. Plasmas* **18** 022505
[10] Yokoyama M *et al* 2006 *Fusion Sci. Technol.* **50** 327
[11] Hirsch M *et al* 2008 *Plasma Phys. Control. Fusion* **50** 053001
[12] Dinklage A *et al* 2013 *Nucl. Fusion* **53** 063022
[13] Dinklage A *et al* 2007 *Nucl. Fusion* **47** 1265
[14] Yamada H *et al* 2005 *Nucl. Fusion* **45** 1684
[15] Beidler C D *et al* 2011 *Nucl. Fusion* **51** 076001
[16] Marushchenko N B *et al* 2012 *EPJ Web Conf.* **32** 01004
[17] Maaßberg H, Lotz W and Nührenberg J 1993 *Phys. Fluids* B **5** 3728
[18] Nührenberg J and Zille R 1988 *Theory of Fusion Plasmas (Varenna 1987)* (Bologna: Editrice Compositori) p 3
[19] Schwab C 1989 *Theory of Fusion Plasmas (Varenna 1988)* (Bologna: Editrice Compositori) p 85
[20] Sengupta A, Geiger J and Mc Carthy P J 2007 *Plasma Phys. Control. Fusion* **49** 649
[21] Gori S, Lotz W and Nührenberg J 1996 *Theory of Fusion Plasmas (Varenna 1996)* (Bologna: Editrice Compositori) p 335
[22] Beidler C D and Maaßberg H 2001 *Plasma Phys. Control. Fusion* **43** 1131
[23] Drevlak M, Geiger J, Helander P and Turkin Yu 2014 *Nucl. Fusion* **54** 073002
[24] Mynick H E, Chu T K and Boozer A H 1982 *Phys. Rev. Lett.* **48** 322
[25] Shaing K C and Hokin S A 1983 *Phys. Fluids* **26** 2136
[26] Helander P *et al* 2012 *Plasma Phys. Control. Fusion* **54** 124009
[27] Lortz D and Nührenberg J 1976 *Z. Naturf.* a **31a** 1277 (http://zfn.mpdl.mpg.de/data/Reihe_A/31/ZNA-1976-31a-1277.pdf)
[28] Kaneko O *et al* 2002 *Phys. Plasmas* **9** 2020
[29] Wakasa A and Murakami S 2008 *Plasma Fusion Res.* **3** S1030
[30] Mynick H E 1985 *Phys. Fluids* **28** 1139
[31] Maaßberg H, Beidler C D and Turkin Y 2009 *Phys. Plasmas* **16** 072504
[32] Marushchenko N B, Turkin Y and Maaßberg H 2014 *Comput. Phys. Commun.* **185** 165
[33] Maaßberg H, Beidler C D and Marushchenko N B 2012 *Phys. Plasmas* **19** 102501
[34] Reiter D 2005 *The EIRENE Code User Manual* www.eirene.de
[35] Feng Y *et al* 2014 *Contrib. Plasma Phys.* **54** 426
[36] Feng Y *et al* 2011 *Plasma Phys. Control. Fusion* **53** 024009
[37] Tanaka K *et al* 2010 *Fusion Sci. Technol.* **58** 70
[38] Pégourié B *et al* 2007 *Nucl. Fusion* **47** 44
[39] Maaßberg H *et al* 2005 *Plasma Phys. Control. Fusion* **47** 1137
[40] Shaing K C and Callen J D 1983 *Phys. Fluids* **26** 3315

IOP Publishing

Plasma Physics and Controlled Fusion

Plasma Phys. Control. Fusion **57** (2015) 014005 (10pp)

doi:10.1088/0741-3335/57/1/014005

Electric propulsion for small satellites

Michael Keidar, Taisen Zhuang, Alexey Shashurin, George Teel, Dereck Chiu, Joseph Lukas, Samudra Haque and Lubos Brieda

Mechanical and Aerospace Engineering, The George Washington University, Washington, DC 20052, USA

E-mail: keidar@gwu.edu

Received 1 July 2014, revised 28 August 2014
Accepted for publication 1 September 2014
Published 28 November 2014

Abstract

Propulsion is required for satellite motion in outer space. The displacement of a satellite in space, orbit transfer and its attitude control are the task of space propulsion, which is carried out by rocket engines. Electric propulsion uses electric energy to energize or accelerate the propellant. The electric propulsion, which uses electrical energy to accelerate propellant in the form of plasma, is known as plasma propulsion. Plasma propulsion utilizes the electric energy to first, ionize the propellant and then, deliver energy to the resulting plasma leading to plasma acceleration. Many types of plasma thrusters have been developed over last 50 years. The variety of these devices can be divided into three main categories dependent on the mechanism of acceleration: (i) electrothermal, (ii) electrostatic and (iii) electromagnetic. Recent trends in space exploration associate with the paradigm shift towards small and efficient satellites, or micro- and nano-satellites. A particular example of microthruster considered in this paper is the micro-cathode arc thruster (μCAT). The μCAT is based on vacuum arc discharge. Thrust is produced when the arc discharge erodes some of the cathode at high velocity and is accelerated out the nozzle by a Lorentz force. The thrust amount is controlled by varying the frequency of pulses with demonstrated range to date of 1–50 Hz producing thrust ranging from 1 μN to 0.05 mN.

Keywords: plasma propulsion, micropropulsion, micro-cathode arc thruster

(Some figures may appear in colour only in the online journal)

1. Introduction

Propulsion is required for satellite motion in outer space. The orbit transfer of a satellite, station-keeping and attitude control are performed by utilizing rocket engines. Rocket engines operate according to the basic principle of action and reaction leading to momentum transfer. A force (thrust) F on the spacecraft is formed by ejecting a jet of gas or plasma in the backward direction. In principle, there are two major types of rocket engines, which are distinguished by the energy source used to accelerate the gas, namely chemical and electric rockets. Electric propulsion uses electric energy to energize or accelerate the propellant. The electric propulsion, which uses electrical energy to accelerate propellant in the form of plasma, is known as plasma propulsion. Plasma propulsion utilizes electric energy to first, ionize the propellant, and then, deliver energy to the resulting plasma leading to plasma acceleration.

An important characteristic of the plasma thruster is the total propulsive action on a spacecraft, which is measured by the velocity change. The term delta-V (ΔV) constitutes the sum of all velocity changes by the spacecraft due to thruster operation. In the proximity of large celestial bodies, ΔV is needed to compensate various ambient forces, including atmospheric drag, radiation pressure, etc. Such orbital maintenance of the satellite (for instance a GEO communication satellite) requires $\Delta V \sim 0.6\,\text{km s}^{-1}$ for a 10-year period. The ΔV of deep-space missions are typically several km s^{-1}, and a ΔV for an orbit transfer is about a few km s^{-1}.

Many types of plasma thrusters have been developed over last 60 years. The variety of these devices can be divided into three main categories dependent on the mechanism of acceleration [1]: (i) electrothermal, (ii) electrostatic and (iii) electromagnetic.

(i) Electrothermal thrusters employ electric energy to heat propellant. In this case the acceleration mechanism is based on the pressure gradient, which is the same acceleration mechanism as the one employed in chemical thrusters.

(ii) Electrostatic thrusters directly employ an electric field to accelerate ions.

(iii) Electromagnetic thrusters wherein combination of both electric and magnetic fields are employed to accelerate the propellant.

Recent trends in space exploration associate with the paradigm shift towards small and efficient satellites, or micro- and nano-satellites. There are many near-future space missions involving science, military, and commercial payloads utilizing micro- and nano-satellite platforms. These platforms require very small levels of thrust for very fine attitude control, for high resolution Earth imaging and astronomy, and very fine positioning requirements, for spacecraft formation flying and interferometry missions. Nowadays many basic components of a spacecraft are being miniaturized allowing micro-satellites and nano-satellites to be designed and built. To satisfy the needs of both the low-thrust missions and the small-scale spacecraft, miniaturized propulsion systems are required.

It should be pointed out that this paper does not attempt to address the entire field of micropropulsion in detail, but rather describe major issues associated with plasma microthrusters. Some of the mature micropropulsion concepts were described in a recent review article [2]. Broader view of electric propulsion can be found in several dedicated books by Jahn [3], Jahn and Choueiri [4], Goebel and Katz [5], and a special issue of *IEEE Transactions of Plasma Science* (vol 36, issue 5, part 1,2008) [6].

The rest of this paper will be organized as follows. In section 2 we will overview various micropropulsion concepts. In section 3 we will describe in detail one particular example of microthruster, the micro-cathode arc thruster.

2. Plasma-based micropropulsion

Micropropulsion devices producing micro-Newton levels of thrust or micro-Newton-second impulse bits are currently under development in many academic, government and industrial laboratories, and yet the field of micropropulsion is still in its infancy [7]. Recall that the most important issue for microthrusters is propellant management and as such issues related to propellant become critical. In the next section we describe various propulsion concepts and we adopt natural division based on the propellant mechanism. In general, ablative thrusters are the simplest devices in terms of functionality and in fact were the first to be flown in space.

2.1. Ablative pulsed plasma thruster

Ablative pulsed plasma thrusters (PPTs) have achieved a high degree of maturity over several decades of research and development, and flight. The first flight of a PPT took place on the Russian Zond 2 spacecraft in 1964. Subsequent flights took place in the US on the geosynchronous MIT Lincoln Lab LES-6 satellite in 1968, 1974 and the US Navy TIP/NOVA satellite in 1981.

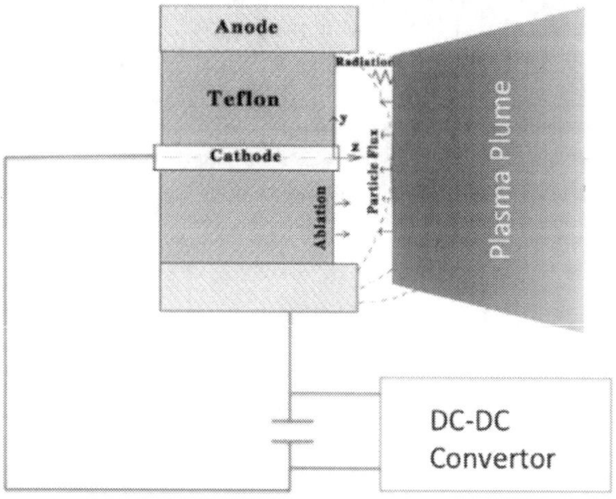

Figure 1. Schematic of the co-axial micro-PPT.

PPTs are currently considered an attractive propulsion option for mass and power limited satellites that require μN s to mN s impulse bits [8–13]. One of the most extensively analyzed propulsion functions has been attitude control. For missions that require precision pointing, the PPT offers a unique advantage over other technologies as it delivers a small impulse 'bit' with high specific impulse (exhaust velocity). Another major functional area is station-keeping, which includes drag compensation and formation flying. These functions require less than 1 mN of thrust. For instance, interferometer missions [7] such as ST-3 require sub-millimeter relative positioning precision. An electromagnetic PPT was successfully operated for pitch axis control on the EO-1 spacecraft [14, 15]. It was shown that PPT can be easily scaled down in power and size. A micro-PPT (μPPT) is the miniature version of the traditional PPT, and has been designed at the Air Force Research Laboratory (AFRL) for delivery of very small impulse bits [16, 17]. AFRL μPPT was developed for a demonstration mission on TechSat21 [18] and was recently employed at FalconSatIII. The μPPT can deliver a thrust in the 10 μN range to provide attitude control and station-keeping for micro-satellites.

With the μPPT, the discharge across the propellant surface ablates a portion of the propellant, ionizes it, and then accelerates it predominantly electromagnetically to generate the thrust (schematically this thruster is shown in figure 1) [19, 20]. It is expected that the use of electromagnetic acceleration to create thrust will also lead to relatively high specific impulse.

Recall that both discharge energy (peak current) and thruster size significantly affect the discharge uniformity in the azimuthal and radial directions. Azimuthal non-uniformity relates to the current constriction and anode spot formation phenomena. This effect occurs when the discharge current or thruster size exceed some critical value. Non-uniformity in discharge associated with the spot mode leads to a much higher localized ablation rate and causes degradation of the specific impulse. On the other hand, small discharge current leads to strong Teflon surface carbonization (charring) and radial non-uniformity, which in turn leads to thruster failure [20]. The primary mechanism of the charring formation was identified and is related to carbon backflux. Thruster size and discharge energy can be optimized by trading between two conflicting requirements of large pulse energy (to prevent charring) and small discharge energy (to prevent current constriction) [21]. Because the charring phenomenon is completely intolerable and leads to thruster failure, the optimal discharge energy should be chosen somewhere near the spot formation limit.

2.2. Micro-laser plasma thruster

Another good example of the ablative type of micro-propulsion system is the micro-laser-ablation plasma thruster (μ-LPT). Some variations of this device have been developed by Phipps and Luke [22] and Gonzalez and Baker [23]. In [22] a Q-switched micro-chip laser, pumped by a cw diode laser, was used to ablate an aluminum target generating thrust in the range of 0.3 nN to 3 μN with power consumption of about 5 W and pulse frequency in the range of 1 Hz to 10 kHz. The wide dynamic range of thrust levels provided by these devices is one of their most attractive features.

The micro-Laser-Ablation Plasma Thruster (μ-LPT) developed by Phipps and Luke [22] uses a 1–10 W, high-brightness diode laser irradiating various absorbing material and substrate combinations (e.g. black ink on paper, black PVC on Kapton™). Laser coupling coefficients on the order of 60 μN W^{-1} and specific impulses on the order of 500 s with a 1 W laser are achieved. One of the major advantages of the μ-LPT is its large dynamic range of impulse bit that can be varied between 0.4 to 16 μN s by simply increasing the laser pulse duration. In addition, selection of the absorber and substrate materials allows the specific impulse and laser characteristics to be tailored for specific mission requirements.

The μ-LPT can be operated in two different modes. In reflection mode (R), the laser is incident on the target and the ablated material ‘reflects’ from the surface. This mode has the potential problem of leading to deposition of plume effluent on the laser optics. In transmission mode (T), the laser passes through a transparent substrate film from the back. The substrate is coated on the other side with an absorbing material that ablates material away from the laser. This approach circumvents the problem of optics contamination found with the R-mode. However, the dynamic range of impulse bit available in T-mode is more restrictive. Coating of the laser optics by plume deposition is one of the major lifetime limitations of the μ-LPT in R-mode. Therefore, there is a certain preference for development of T-mode operation. Computational analysis of the R-mode μ-LPT was performed using hybrid particle-in-cell technique [24]. A lens focuses the laser diode output on a 25 μm diameter spot on the transparent side of a fuel tape. The beam heats an absorbing coating to high temperature, producing a miniature ablation jet. The material that is ablated is usually PVC or Kapton™. Typical parameters of operation are: power of 2 to 14 W, pulse duration of 3–10 ms. The fuel tape thickness is about 185 μm, composed of 125 μm of transparent backing (usually cellulose acetate) and about 60 μm of absorbing coating. Typically Q^* (energy of laser light required to ablate 1 kg of target material) is about 2×10^7 J kg^{-1} and the momentum coupling coefficient is about $C_w = 60 - 100\,\mu$N W^{-1}. Computational analysis of the T-mode μ-LPT was also developed and compared with an experiment [25]. Generally the LPT models have satisfactorily described the main features of the thruster, such as ablated mass and plume expansion.

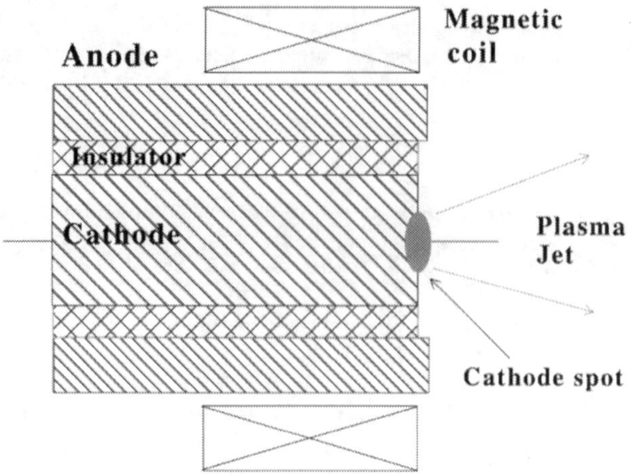

Figure 2. Schematics of the vacuum arc thruster (reproduced with permission from [32]).

2.3. Micro-vacuum arc thruster

One possibility to improve some aspects of PPTs without compromising its efficiency is to use metal as a propellant. Recent success in the development of vacuum arc technology has attracted much attention to this technology [26]. Metal propellant will have the following benefits: lower energy consumption per mass ionized (due to the low ionization potential of metals), high ionization degree, operation with higher repetition rates since the metal melting temperature is higher than that of polymer dielectrics used in some current thrusters, and high pulse-to-pulse stability. For these reasons a vacuum arc source was developed and used for an ion thruster [27]. This work led to development of a co-axial vacuum arc plasma thruster [28]. Schematically this thruster is shown in figure 2. However, the measured efficiency was reported to be only ~1.6% which is strongly different from the estimations, leaving ample room for improvement. On the other hand several problems related to this technology were reported, including lifetime, ion current degradation with cathode recession, etc [29]. Therefore understanding these problems and the efficiency limitations of the vacuum arc thruster technology is an important issue. In addition, basic aspects of vacuum arc thrusters such as thrust mechanism, effects of the magnetic field of the thruster operation and its plume, and droplet generation were not studied in detail. Very recently it was shown that using an axial magnetic field (so-called magnetically enhanced vacuum arc thruster) helps to improve thruster characteristics, such as specific impulse (plasma velocity) and decrease plasma plume divergence [30, 31].

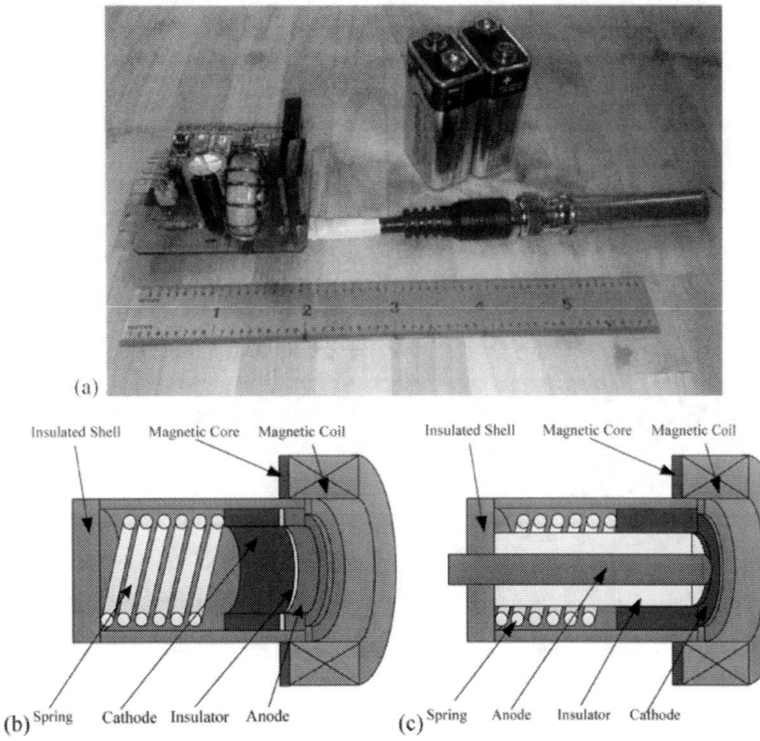

Figure 3. (*a*) μCAT with PPU. (*b*) Schematic design of the ring shape μCAT, (*c*) schematic design of co-axial electrodes μCAT.

With the vacuum arc thruster the thrust is created in the minute spot on the cathode surface called cathode spots (having size of about 10–100 μm) [32]. The high-pressure plasma in the cathode spot creates the reaction force (thrust) on the cathode as schematically shown in figure 2. As earlier as in the 1930s it was found experimentally that the plasma stream exerts a reaction force on the cathode, which is about 17 ± 3 dyn A^{-1} (or ~0.2 mN A^{-1}) [33, 34]. It is very important to note that since the reaction force is proportional to the arc current, the thrust can be controlled relatively easily.

3. Micro-cathode arc thruster

Recently, a new thruster configuration, the micro-cathode arc thruster (μCAT), was developed motivated by need for an efficient low mass propulsion system. This thruster improves on the vacuum arc discharge thruster by applying a specially designed external magnetic field. The unique magnetic field conditions achieved with a vacuum arc offer several potential advantages in these devices [35].

The μCAT is a simple electric propulsion device and combined with a magnetic coil and an inductive energy storage power processing unit (PPU) results in a low mass (<100 g) system. A picture of the μCAT system and two types of geometries of a thruster are shown in figures 3(*a*)–(*c*). Figure 3(*b*) shows the schematic design of the ring electrodes μCAT (RE-μCAT), which consist of an annular titanium cathode and a same diameter annular copper anode with 1 mm width. The annular ceramic insulator tube having same inner and outer diameters and a width of about 1 mm was used as separator between the arc electrodes. Figure 3(*c*) shows the schematic design of the co-axial μCAT. Instead of the ring electrodes, this design employs cylindrically shaped cathode and anode.

Figure 4(*a*) presents a schematic of the thruster and the PPU system. The PPU that has been designed is equipped with an inductive energy storage system as shown in figures 4(*a*) and (*b*). When the trigger pulse is applied to a semiconductor insulated gate bipolar transistor (IGBT) switch, the energy is accumulated in the inductor, while when the trigger pulse ends; a surge voltage with the magnitude proportional to $L\,\mathrm{d}I/\mathrm{d}t$ is generated on the inductor and applied to the electrodes. This leads to a breakdown and initiation of arc discharge between the electrodes. A coil has been applied to the outside of the thruster to produce magnetic field as indicated in figure 4(*a*). The field strength was simulated using Finite Element Method Magnetics (FEMM) magnetic field simulation software. The direction of the magnetic field could be simply reversed by reversing the coil current.

The μCAT operates by producing a fully ionized plasma at the inner surface of the electrode. The plasma is formed in the cathode spots, and expands into the vacuum zone under the applied magnetic field gradient. The plasma then accelerates and expands into vacuum at high velocities, around tens of thousands m s^{-1}, which results in an impulse bit. The effect of the magnetic field on the thruster operation is clearly visible in figure 4(*b*). This figure shows a CCD camera observation of the RE μCAT firing in the vacuum chamber with an added magnetic field ($B = 0.3$ T) and without.

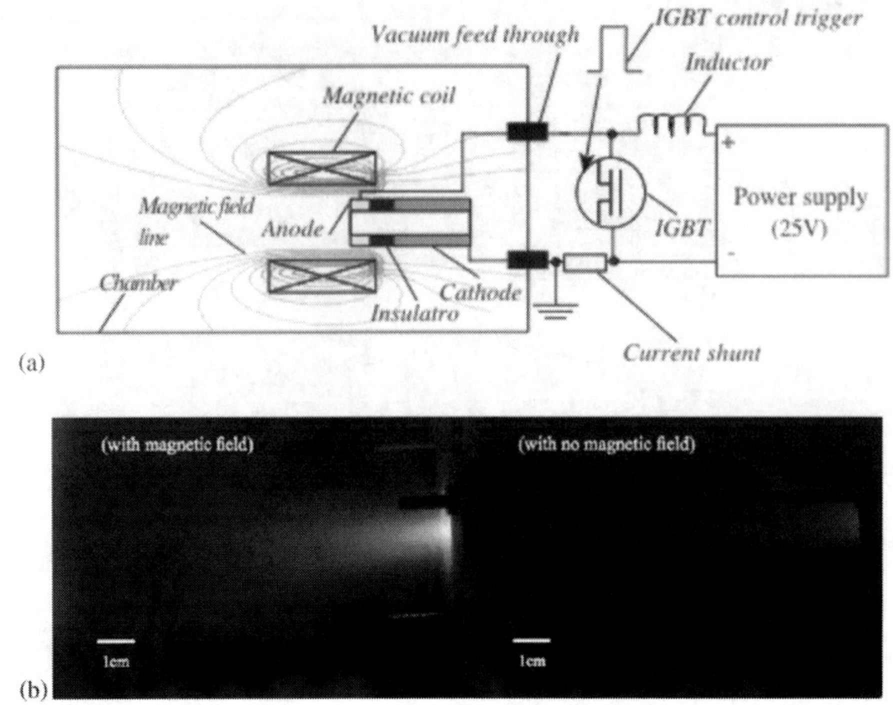

Figure 4. (*a*) Schematic of micro-cathode vacuum arc thruster experimental arrangement. (*b*) CCD camera observation of plasma plume with magnetic field (left image, $B = 300$ mT) and with no magnetic field.

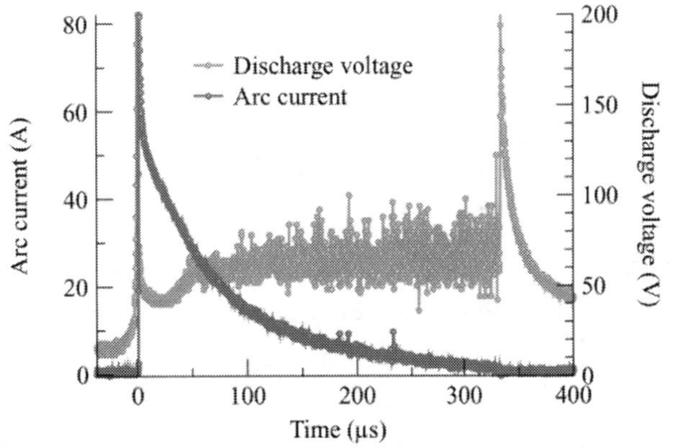

Figure 5. Evolution of μCAT arc current (blue line) and discharge voltage (yellow line) in the case of a magnetic field of about 100 mT.

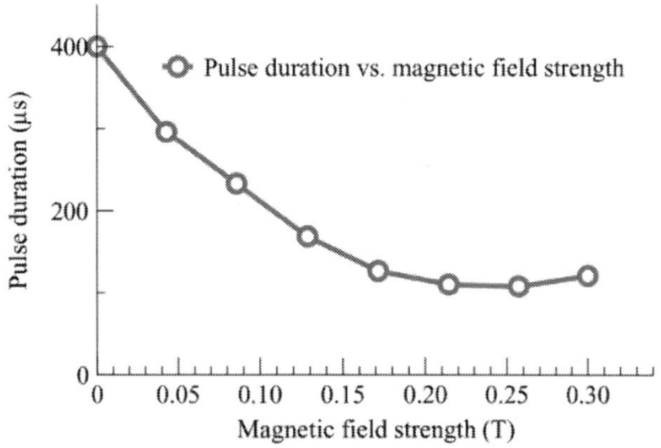

Figure 6. Thruster pulse duration evolution with different magnetic field strength.

3.1. Experimental characterization of the micro-CAT

The arc voltage during the discharge process remains approximately constant, yet it is found that this constant voltage varies with the magnetic field strength. A typical sample of μCAT arc current and discharge voltage evolution is indicated in magnetic field strength of about 100 mT. Typically, during the discharge process, the current that was flowing through the solid-state switch is fully transferred to the cathode spots, as indicated in the figure 5. Consequently, the current dropped down from around 80–100 A to 0 A (for around 100–400 μs varied with magnetic field strength as shown in figure 5). Pulse duration is also affected by a magnetic field as shown in figure 6. A previous work of Keidar *et al* [32] indicated that an axial magnetic field located near the anode region plays an important role, and indicated that the arc voltage increases when a magnetic field is applied. The experimental results agree well with the theoretical predication.

The plasma formed by a vacuum arc is created on the cathode surface spots. Previously optical method were employed and it was observed that the spot consists of either a homogeneous bright region or consists of cells and fragments with a typical total size of about 10–100 μm [36, 37]. The μCAT is equipped with a coil that provides magnetic field strength in the range of 0–300 mT. The observation of cathode spot motion under magnetic field was first studied in the

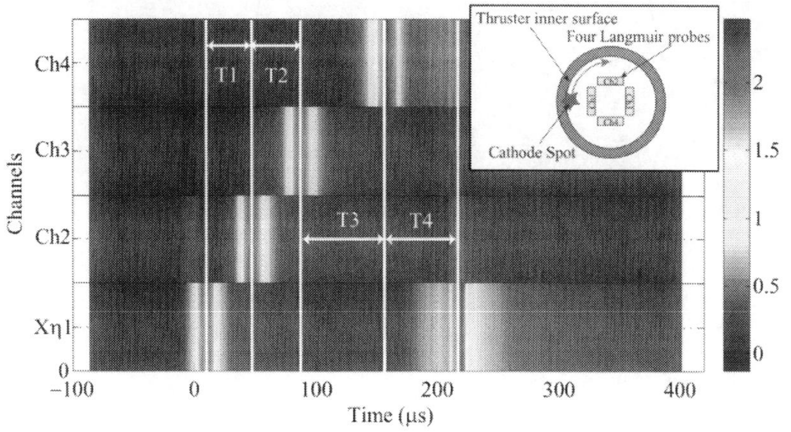

Figure 7. Example of four Langmuir probes method to measure the cathode spot rotation speed.

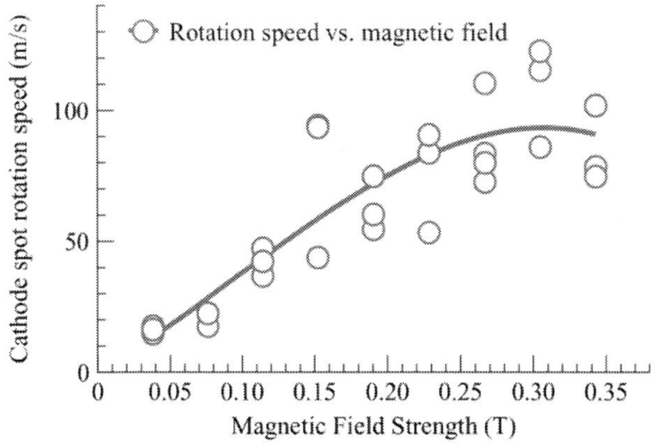

Figure 8. Distribution of μCAT cathode spot rotation speed with different magnetic field strengths [40].

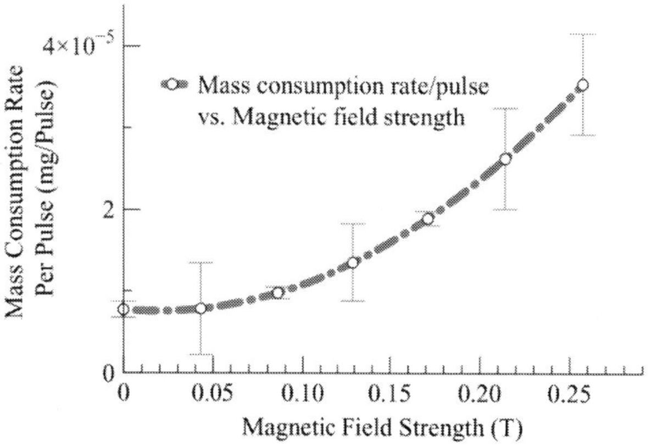

Figure 9. Thruster cathode material erosion rate with different magnetic field strength.

1960s [38, 39]. It is known that the presence of a transverse magnetic field at the cathode surface produces arc rotation in the $-J \times B$ direction. The observed vacuum arc cathode spot rotation has important implications for propulsion. The benefits are observable in that the cathode spot rotation leads to uniform cathode erosion, which is critical for assuring long thruster lifetime. The μCAT cathode spot rotation could be measured by utilizing 4-probe assembly of Langmuir probes. Four single probes were located along the azimuth direction inside the thruster channel, and the four probes ion current measurement results are shown in figure 7. The rotation speed was calculated using a quarter of circumference of the thruster's inner surface, divided by the delay time between each two neighbor peaks. The average rotation speed is shown in figure 8. It was found that the spot rotation speed increased about 5 times (from 20 to 100 m s^{-1}) as the magnet field strength increased (from 0 to 300 mT). More detailed descriptions of this effect can be found in [40].

It was also found that the applied magnetic field increases the thruster erosion rate. In [41] the cathode material erosion rate was reported to increase almost three times when the magnetic field was increased from zero up to 300 mT. The mass consumption rate was measured utilizing a highly accurate mass balance. These results (see figure 9) show that the added magnetic field leads to an increase of mass consumption rate by a factor of 3, from around 1×10^{-5} mg/pulse at zero magnetic field strength to around 3×10^{-5} mg/pulse at 0.25 T magnetic field. There are two reasons for increased mass consumption rate. Firstly, the magnetic field contributes to the transport of the metal plasma out of the thruster channel. Secondly, the rotation of cathode spots promotes erosion efficiency by moving the spot over a larger area of the thruster cathode.

3.2. μCAT plume characterization

In this section we present a 2D spatial and temporal evolution of the plasma plume outside the thruster channel in the presence of an applied magnetic field. The two-dimensional analysis of plasma flow affected by an externally imposed magnetic field was performed by Keidar and co-workers [42–44]. The experimental measurement of μCAT plasma flow was done in [45]. In that work, the spatial distribution of plasma was obtained using a set of axially aligned circular Langmuir probes. Each probe consisted of eight concentric circular planar probes separated by an insulating ring that were aligned with the thruster axis. Figure 10 presents the 2D ion current maps for different magnetic field strengths. One can see that

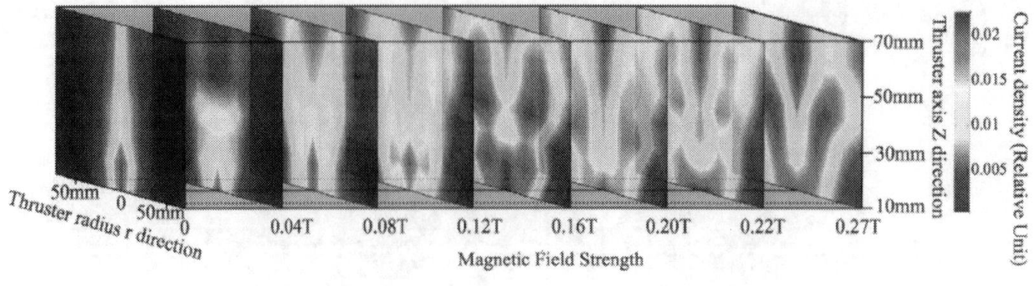

Figure 10. D spatial distribution of ion current with different magnetic field.

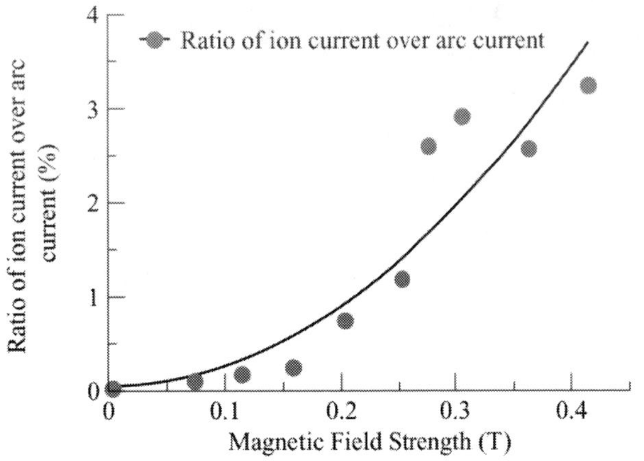

Figure 11. The ratio of thruster total ion current over the arc current with different magnetic field.

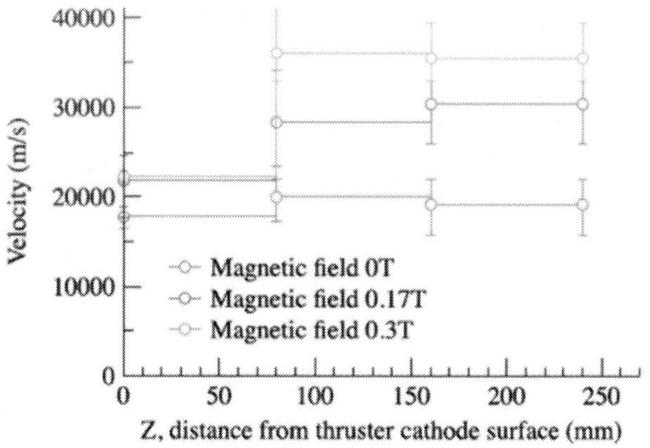

Figure 12. Measured ion velocity along the axis with magnetic field as a parameter. Reprinted with permission from [47]. Copyright 2011, AIP Publishing LLC.

magnetic field leads to a plasma collimation along magnetic field lines as shown in figure 10. In addition, one can notice the hollow shape of the plasma plume consequent by increasing the magnetic field strength. It can be noticed that the external magnetic field of the μCAT leads to an improvement of plasma plume transport efficiency along the magnetic field.

The total ion beam current in the plume is a critical parameter needed to evaluate the thruster efficiency. It is well known that in the case of Ti cathode the maximum ratio of ion current over arc current is about 8% [46]. As indicated in [45], the measured μCAT total ion current output with no magnetic field is about 0.06%, yet, when the magnetic field is applied the aforementioned efficiency increases up to 3.5%, as shown in figure 11. This translates into approximately 45% of transport efficiency.

3.3. Ion velocity in micro-CAT

In this section we will discuss the ion acceleration mechanism in the μCAT. Recall that based on our measurements [47] one can identify three characteristics regions. Approximately up to 80 mm from the cathode is the interelectrode region, in which plasma is being produced and accelerated by a pressure gradient due to electron–ion coupling, i.e. by the gas dynamic mechanism. It should be pointed out that the ion velocity close to the arc source is independent of the magnetic field, which demonstrates that the magnetic field strength utilized in our experiments does not significantly affect the gas dynamic acceleration of ions in vicinity of the cathode spot. The second region, which is about 80–150 mm from the cathode surface is the plasma expansion inside and outside the thruster, where ion acceleration in a magnetic field was observed as shown in figure 12. Let us consider a possible mechanism of plasma acceleration in this region. Plasma flow is subject to the electromagnetic force in the divergent magnetic field region:

$$Mn\frac{\mathrm{d}V}{\mathrm{d}t} = j_\theta B_r \quad (1)$$

where M is the ion mass, n is the plasma density, j_θ is the azimuthal electron current and B_r is the radial component of the magnetic field. One can estimate the azimuthal electron current density as

$$j_\theta = \frac{\omega_e}{\nu_{ei}} j_r \quad (2)$$

where j_r is the electron radial current density. In the plasma jet outside of the interelectrode gap, the global current is zero so that the electron current is equal to the ion current. Thus

$$j_{er} = j_{ir} = enV_r \quad (3)$$

where V_r is the radial component of the ion current. The radial component of the plasma is developed in course of the plasma expansion [42] and could be about 0.3 of the axial component in a large magnetic field.

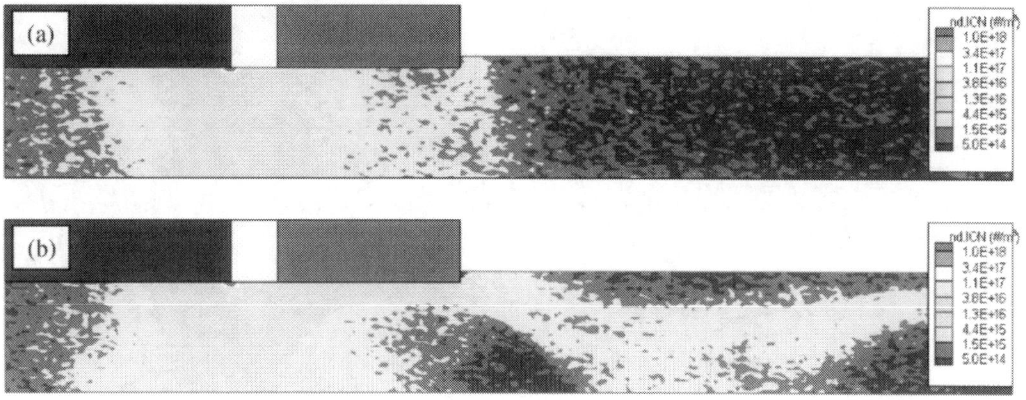

Figure 13. Comparison of ion density without (*a*) and with (*b*) the applied magnetic field. Thruster components correspond to the figure 3.

Based on the above scenario one can estimate the ion velocity change in the axial direction as

$$\Delta V_z \approx \sqrt{\frac{2e^2 B_z B_r V_r \Delta z}{m M \nu_{\mathrm{ei}}}}. \tag{4}$$

Using the following typical parameters of the plasma jet: B (i.e. both B_r and B_z) ~ 0.1 T, $n \sim 10^{20}\,\mathrm{m^{-3}}$, $V_r \sim 10^4\,\mathrm{m\,s^{-1}}$ and $\Delta z \sim 0.01$ m one can estimate that $\Delta V_z \sim 10^4\,\mathrm{m\,s^{-1}}$. One can point out that these results are in good agreement with experimental measurements.

It should be pointed out that cathodic arc plasmas include macroparticles generated in the cathode spot [46]. The size and density of macroparticles depend on the cathode material and arc operation [46]. Typically these are 1–10 μm in diameter spherical droplets [48]. Presence of macroparticle in the plume might raise a concern with respect to thruster contamination. However, it should be pointed out that macroparticle emission generally decreases in the case of a pulsed arc in comparison with dc vacuum arc. In addition, emitted macroparticles interact with high-velocity plasma jet and momentum transfer between plasma and macroparticle leads to macroparticle emission in the direction of jet propagation [49], i.e. away from the satellite. Since no backflux was determined in the case of micro-CAT one can expect no macroparticle backflux and as such negligible contamination risk.

3.4. Simulation of the μCAT

We have also completed an initial numerical analysis of the thruster. The primary goal of this investigation was to demonstrate the feasibility of the numerical approach. We were interested in determining whether a fully-kinetic approach, in which both electrons and ions are treated as particles, is suitable for modeling the turning of the ion beam, and the increased ion velocity observed experimentally. Various kinetic codes were developed for application in electric propulsion [50–52]. As such, we set up the simulation using the 3D electrostatic particle-in-cell code Draco [53]. In this work, two cases are considered. The first case did not contain any magnetic field. In the second case, the background magnetic field was loaded, corresponding to coil current of 0.4 A. In this configuration, the magnetic field lines form an angle of approximately 15° from the cathode surface at the location of the cathode spot. Since Draco is a 3D code, the axisymmetric geometry is represented by modeling a thin slab, in which the periodic boundary condition was applied to the front and back faces. The simulation domain can be seen in figure 13. The cathode is shown by the orange rectangle. The gray rectangle is the insulator, and the blue shape is the anode. Potential difference of 50 V was applied between the electrodes. Electrons and ions were injected from a small control surface located at the junction of the cathode and the insulator, at flow rates corresponding to 0.4 A. We assumed that the beam consisted solely of electrons and doubly charged Ti^{2+} ions [54]. Neutral particles were not modeled directly, but instead we loaded a constant background density corresponding to 6×10^{-5} Torr. Collisions were included through the Monte Carlo collisions (MCC) scheme with cross-sections determined from analytical models. Momentum transfer and Coulomb collisions were considered. The simulation contained $10 \times 50 \times 226$ cells, and approximately 2 million particles. No artificial scaling of permittivity was performed. Figures 13(*a*) and 13 (*b*) show the preliminary results. Comparison of these figures clearly illustrates the influence of the magnetic field. In figure 13(*a*), in which there is no magnetic field, the ions are seen to simply diffuse out from the cathode spot. No significant thrust is expected to be generated in this configuration. However, the addition of the magnetic field, figure 13(*b*), clearly shows the extraction of the ion beam from the thruster as demonstrated experimentally. Although not shown here, the simulation results presented in figure 13(*b*) indicate that with adding a magnetic field, the ion velocity reaches the magnitude of $4 \times 10^4\,\mathrm{m\,s^{-1}}$ which agrees with the experimental results.

3.5. Comments on comparison of micropropulsion technologies

The comparison of micropropulsion technologies operating in the low powers range of <10 W is presented in table 1. Utilization of metal propellant allows the μCAT system to carry larger amounts of propellant in smaller volume in comparison with ion thruster and PPT, which utilize gas and

Table 1. Comparison of micropropulsion technologies.

	μCAT (GWU)	PPT (Clyde Space) [55, 56]	PPT (Busek Co) [57–59]	Electrospray (MIT) [60–62]	Electrospray (Busek Co) [57, 58]	VAT (Alameda) [63]
System mass (g)	200	160	550	45	1150	600
System volume (cm^3)	200	200	500	300	500	200
Propellant	Metal	Teflon	Teflon	Liquid	Liquid	Metal
I_{sp} (s)	3000	590	700	3000	800	1500
Propellant mass (g)	40	10	36	20	75	40
Delta-V (for 4 kg satellite) ($m\,s^{-1}$)	300	15	63	150	151	151
Efficiency (%)	15	4.7	16	71	31	9.4
Thrust-to-mass ratio ($\mu N\,g^{-1}$)	0.63	0.03	0.18	0.5	0.65	0.22
Ionization degree	High	Low	Low	High	High	High
Cost	Low	Low	Low	High	High	Low
Technical readiness level (TRL)	6	7	7	2–3	5	4

Teflon propellants respectively. It is seen that the μCAT subsystem has mass and volume characteristics comparable or exceeding that of other micropropulsion technologies. In addition, the μCAT is providing the highest I_{sp} of about 3500 s, due to the high speed of the plasma produced by the cathodic arc. Both of these factors (large amount of carried propellant and high I_{sp}) ensures leadership of μCAT technology in the Delta-V parameter. The μCAT is characterized by an ionization degree close to 100% in the plasma jets in comparison with several tens of percent for PPT and ion thruster technologies. Low ionization degree is a critical issue, since it causes strong back flux to a satellite and results in contamination issues. Finally, μCAT technology is characterized by very high overall efficiency and thrust-to-mass ratio in comparison with competing technologies.

4. Concluding remarks

In this paper, a comprehensive characterization of the μCAT has been presented. It has been found that the thruster with different magnetic strength exhibits certain regularities in its behavior, which has been observed through experimental measurements. The cathode spot motion was observed and results indicated that the magnetic field leads to cathode spot rotation which is critical for uniform cathode erosion. The plasma plume distribution outside the thruster channel has been presented to analyze the effects of the magnetic field on the plasma plume. A preliminary numerical analysis of the thruster has also been completed, with the simulation results confirming the ion extraction effect of the magnetic field. The μCAT equipped with a magnetic field operates efficiently and gives great flexibility in specific impulse and impulse bit by simply varying the magnetic field strength.

Acknowledgment

This work was supported in part by NASA DC grant.

References

[1] Choueiri E Y 2007 Plasma Propulsion *McGraw-Hill Encyclopedia of Science and Technology* (New York: McGraw-Hill)
[2] Ahedo E 2011 *Plasma Phys. Control. Fusion* **53** 124037
[3] Jahn R 1968 *Physics of Electric Propulsion* (New York: McGraw-Hill)
[4] Jahn R and Choueiri E 2002 Electric propulsion *Encyclopedia of Physical Science and Technology* (New York: Academic) pp 125–41
[5] Goebel D and Katz I 2008 *Fundamentals of Electric Propulsion: Ion and Hall Thrusters* (New York: Wiley)
[6] Keidar M, Gallimore A D, Raitses Y and Bouef J P (ed) 2008 Special Issue on Plasma Propulsion *IEEE Trans. Plasma Science* **36** 1962–6
[7] Keidar M and Beilis I I 2013 *Plasma Engineering: Applications from Aerospace to Bio and Nanotechnology* (New York: Academic)
[8] Spores R A, Cohen R B and Birkan M 1997 The USAF electric propulsion program *Proc. 25th Int. Electric Propulsion Conf. (Worthington, OH, 1997)* vol 1, p 1
[9] Micci M M and Ketsdever A D (ed) 2000 *Micropropulsion for Small Spacecraft* (Reston, VA : AIAA)
[10] Burton R L and Turchi P J 1998 Pulsed plasma thruster *J. Propulsion Power* **14** 716–35
[11] Scharlemann C A, Corey R, Mikellides I G, Turchi P J and Mikellides P G 2000 Pulsed plasma thruster variations for improved mission capabilities *AIAA Paper*-00-3260
[12] Choueiri E Y 1996 System optimization of ablative pulsed plasma thruster for stationkeeping *J. Spacecraft Rockets* **33** 96–100
[13] Keidar M and Boyd I D 2000 Device and plume model for an electrothermal pulsed plasma thruster *AIAA Paper* 2000–3430
[14] Dunning J W, Benson S and Oleson S 2001 NASA's electric propulsion program *27th Int. Electric Propulsion Conf. (Pasadena, CA, October 2001)* IEPC-01-002
[15] Zakrzwsky C, Benson S, Sanneman P and Hoskins A 2002 On-orbit testing of the EO-1 pulsed plasma thruster *38th Joint Propulsion Conf. (Indianapolis, IN, July 2002)* AIAA-2002-3973
[16] Spanjers G G, White D, Schilling J, Bushman S, Lake J and Dulligan M 2001 AFRL microPPT development for the TechSat21 flight *27th Int. Electric Propulsion Conf. (Pasadena, CA 2001)* IEPC Paper 2001-166

[17] Spanjers G G *et al* 2002 AFRL microPPT development for small spacecraft propulsion *38th AIAA Joint Propulsion Conf. (Indianapolis, IN, July 2002)* Paper AIAA-2002-3974
[18] Gulchinski F S *et al* 2000 Micropropulsion research at AFRL *36th AIAA Joint Propulsion Conf. (Huntsville, AL, July 2000)* Paper AIAA-2000-3255
[19] Keidar M, Boyd I D and Beilis I I 2003 Model of an electrothermal pulsed plasma thruster *J. Propulsion Power* **19** 424–30
[20] Keidar M, Boyd I D, Antonsen E L, Gulczinski F S III and Spanjers G G, Propellant charring in pulsed plasma thrusters 2004 *J. Propulsion Power* **20** 978
[21] Keidar M, Boyd I D, Antonsen E L, Burton R L and Spanjers G G 2006 Optimization issues for a micro-pulsed plasma thruster *J. Propulsion Power* **22** 48–55
[22] Phipps C and Luke J 2002 Diode laser-driven microthrusters: a new departure for micropropulsion *AIAA J.* **40** 310–17
[23] Gonzalez D A and Baker R P 2001 Microchip laser propulsion for small satellites *37th AIAA/ASME/SAE/ASEE Joint Propulsion Conf. (Salt Lake City, UT, USA, July 2001) AIAA Paper* 2001-3789
[24] Boyd I D and Keidar M 2002 Simulation of the plume generated by a micro laser-ablation plasma thruster *High-Power Laser Ablation IV* ed C R Phipps *Proc. SPIE* **4760** 852
[25] Keidar M, Boyd I D, Luke J and Phipps C 2004 Plasma generation and plume expansion for a transmission-mode micro-laser plasma thruster *J. Appl. Phys.* **96** 49–56
[26] Polk J *et al* 2001 A theoretical analysis of vacuum arc thruster performance *IEPC Paper* IEPC-2001-211
[27] Qi N, Gensler S, Prasad R R, Krishnan M, Visir A and Brown I G 1998 A pulsed vacuum arc ion thruster for distributed small satellite systems *AIAA Paper* AIAA-98-3663
[28] Qi N, Schein J, Binder R, Krishnan M, Anders A and Polk J 2001 *37th Joint Propulsion Conf. (Salt Lake City, UT, 8–11 July 2001)*
[29] Schein J, Qi N, Binder R, Krishnan M, Ziemer J K, Polk J E and Anders A 2001 Low mass vacuum arc thruster system for station keeping missions *IEPC Paper* 01-228
[30] Au M, Schein J, Gerhan A, Wilson K, Tang B and Krishnan M 2004 *AIAA Paper* 2004-3618
[31] Keidar M and Schein J 2004 Modeling of a magnetically enhanced vacuum arc thruster *40th AIAA Joint Propulsion Conf. (Fort Lauderdale, FL, July 2004)* Paper AIAA-04-4117
[32] Keidar M, Schein J, Wilson K, Gerhan A, Au M, Tang B, Idzkowski L, Krishnan M and Beilis I I 2005 Magnetically enhanced vacuum arc thruster *Plasma Sources Sci. Technol.* **14** 661–9
[33] Tanberg R 1930 *Phys. Rev* **35** 1080
[34] Robertson R 1938 *Phys. Rev.* **53** 578
[35] Beilis I I 2002 *IEEE Trans. Plasma Sci.* **30** 2124–32
[36] Djakov B E and Holmes R 1974 *J. Phys. D: Appl. Phys.* **7** 569
[37] Juttner B 2001 *J. Phys. D: Appl. Phys.* **34** R103
[38] Murphree D 1970 *Phys. Fluids* **13** 1747
[39] Ecker G and Muller K G 1958 *J. Appl. Phys.* **29**1606
[40] Zhuang T, Shashurin A, Keidar M and Beilis I I 2011 Circular periodic motion of plasma produced by a small-scale vacuum arc *Plasma Sources Sci. Technol.* **20** 015009
[41] Zhuang T, Shashurin A, Denz T, Vail P, Pancotti A and Keidar M 2014 Performance characteristics of micro-cathode arc thruster *J. Propulsion Power* **30** 29–34
[42] Keidar M, Beilis I I, Boxman R L and Goldsmith S 1996 *J. Phys. D: Appl. Phys.* **29** 1973–83
[43] Beilis I, Keidar M, Boxman R L and Goldsmith S 1998 *J. Appl. Phys.* **83** 709–17
[44] Keidar M, Beilis I, Boxman R L and Goldsmith S 1997 *IEEE Trans. Plasma Sci.* **25** 580–5
[45] Zhuang T, Shashurin A and Keidar M 2011 *IEEE Trans. Plasma Sci.* **39** 2936–7
[46] Boxman R L, Martin P J and Sanders D M 1995 *Vacuum Arc Science and Technology* (Park Ridge, NJ: Noyes)
[47] Zhuang T, Shashurin A, Beilis I and Keidar M 2011 Ion velocities in a micro-cathode arc thruster *Phys. Plasmas* **19** 063501
[48] Keidar M, Beilis I, Boxman R L and Goldsmith S 1995 Non-stationary macroparticle charging in an arc plasma jet *IEEE Trans. Plasma Sci.* **23** 902–8
[49] Keidar M, Beilis I, Boxman R L and Goldsmith S 1996 Transport of macroparticles in magnetized plasma ducts *IEEE Trans. Plasma Sci.* **24** 226–34
[50] Taccogna F *et al* 2008 *Plasma Sources Sci. Technol.* **17** 024003
[51] Szabo J *et al* 2014 *J. Propulsion Power* **30** 197
[52] Kalentev *et al* 2014 *Contrib. Plasma Phys.* **54** 235
[53] Brieda L, Pierru J, Kafafy R and Wang J 2004 Development of the Draco code for modeling electric propulsion plume interactions *40th Joint Propulsion Conf. (Fort Lauderdale, FL, 2004)* AIAA-2004-3633
[54] Keidar M, Brown I G and Beilis I I 2000 Axial ion charge state distribution in the vacuum arc plasma jet *Rev. Sci. Instrum.* **71** 698–700
[55] Guarducci M and Gabriel S B 2011 Design and testing of a micro pulsed plasma thruster for Cubesat application *32nd Int. Electric Propulsion Conf. (Wiesbaden, Germany, 11–15 September 2011)* IEPC-2011-239
[56] www.clyde-space.com/cubesat_shop/propulsion/303_cubesat-pulse-plasma-thruster
[57] Hruby V 2012 High Is CubeSat propulsion *1st Interplanetary CubeSat Workshop (Massachusetts Institute of Technology, Cambridge, MA, 29–30 May 2012)*
[58] www.busek.com/
[59] Mueller J, Ziemer J, Hofer R, Wirz R and O'Donnell T 2008 A survey of micro-thrust propulsion options for microspacecraft and formation flying missions *Cube Sat—5th Annual Developers Workshop (California Polytechnic State University, San Luis Obispo, CA, 9–11 April 2008)*
[60] Martel F and Lozano P 2012 Ion electrospray thruster assemblies for CubeSats *iCubeSat Workshop (Cambridge, MA, 29–30 May 2012)*
[61] Gassend B L P 2007 Fully microfabricated two-dimensional electrospray array with applications to space propulsion *PhD Thesis* Massachusetts Institute for Technology
[62] Velásquez-García L F, Akinwande A I and Martínez-Sánchez M 2006 Planar array of micro-fabricated electrospray emitters for thruster applications *J. Micromech. Syst.* **15** 1272
[63] ww.aasc.net/micropropulsion

IOP Publishing

Plasma Physics and Controlled Fusion

Plasma Phys. Control. Fusion **57** (2015) 014006 (11pp)

doi:10.1088/0741-3335/57/1/014006

Energetic ion behaviour in MAST

M Cecconello[1], O M Jones[2,3], W U Boeglin[4], R V Perez[4], D S Darrow[5], I Klimek[1], S E Sharapov[2], M Fitzgerald[2], K G McClements[2], D L Keeling[2], S Y Allan[2], C A Michael[6], R J Akers[2], N J Conway[2], R Scannell[2], M Turnyanskiy[7] and G Ericsson[2] and the MAST Team[2]

[1] Department of Physics and Astronomy, Uppsala University, SE-751 05 Uppsala, Sweden
[2] CCFE, Culham Science Centre, Oxon OX14 3DB, UK
[3] Department of Physics, Durham University, Durham DH1 3LE, UK
[4] Department of Physics, Florida International University, 11200 SW, Miami, FL 33199, USA
[5] Princeton Plasma Physics Laboratory, PO Box 451, Princeton, NJ 08543, USA
[6] Australian National University, Canberra, ACT 0200, Australia
[7] ITER Physics Department, EFDA CSU Garching, Boltzmannstrasse 2, D-85748 Garching, Germany

E-mail: marco.cecconello@physics.uu.se

Received 27 June 2014, revised 14 August 2014
Accepted for publication 28 August 2014
Published 28 November 2014

Abstract

Recent studies of fast ion transport resulting from a range of instabilities, including $n = 1$ internal kink modes (fishbones and long-lived modes), toroidal Alfvén eigenmodes and sawteeth have been carried out at MAST. Strong correlations were found between relative changes in magnetic edge coils signals, edge D_α signal a fast ion D_α system, a prototype collimated neutron flux monitor and a recently installed prototype charged fusion product detector array, indicating both redistribution and loss of fast ions. Preliminary interpretation of these observations with a suite of stability, modelling and interpretative codes is discussed.

Keywords: MAST, TAE, fishbones, LLM, sawtooth, fast ion diagnostics

(Some figures may appear in colour only in the online journal)

1. Introduction

Understanding the behaviour of fast particles and their interaction with MHD instabilities is crucial for the operation and control of burning plasmas in devices such as ITER. In present day devices some of the relevant issues can be addressed. For example, on the Mega Ampere Spherical Tokamak (MAST) [1] the super-Alfvénic deuterium neutral beam injection (NBI) makes it possible to study fast ion-driven instabilities that resemble those likely to be driven unstable by α-particles in ITER in plasmas characterized by a large fast particle pressure. The redistribution and loss of fast ions due to such instabilities is detrimental to plasma energy confinement and to the effectiveness of non-inductive current drive, limiting the capability of controlling the current profile. The fast ion population in MAST is studied by measuring uncollimated and collimated neutron fluxes, using a fission chamber and a prototype four-channel neutron camera (NC) respectively, energetic proton yield using a prototype charged fusion product detector (CFPD), and fast ion D_α (FIDA) emission produced by beam ions that have exchanged electrons with beam or thermal neutrals. This set of fast ion diagnostics (FIDs) is suited to measuring the fast ion population with good spatial and temporal resolution, as well as studying the fast ion distribution in the energy E and pitch $\lambda = v_\parallel/v$ space. This work reports on recent studies of fast ion transport resulting from a range of instabilities, including $n = 1$ internal kink modes (fishbones and long-lived modes (LLMs)), toroidal Alfvén eigenmodes (TAEs) and sawteeth. Strong correlations were found between changes in fission chamber, FIDA, NC, CFPD, edge D_α and magnetic edge coil signals. The experiments were carried out with the dual aim of assessing the performance of the prototype diagnostics, and taking advantage of the complementarity of all four FIDs to provide the most complete experimental measurements of the fast ion behaviour on MAST to date, building on and extending previous observations. The next section describes the diagnostics used in this study. Experimental observations and their preliminary interpretation with a suite of stability, modelling and interpretative codes are discussed in section 3. Concluding remarks on the findings, on

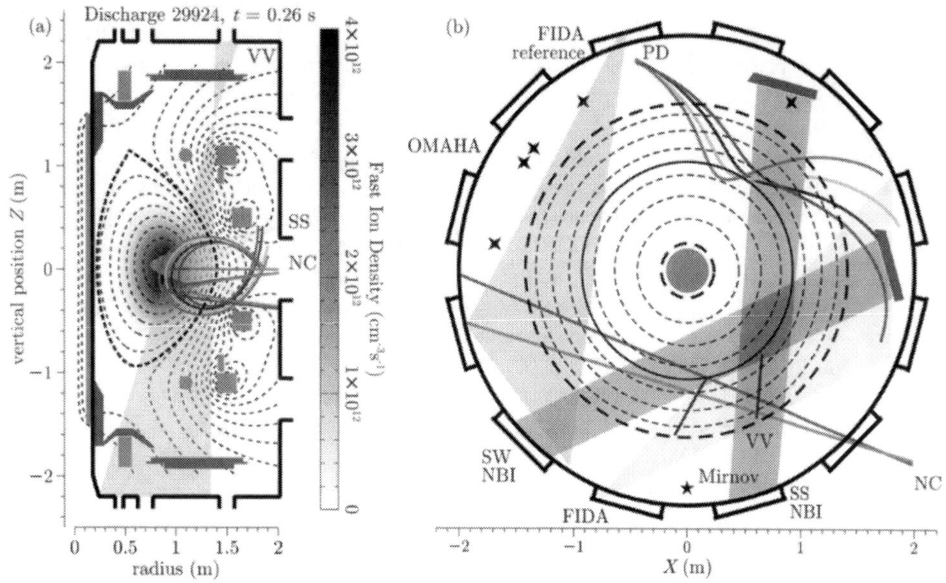

Figure 1. Poloidal and equatorial cross section of MAST showing the FIDs together with magnetic fluctuation coils. (*a*) Vacuum vessel (black), poloidal field coils (grey), divertor and central column first wall (dark magenta), flux surfaces and last closed flux surface (dashed lines) and fast ion density, FIDA vertical and tangential views (grey and green shaded areas), CFPD trajectories (light blue, dark blue, yellow and brown) and NC lines of sight (red and blue lines). (*b*) South–south and south–west neutral beam injectors footprints (shaded blue areas) and beam dumps (dark magenta), FIDA tangential active and reference fields of view (light green and grey shaded area) and vertical active and reference views (dark brown), toroidal array of fast magnetic coils (black crosses) and Mirnov coil (black star) and magnetic flux surfaces (dashed lines) with magnetic axis (solid line).

the prototype diagnostics and on the implications for MAST Upgrade are presented in section 4.

2. Fast ion experiments and diagnostics

MAST is a medium-size ST with a low aspect ratio ($R/a \approx 1.3$ with $R \approx 0.85$ m and $a \approx 0.65$ m), high elongation ($1.5 \leqslant \kappa \leqslant 2.5$) and triangularity ($\delta \leqslant 0.5$), capable of sustaining plasma current I_p up to 1.5 MA in a moderate toroidal field ($B_T \approx 0.6$ T at $R \approx 0.75$ m). Additional heating is provided by two horizontal mid-plane deuterium NBIs with tangency radius of 0.7 m, each delivering 2.5 MW of heating power with full energy up to $E_b \approx 75$ keV. The electron density n_e varies in the range 0.1–1.0×10^{20} m^{-3} and the electron temperature is $T_e \leqslant 2$ keV. An extensive set of plasma diagnostics is used in this study including a Thomson scattering (TS) system, a CXRS diagnostic, soft x-ray (SXR) detectors, a motional Stark effect (MSE) diagnostic, an array of Mirnov coils and an absolutely calibrated ^{235}U fission chamber measuring the global neutron rate Y_n with 10 μs time resolution and an accuracy of 10%. The confined fast ion population arises from the beam ions (Larmor radii r_L up to about 10 cm) and has a kinetic pressure P_f comparable to the thermal pressure. Beam ions dominate the DD fusion reactivity via beam–beam (BB) and beam–thermal (BT) reactions which account for approximately 10% and 85% of the fusion product yield respectively. This implies that the spatial and temporal evolution of the fusion product emission is strongly linked to that of the fast ions and therefore the NC and CFPD can be considered, indirectly, as fast ion diagnostics. An overview of the lines of sight and fields of view of the FIDs is shown in figure 1. The FIDA system's toroidal and vertical views measure spatially resolved fast ion density and distribution in velocity space and are absolutely calibrated [2]. FIDA measurements require NBI to provide D neutrals with which the fast ions can undergo charge exchange (CX). The neutralized fast deuterium atoms emit Balmer-α radiation with a Doppler shift proportional to their line of sight velocity. FIDA intensity is proportional to the product of fast ion and beam neutral density, weighted by the energy-dependent CX reactivity. In the spectral region free from pollution by beam and thermal D_α and impurity line radiation, the FIDA signal is obtained by subtracting bremsstrahlung emission observed by reference views not intersecting the neutral beam. In preparing FIDA data for analysis, close examination was necessary in order to reject data affected by large transient bursts of passive FIDA radiation, during which the background subtraction is prone to systematic errors. Fortunately these transient events typically lasted less than a millisecond, allowing a reliable pre/post-event profile to be obtained by excluding just one or two data points. The magnetic field orientation renders toroidal and vertical views mostly sensitive to passing and trapped particles respectively. The fast ion energy range observable with the FIDA system is 40–70 keV for the toroidal views and 5–70 keV for the vertical views. FIDA spatial resolution along the line of sight is approximately 20 cm, while perpendicular to the line of sight, in the radial direction, the resolution is approximately 2 cm [3]. The maximum temporal resolution of the FIDA system is 0.3 ms. The charged fusion products have large Larmor radius ($r_{L,p,T} \approx 0.5$ m and $r_{L,^3He} \approx 0.2$ m) and are therefore unconfined, leaving the plasma in a single or a few gyro-orbits. These charged fusion products are detected by an absolutely calibrated prototype CFPD [4] consisting of four silicon surface barrier detectors shielded against SXRs by a 0.8 μm thick aluminum foil and mounted at the end of a linear manipulator which can be moved radially on the equatorial plane between pulses. The detectors view the plasma along collimated lines of sight with slightly different orientations, allowing the simultaneous measurement of the emission from four different plasma regions. The line of sight of each detector is defined by the curved trajectory of the charged fusion products and its intersection with the mid-plane provides the nominal spatial location of their emission. Due to the finite acceptance angle of the collimators, the uncertainty in the mid-plane radial position for each line of sight is ± 3 cm. The spatial resolution, also defined along the major radius, is between 5 and 7 cm depending on the magnetic field. Proton and triton count rates of $\approx$200 kHz with a time resolution of 1 ms have recently been reported [5]. The prototype NC [6] measures the neutron emission along four lines of sight, two horizontal views lying on the equatorial plane and two diagonal views slightly inclined in the negative vertical direction. The radial spatial resolution in a single pulse is 20 cm: this can be arbitrarily increased by repeating plasma discharges while rotating the NC around a pivot point to cover tangency radii p in the range 0.2–1.2 m. The vertical positions at which the equatorial and diagonal lines of sight intersect the poloidal plane are fixed at $Z = 0$ m and $Z = -0.2$ m respectively. The neutron count rate time resolution is 1 ms with a counting statistical uncertainty of less than 10%; higher time resolutions can be achieved at the cost of larger statistical fluctuations in the counts. The count rates are proportional to the directional neutron emission within the field of view of each detector, i.e. the volume integral of the neutron flux directed towards the detectors. Absolute calibration of the NC detectors is obtained using the Monte Carlo code NEFF [7].

The plasma equilibrium is modelled using the EFIT++ [8] and HELENA [9] codes. The pre-processor code MC^3 is used to map experimental measurements onto EFIT++ equilibria, with the possibility of including the kinetic fast ion pressure, and to prepare the input files for the TRANSP [10] and NUBEAM [11] codes which are then used for the time dependent analysis of the plasma and the fast ion population. In particular, NUBEAM computes the five-dimensional fast ion distribution $f_i(t, R, Z, E, \lambda)$ and the local neutron emissivity $\varepsilon_n(t, R, Z)$. Sawtooth effects on $f_i(t, R, Z, E, \lambda)$ are treated in NUBEAM using the Kadomtsev magnetic reconnection model. Semi-heuristic models of the effect of fishbones on $f_i(t, R, Z, E, \lambda)$ are also included in NUBEAM based on the suppression of the fast ion population in predefined regions in the (E, λ) space, or on their trapping depth. Finally, the anomalous diffusion of fast ions can be modelled by a time dependent *ad hoc* anomalous diffusion coefficient D_{an} which can be selectively imposed on fast ions of a given energy and orbit-averaged radial position. The level of anomalous diffusion is set by matching the neutron rate measured by the fission chamber with the TRANSP predicted one. Changes in the neutron rate due to non-zero D_{an} are larger than the uncertainty in the TRANSP predicted neutron rate due to the experimental uncertainties in the input TRANSP profiles [12]. The Monte Carlo FIDASIM code [13] is used to model FIDA measurements based on the NUBEAM distribution, while neutron count rates are simulated with the full 3D Monte Carlo solid angle code LINE2 [14]. The non-linear wave–particle interaction code HAGIS [15] is used to compute the resonance condition between the poloidal and toroidal drift motions of the fast ions and $n = 1$ TAEs and kink modes, whose eigenfunctions are computed with the ideal incompressible MHD code MISHKA-1 [16].

Two different plasma scenarios are here studied. Scenario *S1* was used to explore TAEs, fishbones and LLMs, and is characterized by low density, $P_{NBI} = 3$ MW ($E_b = 70$ keV), and plasma current $I_p = 0.8$ MA with a safety factor profile $q(r)$ initially reversed, with $q(0) > 1$, evolving into a monotonic profile with $q \approx 1$ in the plasma core towards the end of the pulse. Scenario *S2* was aimed at the study of the effect of sawteeth on fast ions, and was characterized by $P_{NBI} = 1.5$ MW ($E_b = 50$ keV), $I_p = 0.6$ MA, a high density n_e and a monotonic $q(r)$ with $q(0) \lesssim 1$ during the discharge flat-top. Both *S1* and *S2* are L-mode plasmas in double-null divertor configuration.

3. Observations and analysis

3.1. TAEs and energetic particle modes

Chirping TAEs are observed in *S1* with an initial frequency of approximately 80–90 kHz, which evolves slowly, on the time scale of the background plasma, as shown in figure 2 for pulse 29976. The observed frequency is consistent with the estimated central frequency of the toroidicity-induced gap in the shear Alfvén continuum, which is approximately 90 kHz ($B_T(0) = 0.52$ T, $n_e = 3 \times 10^{19}$ m^3, $q(0) = 1.5$ and $R(0) = 0.9$ m). Both NBIs are producing super-Alfvénic ions with $v_f = 2.0$–2.6×10^6 m s^{-1} ($V_A(0) \approx 1.5 \times 10^6$ m s^{-1}), and can thus excite Alfvén waves via the resonance $v_{f,\parallel} = V_A$. TRANSP estimated beam poloidal beta in the core is such that $\beta_{f,\theta} > \beta_{therm,\theta}$, thus creating the conditions for the excitation of EPMs although in *S1* plasmas, β_f is sufficiently small that beta-induced suppression of core-localized TAEs does not occur [17]. As the q-profile evolves, bursts of EPMs appear in conjunction with TAEs: non-linear interaction of TAEs is inferred from their very fast sweeping down in frequency to approximately 30 kHz, indicative of modes with a substantial non-linear component to their dynamics. Mirnov signals show that both TAEs and EPMs are dominantly $n = 1$ modes. The bursting, frequency-chirping TAEs result in a strong redistribution of the fast ions from the core to the plasma edge accompanied by small bursting losses. Spikes in the divertor D_α emission are interpreted as fast ions being expelled from the plasma core and transported to the plasma edge. Such fast ions losses are not seen so clearly (if at all) by the FIDs. The total neutron yield Y_n and the collimated neutron count rates (for this study the time resolution was pushed to 0.1 ms) for $p = 1.01$ and 1.20 m do not always show clear drops in their amplitude; in some cases, losses are seen at one tangency radius but not at the other suggesting a localized loss

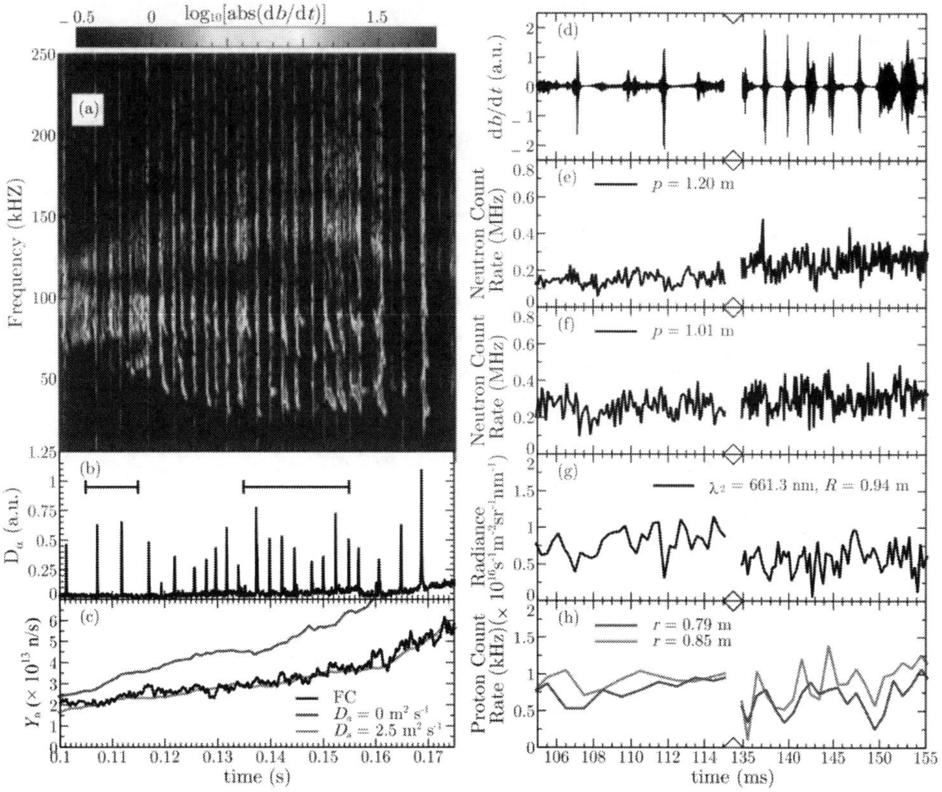

Figure 2. Evolution of MHD activity during the early phase of pulse 29976 and its effect on fast ions: (*a*) spectrogram of an outer Mirnov vertical coil; (*b*) bottom divertor D_α emission; (*c*) experimental (black) and TRANSP-computed total neutron rate with 0 (blue) and $2.5\,m^2\,s^{-1}$ (red) spatially uniform anomalous fast ion diffusion. Detailed view of the two time windows indicated in (*b*): (*d*) time trace of the Mirnov coil signal used to generate the spectrogram in (*a*); (*e*) and (*f*) neutron count rates for NC tangency radius 1.01 and 1.20 m; (*g*) tangential FIDA time trace at $R = 0.94$ m; (*h*) CFPD proton count rates at radial position 0.79 and 0.84 m.

of fast ions. The drop in the collimated neutron count rate for $p = 1.01$ m at $t = 0.108$ s is not seen in Y_n, suggesting that for this event redistribution occurs but the extent of the losses is rather limited. Some evidence of fast ion redistribution is seen in the FIDA signal (time resolution of 0.3 ms), where drops are observed corresponding to some MHD bursts for $R = 0.94$ m. The CFPD, which for this pulse was probing the inboard side of the plasma ($r < 0.9$ m), shows no indication of any redistribution or losses: the 1 ms time resolution of the CFPD might be too slow to follow these fast events. Although fast ion ejection from the plasma due to the bursting TAEs is small, a strong redistribution from the core to the edge is inferred from the need to impose, during the entire interval under study here, a spatially uniform $D_{an} = 2.5\,m^2\,s^{-1}$ to match the measured and TRANSP-computed neutron rates as shown in panel (*c*) of figure 2.

The eigen function of the $n = 1$ TAE at $t = 0.160$ s and the corresponding passing and trapped particle resonances as a function of pitch and energy (resonance map) were computed based on a plasma equilibrium reconstructed with EFIT++ constrained by MSE and TS measurements, including the TRANSP-computed kinetic pressure of the fast ions $(P_{f,\parallel} + P_{f,\perp})/2$, using MISHKA-1 and HAGIS codes. These are shown in figure 3. The computed radial q-profile is reversed in the core, with $q(0) \approx 1.5$ and two $q = 1$ surfaces for $r/a \lesssim 0.6$. For the dominant m harmonics, the modes are radially localized in the outer part of the plasma. In this region, trapped and barely passing fast ions with toroidal and poloidal transit frequencies satisfying the condition $\langle\omega_\phi\rangle + p\langle\omega_\theta\rangle = \omega$, where p is an integer, can resonate with both TAEs and EPMs and therefore be transported by the modes. This is confirmed by the resonance map calculated for $f = 82$ kHz, $n = 1$ mode and for passing fast ions in the energy range 0 to 70 keV (panel (*c*)), which shows how passing beam ions, with normalized perpendicular energy $\mu B_0/E = 0.3$ near $s \approx 0.8$, resonate with the $n = 1$ global TAE for a broad range of energies between 30 and 70 keV. Panel (*d*) shows that, deeper into the plasma in a region with larger fast ion pressure gradient ($R = 1.13$ m, i.e. $s \approx 0.5$), lower energy fast ions in the trapped region of velocity space (small $|\lambda|$) as well as ions of all energies close to the trapped/passing boundary, can also resonate with this mode. The estimated fraction of trapped fast ions at $R = 1.13$ m is approximately 40% of the total fast ion population, explaining the large mode amplitude and strong redistribution observed experimentally. Weakly damped modes, not subject to continuum damping, exist in the toroidicity-induced gaps in the frequency dispersion relation. For $s \geqslant 0.6$, the Landau damping provided by the background ions is negligible and TAEs can be driven unstable by even a small gradient in P_f. Due to the localized nature of these modes, it can be speculated that the NC and FIDA diagnostics, with their line integrated tangential views, are more sensitive

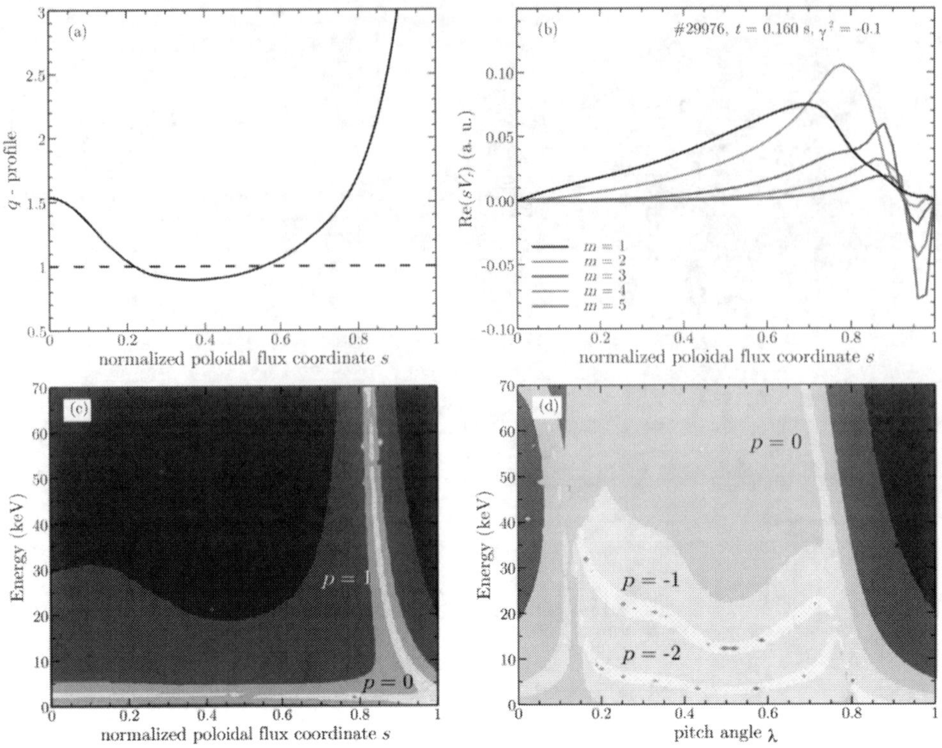

Figure 3. (a) Reversed shear q-profile of MAST shot 29976 at $t = 0.160$ s. (b): Real part of sV_r where V_r is the radial component of the perturbed velocity for $n = 1$ TAE computed with the MISHKA-1 code based on an equilibrium computed with HELENA code (high m modes with lower Re(sV_r) are not displayed, to aid clarity). The mode eigenvalues are given in units of $\gamma^2 = -[\omega R_0/V_A(0)]^2$.
(c), (d) Resonance maps showing $1/\Omega$ (as a function of radius at a fixed $\Lambda = \mu B_0/E = 0.3$, ($c$), and as a function of pitch for $s \approx 0.5$, (d)), where $\Omega = |\omega - <\omega_\phi> - p <\omega_\theta>|$. Note that these latter plots apply to a $n = 1$ TAE at $t = 0.165$ s during a similar shot, #29210, and are computed with the HAGIS code.

to fast ion losses near the plasma edge than the CFPD whose vertical views, although also line integrated, are more sensitive to the plasma inboard region. It is worth mentioning that the NC and CFPD lines of sight probed different plasma regions in similar discharges and, while the NC observations conform to the picture above, the CFPD showed some correlation with MHD bursts when the sight line was more central ($r \approx 0.9$ m).

3.2. Fishbones

A typical example of fishbones in a magnetic spectrogram is shown in figure 4, panel (a), where the characteristic sweep in frequency of the dominantly $n = 1$ mode from approximately 40 kHz down to 20 kHz is observed. Some of these fishbones are accompanied by spikes in the D_α emission from the divertor (panel (c)), indicative of fast ions losses. The amplitude of these spikes steadily decreases over successive fishbones, suggesting that fast ions losses become less severe. A reduction in the global neutron rate Y_n does not always follow a fishbone, and when it does is not always accompanied by a spike in the D_α signal. The redistribution affects a large part of the plasma core in the range 0.85–1.25 m as observed in the FID profiles shown in panels (d)–(f). Since the onset time of fishbone bursts in scenario *S1* varies slightly from pulse to pulse, CFPD and NC radial profiles were obtained by time aligning similar fishbones in pulses 29975 and 29980 to the fishbone at $t_{FB} = 0.214$ s in pulse 29976 (the required time alignment is -9.2 ms for pulse 29975 and -3.0 ms for pulse 29980). Tangential FIDA profile measurements are from pulse 29976 (no vertical FIDA measurements are available for *S1*). The time averaged FID profiles for $t < t_{FB}$ and $t > t_{FB}$ are shown in panels (d)–(f). An overall decrease in the fast ion population is observed by all FIDs across the measured radial positions, while n_e and T_e profiles are unaffected. The average drop across the profile of approximately a third in the tangential FIDA signal for $E_f \geqslant 48$ keV is twice that observed in the NC ($\approx$16%) and CFPD ($\approx$13%) signals, while the average drop in the tangential FIDA signal for $E_f \geqslant 59$ keV is similar to that observed by the CFPD and NC for $R \lesssim 1.05$ m and similar to the drop observed at $E_f \geqslant 48$ keV at larger radii. This suggests that the fishbone burst is efficiently redistributing fast ions with a wide range of energies from at least 48 keV up to the injection energy of $E_b = 75$ keV. The NC and CFPD diagnostics are most sensitive to fast ions with energies closer to E_b due to the strong dependence of the D–D fusion reaction rate on the relative ion velocities, while FIDA is sensitive to a wider energy range of the slowed down distribution function. The radial profiles of the ratio between pre- and post-fishbone profiles for FIDA, CFPD and NC shown in panel (g) are broadly consistent with each other, i.e. they are affected to the same degree by the fast ion redistribution.

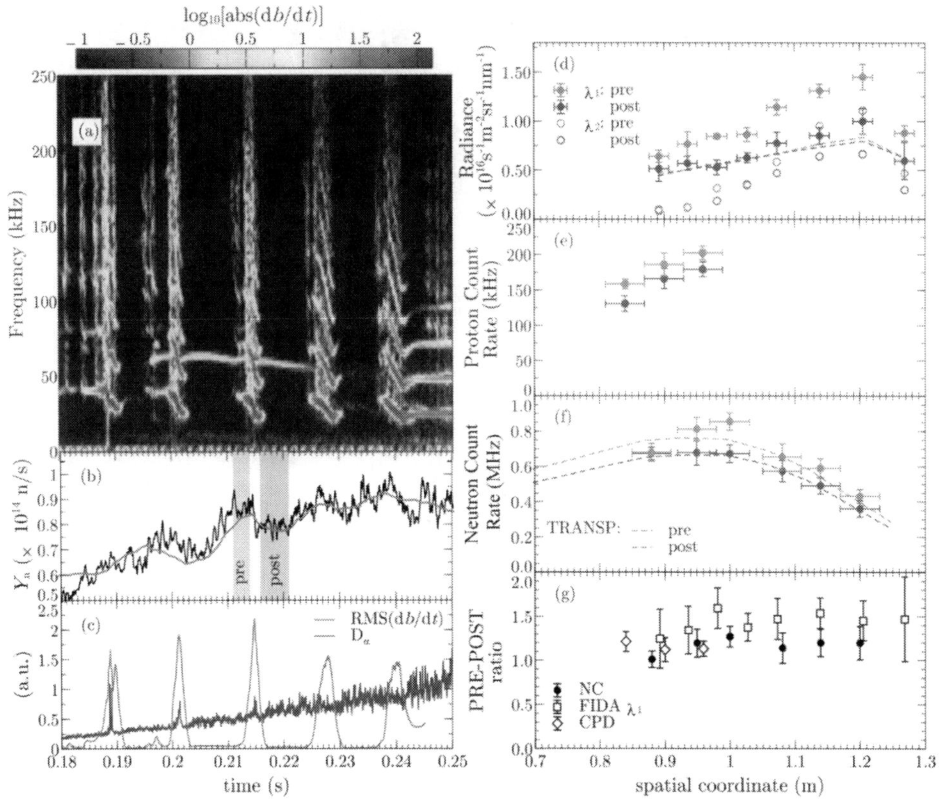

Figure 4. Fishbones and fast ion measurements: (*a*) spectrogram of an outer Mirnov vertical coil; (*b*) neutron rate: experimental (black) and predicted by TRANSP (red) with $D_{an} = 0.5\,m^2\,s^{-1}$ and imposing the loss of all trapped fast ions with energy in the range 50–70 keV (shaded areas indicate the time interval used for the radial profiles pre- and post-fishbone burst); (*c*) RMS of the magnetic fluctuations of (*a*) and D_α light emission from the bottom divertor. Profiles based on pulses 29975, 29976 and 29980: (*d*) FIDA emission profile as a function of beam intersection radius (#29976 only) with TRANSP predicted profiles (dashed lines); (*e*) CFPD proton count rate profile as a function of the major radius (#29980 only); (*f*) NC neutron count rate profile together with TRANSP predicted profiles (dashed lines). Panel (*g*) shows the ratio between pre- and post-fishbone profiles for each FID.

A series of TRANSP/NUBEAM simulations for pulse 29976 during the fishbone phase have been carried out. An inbuilt fishbone loss model, which allows fast ions in TRANSP to be removed from selected regions of velocity space parametrized by E and λ with a user-defined period and characteristic loss time, was used in an attempt to match the drops in Y_n. A systematic scan of the affected portion of (E, λ) space was carried out, and the fast ion distributions from the most promising candidate loss models were compared to the fission chamber, FIDA and NC data using FIDASIM and LINE2 codes. The overall trend of the global neutron rate is readily matched by setting a spatially uniform $D_{an} \approx 0.5\,m^2\,s^{-1}$ while the drops in Y_n at the fishbone bursts have been more difficult to reproduce in a consistent manner. It was found that selectively removing trapped/barely passing ions at high energies reproduced the behaviour of the global neutron rate and NC signal but failed to account for the observed reduction in FIDA signal due to the fishbone bursts, while removing passing fast ions in a narrow region of pitch ($\lambda \in [0.88–0.91]$) at all energies up to the injection energy was required to match the relative size of the drops in FIDA signal, but failed to reproduce the observed drops in NC signal. The best agreement with the fission chamber and NC signals has been obtained by setting the spatially uniform $D_{an} = 0.5\,m^2\,s^{-1}$ in the time interval $t = 0.200–0.235$ s and by forcing the removal of trapped fast ions with energies in the range 50–70 keV regardless of their trapping depth (the results are shown in panels (*b*) and (*f*) of figure 4). The predicted NC count rates for $t < t_{FB}$ and $t > t_{FB}$ agree rather well with the experimental data, especially for $t > t_{FB}$. Before the onset of the fishbone, the NC measurements indicate a more spatially peaked fast ion distribution than that predicted by TRANSP. FIDASIM results for $\lambda = \lambda_1$ and $t < t_{FB}$ and $t > t_{FB}$, using the same TRANSP simulation, are shown in panel (*d*): while good agreement is found between experimental data and predictions is found for $t > t_{FB}$, the pre-fishbone profile is clearly underestimated. The plasma stability for $t = 0.210$ s was studied by computing the eigenfunction of $n = 1$ kink modes with $m = -1$ to 11 with the MISHKA-1 code using an equilibrium reconstructed with EFIT++ and constrained by MSE measurements. The computed radial q-profile is weakly reversed ($q(0) \approx 1.1$) with two $q = 1$ surfaces for $r/a \lesssim 0.6$. The radial and poloidal components of the perturbed velocity are shown in figure 5, panels (*a*) and (*b*): the $m = 1$ mode is by far the dominant component, having the largest amplitude in the plasma core. The passing and trapped particle resonances with a $n = 1$ kink mode with $f = 30$ kHz, matching the frequency of the FB at the beginning of the chirp, were computed as a function

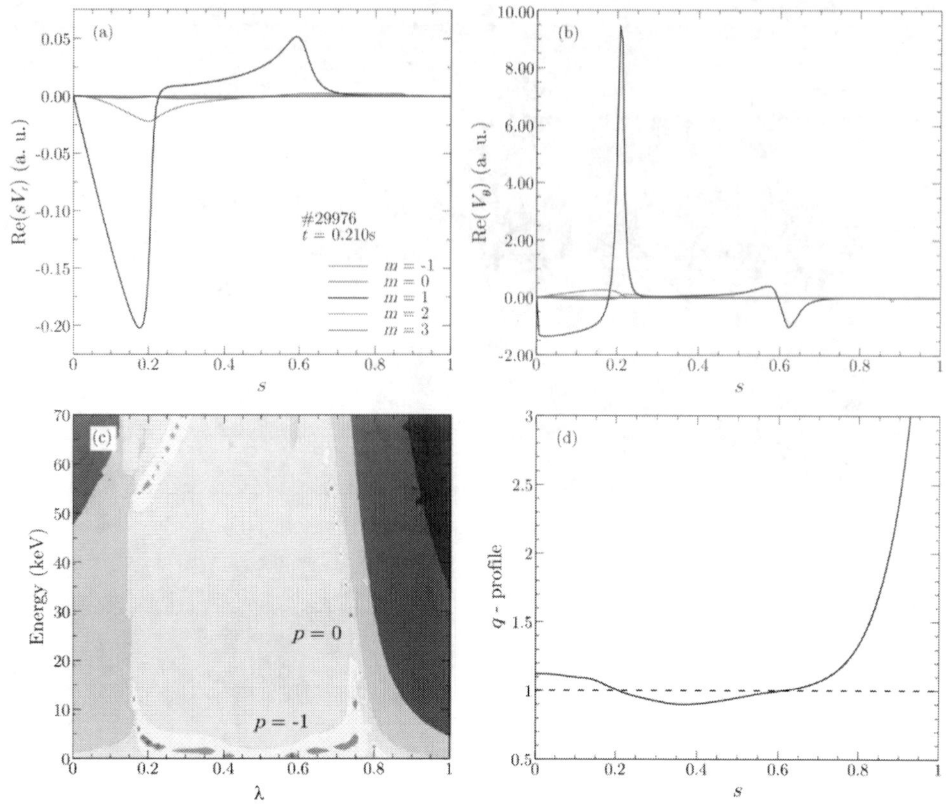

Figure 5. Fishbone modelling for pulse 29976 at $t = 0.210$ s. (*a*), (*b*) Real part of the radial sV_r and poloidal V_θ component of the perturbed velocity for an ideally unstable $n = 1$ kink mode computed with the MISHKA-1 code based on an equilibrium computed with HELENA code (only dominant m modes are displayed) as a function of the normalized poloidal flux coordinate s. (*c*) Fast ion resonance map in the energy-pitch space for $s \approx 0.5$ showing $1/\Omega$ where $\Omega = |\omega - <\omega_\phi> - p <\omega_\theta>|$ for $n = 1$ kink mode computed with HAGIS code. (*d*) q-profile from HELENA code.

of λ and E with the HAGIS code at $R \approx 1.2$ m ($Z = 0$ m) where the fast ion pressure gradient is largest. As may be seen in panel (*c*) of figure 5, the resonance condition is satisfied by low energy trapped or nearly trapped fast ions close to the plasma mid-outer radius, via their precessional toroidal drift frequency, as well as by passing fast ions with energies up to 70 keV through the $p = 1$ resonance.

3.3. Long-lived modes

On MAST, the fishbones typically evolve into long-lived saturated $n = 1$ modes with a frequency close to the plasma core rotation as the current diffuses and q is lowered. These modes persist for a significant portion of the plasma discharge. The LLM phenomenology, stability properties and the role of fast ions have been extensively studied in MAST [18], where it was found that a sufficiently large kink-like displacement of the plasma core could account for the saturation of the modes, the reduction in the energy confinement (reduction in both core n_e and T_e), and the observed enhanced fast ion losses. The latter were inferred from a comparison of TRANSP simulated and measured Y_n: agreement was found throughout the pulse only by doubling the value of D_{an} during the LLM phase. The ensuing reduction in the fast ion pressure gradient was then considered as a possible reason for the absence of fishbones during this phase. The mechanism suggested for the loss of fast ions was the transition to unconfined orbits as fast ions with large r_L radially transported by the kink displacement might easily traverse the last closed flux surface. Numerical simulations of an ideal kink perturbation displacing the core of the plasma column, using realistic 3D equilibrium field geometry and a guiding-centre following code, have indeed been able to reproduce the change in the NC profiles during a LLM on MAST but failed to generate significant fast ion losses [19]. Confirmation of the reduction in the spatial gradient of the fast ion pressure $\nabla_r P_f$ during a LLM is shown in figure 6 for scenario *S1*. The q-profile during the LLM phase is monotonic and approximately equal to unity for $s \lesssim 0.4$, dropping below unity at $s \lesssim 0.6$ towards the end of the pulse. Although a $q = 1$ surface is present in the plasma, no sawteeth are observed. During the LLM, the experimental neutron rate is matched by the TRANSP-computed rate if a spatially uniform $D_{an} \approx 3\,\mathrm{m^2\,s^{-1}}$ is used. The LLM causes significant fast ion redistribution, which is clearly seen by all FID measurements as a reduction of the gradient of the radial profiles, integrated between 0.255 and 0.280 s (shaded grey region in panel (*b*) of the same figure), which are directly linked to the spatial profiles of the fast ion distribution. The reduction of $\nabla_r P_f$ implies a reduction of the drive of the fishbones. The suppression of fishbone in *S1* is however not complete as the LLM coexists

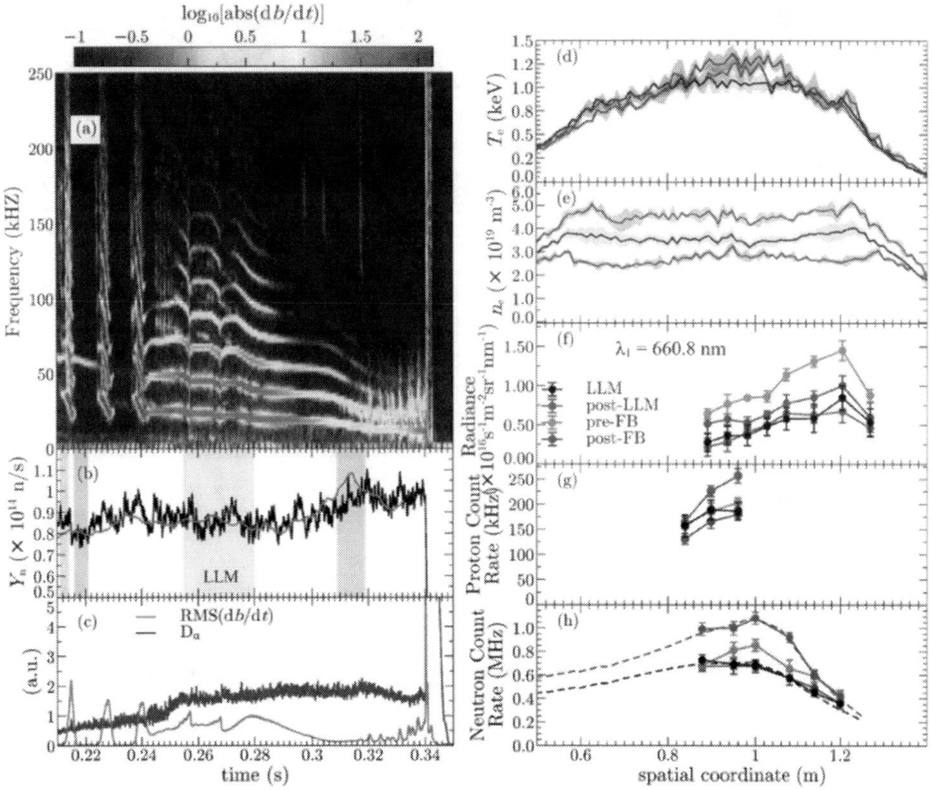

Figure 6. Evolution of the LLM for pulse #29976: (*a*) spectrogram of Mirnov coil signal; (*b*) measured (black) and TRANSP-computed (red) neutron rate (D_{an} is adjusted in the time window shown to match the fission chamber measurements): shaded areas refer to the time intervals used for the radial profiles in (*d*)–(*h*); (*c*) RMS amplitude of the Mirnov coil and edge D_α monitor signal. Radial profile evolution: electron temperature (*d*) and density (*e*); (*f*) tangential FIDA; (*g*) CFPD proton count rate; (*h*) experimental (points) and TRANSP-computed (dashed lines) neutron count rate profiles with $D_{an} \approx 3\,\mathrm{m^2\,s^{-1}}$ (black dashed line) and D_{an} to $\approx 0.7\,\mathrm{m^2\,s^{-1}}$ (purple dashed line).

with very short burst of chirping MHD activity at $t = 0.258$ and 0.268 s, indicating that ∇P_f, however small, is still able to drive fishbone-like instabilities as $q(0) \approx 1$. The same TRANSP simulation provides good agreement with the experimental neutron count rate profile as shown in panel (*h*). The absolute levels of fast ion redistribution due the LLM and the fishbones are comparable according to NC and CFPD measurements, while FIDA seems to indicate only a weak recovery of the fast ion population after the LLM; this difference is explained by the increased beam attenuation due to the higher plasma density later in the discharge, which contributes to the reduction of the FIDA signal. This continuous increase of the density causes the amplitude of the LLM to steadily decrease, reaching a minimum between 0.30 and 0.32 s as shown in panel (*c*); the redistribution of the fast ions is reduced and an increase in the neutron rate is consequently observed. Indeed, it was necessary to reduce D_{an} to $\approx 0.7\,\mathrm{m^2\,s^{-1}}$ in order to match experimental and TRANSP predicted neutron rates and NC profiles at this time. The improved fast ion confinement results in the increase of the spatial gradients of the NC and CFPD profiles. The increase in the proton and neutron beam-thermal $Y_{n,BT}$ emission is not due to the increase in density, as this effect is compensated by the decrease in the fast ions slowing down time of an equal amount. Also the small decrease in Z_{eff} observed later in the pulse cannot account for it. FIDA measurements on the other hand do not show an increase in the fast ion population since the increase in signal is balanced by an increased beam attenuation. Some indication that fast ion redistribution is mitigated after the suppression of the LLM can be inferred from the small FIDA profile differences: during the LLM period, FIDA signal is maximum at $R = 1.15$ m while in the time interval 0.30–0.32 s it tends to peak closer to the plasma core. The steeper gradients that have been built up during the LLM partial suppression are a source of free energy that can drive MHD instabilities. Frequency-sweeping bursts of $n = 1$ modes can be seen starting at around $t = 0.32$ s and lasting to the end of the discharge. Similar back transitions from LLM to chirping modes were observed previously in MAST [20].

3.4. Sawteeth

Investigation of the fast ion behaviour during non-resonant MHD instabilities has focused on the sawtooth. Six similar plasma discharges (#29879–#29882 and #29884–#29885) in scenario *S2* were performed. Analysis of Mirnov coil and SXR signals confirms the sawtooth nature of the observed non-resonant instability and its dominant $n = 1$ nature,

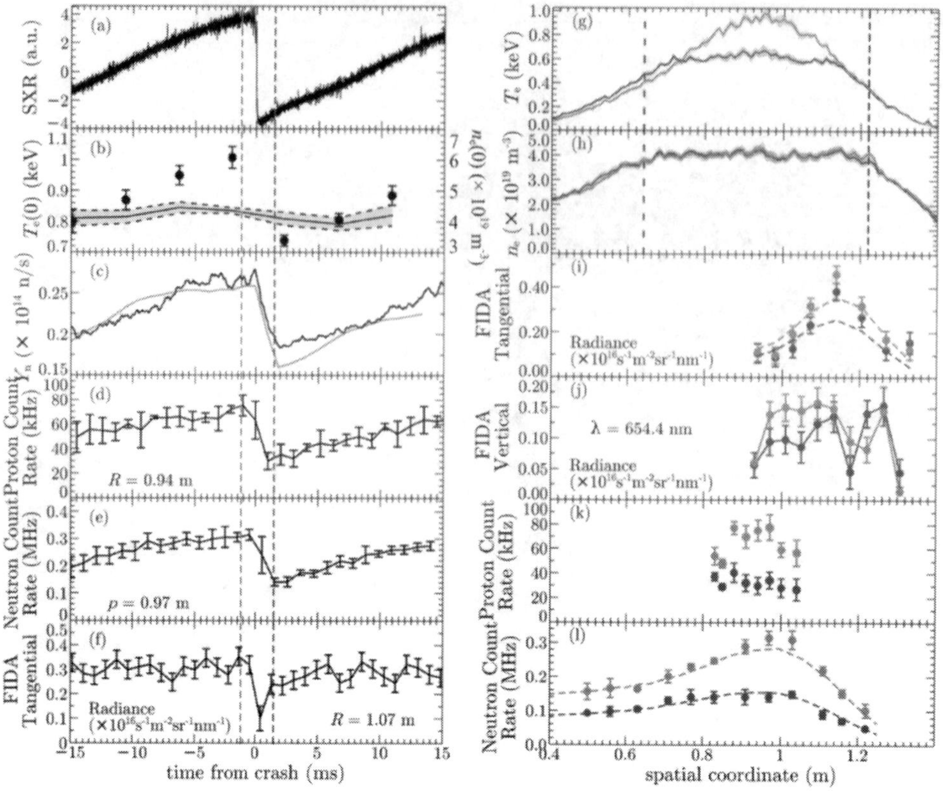

Figure 7. Evolution of the fast ion population during a sawtooth crash: (*a*) tangential SXR signal at $R = 0.9$ m; (*b*) TS core electron temperature (solid circles) and density (solid line); (*c*) measured (black) and TRANSP-computed (magenta) global neutron rate; (*d*), (*e*) CFPD proton and NC time traces; (*f*) FIDA time trace for fast ions with energy $\geqslant 48$ keV. Radial profiles before and after the sawtooth crash: (*g*), (*h*) electron temperature and density (the dashed vertical lines indicate the location of the mixing radius); (*i*) tangential FIDA measurements (points) and FIDASIM profiles (dashed lines); (*j*) vertical FIDA measurements for $E \geqslant 6$ keV; (*k*) CFPD proton count rate profile; (*l*) NC measured and simulated profiles using the Kadomtsev mixing model available in TRANSP/NUBEAM. Uncertainties in the spatial coordinates omitted for clarity.

although there is a significant contribution from higher n harmonics. The evolution of the SXR, $T_e(0)$, $n_e(0)$, Y_n and FID measurements are shown in figure 7 for four sawteeth aligned to a single crash time $t_c = 0$ s and averaged. Drops of approximately 30% in Y_n and up to 50% in the local fast ion population and fusion product yield are observed. Radial profiles for $t < t_c$ and $t > t_c$ are shown in 7, panels (*g*)–(*l*): the core T_e collapses and flattens while no change is seen in $n_e(r)$, indicating an incomplete reconnection [21]. Tangential FIDA profiles show a small redistribution of mainly passing fast ions with $E \geqslant 48$ keV from the mid-outer region of the plasma towards the plasma edge but little or no effect near the magnetic axis (≈0.95 cm). Vertical FIDA radial profiles indicate that some trapped fast ions with $E \geqslant 6$ keV ($\lambda = 654.4$ nm) are redistributed for $0.95\,\text{m} \lesssim R \lesssim 1.10$ m. The NC and CFPD count rates show a reduction in signal as large as ≈60%. This difference between the FIDA and fusion product profiles reflects the fact that different parts of the (E, λ) space are being probed in different regions of the plasma by the three diagnostics. The sawtooth causes fast ion losses observed as a sudden increase in the divertor D_α emission occurring ≈20 μs after each crash, a time scale compatible with fast ions of $E_f \approx E_b$ moving along perturbed field lines (typical connection lengths on MAST are ≈10 m). No information on the region of (E, λ) space from which fast ions are lost is provided by the D_α emission monitor on MAST, however. Modelling of the FID measurements for pulse #29880 has been performed using the Kadomtsev mixing model implemented in TRANSP/NUBEAM, setting $D_{an} = 0\,\text{m}^2\,\text{s}^{-1}$ throughout the pulse. Good agreement with the experimental data is found (figure 7, panels (*c*), (*i*) and (*l*)). The fast ion slowing down time (which drops from ≈60 ms before to ≈30 ms after the crash) is much longer that the characteristic sawtooth crash time, and cannot account for the rapid drop in the neutron or proton rates. TRANSP indicates that the crash causes a large radial redistribution of fast ions from the core to regions of the plasma outside the mixing radius. Outside the mixing radius, the fusion reactivity is greatly diminished, explaining the large net reduction in both NC and CFPD measurements. According to TRANSP simulations, the passing fast ion population is reduced at each crash while the trapped population is increased, indicating not only spatial redistribution but also increased pitch angle scattering. This is consistent with tangential FIDA observations but not with the vertical observations, which reveal that trapped fast ions are affected by the sawteeth. These observations, although affected by a large experimental uncertainty due to the very weak FIDA core signal in this low NBI heating power scenario, are consistent with the findings on DIII-D [22], ASDEX Upgrade [23] and TEXTOR [24]. In particular, the observation of the redistribution of trapped fast ions with the vertical FIDA view is in accordance with ASDEX Upgrade FIDA measurements.

4. Summary and conclusions

A series of experiments has been carried out to assess the effect of resonant and non-resonant MHD instabilities on the fast ions with the comprehensive set of fast ion diagnostics available at MAST: a tangential and vertical FIDA system, a ^{235}U fission chamber, a neutron collimated flux monitor and a charged fusion product detector array. The last two are prototypes, and these experiments were also carried out to assess their performances and suitability in view of MAST Upgrade, and the CFPD in view of NSTX Upgrade. A preliminary attempt has also been made to integrate the observations from all diagnostics with stability, modelling and predictive codes to provide a consistent description of the fast ion behaviour in MAST. Strong fast ion redistribution accompanying resonant and non-resonant instabilities is consistently observed by the FIDs, and where differences in the measurements have been observed these have mostly been explained in term of the characteristic response of each diagnostic to different plasma conditions. While such detectors of confined fast ions provide valuable information on fast ion redistribution, it would also be very useful to have a fast ion loss detector (FILD) with high time resolution (sub-ms) to allow the study of fast ion losses due to MHD instabilities evolving on very fast time scales, such as the chirping TAEs described in section 3.1. FILDs are also capable of inferring the region of energy and pitch space from which the lost fast ions originated, thus complementing the information obtained from the edge D_{α} emission monitors used to detect fast ion losses in this study. A FILD is being designed for MAST Upgrade to provide such information. Complementing the FILD system, the current FIDA system will be adapted to the viewing geometry of MAST Upgrade retaining both the tangential and vertical view. The design of an upgrade of the neutron camera is ongoing with the aim of providing a full radial profile with a spatial resolution of approximately 5 cm and a time resolution of 1 ms in a single pulse. Two sets of lines of sight are envisaged to intercept both the on-axis and off-axis fast ion population [25].

TRANSP modelling of the effect of resonant and non-resonant instabilities has been carried out. While for sawtooth good agreement is found between simulated FIDA and NC drops in the fast ion population, in the case of a fishbone burst, the fast ion loss model included in NUBEAM has so far not provided consistent predictions that could match simultaneously the drops observed with FIDA and the NC. The study and understanding of this discrepancy is the subject of ongoing work. Eigenmode analysis of the TAEs and fishbone observed during these experiments has been performed with the MISHKA-1 code; these two modes are characterized by a different radial extent, with the TAEs localized more towards the mid-outer plasma region and the fishbone a more core-localized mode. The resonance conditions between the fast ion motion and such eigenmodes, with experimentally observed frequencies, have been obtained using HAGIS. According to these calculations the resonance condition is satisfied simultaneously by trapped and passing fast ions in different regions of the energy and pitch space. Self-consistent wave–particle interaction calculations with HAGIS are the subject of ongoing work. These HAGIS calculations will extend a preliminary study of the energy dependence of redistribution of fast ions due to fishbones, in which the mode amplitude and evolution (in time and frequency) was externally imposed, by looking at the pitch-angle dependence as well as by comparing the predictions with the observations reported in this work that were not available in the previous analysis [26].

Acknowledgments

This work was funded by the Swedish Research Council, the RCUK Energy Programme under grant EP/I501045, the European Union's Horizon 2020 research and innovation programme under grant agreement number 633053 and the US Department of Energy Contract Numbers DESC0001157 and DEAC0209CH11466. The views and opinions expressed herein do not necessarily reflect those of the European Commission.

References

[1] Darke A C *et al* 1995 MAST: aMega Amp Spherical Tokamak *Proc. 18th Symp. on Fusion Technology (Karlsruhe, Germany, 22–26 August 1994)* vol 1 (Amsterdam: Elsevier) p 799
[2] Michael C A *et al* 2013 *Plasma Phys. Control. Fusion* **55** 5007
[3] Heidbrink W W 2010 *Rev. Sci. Instrum.* **81** 10D727
[4] Boeglin W U, Perez R V and Darrow D S 2010 *Rev. Sci. Instrum.* **81** 10D301
[5] Perez R V *et al* 2014 *Rev. Sci. Instrum.* **85** 11D701
[6] Cecconello M *et al* 2014 *Nucl. Instrum. Methods* A **753** 72
[7] Dietze G *et al* 1982 NRESP4 and NEFF4: Monte Carlo codes for the calculation of neutron response functions and detection efficiencies for NE213 scintillation detectors *PTB Report* PTB-ND-22, Braunschweig
[8] Appel L *et al* 2006 *33rd EPS Conf. on Plasma Physics (Rome, Italy, 19–23 June 2006)*
[9] Huysmans G T A, Goeldbloed J P and Kerner W 1997 *Proc. CP90 Conf. (Amsterdam, The Netherlands, 1990)* (Singapore: World Scientific) *Plasma Phys. Rep.* **23** 371
[10] Goldston R J *et al* 1981 *J. Comput. Phys.* **43** 61
[11] Pankin A, McCune D, Andre R, Bateman G and Kritz A 2004 *Comput. Phys. Commun.* **159** 157
[12] Klimek I *et al* 2014 submitted to *Nucl. Fusion*
[13] Heidbrink W W, Liu D, Luo Y, Ruskov E and Geiger B 2011 *Commun. Comput. Phys.* **10** 716
[14] Conroy S 2009 private communication
[15] Pinches S *et al* 1998 *Comput. Phys. Commun.* **111** 133
[16] Mikhailovskii A B *et al* 1997 *Plasma Phys. Rep.* **23** 844
[17] Gryaznevich M P and Sharapov S E 2004 *Plasma Phys. Control. Fusion* **46** S15
[18] Chapman I T, Hua M D, Pinches S D, Akers R J, Field A R, Graves J P, Hastie R J, Michael C A and the MAST Team 2010 *Nucl. Fusion* **50** 045007
[19] Pfefferle D, Graves J P, Cooper W A, Misev C, Chapman I T, Turnyanskiy M and Sangaroon S 2014 *Nucl. Fusion* **54** 064020
[20] Gryaznevich M P *et al* 2008 *Nucl. Fusion* **48** 084403

[21] Chapman I T, Scannell R, Cooper W A, Graves J P, Hastie R J, Naylor G and Zocco A 2010 *Phys. Rev. Lett.* **105** 255002
[22] Muscatello C M, Heidbrink W W, Kolesnichenko Y I, Lutsenko V V, Van Zeeland M A and Yakovenko Y V 2012 *Plasma Phys. Control. Fusion* **54** 025006
[23] Geiger B *et al* 2014 *Nucl. Fusion* **54** 022005
[24] Nielsen S *et al* 2011 *Nucl. Fusion* **51** 063014
[25] Wesizflog M *et al* 2014 *Rev. Sci. Instrum.* **85** 11E121
[26] Lake R 2013 Consequences of fast ion driven modes in MAST *PhD Thesis* Warwick University

IOP Publishing
Plasma Physics and Controlled Fusion
Plasma Phys. Control. Fusion 57 (2015) 014007 (11pp)
doi:10.1088/0741-3335/57/1/014007

Enhanced efficiency of plasma acceleration in the laser-induced cavity pressure acceleration scheme

J Badziak[1], M Rosiński[1], S Jabłoński[1], T Pisarczyk[1], T Chodukowski[1], P Parys[1], P Rączka[1], E Krousky[2], J Ullschmied[3], R Liska[4] and M Kucharik[4]

[1] Institute of Plasma Physics and Laser Microfusion, Warsaw, Poland
[2] Institute of Physics AS CR, Prague, Czech Republic
[3] Institute of Plasma Physics AS CR, Prague, Czech Republic
[4] Czech Technical University, FNSPE, Prague, Czech Republic

E-mail: jan.badziak@ipplm.pl

Received 20 June 2014, revised 17 August 2014
Accepted for publication 5 September 2014
Published 28 November 2014

Abstract

Among various methods for the acceleration of dense plasmas the mechanism called laser-induced cavity pressure acceleration (LICPA) is capable of achieving the highest energetic efficiency. In the LICPA scheme, a projectile placed in a cavity is accelerated along a guiding channel by the laser-induced thermal plasma pressure or by the radiation pressure of an intense laser radiation trapped in the cavity. This arrangement leads to a significant enhancement of the hydrodynamic or electromagnetic forces driving the projectile, relative to standard laser acceleration schemes. The aim of this paper is to review recent experimental and numerical works on LICPA with the emphasis on the acceleration of heavy plasma macroparticles and dense ion beams. The main experimental part concerns the research carried out at the kilojoule sub-nanosecond PALS laser facility in Prague. Our measurements performed at this facility, supported by advanced two-dimensional hydrodynamic simulations, have demonstrated that the LICPA accelerator working in the long-pulse hydrodynamic regime can be a highly efficient tool for the acceleration of heavy plasma macroparticles to hyper-velocities and the generation of ultra-high-pressure (>100 Mbar) shocks through the collision of the macroparticle with a solid target. The energetic efficiency of the macroparticle acceleration and the shock generation has been found to be significantly higher than that for other laser-based methods used so far. Using particle-in-cell simulations it is shown that the LICPA scheme is highly efficient also in the short-pulse high-intensity regime and, in particular, may be used for production of intense ion beams of multi-MeV to GeV ion energies with the energetic efficiency of tens of per cent, much higher than for conventional laser acceleration schemes.

Keywords: laser ion acceleration, laser plasma, fast ignition, ion diagnostics, LICPA

(Some figures may appear in colour only in the online journal)

1. Introduction

The generation of intense fluxes of plasma in various forms (fast macroparticles, ion beams, plasma jets) is currently the topic of high relevance for contemporary research in high-energy-density physics (HEDP) [1], nuclear physics [2], inertial confinement fusion (ICF) [3], space research [4] and laboratory astrophysics [1, 5] as well as in medicine [6] and technology [7]. To achieve plasma fluxes relevant for the above mentioned applications various schemes of acceleration of dense plasma have been proposed and demonstrated. In particular, these include Z-pinch machines [8], coaxial plasma guns [9], and laser-driven accelerators [3, 10–13]. However, only laser-driven acceleration is presently capable of accelerating dense plasma projectiles to velocities in the range of 100–1000 km s^{-1} [3, 10, 14] and has the potential to achieve

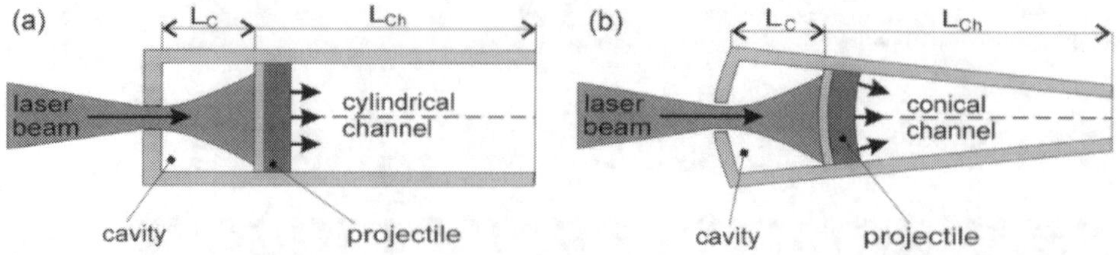

Figure 1. The LICPA accelerator in two geometries: (*a*) the cylindrical accelerator and (*b*) the conical accelerator.

much higher velocities [11–13]. Due to their high density and very high velocity, the laser-driven plasma projectiles could therefore achieve very high-energy fluence or the plasma flux intensity.

The most commonly used scheme of laser-driven acceleration of dense plasma is the so-called ablative acceleration (AA) [3, 10, 14]. In the AA scheme the surface of a solid target is irradiated either directly by the laser beam or indirectly by secondary radiation induced by the laser, which leads to the creation of a hot plasma expanding backwards, thus accelerating the remaining denser part of the target (the projectile) in the forward direction via the 'rocket effect' [3]. In particular, the AA scheme is used to drive and compress the DT fuel in fusion targets [3, 14] or to accelerate a macroparticle to ignite the compressed fuel by the impact [10, 15, 16]. Unfortunately, this scheme has a relatively low energetic efficiency of the acceleration $\eta_{acc} = E_p/E_L$, where E_p is the kinetic energy of the projectile and E_L is the laser energy, which in practice is well below 10% [3, 14, 16]. Moreover, the maximum attainable velocity of the projectile driven by the AA is limited due to the Rayleigh–Taylor (RT) instability to about $10^8\,\mathrm{cm\,s^{-1}}$. This limit for the projectile velocity may be shifted up to relativistic velocities by using the radiation pressure acceleration (RPA) [11, 13, 17–20] (also referred to as skin-layer ponderomotive acceleration—SLPA [21, 22]). In the RPA scheme a plasma bunch is driven by the radiation pressure of a short (ps or sub-ps) laser pulse of very high intensity ($I_L \gg 10^{18}\,\mathrm{W\,cm^{-2}}$). The efficiency of these schemes may reach tens of per cent [13, 17, 18] in the relativistic regime, which is attainable however either at very small masses of the bunch or at very high laser energies. Unfortunately, at sub-relativistic velocities, which are important for many applications (e.g. for ICF, HEDP), the acceleration efficiency in the RPA scheme is lower, ~10% or less [13, 19].

Recently, Badziak *et al* have proposed [23, 24] a new scheme of acceleration of dense matter which combines the old idea of the conventional gun, using the pressure of combustion gasses, with the idea of laser acceleration using the hydrodynamic pressure of hot plasma (like AA) or ponderomotive (radiation) pressure (like RPA/SLPA). Apart from high energetic efficiency, the proposed scheme has a potential to accelerate dense plasma projectiles up to relativistic velocities. This scheme has been referred to as laser-induced cavity pressure acceleration (LICPA) [23, 24].

In the LICPA scheme (figure 1), a projectile (e.g. foil target) placed in a cavity is irradiated by a laser beam introduced into the cavity through a hole; the projectile is then accelerated in a guiding channel (cylindrical or conical) by the pressure created in the cavity by the laser-produced hot plasma expanding from the irradiated side of the projectile and from the cavity walls, or by the photon (radiation) pressure of the ultra-intense laser pulse trapped in the cavity. Depending on initial parameters of the projectile and the laser beam, the accelerated projectile can have a form of a macroparticle (i.e. a compact object of the density ρ comparable to or higher than the solid density ρ_s) or an ion beam or a plasma jet (i.e. an extended in space, highly directional plasma outflow). An important part of the scheme is the guiding channel, which plays a role similar to that of a barrel in a conventional gun, allowing the accelerating forces to act for an extended period of time, as well as collimating and compressing the accelerated matter.

Depending on the laser intensity and the pulse length, different regimes of the LICPA accelerator operation can be distinguished [24]. At relatively low laser intensities and moderately long laser pulses (say, at $I_L < 10^{17}\,\mathrm{W\,cm^{-2}}$ and $\tau_L > 10^{-11}$ s) we can speak of a pure hydrodynamic LICPA regime, as the dominating force driving the projectile is due to the hydrodynamic pressure of hot plasma produced and confined in the accelerator cavity. On the other hand, at very high laser intensities and short laser pulses of circular polarization (say, at $I_L \geqslant 10^{20}\,\mathrm{W\,cm^{-2}}$, $\tau_L \leqslant 10^{-11}$ s), the photon pressure LICPA regime occurs, as the acceleration of the projectile is predominantly due to the photon (radiation) pressure of the laser pulse trapped in the cavity. For intermediate laser intensities (say, $10^{17}\,\mathrm{W\,cm^{-2}} \leqslant I_L < 10^{20}\,\mathrm{W\,cm^{-2}}$) we may speak of a mixed LICPA regime, as in this case both hydrodynamic and photon pressure can contribute to the acceleration process and a large amount of non-thermal hot electrons is produced, which significantly influence the process.

This paper summarizes briefly the main results of our recent LICPA studies, in particular, experimental investigations of the ion beam generation and the heavy plasma macroparticle acceleration in the hydrodynamic LICPA accelerator, as well as numerical, particle-in-cell (PIC) studies of acceleration of dense plasma (ion) bunches in the LICPA accelerator working in the photon pressure or mixed regime. Preliminary results on the generation of ultra-high-pressure (>100 Mbar) shocks through the collision of the LICPA-driven plasma macroparticle with a solid target are also presented.

2. The hydrodynamic LICPA accelerator

In the LICPA accelerator working in the hydrodynamic regime a projectile is driven by the hydrodynamic pressure of hot plasma produced in the accelerator cavity by a sub-ns or ns laser pulse of moderate intensity ($\sim 10^{10}$–10^{17} W cm^{-2}). In this accelerator, the energy of the ablating hot plasma, which is lost in the AA (open) scheme, is confined in the cavity and most of this energy can be converted into the kinetic energy of the projectile. Since the ablating plasma carriers up to 80–90% of the absorbed laser energy, the energetic efficiency of acceleration for LICPA can be much higher than that for AA and can reach tens of per cent [24]. In this section we will demonstrate the possibility of using the hydrodynamic LICPA accelerator for an efficient acceleration of heavy plasma macroparticles or production of high-current-density moderate-energy ion beams as well as for generation of ultra-high-pressure shocks by collision of the accelerated macroparticle with a solid target.

2.1. Acceleration of a heavy plasma macroparticle

Acceleration of heavy macroparticles (i.e. dense and compact objects of the mass $m_p \sim 1$–$100\,\mu$g or bigger) to hyper-velocities ~ 100–1000 km s^{-1} is a topic of high relevance for inertial fusion [3, 15, 16] and HEDP [1, 5, 25] and has been studied for a long time. So far, only laser-based methods using laser-induced ablative acceleration (AA) [3] have been capable of accelerating a heavy macroparticle to such velocities [3, 14, 16, 25]. However, due to the low energetic acceleration efficiency of AA big multi-kJ laser drivers had to be used to this end [14, 16, 25]. The LICPA accelerator could potentially reduce the required laser driver energy by an order of magnitude. To check the efficiency of the LICPA accelerator we performed an experiment at the kilojoule PALS laser facility [26] in Prague.

In this experiment an aluminum disc covered with a 5 μm thick CH ablator was irradiated by an 0.3 ns, 3ω PALS beam of energy 200 J in free space (the AA scheme) or inside the cavity of the cylindrical LICPA accelerator (figure 1(*a*)) characterized by the parameters $L_c = 0.4$ mm, $L_{Ch} = 2$ mm and the channel diameter $D_{Ch} = 0.3$ mm. In both schemes the beam diameter ($\sim 300\,\mu$m) and the laser intensity ($\sim 10^{15}$ W cm^{-2}) at the CH/Al target surface were the same. We used Al discs of the thickness 10 μm, 20 μm, 50 μm and 75 μm with masses m_p equal to 2 μg, 4 μg, 10 μg and 15 μg, respectively. The macroparticle velocity was measured using the time-of-flight (TOF) method employing the ion collector placed on the laser beam axis at the distance of 25 cm from the irradiated target (figure 2). Additionally, we recorded the density distribution of plasma leaving the accelerator channel using three-frame interferometry [24, 27].

Figure 3 presents exemplary results of interferometric measurements obtained for the plasma macroparticle with $m_p = 4\,\mu$g, driven according to the LICPA and AA schemes. To calculate the electron density distributions shown in the figure, phase shifts were extracted from the interferograms using the maximum-of-the-fringe method [27]. To obtain the

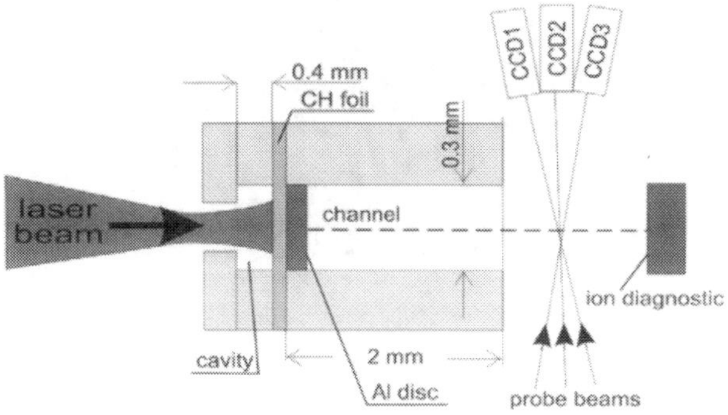

Figure 2. A scheme of the experimental set-up for measurements of the macroparticle velocity and the density profiles of plasma flowing out of the accelerator channel.

Abel equation solutions, the symmetrized phase distributions were expanded into Fourier series with the use of the Fast Fourier Transform method [28]. In the case of AA, the CH foil with the attached Al disc was mounted on the entrance of a large cylindrical channel of the diameter and the length equal to 2 mm. It can be seen from the electron density distributions presented in the figure that in the LICPA scheme a dense and collimated plasma flux reaches the distance ~ 1 mm from the channel exit after the time lapse of 30 ns from the moment the CH/Al target irradiation was started. This corresponds to the average flux velocity in the channel ~ 100 km s^{-1} (the flux velocity at the channel exit is a factor ~ 1.5–2 higher than the average one since the initial flux velocity is equal to 0). This is to be contrasted with the case of AA, when only a low density CH plasma sliding on the large channel wall leaves the channel, while the denser part of plasma originating from the Al disc is still inside the channel. Results obtained for macroparticles with other masses were qualitatively similar.

A more accurate—although still approximate—method to measure the macroparticle velocity is the TOF method relying on the use of ion charge collectors [29]. This method is feasible in the case investigated here since the macroparticle is actually a rather compact plasma bunch in which the charge density is roughly proportional to the plasma density. As we have learned from our two-dimensional (2D) hydrodynamic simulations of the macroparticle acceleration, the velocity distribution across the plasma bunch is inhomogeneous and in the LICPA accelerator plasma achieves highest velocities in regions off the accelerator axis (see below). Moreover, when the bunch leaves the accelerator channel, the plasma moves not only along the accelerator and the laser beam axis (the main direction) but also in other directions, which is due to internal pressure inside plasma. As a result, the spatial distribution of plasma velocity is rather complex, so in order to estimate the average velocity of the bunch v_p and its kinetic energy E_p many ion collectors placed at various distances from the axis are needed. In the preliminary measurements performed at PALS we used only a single ion collector situated on the laser beam axis. Thus our determination of the macroparticle velocity is very approximate and most likely underestimates the true value.

Figure 4 presents the TOF signals for the 10 μg plasma macroparticle accelerated in the LICPA and AA schemes. The

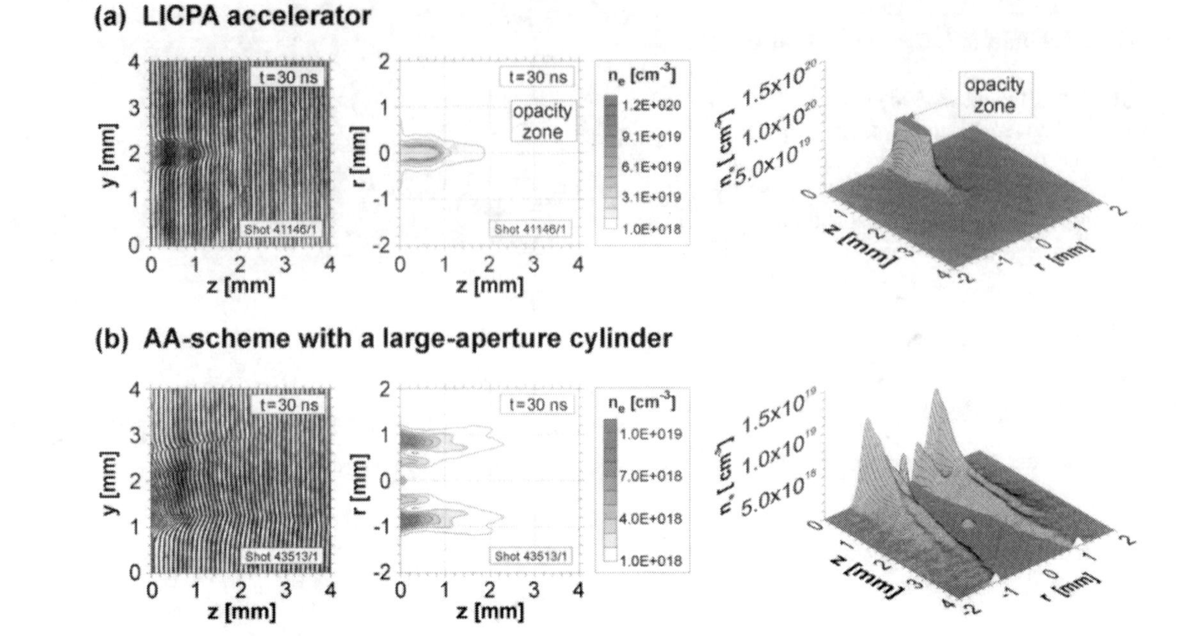

Figure 3. The interferograms (left), the electron isodensitograms (middle) and the space profiles of electron distributions (right) for the plasma flowing out of the channel for the LICPA and AA schemes recorded 30 ns after the CH/Al20 μm target irradiation. $E_L \approx 200$ J.

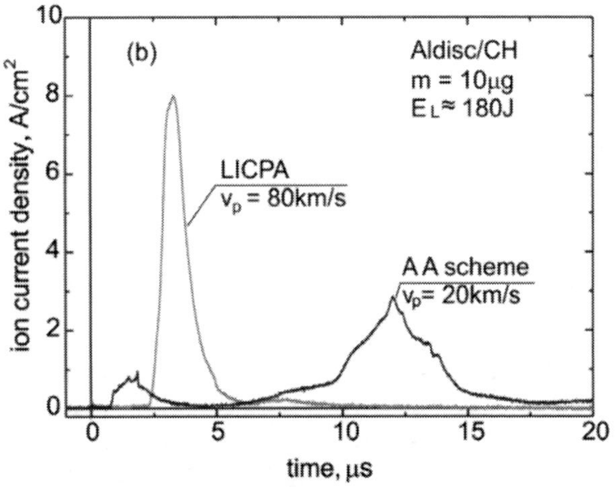

Figure 4. The TOF ion signals for the macroparticle driven by LICPA or AA recorded by the ion collector placed on the laser beam axis at the distance of 25 cm from the irradiated target.

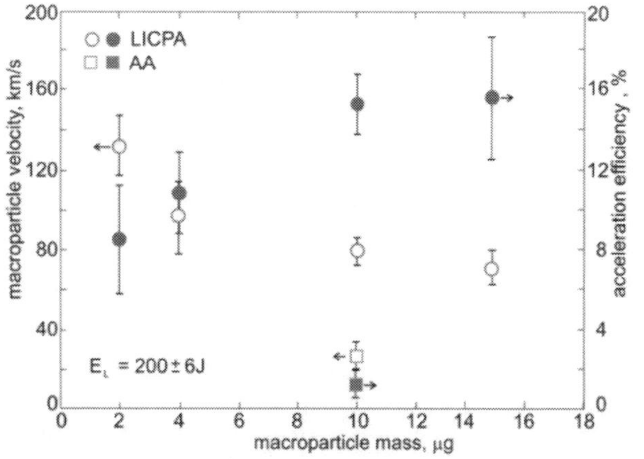

Figure 5. The macroparticle velocity and the energetic efficiency of acceleration as a function of the macroparticle mass.

TOF signal for LICPA is significantly stronger and appears with a much shorter delay than the one for AA and the macroparticle velocity estimated from this signal corresponds fairly well to the average velocity in the accelerator channel deduced from the interferometric measurements. Figure 5 shows the macroparticle velocity v_p and the macroparticle acceleration efficiency $\eta_{acc} = E_p/E_L$ as a function of the macroparticle mass (E_p was calculated under the assumption that the whole mass of the macroparticle moves with the velocity v_p). For comparison, the values of v_p and η_{acc} for the 10 μg plasma macroparticle driven by the ablative acceleration are also shown in this figure. It can be seen that the velocity of a macroparticle driven by LICPA decreases from $\sim$130 to $\sim$70 km s^{-1} and the acceleration efficiency increases from $\sim$9% to $\sim$15% with an increase in m_p (for AA η_{acc} is only $\sim$1%). The lower acceleration efficiency for lighter macroparticles is somewhat unexpected. It may result from the fact that velocities of these macroparticles have lower axial and higher radial components outside the channel (due to the higher temperature of the thinner Al discs), but this effect cannot be revealed with a single ion collector.

To have a deeper insight into the process of macroparticle acceleration in the LICPA scheme we carried out numerical simulations using 2D hydrodynamic PALE (Prague Arbitrary Lagrangian Eulerian) code [30, 31]. Simulations were performed for the same parameters of the targets, of the accelerator (L_C, L_{Ch}, D_{Ch}, the cavity hole diameter d_h) and the laser beam as in the experiment (the spatial intensity

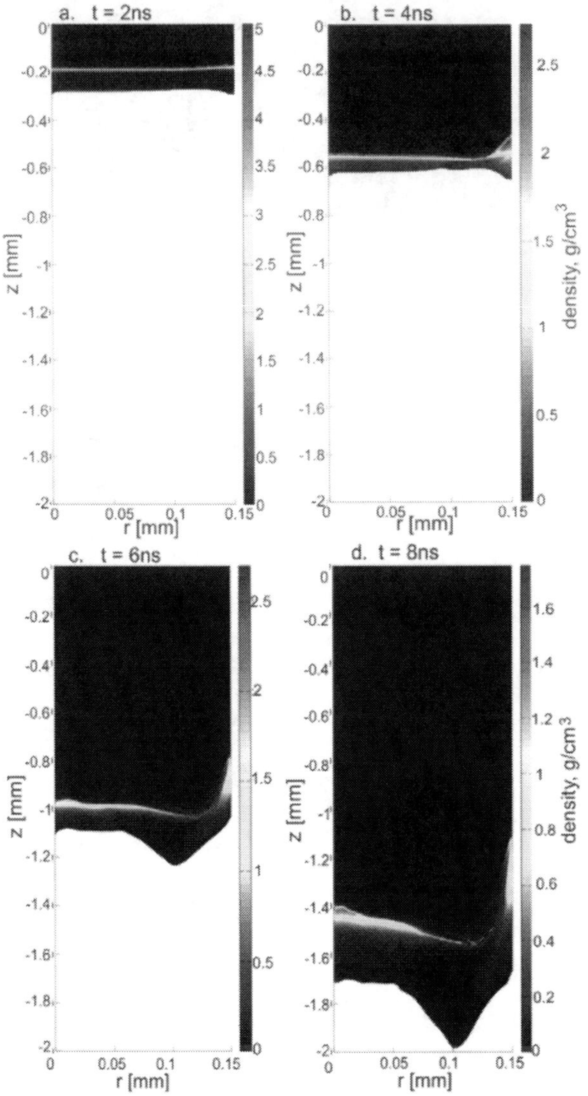

Figure 6. 2D spatial profiles of density in the LICPA accelerator channel for various times of acceleration of the macroparticle of the mass of 4 μg. The CH/Al20 μm target placed at $z = 0$ is irradiated by the laser beam directed along the channel axis ($r = 0$). The accelerator cavity (not shown) of the length $L_c = 0.4$ mm is situated in the region with $z > 0$. Time $t = 0$ corresponds to the moment when the target is irradiated at the maximum beam intensity. The result of 2 D hydrodynamic simulations at $E_L \approx 200$ J.

distribution of the laser beam was assumed to be super-Gaussian with the shape defined by $\exp[-(r/R)^6]$ with $R = 170\,\mu$m). Simulations of the acceleration phase were conducted in the cylindrical r, z geometry and in Lagrangian coordinates employing mesh moving with the plasma. Figure 6 presents the spatial 2D distributions of the plasma macroparticle density for $m_p = 4\,\mu$g (CH/Al 20 μm target) at various moments of acceleration in the accelerator channel. It can be seen that the macroparticle remains a compact and dense (>0.5 g cm^{-3}) plasma object for all phases of acceleration in the channel. The transverse plasma density distribution is relatively flat with the maximum density situated close to the channel axis. On the other hand, in the final phase of acceleration the plasma attains the highest velocity away from the channel axis. At this stage the most dense part of the plasma bunch (close to the axis) propagates with the velocity $\sim 2 \times 10^7$ cm s^{-1}. The figure shows the most significant difference between the plasma macroparticle driven by LICPA and the one accelerated in the AA scheme, apart from higher velocity: the LICPA macroparticle is a much more compact object, with much higher density. The density of a plasma macroparticle driven by the AA in free space may be even two orders of magnitude lower than that of the LICPA macroparticle, due to both longitudinal and radial expansion of plasma, while for the AA with a channel the difference between LICPA and AA is by one order of magnitude. The difference in the density of LICPA-driven macroparticles and AA-driven macroparticles is very important for their interactions with solid targets (see the next sub-section).

To conclude, we have shown that with the LICPA accelerator it is possible to accelerate a plasma macroparticle with the mass $\sim 10\,\mu$g to the velocity ~ 100 km s^{-1} with the laser driver of energy of a few hundred J. So far, multi-μg macroparticles could be accelerated to such velocities only with the use of multi-kJ lasers [25]. The energetic acceleration efficiency η_{acc} for LICPA is by an order of magnitude higher than that for AA and can reach values $\geqslant 15\%$. The attained value of η_{acc} is probably the highest achieved so far for heavy (>1 μg) macroparticles.

2.2. Generation of ultra-high-pressure shocks

Ultra-high pressures from tens of Mbar (10^6 bar) to Gbar (10^9 bar) and beyond are common in the cores of astrophysical objects like stars and planets [1], and are predicted to exist in laboratory, for example in highly compressed matter produced in inertial confinement fusion experiments [3]. Pressures of this magnitude are also important for material science, as they enable us to study the properties of materials in extreme conditions, to understand their phase transition and contribute to the knowledge necessary to produce principally new kinds of materials.

The pressures up to $\sim$1–2 Mbar can be produced in the laboratory in a steady-state regime using the diamond anvil techniques [32]. Higher pressures may be achieved by the use of dynamic shocks (shock waves) which can be generated in the condensed matter using chemical [33] or nuclear [34] explosions, light-gas guns [35], Z-pinch machines [8, 36] or laser-based methods [1, 3, 25]. The highest pressures, from tens of Mbar to Gbar and possibly beyond can be produced by nuclear explosions and laser-based methods. For precise determination of the shock parameters and the equation of state (EOS) of the shock-compressed material the quality of the shock is very important, and the material being compressed should have well defined initial parameters. These requirements impose significant constraints on the maximum shock pressures which can be attained in laboratory experiments.

The generation of strong shocks using lasers is possible by a direct irradiation of a solid target by the laser beam

(a) [1, 3, 37, 41, 42] or the soft x-ray flux produced with a laser (b) [3, 39, 40] as well as by an impact of a macroparticle driven by the laser beam (c) [3, 38] or the x-ray flux (d) [3, 25]. A common drawback of all these laser-based methods is their low energetic efficiency $\eta_s = E_s/E_L$ (E_s is the shock energy and E_L is the energy of the laser beam which generates the shock or produces x-rays or accelerates the projectile). In the case of a short-wavelength (UV) laser driver this efficiency is of the order of ~1–3% [3] for the method (a) and is below 1% for the method (d) [3, 25], achieving still lower values when a long-wavelength (~1 μm) laser driver is applied. Consequently, in order to obtain the quasi-planar shock pressure in the 0.1–1 Gbar range, a multi-kJ short-wavelength laser-driver has to be used [25]. Below we will show that using LICPA accelerator the shock pressures above 0.1 Gbar can be produced with a laser driver of energy only 200 J.

In the experiment performed at PALS, a heavy plasma macroparticle (CH/Al disc) accelerated in the LICPA accelerator like the one presented in figure 2 collided with a massive Al target placed at the accelerator channel exit. We measured dimensions of the crater produced in the massive target by the macroparticle impact, since the crater volume is a measure of the energy of the shock generated in the target by the impact and it is a useful source of information about the shock pressure. For comparison we also measured craters produced in the massive target by the impact of a plasma macroparticle (CH/Al disc) driven by the AA, as well as craters created by the direct irradiation of the massive target by the laser beam. In all three cases, parameters of the laser beam at the irradiated target were the same and CH/Al targets for LICPA and AA were identical. The result of this comparison is shown in figure 7 where replicas of craters produced with the use of LICPA or AA with CH/Al20 μm target as well as by the direct irradiation are presented. The volumes of craters produced by the LICPA-driven macroparticle are by more than order of magnitude larger than craters produced using AA or the direct irradiation. This is a strong indication that the energy and the pressure of the shock generated with the use of the LICPA accelerator are also much higher than those achieved with AA and the direct irradiation.

To estimate the pressure of the shock generated in a massive target by the impact of a LICPA-driven plasma macroparticle we performed detailed numerical simulations of the macroparticle acceleration and formation of craters by the macroparticle impact using the 2D PALE code [30, 31]. As the computational domain in the impact phase does not change, we performed the simulation of this phase in Eulerian coordinates with static computational mesh. A comparison of the measured and computed volumes of craters produced in the massive Al target by the impact of CH/Al plasma macroparticles of various masses is presented in figure 8, together with volumes of craters produced with the use of the AA scheme. We see that numerical results are in reasonable agreement with experimental data, which indicates that our estimates of the parameters of the shock are reliable [44].

A structure of the shock generated in the massive target by the impact of the 4 μg plasma macroparticle is shown in figure 9. The figure presents 2D spatial profiles of density and pressure inside the target at the moment when the pressure attains a maximum value. The shock front is flat over the diameter $2r \approx 200\,\mu$m, the maximum density in the shock is around 14.3 g cm^{-3}, and the shock velocity is estimated to be 80 km s^{-1}. The pressure at the shock front reaches 144 Mbar and the maximum pressure behind the front approaches 290 Mbar. All these parameters are much higher than those achieved for the plasma macroparticle driven by the AA or for the case of a direct laser irradiation, as could be expected from the comparison of craters. It should be noted that despite the fact that main reasons for such high parameters of the generated shocks are high density and high velocity of the plasma macroparticle efficiently accelerated in the LICPA accelerator, there is an additional factor which enhances the shock velocity and pressure, namely the additional force acting on the macroparticle during and after the collision, originating from the pressure of plasma still remaining in the accelerator channel.

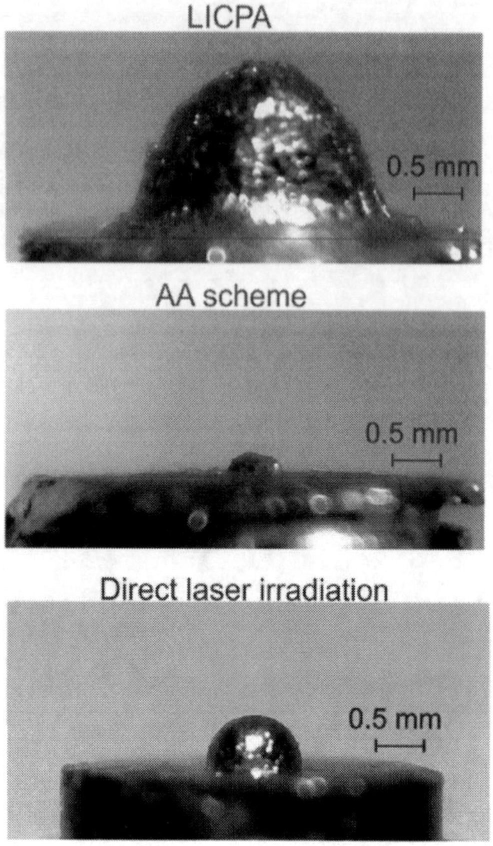

Figure 7. The pictures of replicas of craters produced in the massive Al target by the impact of 4 μg plasma macroparticle accelerated in the LICPA scheme or the AA scheme or by the direct irradiation of the target by laser beam. The distance from the CH/Al20 μm target to the massive target for LICPA and AA is the same and equal to $L_{Ch} = 2$ mm. $E_L \approx 200$ J.

To conclude, it has been shown that hydrodynamic LICPA accelerator can be a highly efficient tool for generating high-pressure shock by collision of the LICPA-driven plasma macroparticle with a solid target. In particular, using such

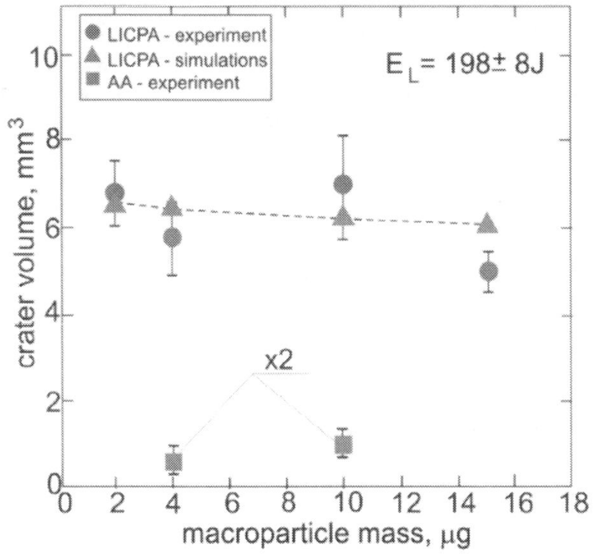

Figure 8. The volumes of craters produced in the massive Al target by the plasma macroparticle driven by LICPA or AA as a function of the macroparticle mass. The red circles and the blue squares—results of measurements, the green triangles—results of numerical simulations. Note that the crater volumes for AA are magnified by the factor 2 in the figure. A procedure of calculation of the crater volume from measurements was described in [43].

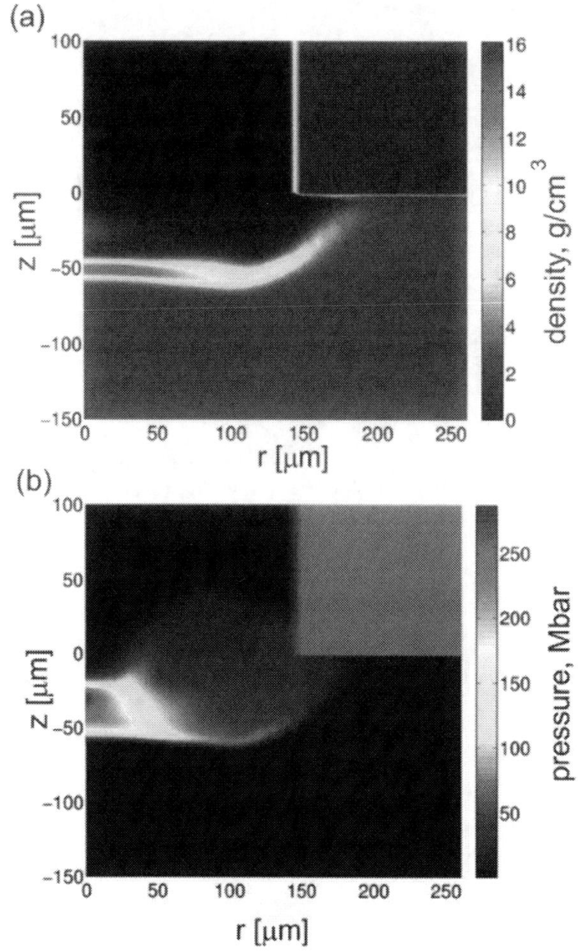

Figure 9. 2D spatial profiles of density (*a*) and pressure (*b*) inside the massive Al target ($z \leqslant 0$, $0 \leqslant r \leqslant 250\,\mu$m) after the macroparticle impact at the moment when the pressure attains a maximum value. The 4 μg plasma macroparticle (CH/Al20 μg target) is accelerated in the channel situated in the region $z > 0$, $0 \leqslant r \leqslant 150\,\mu$m. There is a free space in the region $z > 0$, $150\,\mu$m $< r \leqslant 250\,\mu$m. The result of 2D hydrodynamic simulations at $E_L = 200$ J.

an accelerator enables us to generate quasi-planar shocks of the pressure above 100 Mbar with the laser driver of energy as low as 200 J.

2.3. Generation of high-current-density ion beams

The hydrodynamic LICPA accelerator has also a potential to be an efficient source of intense ion beams of moderate ion energies ranging from keV to multi-MeV.

In the experiment performed on the PALS laser facility, ion beams produced from a 25 μm CD_2 foil target in the cylindrical (figure 1(*a*)) or conical (figure 1(*b*)) LICPA accelerator were compared with the ones generated from the CD_2 foil in the conventional laser–planar target interaction scheme (the AA scheme) and the scheme where the foil was placed at the entrance of the cylindrical or conical channel without a cavity (the AA-channel scheme). Parameters of ion fluxes emitted in the forward direction were measured by the time-of-fight (TOF) method using two ion collectors [45, 46] placed 30 cm from the CD_2 target at angles 0° and 5°, respectively, with respect to the target normal and the laser beam axis. The CD_2 target was irradiated by the 3ω ($\lambda = 0.438\,\mu$m) PALS laser beam of 0.3 ns duration and energy up to ~300 J. The parameters of ion beams produced in different acceleration schemes were compared for the same laser intensity on the CD_2 target, which was varied in the range 10^{14}–10^{15} W cm^{-2}.

Figure 10 presents an exemplary TOF ion signal, recorded by the ion collector situated on the laser beam axis, for ions accelerated in the cylindrical LICPA accelerator and ions originating from a planar target irradiated in the AA scheme. The presented signals reflect only the thermal (slow) component of the ion beam of keV ion energies, since the number of fast ions is too small here to produce a fast component visible in the scale presented in the figure. The difference between parameters of ion beams produced in the LICPA accelerator and those generated via the AA scheme is very pronounced: for LICPA the peak current density of thermal ions is by more than an order of magnitude higher and the mean ion energy is by a factor ~4 higher than in the AA scheme.

Similarly, higher current densities and ion energies (both the thermal and the fast ions) were observed when the LICPA scheme was compared with other AA schemes (with a conical channel or without the channel). Further results of the measurements, including the energy dependence of ion beam parameters obtained in different schemes, were presented in [29]. It should be noted that the current density of 10–150 keV deuteron beam leaving the conical LICPA accelerator achieved an extremely high value ~1 GA cm^2 sr^{-1}, which made it

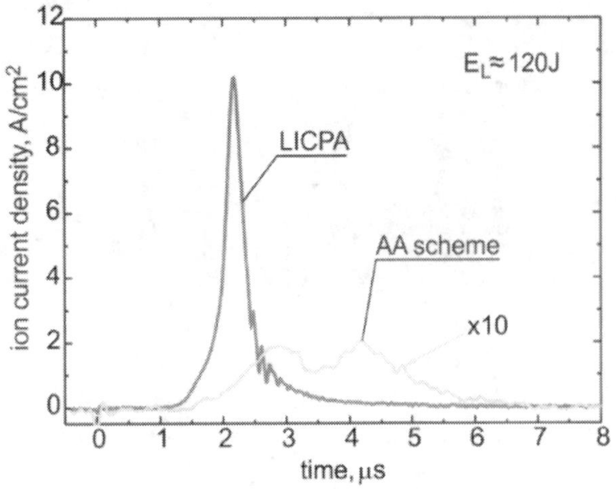

Figure 10. The TOF ion signals for the ion beam driven by LICPA or AA recorded on the laser beam axis at the distance of 30 cm from the irradiated CD_2 target.

possible to produce fusion neutrons from the interaction of the beam with a CD_2 target [29]. Parameters of ion beams produced in the LICPA accelerator, in particular the ion energies, can be further enhanced by a careful selection of the target thickness and the optimization of the accelerator geometry, so that efficient generation of intense ion beams with MeV ion energies at laser intensities on the target $\sim 10^{16}$ W cm^{-2} seems to be feasible.

3. The LICPA accelerator working in the mixed regime

When the intensity of a laser driver of the LICPA accelerator reaches values between 10^{17} W cm^{-2} and a few 10^{20} W cm^{-2}—which is the intensity range of the majority of contemporary laser–plasma experiments with short-pulse lasers—we can speak about the mixed LICPA regime. In this regime, a large amount of hot (non-thermal) electrons is produced in the plasma confined inside the accelerator, and electromagnetic forces (electrostatic, ponderomotive, others) generated in the plasma clearly dominate over the hydrodynamic forces. The plasma ions are accelerated predominantly by these forces and various mechanism of ion (plasma) acceleration can contribute to the process including TNSA (target normal sheath acceleration) [47, 48], RPA [17–20], BOA (laser break-out afterburner) [49, 50] or others. A contribution of each of them to the acceleration process is determined by the laser beam and the target parameters and, in part, by the accelerator configurations.

To examine properties and efficiency of ion beam generation in the LICPA accelerator working in the mixed regime, we performed numerical simulations using a fully relativistic 1D particle-in-cell code (PIC1D) [51] based on similar physics and numerical concept as in the popular LPIC++ code. Results of simulations for LICPA were compared with the ones for the conventional acceleration scheme i.e. when a laser beam interacts with a planar target in the free space. Parameters of the laser beam used in the simulations were typical for those of current Ti : sapphire petawatt lasers (e.g. the CETAL laser [52]). The laser pulse was assumed to be super-Gaussian with duration (FWHM) 25 fs and intensity ranged from 2×10^{19} to 2×10^{20} W cm^{-2}. The laser pulse interacted with a fully ionized carbon target (ion density 10^{23} cm^{-3}) of the thickness 20, 50 and 100 nm placed in the cavity of the length L_c from 50 to 150 μm. The internal cavity wall was represented by a reflective boundary for the electromagnetic waves circulating between the plasma target and the wall (the laser pulse was initialized on the grid with plasma in front of it). The (effective) reflection coefficient of the wall R_c included a loss of radiation reflected from the target in the cavity hole (it was calculated as $(d_h/d_L)^2$ where d_h and d_L are diameters of the hole and the laser beam on the target, respectively) as well as the loss due to imperfectness of the cavity wall as a mirror (the last one was calculated with the 1D PIC code taking into account plasma production on the wall by the radiation circulating in the cavity). In the simulations presented here we assumed $R_c = 70\%$ in accordance with numerical calculations performed for the cavity wall made of polished Au and the assumption that the loss in the hole is equal to 20%.

The presented 1D model of the LICPA accelerator has some limitations which must be taken into account when results obtained using this model are assessed. An obvious limitation of the model is impossibility of inclusion in the calculations transverse instabilities in the acceleration process and losses of energy and number of accelerated ions due to their interaction with the accelerator channel wall. Other problem is a loss of part of the hot electron population due to interaction of the electrons with the cavity side wall. This problem can be minimized, if the target and the cavity parameters are properly selected—the target is made of insulator (e.g. C), its thickness is small (in the sub-μm range) and the cavity diameter is much larger than the target thickness—and, moreover, the laser pulse contrast is relatively high (the preplasma thickness is in the sub-μm–μm range) and the pulse is not too long (fs–sub-ps range). In such a case only relatively small part of the electron population remains inside the cavity side wall and is lost for the acceleration process. It is due to the fact that in such conditions the Debye length is short, most of electrons is strongly coupled to ions (especially, if the ion charge state $z \gg 1$, as for C) and the transverse transport of electrons in the non-ionized part of the (insulator) target is inhibited. Thus, only hot electrons produced in the vicinity of the cavity side wall are lost in the wall while majority of them take part in the acceleration process. The above conditions, though seem to be idealized, are fully realizable in current experiments. Just such conditions were assumed for demonstration of ion acceleration in the mixed LICPA regime using 1D PIC simulations.

The simulations were performed for both the linear and the circular polarization of the laser pulse. As could be expected, both for the LICPA scheme and the conventional scheme the energy spectrum of the accelerated ions was broad with the Maxwellian-like shape. However, the ion energies and the laser-to-ion energy conversion efficiency for both schemes

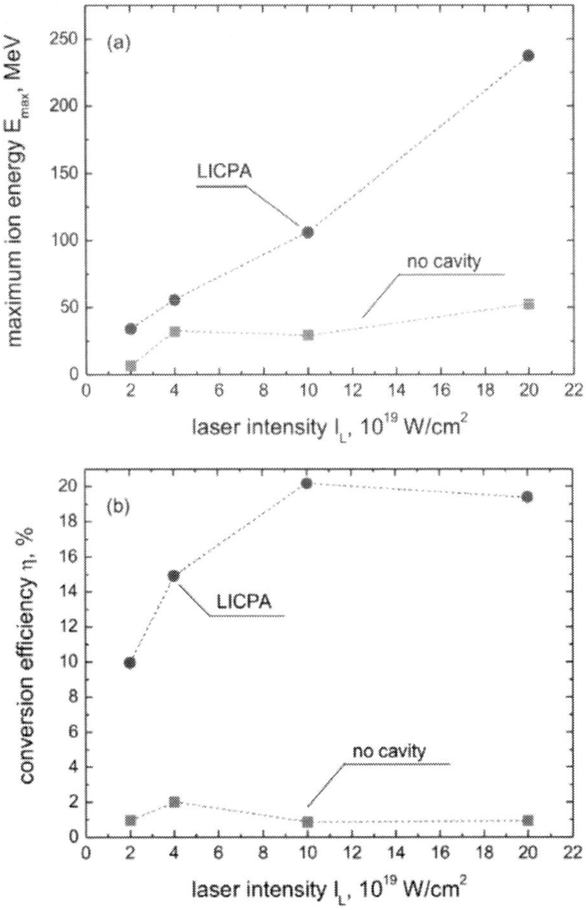

Figure 11. The maximum ion energy (*a*) and the laser-ions energy conversion efficiency (*b*) for ion beams driven in the LICPA accelerator or the conventional scheme (no cavity) as a function of laser intensity on the carbon target. The result of 1 D PIC simulations for $L_T = 50$ nm, $L_c = 100\,\mu$m and the linear light polarization.

were essentially different. For the cases presented in figure 11, the conversion efficiency for LICPA reaches approximately 20% and is by a factor ∼10 higher than that for the conventional scheme. The maximum ion energies are higher by a factor of 2–5 and the number of high-energy ions (of $E > 5$ MeV) was found to be higher by more than an order of magnitude relative to the conventional scheme. However, it should underline, that the comparison of the LICPA and the conventional scheme was carried out for the laser and target parameters the same for both the scheme. In the case when these parameters are individually matched to the optimal ones for each the scheme, the differences in output parameters of these schemes can be smaller.

Although results of simulations presented above were obtained for an idealized 1D case and, as a result, the absolute values of the calculated ion beam parameters can be overestimated, they clearly demonstrate the potential of the LICPA accelerator driven by currently attainable short-pulse lasers as a highly efficient tool for generation of high-energy ion beams with ion energies approaching hundreds of MeV.

4. The photon pressure-driven LICPA accelerator

The LICPA accelerator working in the photon pressure regime requires very high laser intensities ($>10^{20}$ W cm^{-2}) and this regime seems to be especially interesting for the generation of ultra-intense ion beams or acceleration of macroparticles to ultra-high velocities for future applications in HEDP or ICF. We tried to check the possibility of using the LICPA accelerator for these purposes by numerical simulations [24, 53, 54]. Below, we briefly summarize our numerical works related to generation of carbon ion beams for the ion fast ignition of a fusion target [53] and acceleration of a gold macroparticle for application in impact ignition fusion [54].

4.1. Generation of ultra-intense ion beams for ion fast ignition

In the ion fast ignition scenario of ICF [50, 55–57], a short-pulse (5–20 ps) intense beam of light ions of the mean energy $\bar{E}_i \sim$10–50 MeV amu^{-1} [57] ignites a DT fuel compressed to the density ∼300–500 g cm^{-3}. The ion energies are relatively low, however the intensity $I_i \geqslant 10^{20}$ W cm^{-2}, the energy fluence $F_i \geqslant 2$ GJ cm^{-2}, and the total beam energy $E_{ib} \geqslant 20$ kJ of the beam are extremely high [50, 56, 57]. In addition, the laser-to-ions energy conversion efficiency η should be sufficiently high ($\eta \geqslant 15\%$) to limit the laser driver energy and cost to a reasonable value.

For a quantitative estimation of the possibility of achieving the ion beam parameters required for the fast ignition in the LICPA and the conventional RPA schemes we performed simulations of acceleration of carbon ions using the relativistic PIC1D code [53]. In the simulations, a laser beam interacted with a fully ionized carbon plasma target of the thickness $L_T = 2$ or $8\,\mu$m and the ion density $n_i = 10^{23}$ cm^{-3}. We assumed that a circularly polarized laser pulse has wavelength of 1.05 μm, a super-Gaussian temporal profile with the peak intensity $I_L = 2.5 \times 10^{21}$ W cm^{-2}, 5×10^{21} W cm^{-2} or 8×10^{21} W cm^{-2} and duration (FWHM) equal to 2 ps. In the LICPA scheme, $R_c = 0.64$ and $L_c = 120\,\mu$m were assumed.

The simulations showed [53] that in the considered range of laser intensities (2.5–8) $\times 10^{21}$ W cm^{-2}, both LICPA and RPA produces quasi-monoenergetic ion beams of similar energy dispersions $\Delta E_i/\bar{E}_i \sim$ 0.3–0.4, but of significantly different mean ion energies $\bar{E}_i$: from 7.2 to 41 MeV amu^{-1} for LICPA and from 2.2 to 18 MeV amu^{-1} for RPA. In turn, the energy conversion efficiency η changes from 23% to 41% for LICPA and from 7% to 18% for RPA in this intensity range. As a result, for LICPA all the ion beam parameters are well above the threshold for the ignition at the laser intensity of 5×10^{21} W cm^{-2}, which corresponds to 100 kJ of laser energy at the laser focal spot on the target of 35 μm. To reach the ignition threshold in the case when RPA is used, the laser intensity and energy should be by a factor 2–3 higher.

4.2. Acceleration of a gold plasma macroparticle for impact ignition fusion

In the impact ignition scenario of inertial fusion, an ignition of the compressed DT fuel is the result of collision of a macroparticle of the mass $m_p \sim 10^{-6}$–10^{-4} g and the velocity above 10^8 cm s^{-1} with the fuel [15, 58]. In the ignition scheme proposed by Caruso and Pais [58], a $\sim 1\,\mu$g macroparticle made of Au and accelerated to $\sim 5 \times 10^8$ cm s^{-1} collides with the DT fuel compressed to the density $\gtrsim$200 g cm^{-3}. Due to the collision, the macroparticle is rapidly collapsing to high densities ($>$1000 g cm^{-3}) and a large fraction of its energy is transferred to the fuel in very short time ($\sim 10^{-11}$ s). To reach the fuel ignition, the kinetic energy and the energy fluence of the macroparticle should be, respectively, $\sim$10–20 kJ and $\sim$500 MJ cm^{-2} [58].

We investigated numerically [54] the possibility of accelerating a gold plasma macroparticle to the required velocities and energies in the LICPA and RPA schemes using the same PIC1D code as the one used for simulations of acceleration of carbon ions. The gold (Au^{10+}) plasma target of the thickness $L_T = 5\,\mu$m was irradiated by a circularly polarized super-Gaussion laser pulse of duration $\tau_L = 10$, 5 or 2.5 ps (FWHM), $\lambda = 1.05\,\mu$m and the focal spot diameter $d_L = 80\,\mu$m (which corresponds to the target mass of $\approx$0.5 μg). The laser pulse energy E_L was assumed to be of 100, 200 or 300 kJ. For LICPA we assumed $L_c = 300\,\mu$m and $R_c = 0.75$.

We have found that for LICPA the macroparticle energy fluence reaches the threshold value for ignition at the laser energy $\sim$150 kJ, while for RPA the energy fluence is well below the threshold even at $E_L = 300$ kJ. In the considered laser energy range of 100–300 kJ and at $\tau_L = 10$ ps, the macroparticle velocity v_p and the acceleration efficiency η_{acc} reach for LICPA the values from 7.6×10^8 cm s^{-1} to 1.6×10^9 cm s^{-1} and from 12% to 21%, respectively, while for RPA v_p and η_{acc} are in the range of (2.4–6.9) $\times 10^8$ cm s^{-1} and 1.2–3.3%, respectively. Thus, at the laser energy $\sim$150 kJ, the velocity and the energy fluence as well as the total kinetic energy of the macroparticle driven by LICPA achieve the values required for the ignition. In the case of using RPA, the ignition seems to be possible only at laser energies at least 3–4 times higher.

The results obtained for LICPA are highly promising for both cases considered in the sections 4.1 and 4.2. However, since the simulations were performed for very idealized conditions, they should be treated as a first step to more comprehensive studies of the photon pressure-driven LICPA accelerator employing, in particular, 2D codes and a more detailed model of laser–plasma interaction in the accelerator cavity.

5. Summary

Our measurements performed at the kilojoule PALS laser facility, supported by advanced 2D simulations, have demonstrated that the hydrodynamic LICPA accelerator can be a highly efficient tool for acceleration of heavy plasma macroparticles to hyper velocities and generation of ultra-high-pressure ($>$100 Mbar) shocks by collision of the macroparticle with a solid target. Parameters of macroparticles driven by LICPA and the generated shocks can be much higher than the ones achieved at similar laser energies with other laser-based methods known so far.

Our 1D PIC simulations have shown that LICPA accelerator working in the photon pressure or the mixed regime can be highly efficient like the hydrodynamic one and, in particular, can produce intense ion beams of multi-MeV to GeV ion energies with the energetic efficiency of tens of per cent, much higher than the one for the conventional RPA or TNSA schemes. The photon pressure-driven LICPA accelerator can also efficiently accelerate heavy plasma macroparticles to velocities required for impact ignition fusion and the efficiency of macroparticle acceleration is significantly higher than that in the RPA scheme.

The LICPA accelerator can, potentially, be driven by lasers covering a very broad range of laser energies (from 1 J to 1 MJ) and intensities (from 10^{10} W cm^{-2} to 10^{23} W cm^{-2}) as well as pulse lengths (from ns to fs) and laser wavelengths (from UV to IR). As a result, the accelerator can produce dense, fast or ultrafast projectiles of a wide variety of parameters. The unique efficiency of the LICPA accelerator coupled with its flexibility in using various laser drivers and in producing projectiles of various parameters create a potential for a broad range of the accelerator applications, in particular in high-energy density physics, ICF research, nuclear physics or material science.

Acknowledgments

This work was supported in part by the funds related to the IFE Keep-in-Touch activity within the Eurofusion consortium and the National Centre for Science (NCN), Poland under grant no. 2012/04/M/ST2/00452. The experiments were performed within the Access to Research Infrastructure activity in the 7th Framework Programme of EU (contract no. 212025, Laserlab Europe-Continuation) at the PALS Research Infrastructure in Prague supported by the Ministry of Education, Youth and Sport of CR (project LC528 and LM2010014).

References

[1] Drake R P 2006 *High-Energy-Density Physics* (Berlin: Springer)
[2] Ledingham K W D and Galster W 2010 *New J. Phys.* **12** 045005
[3] Atzeni S and Meyer-ter-Vehn J 2004 *The Physics of Inertial Fusion* (Oxford: Clarendon)
[4] Thornhill T F *et al* 2006 *Int. J. Impact Eng.* **33** 799
[5] Foster J M *et al* 2005 *Astrophys. J.* **634** L77
[6] Khoroshkov V S and Minakova E I 1998 *Eur. J. Phys.* **19** 523
[7] Rosiński M *et al 2005 Vacuum* **78** 435
[8] Lemke R W *et al* 2011 *Int. J. Impact Eng.* **38** 480
[9] Degnan J *et al* 1993 *Phys. Fluids* B **5** 2938
[10] Karasik M *et al* 2010 *Phys. Plasmas* **17** 056317
[11] Macchi A *et al* 2005 *Phys. Rev. Lett.* **94** 165003
[12] Yu W *et al* 2005 *Phys. Rev.* E **72** 046401
[13] Badziak J and Jabłoński S 2011 *Appl. Phys. Lett.* **99** 071502
[14] Glenzer S H *et al* 2010 *Science* **327** 1228
[15] Murakami M *et al* 2006 *Nucl. Fusion* **99** 99
[16] Azechi H *et al* 2009 *Phys. Rev. Lett.* **102** 235002
[17] Esirkepov T *et al* 2004 *Phys. Rev. Lett.* **92** 175003
[18] Robinson A P L *et al* 2008 *New J. Phys.* **10** 013021
[19] Liseykina T V *et al* 2008 *Plasma Phys. Control. Fusion* **50** 124033
[20] Macchi A *et al 2009 Phys. Rev. Lett.* **103** 085003
[21] Badziak J *et al* 2004 *Plasma Phys. Control. Fusion* **46** B541

[22] Badziak J *et al* 2008 *J. Appl. Phys.* **104** 063310
[23] Badziak J *et al* 2010 *Appl. Phys. Lett.* **96** 251502
[24] Badziak J *et al* 2012 *Phys. Plasmas* **19** 053105
[25] Cauble R *et al* 1993 *Phys. Rev. Lett.* **70** 2102
[26] Jungwirth K *et al* 2001 *Phys. Plasmas* **8** 2495
[27] Kasperczuk A and Pisarczyk T 2001 *Opt. Appl.* **31** 571
[28] Kalal M and Nugent A 1988 *Appl. Opt.* **27** 1956
[29] Badziak J *et al* 2013 *Appl. Phys. Lett.* **103** 124104
[30] Liska R *et al* 2011 *Finite Volumes for Complex Applications VI, Problems and Perspectives* vol 2, ed J Fort *et al* (Berlin: Springer) pp 57–73
[31] Kapin T *et al* 2008 *Int. J. Numer. Methods Fluids* **56** 1337
[32] Hemley R J and Mao H K 1988 *Phys. Rev. Lett.* **61** 857
[33] Altshuler L V *et al* 1960 *Sov. Phys.—JETP* **11** 573
[34] Trunin R F 1994 *Phys.—Usp.* **37** 1123
[35] Mitchell A C *et al* 1991 *J. Appl. Phys.* **69** 2981
[36] Knudson M D *et al* 2009 *Phys. Rev. Lett.* **103** 225501
[37] Veeser L R and Solem S C 1978 *Phys. Rev. Lett.* **40** 1391
[38] Fabbro R *et al* 1986 *Laser Part. Beams* **4** 413
[39] Löwer Th *et al* 1994 *Phys. Rev. Lett.* **72** 3186
[40] Benuzzi A *et al* 1996 *Phys. Rev.* E **54** 2162
[41] Batani D *et al* 2002 *Phys. Rev. Lett.* **88** 235502
[42] Ozaki N *et al* 2004 *Phys. Plasmas* **11** 1600
[43] Pisarczyk T *et al* 2014 *Phys. Plasmas* **21** 012708
[44] Koester P *et al* 2013 *Plasma Phys. Control. Fusion* **55** 124045
[45] Laska L *et al* 2003 *Plasma Phys. Control. Fusion* **45** 585
[46] J Badziak *et al* 2001 *J. Phys. D: Appl. Phys.* **34** 1885
[47] Wilks S C *et al* 2001 *Phys. Plasmas* **8** 542
[48] Borghesi M *et al* 2006 *Fusion Sci. Technol.* **49** 412
[49] Yin L *et al* 2007 *Phys. Plasmas* **14** 056706
[50] Fernandez J C *et al* 2009 *Nucl. Fusion* **49** 065004
[51] Badziak J and Jabłoński S 2010 *Phys. Plasmas* **17** 073106
[52] www.cetal.inflpr.ro
[53] Badziak J *et al* 2012 *Appl. Phys. Lett.* **101** 084102
[54] Badziak J and Jabłoński S 2014 *Phys. Scr.* **T161** 014031
[55] Roth M *et al* 2001 *Phys. Rev. Lett.* **86** 436
[56] Temporal M *et al* 2002 *Phys. Plasmas* **9** 3098
[57] Davis J *et al* 2011 *Plasma Phys. Control. Fusion* **53** 045013
[58] Caruso A and Pais V A 1996 *Nucl. Fusion* **36** 745

IOP Publishing

Plasma Physics and Controlled Fusion

Plasma Phys. Control. Fusion **57** (2015) 014008 (5pp)

doi:10.1088/0741-3335/57/1/014008

Symbolic regression via genetic programming for data driven derivation of confinement scaling laws without any assumption on their mathematical form

A Murari[1], E Peluso[2], M Gelfusa[2], I Lupelli[2,3], M Lungaroni[2] and P Gaudio[2]

[1] Consorzio RFX-Associazione EURATOM-ENEA per la Fusione, I-35127 Padova, Italy
[2] Associazione EURATOM-ENEA, University of Rome 'Tor Vergata', Roma, Italy
[3] Culham Centre for Fusion Energy, Abingdon, Oxfordshire, OX14 3DB, UK

E-mail: Andrea.Murari@igi.cnr.it

Received 28 August 2014, revised 8 September 2014
Accepted for publication 9 September 2014
Published 28 November 2014

Abstract

Many measurements are required to control thermonuclear plasmas and to fully exploit them scientifically. In the last years JET has shown the potential to generate about 50 GB of data per shot. These amounts of data require more sophisticated data analysis methodologies to perform correct inference and various techniques have been recently developed in this respect. The present paper covers a new methodology to extract mathematical models directly from the data without any *a priori* assumption about their expression. The approach, based on symbolic regression via genetic programming, is exemplified using the data of the International Tokamak Physics Activity database for the energy confinement time. The best obtained scaling laws are not in power law form and suggest a revisiting of the extrapolation to ITER. Indeed the best non-power law scalings predict confinement times in ITER approximately between 2 and 3 s. On the other hand, more comprehensive and better databases are required to fully profit from the power of these new methods and to discriminate between the hundreds of thousands of models that they can generate.

Keywords: symbolic regression, genetic programming, scaling laws

(Some figures may appear in colour only in the online journal)

1. Introduction to the data analysis problem in fusion

Thermonuclear plasmas are nonlinear, open systems, kept well out of equilibrium to maximize their efficiency and rate of energy production. They therefore present all the problems typical of open systems and living organisms, from the need of adequate inputs of energy and matter to stringent requirements in terms of exhaust, control and purity. They are also characterized by a very high level of complexity, which practically prevents the formulations of theories from basic principles. This leads to a hierarchy of descriptions of thermonuclear plasmas (particle, kinetic, fluid) and to a plethora of ad hoc models of limited applicability (see for example the L-H transition, for which an undisputed control parameter has not been found yet, or the magnetohydrodynamic (MHD) treatment with its various stability regions and the variety of instabilities). This plurality of models is not a fault per se but more a specific characteristic of complex systems; on the other hand, robust statistical techniques are required to assess the quality of the various alternative descriptions of the phenomena.

These challenges are to be tackled first by analysing the large amounts of data produced by present day diagnostics. In JET for example, about 50 GB of data can be generated in a well diagnosed discharge and the whole database (DB)

now exceeds 250 TB. On the other hand, given the lack of a unifying theory and the large amounts of data available, sometimes important information remains hidden in the DBs and it can be difficult to identify it manually or with traditional methods. To overcome these difficulties, in the last years a series of new data analysis tools have been developed to increase the physics output that can be derived from the measurements. These data mining tools consist of sophisticated techniques of correlation and pattern recognition to discover within the DBs useful and understandable knowledge that was previously unknown. Their application ranges from exploration of the physics to pattern recognition and to prediction [1–5]. To illustrate the potential of the tools for the extraction of mathematical models directly from the data, in this paper the subject of the determination of scaling laws has been chosen. New regression tools, indicated collectively by the term symbolic regression, have been developed to determine the mathematical form of various scaling laws for the energy confinement time in Tokamaks. These methods exploit genetic algorithms and do not assume *a priori* that the scaling laws are in power law (PL) monomial form. With regard to the structure of the paper, in the next section 2 the statistical tools developed to obtain the results of the paper are introduced. Section 3 describes the main features of the International Tokamak Physics Activity (ITPA) DB analysed. In section 4 the main results about the confinement time in terms of dimensional quantities are described and section 5 contains the conclusions.

2. Symbolic regression via genetic programming

As mentioned in the previous section, this paper describes the application of advanced knowledge discovery techniques to the problem of deriving scaling laws for the confinement time in Tokamaks. The main advantage of the proposed approach consists of practically eliminating any assumption about the form of the scaling laws. The methods developed indeed allow identifying the most appropriate mathematical expressions for the regression equations and to demonstrate that they have the potential to better interpret the experimental data for the confinement time in comparison with PLs.

The goal of the method consists of testing various mathematical expressions to fit a given DB. The main steps to perform such a task are reported in figure 1. First of all, the various candidate formulas are expressed as trees, composed of functions and terminal nodes. The function nodes can be standard arithmetic operations and/or any mathematical functions, squashing terms (terms of the form $1/(1 + e^{-x})$) as well as user-defined operators [6, 7]. This representation of the formulas allows an easy implementation of the next step, symbolic regression with genetic programming. Genetic programs (GPs) are computational methods able to solve complex optimization problems [6, 7]. They have been inspired by the genetic processes of living organisms. They work with a population of individuals, e.g. mathematical expressions

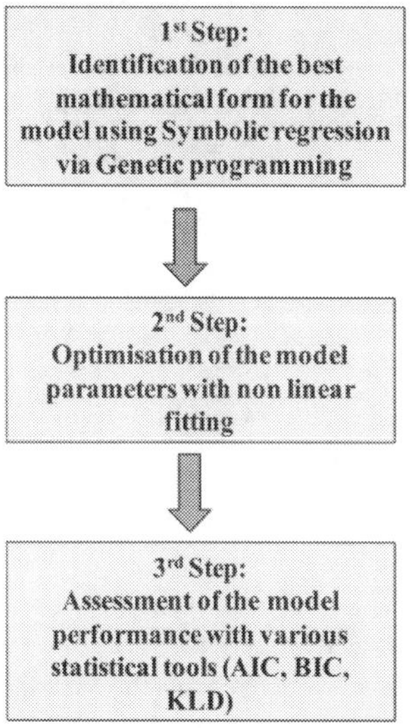

Figure 1. The main steps of the proposed methodology to identify the best scaling laws without assumption on their mathematical form.

in our case. Each individual represents a possible solution, a potential scaling law in the present application. A fitness function (FF) is used to measure how good an individual is with respect to the DB. A higher probability to have descendants is assigned to those individuals with better FF. Therefore, the better the adaptation (the value of the FF) of an individual to a problem, the higher is the probability that its genes can be passed to its descendants.

The first step of the method is the generation of the initial population of computer programmes (CPs), formulas for scaling laws in our case, and then the algorithm finds out how well an element of the population works, evaluating its behaviour with some appropriate metrics. Then, as with most evolutionary algorithms, genetic operators (reproduction, crossover and mutation) are applied to individuals that are probabilistically selected on the basis of the FF, in order to generate the new population. That is, better individuals are more likely to have more child elements than inferior individuals. When a stable and acceptable solution, in terms of complexity, is found or some other stopping condition is met (e.g. a maximum number of generations or acceptable error limits are reached), the algorithm provides the solution with best performance in terms of the FF.

In this work, CPs are composed of functions and terminal nodes and can be represented as a combination of syntax trees. The function nodes can be standard arithmetic operations and/or any mathematical functions, squashing terms as well as user-defined operators. The function nodes included in the analysis performed in this paper are reported in table 1.

The FF is a crucial element of the genetic programming approach and it can be implemented in many ways. To derive the results presented in this paper, the AIC (Akaike information criterion) has been adopted [8] for the FF. The AIC form used is:

$$\mathrm{AIC} = 2k + n \cdot \ln(\mathrm{RMSE}/n) . \quad (1)$$

Table 1. Types of function nodes included in the symbolic regression used to derive the results presented in this paper, x_i and x_j are the generic independent variables.

Function class	List
Arithmetic	c (real and integer constants), +, −, ×, /
Exponential	$\exp(x_i)$, $\log(x_i)$, power(x_i, x_j), power(x_i, c)
Squashing	logistic(x_i), step(x_i), sign(x_i), gauss(x_i), tanh(x_i), erf(x_i), erfc(x_i)

In equation (1), RMSE is the root mean square error, k is the number of parameters used in the model plus one and n is the number of y_{data} provided, so the number of entries in the DB (DB). The FF parameterized above allows considering the goodness of the models, thanks to the RMSE, and at the same time their complexity is penalized by the dependence on the number of nodes. Therefore the lower the AIC the better he model. It is also worth mentioning that the choice of the RMSE for the FF is driven by the fact that the RMSE satisfies very general statistical properties. In particular, it can be easily demonstrated that the RMSE is an unbiased estimator that maximizes the likelihood of the results. This is the main reason why the RMSE is so widely used in the statistical community; moreover its unbiased character makes the RMSE the best candidate to the present application, meant at verifying what is the most suitable mathematical expression for the scaling laws.

To assess the quality of the final models the well-known criteria of BIC (Bayesian information criterion) and Kullback–Leibler divergence (KLD) have been used. The BIC criterion is defined as:

$$\text{BIC} = n \cdot \ln\left(\sigma^2_{(\varepsilon)}\right) + k \cdot \ln(n), \tag{2}$$

where $\varepsilon = y_{\text{data}} - y_{\text{model}}$ are the residuals, $\sigma^2_{(\varepsilon)}$ their variance and the others symbols are defined in analogy with the AIC expression. Again the better the model, the lower its BIC.

Then the aim of the KLD is to quantify the information lost when $p\left(\vec{y}_{\text{model}}(\vec{x})\right)$ is used to approximate $q\left(\vec{y}_{\text{data}}(\vec{x})\right)$ [8]. The KDL is defined as:

$$\text{KLD}(PQ) = \int p(x) \cdot \ln(p(x)/q(x))\text{d}x, \tag{3}$$

where the symbols have been defined as above. The KLD assumes positive values and is zero only when the two probability distribution functions (pdfs), p and q, are exactly the same. Therefore the smaller the KLD is, the better the model approximates the data, i.e. the less information is lost by representing the data with the model.

3. The main characteristics of the ITPA databaseDB3v13f

The term $h(n)$ is:

$$h(n) = n^{0.100^{0.109}_{0.091}} \cdot \left(1 + \text{e}^{-0.408^{-0.390}_{-0.426} \cdot n^{1.036^{1.108}_{0.996}}}\right)^{-1}.$$

Table 2. PLs and non-PL model. PL1 is the IPB98(y, 2) scaling while PL2 is (EIV) [10].

PL1	$5.62 \times 10^{-2} I^{0.93} B^{0.15} n^{0.41} M^{0.19} R^{1.97} \varepsilon^{0.58} \kappa_{\text{a}}^{0.78} P^{-069}$
PL2	$5.55 \times 10^{-2} I^{0.75} B^{0.32} n^{0.35} M^{0.06} R^{2.0} \varepsilon^{0.76} \kappa_{\text{a}}^{1.14} P^{-062}$
NPL1	$0.070^{0.071}_{0.069} \cdot I^{1.071\|^{1.079}_{1.062}} R^{1.706\|^{1.727}_{1.685}} \kappa_{\text{a}}^{1.250\|^{1.290}_{1.211}} P^{-0.715^{-0.707}_{-0.723}} h(n)$

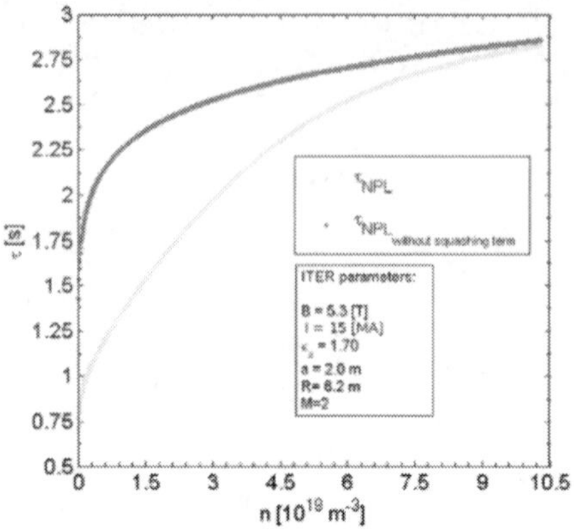

Figure 2. Comparison of the NPL scaling law behaviour with the plasma density at the parameters of ITER with or without the squashing term.

Table 3. Statistical estimators used to qualify the scaling reported in table 2. The KLD has been computed in a range of $\pm 6\sigma$ around the mean value of the data.

	k	AIC	BIC	MSE (s^2)	KLD
PL1	10	−19416.86	−19362.86	1.866×10^{-3}	0.0337
PL2	10	−19084.36	−19203.68	2.077×10^{-3}	0.0802
NPL	9	−19610.81	−19556.55	1.753×10^{-3}	0.0255

To maximize the generality of the results obtained with the methodology described in the previous sections, an international DB has been considered [9]. This DB was explicitly conceived to support advanced studies of the confinement time and includes validated signals from the vast majority of the most relevant Tokamak machines ever operated in the world. In line with the previous literature on the subject [10], the following quantities have been considered good candidate regressors in the present work: B (T), I (MA), n ($10^{19}\,\text{m}^{-3}$), R (m), M, ε, k_{a}; P (MW), q_{95}. In the previous lists, k_{a} indicates the volume elongation measurement, ε the inverse aspect ratio, q_{95} the plasma safety factor evaluated at the flux surface enclosing the 95% of the poloidal flux, M the effective atomic mass in amu, n the central line average plasma density, B the toroidal magnetic field, R the plasma major radius, I the plasma current and finally P the estimated lost power [10]. All the selected quantities are generally known with accuracy better than ±20% and they are routinely available in all the major Tokamaks, providing enough data for an acceptable statistical analysis. The entries of the DB, for which all the variables used are available or computable, have been considered, resulting in a total of 3093 entries, corresponding to the DB3 dataset [9].

Table 4. PLs and NPL model fitted on the subset of data without the JET's entries.

PLs	$5.35_{5.30}^{5.39} \times 10^{-2} I^{0.67_{0.62}^{0.73}} B^{0.12_{0.07}^{0.17}} n^{0.38_{0.35}^{0.41}} M^{0.43_{0.35}^{0.51}} R^{1.69_{1.59}^{1.78}} \epsilon^{0.53_{0.44}^{0.63}} \kappa_a^{0.39_{0.29}^{0.48}} P^{-0.54_{-0.57}^{-0.52}}$
NPL	$0.074_{0.073}^{0.075} \cdot I^{0.954_{0.927}^{0.981}} R^{1.369_{1.323}^{1.415}} \kappa_a^{0.219_{0.202}^{0.236}} P^{-0.514_{-0.539}^{-0.489}} h\,(n)$

4. Results in terms of dimensional quantities

The best models found with symbolic regression via genetic programming in general are not in PL form but they present additional terms. In particular, the most performing models include multiplicative squashing terms in the plasma current, the plasma density and the plasma density divided by the magnetic field. It is interesting to note that the method by itself selects these quantities as the ones responsible for the saturation of the confinement time. The effect of the plasma volume, represented by the major radius R, is not found to be affected as intuitively expected.

One of the best functional forms for the energy confinement time is reported in table 2. Also the power laws (PL1 and PL2) typically used as reference by the community are reported: PL1 is IPB98(y, 2) and PL2 is EIV of [10]. The most important aspect of the non-power law (NPL) functional form is the presence of a squashing term for the density whose effect can be clearly seen in figure 2.

The equations of the scaling laws have been used to generate their estimate corresponding to the values of the DB and their pdfs have been compared with the experimental ones. The KLD, the MSE and the BIC and AIC criteria all show that the NPL scaling is better than the PLs in interpreting the experimental data available. The comparison between the traditional PL and the NPL scalings, in terms of statistical indicators, is summarized in table 3, proving the best quality of the NPL regression.

Even if the superior properties of the NPL scaling are quite consolidated in statistical terms, its extrapolation capability could still be questioned and has therefore been checked with additional tests. A first test has been performed by fitting the NPL model on the DB for currents below 2.5 MA and then extrapolating this model to higher current data; all the criteria agree that the NPL scaling performs better than the PLs ($KL_{NPL} = 0.2347$, while $KL_{PL1} = 0.3210$ or $KL_{PL2} = 0.7478$).

The term $h(n)$ is:

$$h\,(n) = n^{0.139_{0.120}^{0.158}} \cdot \left(1 + e^{-0.350_{-0.378}^{-0.322} \cdot n^{1.363_{1.245}^{1.480}}}\right)^{-1}.$$

A second more relevant test has been performed by non-linearly fitting the various models, using data of smaller devices, and testing the results with JET data. Since the PL1,2 models have the same variables, only one fit has been performed and labelled as PLs. The non linear fits for the PLs and the NPL have been performed using the weights obtained by the percentiles method. This technique allows weighting data depending on their distribution in order to give more importance to the data falling in their tails since they carry more relevant of information for the determination of scaling laws (e.g. high plasma current values ($I > 2.5$ MA) of JET are extremely valuable since they are the closest to ITER operational region of 15 MA). For each physical quantity, five percentiles have been computed in order to divide the distribution function in six partitions, which can be independently weighted. In our case we have chosen the inverse of the cumulative probability, defining the percentiles themselves, as the weight for data falling in each different partition. This can be repeated for all the selected physical quantities to obtain a final weight for each entry of the DB. The PLs perform slightly better on the small machines ($MSE_{PLs} = 3.322 \times 10^{-4}\,s^2$ and $KLD_{PLs} = 0.0516$; while $MSE_{NPL} = 3.554 \times 10^{-4}\,s^2$ and $KLD_{NPL} = 0.0630$, but when the scaling are applied to JET data, the superiority of the NPL model in terms of the indicators considered can be clearly seen; the fitted models are reported in table 4 and their statistical estimators for JET data in table 5.

Table 5. Statistical estimators used to qualify the extrapolations to JET data using the equations derived for the smaller devices.

	k	AIC	BIC	MSE (s^2)	KLD
PLs	10	−5842.20	−6720.79	1.578×10^{-2}	5.895
NPL	9	−6052.17	−6758.98	1.360×10^{-2}	2.831

Table 6. PLs and NPL. PL1 is the IPB98(y, 2) scaling while PL2 is (EIV) [10].

PL1	$5.62 \times 10^{-2} I^{0.93} B^{0.15} n^{0.41} M^{0.19} R^{1.97} \varepsilon^{0.58} \kappa_a^{0.78} P^{-069}$
PL2	$5.55 \times 10^{-2} I^{0.75} B^{0.32} n^{0.35} M^{0.06} R^{2.0} \varepsilon^{0.76} \kappa_a^{1.14} P^{-062}$
NPL	$0.101_{0.099}^{0.102} \cdot R^{1.666_{1.621}^{1.712}} \kappa_a^{1.270_{1.192}^{1.346}} P^{-0.716_{-0.732}^{-0.700}} h\,(n)\, g(I)$

Table 7. Statistical estimators used to qualify the scaling reported in table 6.

	k	AIC	BIC	MSE	KLD
PL1	10	−19416.86	−19362.86	1.866×10^{-3}	0.0037
PL2	10	−19084.36	−19203.68.	2.077×10^{-3}	0.0802
NPL	11	−19719.08	−19653.64	1.691×10^{-3}	0.0219

The two terms $h(n)$ and $g(n)$ are:

$$h\,(n) = n^{0.130_{0.114}^{0.147}} \cdot \left(1 + e^{-0.501_{-0.535}^{-0.468} \cdot n^{1.055_{0.972}^{1.139}}}\right)^{-1}$$

$$g\,(I) = I^{0.760_{0.743}^{0.776}} \cdot \left(1 + e^{-0.450_{-0.480}^{-0.420} \cdot I^{1.701_{1.615}^{1.787}}}\right)^{-1}.$$

The higher extrapolation capability of the scaling laws in NPL form motivates a revision of the expected performance in terms of confinement time for ITER. Using the equations of table 2, the predicted value of the confinement time for ITER ($n_e = 1.03 \times 10^{20}\,m^{-3}$, $\kappa_a = 1.70$, $I_p = 15$ MA, $R = 6.2$ m, $P = 87$ MW) is $2.83_{2.42}^{3.31}$ s to be compared to the $3.6_{4.14}^{3.13}$ s of the traditional extrapolations obtained with PL scalings. The expected confinement time of the NPL scaling is therefore significantly lower than the one of the IPB98 PL, even if there

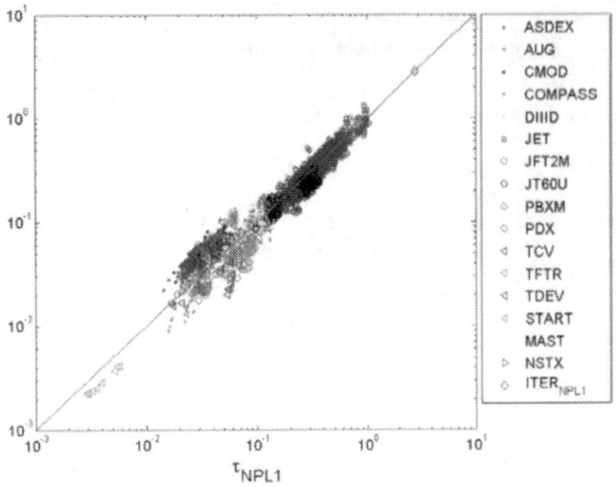

Figure 3. Plots of the experimental points versus the NPL1 scaling law on a log-log scale. The top diamond is the 2.83 s estimated value of the confinement time on ITER according to the NPL1 equation.

is an overlap of the 95% confidence intervals between the two estimates. On the other hand, the solution of table 2 is not the only candidate. Other models have quite good statistical performance and similar physics substance and credibility. One example is reported in table 6. This time two squashing terms are present. The high quality of this model can be seen in table 7 which summarizes its statistical performance. The extrapolation to ITER of this model would give more pessimistic estimate for the confinement time ($1.8^{2.4}_{1.3}$ s).

5. Conclusions and further developments

In this paper, symbolic regression via genetic programming has been applied to the derivation of empirical scaling laws for the energy confinement time in Tokamaks. The analysis has been particularized for the H-mode of confinement. The examples used for the present investigation belong to the ITPA international DB, which include all of the most relevant machines in the world.

Contrary to the main results reported in the literature, the obtained empirical scalings are not in the form of PLs. Indeed they present multiplicative squashing terms. The superior quality of these new scalings, compared to the traditional PLs, has been demonstrated first of all with the help of a series of statistic indicators. Also a visual inspection of the traditional log-log plots confirms the quality of the results. A clear example is provided by figure 3, for the scaling NPL1 of table 2. The flexibility provided by the squashing term can be appreciated by this plot. Indeed, even analysing visually the available DB, it appears very clearly that the smaller and larger devices present different trends. The PL scalings are too rigid and converge on exponents, which are strongly weighted in favour of the smaller devices. Our approach, being much more flexible, allow separating the two trends and the final scaling law is dominated by the larger devices at high values of τ, which seems to be only appropriate for the case of extrapolations to ITER. The trend of these larger devices is less favourable and results in shorter confinement times for ITER.

In any case, to complement the analysis, the extrapolation capability of the new scalings has been verified by dedicated investigations of different groups of devices (small and large machines), confirming the superior quality of the NPL equations. On the other hand, a specific value of the confinement time cannot be given now since the DB is not of enough quality to discriminate between various equally satisfactory models, presenting different quashing terms. In any case, the new found scalings predict a confinement time in ITER between about 2 and 3, significantly lower than the 3.66 s foreseen by the PL scaling IPB98(y, 2). This is due to the excess rigidity of the PL scalings, which probably tend to overestimate the confinement time in the case of large extrapolations. Better DBs and/or specific experiments will have to be considered in order to narrow down the best estimate for the confinement time in ITER.

References

[1] Murari A *et al* 2008 *Nucl. Fusion* **48** 035010
[2] Dormido-Canto S *et al* 2013 *Nucl. Fusion* **53** 113001
[3] Murari A *et al* 2008 *Nucl. Fusion* **48** 035010
[4] Murari A *et al* 2013 *Nucl. Fusion* **53** 033006
[5] Murari A *et al* 2012 *Nucl. Fusion* **52** 063016
[6] Schmid M and Lipson H 2009 *Science* **324** 81–5
[7] Koza J R 1992 *Genetic Programming: On the Programming of Computers by Means of Natural Selection* (Cambridge, MA: MIT Press)
[8] Burnham K P and Anderson D R 2002 *Model Selection and Multi-Model Inference: A Practical Information-Theoretic Approach* 2nd edn (Berlin: Springer)
[9] http://efdasql.ipp.mpg.de/hmodepublic/DataDocumentation/Datainfo/DB3v13/db3v13.html
[10] McDonald D C *et al* 2007 *Nucl. Fusion* **47** 147–74

IOP Publishing
Plasma Physics and Controlled Fusion
Plasma Phys. Control. Fusion **57** (2015) 014009 (15pp)
doi:10.1088/0741-3335/57/1/014009

Excited-state kinetics and radiation transport in low-temperature plasmas

G Colonna[1], G D'Ammando[2], L D Pietanza[1] and M Capitelli[1,2]

[1] CNR-IMIP, Via Amendola 122/D, 70126 Bari, Italy
[2] Dipartimento di Chimica, Universitá degli studi di Bari, Via Orabona 4, 70125 Bari, Italy

E-mail: gianpiero.colonna@imip.cnr.it

Received 1 July 2014, revised 5 September 2014
Accepted for publication 15 September 2014
Published 28 November 2014

Abstract

An advanced self-consistent plasma physics model including non-equilibrium vibrational kinetics, a collisional radiative model for atomic species, a Boltzmann solver for the electron energy distribution function, a radiation transport module coupled to a steady inviscid flow solver and, has been applied to study non-equilibrium in high enthalpy flows for Jupiter's atmosphere. Two systems have been considered, a hypersonic shock tube and nozzle expansion, emphasizing the role of radiation reabsorption on macroscopic and microscopic flow properties. Large differences are found between thin and thick plasma conditions not only for the distributions, but also for the macroscopic quantities. In particular, in the nozzle expansion case, the electron energy distribution functions are characterized by a rich structure induced by superelastic collisions between excited species and cold electrons.

Keywords: non-equilibrium plasma kinetics, excited states, radiation transport, electron energy distribution function

(Some figures may appear in colour only in the online journal)

1. Introduction

Non-equilibrium plasma kinetics is a topic of large interest due to its wide use in different technological applications including negative ion sources for fusion [1, 2], microwave electrical discharges for diamond film formation [3, 4], parallel plates capacitive RF discharges for microelectronics [5–7], infrared [8] and excimer laser kinetics [9] and laser plasma interaction [10–12].

In the last 20 years there has been a growing interest in applying advanced plasma kinetics models to study the environment encountered by a vehicle during high-speed (re) entry in planetary atmospheres, including boundary layer flows [13–15], shock waves [16] and nozzle expansion flows [17, 18]. In these applications special attention is devoted to non-equilibrium vibrational kinetics and its coupling to electron energy distribution function (EEDF) [19–23] . Under many circumstances however the interaction vehicle-planetary atmosphere occurs at very large Mach numbers activating, in addition to the dissociation of molecules, the processes of ionization/recombination. This is for example the case of vehicles entering Jupiter atmosphere, mainly composed by H_2 and He, at supersonic/hypersonic speed typically in the range 25–50 km s^{-1}. Under these conditions, H_2 molecules are almost completely dissociated and the formed atoms are also extensively ionized.

During atmospheric entry, the vehicle undergoes shock wave interaction at its front, while a plasma expansion region appears in the rear, where the radiative heating plays a relevant role [24, 25]. These two situations are very different from the point of view of plasma kinetics. During the shock wave formation there is an abrupt increase of translational temperature of heavy particles followed by an increase of both the electron temperature and concentration, a condition which can be described with good accuracy by a multi-temperature approach. The expansion regime, on the other hand, is characterized by a sudden decrease of the translational temperature of heavy particles and electrons, resulting in a recombining plasma. This condition can not be described by a multi-temperature model, since the latter completely disregards the possibility of forming non-equilibrium distributions of atomic/molecular excited states and free electrons as well, and a fully non-equilibrium state-to-state (STS) treatment of the plasma kinetics is mandatory to adequately describe the complex

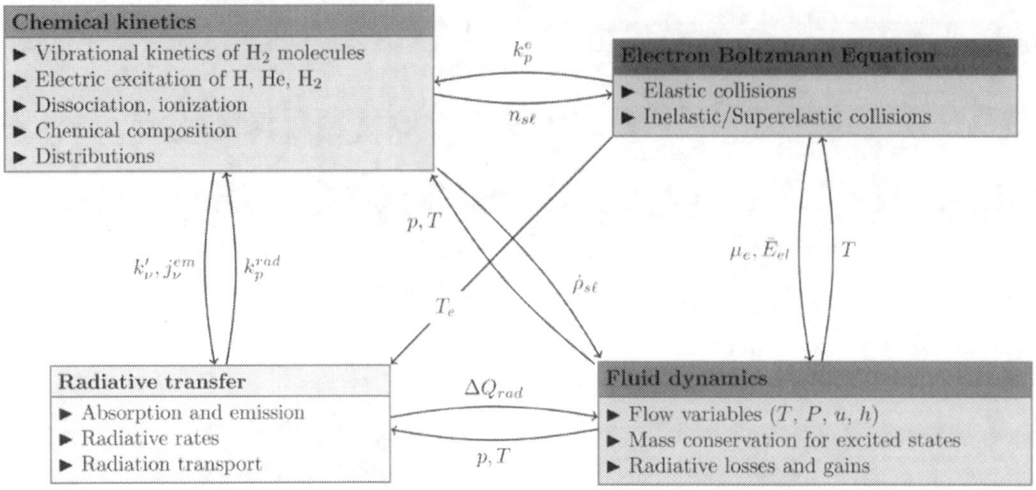

Figure 1. A schematic representation of the modules included in the numerical model.

phenomenology. In particular the role of superelastic collisions involving electronically excited states and low energy electrons can play a major role in creating structures in the electron energy distribution function, which in turn affects the electron impact rate coefficients [11, 26]. The effectiveness of superelastic collisions is largely determined by the concentration of electronically excited states which in turn depends on the degree of reabsorption of photons emitted by atoms. The effect of reabsorption can be parametrically estimated reducing the Einstein coefficients for spontaneous emission between atomic levels i, j by the escape factor $0 \leqslant \lambda_{ij} \leqslant 1$ to account for reabsorption in radiative transitions. A fully consistent but costly approach requires an explicit modeling of radiation transfer in the gas and the self-consistent calculation of the rate coefficients of radiative processes.

Non-equilibrium effects on the EEDF are investigated solving a Boltzmann transport equation including second kind (superelastic) collisions as well as electron–electron Coulomb collisions, coupled to the kinetics described by a system of master equations for the distribution of atomic and molecular excited states. The radiation model plays a significant indirect control on the EEDF through its influence on the concentration of excited states, which in turns exchange energy with the free electrons through inelastic and superelastic collisions. Using the escape factor method, upper and lower limits to the influence of excited states on EEDF are obtained, corresponding to thin and thick plasma conditions.

In this work the interplay between radiation and EEDF is assessed using the Jupiter atmosphere kinetic model under shock wave and plasma expansion conditions. In order to obtain a reasonably complete description of the problem at hand, it is necessary to include (*a*) the vibrational kinetics of H_2, (*b*) collisional radiative model (CRM) for both H and He atoms, (*c*) a plasma chemistry module for the dissociation and ionization processes and their reverse ones, (*d*) a radiative transfer module and (*e*) an Euler code to describe the fluid dynamics of shock wave and expansion regimes. This model is being developed in our laboratory and used in the present paper to emphasize the role of non-equilibrium plasma kinetics (in particular second kind collisions) in affecting shock waves and plasma expansion in the Jupiter atmosphere submitted to the interaction of a hypersonic vehicle entering in the planetary atmosphere. In particular we want to show the basic differences between the two cases (shock wave against plasma expansion) presenting dedicated numerical experiments. The paper is divided into four sections. After the introduction, section 2 presents a description of the numerical model, while sections 3.1 and 3.2 respectively report simulation results for the shock tube and nozzle expansion conditions in an H_2/He plasma. Conclusions and perspectives are given in section 4.

2. Model description

The model used to obtain the numerical results presented in this paper is represented in figure 1. It comprises a collisional–radiative model for a plasma of $H_2(v = 0-14)$, $H_2^*(i)$ (9 levels), $H_2{}^+$, $H(i = 1-25)$, H^+, $He(i = 0-58)$, He^+ and e^-, where $H_2(v)$ is the hydrogen molecule in its ground electronic state $X^1\Sigma_g^+$, $H_2^*(i)$ is an electronically excited state of the hydrogen molecule, $H(i)$ is atomic hydrogen in an electronic state with principal quantum number i, and $He(i)$ is atomic helium in an electronic state indexed by i. The energy level scheme of He is constructed by grouping closely spaced atomic levels of the real atom in different ways according to the principal quantum number n of the optical electron and total angular momentum quantum number L of the atomic term as follows [27] (i) for $n \leqslant 4$, each atomic term is a level in the atom, with energy calculated averaging over fine structure (ii) for $n = 5-7$ individual terms are retained for $L < 3$, while the remaining terms with $L \geqslant 3$ are grouped in a single level (iii) for $n = 8-10$ atomic levels within each n are grouped in a single level keeping singlet an triplet terms separate (iv) for $10 < n \leqslant 20$ the model considers a level for each n with hydrogen-like energy and degeneracy. The resulting energy level scheme has 59 levels.

The generalized CRM considered in this paper, represented in figure 2, takes into account a comprehensive kinetic

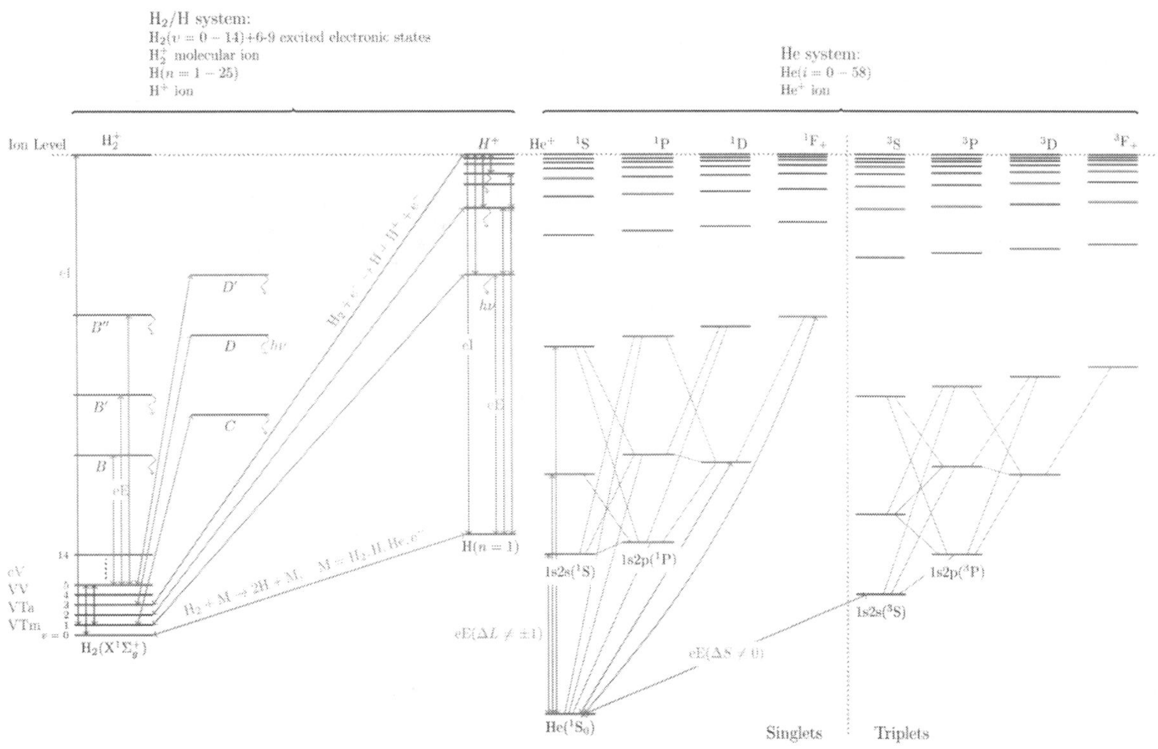

Figure 2. A schematic representation of the H_2/He collisional–radiative model used in this study.

scheme for the vibrational kinetics of $H_2\left(X^1\Sigma_g^+\right)$, including V–V mono-quantum transitions [28] (υ and ω are vibrational quantum numbers)

$$H_2(\upsilon+1)+H_2(\omega) \rightleftharpoons H_2(\upsilon)+H_2(\omega+1) \quad (1)$$

V–T energy transfer and dissociation

$$H_2(\upsilon)+\mathrm{M} \rightleftharpoons H_2(\upsilon \pm \Delta\upsilon)+\mathrm{M} \quad (2)$$

$$H_2(\upsilon)+\mathrm{M} \rightleftharpoons 2\,\mathrm{H}(n=1)+\mathrm{M} \quad (3)$$

The rate coefficients for M = H, He have been calculated by Esposito *et al* in [29–32] using a quasi-classical trajectory method and the reader is referred to those papers and [33] for an in-depth discussion on their range of validity. Rate coefficients for the H_2 + H_2 V–T energy transfer have been estimated rescaling the vibrationally specific H_2 + He rate coefficients by a temperature-dependent factor obtained comparing with the thermal rate coefficient of [34] due to the lack of level specific data at high temperature (5000–20 000 K) for the above process in literature.

The atom–atom excitation and ionization of A = H, He atoms

$$\mathrm{A}(i)+\mathrm{A}^0 \rightleftharpoons \mathrm{A}(j)+\mathrm{A}^0 \quad (4)$$

$$\mathrm{A}(i)+\mathrm{A}^0 \rightleftharpoons \mathrm{A}^+ +\mathrm{A}^0+e \quad (5)$$

is very important under shock wave conditions and the relevant rate coefficients have been calculated using analytical expressions proposed by Drawin [35, 36].

Spontaneous radiative decay of electronically excited atoms

$$\mathrm{A}(j) \rightarrow \mathrm{A}(i)+h\nu_{ij} \quad (6)$$

with A = H, He has been included using the Einstein coefficients for spontaneous emission of [37, 38] for H transitions and NIST's data [39] for He ones.

Electron impact processes include resonant vibrational excitation of $H_2\left(X^1\Sigma_g^+\right)$ molecule and dissociative excitation through singlet and triplet molecular electronic states [40–43]

$$H_2(X^1\Sigma_g^+,\upsilon)+e \rightleftharpoons H_2^-(^2\Sigma_u^+) \rightleftharpoons H_2(X^1\Sigma_g^+,\omega)+e \quad (7)$$

$$H_2\left(X^1\Sigma_g^+,\upsilon\right)+e \rightleftharpoons H_2^*\left(a^3\Sigma_g^+,b^3\Sigma_u^+,c^3\Pi_u,e^3\Sigma_u^+\right)+e \rightleftharpoons 2\,\mathrm{H}(1)+e \quad (8)$$

$$H_2(X^1\Sigma_g^+,\upsilon)+e \rightleftharpoons H_2^*(B^1\Sigma_u^+,C^1\Pi_u,D^1\Pi_u,B'^1\Sigma_u^+,D'^1\Pi_u,B''^1\Sigma_u^+)+e \rightleftharpoons \mathrm{H}(1)+\mathrm{H}(2,3,4)+e \quad (9)$$

molecular ionization and electron-impact processes involving the molecular ion [44]

$$H_2\left(X^1\Sigma_g^+,\upsilon\right)+e \rightleftharpoons H_2^+ +2e \quad (10)$$

$$H_2\left(X^1\Sigma_g^+,\upsilon\right)+e \rightleftharpoons H_2^+\left(X^2\Sigma_g^+,{}^2\Sigma_u^+\right) \rightarrow \mathrm{H}(1)+\mathrm{H}^+ +2e \quad (11)$$

$$H_2^+ +e \rightleftharpoons 2\mathrm{H} \quad (12)$$

$$H_2^+ +e \rightleftharpoons 2\mathrm{H}^+ +2e \quad (13)$$

$$H_2^+ +e \rightleftharpoons \mathrm{H}+\mathrm{H}^+ +e \quad (14)$$

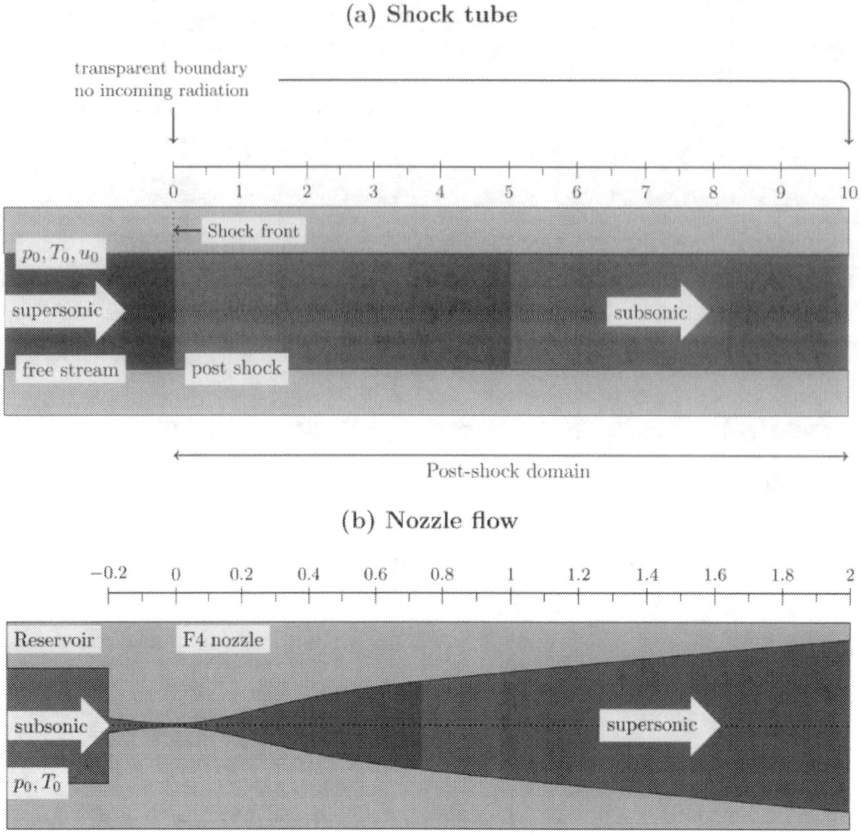

Figure 3. Schemes of the two space geometries considered in this paper (*a*) the shock tube and (*b*) the F4 nozzle flow.

as well as electron impact excitation, ionization of atoms and radiative recombination of atomic ions [45–48]

$$A(i) + e \rightleftharpoons A(j) + e \tag{15}$$

$$A(i) + e \rightleftharpoons A^+ + 2e \tag{16}$$

$$A^+ + e \rightarrow A(i) + h\nu \tag{17}$$

with A = H, He.

The electron impact rate coefficients needed in the CRM are calculated from the EEDF as

$$n_e k_p^e = \int_{\varepsilon^{th}}^{\infty} n(\varepsilon)\,\sigma(\varepsilon)\,v_e(\varepsilon)\,d\varepsilon \tag{18}$$

and the EEDF is obtained solving the electron Boltzmann equation in compact form

$$\frac{\partial n(\varepsilon, t)}{\partial t} = -\frac{\partial (J_E + J_{el} + J_{e-e})}{\partial \varepsilon} + S_{in} + S_{sup} \tag{19}$$

where $n(\varepsilon, t)d\varepsilon$ is the number of electrons in the energy range $(\varepsilon, \varepsilon + d\varepsilon)$ at time t, $\frac{\partial J_E}{\partial \varepsilon}$ is the flux of electrons due to an the electric field, $\frac{\partial J_{el}}{\partial \varepsilon}$ is flux of electrons due to elastic collisions with heavy particles, $\frac{\partial J_{ee}}{\partial \varepsilon}$ is the flux of electrons due to e–e collisions and S_{in}, S_{sup} are inelastic/superelastic collisions source terms. The Boltzmann equation is numerically solved using the formulation proposed by Rockwood [49] and the reader is referred to our previous works [17, 50, 51] for the full implementation details.

The complete STS kinetics and the Boltzmann equation are coupled to a quasi 1D steady state code [52] for solving the mass, momentum and energy conservation equations along the nozzle or shock tube (constant section) coordinate

$$\begin{aligned} \rho u A &= \rho_0 u_0 A_0 \\ (\rho u^2 + p)\,A &= \left(\rho_0 u_0^2 + p_0\right) A_0 + \int_{x_0}^{x} p\,(y)\,\frac{\partial A}{\partial y}\,dy \\ \frac{1}{2}u^2 + h &= \frac{1}{2}u_0^2 + h_0 + \Delta Q_{rad} \end{aligned} \tag{20}$$

where ρ, u and h are the mass density, velocity and enthalpy per unit mass, p is the total pressure, A is the nozzle section and ΔQ_{rad} accounts for radiative energy transfer or other external energy sources such as Joule heating. The shock tube and nozzle geometries considered in the present paper are represented in figure 3. Unsubscripted quantities in equation (20) refer to position x along the shock tube or nozzle axis, while subscripted quantities refer to the initial condition in the free stream (shock tube) or reservoir (nozzle) respectively. The gas pressure is determined using the ideal gas equation of state

$$p = \rho \frac{R}{\overline{m}} T \tag{21}$$

where R is the ideal gas constant, $\overline{m}$ is the average molar weight of the mixture and T the gas temperature. The enthalpy per unit mass is composed by a translational contribution which also includes the heat capacity of degrees of freedom in equilibrium with translation and an internal contribution resulting from the degrees of freedom that are not assumed in thermal equilibrium

$$h = h_{\rm tr} + h_{\rm int} \tag{22}$$

$$h_{\rm tr} = \alpha \frac{R}{\overline{m}} T \tag{23}$$

$$h_{\rm int} = \frac{1}{\overline{m}} \sum_s x_s \left(h_{fs} + \sum_i \varepsilon_{si} f_{si} \right) \tag{24}$$

where x_s is the molar fraction of the s-th species, $h_{f,s}$ the formation energy per particle of species s, $\varepsilon_{s,i}$ the energy of the i-th energetic level of species s referred to its ground state and f_{si} the faction of s-particles in their i-th energy level. The quantity α is given by

$$\alpha = \sum_s x_s \alpha_s \tag{25}$$

where $\alpha_s = \frac{5}{2}$ for monoatomic species and the electron and $\alpha_s = \frac{7}{2}$ for diatomic species.

Transport of radiation is described using the 1D radiative transfer equation (RTE) in slab geometry [53]

$$\mu \frac{\mathrm{d}I_\nu(x,\mu)}{\mathrm{d}x} = j_\nu^e(x) - \kappa_\nu(x) I_\nu(x,\mu) \tag{26}$$

where $I_\nu(x, \mu)$ is the radiation intensity propagating along a ray with director cosine μ with respect to the x-axis. The term $j_\nu^e(x)$ is the spectral emissivity, $\kappa_\nu(x)$ is the absorption coefficient corrected for stimulated emission. These spectral coefficients are calculated taking into account the bound–bound and bound-free radiative processes of atomic species as well as the free–free e-H^+ radiation, neglecting the contribution of He^+ having a very low concentration. The coupling between the RTE and chemical kinetics is due to the possibility of radiation reabsorption in atomic transitions, so that the net rate of a radiative transition between atomic levels i and j is given by

$$\left(\frac{\mathrm{d}n_j}{\mathrm{d}t} \right)_{ij} = -\left[A_{ji} + B_{ji} \overline{J}_{ji} \right] n_j + B_{ij} \overline{J}_{ji} n_i \tag{27}$$

where A_{ji}, B_{ji} and B_{ij} are respectively the spontaneous emission, stimulated emission and absorption Einstein coefficients associated to the $i \leftrightarrow j$ transition and

$$\overline{J}_{ij} = \overline{J}_{ji} = \int_0^\infty J_\nu(x)\, \phi_{ij}(x,\nu)\, \mathrm{d}\nu \tag{28}$$

where ϕ_{ij} is the normalized line shape of the $i \leftrightarrow j$ transition, which has been modeled using an approximate Voigt function [54], taking into account the main line broadening mechanisms such as Doppler broadening, characterized by a Gaussian profile, as well as different kinds of pressure broadening [38] approximated by a Lorentzian profile. The average spectral intensity $J_\nu(x)$ needed in equation (28) is directly calculated from the solution of the RTE, equation (26), as follows [53]

$$J_\nu(x) = \frac{1}{2} \int_{-1}^{1} I_\nu(x,\mu)\, \mathrm{d}\mu \tag{29}$$

It should be remarked that radiation transport significantly adds to the computational time, due to the need of iterating to self-consistency the system of kinetic master equations for the level populations depending the rates in equation (27) and the spectral intensity obtained solving equation (26), which depends non-linearly on the level populations through the emissivity j_ν^e and absorption coefficient κ_ν. For the shock wave results reported in the following section, the calculation time for a serial run on an Intel 3.2 GHs i3 processor is about three days, to be compared with about 20 min required by a run neglecting radiation transport.

3. Results and discussion

3.1. Shock wave

In this section we apply the model described in section 2 to investigate the non-equilibrium region behind a strong shock, under typical entry conditions in Jupiter's atmosphere. The geometry investigated in this section, corresponding to an 1D shock tube experiment, is represented in figure 3(a). In the free stream, the initial temperature and pressure are $p_0 = 700$ Pa, $T_0 = 160$ K and a free stream gas speed $u_0 = 28441.85$ m s^{-1}, corresponding to a mach number $M_0 = u_0/c \approx 31$. The cold gas is an H_2/He mixture in molar ratio 89:11. The initial conditions have been chosen to reproduce the trajectory point at about 100 km of altitude from the planetary surface reported in the numerical rebuilding of Galileo's probe entry in Jupiter atmosphere of [55]. The shock wave problem is characterized by a sudden rise of temperature ($T_1 \approx 31000$ K) and pressure ($p_1 \approx 7.7$ atm) while the gas speed becomes subsonic ($u_1 \approx 4950$ m s^{-1}) just behind the shock wave.

The increase of translational temperature triggers the heavy particle impact processes (1)–(5). Due to the high impact speed, the gas is extensively ionized, with first electron generated by the atom–atom processes (5) and (12). When a sufficient electron number density is reached, the electron impact processes (7)–(17) are also activated and start to dominate the chemical kinetics behind the shock. Radiation is mostly generated by radiative decay of electronically excited atomic states, see equation (6) and provides a further energy transfer mechanism that is accounted for solving equation (26) through the plasma slab. In presenting the following results, we consider both the case when radiation transfer is fully coupled to the kinetics of excited electronic states of atoms through the self-consistent radiative rate coefficients given in equation (27) and the optically thin model which is obtained neglecting terms proportional to J_{ij} in the same equation. The thick plasma case, obtained assuming that the radiative rates given by equation (27) are null, gives results which are very close to the those obtained in the self-consistent model under the present free stream conditions.

Figure 4 reports the composition profiles in the post-shock region as a function of the distance from the shock. The high

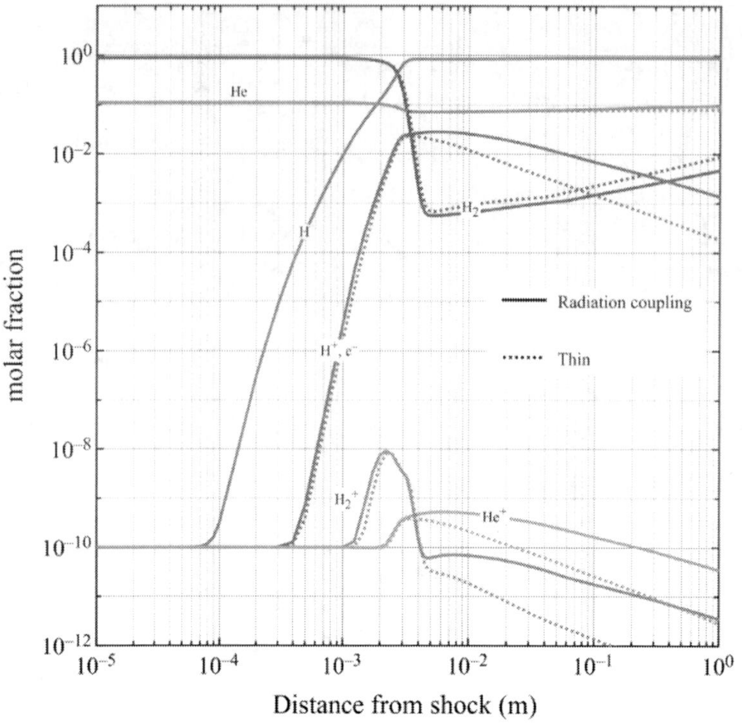

Figure 4. Post-shock molar fraction and profiles as a function of the distance from the shock front, calculated at the free stream conditions of section 3.1, using the radiation coupled model (solid lines) and in an optically thin plasma (dotted lines).

heavy particles translational temperature just behind the shock front promotes H_2 dissociation and ionization of the formed H atoms. Heavy particle impact processes, see equations (1)–(3) populate the excited vibrational levels of H_2, which are successively dissociated, building up a large concentration of H atoms at about 0.1 mm from the shock front. H atom ionization starts at about 0.3–0.4 mm from the shock and is initially due to the atom–atom process (5) only, until a sufficient electron number density builds up, that increases the rates of electron impact ionization processes (16). Figure 4 shows that the optically thin composition is almost identical to the one obtained using the full model, except that for distances larger than 3 mm the concentration of charged species, mostly e^- and H^+ ions, is 2–3 times smaller of the corresponding one in the radiation coupled case. In this case the radiative losses are very efficient in cooling the H atom level distribution, providing an indirect mechanism of dissipation of the free electron translational energy. As a result, both the electron concentration and temperature are lower in an optically thin plasma compared to the radiation coupled model.

Figure 5 shows the evolution of the translational gas temperature, together with the internal temperatures of both H and He atoms, comparing the radiation coupled model with both the plasma thin and plasma tick approximations. The internal temperatures of atomic species are calculated as the temperature of a Boltzmann distribution interpolating the population of the ground and first electronically excited state of the relevant atom. At shock distances $x > 6{\cdot}10^{-4}$ m the optically thick plasma approximation closely reproduces the results of the radiation coupled calculations, while at lower distances radiation reabsorption increases the population of electronically excited states in the full model. The optically thin approximation, on the other hand, overestimates radiation losses effects resulting in lower excited states populations.

Figure 6 reports the normalized internal state distribution (ISD) of H atom, obtained dividing the molar fraction $N_{H(i)}/N_H$(N_H is the total number density of H atoms) of the i-th atomic state H(i) by its statistical weight g_i, at increasing distances from the shock front. At distances larger than about 5 mm the distributions are equilibrium. At smaller distances the distributions are characterized by underpopulated excited states due to the ionization process and a substantially larger population of the H(n = 2) and H(n = 3) excited states in the radiation coupled model compared to the plasma thin case.

Figure 7 compares the EEDF calculated by the radiation coupled model and the optically thin approximation. For distances lower than about 1 mm the optically thin distributions are underpopulated compared to the radiation coupled ones at electron energy values larger than about 10 eV due to a lesser effect of superelastic collisions involving the H(n = 1) resulting from the lower concentration of excited states in the thin case. On the other hand, for distances larger than about 1 cm, the radiation coupled distributions are essentially Maxwellian at the gas temperature, while in the optically thin case the decrease in the electron density concentration is such to avoid the thermalization of the EEDF by electron–electron collisions, while the concentration of excited states is not large enough to create significant superelastic structures on the EEDF induced by superelastic collisions. It is worth noting that in the present implementation,

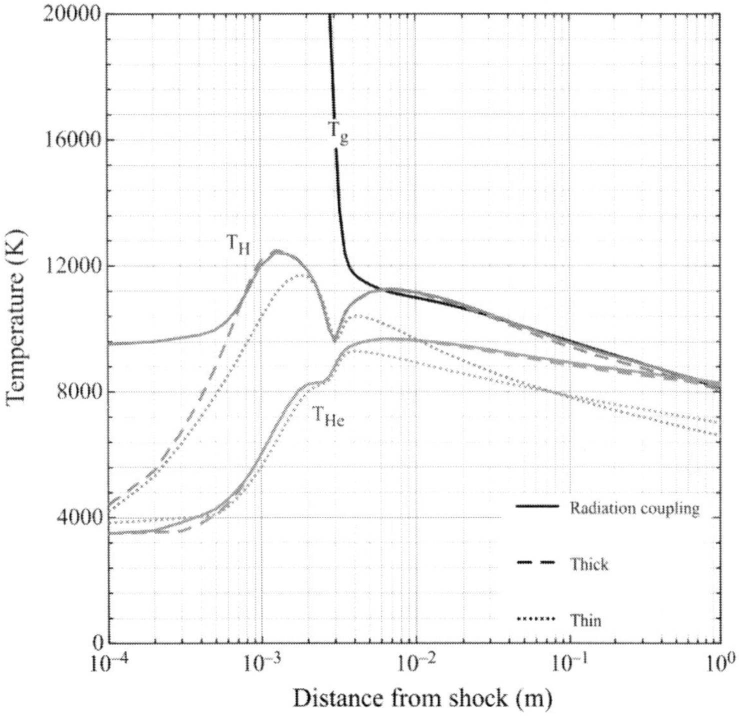

Figure 5. Translational and internal atom temperature profiles calculated at the free stream conditions of section 3.1, using the radiation coupled model (solid lines) or under plasma thin (dotted lines) and plasma thick conditions (dashed lines).

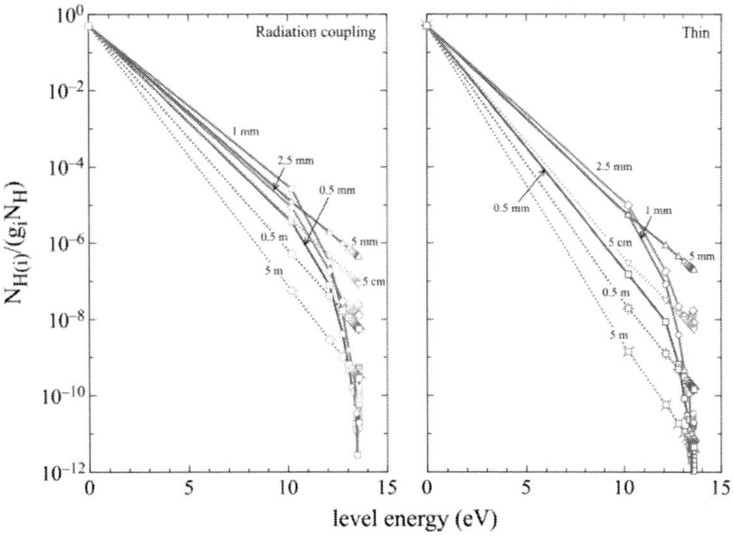

Figure 6. Internal distribution of H atoms at different distances from the shock front, calculated at the free stream conditions of section 3.1, using the radiation coupled model (left) and in an optically thin plasma (right).

the coupling between radiation and the EEDF is indirect, through the inelastic and superelastic terms in equation (19). In particular, S_{sup} is affected by the concentration of excited states which is in turn strongly dependent on the assumed radiation model.

Figure 8 reports the concentration profiles of the H(n = 2) and H(n = 3) excited states and the evolution of the average electron energy computed from the non-equilibrium EEDF. Reabsorption effects in the radiation coupled model tend to increase the concentration of atomic excited states at distances lower than about $6 \cdot 10^{-4}$ m compared to both the plasma thin and thick assumption, while the average electron energy is independent on the optical model due to the low excited state and electron concentrations, which makes inelastic and superelastic collision processes negligible. This situation lasts up to about $3 \cdot 10^{-3}$ m, where superelastic collisions between

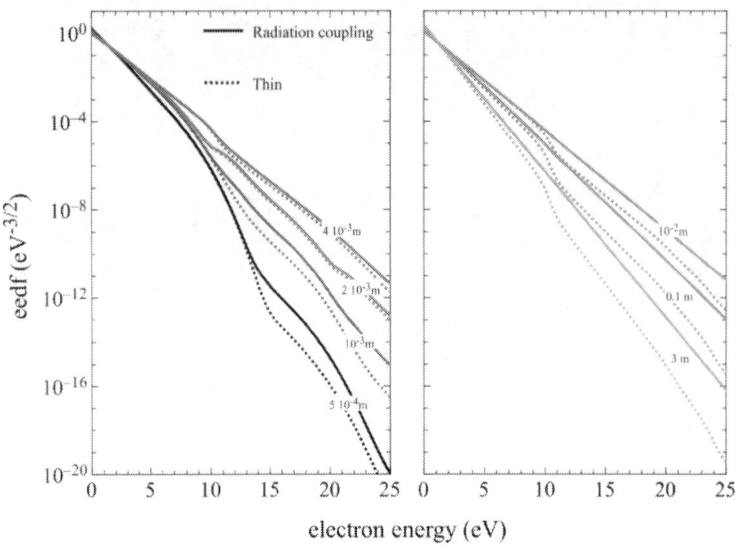

Figure 7. The EEDF at different distances from the shock front, calculated at the free stream conditions of section 3.1, using the radiation coupled model (solid lines) and in an optically thin plasma (dotted lines).

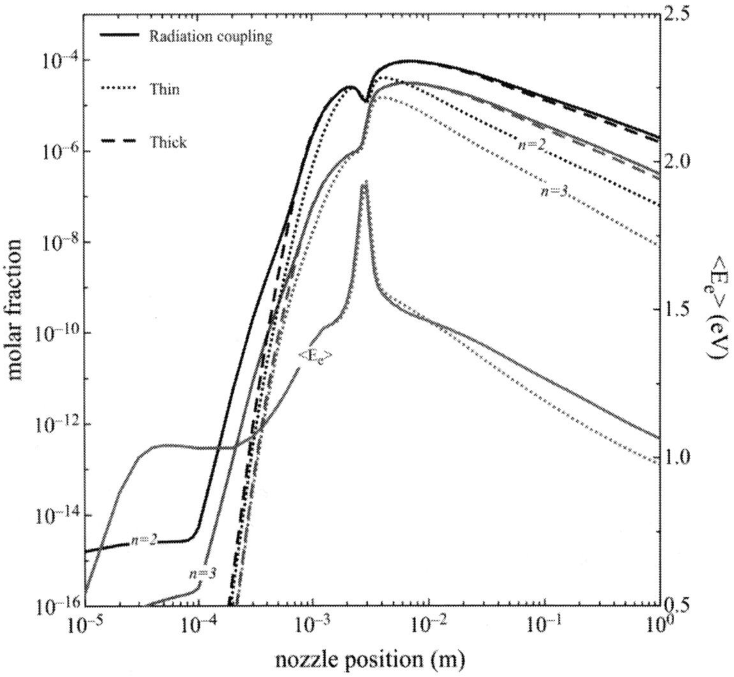

Figure 8. Molar fraction profiles of the H(n = 2) and H(n = 3) and average electron energy $\langle E_e \rangle$ as a function of the distance from the shock front, calculated at the free stream conditions of section 3.1, using the radiation coupled model (solid lines) or under plasma thin (dotted lines) and plasma thick conditions (dashed lines).

electrons and atomic excited states start to be effective, resulting in a larger average electron energy in the radiation coupled and plasma thick cases compared to the thin one, as a consequence of the larger excited state concentration in the first two cases.

The larger concentration of electronically excited states in the radiation coupled and plasma thick cases, coupled to the larger concentration of high energy electrons, see figure 7, results in a much larger ionization degree in the first two cases, with a peak electron concentration which is about 50% higher compared to the plasma thin assumption as shown in figure 9.

3.2. Nozzle expansion

Let us now examine the behavior of the EEDF in the nozzle case for both thin and thick plasma situations. The nozzle

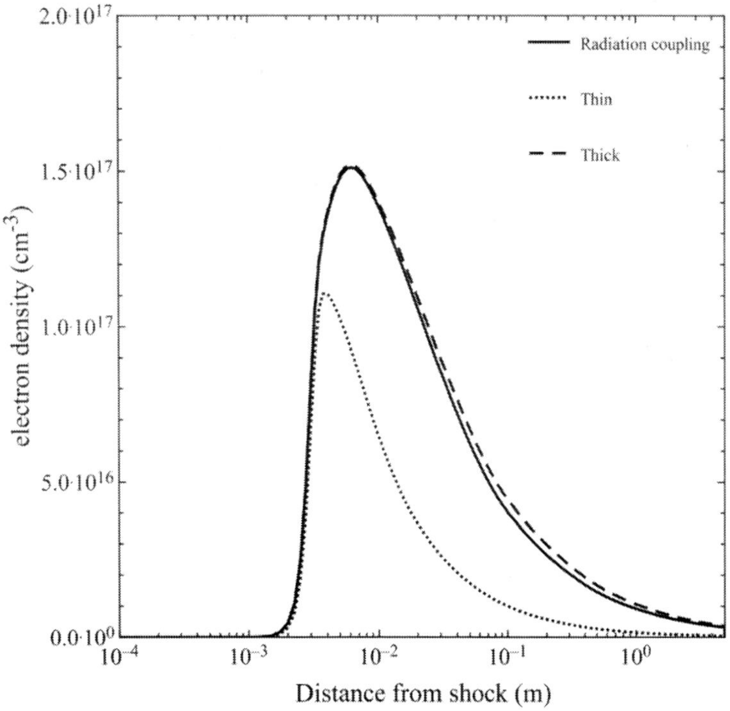

Figure 9. Electron number density profiles as a function of the distance from the shock front, calculated at the free stream conditions of section 3.1, using the radiation coupled model (solid lines) or under plasma thin (dotted lines) and plasma thick conditions (dashed lines).

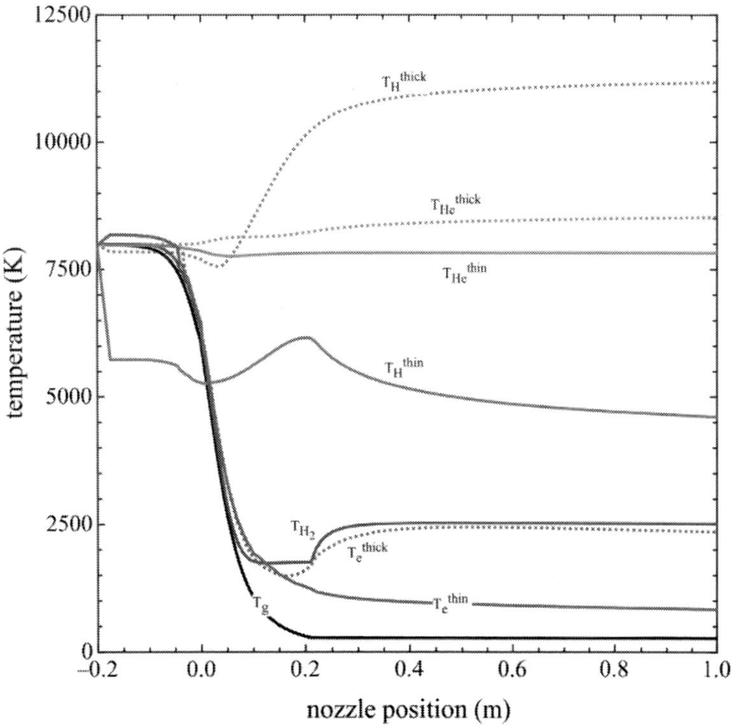

Figure 10. Evolution of internal and translational temperatures along the nozzle axis obtained imposing p_0 = 1 atm and T_0 = 8000 K in the reservoir, under plasma thin (solid lines) and plasma thick conditions (dotted lines).

geometry has been described in a previous work [56] and is schematically reproduced in figure 3(*b*). The initial temperature and pressure at the reservoir are T_0 = 8000 K and p_0 = 1 atm respectively, the chemical composition is obtained running the CRM to steady state while neglecting radiative processes. Moreover, the initial EEDF and distributions of

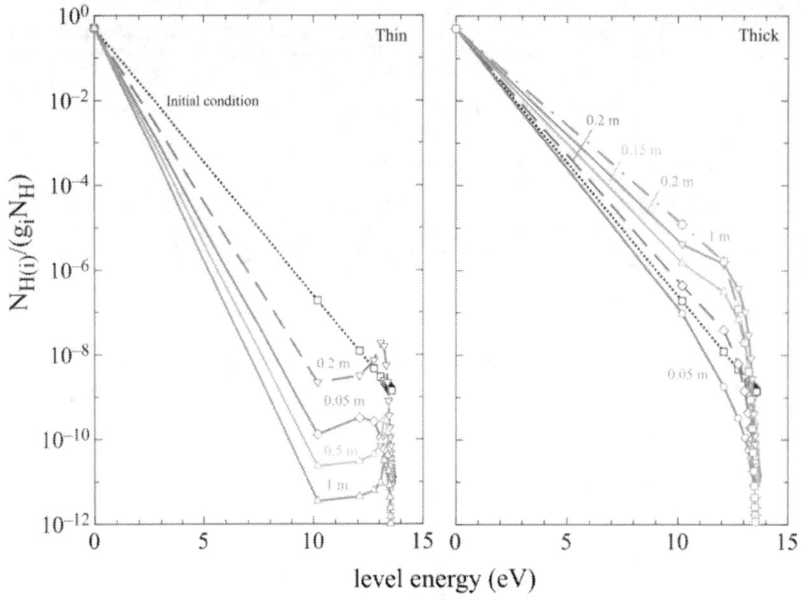

Figure 11. Evolution of the H internal electronic distribution at different positions along the nozzle axis obtained imposing $p_0 = 1$ atm and $T_0 = 8000$ K in the reservoir, under plasma thin (left) and plasma thick conditions (right).

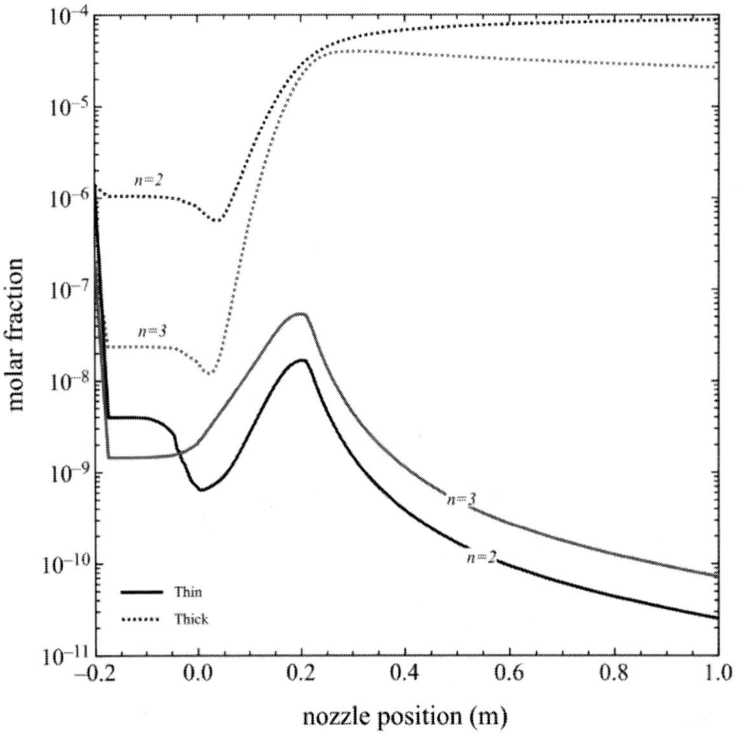

Figure 12. Evolution of the molar fraction of H(n = 2) and H(n = 3) along the nozzle axis obtained imposing $p_0 = 1$ atm and $T_0 = 8000$ K in the reservoir, under plasma thin (left) and plasma thick conditions (right).

excited H, He and H_2 species are respectively Maxwell and Boltzmann at $T_0 = 8000$ K.

Figure 10 reports the evolution of different translational (T and T_e) and internal (T_{H_2}, T_H and T_{He}) temperatures on the nozzle axis. In both the plasma thin and thick conditions, a strong decrease of T and T_e along the nozzle axis is observed, caused by the supersonic expansion of the gas through the throat. The electron temperature in the thin plasma reaches a final value of about 800 K at a distance of 0.4 m from the nozzle throat, to be compared with a final value of about

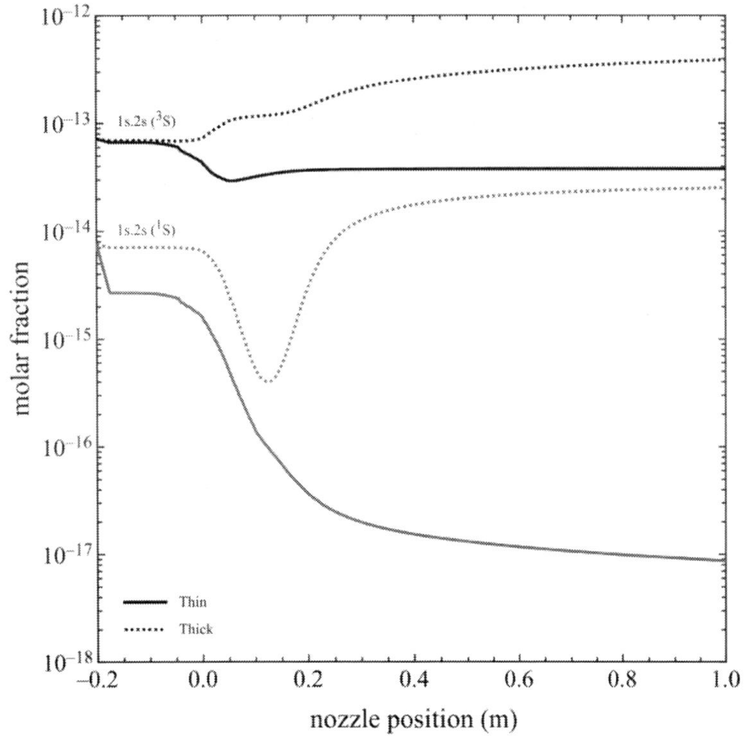

Figure 13. Evolution of the molar fraction of He[1*s*.2*s*.(^{3}S)] and He[1*s*.2*s*.(^{1}S)] along the nozzle axis obtained imposing $p_0 = 1$ atm and $T_0 = 8000$ K in the reservoir, under plasma thin (left) and plasma thick conditions (right).

2300 K in the thick case. The low values of T_e in the thin case, associated to an appreciable density of electronically excited atomic states represent an ideal situation for enhancing the role of second kind collisions in affecting the EEDF. In particular we can expect the coupling of EEDF and excited states through the following processes

$$e\,(\varepsilon \approx 0) + \mathrm{H}(n = 2) \rightarrow e\,(\varepsilon = 10.2\ \mathrm{eV}) + \mathrm{H}(n = 1) \qquad (30)$$

$$e\,(\varepsilon = 10.2\ \mathrm{eV}) + \mathrm{H}(n = 2) \rightarrow e\,(\varepsilon = 20.4\ \mathrm{eV}) + \mathrm{H}(n = 1) \qquad (31)$$

$$e\,(\varepsilon \approx 0) + \mathrm{H}(n = 3) \rightarrow e\,(\varepsilon = 12.3\ \mathrm{eV}) + \mathrm{H}(n = 1) \qquad (32)$$

Figure 11 reports the evolution of the H ISD along the nozzle axis. In the thin plasma case the H($n \geqslant 3$) are overpopulated compared to a Boltzmann distribution at T_{H} due to the efficient depletion of the H($n = 2$) level by spontaneous emission, which dissipates the excitation energy leading to a progressive decay of the entire ISD. A completely different behavior is observed in an optically thick plasma, where the ISD are enhanced due to an interplay between recombination and electron impact excitation as explained later in the paper.

The evolution of molar fractions of H($n = 2$) and H($n = 3$) excited states along the nozzle axis (the H($n = 2$) concentration is directly related to the evolution of T_{H} shown in figure 10) are reported in figure 12 for both thin and thick situations. The thick case is characterized by concentrations of H($n = 2$) and H($n = 3$) excited states above 0.2 m about about 4–6 orders of magnitude larger compared to the thin case. Furthermore, the qualitative trend above 0.2 m is also completely different and can be explained as follows. The e^--H^+ three body recombination process preferentially populates excited levels which are subsequently de-excited through superelastic collisions towards lower excited states, leading to an increase of both the concentration of the H($n = 2$) and H($n = 3$). In the plasma thick case excited states can be de-excited mainly through superelastic collisions (equations (30)–(32)), enhancing the EEDF around the relevant threshold energy, which in turn promotes electron impact excitation from the ground state. As a result of this interaction, the concentration of electronically excited atomic levels keeps an approximately constant value in the expanding thick plasma. On the other hand the thin case is characterized by an exponential depletion of excited states due to the spontaneous radiative decay, which provides an efficient dissipation mechanism for the energy stored in atomic excited states. The evolution of He the two lowest He excited states, the triplet state 1*s*.2*s* (^3S) and the singlet 1*s*.2*s* (^1S) reported in figure 13 has a similar behavior. The initial increase of concentration of excited states is absent in this case, due to the almost negligible ionization degree of He under the present conditions, preventing extensive recombination. Radiative decay of the triplet He state, implying a spin-forbidden transition, is also absent in the thin case.

The EEDF for the thin case are reported in figure 14 at different positions along the nozzle axis. Very structured EEDF's are created as a result of processes described by equations (30)–(32). Second kind collisions from process (30) form a peak at 10.2 eV which is subsequently spread in a plateau by elastic collisions, while process (31) forms a new plateau

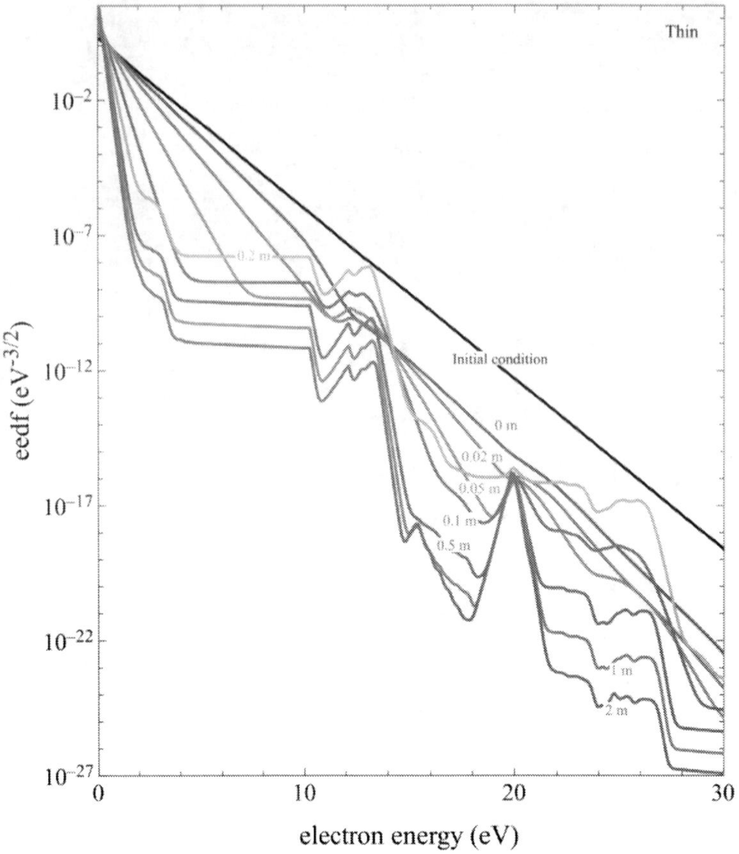

Figure 14. Evolution of the EEDF at different positions along the nozzle axis obtained imposing $p_0 = 1$ atm and $T_0 = 8000$ K in the reservoir in the plasma thin case.

starting at 20.4 eV, due to two successive superelastic collisions. Moreover process (32) forms a hump at 12.3 eV, while different inelastic processes including collisions with H_2 molecules are responsible of additional smaller structures present in the EEDF. In the thin case, the EEDF's also show an hump at 19.6 eV created by the second kind collisions involving a metastable triplet state of He

$$e(\varepsilon = 0) + \mathrm{He}(^3\mathrm{S}) \rightarrow e(\varepsilon = 19.6\,\mathrm{eV}) + \mathrm{He}(^1\mathrm{S}) \quad (33)$$

The height of this peak is much smaller compared the other maxima due the very low concentration of the He metastable triplet state, see figure 13. The EEDF's in the thick case, reported in figure 15 present a similar qualitative behavior, but the plateau formed by processes (30) and (31) are orders of magnitude higher compared to the thin case. Moreover, an additional third plateau formed by process

$$e\,(\varepsilon = 20.4) + \mathrm{H}(n = 2) \rightarrow e\,(\varepsilon = 30.6\,\mathrm{eV}) + \mathrm{H}(n = 1) \quad (34)$$

is well evident in the figure, the peak due to He $1s.2s$ (^{3}S) disappears, while the peak at 12.3 eV is still present but less structured compared to the thin case.

4. Conclusion and perspectives

An advanced state-to-state kinetic code for heavy particles and electrons, coupled to a steady state 1D flow solver and a radiative transfer equation has been applied to investigate typical conditions encountered during high speed entry in a planetary atmosphere, i.e. shock wave and plasma expansion regimes similar to those created in hypersonic wind tunnels. A comprehensive kinetic scheme for a H_2/He plasma, recently developed in our laboratory to model Jupiter's atmosphere has been used to perform numerical simulations of a shock tube and nozzle expansion geometries.

The results highlight a significant coupling between the distribution of electronically excited states of atoms and the EEDF. The optical thickness of the plasma has an important effect on the concentration of excited states, which in turn influence the EEDF through the action of superelastic collisions. For the shock wave simulations three optical models have been considered, i.e. thin and thick plasma as well as a self-consistent treatment of radiation reabsorption through the solution of a radiative transfer equation. The results show that under hypersonic speed conditions the plasma is well described by the plasma thick approximation, except for a significant increase of the internal H atom temperature at short distance from the shock front ($d \leqslant 1$ mm) due to the formation of electronically excited states by radiation reabsorption. The plasma optical model has a major effect on the ionization degree of the plasma at large distances from the shock, which is much lower in the plasma thin case, resulting in incomplete thermalization of the EEDF.

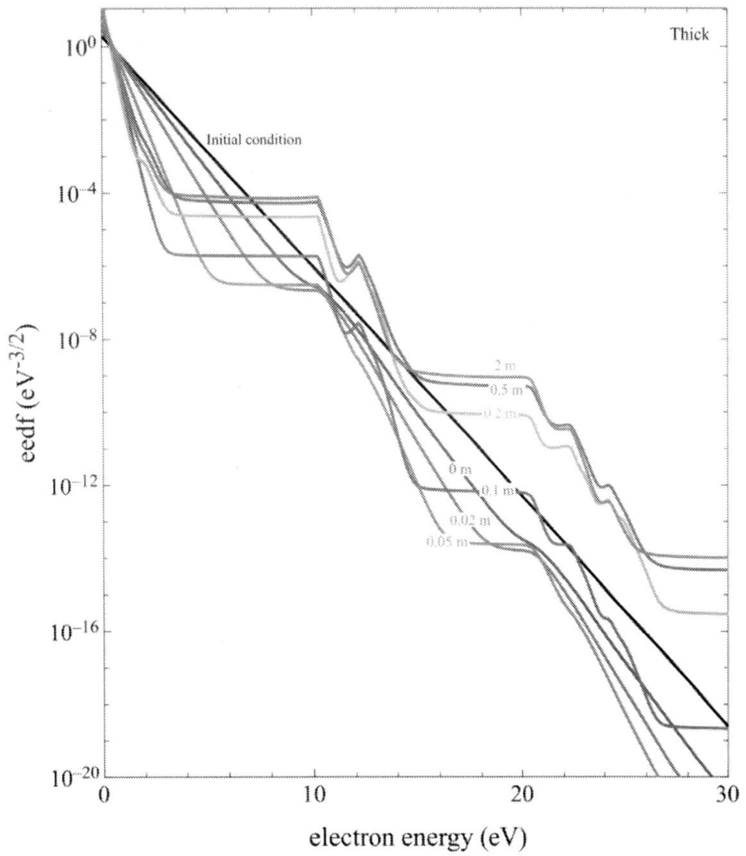

Figure 15. Evolution of the EEDF at different positions along the nozzle axis obtained imposing $p_0 = 1$ atm and $T_0 = 8000$ K in the reservoir in the plasma thick case.

For nozzle expansion, the cases of an optically thick and thin plasma are considered, which evolve along the nozzle axis starting from the same initial condition. Structured EEDF are created, showing long plateaux and humps due to superelastic collisions of low energy electrons and electronically excited atomic hydrogen. The magnitudes of the plateaux are much higher in the thick opposed to the thin plasma case. Because of their low concentration, He excited states have a role only in the plasma thin case, thanks to the low electron energy value and electron number density. This situation slows down the thermalization of EEDF through electron–electron collisions, preserving the complex structure of EEDF created by superelastic collisions.

Acknowledgments

This work received funding from the European Community's Seventh Framework Programme (FP7/2007-2013) under Grant agreement no. 242311 and from the project 'Apulia Space', PON 03PE-00067.6 from DTA Brindisi.

References

[1] Capitelli M *et al* 2006 Vibrational kinetics, electron dynamics and elementary processes in H_2 and D_2 plasmas for negative ion production: modelling aspects *Nucl. Fusion* **46** S260–74
[2] Hassouni K, Gicquel A and Capitelli M 1998 The role of dissociative attachment from Rydberg states in enhancing H^- concentration in moderate- and low-pressure H_2 plasma sources *Chem. Phys. Lett.* **290** 502–8
[3] Hassouni K, Gicquel A, Capitelli M and Loureiro J 1999 Chemical kinetics and energy transfer in moderate pressure H_2 plasmas used in diamond MPACVD processes *Plasma Sources Sci. Technol.* **8** 494–512
[4] Hassouni K, Silva F and Gicquel A 2010 Modelling of diamond deposition microwave cavity generated plasmas *J. Phys. D: Appl. Phys.* **43** 153001
[5] Longo S and Boyd I 1998 Nonequilibrium dissociation of hydrogen in a parallel-plate radio frequency discharge *Chem. Phys.* **238** 445–53
[6] Longo S, Diomede P and Capitelli M 2008 Vibrational kinetics of diatomic molecules in parallel plate RF plasmas *Phys. Scr.* **78** 058107
[7] Longo S and Diomede P 2009 Modeling of capacitively coupled RF plasmas in H_2 *Plasma Process. Polym.* **6** 370–9
[8] Roettgen A, Adamovich I V and Lempert W R 2013 Measurements of N_2 vibrational distribution function in pulsed nanosecond nonequilibrium discharge by spontanous Raman scattering *51st AIAA Aerospace Sciences Mtg (7–10 January 2013, Grapevine, TX)* paper 2013-0576
[9] Longo S, Gorse C and Capitelli M 1991 Open problems in the XeCl laser physics *IEEE Trans. Plasma Sci.* **19** 379–86

[10] Pietanza L D, Colonna G, De Giacomo A and Capitelli M 2010 Kinetic processes for laser induced plasma diagnostic: a collisional–radiative model approach *Spectrochim. Acta* B **65** 616–26
[11] Colonna G, Pietanza L D and Capitelli M 2001 Coupled solution of a time-dependent collisional–radiative model and Boltzmann equation for atomic hydrogen plasmas: possible implications with LIBS plasmas *Spectrochim. Acta* B **56** 587–98
[12] Colonna G, Casavola A and Capitelli M 2001 Modelling of LIBS plasma expansion *Spectrochim. Acta* B **56** 567–86
[13] Armenise I, Capitelli M and Gorse C 1996 Nonequilibrium vibrational kinetics in the boundary layer of re-entering bodies *J. Thermophys. Heat Transfer* **10** 397–405
[14] Armenise I, Capitelli M and Longo S 2009 Fourier and diffusive heat transfer in hypersonic nitrogen flows: state-to-state approach *J. Thermophys. Heat Transfer* **23** 674–83
[15] Armenise I A and Kustova E V B 2013 State-to-state models for CO_2 molecules: from the theory to an application to hypersonic boundary layers *Chem. Phys.* **415** 269–81
[16] Colonna G, Pietanza L D and D'Ammando G 2012 Self-consistent collisional–radiative model for hydrogen atoms: atom–atom interaction and radiation transport *Chem. Phys.* **398** 37–45
[17] Colonna G and Capitelli M 2001 Self-consistent model of chemical, vibrational, electron kinetics in nozzle expansion *J. Thermophys. Heat Transfer* **15** 308–16
[18] Colonna G and Capitelli M 2001 The influence of atomic and molecular metastable states in high-enthalpy nozzle expansion nitrogen flows *J. Phys. D: Appl. Phys.* **34** 1812
[19] Kadochnikov I N, Loukhovitski B I and Starik A M 2013 Kinetics of plasmachemical processes in the expanding flow of nitrogen plasma *Phys. Scr.* **88** 058306
[20] Panesi M, Magin T, Bourdon A, Bultel A and Chazot O 2009 Fire II flight experiment analysis by means of a collisional–radiative model *J. Thermophys. Heat Transfer* **23** 236–48
[21] Guy A, Bourdon A and Perrin M -Y 2013 Consistent multi-internal-temperatures models for nonequilibrium nozzle flows *Chem. Phys.* **420** 15–24
[22] Munafò A, Lani A, Bultel A and Panesi M 2013 Modeling of non-equilibrium phenomena in expanding flows by means of a collisional–radiative model *Phys. Plasmas* **20** 073501
[23] Colonna G and Capitelli M 1996 Electron and vibrational kinetics in the boundary layer of hypersonic flow *J. Thermophys. Heat Transfer* **10** 406–12
[24] Surzhikov S 2012 Coupled radiative gasdynamic interaction and non-equilibrium dissociation for large-scale returned space vehicles *Chemi. Phys.* **398** 56–63
[25] Surzhikov S T A and Shang J S B 2012 Coupled radiation-gasdynamic model for stardust earth entry simulation *J. Spacecr. Rockets* **49** 875–88
[26] Capitelli M, Colonna G, Pietanza L D and D'Ammando G 2013 Coupling of radiation, excited states and electron energy distribution function in non-equilibrium hydrogen plasmas *Spectrochim. Acta* B **83–4** 1–13
[27] Goto M 2003 Collisional–radiative model for neutral helium in plasma revisited *J. Quant. Spectrosc. Radiat. Transfer* **76** 331–44
[28] Matveyev A A and Silakov V P 1995 Kinetic processes in a highly-ionized non-equilibrium hydrogen plasma *Plasma Sources Sci. Technol.* **4** 606–17
[29] Esposito F, Gorse C and Capitelli M 1999 Quasi-classical dynamics calculations and state-selected rate coefficients for $H + H_2 (v, j) \rightarrow 3H$ processes: application to the global dissociation rate under thermal conditions *Chem. Phys. Lett.* **303** 636–40
[30] Esposito F and Capitelli M 2001 Dynamical calculations of state to state and dissociation cross-sections for atom-molecule collision processes in hydrogen *At. Plasma-Mater. Interact. Data Fusion* **9** 65
[31] Esposito F and Capitelli M 2009 Selective vibrational pumping of molecular hydrogen via gas phase atomic recombination *J. Phys. Chem.* A **113** 15307–14
[32] Capitelli M, Colonna G, Esposito F, Hassouni K, Laricchiuta A and Longo S 2015 *Fundamental Aspects of Plasma Chemical Physics: Kinetics* (New York: Springer)
[33] D'Ammando G, Capitelli M, Esposito F, Laricchiuta A, Pietanza L D and Colonna G 2014 The role of radiative reabsorption on the electron energy distribution functions in H_2/He plasma expansion through a tapered nozzle *Phys. Plasmas* **21** 093503
[34] Park C 2012 Nonequilibrium ionization and radiation in hydrogen–helium mixtures *J. Thermophys. Heat Transfer* **26** 231–43
[35] Drawin H W 1969 Influence of atom–atom collisions on the collisional–radiative ionization and recombination coefficients of hydrogen plasmas *Z. Phys.* **225** 483–93
[36] Drawin H W and Emard F 1973 Atom-atom excitation and ionization in shock waves of the noble gases *Phys. Lett.* A **43** 333–5
[37] Hey J D 2006 On the determination of radial matrix elements for high-*n* transitions in hydrogenic atoms and ions *J. Phys. B: At. Mol. Opt. Phys.* **39** 2641
[38] D'Ammando G, Pietanza L D, Colonna G, Longo S and Capitelli M 2010 Modelling spectral properties of nonequilibrium atomic hydrogen plasma *Spectrochim. Acta* B **65** 120–9
[39] Kramida A, Ralchenko Y, Reader J and NIST ASD Team 2012 *NIST Atomic Spectra Database* version 5.0 (Gaithersburg, MD: National Institute of Standards and Technology)
[40] Celiberto R, Cacciatore M and Capitelli M 1990 Electron impact direct dissociative-ionization cross sections from vibrationally excited H_2 molecules and translational energy distribution functions of protons *Chem. Phys.* **140** 209–15
[41] Cacciatore M, Capitelli M, Celiberto R, Cives P and Gorse C 1990 Coupled solution of Boltzmann equation and non equilibrium vibrational kinetics in H_2 volume discharges: an analysis of the input data *AIP Conf. Proc.* **210** 74
[42] Celiberto R, Janev R K, Laricchiuta A, Capitelli M, Wadehra J M and Atems D E 2001 Cross section data for electron-impact inelastic processes of vibrationally excited molecules of hydrogen and its isotopes *At. Data Nucl. Data* **77** 161–213
[43] Celiberto R, Janev R K, Wadehra J M and Laricchiuta A 2008 Cross sections for 11 − 14 eV $e - H_2$ resonant collisions: vibrational excitation *Phys. Rev.* A **77** 012714
[44] Gorse C, Celiberto R, Cacciatore M, Laganà A and Capitelli M 1992 From dynamics to modeling of plasma complex systems: negative ion (H^-) sources *Chem. Phys.* **161** 211–27
[45] IAEA 2009 ALADDIN database (https://www-amdis.iaea.org/ALADDIN)
[46] The Opacity Project Team 2012 Opacity Project database (http://cdsweb.u-strasbg.fr/topbase/topbase.html)
[47] Nahar S N 2010 Photoionization and electron–ion recombination of He *New Astron.* **15** 417
[48] Ralchenko Y, Janev R K, Kato T, Fursa D V, Bray I and de Heer F J 2008 Electron-impact excitation and ionization cross sections for ground state and excited helium atoms *At. Data Nucl. Data* **94** 603–22
[49] Rockwood S D 1973 Elastic and inelastic cross sections for electron-Hg scattering from hg transport data *Phys. Rev. Lett.* **8** 2348–58
[50] Colonna G and Capitelli M 2008 Boltzmann and master equations for MHD in weakly ionized gases *J. Thermophys. Heat Transfer* **22** 414–23

[51] D'Angola A, Coppa G, Capitelli M, Gorse C and Colonna G 2010 An efficient energy-conserving numerical model for the electron energy distribution function in the presence of electron–electron collisions *Comput. Phys. Commun.* **181** 1204–11

[52] Colonna G, Tuttafesta M and Giordano D 2001c Numerical methods to solve Euler equations in 1D steady nozzle flow *Comput. Phys. Commun.* **138** 213–21

[53] Mihalas D and Mihalas B W 1999 *Foundations of Radiation Hydrodynamics* (New York: Dover)

[54] Wells R J 1999 Rapid approximation to the Voigt/Faddeeva function and its derivatives *J. Quant. Spectrosc. Radiat. Transfer* **62** 29–48

[55] Matsuyama S, Ohnishi N, Akihiro Sasoh and Sawada K 2005 Numerical simulation of Galileo probe entry flowfield with radiation and ablation *J. Thermophys. Heat Transfer* **19** 28–35

[56] Giordano D, Colonna G, Capitelli M and Tuttafesta M 1999 Non-Arrhenius No formation rate in 1D nozzle airflow *J. Thermophys. Heat Transfer* **13** 372–5

IOP Publishing | Plasma Physics and Controlled Fusion

Plasma Phys. Control. Fusion **57** (2015) 014010 (9pp) | doi:10.1088/0741-3335/57/1/014010

Industrial plasmas in academia

Ch Hollenstein[1], AA Howling[2], Ph Guittienne[3] and I Furno[2]

[1] Avenue William 46, CH-1095 Lutry, Switzerland
[2] Ecole Polytechnique Fédérale de Lausanne, Centre de Recherches en Physique des Plasmas CH-1015 Lausanne, Switzerland
[3] Helyssen Sàrl, Route de la Louche 31, CH-1092 Belmont, Switzerland

E-mail: christophe.hollenstein@epfl.ch

Received 1 July 2014, revised 10 September 2014
Accepted for publication 16 September 2014
Published 28 November 2014

Abstract

The present review, written at the occasion of the 2014 EPS Innovation award, will give a short overview of the research and development of industrial plasmas within the last 30 years and will also provide a first glimpse into future developments of this important topic of plasma physics and plasma chemistry. In the present contribution, some of the industrial plasmas studied at the CRPP/EPFL at Lausanne are highlighted and their influence on modern plasma physics and also discharge physics is discussed. One of the most important problems is the treatment of large surfaces, such as that used in solar cells, but also in more daily applications, such as the packaging industry. In this contribution, the advantages and disadvantages of some of the most prominent plasmas such as capacitively- and inductively-coupled plasmas are discussed. Electromagnetic problems due to the related radio frequency and its consequences on the plasma reactor performance, and also dust formation due to chemical reactions in plasma, are highlighted. Arcing and parasitic discharges occurring in plasma reactors can lead to plasma reactor damages. Some specific problems, such as the gas supply of a large area reactor, are discussed in more detail. Other topics of interest have been dc discharges such as those used in plasma spraying where thermal plasmas are applied for advanced material processing. Modern plasma diagnostics make it possible to investigate sparks in electrical discharge machining, which surprisingly show properties of weakly-coupled plasmas. Nanosecond dielectric barrier discharge plasmas have been applied to more speculative topics such as applications in aerodynamics and will surely be important in the future for ignition and combustion.

Most of the commonly-used plasma sources have been shown to be limited in their performance. Therefore new, more effective plasma sources are urgently required. With the recent development of novel resonant network antennas for new advanced large area or large volume plasma sources, an important step towards high performance plasmas and new fast processes is made.

Keywords: plasmas, industry, academia

(Some figures may appear in colour only in the online journal)

1. Introduction

Plasma processing has proven to be an important tool in industry. Many of our present-day technologies and applications rely strongly on plasma processes. Without any doubt, plasmas have been at the origin of the overwhelming success of semiconductors and their application in electronics and communications. The upcoming very diverse industrial applications led to the use of quite a large variety of plasmas. The classes of plasmas used in industrial applications differ substantially from plasmas such as fusion plasmas and astrophysical plasmas as typically investigated in academia. Thermal and in most cases, non-thermal plasmas, are applied for processing, running mostly in a chemically reactive environment. In particular, non-thermal plasmas are often used, since they allow low temperature processing where normally high temperatures are needed. Efforts to understand the complex plasma physics and also plasma chemistry, together with

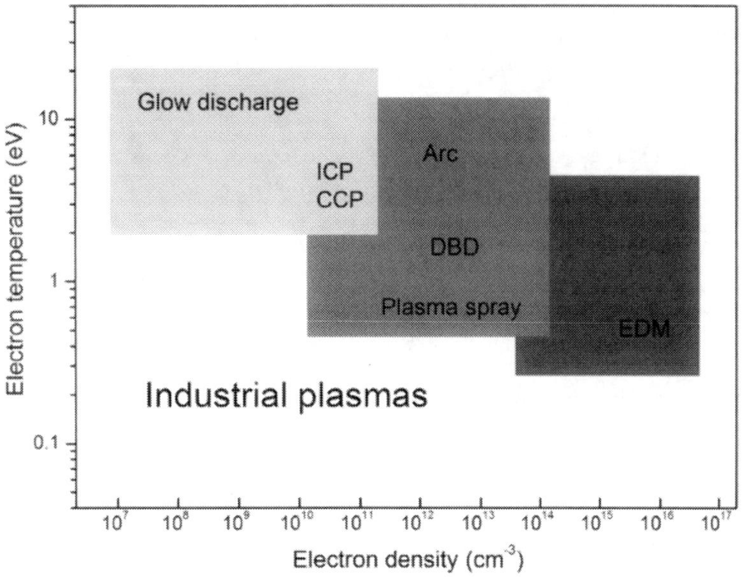

Figure 1. Electron temperatures and electron densities of industrial plasmas. The parameters of processing plasmas discussed in this paper are indicated.

investigations of the surface physics taking place during processing, have been made in the recent years in academia, but also to a certain extent in industry.

Interestingly, many of the problems encountered related to industrial plasma processing also lead to new fundamental topics in plasma research. One of the best-known examples of basic plasma physics research triggered by applications is the complex (or dusty) plasma, since dust particles were found to influence the properties of the processing plasma and of the deposited films. Such (or similar) dusty or complex plasmas are also found in astrophysics and are therefore of fundamental scientific interest as well.

Very important industrial applications are the production of thin film solar cells, flat displays, architectural glass and packaging, which require large area plasmas for deposition. Therefore, large area plasmas have been investigated in strong relation to their specific industrial exploitation. Basic investigations and development of large area plasmas are rarely reported in the literature. Large-scale plasmas often show a different behavior with respect to small laboratory plasmas, which are mostly edge dominated. Therefore, the necessary upgrade is in most cases a cumbersome task, since reactor and plasma parameters cannot simply be scaled up. In addition, process-specific parameters such as deposition rate and material properties also require highly uniform plasmas. An additional industrial demand is the reduction of individual processing time leading to plasmas with higher and higher radical densities. Therefore in future, versatile, high-density plasma sources will be in strong demand for modern plasma processing.

Many problems arising in large area plasmas are related to the large substrate sizes and the high electrical powers involved in driving the plasma. Electrical and fluid dynamic problems of the reactor and the complicated plasma chemistry contribute to the complexity of large area plasma reactor use.

The aspect of large area industrial plasmas is contrasted by other potential plasma applications, such as the use of small, but very dense, plasmas for mechanical small-scale machining and the application of dc plasmas used in thermal plasma spraying for fast deposition of dense, thin films. Recently, plasmas have also entered into new applications and challenges, as will shortly be demonstrated by some insight into the development of plasma for aerodynamics and the important field of combustion and ignition. Novel studies on resonant network antennas demonstrate the wide future potential of this new type of radio frequency (RF) antenna for advanced applications in industrial processing.

In this paper, the different plasmas used for some typical industrial applications are presented and their properties are discussed. The most important problems in plasma processing investigated in academia will then be highlighted and some specific problems such as the working gas supply in large area reactors and its solution will be discussed in more detail. Future developments in industrial plasmas and related topics will also be presented.

2. Industrial plasma sources

The choice of the plasma source is mainly dictated by the specific application. The use of plasmas in widespread applications leads to a large number of quite different methods of producing plasmas with parameters and properties demanded by the material processing. As shown in figure 1, industrial plasma sources operate over a wide plasma parameter range. These plasmas are driven in gas pressures ranging from a few Pa up to atmospheric pressure. As a consequence, all different types of non-equilibrium plasmas, but often also thermal plasmas, are found in industry. Industrial applications also lead to the use of plasmas produced in a wide variety of gases, or even produced in liquids and solids. Besides noble gases, many types of molecular gases and even complex metal-organics, are nowadays employed for plasma treatments. With the use of such different gases, the number of possible plasma constituents is considerably increased. Plasmas sources are mainly conceived to optimize particular plasma species such as negative ions, radical or excited species, depending on the process. Such plasma optimization needs a detailed understanding of the ongoing plasma physics and importantly also of the plasma chemistry.

One very specific application of plasmas in industry is thin film deposition by plasmas running in reactive gases. The deposition in general needs high throughput, homogeneous plasmas for uniform coating and a high degree of reproducibility. A very specific case is large area thin film deposition as used in solar cells, flat displays or in the packaging industry. The specifications for deposition plasma in the semiconductor domain will have quite different demands concerning impurity content from plasmas used for barrier coating in packaging. Therefore, the type of plasma reactor might imply quite different design and construction depending on the specific applications.

In the following, some of the problems and some solutions in large area depositions are discussed. Presently, various plasma sources for large area industrial applications are given in the literature [1]. For large area applications besides capacitively- and inductively-coupled RF plasmas, magnetron discharges are often applied [2]. In particular, magnetron plasmas are widely used in industrial roll-to-roll processing. Magnetron discharges, since they are also used for manufacturing of a large palette of products, are one of the best-investigated industrial plasmas.

Capacitively- and inductively-coupled plasmas used in the so-called batch mode have been upscaled from laboratory-sized plasmas to large areas of the order of several square meters [3]. Various modifications to the plasma source during upscaling, such as the use of lens-shaped electrodes in capacitively-coupled plasma reactors [4], have been necessary to obtain sufficient industrial performance.

One of the problems arising from upscaling of capacitive plasma sources to large sizes is due to the change in impedance with size: increasing the size leads to an increased capacity of the reactor and so the reactor presents a lower and lower impedance. Therefore, in the case of a large area capacitively-coupled plasma, a strong RF current flows in the input lines, requiring the construction of ruggedized RF transmission lines and RF feedthroughs. This tendency can give rise to further problems, since parasitic discharges can be triggered by faulty design of RF components if no precautions are taken.

On the other hand, upscaling of inductively-coupled plasmas leads to high voltages on the inductive device. Therefore, the question of whether the plasma production is being more capacitively driven than inductively driven has to be considered. The installation of electrical screens to avoid parasitic capacitive coupling leads to further design complications of conventional inductive plasma sources.

In order to decrease the processing time, high plasma densities are generally demanded, since high rate processing is closely related to rapid production of radicals. Various efforts have been made to increase the overall efficiency of large area capacitively-coupled plasma reactors. However, it is believed that substantially higher deposition rates than presently achieved cannot be obtained with such a device. Higher particle densities can be obtained in inductively-coupled plasma and even higher densities, only in wave-excited plasmas such as in helicon devices. Recently, great efforts have been made in order to advance these plasma sources for high deposition rate processing.

In many different applications of the growing layer or for the treated surface, the resulting ion bombardment from the plasma is of great importance in terms of obtaining the desired material properties. For instance, this is the case in the production of thin film silicon layers for solar cells where, most probably, a modest ion bombardment of the growing layer is necessary for device quality material. Various modifications of plasma reactors or processing methods have been developed to control the ion bombardment energy. One way to obtain a controllable ion bombardment is by applying different RF frequencies in a capactively-coupled device [5]. In inductively-coupled plasmas a control of the ion energy is obtained by biasing an additional electrode by means of a dc or an additional RF power.

Sources which are completely different from the discussed inductively-coupled or capacitively-coupled plasmas are used for applications in aerodynamics and for plasma assisted combustion. For these applications, ac dielectric barrier discharges (DBD) which run at ambient atmosphere are used and recently the more promising nanosecond DBD have been tested [6]. Plasma actuators for flow control made of such dielectric barrier discharges have the advantage with respect to conventional flow actuators of a simple design, low mass, large frequency bandwidth, the absence of moving parts and real–time feedback control. The control mechanisms of the ac DBD and of the nanosecond DBD differ fundamentally. In the case of the ac DBD, the ion wind produced interacts with the boundary layer. This ion wind is presently limited to about $10\,m\,s^{-1}$, a value too low for real aerodynamic applications. The nanosecond DBD are known to excite a compression wave. At present, it is thought that the control capability of this plasma actuator does not originate from electro-hydrodynamic flow acceleration, but from the excitation of flow instabilities by the compression wave generated by the Joule heating of the DBD. Depending on the application, control of the leading edge separation on an airfoil could be obtained up to about $60\,m\,s^{-1}$. Flow control might be possible in the future after more intense investigations and development of the interaction of the plasma with the high velocity flow. It has to be mentioned that, for the nanosecond pulsed DBD, fast high voltage power supplies are necessary. For control tests, such power supplies small enough to fit into a model aircraft have been constructed and by carefully avoiding electrical disturbances, used for flow control. The development of small and light high voltage pulse generators is also important for other applications of atmospheric DBD such as in medicine and in particular also for the widespread applications in combustion and ignition [7].

All the previously described plasmas belong to the group of non-thermal plasmas. A very typical application for thermal plasma is the well-established plasma spraying. Classical plasma spray is performed at atmospheric pressure or in a low vacuum environment with the aim of reducing oxidation of the coating. Plasma spraying finds great success in applications such as automotive and avionics. In plasma spraying, small powder particles are introduced into the plasma, which melt and finally splash onto the substrate forming a mostly porous, thick layer. At high operation pressures essentially solid precursors are used, whereas for liquid and gaseous precursors the plasma torch parameters have been optimized at intermediate or low pressure. In this low-pressure range and with liquid or gaseous precursors, the chemistry within the plasma starts to play an essential role [8]. This latter process has to be distinguished from the so-called reactive plasma spraying where reactive gases are added in order to modify the powder particles. In reactive thermal plasma chemical vapor deposition (CVD), the precursors are evaporated and the gaseous components contribute to the plasma chemistry and coating. With this new low pressure process the high porosity of the plasma spray coatings has been overcome and dense layers at high deposition rate, similar to the films of non-thermal plasmas, have been obtained. Due to the high enthalpy in the plasma, the thermal plasma CVD is more efficient than the widespread RF plasma processing. In particular, due to liquid and solid precursors, higher throughput and higher deposition rates than in RF plasmas are obtained [9]. With this process development, it is possible to obtain a variety of coating qualities at high rate using the same plasma tool.

3. Problems of industrial plasmas

The problems encountered in industrial plasma devices are related to different topics in science and engineering. The large variety of problems found in the case of large area capacitively-coupled RF plasma reactors [1] are summarized in figure 2. These, or similar problems, are also found in other types of plasmas used in industry.

Unfortunately, many of the problems considered arise from improperly designed hardware such as the geometry of the plasma reactor, the incorrect implantation of the RF power connection(s) or poor vacuum-compatible reactor design. The most well-known difficulties arise from electrical problems and are related to RF. Additional difficulties can arise from substrate charging during processing, making substrate handling more difficult since electrostatic discharges have to be avoided [10]. The gas supply and pumping design can have a significant influence on the properties of the obtained layers, whereas various plasma properties such as chemical reactions in many reactive plasmas can lead to dust formation, causing several problems for the substrates and can lead to strong contamination of the reactor. In the following, some of these topics shall be discussed in more detail.

The standing wave effect in large area capacitively-coupled plasmas leads, as is widely known, to electric field non-uniformity and consequently inhomogeneous deposition. This effect has been corrected by a more complex electrical design of the plasma reactor. Lens shaping of the RF electrode is one of such designs to compensate for the size-induced standing wave effect [3, 4]. Further electromagnetic effects with asymmetric electrodes lead to inhomogeneity at the plasma edge, thus reducing the effective usable size of the substrate [11, 12]. As it turns out, the details of the RF feeding of the electrode are also important and also influence the deposition. To demonstrate the importance of design details of the plasma reactor, it has to be mentioned that small changes in the reactor capacity due to a small hole in an electrode already lead to visible inhomogeneity in the coating.

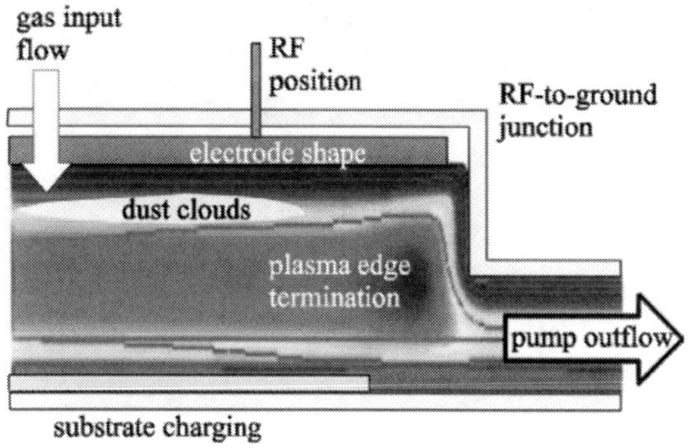

Figure 2. Schematic of a generalized, large area capacitively-coupled RF reactor. The labels show some possible design difficulties mainly associated with large area plasmas.

The large area of the reactor might also strongly influence the contamination of the growing layers. This is particularly important in the case of semiconductor production, such as for thin film silicon deposition for solar cells. The contamination problem can be solved in various ways by the design of the plasma reactor, for example, with the differentially pumped plasma box concept [3].

A significant problem in deposition reactors is the gas supply and the pumping of the plasma reactor [13]. This problem will be treated here in more detail.

Figure 3 shows a schematic of a rectangular RF capacitive reactor with lateral pumping on two opposite sides and a uniform showerhead gas inflow through one of the electrodes. The pressure is uniform over the whole reactor, to a good approximation, provided that the electrode gap is not too small because the pressure gradient is proportional to the inverse cube of the gap distance. For these conditions, the gas transport is described by finite-gap stagnation point flow [13] whose similar solution exhibits lateral uniformity. In the approximation of creeping flow, relevant to most plasma deposition processes, the vertical flow velocity has a cubic vertical profile, but is independent of lateral position, whereas the horizontal flow velocity is proportional to distance from the mid-point and has a parabolic vertical profile. The resulting velocity field is shown by the streamlines in figure 3 for this example case of plasma deposition; the deposition rate onto the growing film surface is represented by a non-zero fluid velocity into the walls known as Stefan flow.

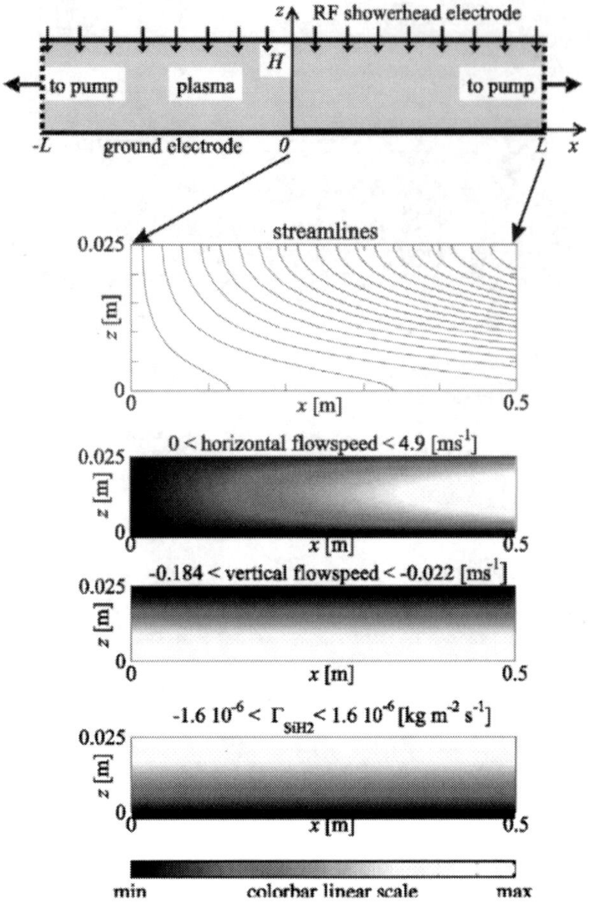

Figure 3. Schematic of a rectangular showerhead reactor, showing the numerical solution of the gas flow streamlines in the right half. The surface plots show the horizontal and vertical flow speeds and the diffusive flux of SiH_2 for an example case of plasma deposition.

The key property of stagnation point flow is that the fluid vertical flow velocity and the concentrations of all species are independent of lateral position.

This implies that the convective and diffusive flux and therefore the net flux of each species, is uniform over the whole substrate area, as shown by the lateral uniformity in the numerical solutions of figure 3. This results in the desired uniform deposition rate on the substrate. This stagnation flow solution holds for rectangular reactors with lateral pumping through one entire side, or two opposed entire sides and for cylindrical reactors with uniform circumferential pumping.

Another important problem in reactive plasmas is the formation of nanoparticles. Besides the damage that particles can cause in the thin film, dust can also lead to inhomogeneity of the deposited layer. The localised powder cloud changes the local RF power density in the plasma and consequently inhomogeneous films originate. Dust reduction or elimination is a complex problem involving plasma chemistry as well as reactor design. To solve this problem, the dust formation due to volume reactions in reactive plasmas had to be understood, leading to the opening of a new topic: complex or dusty plasmas.

The presence of nanometre particles was first observed in etching devices [14], triggering research into the origin and behavior of particles in low-pressure plasmas. Later, abundant particle formation was found in hydrocarbon and silane plasmas, which are used for many important industrial applications. The powder formation in capacitively-coupled RF silane plasmas was extensively investigated in order to find the powder precursors and the responsible gas phase reactions leading from the SiH_4 gas to nano- or micrometre sized particles [15]. Mass spectrometry in these capacitively-coupled RF silane plasmas revealed no high mass neutrals or positive ions, but high mass anions are produced. Anions up to a mass of 1600 amu containing up to 60 silicon atoms have been found in these plasmas [16]. In the case of RF silane plasmas, anions play an important role as precursors for nucleation, since they have a long dwell time in the plasma due to trapping in the plasma sheath [17]. During the cluster growth or nucleation, the charge of the small particles fluctuates between neutral or singly-charged anion clusters. This charge fluctuation allows agglomeration and coagulation of the precursors into macroscopic particles bearing multiple negative charges, which in turn are trapped within the plasma. These particles can increase in size by neutral particle accretion to become micrometre-sized particles which behave in a similar way, as has been shown in numerous so-called dusty plasmas experiments.

Powder formation in other reactive plasmas containing metal- or silicon-organic precursors, such as for the deposition of SiO_x (for packaging) and diamond-like carbon, are also found and can represent serious restrictions to the process.

Finally, a very serious and little investigated problem in plasma processing is the reproducibility and plasma parameter variations from device to device. One other important problem in industrial plasmas is related to uncontrolled arcing and the formation of parasitic discharges. The latter problem can have expensive consequences, since in plasma reactors RF power levels of several kilowatts can be involved. A small fraction might go undetected into an arc or other parasitic discharge, which in turn could lead to serious damage of the reactor. Many different types of parasitic discharges and arcing can occur in an industrial reactor. The occurrence of these discharges is strongly dependent on the design or geometry of the plasma reactor and depends also on the process parameters such as pressure and power. Precise monitoring of the input power and the matching condition and a consequent power limiting control, can suppress arcing events; however a good understanding of discharge physics is necessary to avoid, or at least retard, arcing. For these reasons, a few of the typical situations often encountered in large area plasma reactors have been investigated in experiments simulating the situation in real industrial reactors [18]. Such investigations turned out to be highly efficient in terms of improving process reliability. Another phenomenon related to parasitic discharges, so-called plasmoids, was recently investigated in detail. When a plasmoid forms in the hole of an electrode, it transports electrons through to the reverse side, for instance, of a pumping grid in a large area reactor, which is meant to contain the plasma. It was found that plasmoids are actually sustained by the radio–frequency plasma potential, forming a positive plasma potential on both sides of a grid in the reactor. It is the funneling of an intense electrical current that can cause strong heating of the plasmoid and subsequent damage to the reactor components [19].

Besides arcing in plasma reactors, arcing is a serious issue in many technical devices. Particularly in the miniaturisation of equipment and of electronic device, the small dimensions can lead to electrical breakdown and finally to discharges which can be dangerous for the devices. Arcing in industrial devices is a serious problem for the safety of personnel and equipment. One example that has been investigated in collaboration with industry and ESA is arcing phenomena in spacecraft equipment. Arcing in spacecraft can put an entire costly mission at risk. Therefore arcing - in particular in the power supply section - must be avoided by all means. The approach is to study experimentally and theoretically the susceptibility of specific components to electrical breakdown [20]. Breakdown experiments are performed on the original piece or importantly also using models of principal parts of the equipment under test. Numerical simulations of breakdown of these complex 3D geometries have been made in order to understand electrical breakdown in complex geometries. A basic understanding of breakdown in complex geometries can be obtained from such investigations and safer equipment can be designed. Such new physical rules are complementary to the well-known Paschen law, which applies only to the ideal case of breakdown between plane parallel plates.

4. Plasma diagnostics and simulation in industrial plasmas

Diagnostics in large area plasmas, or in industrial plasmas, has several aims. Firstly, plasma diagnostics serve as process monitoring of the most important parameters of the plasma and the reactor. Furthermore, plasma diagnostics are necessary to monitor reproducibility of the process, but also for transferring plasma parameters from reactor to reactor in equipment manufacturing. The application of plasma diagnostics in an industrial environment also requires efficient, cheap, robust and simple diagnostics, which can withstand high temperatures, vacuum and chemical attack.

It is apparent that, in industry, only a few plasma diagnostics are available, not only due to the cost, but also because of the complexity of using diagnostics in an industrial environment. However, in the literature only a few contributions concerning specific diagnostic of industrial plasma can be found [21]. Many sophisticated plasma diagnostics are available in academia, including trained personnel for interpretation, who are of great help to the research undertaken in industry.

Most recent developments have been performed in testing the coatings resulting from the plasma. Layer thickness, or the important deposition rate, material properties, and also the homogeneity of the film, or of the required functionality produced by the plasma, are determined by various, often time consuming, ex-situ measurements. In-depth diagnostics of large area plasmas are only rarely performed due to the poor access to the large area plasma and the few existing plasma diagnostics adapted to large area plasma reactors.

The most common diagnostics are measurements of process relevant parameters such as gas pressure and RF parameters. Electrical parameter measurements are necessary to monitor the process and to detect parasitic discharges to protect the equipment from dangerous electrical arcing.

Various types of spectroscopy such as emission spectroscopy and imaging using CCD cameras are applied to investigate large area plasmas. Fourier Transform Infrared (FTIR) absorption spectroscopy, often applied in the exhaust line, can be used for depletion measurements. Depletion measurements are useful to estimate the efficiency of the plasma for radical production. Absorption measurements are also particularly useful for determining the plasma composition. Other plasma diagnostics, such as mass spectrometry of neutral, positive and negative ions, various types of electrical probes and active and passive laser diagnostics or Mie scattering for detecting particles within the plasmas, are mostly used only in the research and development phase, or in collaboration with academia. Since the diagnostics are difficult to implant on industrial devices, academic research is necessary to improve the understanding of the plasma process, to control the plasma process and to conduct further process development for industry.

Recent advances in the understanding of electrical discharge machining (EDM) show that modern diagnostics also lead to advances in industrial plasmas. In the case of EDM, fast time-resolved emission spectroscopy gives a first insight into the evolution of the plasma parameters during a single spark event. Emission spectroscopy shows the plasma composition and relative line intensity ratios are useful for measuring the electron temperature within the small discharge [22]. Stark broadening, in particular of hydrogen lines, can be used to determine the electron density which, in the case of the EDM discharge, is between 10^{16} to $10^{17}\,cm^{-3}$. The Coulomb coupling coefficient can be calculated from these measurements, which astonishingly shows that the EDM spark is a weakly coupled plasma. Further diagnostics of this weakly coupled plasma is a future challenge. The investigations on the EDM plasma also give rise to new issues, which might be of broad interest [23, 24]. One of these issues is the breakdown in liquids. As has been shown, the electrical breakdown in water behaves differently from oil, giving rise to new fundamental investigations. Breakdown in liquids is an important topic for the application of plasmas in biology and medicine. Breakdown and abrasion of electrodes are still important topics, since the increased power consumption in vehicles demands higher supply voltages, leading to redesign of part of the existing electrical equipment such as electrical relays in future.

Recently, very important advances have been made in the numerical simulation of applied and industrial plasmas. In particular, advances in modelling and simulation of different industrial plasma reactors, such as capacitivly and inductively coupled devices [25–27] and various plasma processes, have been made. Without any doubt, such theoretical efforts lead to quite substantial advances in plasma physics and chemistry and also to more efficient and economic industrial plasmas.

Significant efforts in the development of simplification of software for calculation and numerical simulation have been made. In particular, finite element calculations have been very much simplified by a wide choice of algorithms and solutions offered by many commercial software programs. This progress reduces the problems involved with numerical simulation programming, but does not solve the overall and often underestimated, complexity of the numerical simulation of plasmas. The inherent multi-scale and the multi-timescale problems in plasmas are still to be overcome in the calculation of plasmas. Also, useful insight from simple models can still be obtained [28] by judicious choice of the relevant physical phenomena which are well adapted to the real situation, without blindly using the full program complexity, thus avoiding long calculation times and large computer memories.

However, it has to be mentioned that not only simulation of the processing plasmas are important, but also many technical aspects of the design of the plasma reactor. In fact, many problems related to the desired outcome of the product are linked to a correct design of the plasma reactor. Heat conduction problems, fluid problems for gas supply and many others have to be solved in order to obtain an efficient, reproducible plasma reactor.

5. Future plasmas for industrial applications

A great industrial potential for the future lies in atmospheric plasmas. New and less known applications, such as those used in biology and medicine and also in technological applications such as aerodynamics, combustion and ignition, need atmospheric plasmas. The development of the technology and control of atmospheric plasma is an important future field of development and research in industry and academia. In particular, the well-known heating of neutrals observed in many high density plasmas has to be understood and controlled for applications in deposition, functionalization and many other fields.

New materials or advanced material processing and applications have to be explored for plasma treatment and the plasma processing itself must be made more reliable and faster. At present, new applications such as various topics in medicine, have found increasing interest in the industrial plasma community. One example is the promising application of fast high power pulsed magnetron sources (HIPIMS) for hard material processing, whereas nanosecond pulsed atmospheric discharges find application in advanced combustion, ignition and eventually in aerodynamics.

High rate plasma processing is expected by the use of the modular resonant network antenna. Recent design studies of resonant RF network antennas show that closed (birdcage) and open (planar) configurations of the antenna can be used for plasma production [29, 30]. Both are composed of similar electrical meshes, as shown in figure 4. It has been shown that it is possible to construct complex RF antennas out of the basic network element which can be adapted to their application and which are therefore very user friendly. Resonant modes arise for both closed and open antennas. Each mode is characterized by a resonance frequency determined by the L and C components and the mutual inductances and has specific current and voltage distributions in the antenna legs [29]. When driven at one of its resonant frequencies, very high currents are generated within the antenna structure. A resonant RF network antenna can therefore be used as a source for inductively-coupled plasma. There are a large number of different RF antenna arrangements possible in view of geometry and also in view of the RF operation and of the plasma obtained.

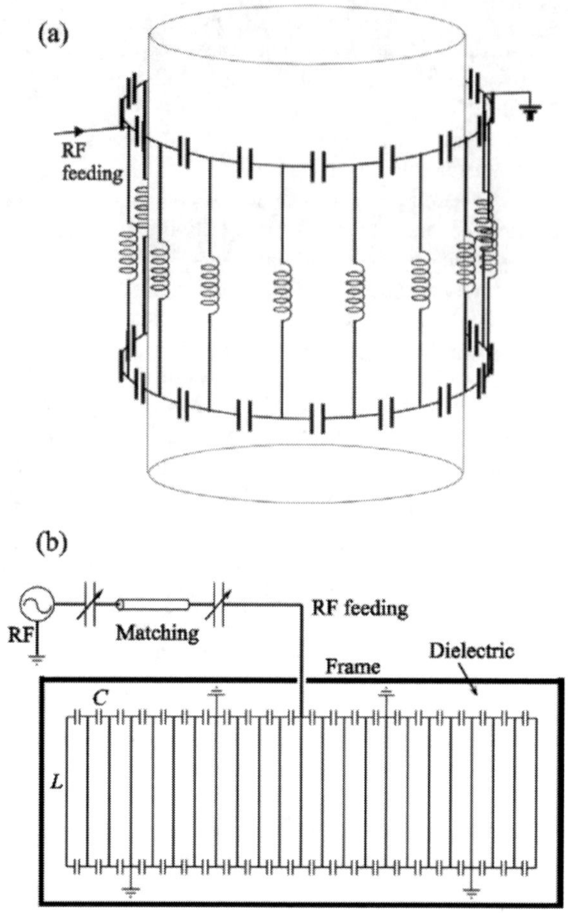

Figure 4. (*a*) Schematic of a birdcage RF antenna, showing the equivalent electrical circuit and (*b*) schematic of a planar RF antenna showing the antenna bars (inductance L), the capacitors (C) and the matching network.

Resonant network antennas can be upscaled to run as large area or large volume plasma sources [30] with much fewer difficulties than for the commonly used capacitive and inductive plasma sources. They can be adapted to the size and geometry of the optimal plasma volume for processing. At a resonance frequency, the antenna presents a real input impedance, thus facilitating matching to RF power generators. They show many advantages compared to the conventional capacitively- or inductively-coupled plasma devices and could therefore replace these plasma reactors for certain industrial applications.

One of the many promising applications is as a plasma source in future neutral beam injectors for fusion research, but they could also be considered as a replacement for the large area capacitive plasma sources for high rate deposition in batch and roll-to-roll processes. Helicon wave excited discharge produced by the resonant network antenna is thought to be an interesting source for future ion thrusters and related devices. Also, wave excited large surface and volume plasmas sources will be possible, as demonstrated by preliminary experiments with whistler-wave heated plasmas [31].

An important point for the development of future plasma reactors is a better adapted design and construction. Many of the present day plasma reactors are not designed to have optimized conditions for the plasmas and the desired process. Such topics include electrical installations such as RF, electrical perturbations, gas flow and gas supply of the plasma and mechanical construction to inhibit arcing and parasitic discharges. All such matters have to be taken into consideration for a successful, reproducible application of plasmas in industry.

6. Conclusion

In the past, plasma processes in industry have been very successful in opening the way for semiconductors and other related technologies. Also at present, plasma technology plays an important role in producing many daily and high technology products. In future, larger, flexible and improved performance high-density plasmas sources will be demanded. A far better knowledge of the underlying plasma and processes will trigger new impulses in advanced plasma source development, including processing of new materials. A strong interaction of research in industry and in academia can lead to the development of new topics in plasma physics and give rise to innovative plasma processing involving advanced materials.

Industrial plasmas are an important source for new fundamental topics in thermal and non-thermal plasma physics. New developments might lead to new types of processing and to original applications of plasma in modern technologies.

Acknowledgments

The authors would like to thank the Centre de Recherches en Physique des Plasmas (CRPP) of the Ecole Polytechnique Fédérale de Lausanne (EPFL) for the great support over the years. Much of the interesting research within the last 30 years would not have been possible without the generous support of the EPFL, KTI, ESA, SNF and the European FP7 project PLASMAERO. We would also like to thank the numerous industries such as Tetra Pak Romont, TEL Solar (formerly Oerlikon Solar), Oerlikon Metco (formerly Sulzer Metco), RUAG, Charmilles/Agie and Helyssen and their competent collaborators for their confidence and their close and excellent collaborations.

References

[1] Anders A 2005 Plasma and ion sources in large area coatings: a review *Surf. Coat. Technol.* **200** 1893

[2] Musil J, Vlcek J and Baroch P 2006 Magnetron discharges for thin films plasma processing in Y Pauleau (Editor) *Materials Surface Processing by Direct Energy Techniques* Chapter 3 (Oxford: Elsevier Limited)

[3] Perrin J, Schmitt J, Hollenstein Ch, Howling A A and Sansonnens L 2000 The physics of plasma-enhanced chemical vapour deposition for large-area coating: industrial application to flat panel displays and solar cells *Plasma Phys. Control. Fusion* **42** B353

[4] Schmidt H, Sansonnens L, Howling A A, Hollenstein Ch, Elyaakoubi M and Schmitt J P M 2004 Improving plasma uniformity using lens-shaped electrodes in a large area very high frequency reactor *J. Appl. Phys.* **95** 4559

[5] Schüngel E, Eremin D, Schulze J, Mussenbrock T and Czarnetzki U 2012 The electrical asymmetry effect in geometrically asymmetric capacitive radio frequency plasmas *J. Appl. Phys.* **112** 053302

[6] Benard N, Moreau E, Zouzou N, Rabat H, Pons J, Hong D, Leroy-Chesneau A, Peschke P and Hollenstein C 2013 Nanosecond pulsed plasma actuators *ERCOFTAC Bulletin No.* 94 March

[7] Starikovskaia S M 2006 Plasma assisted ignition and combustion *J. Phys. D: Appl. Phys.* **39** R265–99

[8] Dorier J L, Guittienne Ph, Hollenstein Ch, Gindrat M and Refke A 2009 Mechanisms of films and coatings formation from gaseous and liquid precursors with low pressure plasma spray equipment *Surf. Coat. Technol.* **203** 2125–30

[9] Gindrat M, Höhle H M, von Niessen K, Guittienne Ph, Grange D and Hollenstein Ch 2011 Plasma Spray-CVD: a new thermal spray process to produce thin films from liquid or gaseous precursors *J. Thermal Spray Technol.* **20** 882–7

[10] Howling A A, Belinger A, Bulkin P, Delaunay L, Elyaakoubi M, Hollenstein Ch, Perrin J, Sansonnens L, Schmitt J and Turlot E 2001 Mechanism of substrate charging after plasma processing *Proc. 15th Int. Symposium on Plasma Chemistry (Orléans, France, 9–13 July 2001)* **1** 33

[11] Chen Z, Rauf S and Collins K 2010 Self-consistent electrodynamics of large-area high-frequency capacitive plasma discharge *J. Appl. Phys.* **108** 073301

[12] Sansonnens L, Howling A A and Hollenstein Ch 2006 Electromagnetic field nonuniformities in large area, high-frequency capacitive plasma reactors, including electrode asymmetry effects *Plasma Sources Sci. Technol.* **15** 302

[13] Howling A A, Legradic B, Chesaux M and Hollenstein Ch 2012 Plasma deposition in an ideal showerhead reactor: a 2D analytical solution *Plasma Sources Sci. Technol.* **21** 015005

[14] Selwyn G S, Singh J and Bennett R S 1989 *In-situ* laser diagnostics studies of plasma-generated particle contamination *J. Vac. Sci. Technol.* **A7** 2758

[15] Bouchoule A 1999 Technological impact of dusty plasmas in *Dusty plasmas Physics, Chemistry and Technological Impacts in Plasma Processing* ed A Bouchoule (New York: Wiley)

[16] Hollenstein Ch, Howling A A, Courteille C, Dorier J L, Sansonnens L, Magni D and Mueller H 1998 Dust particle diagnostics in rf plasma deposition of silicon and silicon oxide films *MRS Symp. Proc. Vol. 507 Amorphous and Microcrystalline Silicon Technology* pp 547–57

[17] Perrin J and Hollenstein Ch 1999 Sources and growth of particles in *Dusty plasmas Physics, Chemistry and Technological Impacts in Plasma Processing* ed A Bouchoule (New York: Wiley)

[18] Legradic B, Howling A A and Hollenstein Ch 2010 Radio frequency breakdown between structured parallel plate electrodes with a millimetric gap in low pressure gases *Phys. Plasmas* **17** 102111

[19] Chesaux M, Howling A A and Hollenstein Ch 2013 Funnelling of rf current via a plasmoid through a grid hole in an rf capacitive plasma reactor *Plasma Sources Sci. Technol.* **22** 055006

[20] Schnyder R, Howling A A, Bommottet D and Hollenstein Ch 2013 Direct current breakdown in gases for complex geometries from high vacuum to atmospheric pressure *J. Phys. D: Appl. Phys.* **46** 285205
[21] Hopkins M B and Lawler J F2000 Plasma diagnostics in industry *Plasma Phys. Control. Fusion* **42** B189–97
[22] Descoeudres A, Hollenstein Ch, Wälder G, Demellayer R and Perez R 2008 Time- and spatially-resolved characterization of electrical discharge machining plasma *Plasma Sources Sci. Technol.* **17** 024008
[23] Maradia U, Boccadoro M, Stirnimann J, Beltrami I, Kuster F and Wegener K 2012 Die-sink EDM in meso-micro machining *Procedia CIRP* **1** 166–71
[24] Roth R, Balzer H, Kuster F and Wegener K 2012 Influence of the anode material on the breakdown behavior in dry electrical discharge machining *Procedia CIRP* **1** 639–44
[25] Logue M D, Shin H, Zhu W, Xu L, Donnelly V M, Economou D J and Kushner M J 2012 Ion energy distributions in inductively coupled plasmas having a biased boundary electrode *Plasma Sources Sci. Technol.* **21** 065009
[26] Turner M M, Derzsi A, Donko Z, Eremin D, Kelly S J, Lafleur T and Mussenbrock T 2013 Simulation benchmarks for low pressure plasmas: capacitive discharges *Phys. Plasmas* **20** 013507
[27] Eremin D, Hemke T, Brinkmann R P and Mussenbrock T 2013 Electromagnetic simulation of capacitively coupled discharges *J. Phys. D: Appl. Phys.* **46** 084017
[28] Hollenstein Ch and Howling A A 2008 Plasma edge simulations by finite elements using COMSOL *Proc. European COMSOL Conf.* (*Hannover, Germany, 4–6 November 2008*)
[29] Guittienne Ph, Lecoultre S, Fayet P, Larrieu J, Howling A A and Hollenstein Ch 2012 Resonant planar antenna as an inductive plasma source *J. Appl. Phys.* **111** 083305
[30] Hollenstein Ch, Guittienne Ph and Howling A A 2013 Resonant RF network antennas for large-area and large-volume inductively coupled plasma sources *Plasma Sources Sci. Technol.* **22** 055021
[31] Guittienne Ph, Howling A A and Hollenstein Ch 2013 Generation of whistler wave heated discharges with planar resonant RF networks *Phys. Rev. Lett.* **111** 125005

Plasma Phys. Control. Fusion **57** (2015) 014011 (7pp) doi:10.1088/0741-3335/57/1/014011

Alfven wave acceleration of particles in the aurora

F Mottez

Laboratoire Univers et THéories (LUTH), Observatoire de Paris-Meudon, CNRS, Université Paris Diderot, Meudon, France

E-mail: fabrice.mottez@obspm.fr

Received 18 June 2014, revised 11 September 2014
Accepted for publication 12 September 2014
Published 28 November 2014

Abstract

This paper is an introduction to the acceleration of the electrons that cause polar auroras. Emphasis is placed on acceleration processes involving Alfvén waves and their role in the global auroral context.

Keywords: acceleration, magnetosphere, Alfvén waves, auroral zone

(Some figures may appear in colour only in the online journal)

1. Introduction

The Earth plasma environment is constrained externally by the solar wind, a low density (~ 5 cm^{-3}) and high velocity ($\sim$ 300–900 km s^{-1}) collisionless flow of plasma of solar origin. Internally, it is bounded by the ionosphere, a mixture of neutral and ionized gas mostly in the collisional regime. Because of the Earth's magnetic field, these two areas are separated by a huge and rather complex region filled with a low density, magnetized and collisionless plasma, called the magnetosphere. Figure 1 shows a sketch of the magnetosphere where a few regions relevant for auroral physics are mentioned. The magnetosphere is behind a shock where the solar wind becomes sub-Aflvénic. The magnetosphere boundary behind this shock is the *magnetopause*. Within the magnetopause, all the magnetic field lines are connected to the Earth, at least on one side. The *magnetotail* is opposite to the Sun direction. It is on the right-hand side of figure 1. It contains two vast and almost empty regions, the *lobes*, separated by a magnetic reversal region, the *current sheet* or current layer, also called the *plasma sheet* because the plasma density is much higher than in the lobes. The direction of the electric current is marked with arrows. The electric current of the current sheet is closed along the magnetopause [1].

The magnetosphere is a dynamic region that almost never reaches a stationary regime. Instead, it endures several reconfigurations of the magnetic field and of the associated electric currents. Some of these reconfiguration progress for hours, like the increase of the cross tail electric current that tends to make the magnetotail thinner and more magnetized [2]; other reconfigurations are rather sudden. The most remarkable sudden reconfigurations of the magnetosphere are local and global reductions of the cross tail electric current [3]. When they are local they are called 'pseudo-break ups', when they encompass the whole current sheet we call them 'substorms' [4]. Typically, substorms or pseudo-breakups happen several times a day. They are more frequent when the Sun is active, because of the increased variability of the solar wind.

The magnetosphere is mostly composed of out of local equilibrium electrons, protons and oxygen ions of various energies ranging from a few eV to tens of keV. They originate from the solar wind and also (oxygen ions and low energy electrons) from the ionosphere.

When fluxes of magnetospheric electrons are directed towards the Earth, the ionosphere is like a screen on which the electrons give rise, when they carry enough energy, to spectacular visual displays called the *polar auroras*. There is a wide variety of auroras [5], characterized by various shapes and time evolution. The *diffuse auroras* are very common and hardly noticeable. *Discrete auroras* can be seen in visible light and in UV and x rays from space. The brightest ones appear mainly on the night-side (22 : 00–2 : 00 local time) and their appearance is correlated with substorms and pseudo-breakups [4]. Substorms (i) start progressively with an onset phase and quiet auroras, (ii) are followed by a break-up and the development of new patterns of discrete auroras, (iii) become global during expansion phase, accompanied by the most dynamic and bright auroral arcs, (iv) and end with a recovery phase during which things progressively calm down [4, 5].

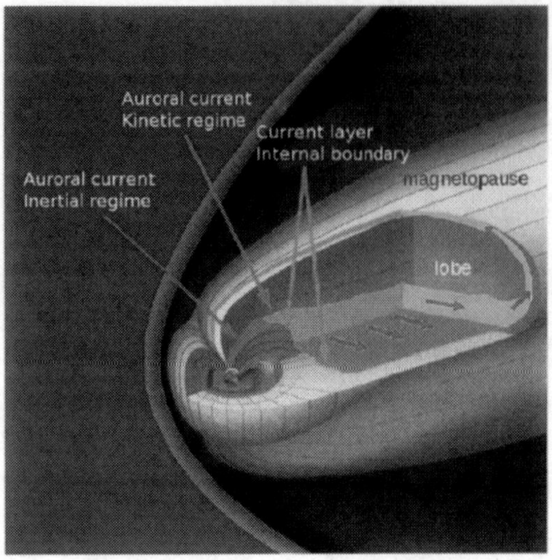

Figure 1. Schematic view of the magnetosphere, the main currents and the auroral zone, shown in this figure through the auroral currents. The sun direction is on the left-hand side.

Discrete auroras are visible at altitudes between 80 km and 400 km, in two ovals around the northern and the southern magnetic poles, for magnetic latitudes in the range 60–80°. The radiative processes that underlie them are well understood [6]. The light from auroras is emitted when ionospheric atoms and molecules (mostly O, N, O_2 and N_2) are de-excited after collisions with electrons precipitated from the magnetosphere with energies ~ few keV. The energy per particle is much larger than the thermal energy of the ionospheric particles (~ 1 eV) and of the solar wind (~ 100 eV). This means that the auroral electrons have been accelerated in the magnetosphere.

Our purpose is to introduce a series of theoretical works about the cause of the acceleration of of the most energetic auroral electrons, that we now call '*auroral acceleration*'. We underline the role of Alfvén waves (AW) in auroral acceleration.

2. Auroras as markers of energy fluxes in the magnetosphere

The overall flux of the solar wind flowing through the magnetosphere section is 1.3×10^{13} W from which about one percent is captured by the magnetosphere.The most energetic process in the magnetosphere is the transverse electric current across the magnetotail that drives the cross tail current (marked by the arrows between the lobes on figure 1), requiring a power of 3×10^{11} W. A large part of the energy is accumulated into the magnetotail when the magnetosphere is quiet, while a flux of 1.3×10^{11} W is continuously carried by the plasma towards the Earth, across the inner boundary of the magnetotail current layer. During substorms, a more substantial fraction of it (3×10^{11} W) is released toward the Earth. This energy corresponds to the liberation of the magnetic energy accumulated into the magnetotail during the quiet times and it corresponds also to a full transfer of the energy that transits through the magnetotail current [7].

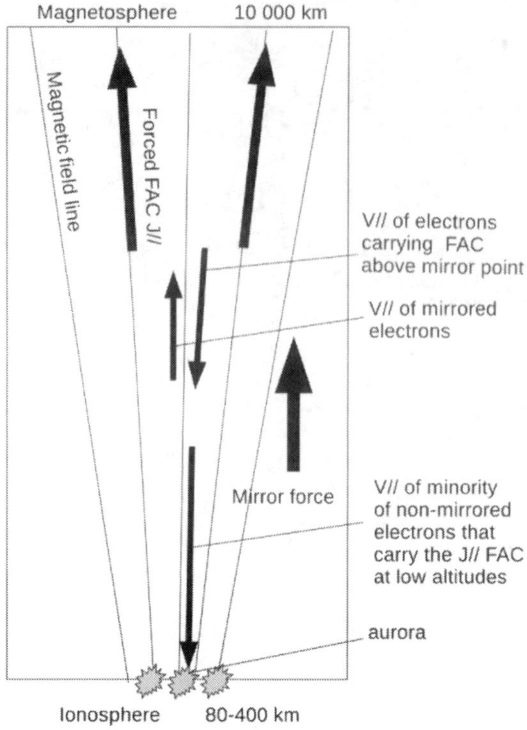

Figure 2. Schematic representation of the auroral zone and of the Knight effect.

Most of the energy directed from the magnetotail towards the Earth is conveyed through the *field aligned currents* (FAC), also called Birkeland currents, which are electric currents in the auroral zone propagating along the magnetic field lines. The energy flux along the FAC is 1.4×10^{11} W during quiet times and 2.7×10^{11} W during substorms. Most of the energy flowing Earthward from the magnetotail is dissipated in the auroral region along the FAC. Most of the auroral acceleration occur along the FAC and the polar auroras are the visible markers of this energy transfer [7].

Because the FAC are carried at low altitude by fast electrons and because the auroras are caused by down-going electrons, the relevant FAC for auroras correspond to upward currents.

2.1. The role of forced electric currents in auroral acceleration

Models of auroral acceleration do not include the whole magnetosphere. They are developed in a box with an upward boundary on the magnetospheric side and a downward boundary on the ionosphere, as shown in figure 2. As with electric circuits, we are concerned with a (field aligned) electric current and an (accelerating) electric field. With electric circuits, the boundary conditions can be set equivalently as a voltage generator and an impedance in series (Thevenin circuit), or a current generator mounted in parallel with an impedance (Norton circuit). The two systems are equivalent from a computational point of view, but the physics behind them is different. In the auroral zone, a 'normal conductivity' is defined by $\vec{j}_{\parallel} = \sigma_{\parallel}\vec{E}_{\parallel}$ and $\sigma_{\parallel} \sim n_e e^2/(m_e\, \nu_{en}) \to \infty$. Because $\sigma_{\parallel} \to \infty$, it is easy to create a current and difficult to set an electric field. To prove acceleration, one needs to show the rise of parallel electric field and a kind of *finite anomalous resistivity*. Thus, a forced current is to be considered and the question is how it could trigger an accelerating electric field. We must be aware that this resistivity imposes a deceleration of *most* of the plasma and the efficient acceleration of a *minority* of particles that carry the (forced) parallel current.

First derived by Knight [8] and sketched in figure 2, the most important example in the auroral zone is based on the increase of the magnetic field amplitude at low altitude. When the particles carrying the current approach the ionosphere, the mirror effect comes into action: the particles with pitch angle higher than a critical value bounce back and cease to carry the field aligned current. Because the current is forced, an electric field is created to counteract the mirror effect. It is efficient only on the low pitch angle passing particles. These particles are accelerated and they maintain the current at low altitude below the mirror points. The relation between the electric potential drop $\Delta\,\Phi$ and the parallel to $\vec{B}$ current $j_{\|}$ (obtained from the Vlasov–Poisson equation) is

$$\text{for } 1 << \frac{e\Delta\Phi}{T_{\rm I}} \text{ and } \frac{e\Delta\Phi}{T_{\rm M}} << \frac{B_{\rm I}}{B_{\rm M}}, j_{\|} \sim -eN_{\rm M}\left(\frac{T_{\rm M}}{2\pi m_e}\right)^{1/2}\frac{e\Delta\Phi}{T_{\rm M}}, \tag{1}$$

where $B_{\rm I}$ and $B_{\rm M}$ are magnetic field amplitudes on the ionospheric and magnetospheric boundaries, $T_{\rm I}$ and $T_{\rm M}$ are the temperatures, and $N_{\rm M}$ is the magnetospheric number densities. We have there a current-potential relation that shows the existence of a finite anomalous resistivity given by the ratio $\Delta\,\Phi/j_{\|}$, or equivalently, a field aligned conductance $K = j_{\|}/\Delta\,\Phi$. The anomalous resistivity implies that the low pitch-angle particles have been accelerated and Knight suggested that it would be the cause of the auroral energetic electrons. In an example derived from standard figures, Knight showed that a $V = 6$ keV potential drop would correspond to a current density $j_{\|} \sim 2.\ 10^{-5}$ A.m^{-2}; this corresponds to $K \sim 3.\ 10^{-9}$ S.m^{-2}.

2.2. Acceleration by double layers and by waves

Data from rocket flights was used to show that, consistently with Knight's theory, the current-voltage relation is linear near and over auroral forms [9]. Measurements from Fast satellite confirmed the linear relation for upward currents (associated with auroral arcs), while it is non-linear for downward currents [10]. (Downward currents are associated with a lack of precipitated electrons and of aurora.) A quantitative comparison of the field aligned conductances K derived from Freija satellite data and of previously published values [11] was done, using three different methods of evaluation of the current density. With quiet auroral arcs, experimental values of $K \sim 10^{-9}$ S.m^{-2} fit reasonably well the value found by Knight. During substorm events, K is lower by one or two orders of magnitude. Therefore, the convergence of the magnetic field lines is favourable but not sufficient to explain the observed auroral electron energies associated with bright auroras.

Actually, Knight theory does not include the reaction of the plasma on the electric field. Experimental [12] and theoretical [13–16] works showed that when an electric potential drop appears in a plasma, it tends to concentrate over a few Debye lengths. It forms a coherent electrostatic structure, called *double layer* (DL) which is said to be *strong* when the potential drop exceeds the plasma thermal energy and *weak* otherwise. In space, the analysis of accelerated particles showed that strong and weak DL exist in the auroral zone. Strong DL support most of the potential drop; they are located at a few thousands of km above the ionosphere [17, 18]. Weak DL develop rather after acceleration, as a *consequence* of the evolution of the beams of accelerated particles [19].

Observations and models show that strong DL produce quasi-mono-energetic beams of particles [20]. On the contrary, particle acceleration by waves (see next sections) produce broadband particle distributions [21]. This is used with *in-situ* particle measurements to evaluate the proportion of acceleration by DL or by other processes. Newell *et al* [22] showed that the most significant input of energy per precipitated particles is (surprisingly) due to the diffuse auroras, that are not specifically addressed by the present paper. The second contribution to the energy flux comes from the quasi-mono-energetic beams of particles accelerated by strong DL (10% of the energy flux during quiet conditions to 15% during moderately active periods). Broadband acceleration is associated with another type of process, that is only 6% during quiet periods and rises to 13% (almost as much as DL acceleration) during active periods. Broadband acceleration by waves accounts for 28% of the particle number flux and is exceeded only by the mechanism responsible for diffuse auroras (48%).

3. Acceleration by Alfvén waves

Alfvén waves (AW) carry an oscillating FAC with an amplitude $j_{\|} = -\,i\delta B_{\perp}/\mu_0$ and a Poynting flux $S_{\|} = \delta B_{\perp}^2 V_{\rm A}\,/\,\mu_0$. Several measurements in the auroral zone identify Alfvén waves with a Poynting flux large enough to allow for the acceleration of the electrons, even for an acceleration process with an energetic efficiency of only a few percent [23–27].

But according to the linear MHD theory of shear Alfvén and magnetosonic waves the parallel electric field is null. Therefore, to see how AW can create a parallel electric field, it is necessary to consider a non homogeneous medium, or to go beyond linear MHD, or both.

Various time-scales observed in the magnetosphere are associated with AW: (i) long wavelength AW: magnetic field line resonance (FLR). Period > 100 s, associated with substorm onsets [28, 29]. In this case, the medium is clearly non-homogeneous. (ii) meso-scale AW, period < 40 s, often 1 s. They are the resonance of a limited part of a magnetic field line. They are associated with the Ionospheric Alfvénic Resonator (IAR, see below) that also invoke a non uniform medium [30, 31]. (iii) Bursty wave packets. Duration < 1 s [32–34]. (iv) Wave packets named *Solitary kinetic Alfvén Waves* (SKAW), with a duration of about 10 ms by wave packet [24]. In cases (iii) and (iv), the wavelengths are small in regards of the auroral zone dimensions and non-MHD or non-linear effects are necessary to understand how they can accelerate auroral electrons.

4. Acceleration by long wavelength AW

Linear MHD equations of the shear (no compression) oscillations of dipole magnetic field lines were derived in a MHD context [30, 31]. Plasma compression effects were taken into account [32] and Field Line Resonant solutions (FLR) were found for the fundamental mode and its harmonics [33, 34]. Ionosphere's electrodynamic properties [35, 36] and finite electron inertia [37] were included in models that can reproduce the magnetic fluctuations observed from the ground and from space [38]. The ability of the FLR to accelerate the electrons was investigated by Tikhonchuk and Rankin [39]. Their model revealed electric fields that result from the Knight effect combined with the MHD electromagnetic field of the field line oscillation. It is suggested that auroras associated with FLR could appear during the onset of substorms, when the magnetic field lines in the magnetotail are stretched by the intensification of the magnetotail transverse current [38]. These auroras, called *quiet auroras*, are static and of moderate amplitude. Later, during and after substorm break-up, smaller scale AW are involved.

5. Alfvén waves beyond MHD

In the low altitude auroral zone (see *inertial regime* in figure 1) the plasma beta is $\beta \ll m_e/m_i$. The motion of the particles in the strong magnetic field is given by the guiding-centre approximation,

$$\frac{d\vec{v}_\parallel}{dt} = +\frac{e}{m}\vec{E}_\parallel \text{ and } \vec{v}_\perp = \vec{E}\times\frac{\vec{B}}{B^2} + \frac{m}{qB^2}\frac{d\vec{E}_\perp}{dt}. \qquad (2)$$

Because of their low mass, the electrons have a high mobility along the magnetic field lines, while the ions are sensitive to the polarization drift (last term of the second equation). When the ions see the perpendicular electric field of the AW, they move accordingly across the magnetic field lines. The much lighter electrons cannot do so and a charge density establishes and causes a parallel electric field $E_\parallel$. Therefore, the electrons are accelerated along the magnetic field direction. Then [40],

$$\omega^2 = \frac{k_\parallel^2 V_A^2}{1 + k_\perp^2 c^2/\omega_p^2}, \quad \frac{E_\parallel}{E_\perp} = \frac{k_\parallel}{k_\perp}\frac{k_\perp^2 c^2/\omega_p^2}{1 + k_\perp^2 c^2/\omega_p^2}. \qquad (3)$$

At higher altitudes (see *kinetic regime* in figure 1) $1 \gg \beta \gg m_e/m_i$, the inertial effects are replaced by kinetic effects linked to finite temperature [41],

$$\omega^2 = k_\parallel^2 V_A^2\left(1 + k_\perp^2\rho_i^2\left(\frac{3}{4} + \frac{T_e}{T_i}\right)\right), \quad \frac{E_\parallel}{E_\perp} = \frac{k_\parallel}{k_\perp}\frac{T_e}{T_i}k_\perp^2\rho_i^2, \qquad (4)$$

where ρ_i is the ion Larmor radius. In summary, a parallel electric field is set as soon as $\beta < m_e/m_i$ and $k_\perp^2 c^2/\omega_p^2 \sim 1$ or $m_e/m_i < \beta << 1$ and $k_\perp^2\rho_i^2 \sim 1$. This is illustrated in figure 3. Oblique propagation is the building block of most theories of auroral acceleration by AW.

6. The ionospheric Alfvénic resonator

Lysak and Song [42] developed a theory of the resonance of a portion of field line where an AW with a given frequency from 0.1 Hz to 1 Hz and a $k_\perp$ in the inertial regime are imposed. This resonance is based notably on the fact that the Alfvén velocity has a strong vertical gradient at the ionosphere, reaches a maximum and then decreases at altitudes of the order of 1 Earth radius (R_E). This is illustrated in figure 4. The authors show that the absorption of the wave energy by the plasma is important above 6000 km: this is where particle acceleration must be expected. At lower altitudes, the wave is partly reflected and partly dissipated in the ionosphere by Joule heating. If the electrons in the magnetospheric side have a temperature exceeding a few hundreds eV, a parallel electric field of a few mV/m can be produced over large distances (about 1000 km). This model is called the ionospheric Alfvénic resonator (IAR). Contrary to FLR's, IAR's can be applied to open field lines. Indeed, satellite and ground measurements have shown that Alfvénic auroral arcs can originate in the polar cap where magnetic field lines are open [53].

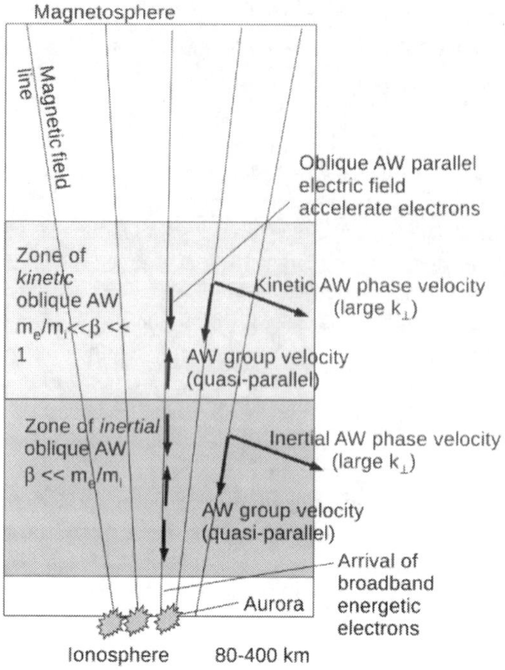

Figure 3. Schematic representation of the geometry of inertial and kinetic oblique AW in the auroral region.

Many variants of the IAR model were developed [43–46]. The AW Poynting flux transferred to the particles reaches 10% and this is compatible with the development of bright auroras [43]. Some models place the IAR at lower altitudes, below 5000 km [47]. Observational evidence of the IAR are mostly based on the detection with ground magnetometers of the resonant frequencies computed in the models [48, 49]. IAR's resonant waves are measured almost daily during local night-time and disappear during the daylight hours [50]. Measurements of IAR resonant frequencies from space have been performed, for instance, at altitudes 1700 km [51] and 500 km [52]. They become increasingly difficult with altitude; this is consistent with theoretical models that predict a decreasing wave amplitude with altitude.

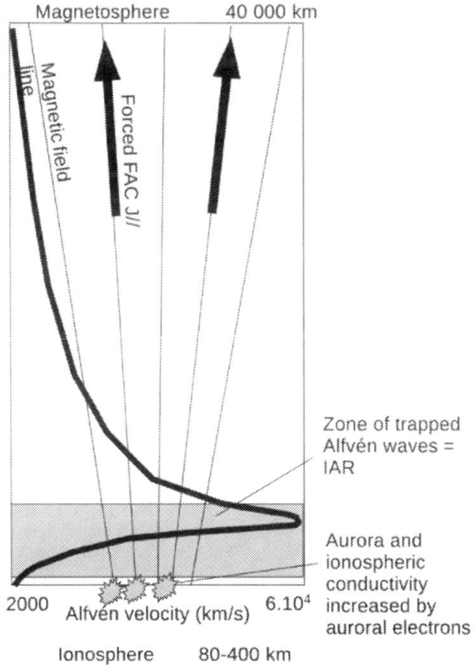

Figure 4. Schematic representation of the Ionospheric Alfvénic Resonator (IAR). The thick line represent the Alfvén velocity (horizontal axis) as a function of the altitude (vertical axis).

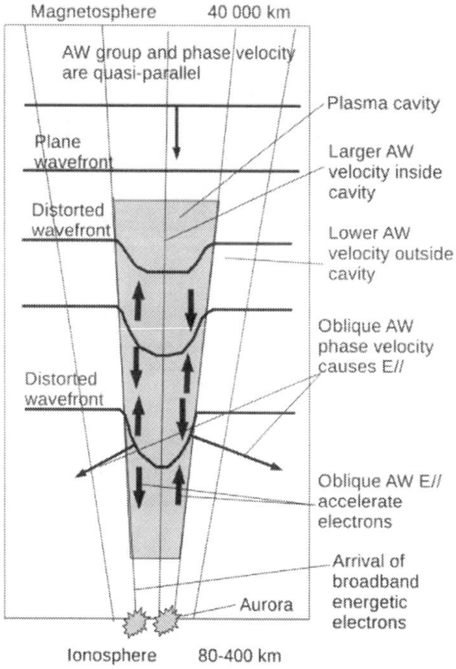

Figure 5. Schematic representation of acceleration by an AW along the density gradients of an auroral cavity.

The ionosphere is not passive in that context. The accelerated particles increase the ionospheric ionisation when they reach it. They subsequently change the current forcing conditions. An instability called the *ionospheric feedback instability* (IFI), can develop if the FAC is time dependent [54]. Because the FAC varies when it is modulated by an AW [36], the ionospheric resonant cavity was adapted to take the IFI into account [55]. The trapped AW was more amplified with the IFI. The ponderomotive force associated with the AW electric field can dig plasma cavities that reproduce what is observed in the auroral region [56]. If the ionospheric conductivity is larger that the Alfvén wave conductivity, the FAC splits in 10 km wide filaments. The filamentary structure of the FAC and subsequently of the associated accelerated particle flux is compatible with the auroral fine structures appearing during substorm expansion phase.

6.1. Acceleration by small AW packets

The above mentioned plasma cavities (they are stable [57]) can also have a role in the creation of small AW perpendicular wavelengths $k_\perp$.

Localized wave packets called *Solitary Kinetic Alfvén Wave* (SKAW) were observed in the auroral zone. They are isolated electromagnetic perturbation (observed typically during 10 ms) with the characteristic polarisation of small $k_\perp$ AW in the inertial regime $\beta < m_e/m_i$) [24, 26, 27]. Some SKAW were observed propagating along a plasma cavity with transverse (horizontal) gradients in the presence of energetic electrons. It is suggested that these electrons have been accelerated by the SKAW, as illustrated in figure 5. Let us consider a SKAW initially set with $k_\perp = 0$, propagating along the magnetic field direction. Initially a wave front is a horizontal surface. But in an auroral cavity, the magnetic field is hardly modified when the density can reach a very deep depletion [57, 59]. The transverse gradients can be very sharp, with a typical size comparable to the electron inertial length. Therefore, the Alfvén velocity $V_A = B/(\mu_0 \rho_M)^{1/2}$ where ρ_M is the mass density, increases in the low plasma density region. Then, the wave propagates faster in the cavity than outside and the wave front is distorted. It becomes oblique to the magnetic field along the transverse density gradients, with the development of small scales of the order of the electron inertial length [60]. As was shown with equation (3), when $k_\perp$ reaches this scale, the AW develops a parallel electric field and acceleration becomes possible [58]. Numerical simulations confirmed that parallel electric fields with the same wavelength as the AW develop along the density gradients. These electric fields, in the regions where they are strong, heat the electrons and some of them reach velocities greater than V_A. Then, the most energetic electrons escape from the high parallel electric field regions, forming beams [61, 62]. The (finite size) electron beams quickly generate charge separation that result in electrostatic instabilities that themselves cause a complete re-organization of the electric field with localized electric structures of high intensity. These structures were identified as weak double layers and electric solitary waves [19, 63]. They are analogous to the weak DL observed in the auroral zone in the vicinity of the auroral plasma cavities. The AW acceleration increases by one decade the flux of kinetic energy of the accelerated electrons. This flux characterizes the ability of the energetic electrons to trigger the auroral visual display, therefore the interaction of auroral cavities and SKAWs is favourable to bright polar auroras [64].

All these works are based on the existence or on the creation of small $k_\perp$ AW. Nevertheless, it was shown more recently with simulations and analytical computations that two counter propagating parallel AW packets ($k = k_\|$) with the same circular polarization crossing each other, generate a magnetic field gradient and a parallel electric field [65] that can trigger particle acceleration. Ion beams are generated as well as broadband energetic electron distributions [66]. In the auroral context, these opposite wave packets could be SKAWs before and after reflection on the ionosphere or in the IAR.

6.2. Perspectives

This introduction to auroral acceleration should convince that there is not a single cause of energetic electrons, but many. They belong to two main families: quasi-static structures and acceleration by AW waves. Concerning Earth, recent works are devoted to the relation between them. Some observations show that the ion beams that have been accelerated by strong double layers can trigger AW instabilities that in turn could generate Alfvénic electron acceleration [67–69]. On the contrary, other observations show that Alfvénic acceleration acts as a precursor of the acceleration by strong double layers [20, 70, 71].

A general overview of questions concerning observational auroral physics can be found in [72]. Review on observations and theory can be found about auroras [5], about acceleration processes in general [73] and about the role of Alfvén waves [74, 75].

Auroral acceleration is part the of science case of the 'Alfvén +' proposal to the next ESA proposal for M4 space missions. If accepted, it would be a multi-spacecraft mission launched in the auroral zone with fast and sensitive instrumentation. The last auroral mission of that type was the two spacecraft Dynamic Explorer mission, launched in 1981 and terminated more than 23 years ago.

The processes studied *in-situ* for the Earth can be relevant in other astrophysical contexts. Auroral acceleration on other magnetized planets is already investigated. Because the Jupiter (Galileo) and Saturn (Cassini) orbiters did not go across the auroral zone, the studies are based on IR to UV images [76, 77] and radio emissions [78]. For Jupiter, the decametric radio emission from the flux tube that connects Jupiter to its closest big satellite Io are interpreted as sources emitting at the local electron gyrofrequency triggered by bunches of energetic electrons [79, 80]. The characteristics of the dynamic spectrum of these emissions can be explained by the presence of quasi-monochromatic AW in the auroral region [81–83] that would accelerate electrons [84–87]. The *in-situ* exploration of the auroral zone of Jupiter with Juno spacecraft will soon be possible [88], and theoretical models already allow to speculate on the waves and acceleration processes that would be observed there [89].

References

[1] Atkinson G 1978 *J. Geophys. Res.* **83** 1089–103
[2] Sergeev V A, Mitchell D G, Russell C T and Williams D J 1993 *J. Geophys. Res.* **98** 17345–66
[3] Lui A T Y, Mankofsky A, Chang C-L, Papadopoulos K and Wu C S 1990 *Geophys. Res. Lett.* **17** 745–8
[4] Akasofu S-I and Meng C-I 1969 *J. Geophys. Res.* **74** 293–313
[5] Elphinstone R D, Murphree J S and Cogger L L 1996 *Rev. Geophys.* **34** 169–232
[6] Jones A V 1975 *Aurora* (Dordrecht: D Reidel)
Jones A V 1975 *Geophys. Astrophys. Monogr.* **9** 315
[7] Stern D P 1984 *Space Sci. Rev.* **39** 193–213
[8] Knight S 1973 *Planet. Space Sci.* **21** 741–50
[9] Lyons L R, Evans D S and Lundin R 1979 *J. Geophys. Res.* **84** 457–61
[10] Elphic R C *et al* 1998 *Geophys. Res. Lett.* **25** 2033–6
[11] Olsson A *et al* 1998 *J. Geophys. Res.* **103** 4285–302
[12] Andersson D 1981 *J. Phys.: D Appl. Phys.* **14** 1403–18
[13] Bernstein I B, Greene J M and Kruskal M D 1957 *Phys. Rev.* **108** 546–50
[14] Block L P 1972 *Earth's Magnetospheric Processes Astrophysics and Space Science Library* vol 32) (Dordrecht: D Reidel) p 258
[15] Block L P 1978 *Astrophys. Space Sci.* **55** 59–83
[16] Raadu M A 1989 *Phys. Rep.* **178** 25–97
[17] Block L P and Falthammar C-G 1990 *J. Geophys. Res.* **95** 5877–88
[18] Bruning K, Block L P, Marklund G T, Eliasson L and Pottelette R 1990 *J. Geophys. Res.* **95** 6039–49
[19] Génot V, Louarn P and Mottez F 2004 *Ann. Geophys.* **6** 2081–96
[20] Hull A J, Wilber M, Chaston C C, Bonnell J W, McFadden J P, Mozer F S, Fillingim M and Goldstein M L 2010 *J. Geophys. Res.* **115** 6211
[21] Chaston C C, Bonnell J W, Peticolas L M, Carlson C W, McFadden J P and Ergun R E 2002 *Geophys. Res. Lett.* **29** 1535
[22] Newell P T, Sotirelis T and Wing S 2009 *J. Geophys. Res.* **114** A09207
[23] Mozer F S, Cattell C A, Hudson M K, Lysak R L, Temerin M and Torbert R B 1980 *Space Sci. Rev.* **27** 155–213
[24] Louarn P, Wahlund J E, Chust T, de Feraudy H, Roux A, Holback B, Dovner P O, Eriksson A I and Holmgren G 1994 *Geophys. Res. Lett.* **21** 1847
[25] Wygant J R 2000 *J. Geophys. Res.* **105** 18675–92
[26] Volwerk M, Louarn P, Chust T, Roux A, de Feraudy H and Holback B 1996 *J. Geophys. Res.* **101** 13335–44
[27] Keiling A, Wygant J R, Cattell C, Temerin M, Mozer F S, Kletzing C A, Scudder J, Russell C T, Lotko W and Streltsov A V 2000 *Geophys. Res. Lett.* **27** 3169–72
[28] Baumjohann W and Glassmeier K-H 1984 *Planet. Space Sci.* **32** 1361–70
[29] Lessard M R, Lund E J, Kim H M, Engebretson M J and Hayashi K 2011 *J. Geophys. Res.* **116** A06203
[30] Radoski H R 1967 *J. Geophys. Res.* **72** 418
[31] Cummings W D, O'Sullivan R J and Coleman P J 1969 *J. Geophys. Res.* **74** 778–93
[32] Lee D-H and Lysak R L 1989 *J. Geophys. Res.* **941** 17097–103
[33] Southwood D J and Kivelson M G 1990 *J. Geophys. Res.* **95** 2301–9
[34] Streltsov A and Lotko W 1995 *J. Geophys. Res.* **100** 19457–72
[35] Maltsev Y P 1977 *Geomagn. Aeron.* **17** 1008–11
[36] Trakhtengertz V Y and Feldstein A Y 1984 *Planet. Space Sci.* **32** 127–34
[37] Lu J Y, Rankin R, Marchand R, Tikhonchuk V T and Wanliss J 2003 *J. Geophys. Res.* **108** 5–1

[38] Samson J C, Rankin R and Tikhonchuk V T 2003 *Ann. Geophys.* **21** 933–45
[39] Tikhonchuk V T and Rankin R 2000 *Phys. Plasmas* **7** 2630–45
[40] Goertz C K 1984 *Planet. Space Sci.* **32** 1387–92
[41] Hasegawa A and Mima K 1978 *J. Geophys. Res.* **83** 1117–23
[42] Lysak R L and Song Y 2003 *J. Geophys. Res.* **108** 9–1
[43] Lysak R L and Song Y 2005 *J. Geophys. Res.* **110** 10
[44] Clark A E and Seyler C E 1999 *J. Geophys. Res.* **104** 17233–50
[45] Watt C E J and Rankin R 2007 *Planet. Space Sci.* **55** 714–21
[46] Watt C E J and Rankin R 2007 *J. Geophys. Res.* **112** 4214
[47] Chaston C C, Bonnell J W, Carlson C W, McFadden J P, Ergun R E, Strangeway R J and Lund E J 2004 *J. Geophys. Res.* **109** 4205
[48] Belyaev P P, Bösinger T, Isaev S V and Kangas J 1999 *J. Geophys. Res.* **104** 4305–18
[49] Parent A, Mann I R and Rae I J 2010 *J. Geophys. Res.* **115** 2312
[50] Beggan C D 2014 *Ann. Geophys.* **32** 951–8
[51] Grzesiak M 2000 *Geophys. Res. Lett.* **27** 923–6
[52] Dudkin D, Pilipenko V, Korepanov V, Klimov S and Holzworth R 2014 *J. At. Sol.—Terr. Phys.* **117** 81–7
[53] Haerendel G and Frey H U 2014 *J. Geophys. Res.* **119** 2945–62
[54] Atkinson G 1970 *J. Geophys. Res.* **75** 4746–55
[55] Lysak R L 1990 *Space Sci. Rev.* **52** 33–87
[56] Lu J Y, Rankin R, Marchand R, Rae I J, Wang W, Solomon S C and Lei J 2007 *J. Geophys. Res.* **112** 10219
[57] Hilgers A, Holback B, Holmgren G and Bostrom R 1992 *J. Geophys. Res.* **97** 8631–41
[58] Génot V, Louarn P and Le Quéau D 1999 *J. Geophys. Res.* **104** 22649–56
[59] Mottez F 2003 *Phys. Plasmas* **10** 2501
[60] Heyvaerts J and Priest E R 1983 *Astron. Astrophys.* **117** 220–34
[61] Génot V, Louarn P and Mottez F 2000 *J. Geophys. Res.* **105** 27611–20
[62] Génot V, Mottez F and Louarn P 2001 *Phys. Chem. Earth* C **26** 219–22
[63] Génot V, Louarn P and Mottez F 2001 *J. Geophys. Res.* **106** 29633–44
[64] Mottez F and Génot V 2011 *J. Geophys. Res.* **116** A00K15
[65] Mottez F 2011 *Ann. Geophys.* **30** 81–95
[66] Mottez F 2014 *J. Plasma Phys.* at press
[67] Klatt E M, Kintner P M, Seyler C E, Liu K, MacDonald E A and Lynch K A 2005 *J. Geophys. Res.* **110** 10
[68] Chen L, Kletzing C A, Hu S and Bounds S R 2005 *J. Geophys. Res.* **110** 10
[69] Asamura K *et al* 2009 *Geophys. Res. Lett.* **36** 5105
[70] Newell P T, Lee A R, Liou K, Ohtani S, Sotirelis T and Wing S 2010 *J. Geophys. Res.* **115** 9226
[71] Zou S *et al* 2010 *J. Geophys. Res.* **115** 12309
[72] Berthomier M *et al* 2012 *Exp. Astron.* **33** 445–89
[73] Paschmann G, Haaland S and Treumann R 2002 *Space Sci. Rev.* **103** 169–232
[74] Keiling A 2009 *Space Sci. Rev.* **142** 73–156
[75] Mottez F 2012 *Proc. of Waves and Instabilities in Space and Astrophysical Plasmas (Eilat, Israel, 19–24 June 2011)*
[76] Prangé R, Rego D, Southwood D, Zarka P, Miller S and Ip W 1996 *Nature* **379** 323–5
[77] Bonfond B, Grodent D, Gérard J-C, Radioti A, Saur J and Jacobsen S 2008 *Geophys. Res. Lett.* **35** 5107
[78] Zarka P and Kurth W S 2005 *Space Sci. Rev.* **116** 371–97
[79] Ellis G R A 1965 *Radio Sci.* D **69** 1513–30
[80] Zarka P, Farges T, Ryabov B P, Abada-Simon M and Denis L 1996 *Geophys. Res. Lett.* **23** 125–8
[81] Ergun R E, Su Y-J, Andersson L, Bagenal F, Delemere P A, Lysak R L and Strangeway R J 2006 *J. Geophys. Res.* **111** 6212
[82] Su Y-J, Jones S T, Ergun R E, Bagenal F, Parker S E, Delamere P A and Lysak R L 2006 *J. Geophys. Res.* **111** 6211
[83] Arkhypov O V and Rucker H O 2011 *Icarus* **212** 714–8
[84] Hess S, Zarka P and Mottez F 2007 *Planet. Space Sci.* **55** 89–99
[85] Hess S, Mottez F and Zarka P 2009 *Geophys. Res. Lett.* **36** 14101
[86] Hess S, Zarka P, Mottez F and Ryabov V B 2009 *Planet. Space Sci.* **57** 23–33
[87] Hess S L G, Delamere P, Dols V, Bonfond B and Swift D 2010 *J. Geophys. Res.* **115** 6205
[88] Matousek S 2007 *Acta Astronaut.* **61** 932–9
[89] Mottez F, Hess S and Zarka P 2010 *Planet. Space Sci.* **58** 1414–22

IOP Publishing | Plasma Physics and Controlled Fusion

Plasma Phys. Control. Fusion **57** (2015) 014012 (6pp) | doi:10.1088/0741-3335/57/1/014012

The influence of finite ion temperature on plasma blob dynamics

P Manz[1,2], G Birkenmeier[1,2], D Carralero[2], G Fuchert[3], H W Müller[2], S H Müller[4], B D Scott[2], U Stroth[1,2], T T Ribeiro[2], E Wolfrum[2] and the ASDEX Upgrade Team

[1] Physik-Department E28, Technische Universität München, James-Franck-Str.1, D-85748 Garching, Germany
[2] Max-Planck-Institut für Plasmaphysik, Boltzmannstr.2, D-85748 Garching, Germany
[3] IJL, Université de Lorraine, CNRS (UMR 7198), BP 40239, Vandoeuvre-lès-Nancy, France
[4] Center for Momentum Transport and Flow Organization, University of California at San Diego, San Diego, USA

E-mail: peter.manz@ipp.mpg.de

Received 19 June 2014, revised 28 August 2014
Accepted for publication 12 September 2014
Published 28 November 2014

Abstract

In the scrape-off layer of magnetically confined fusion devices, the ion temperature is at least as high as the electron temperature and usually even much higher. The effects of the finite ion temperature enhance the blob drive and modify the vorticity. Recently developed scaling laws for blob velocity independent of its size, based on the full drift-interchange-Alfvén fluid equations are compared with recent experiments on the ASDEX Upgrade tokamak and gyrofluid simulations, showing remarkable agreement for the blob sizes and reasonable agreement for the blob velocities.

Keywords: scrape-off layer, transport, blobs, ion temperature, finite Lamor radius effects

(Some figures may appear in colour only in the online journal)

1. Introduction

One of the critical issues of magnetically confined fusion devices is the exhaustion of particles and heat without seriously damaging the vessel walls of the device. The transport in the region of open field lines beyond the confined region, called the scape-off layer (SOL), is dominated by filamentary structures also called blobs, which are elongated along the magnetic field lines and localized in the drift plane perpendicular to them. Blob motion has been and is still extensively studied in fusion devices. However, most of the blob theories and simulations invoke cold ion models, which are realistic for most basic plasma physics experiments [1–5]. Here, theory and experiments with plasma blobs seem to converge [6]. But the cold ion case is not realistic for the tokamak scrape-off layer, where typically the ion temperature T_i exceeds that of the electrons T_e [7–10]. In the divertor it can be assumed that the ion and electron temperatures are thermally coupled resulting in $\tau_i = T_i/T_e \approx 1$ in agreement with the experiments, at least in the near SOL [9]. Upstream in the divertor, in the case of a sheath connected plasma the parallel heat transport is dominated by conduction and the ion to electron temperature ratio can be roughly estimated by $\tau_i \sim (\kappa_e/\kappa_D)^{2/7} \sim 3$ [11] in agreement with the experiments [7–9]. Here $\kappa_{e,\,D}$ is the Spitzer heat conduction coefficient for electrons and ions, respectively. Towards the wall, in the far SOL, the perpendicular transport by blobs becomes more important, reducing the parallel transport by heat conduction and therefore modifying τ_i. In the far SOL $6 \leqslant \tau_i \leqslant 12$ is observed [10]. Therefore, the relevance of warm ions for the blob dynamics increases with the relevance of the blobs themselves. However, the SOL physics in the case of warm ions relevant for fusion experiments is not yet understood and investigations of this are still rare [12–15].

2. Inertial and sheath limited regimes

2.1. Analytical theory

By magnitude estimates based on the DALF model [16, 17] which describes drift-Alfvén turbulence in toroidal geometry and therefore also considers interchange and MHD instabilities, the blob velocity can be estimated analytically independent of its size. The scaling laws have been derived from the evolution of the polarization equation, which in the sheath limited regime reads as [18]

$$\frac{\mathrm{d}\nabla_\perp^2\left(\widetilde{\phi}+\tau_i\widetilde{p}_e\right)}{\mathrm{d}t}=\frac{L_\perp}{L_\parallel}\widetilde{\phi}-(1+\tau_i)\frac{2L_\perp}{R}\frac{\partial}{\partial y}\widetilde{p}_e, \quad (1)$$

where $\widetilde{\phi}$ and $\widetilde{p}_e$ are the plasma potential and electron pressure perturbations, normalized to the background electron temperature to elementary charge ratio T_e/e and background pressure, respectively. The mean profile scale length is $L_\perp$, the parallel connection length is $L_\parallel$ and R is the curvature radius. The blob speed will be given in the sound speed $c_s=\sqrt{T_e/m_i}$, with ion mass m_i and the blob size will be given in $\rho_s=\sqrt{T_e m_i}/eB$, with magnetic field strength B. The term on the left hand side of (1) is the divergence of the polarization current and gives the evolution of the ion stream function or potential vorticity $\nabla_\perp^2\left(\widetilde{\phi}+\tau_i\widetilde{p}_e\right)$. The first and second terms on the right hand side of (1) are the sheath dissipation and the interchange forcing. The effects of finite ion temperatures are also called finite Lamor radius (FLR) effects as the thermal ion gyroradius is given by $\rho_i=\sqrt{\tau_i}\rho_s$ [13].

Both contributions $\nabla_\perp^2\widetilde{\phi}$ and $\tau_i\nabla_\perp^2\widetilde{p}_e$ to the vorticity are taken into account. By applying dimensional analysis ($\mathrm{d}/\mathrm{d}t\to \mathrm{i}\omega_b=\mathrm{i}v_b/\delta_b$) with the blob correspondence principle [6, 19] the linear instability of these fluid equations is related to the radial blob velocity and scale size by $\omega_b\to\frac{v_b}{\delta_b}$, $k_\perp\to\frac{1}{\delta_b}$ and $k_\parallel\to\frac{1}{L_\parallel}$ where ω_b is the characteristic blob frequency and $k_\perp$ and $k_\parallel$ its perpendicular and parallel wavenumber, respectively. The radial velocity is given by the $E\times B$ drift, hence $\widetilde{\phi}=-\mathrm{i}(\delta_b v_b)/(\rho_s c_s)$. After completing the square and matching the real and imaginary parts the blob velocity is given by [18]

$$\left|\frac{v_b}{c_s}\right|=\sqrt{\frac{\sqrt{f^2+g^2}-f}{2}} \quad (2)$$

where

$$g=(1+\tau_i)\frac{2\delta_b}{R}\widetilde{p}_e \quad (3)$$

is the effective gravity due the interchange forcing and

$$f=\frac{1}{4}\left(\frac{\tau_i\rho_s}{\delta_b}\widetilde{p}_e+\frac{\delta_b^3}{\rho_s^2L_\parallel}\right)^2. \quad (4)$$

The balance between the two terms gives the boundary

$$(\delta_b/\rho_s)=\sqrt[4]{\tau_i(L_\parallel/\rho_s)\widetilde{p}_e} \quad (5)$$

above which the sheath effects become important. Blobs smaller than $(\delta_b/\rho_s)^4<\tau_i(L_\parallel/\rho_s)\widetilde{p}_e$ are in the ion pressure dominated resistive ballooning (iRB) inertial regime which can also be obtained by considering the following limit case. For a moderate to high τ_i and/or strong pressure fluctuations, which certainly seems justified for a blob, the ion flow stream function is determined by the ion pressure contribution $W=\phi+\tau_ip_e\approx\tau_ip_e$ and a scaling law for blob propagation can be derived [18]

$$\frac{v_b}{c_s}=2\frac{(1+\tau_i)}{\tau_i}\left(\frac{\delta_b}{\rho_s}\right)^2\frac{\rho_s}{R} \quad (6)$$

which scales with the square of the blob size in contrast to the standard scalings [20, 21].

If the sheath dissipation is stronger than the ion diamagnetic contribution to the polarization, then $f=\left(\left(1/2L_\parallel\right)\left(\delta_b^3/\rho_s^2\right)\right)^2$. There are two subregimes of the sheath affected regime. The boundary is given by $|g|=|f|$ or

$$(\delta_b/\rho_s)\approx\sqrt[5]{8(1+\tau_i)\widetilde{p}_eL_\parallel^2/(\rho_sR)}. \quad (7)$$

Commonly this boundary is identified with the typical blob size, where an additional factor of $\sqrt[5]{8(1+\tau_i)}$ (between 1.5 for $\tau_i=1$ and 2.4 for $\tau_i=10$) compared to the standard blob size given by [19]

$$(\delta_b/\rho_s)\approx\sqrt[5]{\widetilde{p}_eL_\parallel^2/(\rho_sR)} \quad (8)$$

is obtained. Recent investigations in ALCATOR C-Mod show that the typical blob size is displaced by this factor [22]. For $|g|\gg|f|$, which holds for large blobs $(\delta_b/\rho_s)^3\gg\tau_i^2R\widetilde{p}_e/(8(1+\tau_i)\rho_s)$, besides a factor of $\sqrt{(1+\tau_i)/2}$, the resistive ballooning (RB) inertial scaling [21] is obtained

$$\left|\frac{v_b}{c_s}\right|=\sqrt{(1+\tau_i)\frac{\delta_b}{R}\widetilde{p}_e}. \quad (9)$$

The factor of $\sqrt{1+\tau_i}$ is consistent with recent numerical investigations [13]. Larger blobs will follow the Krasheninnikov scaling [20]

$$\left|\frac{v_b}{c_s}\right|=(1+\tau_i)\left(\frac{L_\parallel}{R}\right)\left(\frac{\rho_s}{\delta_b}\right)^2\widetilde{p}_e, \quad (10)$$

beside the additional drive by the factor of $(1+\tau_i)/2$. Here, this regime will be called the sheath connected (SC) regime. Therefore under the consideration of sheath dissipation, we observe three regimes: the sheath non-effected iRB (6), the interchange dominated conventional RB regime (9) and the sheath connected (SC) regime (10).

2.2. Results from GEMR simulations

Until now a few studies of the effects of finite ion temperature on blob dynamics have been carried out [12, 13, 15]. Primarily the dynamics of seeded blobs have been studied, where the blob parameters (for example size, shape, amplitude) are predefined. We will quickly summarize the results: As the ion pressure adds to the electron pressure of the interchange drive, which is responsible for charge separation, propagation and therefore for the transport capabilities of the blobs, the blobs become faster [13]. The ion temperature induces polarization currents and thus adds to the vorticity. The dipolar vortex induced by the interchange drive is due to the ion pressure. This dipolar perturbation adds to the blob pressure inducing an asymmetry in the poloidal direction, a feature also seen in simulations [12, 15]. Due to the asymmetry additional dipoles may form [15]. As the interchange drive will be stronger at the steeper flank, this part of the blob will move faster resulting in a tilt of the blob. Due to this tilt the blob velocity is no longer just in the radial direction and a large fraction of the blob velocity may be in the poloidal direction. This has also been seen in simulations [12, 13, 15].

We choose a different approach, where blobs are generated self-consistently and are not seeded. Such a simulation is provided by the 3D gyrofluid electromagnetic turbulence model GEMR [23] used here. Within this framework, experimental and modelling results have already been compared with reasonably good agreement [8, 24, 25]. A circular plasma cross-section with toroidal axisymmetry is assumed. The coordinate system is aligned with the equilibrium magnetic field. The set of grid points with a constant parallel coordinate represents the drift plane, which is perpendicular to the magnetic field. The open field lines corresponding to the SOL are implemented by the boundary conditions via a perturbed Debye sheath current [26]. The actual boundary conditions used in the simulation, which include the temperature dynamics, can be found in [27]. As GEMR is a 3D code no parallel closure is preassigned.

Although being of δ-f limited code the gradients evolve freely, as required by the strength of fluctuating dynamics in this plasma region. The profiles are included in the dependent variables and also in the polarization, therefore GEMR is a global model. Details on the self-consistent treatment of the profiles and MHD equilibrium can be found in [23]. However, the deviation from the preset background values must be small. The main consequence of the presented simulation is that the gradient lengths in the confined region and the SOL cannot differ strongly. Therefore, realistic gradients in the confined region lead to stronger gradients as well as to higher background values in the SOL. Therefore, in the present implementation, GEMR is restricted to near SOL physics when including realistic gradients in the confined region. Recently a full-f gyrofluid model was developed [28] and a first investigation of seeded blob simulation comparing global and local simulations shows higher blob amplitudes in the global simulations [29].

By means of an object recognition technique [30] blob events are tracked over several images. The radial velocity of the trajectory is the blob velocity. The blob size is the diameter of a circle of the corresponding blob area, which is not circular in general. In this presentation, the blob size and velocity of these trajectories are shown from the region from 1 cm to 2 cm outside the LCFS. The local plasma parameters are $\rho_s \approx 3.5 \cdot 10^{-4}$ m and $c_s \approx 43$ km s^{-1}, the connection length is about $L_{\|} \approx 15$ m. For a better comparison with the experiment the blob-averaged fluctuation level of the density (of about $\tilde{n}_e / \langle n \rangle (\tau_i = 0.1) = 0.22$ and $\tilde{n}_e / \langle n \rangle (\tau_i = 3) = 0.48$) is taken as the blob amplitude. The normalization is performed with the local background density $\langle n \rangle (r)$ as typically done in the experiments not with one density n_0 for all radial positions as typically done in the simulations. Taking n_0 as the LCFS value, local large fluctuations can still be small in the simulation $\tilde{n}_e / n_0 < \tilde{n}_e / \langle n \rangle$ as $n_0 > \langle n \rangle (r)$. The blob amplitudes strongly exceed the total fluctuation level of about $\tilde{n}_e / \langle n \rangle = 0.022$. For both the cold ion (figure 1) and warm ion (figure 2) cases, the simulated blob size-velocity distribution is within the limits, as shown by the dotted (iRB), dash–dotted (RB)and dotted (SC) lines. For cold ions the predicted blob size-velocity dependence agrees very well with the simulated distribution. As the ion to electron ratio increases the density gradient in the SOL becomes flatter. We

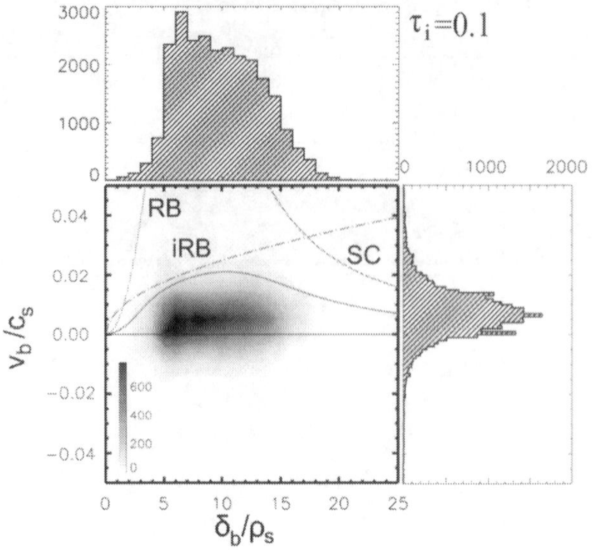

Figure 1. The centre plot shows the distribution of blob sizes $\delta_b\rho_s$ and blob velocities v_b/c_s estimated from blob trajectories in a GEMR simulation with cold ions $\tau_i = 0.1$. The prediction (equation (2) solid line) as well as the limiting regimes (iRB dotted, RB dash–dotted, SC dotted line) overlay the centre plot. At the top, the average blob size distribution is shown and on the right-hand side the average blob velocity distribution is shown.

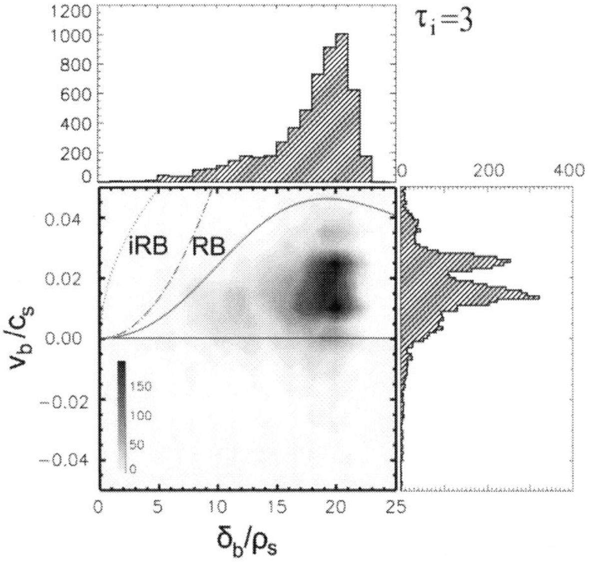

Figure 2. Same as figure 1 for warm ions $\tau_i = 3$.

observe fewer but larger and faster blobs (figure 2). As the finite ion temperature effects become important the blob size distribution changes from positive skewed to negative skewed and a bimodal velocity distribution develops. In the simulation the mean blob size changes from $\delta_b/\rho_s(\tau_i = 0.1) \approx 10$ to $\delta_b/\rho_s(\tau_i = 3) \approx 18$, where the cold ion model (8) predicts $\delta_b/\rho_s \approx 18$ for $\tau_i = 0.1$ and $\delta_b/\rho_s \approx 21$ for $\tau_i = 3$. The changes are mainly due to changes in the blob amplitudes. The warm ion predicts the blobs appearing in the RB inertial regime [18]. This is between the maximum growth rate given by $\gamma_b = v_b/\delta_b$ [18] (which is $\delta_b/\rho_s(\tau_i = 0.1) \approx 4$ and $\delta_b/\rho_s(\tau_i = 3) \approx 14$) and the boundary to the SC regime given by equation (7) (which is $\delta_b/\rho_s(\tau_i = 0.1) \approx 27$ and $\delta_b/\rho_s(\tau_i = 3) \approx 41$, respectively). The present simulations are in the inertial RB regime. For finite ion temperatures the blob size distribution is limited to a very narrow region of only a few ρ_s. This feature is often observed in experiments [22] and is a further indication that the effects of finite ion temperature play a significant role in the blob dynamics of high temperature fusion experiments.

Previously, GEMR simulations have been carried out in comparison to Alcator C-Mod, which also did not show any magnetic field dependence in the simulation or the experiments [24].

2.3. Results from ASDEX upgrade experiments

Blob velocities and sizes have been measured in low density *L*-mode discharges in ASDEX Upgrade (AUG) using lithium beam emission spectroscopy (Li-BES) [31] and in *L*- and *H*-mode with gas-puff imaging (GPI) [32]. The Li-BES measurements were performed for a wide range of B in order to investigate the ρ_s dependence. The Li-BES signal is dominantly sensitive to the electron density. The GPI can measure the spatio-temporal behaviour in more detail compared to the Li-BES with the penalty that the intensity depends on a combination of density and temperature. The results from the LiBES analysis are presented from around $\rho_{pol} = 1.044$. The maximum intensity of the gas puff was located close to $\rho_{pol} = 1.065$. The experimental results are from the far SOL. The electron temperature was measured using Thomson scattering yielding for all discharge values in the range of $T_e = 12 \pm 8$ eV at the normalized poloidal flux coordinate position chosen as the reference for the analysis of the Li-BES data [31]. Due to the lack of T_e profiles from probe measurements, T_e had to be assumed for the comparison of the GPI data with the analytical model. Here, in *L*-mode $T_e = 15 \pm 5$ eV and $T_e = 20 \pm 5$ eV in *H*-mode have been assumed [32].

To determine the blob size with Li-BES the radial HWHM of the density perturbations is taken for the blob size. Due to the finite lifetime of the Li_{2p} state the blobs appear to be smeared out in the emission response [33]. For the detailed translation to the actual blob size we refer to [31]. We compare the cold ion case ($\tau_i = 0$) to the case with $\tau_i = 3$, which agrees with previous measurements in AUG [8] and simple theoretical considerations [11]. Assuming $\tau_i = 3$ the warm ion scaling predicts the blob size (figure 3(*a*)) very well. From the GPI the blob size δ_b is determined by the poloidal semi axis of a fitted ellipse. It is also the poloidal size, which accelerates the blob. The

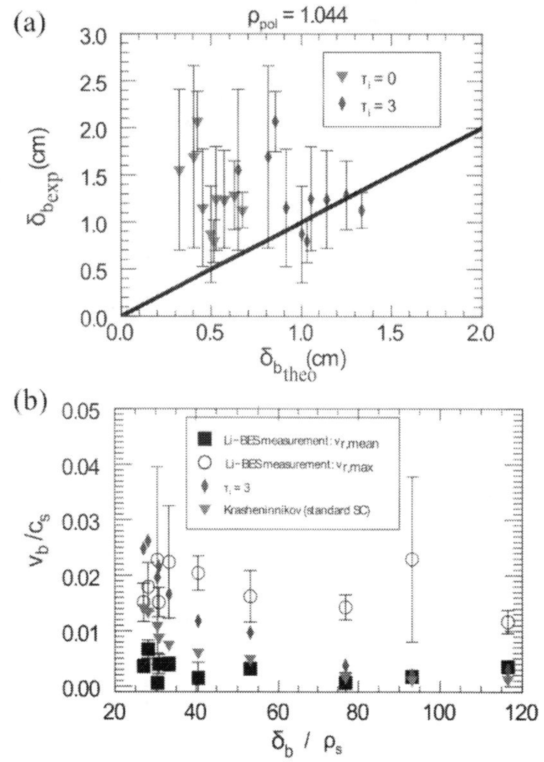

Figure 3. Characteristic blob size (*a*) and velocity (*b*) from LiBES compared with cold and warm-ion models (adapted from [31]).

GPI measurements show a better accordance with respect to the cold-ion approximation (figure 4(*a*)). For the Li-BES data the radial velocity is calculated after conditionally averaging, where the mean and maximum radial velocities are shown in figure 3(*b*). As shown in figure 3(*b*) the velocity decreases with increasing blob size indicating the sheath connected regime. The absolute values from the scaling formula agree well with the measured mean velocities for larger blob sizes ($\delta_b/\rho_s > 50$). The measured maximum velocities generally exceed the cold ion sheath-connected scaling velocities especially for large blob sizes. For the smallest blob sizes ($\delta_b/\rho_s \approx 30$), however, the scaling agrees well with the maximum velocities. Since the scaling formulae should be related to the maximum velocities rather than to the mean velocities as explained in [34], the greater agreement of the warm ion case with the measurements again points to a possible influence of finite ion temperatures on the blob dynamics. Similar to the treatment of the simulated data an object recognition method was used for the GPI data, where the blobs are tracked over several images and their radial velocity is estimated by their trajectories. In both the Li-BES and GPI the measured velocities are substantially smaller compared to the model predictions (figures 3(*b*) and 4(*b*)). Regarding the effect of uncertainties on the predictions by the model, the multiple dependencies on the electron temperature (c_s, ρ_s, τ_i) are most challenging.

Changing the toroidal magnetic field does not reveal any dependence on the blob size or velocity in the analysis of the Li-BES data, which contradicts the explicit dependence in the model. Also a strong increase in the radial velocity from *L*- to

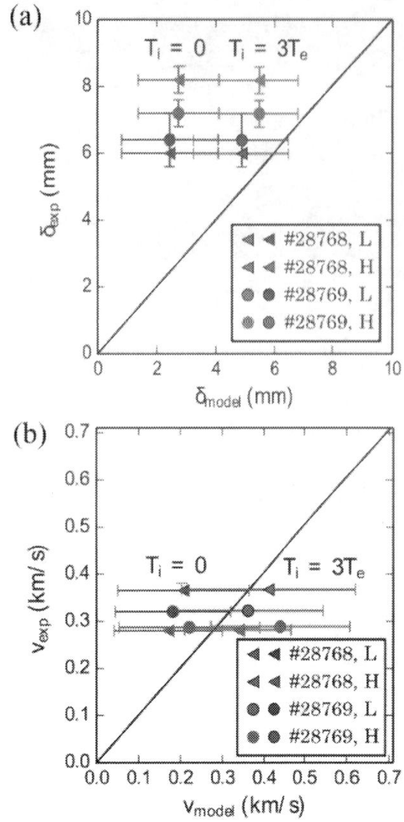

Figure 4. Characteristic blob size (*a*) and velocity (*b*) from GPI compared with cold and warm-ion models (adapted from [32]).

inter-ELM *H*-mode is expected due to the increase in the electron temperature and therefore ρ_s. This is also not observed in the GPI data.

3. Transition to other regimes

With increasing background density a transition in SOL transport can occur, manifesting itself in the development of a shoulder in the radial density profile. As proposed by Myra *et al* [35] recent experiments in ASDEX shows that this transition occurs when the collisionality $\Lambda = (\nu_{ei} L_{\parallel})/(\omega_{ce}\rho_s) > 1$ [36]. Here ω_{ce} is the electron cyclotron frequency and ν_{ei} is the electron-ion collisionality. As collisions become dominant the divertor begins to detach and an increase in both blob size and velocity is observed [36]. The dissipation mechanism does not modify the inertial regime, but in the collisional dominated regime ($\Lambda > 1$)the boundary between inertial and dissipative regime changes to [18]

$$(\delta_b/\rho_s) \approx \sqrt[5]{8(1+\tau_i)\Lambda\,\tilde{p}_e L_{\parallel}^2/(\rho_s R)}, \qquad (11)$$

which can be used to determine the blob size, which increases with Λ. Blobs larger than this limit, are in the collisional dissipative regime, where they fulfil the scaling [18]

$$\left|\frac{v_b}{c_s}\right| = (1+\tau_i)\Lambda\left(\frac{L_{\parallel}}{R}\right)\left(\frac{\rho_s}{\delta_b}\right)^2 \tilde{p}_e, \qquad (12)$$

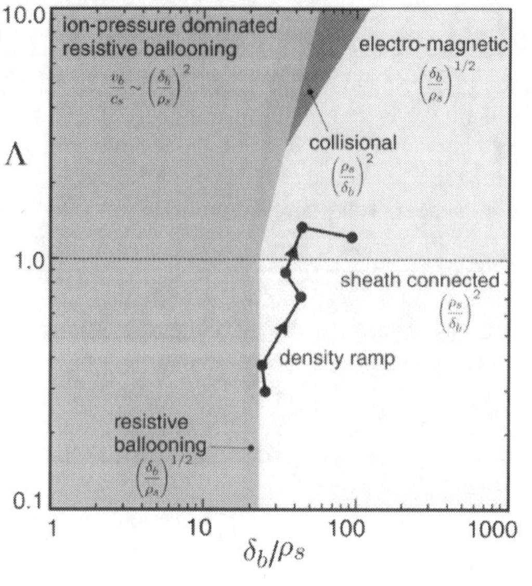

Figure 5. Limiting regimes for finite-T_i blob model as a function of collisionality Λ and blob size δ_b/ρ_s with the corresponding scaling dependences and a trace (solid line) of parameters estimated as the density increases. Plasma parameters used for this example are $\tau_i = 6$, $\rho_s = 6\cdot 10^{-4}$ m, $\tilde{p}_e = 1.5$, $\beta = 0.01$ and $L_{\parallel} = 10$ m.

also increasing with Λ. Blob sizes and collisionalities have been estimated from the midplane values of the experiments in [36]. Figure 5 shows a trace of the parameters with increasing density. A transition from the sheath-connected to the electromagnetic regime is indicated.

The electromagnetic regime is valid in finite beta turbulence ($\beta > m_e/m_i$), for blobs exceeding the velocity of $(v_b/c_s) > (R\rho_s)/(2\beta\, L_{\parallel}^2)$ with a size above $(\delta_b/\rho_s) > 2\Lambda\, L_{\parallel}/R$. As the electromagnetic terms dominate the dissipative terms the electromagnetic regime represents an inertial regime with stronger effective gravity due to the electromagnetic effects [18]

$$\left|\frac{v_b}{c_s}\right| = \sqrt{\frac{\omega_B}{2}\frac{\rho_s}{L_\perp}\left(1+\tau_i+\frac{\omega_A^2}{\omega_B^2}\right)\left(\frac{\delta_b}{\rho_s}\right)}, \qquad (13)$$

where $\omega_A^2/\omega_B^2 = R^2/\left(4\beta L_{\parallel}^2\right)$, with $\beta = (c_s/v_A)^2$ and v_A the Alfvén speed. Therefore the large blob sizes observed in the experiments at high background densities and collisionalities [36] indicate that electromagnetic effects may be important for the SOL dynamics in this regime.

4. Summary and conclusion

In response to measurements in an ASDEX Upgrade, which revealed ion temperatures close to limiters of up to 100 eV [8], an analytic model including finite ion temperature effects has been developed [18]. This has been compared in detail with experiments in the ASDEX Upgrade [31, 32] and GEMR simulations.

A direct comparison of the simulation and the experiment is difficult, as the experiments were carried out in the far SOL and the simulation was performed in the near SOL. However, both can be compared to the analytical model. By including the ion diamagnetic contribution to the polarization, increased interchange forcing and sheath dissipation, equation (2) presents a suitable boundary condition for the blob velocity independent of its size, self-consistently simulated for near SOL conditions with GEMR. The simulated data appears to be in the resistive ballooning (RB) inertial range. The analytical model can reflect the changes in the simulation, varying the ion to electron temperature ratio.

The experiments in the ASDEX Upgrade carried out with Li-BES and GPI show the same trends:

(a) The predicted blob size agrees very well taking finite ion temperatures into account. This is remarkable since rough approximations were entered into the derivation of the scaling laws and it shows the ability of the scalings to reasonably estimate the blob size.
(b) On the other hand, the velocity is overestimated. The inertial scaling appears to be an upper boundary for the radial velocity. Deviation from a pure interchange cross-phase between radial velocity and pressure fluctuations naturally develops in the experiments and simulations, which reduces the radial velocity of the blobs compared to the model predictions. An improved agreement can be achieved by including the cross-phase $v_b \rightarrow v_b \sin\left(\alpha^{\tilde{\phi},\tilde{p}_e}\right)$ (with cross-phase $\alpha^{\tilde{\phi},\tilde{p}_e}$ between potential and pressure fluctuations) as has been shown in [5].
(c) The explicit ρ_s dependence of the model does not apply to experiments nor to GEMR simulations.

According to the presented analytical model, electromagnetic terms may become dominant after the transition at high densities [36]. The simplified analytical model presented here does not take into account the real geometry of the divertor. Near the density limit at high collisionalities, the large increase in perpendicular transport as observed in [36] can lead to a thermal collapse of the SOL as predicted in the resistive X-point regime [37, 38]. In the next step, heuristic X-point boundary conditions [39] can be used to include divertor leg instabilities [39] and resistive X-point modes [39]. As the transition at high collisionalities is accompanied by detachment parallel losses by atomic processes as ionization and recombination should be taken additionally into account.

This project received funding from the Euratom research and training programme 2014–2018

References

[1] Theiler C, Furno I, Ricci P, Fasoli A, Labit B, Müller S H and Plyushev G 2009 *Phys. Rev. Lett.* **103** 065001
[2] Happel T, Greiner F, Mahdizadeh N, Nold B, Ramisch M and Stroth U 2009 *Phys. Rev. Lett.* **102** 255001
[3] Müller S H, Theiler C, Fasoli A, Furno I, Labit B, Tynan G R, Xu M, Yan Z and Yu J H 2009 *Plasma Phys. Control. Fusion* **51** 055020
[4] Furno I, Spolaore M, Theiler C, Vianello V, Cavazzana R and Fasoli A 2011 *Phys. Rev. Lett.* **106** 245001
[5] Fuchert G, Birkenmeier G, Nold B, Ramisch M and Stroth U 2013 *Plasma Phys. Control. Fusion* **55** 125002
[6] D'Ippolito D A, Myra J R and Zweben S J 2011 *Phys. Plasmas* **18** 060501
[7] Kočan M and Gunn J P 2010 *Plasma Phys. Control. Fusion* **52** 045010
[8] Kočan M, Gennrich F P, Kendl A, Müller H W and ASDEX Upgrade Team 2012 *Plasma Phys. Control. Fusion* **54** 085009
[9] Elmore S, Allan S Y, Kirk A, Fishpool G, Harrison J, Tamain P, Kočan M, Gaffka R, Stephen R and Bradley W 2012 *Plasma Phys. Control. Fusion* **54** 065001
[10] Schmid A, Herrmann A, Müller H W and the ASDEX Upgrade Team 2008 *Plasma Phys. Control. Fusion* **50** 045007
[11] Stangeby P C and Leonard A W 2011 *Nucl. Fusion* **51** 063001
[12] Jovanovic D, Shukla P K and Pegorano F 2008 *Phys. Plasmas* **15** 112305
[13] Madsen J, Garcia O E, Larsen J S, Naulin V, Nielsen A H and Rasmussen J J 2011 *Phys. Plasmas* **18** 112504
[14] Bisai N, Singh R and Kaw P K 2012 *Phys. Plasmas* **19** 052509
[15] Bisai N and Kaw P K 2013 *Phys. Plasmas* **20** 042509
[16] Scott B D 1998 *Contrib. Plasma Phys.* **38** 171
[17] Scott B 1998 *Plasma Phys. Control. Fusion* **40** 823
[18] Manz P, Carralero D, Birkenmeier G, Müller H W, Müller S H, Fuchert G, Scott B D and Stroth U 2013 *Phys. Plasmas* **20** 102307
[19] Krasheninnikov S and Smolyakov A 2008 *Phys. Plasmas* **15** 055909
[20] Krasheninnikov S 2001 *Phys. Lett.* A **283** 368
[21] Garcia O E, Bian N H and Fundamenski W 2006 *Phys. Plasmas* **13** 082309
[22] Kube R, Garcia O E, LaBombard B, Terry J L and Zweben S J 2013 *J. Nucl. Mater.* **438** S505–8
[23] Scott B D 2006 *Contrib. Plasma Phys.* **46** 714
[24] Zweben S J, Scott B D, Terry J L, LaBombard B, Hughes J W and Stotler D P 2009 *Phys. Plasmas* **16** 082505
[25] Nold B, Ribeiro T T, Ramisch M, Huang Z, Müller H W, Scott B D, Stroth U and ASDEX Upgrade Team 2012 *New. J. Phys.* **14** 063022
[26] Ribeiro T T and Scott B 2005 *Plasma Phys. Control. Fusion* **47** 1657
[27] Ribeiro T T and Scott B 2008 *Plasma Phys. Control. Fusion* **50** 055007
[28] Madsen J 2013 *Phys. Plasmas* **20** 072301
[29] Wiesenberger M, Madsen J and Kendl A 2014 *Phys. Plasmas* **21** 092301
[30] Crocker J C and Grier D G 1996 *J. Colloid Interface Sci.* **179** 298
[31] Birkenmeier G *et al* 2014 *Plasma Phys. Control. Fusion* **56** 075019
[32] Fuchert G *et al* 2014 *Plasma Phys. Control. Fusion* **56** 125001
[33] Willensdorfer M *et al* 2014 *Plasma Phys. Control. Fusion* **56** 025008
[34] Garcia O E, Horacek J, Pitts R A, Nielsen A H, Fundamenski W, Graves J P, Naulin V and Rasmussen J J 2006 *Plasma Phys. Control. Fusion* **48** L1
[35] Myra J R, Russell D A and D'Ippolito D A 2006 *Phys. Plasmas* **13** 112502
[36] Carralero D *et al* 2014 *Nucl. Fusion* **54** 123005
[37] D'Ippolito D A and Myra J R 2006 *Phys. Plasmas* **13** 062503
[38] Myra J R, D'Ippolito D A, Xu X Q and Cohen R H 2000 *Phys. Plasmas* **7** 4622
[39] Ryutov D D and Cohen R 2004 *Contrib. Plasma Phys.* **44** 168

IOP Publishing | Plasma Physics and Controlled Fusion

Plasma Phys. Control. Fusion **57** (2015) 014013 (7pp) doi:10.1088/0741-3335/57/1/014013

Kinetics of general electromagnetic fluctuations in unmagnetized plasmas: aperiodic thermal noise

R Schlickeiser[1] **and P H Yoon**[2,3]

[1] Institut für Theoretische Physik, Lehrstuhl IV: Weltraum- und Astrophysik, Ruhr-Universität Bochum, 44780 Bochum, Germany

[2] IPST, University of Maryland, College Park, MD 20742-2431, USA

[3] School of Space Research, Kyung Hee University, Yongin-Si, Gyeonggi-Do, 446-701, Korea

E-mail: rsch@tp4.rub.de

Received 18 June 2014, revised 7 September 2014
Accepted for publication 16 September 2014
Published 28 November 2014

Abstract

Any fully-ionized collisionless plasma with finite random particle velocities contains electric and magnetic field fluctuations which are of three different types: weakly damped, weakly propagating or aperiodic. The kinetics of these fluctuations in general unmagnetized plasmas is governed by the competition of spontaneous emission, absorption and stimulated emission processes. The generalized Kirchhoff laws for both collective and non-collective fluctuations are derived, which in stationary plasmas provide the equilibrium energy densities of electromagnetic fluctuations by the ratio of the respective spontaneous emission and true absorption coefficients. The equilibrium energy densities of aperiodic transverse collective electric and magnetic fluctuations in an isotropic thermal electron–proton plasma of density n_e is calculated as $|\delta B| = \sqrt{(\delta B)^2} = 2.8(n_e m_e c^2)^{1/2} g^{1/2} \beta_e^{7/4}$ and $|\delta E| = \sqrt{(\delta E)^2} = 3.2(n_e m_e c^2)^{1/2} g^{1/3} \beta_e^2$, where g and β_e denote the plasma parameter and the thermal electron velocity in units of the speed of light. For densities and temperatures of the reionized early intergalactic medium $|\delta B| = 6 \cdot 10^{-18}$ G and $|\delta E| = 2 \cdot 10^{-16}$ G result.

Keywords: waves, oscillations, instabilities in plasmas and intense beams, electromagnetic waves, emission, absorption, scattering of electromagnetic radiation

(Some figures may appear in colour only in the online journal)

1. Introduction

All plasmas, including unmagnetized and those in thermal equilibrium, have fluctuations so that their state variables such as density, pressure and electromagnetic fields fluctuate in position and time.

Because of the large sizes of astrophysical systems compared to the plasma Debye length, the fluctuations are descibed by real wave vectors $(\vec{k})$ and complex frequencies $\omega(\vec{k}) = \omega_R(\vec{k}) + i\gamma(\vec{k})$, implying for the space- and time-dependence of e.g. magnetic fluctuations the superposition of $\delta\vec{B}(\vec{x}, t) \propto \exp[i(\vec{k} \cdot \vec{x} - \omega_R t) + \gamma t]$.

One distinguishes between collective modes with a fixed dispersion relation $\omega = \omega(\vec{k})$, e.g. electromagnetic waves in vacuum $\omega_R^2 = c^2k^2$ and $\gamma = 0$, and non-collective modes with no dispersion relation $\omega = \omega(\vec{k})$. Regarding the real (ω_R) and imaginary (γ) part of the frequency, they can be characterized either as *weakly damped/amplified* wave-like modes with $|\gamma| \ll |\omega_R|$, including as limiting case undamped waves, or *weakly propagating* modes with $|\omega_R| \ll |\gamma|$ sketched in figure 1, including aperiodic fluctuations ($\omega_R = 0$), sketched in figure 2. Aperiodic modes fluctuate only in space, do not propagate as $\omega_R = 0$, but permanently grow or decrease in time depending on the sign of γ.

Because of their comparably low gas densities, all cosmic fully and partially ionized non-stellar plasmas are collision-poor, as indicated by the very small values of the plasma parameter

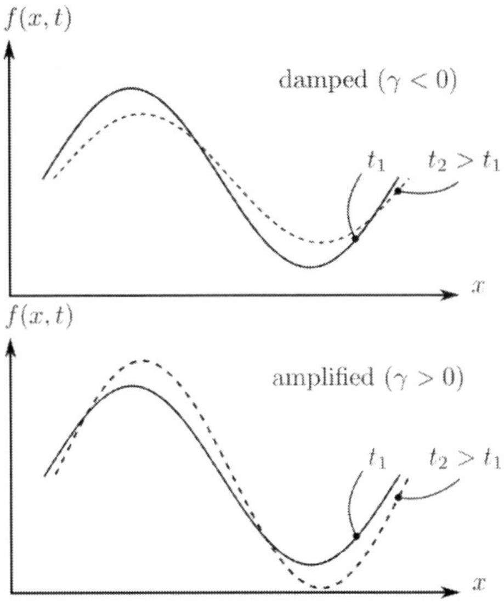

Figure 1. Sketch of weakly propagating damped (γ <0, top) and weakly propagating amplified (γ <0, bottom) fluctuation ($\omega_R \ll |\gamma|$).

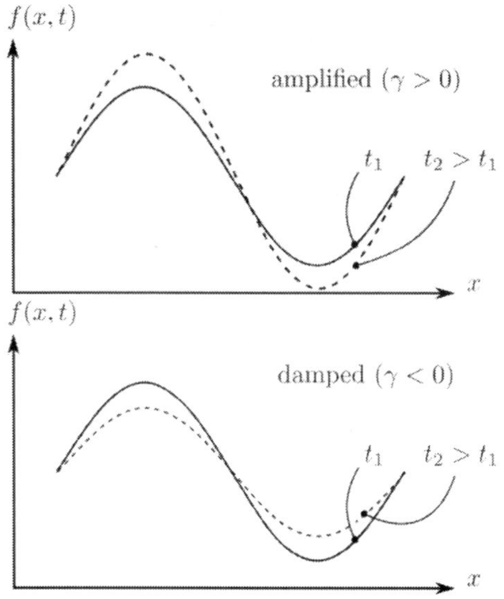

Figure 2. Sketch of aperiodic ($\omega_R = 0$) growing (γ >0, top) and aperiodic damped (γ <0, bottom) fluctuation.

$$g = \nu_{ee} / \omega_{p,e} = 7.3 \cdot 10^{-4} (n_e / \mathrm{cm}^{-3})^{1/2} (T_e / \mathrm{K})^{-3/2} \leqslant \mathcal{O}(10^{-10}), \tag{1}$$

given by the ratio of the electron–electron Coulomb collision frequency ν_{ee} to the electron plasma frequency $\omega_{p,e}$, characterizing interactions with electromagnetic fields, so that fully kinetic plasma descriptions are necessary.

Unlike for weakly amplified/damped modes, however, for aperiodic fluctuations the expected fluctuation level has never been calculated quantitatively. Only recently general expressions for the electromagnetic fluctuation spectra (electric and magnetic field, charge and current densities) from uncorrelated plasma particles in unmagnetized plasmas for arbitrary frequencies have been derived [1, 2] using the system of the Klimontovich and Maxwell equations. These general expressions are covariantly correct within the theory of special relativity, and hold for arbitrary momentum dependences of the plasma particle distribution functions and for collective and non-collective fluctuations.

2. Fluctuation theory

2.1. Basic equations

We linearize the system of Maxwell's equations and the Klimontovich equation by investigating small deviations

$$f_a(\vec{x}, \vec{p}, t) = f_{a0}(\vec{p}) + \delta f_a(\vec{x}, \vec{p}, t), \vec{B}(\vec{x}, t) = \delta\vec{B}(\vec{x}, t), \vec{E}(\vec{x}, t) = \delta\vec{E}(\vec{x}, t) \tag{2}$$

around the adopted 'equilibrium' state of a spatially uniform and stationary plasma $\left(f_{a0}(\vec{p})\right)$ with vanishing ordered magnetic and electric field ($\vec{B_0} = 0, \vec{E_0} = 0$). Small fluctuations mean that $|\delta f_a| \ll f_{a0}$. This results in the linearized equations e.g.

$$\nabla \times \delta\vec{B}(\vec{x}, t) - \frac{1}{c}\frac{\partial}{\partial t}\delta\vec{E}(\vec{x}, t) = \frac{4\pi}{c}\sum_a e_a \int \mathrm{d}^3p\, \vec{v}\, \delta f_a(\vec{x}, \vec{p}, t), \tag{3}$$

$$\left[\frac{\partial}{\partial t} + \vec{v}\cdot\frac{\partial}{\partial\vec{x}}\right]\left[\delta f_a(\vec{x}, \vec{p}, t) - \delta N_a^0(\vec{x}, \vec{p}, t)\right] = -e_a\left[\delta\vec{E} + \frac{\vec{v} \times \delta\vec{B}}{c}\right]\cdot\frac{\partial f_{a0}(\vec{p})}{\partial\vec{p}}, \tag{4}$$

where the Klimontovich perturbation $\delta N_a^0(\vec{x}, \vec{p}, t)$ accounts for the near interactions of neighbouring uncorrelated charged particles in the plasma.

We introduce the Fourier–Laplace transforms of the fluctuations e.g.

$$\delta f_a(\vec{x}, \vec{p}, t) = \frac{1}{(2\pi)^4}\int \mathrm{d}^3k \int \mathrm{d}\omega\, \delta f_a(\vec{k}, \omega, \vec{p}, t) \times \exp\left[\mathrm{i}(\vec{k}\cdot\vec{x} - \omega t)\right] \tag{5}$$

with the restriction $\Im\omega = \gamma > 0$ for convergence reasons. The temporal dependence associated with the spectral amplitude, $\delta f_a(\mathbf{k}, \omega, \vec{p}, t)$, is supposed to be weakly adiabatic. Strictly speaking, the amplitude must be independent of t, but we retain the adiabatic slow-time dependence for reasons to be explained later. This prescription is a shortcut approach to the more rigorous multiple-time scale analysis [3]. With this heuristic prescription

$$\left(\frac{\partial}{\partial t} - \mathrm{i}\omega + \mathrm{i}\mathbf{k}\cdot\mathbf{v}\right)[\delta N^a_{\mathbf{k},\omega}(\mathbf{p}, t) - \delta N^{a0}_{\mathbf{k},\omega}(\mathbf{p}, t)] = -\mathrm{e}_a n_a\left(\delta\mathbf{E}_{\mathbf{k},\omega}(t) + \frac{\mathbf{v}}{c} \times \delta\mathbf{B}_{\mathbf{k},\omega}(t)\right)\cdot\frac{\partial f_{a0}(\mathbf{p})}{\partial\mathbf{p}},$$

$$\mathrm{i}\mathbf{k} \times \delta\mathbf{B}_{\mathbf{k},\omega}(t) - \frac{1}{c}\left(\frac{\partial}{\partial t} - \mathrm{i}\omega\right)\delta\mathbf{E}_{\mathbf{k},\omega}(t) = \sum_a \frac{4\pi e_a}{c}\int \mathrm{d}^3p\, \mathbf{v}\, \delta N^a_{\mathbf{k},\omega}(\mathbf{p}, t),$$

$$\mathrm{i}\mathbf{k} \times \delta\mathbf{E}_{\mathbf{k},\omega}(t) + \frac{1}{c}\left(\frac{\partial}{\partial t} - \mathrm{i}\omega\right)\delta\mathbf{B}_{\mathbf{k},\omega}(t) = 0 \tag{6}$$

We absorb the slow-time derivative into the definition for the generalized angular frequency, $\omega \to \omega + i(\partial/\partial t)$, and shall reintroduce the slow-time derivative later. If we permit such an approach, then equations (6) reduce to a set of algebraic equations,

$$\delta \mathbf{E}^{\parallel}_{\mathbf{k},\omega} = - \sum_a \frac{4\pi i e_a}{\omega} \int d^3p \frac{\mathbf{k}(\mathbf{k}\cdot\mathbf{v})}{k^2} \delta N^a_{\mathbf{k},\omega},$$

$$\left(1 - \frac{c^2k^2}{\omega^2}\right)\delta \mathbf{E}^{\perp}_{\mathbf{k},\omega} = - \sum_a \frac{4\pi i e_a}{\omega} \int d^3p \frac{(\mathbf{k}\times\mathbf{v})\times\mathbf{k}}{k^2} \delta N^a_{\mathbf{k},\omega},$$

$$\delta N^a_{\mathbf{k},\omega} = N^{a0}_{\mathbf{k},\omega} - \frac{i e_a n_a}{\omega - \mathbf{k}\cdot\mathbf{v}} \left(\delta \mathbf{E}_{\mathbf{k},\omega} + \frac{\mathbf{v}}{\omega} \times \mathbf{k} \times \delta \mathbf{E}_{\mathbf{k},\omega}\right)\cdot \frac{\partial f_{a0}}{\partial \mathbf{p}}, \tag{7}$$

where we separated the electric and magnetic field into spectral components parallel and perpendicular to the $\mathbf{k}$-vector. Then after some straightforward algebraic manipulations under the assumption of the vanishing current density condition and gyrotropic distribution functions $f_{a0}(p_\perp, p_\parallel)$, we obtain for the ensemble averaged fluctuation spectra

$$< \delta E^2_{\parallel} >_{\mathbf{k},\omega} = \frac{<\delta E^2_{\parallel}>^0_{\mathbf{k},\omega}}{|\epsilon_\parallel(\mathbf{k},\omega)|^2}, \quad < \delta E^2_{\perp} >_{\mathbf{k},\omega} = \frac{|\omega|^4 < \delta E^2_{\perp}>^0_{\mathbf{k},\omega}}{|\omega^2\epsilon_\perp(\mathbf{k},\omega) - c^2k^2|^2},$$

$$< \delta B^2 >_{\mathbf{k},\omega} = \frac{c^2k^2|\omega|^2 < \delta E^2_{\perp}>^0_{\mathbf{k},\omega}}{|\omega^2\epsilon_\perp(\mathbf{k},\omega) - c^2k^2|^2}, \tag{8}$$

with the source terms

$$< \delta E^2_{\parallel} >^0_{\mathbf{k},\omega} = \sum_a\sum_b \frac{(4\pi)^2 e_a e_b}{|\omega|^2} \int d^3p \int d^3p' \frac{(\mathbf{k}\cdot\mathbf{v})(\mathbf{k}\cdot\mathbf{v}')}{k^2} < \delta N^0_a(\mathbf{p})\delta N^0_b(\mathbf{p}') >_{\mathbf{k},\omega},$$

$$< \delta E^2_{\perp} >^0_{\mathbf{k},\omega} = \sum_a\sum_b \frac{(4\pi)^2 e_a e_b}{|\omega|^2} \int d^3p \int d^3p' \frac{[(\mathbf{k}\times\mathbf{v})\times\mathbf{k}]}{k^2}\cdot\frac{[(\mathbf{k}\times\mathbf{v}')\times\mathbf{k}]}{k^2} <\delta N^0_a(\mathbf{p})\, \delta N^0_b(\mathbf{p}') >_{\mathbf{k},\omega}. \tag{9}$$

determined by the fluctuation of uncorrelated particles in unmagnetized plasmas $<\delta N^0_a(\mathbf{p})\ \delta N^0_b(\mathbf{p}')>_{\mathbf{k},\omega}$.

To compute the latter we consider the source fluctuation associated with the perturbed Klimontovich phase space distributions, $<\delta N^0_a(\mathbf{r}, \mathbf{p}, t)\,\delta N^0_b(\mathbf{r}', \mathbf{p}', t')> = \delta_{ab}\, n_a\, \delta[\mathbf{r} - \mathbf{r}' - \mathbf{v}(t-t')]\ \delta(\mathbf{p}-\mathbf{p}')\, f_{a0}(\mathbf{p})$. Its Fourier–Laplace transformation is

$$< \delta N^0_a(\mathbf{p}\, \delta N^0_b(\mathbf{p}') >_{\mathbf{k},\omega} = \frac{\delta_{ab}\, n_a\, \delta(\mathbf{p} - \mathbf{p}')}{(2\pi)^4} \int d(t - t') f_{a0}(\mathbf{p}) e^{-i\mathbf{k}\cdot\mathbf{v}(t-t') + i\omega(t-t')}. \tag{10}$$

If $\omega = \omega_R$ is purely real, then the range of temporal integration can be taken over $(-\infty, \infty)$, and we have the customary result, $\int_{-\infty}^{\infty} d(t-t')\, e^{-i\mathbf{k}\cdot\mathbf{v}(t-t') + i\omega(t-t')} = 2\pi\, \delta(\omega_R - \mathbf{k}\cdot\mathbf{v})$.

However, for general complex $\omega = \omega_R + i\gamma$, one must restrict the range of integration by

$$\int_0^{\infty} d(t - t') e^{-i\mathbf{k}\cdot\mathbf{v}(t-t') + i\omega(t-t')} = \frac{i}{\omega - \mathbf{k}\cdot\mathbf{v}} = \frac{i}{\omega_R + i\gamma - \mathbf{k}\cdot\mathbf{v}}, \quad (\gamma > 0), \tag{11}$$

and

$$\int_{-\infty}^{0} d(t - t') e^{-i\mathbf{k}\cdot\mathbf{v}(t-t') + i\omega(t-t')} = \frac{-i}{\omega - \mathbf{k}\cdot\mathbf{v}} = \frac{-i}{\omega_R + i\gamma - \mathbf{k}\cdot\mathbf{v}}, \quad (\gamma < 0). \tag{12}$$

Consequently, the source particle phase space density fluctuation generally becomes

$$<\delta N^0_a(\mathbf{p}\, \delta N^0_b(\mathbf{p}') >_{\mathbf{k},\omega} = \frac{i\sigma\, \delta_{ab}\, n_a\, \delta(\mathbf{p} - \mathbf{p}')\, f_{a0}(\mathbf{p})}{(2\pi)^4\, (\omega - \mathbf{k}\cdot\mathbf{v})}, \tag{13}$$

with $\sigma = \pm 1$ for $\gamma > 0$ and $\gamma < 0$, respectively. Then the source electric field fluctuation spectra (9) reduce to

$$< \delta E^2_{\parallel} >^0_{\mathbf{k},\omega} = - \text{Im} \sum_a \frac{\omega^2_{pa} m_a}{4\pi^3|\Omega|^2} \int d^3p \frac{(\mathbf{k}\cdot\mathbf{v})^2}{k^2} \frac{f_{a0}(\mathbf{p})}{\omega - \mathbf{k}\cdot\mathbf{v}},$$

$$< \delta E^2_{\perp} >^0_{\mathbf{k},\omega} = - \text{Im} \sum_a \frac{\omega^2_{pa} m_a}{4\pi^3|\Omega|^2} \int d^3p \frac{[(\mathbf{k}\times\mathbf{v})\times\mathbf{k}]^2}{k^4} \frac{f_{a0}(\mathbf{p})}{\omega - \mathbf{k}\cdot\mathbf{v}}. \tag{14}$$

These forms are particularly useful when the denominators do not become close to zero, i.e. if we consider non-collective fluctuations. However, if we consider collective longitudinal modes, defined by $\epsilon_\parallel(\mathbf{k}, \omega) = 0$, or collective transverse modes, defined by $\omega^2\, \epsilon_\perp(\mathbf{k}, \omega) - c^2k^2 = 0$, a different approach is appropriate.

3. Collective plasma eigenmodes

Suppose that the collective plasma eigenmodes are denoted by $\varpi\_k^{\parallel} = \omega\_k^{\parallel} + i\gamma\_k^{\parallel}$ and $\varpi\_k^{\perp} = \omega\_k^{\perp} + i\gamma\_k^{\perp}$, respectively, where $\epsilon_\parallel(\mathbf{k}, \varpi\_k^{\parallel}) = 0$ and $(\varpi\_k^{\perp})^2\epsilon_\perp(\mathbf{k}, \varpi\_k^{\perp}) - c^2k^2 = 0$. Strictly speaking the notion of plasma eigenmode is an approximation when one ignores the spontaneous emission. Under such an approximation, the amplitudes $\delta\mathbf{E}^{\parallel}_{\mathbf{k},\omega}$ and $\delta\mathbf{E}^{\perp}_{\mathbf{k},\omega}$ become the eigenfunctions,

$$\delta\mathbf{E}^{\parallel}_{\mathbf{k},\omega} = \delta\mathbf{E}^{\parallel}_{\mathbf{k}}\delta\left(\omega - \varpi^{\parallel}_{\mathbf{k}}\right), \quad \delta\mathbf{E}^{\perp}_{\mathbf{k},\omega} = \delta\mathbf{E}^{\perp}_{\mathbf{k}}\delta\left(\omega - \varpi^{\perp}_{\mathbf{k}}\right). \tag{15}$$

From this, the spectral wave energy densities for the longitudinal and transverse modes can be computed,

$$< \delta\mathbf{E}^{\parallel}_{\mathbf{k}}\cdot\delta\mathbf{E}^{\parallel *}_{\mathbf{k}} > = \int d\omega \int d\omega' e^{i(\omega-\omega')t} < \delta\mathbf{E}^{\parallel}_{\mathbf{k},\omega'}\cdot\delta\mathbf{E}^{\parallel *}_{\mathbf{k},\omega} > = I^{\parallel}_{\mathbf{k}}\, e^{2\gamma^{\parallel}_{\mathbf{k}} t},$$

$$< \delta\mathbf{E}^{\perp}_{\mathbf{k}}\cdot\delta\mathbf{E}^{\perp *}_{\mathbf{k}} > = \int d\omega \int d\omega' e^{i(\omega-\omega')t} < \delta\mathbf{E}^{\perp}_{\mathbf{k},\omega'}\cdot\delta\mathbf{E}^{\perp *}_{\mathbf{k},\omega} > = I^{\perp}_{\mathbf{k}}\, e^{2\gamma^{\perp}_{\mathbf{k}} t}, \tag{16}$$

From equations (16) it readily follows that

$$\left[\frac{\partial I^{\parallel}_{\mathbf{k}}}{\partial t}\right]_{\text{collective}} = 2\gamma^{\parallel}_{\mathbf{k}} I^{\parallel}_{\mathbf{k}}, \quad \left[\frac{\partial I^{\perp}_{\mathbf{k}}}{\partial t}\right]_{\text{collective}} = 2\gamma^{\perp}_{\mathbf{k}} I^{\perp}_{\mathbf{k}}. \tag{17}$$

These are the customary quasilinear wave kinetic equations in which the effects of spontaneous emission are ignored.

3.1. Spontaneous emission

In order to incorporate the effects of spontaneous emission, we first rewrite equation (8) as

$$\epsilon_{\|}(\mathbf{k},\omega) < \delta E_{\|}^2 >_{\mathbf{k},\omega} = \frac{< \delta E_{\|}^2 >^0_{\mathbf{k},\omega}}{\epsilon_{\|}^*(\mathbf{k},\omega)},$$

$$[\omega^2\epsilon_{\perp}(\mathbf{k},\omega) - c^2k^2] < \delta E_{\perp}^2 >_{\mathbf{k},\omega} = \frac{< \delta E_{\perp}^2 >^0_{\mathbf{k},\omega}}{(\omega^*)^2\epsilon_{\perp}^*(\mathbf{k},\omega) - c^2k^2}, \tag{18}$$

and reintroduce the slow-time derivative within the generalized complex angular frequency,

$$\epsilon_{\|}\left(\mathbf{k},\omega + \frac{\mathrm{i}}{2}\frac{\partial}{\partial t}\right) < \delta E_{\|}^2 >_{\mathbf{k},\omega} \simeq \epsilon_{\|}(\mathbf{k},\omega) < \delta E_{\|}^2 >_{\mathbf{k},\omega}$$
$$+ \frac{\mathrm{i}}{2}\frac{\partial\epsilon_{\|}(\mathbf{k},\omega)}{\partial\omega}\frac{\partial < \delta E_{\|}^2 >_{\mathbf{k},\omega}}{\partial t} = \frac{< \delta E_{\|}^2 >^0_{\mathbf{k},\omega}}{\epsilon_{\|}^*(\mathbf{k},\omega)}, \tag{19}$$

and

$$\left[\omega^2\,\epsilon_{\perp}\left(\mathbf{k},\omega + \tfrac{\mathrm{i}}{2}\tfrac{\partial}{\partial t}\right) - c^2k^2\right] < \delta E_{\perp}^2 >_{\mathbf{k},\omega} \simeq [\omega^2\epsilon_{\perp}(\mathbf{k},\omega)$$
$$- c^2k^2] < \delta E_{\perp}^2 >_{\mathbf{k},\omega} + \tfrac{\mathrm{i}}{2}\tfrac{\partial\omega^2\epsilon_{\perp}(\mathbf{k},\omega)}{\partial\omega}\tfrac{\partial < \delta E_{\perp}^2 >_{\mathbf{k},\omega}}{\partial t}$$
$$= \frac{|\,\omega\,|^4 < \delta E_{\perp}^2 >^0_{\mathbf{k},\omega}}{(\omega^*)^2\,\epsilon_{\perp}^*(\mathbf{k},\omega) - c^2k^2}. \tag{20}$$

Apparently, in these two equations we replaced the complex frequency by the operator only on their left-hand sides, but we ignored the slow-adiabatic time derivative in the denominator on the respective right-hand sides. This essential two-time step approximation bypasses the mathematical rigor in the full multiple-time scale analysis [3].

Since we are only interested in the collective plasma eigenmodes, we should write the wave spectral intensities as $< \delta E_{\|}^2 >_{\mathbf{k},\omega} = I_{\mathbf{k}}^{\|}\delta\left(\omega - \varpi_{\mathbf{k}}^{\|}\right)$ and $< \delta E_{\perp}^2 >_{\mathbf{k},\omega} = I_{\mathbf{k}}^{\perp}\delta\left(\omega - \varpi_{\mathbf{k}}^{\perp}\right)$. As the eigenmodes exactly satisfy the dispersion relations, terms associated with the linear responses in equations (19) and (20) disappear, leaving

$$\left[\frac{\partial I_{\mathbf{k}}^{\|}}{\partial t}\right]_{\text{spont}}\delta(\omega - \varpi_{\mathbf{k}}^{\|}) = \mathrm{Im}\frac{2\sigma}{\partial\epsilon_{\|}(\mathbf{k},\omega)/\partial\omega}\frac{<\delta E_{\|}^2>^0_{\mathbf{k},\omega}}{\epsilon_{\|}^*(\mathbf{k},\omega)},$$

$$\left[\frac{\partial I_{\mathbf{k}}^{\perp}}{\partial t}\right]_{\text{spont}}\delta(\omega - \varpi_{\mathbf{k}}^{\perp}) = \mathrm{Im}\frac{2\sigma}{\partial\omega^2\epsilon_{\perp}(\mathbf{k},\omega)/\partial\omega}\frac{|\omega|^4 < \delta E_{\perp}^2>^0_{\mathbf{k},\omega}}{(\omega^*)^2\epsilon_{\perp}^*(\mathbf{k},\omega) - c^2k^2}. \tag{21}$$

Since we already know that the wave intensity associated with the collective modes evolves according to equation (18), the above result must be interpreted as the additional source term that comes from the effects of spontaneous emission, hence, the designation on the left-hand side. The net change in the wave intensity is the sum of collective terms in equation (18) and the contribution from the spontaneously emitted power given by equation (21), $\partial I_{\vec{k}} / \partial t = \partial I_{\vec{k}} / \partial t_{\text{collective}} + \partial I_{\vec{k}} / \partial t_{\text{spont}}$.

Note that the denominators associated with the source terms in equation (21) contain linear response terms. For collective modes, with $\epsilon_{\|}(\mathbf{k},\omega) = 0$ and $\omega^2\epsilon_{\perp}(\mathbf{k},\omega) - c^2k^2 = 0$, we expand the linear response functions as

$$\epsilon_{\|}(\mathbf{k},\omega) \simeq \left(\omega - \varpi_{\mathbf{k}}^{\|}\right)\frac{\partial\epsilon_{\|}\left(\mathbf{k},\varpi_{\mathbf{k}}^{\|}\right)}{\partial\varpi_{\mathbf{k}}^{\|}},$$

$$\omega^2\epsilon_{\perp}(\mathbf{k},\omega) - c^2k^2 \simeq \left(\omega - \varpi_{\mathbf{k}}^{\perp}\right)\frac{\partial\left(\varpi_{\mathbf{k}}^{\perp}\right)^2\epsilon\left(\mathbf{k},\varpi_{\mathbf{k}}^{\perp}\right)}{\partial\varpi_{\mathbf{k}}^{\perp}}. \tag{22}$$

After inserting in equation (21) we integrate over all frequencies ω with residuum integration.

3.2. *Full kinetic equations for fluctuations*

Adding the two contributions from both the collective and spontaneous emission terms, we finally obtain the desired wave kinetic equation for collective plasma eigenmodes that includes the effects of spontaneous emission,

$$\frac{\partial I_{\mathbf{k}}^{\|}}{\partial t} = \alpha_{\|}(\mathbf{k}) - \mu_{\|}(\mathbf{k})\,I_{\mathbf{k}}^{\|},\ \frac{\partial I_{\mathbf{k}}^{\perp}}{\partial t} = \alpha_{\perp}(\mathbf{k}) - \mu_{\perp}(\mathbf{k})\,I_{\mathbf{k}}^{\perp}, \tag{23}$$

where $\alpha_{\|}(\mathbf{k})$ and $\alpha_{\perp}(\mathbf{k})$ are the spontaneous emission coefficient, while $\mu_{\|}(\mathbf{k})$ and $\mu_{\perp}(\mathbf{k})$ stand for absorption coefficients for longitudinal and transverse modes,

$$\alpha_{\|}(\mathbf{k}) = \frac{4\pi\left\langle\delta E_{\|}^2\right\rangle^0_{\mathbf{k}}}{|\partial\epsilon_{\|}(\mathbf{k},\varpi_{\mathbf{k}}^{\|})/\partial\varpi_{\mathbf{k}}^{\|}|^2},\ \alpha_{\perp}(\mathbf{k}) = \frac{4\pi\left|\varpi_{\mathbf{k}}^{\perp}\right|^4\left\langle\delta E_{\perp}^2\right\rangle^0_{\mathbf{k}}}{|\partial(\varpi_{\mathbf{k}}^{\perp})^2\epsilon_{\perp}(\mathbf{k},\varpi_{\mathbf{k}}^{\perp})/\partial\varpi_{\mathbf{k}}^{\perp}|^2}$$
$$\mu_{\|}(\mathbf{k}) = -2\gamma_{\mathbf{k}}^{\|},\ \mu_{\perp}(\mathbf{k}) = -2\gamma_{\mathbf{k}}^{\perp}, \tag{24}$$

where $\varpi\_\mathrm{k}^{\|} = \omega\_\mathrm{k}^{\|} + \mathrm{i}\gamma\_\mathrm{k}^{\|}$ and $\varpi\_\mathrm{k}^{\perp} = \omega\_\mathrm{k}^{\perp} + \mathrm{i}\gamma\_\mathrm{k}^{\perp}$ are the complex roots of the respective dispersion relations for longitudinal and transverse modes.

Equations (23) and (24) are the desired kinetic equations for collective eigenmodes in unmagnetized plasmas, which include the spontaneously emission of plasma waves and their absorption.

3.3. *Generalized Kirchhoff laws*

For stationary plasmas, with distribution functions $f_{a0}(\mathbf{p})$ independent of time, the kinetic equations (23) are solved as

$$I_{\mathrm{K}}^{\|}(\mathbf{k},t) = \frac{\alpha_{\|}(\mathbf{k})}{\mu_{\|}(\mathbf{k})}\left[1 - \mathrm{e}^{-\mu_{\|}(\mathbf{k})t}\right],\ I_{\mathrm{K}}^{\perp}(\mathbf{k},t) = \frac{\alpha_{\perp}(\mathbf{k})}{\mu_{\perp}(\mathbf{k})}\left[1 - \mathrm{e}^{-\mu_{\perp}(\mathbf{k})t}\right] \tag{25}$$

which for large times $t \gg \mu_{\|,\perp}^{-1}(\mathbf{k})$ or in steady-state situations provide the corresponding generalized Kirchhoff laws, that take into account the collective eigenmodes only, as

$$I_{\mathrm{K}}^{\|}(\mathbf{k}) = \frac{\alpha_{\|}(\mathbf{k})}{\mu_{\|}(\mathbf{k})},\ I_{\mathrm{K}}^{\perp}(\mathbf{k}) = \frac{\alpha_{\perp}(\mathbf{k})}{\mu_{\perp}(\mathbf{k})},\ I_{\mathrm{K}}^{B}(\mathbf{k}) = \frac{c^2k^2\alpha_{\perp}(\mathbf{k})}{|\varpi_{\mathbf{k}}^{\perp}|^2\mu_{\perp}(\mathbf{k})}. \tag{26}$$

3.4. *Weakly damped modes*

In the case of weakly damped modes we assume $|\omega\_\mathrm{k}| \gg |\gamma\_\mathrm{k}|$ to approximate resonant denominators by

$$\frac{1}{\varpi_{\mathbf{k}} - \mathbf{k}\cdot\mathbf{v}} \to \lim_{\gamma_{\mathbf{k}}\to 0^+}\frac{1}{\omega_{\mathbf{k}} - \mathbf{k}\cdot\mathbf{v} + \mathrm{i}\gamma_{\mathbf{k}}} = -\mathrm{i}\pi\,\delta\left(\omega_{\mathbf{k}} - \mathbf{k}\cdot\mathbf{v}\right), \tag{27}$$

and ignore all γ_k's elsewhere, except in the absorption coefficient.

As there are no weakly damped electromagnetic plasma normal modes in unmagnetized plasmas, we reproduce the customary textbook results for weakly damped longitudinal modes

$$\frac{\partial I^{\parallel}_{\mathbf{k}}}{\partial t} = -2\gamma^{\parallel}_{\mathbf{k}} I^{\parallel}_{\mathbf{k}} + \frac{4\pi < \delta E^2_{\parallel} >^0_{\mathbf{k}}}{|\partial \mathrm{Re}\, \epsilon_{\parallel}(\mathbf{k}, \omega^{\parallel}_{\mathbf{k}})/\partial\omega^{\parallel}_{\mathbf{k}}|^2}; \; \gamma^{\parallel}_{\mathbf{k}} = -\frac{\mathrm{Im}\, \epsilon_{\parallel}(\mathbf{k}, \omega^{\parallel}_{\mathbf{k}})}{\partial \mathrm{Re}\, \epsilon_{\parallel}(\mathbf{k}, \omega^{\parallel}_{\mathbf{k}})/\partial\omega^{\parallel}_{\mathbf{k}}},$$

$$\mathrm{Im}\, \epsilon_{\parallel}(\mathbf{k}, \omega^{\parallel}_{\mathbf{k}}) = -\sum_a \frac{\pi\omega^2_{pa} m_a}{k^2} \int d^3p\, \delta(\omega^{\parallel}_{\mathbf{k}} - \mathbf{k}\cdot\mathbf{v})\, \mathbf{k}\cdot\frac{\partial f_{a0}}{\partial \mathbf{p}},$$

$$\mathrm{Re}\, \epsilon_{\parallel}(\mathbf{k}, \omega^{\parallel}_{\mathbf{k}}) = 1 + \sum_a \frac{\omega^2_{pa} m_a}{k^2} P \int d^3p \frac{\mathbf{k}\cdot\partial f_{a0}/\partial\mathbf{p}}{\omega^{\parallel}_{\mathbf{k}} - \mathbf{k}\cdot\mathbf{v}},$$

$$\left\langle \delta E^2_{\parallel} \right\rangle^0_{\mathbf{k}} = \sum_a \frac{\omega^2_{pa} m_a}{4\pi^2 |\omega^{\parallel}_{\mathbf{k}}|^2 k^2} \int d^3p (\mathbf{k}\cdot\mathbf{v})^2\, \delta(\omega^{\parallel}_{\mathbf{k}} - \mathbf{k}\cdot\mathbf{v}) f_{a0}(\mathbf{p}), \tag{28}$$

where $\mathcal{P}$ denotes the principal part.

3.5. Aperiodic electromagnetic modes

For aperiodic modes we have $\omega\_\mathbf{k} = 0$ and and $\gamma\_\mathbf{k} \neq 0$. For aperiodic electromagnetic modes we obtain

$$\frac{\partial I^{\perp}_{\mathbf{k}}}{\partial t} = \alpha_{\perp}(\mathbf{k}) - \mu_{\perp}(\mathbf{k})\, I^{\perp}_{\mathbf{k}},\ \mu_{\perp}(\mathbf{k}) = -2\gamma^{\perp}_{\mathbf{k}},$$

$$\alpha_{\perp}(\mathbf{k}) = \frac{|\gamma^{\perp}_{\mathbf{k}}|^3}{|\partial(\gamma^{\perp}_{\mathbf{k}})^2 \epsilon_{\perp}(\mathbf{k}, i\gamma^{\perp}_{\mathbf{k}})/\partial\gamma^{\perp}_{\mathbf{k}}|^2} \sum_a \frac{\omega^2_{pa} m_a}{\pi^2} \int d^3p \frac{v^2_{\perp}}{\gamma^{\perp 2}_{\mathbf{k}} + k^2 v^2_{\parallel}} f_{a,0}(\mathbf{p}), \tag{29}$$

where $\mathbf{v}_{\perp} = (\mathbf{k} \times \mathbf{v}) \times \mathbf{k}/k^2, \mathbf{v}_{\parallel} = \mathbf{k}(\mathbf{k}\cdot\mathbf{v})/k^2$, and the dispersion relation for the aperiodic mode is given by

$$0 = \gamma^{\perp 2}_{\mathbf{k}} + c^2k^2 - \sum_a \omega^2_{pa} m_a \int d^3p \frac{v_{\perp}}{2}\left(\frac{\partial f_{a0}}{\partial p_{\perp}} + \frac{k v_{\perp}}{i\gamma^{\perp}_{\mathbf{k}} - k v_{\parallel}} \frac{\partial f_{a0}}{\partial p_{\parallel}}\right). \tag{30}$$

4. Thermal aperiodic noise in the IGM

The most important unmagnetized cosmic plasma is the early intergalactic medium (IGM), photoionized from the earliest generation of stars at redshift $z \simeq 20$, which transformed the Universe from darkness after recombination to light. Modeling the photoionization by the first forming stars [4, 5] indicates IGM temperatures of about $T_e = T_p = T = 10^4 T_4$ K and ionized gas densities of $n_e = 10^{-7} n_{-7}$ cm^{-3} at redshift $z = 4$[4].

As an application we calculate the electric and magnetic equilibrium spectra of aperiodic fluctuations in a thermal isotropic electron–proton plasma using the generalized Kirchhoff laws (29), accounting self-consistently for the simultaneous competition of spontaneous emission and absorption processes. The spontaneously emitted aperiodic transverse magnetic and electric field fluctuation spectra over the whole complex frequency plane for nonrelativistic plasma temperatures $u_a \ll c$ are dominated by a damped mode with negative values of γ [6]. These guaranteed magnetic fields in the form of randomly distributed fluctuations, produced by the spontaneous emission of the isotropic thermal IGM plasma, may serve as seed fields for possible amplification by later possible plasma instabilities from anisotropic plasma particle distribution functions, MHD instabilities and/or the MHD dynamo process [7].

Equation (29) readily yields at equilibrium with $\gamma = \gamma\_\mathbf{k}^{\perp}$,

$$I_{\perp}(k) = \frac{\alpha_{\perp}(\mathbf{k})}{\mu_{\perp}(\mathbf{k})} = -\frac{|\gamma|^3}{\pi^2 \gamma |M(k,\gamma)|^2} \sum_a \omega^2_{p,a} m_a \int d^3p \frac{v^2_{\perp} f_{a0}(\mathbf{p})}{\gamma^2 + k^2 v^2_{\parallel}}, \tag{31}$$

where

$$M(k,\gamma) = \frac{\partial[\gamma^2 \Lambda(k,\gamma)]}{\partial\gamma}, \Lambda(k,\gamma) = \epsilon_{\perp}(\mathbf{k}, i\gamma) + \frac{c^2k^2}{\gamma^2}. \tag{32}$$

In an isotropic thermal plasma the plasma particle distribution function is given by the Jüttner distribution

$$f_{a0}(\mathbf{p}) = \frac{\mu_a e^{-\mu_a E}}{4\pi (m_a c)^3 K_2(\mu_a)}, \tag{33}$$

where $\mu_a = m_a c^2/(k_B T_a)$, k_B being the Boltzmann constant and T_a being the kinetic temperature; $E = (m_a c)^{-1} \sqrt{m_a^2 c^2 + p_{\parallel}^2 + p_{\perp}^2}$ is the Lorentz factor; and $K_2(\mu_a)$ denotes the modified Bessel function of the second kind of order 2. For nonrelativistic plasma temperatures $\mu_a \gg 1$ the aperiodic fluctuations in an electron–proton plasma are dominated by a damped mode with

$$\gamma(\kappa) = -\eta\kappa y(\kappa),\ y(\kappa) = \frac{\kappa^2 (1 + \kappa^{4/3})}{\frac{\beta_e^2}{2\chi} + \kappa^{4/3}}, \tag{34}$$

with the dimensionless electron thermal velocity $\beta_e = u_e / c = \sqrt{2/\mu_e}$, the mass ratio $\chi^2 = m_p/m_e = 1836$, $\eta = 12.3\omega_{p,e} c / (\beta_e^{1/2} u_e)$. and the normalized wavenumber $\kappa = kc/\eta$. For this mode equation (2) becomes

$$M(\kappa) = \eta \frac{y^4 + \left(3 - \frac{\beta_e^2}{2\chi}\right) y^2 + \frac{\beta_e^2}{2\chi}}{\kappa (y^2 + 1)}, \tag{35}$$

Performing the integrations in equation (1) we obtain for the electric and magnetic fluctuation spectra

$$I_{\perp}(\kappa) \simeq \frac{\omega^2_{p,e} m_e c^2\, y^2 (1 + y^2)}{2\pi^{3/2} M^2(\kappa)} \frac{D\left(\frac{y}{\beta_e}\right)}{\frac{y}{\beta_e}}, I_B(\kappa) = \frac{I_{\perp}(\kappa)}{y^2} \tag{36}$$

where $D(x) = \exp(x^2)\mathrm{erfc}\,(x)$, given in terms of the complementary error function.

The equilibrium fluctuation spectra (36) hold for times $t > \mu_{\perp}(k)$ which for our case of aperiodic fluctuations corresponds to times

$$t > (2|\gamma(\kappa)|)^{-1} = \frac{\beta_e^2}{4\chi\eta\kappa^3}, \tag{37}$$

where we use the dispersion relation (34) at small wavenumbers. Hence during typical cosmological time scales of $t \simeq t_c = 10^{10}$ yrs magnetic fluctuations with wavenumbers

$$\kappa > \kappa_{\min} = \left(\frac{\beta_e^2}{4\chi\eta t_c}\right)^{1/3} = 2.9\cdot 10^{-11} T_4^{7/12} n_{-7}^{-1/6} \tag{38}$$

are generated, corresponding to maximum spatial magnetic turbulence scales of

$$\lambda < \frac{2\pi}{k_c \kappa_{\min}} < 2.4\cdot 10^{15} T_4^{1/6} n_{-7}^{-1/3} \mathrm{cm} \tag{39}$$

[4] We scale physical quantities as $H = H_x 10^x$.

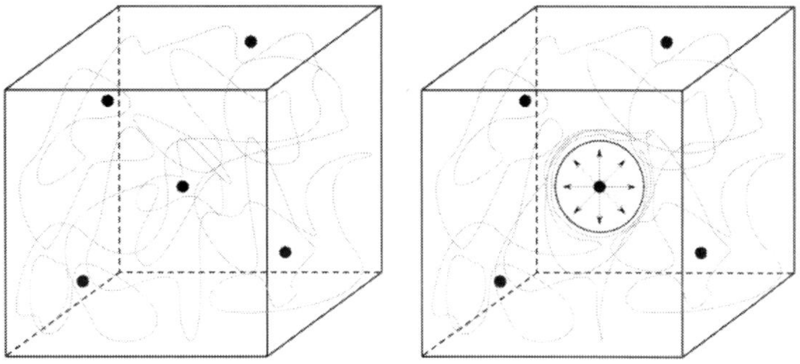

Figure 3. Illustration of the hydrodynamical stretching and ordering of cosmic magnetic fields. On the left figure a turbulent random magnetic field pervades the medium between five protostars. The right figure shows the ordering and stretching of the magnetic field as one of the stars explodes as a supernova. The outgoing shock wave compresses and orders the magnetic field in its vicinity.

By integrating equation (36) over all wavenumber values we find for the total magnetic and electric field strengths

$$|\delta B| = \sqrt{(\delta B)^2} = 2.8(n_e m_e c^2)^{1/2}\, g^{1/2} \beta_e^{7/4} = 6.3 \cdot 10^{-18} n_{-7}^{3/4} T_4^{1/8}\mathrm{G},$$
$$|\delta E| = \sqrt{(\delta E)^2} = 3.2(n_e m_e c^2)^{1/2}\, g^{1/3} \beta_e^{2} = 1.9 \cdot 10^{-16} n_{-7}^{2/3} T_4^{1/2}\mathrm{G} \quad (40)$$

These guaranteed magnetic and electric fields in the form of randomly distributed fluctuations, produced by the spontaneous emission of the isotropic thermal IGM plasma, serve as seed fields for possible amplification by later possible plasma instabilities from anisotropic plasma particle distribution functions, MHD instabilities and/or the MHD dynamo process [7]. The strength of the guaranteed spontaneously emitted magnetic seed fields (40) is significantly larger than the seed fields from the Biermann (1950) battery process [8] and cosmological phase transitions [9] (10^{-20} G). They also have a nearly 100 percent volume filling factor. These spontaneously emitted fluctuations have typical plasma scale lengths $\leqslant 10^{15} n_{-7}^{-1/3} T_4^{1/6}$ cm, but as argued below, the first hydrodynamical compression generates considerably longer correlation lengths of the compressed magnetic fields determined by the spatial scale of the compressor.

5. From disordered to ordered magnetic field structures

The unmagnetized, isotropic, thermal and steady IGM plasma by spontaneous emission generates steady tangled fields, isotropically distributed in direction, on small spatial scales $\leqslant 10^{15}$ cm. Because of its ultrahigh turbulent plasma beta value $\left(\beta_t = 10^{17} n_{-7}^{-1/2} T_4^{3/4}\right)$, these seed fields are too weak to affect the dynamics of the IGM plasma, but are tied passively to the highly conducting IGM plasma as frozen-in magnetic fluxes.

Earlier analytical considerations and numerical simulations [10–12] showed that any shear and/or compression of the IGM medium enormously amplify these seed magnetic fields and make them anisotropic. Depending on the specific exerted compression and/or shear, even 1D ordered magnetic field structures can be generated out of the original isotropically tangled field configuration. Hydrodynamical compression or shearing of the IGM medium arises from the shock waves of the supernova explosions of the first stars at the end of their lifetime, or from supersonic stellar and galactic winds. Figure 3 sketches the basic physical process.

Obviously, this magnetic field stretching and ordering occurs only in gas regions overrun frequently by shocks and winds. Each individual shock or wind (with speed V_s) compression orders the field on spatial scales R on time scales given by the short shock crossing time R/V_s, but signifant amplification requires multiple compressions. The ordered magnetic field filling factor is determined by the shock's and wind's filling factors which are large (80 percent) in the coronal phase of interstellar media [13] and near shock waves in large-scale cosmic structures [14]. In cosmic regions with high shock/wind activity, this passive hydrodynamical amplification and stretching of magnetic fields continues until the magnetic restoring forces affect the gas dynamics, i.e. at ordered plasma betas near unity. As a consequence, magnetic fields with equipartition strength are not generated uniformly over the whole universe by this process, but only in localized cosmic regions with high shock/wind activity. It continues until the magnetic restoring forces affect the gas dynamics, i.e. at ordered plasma betas near unity.

6. Summary and conclusions

Any fully-ionized collisionless plasma with finite random particle velocities contains electric and magnetic field fluctuations. The fluctuations are of three different types: weakly damped, weakly propagating or aperiodic. The kinetics of these fluctuations in general unmagnetized plasmas is governed by the competition of spontaneous emission, absorption and stimulated emission processes. The generalized Kirchhoff laws for both collective and non-collective fluctuations are derived, which in stationary plasmas provide the equilibrium energy densities of electromagnetic fluctuations by the ratio of the respective spontaneous emission and true absorption coefficients. The equilibrium energy densities of aperiodic transverse collective electric and magnetic fluctuations in an isotropic thermal electron–proton plasma of density n_e is calculated as $|\delta B| = \sqrt{(\delta B)^2} = 2.8(n_e m_e c^2)^{1/2}\, g^{1/2} \beta_e^{7/4}$

and $|\delta E| = \sqrt{(\delta E)^2} = 3.2(n_e m_e c^2)^{1/2} g^{1/3} \beta_e^2$, where g and β_e denote the plasma parameter and the thermal electron velocity in units of the speed of light. For densities and temperatures of the reionized early intergalactic medium $|\delta B| = 6 \cdot 10^{-18}$ G and $|\delta E| = 2 \cdot 10^{-16}$ G result. The electromagnetic fields serve well as cosmological seed fields for further amplification processes once additional kinetic free energy becomes available in these plasmas. The shear and/or compression of the intergalactic medium exerted by the first supernova explosions amplify these seed fields and make them anisotropic, until the magnetic restoring forces affect the gas dynamics at ordered plasma betas near unity.

Acknowledgments

R S acknowledges support by the Deutsche Forschungsgemeinschaft through grant Schl 201/25-1 within the priority programme 1573: Physics of the Interstellar Medium of the Deutsche Forschungsgemeinschaft. PHY acknowledges the support by the BK21 plus program through the National Research Foundation (NRF) funded by the Ministry of Education of Korea, and NSF grant AGS1138720 to the University of Maryland.

References

[1] Schlickeiser R and Yoon P H 2012 *Phys. Plasmas* **19** 022105
[2] Yoon P H, Schlickeiser R and Kolberg U 2014 *Phys. Plasmas* **21** 032109
[3] Davidson R C 1972 *Methods Nonlinear Plasma Theory* (New York: Academic)
[4] Hui L and Gnedin N Y 1997 *Mon. Not. R. Astron. Soc.* **292** 27
[5] Hui L and Haiman Z 2003 *Astrophys. J.* **596** 9
[6] Felten T, Schlickeiser R, Yoon P H and Lazar M 2013 *Phys. Plasmas* **20** 052113
[7] Schlickeiser R 2012 *Phys. Rev. Lett.* **109** 261101
[8] Biermann L 1950 *Z. Naturf.* A **5** 65
[9] Sigl G, Olinto A V and Jedamzik K 1997 *Phys. Rev.* D **55** 4582
[10] Laing R A 1980 *Mon. Not. R. Astron. Soc.* **193** 439
[11] Hughes P A, Aller H D and Aller M F 1985 *Astrophys. J.* **298** 301
[12] Matthews A P and Scheuer P A G 1990 *Mon. Not. R. Astron. Soc.* **242** 616
[13] McKee C F and Ostriker J P 1977 *Astrophys. J.* **218** 148
[14] Miniati F, Ryu D, Kang H, Jones T W, Cen R and Ostriker J P 2000 *Astrophys. J.* **542** 608

IOP Publishing Plasma Physics and Controlled Fusion

Plasma Phys. Control. Fusion 57 (2015) 014014 (7pp) doi:10.1088/0741-3335/57/1/014014

Flow damping in stellarators close to quasisymmetry

Iván Calvo[1], **Felix I Parra**[2,3], **José Luis Velasco**[1] **and J Arturo Alonso**[1]

[1] Laboratorio Nacional de Fusión, CIEMAT, 28040 Madrid, Spain
[2] Rudolf Peierls Centre for Theoretical Physics, University of Oxford, Oxford, OX1 3NP, UK
[3] Culham Centre for Fusion Energy, Abingdon, OX14 3DB, UK

E-mail: ivan.calvo@ciemat.es, felix.parradiaz@physics.ox.ac.uk, joseluis.velasco@ciemat.es and arturo.alonso@externos.ciemat.es

Received 27 June 2014, revised 15 September 2014
Accepted for publication 17 September 2014
Published 28 November 2014

Abstract

Quasisymmetric stellarators are a type of optimized stellarators for which flows are undamped to lowest order in an expansion in the normalized Larmor radius. However, perfect quasisymmetry is impossible. Since large flows may be desirable as a means to reduce turbulent transport, it is important to know when a stellarator can be considered to be sufficiently close to quasisymmetry. The answer to this question depends strongly on the size of the spatial gradients of the deviation from quasisymmetry and on the collisionality regime. Recently, criteria for closeness to quasisymmetry have been derived in a variety of situations. In particular, the case of deviations with large gradients was solved in the $1/\nu$ regime. Denoting by α a parameter that gives the size of the deviation from quasisymmetry, it was proven that particle fluxes do not scale with $\alpha^{3/2}$, as typically claimed, but with α. It was also shown that ripple wells are not the main cause of transport. This paper reviews those works and presents a new result in another collisionality regime, in which particles trapped in ripple wells are collisional and the rest are collisionless.

Keywords: stellarator, quasisymmetry, flow damping

(Some figures may appear in colour only in the online journal)

1. Introduction

Microturbulence is the main cause of transport in tokamaks and in the outer region of stellarator plasmas [1]. Since flow shear reduces microturbulent transport [2], it is desirable to understand under which conditions large flow shear can be achieved. Even if not the only one, a reasonable route to large flow shear consists of designing magnetic confinement devices that admit large flows or, equivalently, that have weak flow damping.

Whereas a tokamak has inherently weak damping of flows in the toroidal direction due to axisymmetry, this does not need to be the case for a stellarator, even if its magnetic configuration has been neoclassically optimized, i.e. even if it is such that collisionless trajectories are confined. A stellarator optimized for neoclassical transport is called omnigenous and its defining property is the vanishing of the average of the radial magnetic drift along the lowest order trapped orbits. The stellarator Wendelstein 7-X has been designed to be omnigenous [3]. Whereas omnigeneity is a necessary condition for weak flow damping, it is not sufficient. It has been proven [4] that there exists a direction in which flows are undamped if and only if the stellarator is quasisymmetric. A magnetic field is quasisymmetric [5, 6] when, in Boozer coordinates (see below), its magnitude depends only on a single helicity; that is, on a single linear combination of the Boozer poloidal and toroidal angles. In a quasisymmetric stellarator, the guiding-center equations are equivalent to those of a tokamak to lowest order. They possess a symmetry direction, and flows are undamped along that direction.

Technological issues place relevant bounds to how omnigeneous a real stellarator can be, but there is no theoretical reason to believe that omnigenous stellarators cannot be built. This is not the case for quasisymmetric magnetic fields. By means of an expansion in the inverse aspect ratio, Garren and Boozer proved [7] that quasisymmetry can be satisfied up to second order, but it is necessarily violated to third order in the expansion parameter. Then, it only makes sense to talk about stellarators close to quasisymmetry. It is actually possible to be close to quasisymmetry, as the existence of the HSX stellarator [8] shows. An experimental study of flow damping in HSX can be found in [9]. Let $\mathbf{B} = \mathbf{B}_0 + \alpha\, \mathbf{B}_1$ be the magnetic field, where $\mathbf{B}_0$ is quasisymmetric, $\alpha\, \mathbf{B}_1$ is the deviation from quasisymmetry and $0 \leqslant \alpha \ll 1$ is a small parameter. The theoretical question is to understand when α is sufficiently small for the stellarator to be considered quasisymmetric in practice.

The systematic approach to this question was started in references [10 and 11] by exploiting the equivalence of quasisymmetry and intrinsic ambipolarity [4], i.e. the fact that

$$\langle \mathbf{J}_n \cdot \nabla \psi \rangle_\psi = 0 \tag{1}$$

for any density, temperature and radial electric field profiles is equivalent to having no flow damping. Here, $\mathbf{J}_n$ is the neoclassical electric current density and ψ is a flux-label coordinate. In the previous equation $\langle \cdot \rangle_\psi$ denotes the flux surface average, defined as

$$\langle f \rangle_\psi = \frac{1}{V'} \int_0^{2\pi} \int_0^{2\pi} \sqrt{g} f(\psi, \Theta, \zeta)\, \mathrm{d}\Theta \mathrm{d}\zeta, \tag{2}$$

where Θ is a poloidal angle, ζ a toroidal angle, $\sqrt{g}$ is the square root of the metric determinant, $V(\psi)$ is the plasma volume enclosed by the surface labeled by ψ, and its derivative is given by

$$V'(\psi) = \int_0^{2\pi} \int_0^{2\pi} \sqrt{g}\, \mathrm{d}\Theta \mathrm{d}\zeta. \tag{3}$$

In [4] it is shown that in a stellarator far from quasisymmetry flow damping is given precisely by $\langle \mathbf{J}_n \cdot \nabla \psi \rangle_\psi$. Thus, this quantity vanishes identically for a quasisymmetric magnetic field, i.e. for $\alpha = 0$. Therefore, it is the scaling of $\langle \mathbf{J}_n \cdot \nabla \psi \rangle_\psi$ with α that needs to be determined in order to derive criteria for closeness to quasisymmetry. How the determination of the scaling allows to obtain the criterion is explained in section 2.

The calculation of the scalings, and the scalings themselves, have revealed to be quite different depending on the size of the gradients of the perturbation, and so will be the criteria derived from them. Let us give a summary of some of the results contained in [10 and 11]. From now on, we assume that (ψ, Θ, ζ) are Boozer coordinates [12]. They exist as long as $\mathbf{J} \cdot \nabla \psi = 0$, where $\mathbf{J}$ is the total electric current density, and are defined by requiring that the magnetic field can be simultaneously written as

$$\mathbf{B} = -\tilde{\eta}\, \nabla \psi + \frac{I_t(\psi)}{2\pi} \nabla \Theta + \frac{I_p(\psi)}{2\pi} \nabla \zeta \tag{4}$$

and as

$$\mathbf{B} = \frac{\Psi'_p(\psi)}{2\pi} \nabla \zeta \times \nabla \psi + \frac{\Psi'_t(\psi)}{2\pi} \nabla \psi \times \nabla \Theta. \tag{5}$$

Here, I_t and I_p are flux functions, Ψ_t is the toroidal flux, Ψ_p is the poloidal flux, $\tilde{\eta}(\psi, \Theta, \zeta)$ is a singly-valued function and primes denote differentiation with respect to ψ. Two properties of Boozer coordinates will be relevant for us. The first one is that $\sqrt{g}$ takes the form

$$\sqrt{g} = \frac{V' \langle B^2 \rangle_\psi}{4\pi^2 B^2}. \tag{6}$$

The second one is that if $\mathbf{B}$ is quasisymmetric, then its magnitude B depends on a single linear combination, or helicity, of the Boozer angles.

With the above notation, B_0 depends only on one helicity, $B_0 \equiv B_0(\psi, M\Theta - N\zeta)$, and we can take $B_1(\psi, \Theta, \zeta)$ such that it does not contain the helicity $M\Theta - N\zeta$. Then, we look at $|\partial_\Theta B_1| / |\partial_\Theta B_0|$ and $|\partial_\zeta B_1| / |\partial_\zeta B_0|$. As shown in reference [10], if

$$\frac{|\alpha \partial_\Theta B_1|}{|\partial_\Theta B_0|} \sim \alpha, \tag{7}$$

$$\frac{|\alpha \partial_\zeta B_1|}{|\partial_\zeta B_0|} \sim \alpha,$$

then

$$\langle \mathbf{J}_n \cdot \nabla \psi \rangle_\psi \sim \alpha^2 k \tag{8}$$

for any value of the collisionality (the quadratic scaling was obtained in [13] for high collisionality). The scaling of the function k with the collision frequency depends on the collisionality regime. If, on the contrary,

$$\frac{|\alpha \partial_\Theta B_1|}{|\partial_\Theta B_0|} \sim 1, \tag{9}$$

$$\frac{|\alpha \partial_\zeta B_1|}{|\partial_\zeta B_0|} \sim 1, \tag{10}$$

without further assumptions, then

$$\langle \mathbf{J}_n \cdot \nabla \psi \rangle_\psi = O(\alpha^0) \tag{11}$$

and the quasisymmetric properties of $\mathbf{B}_0$ have been completely destroyed by the perturbation. However, as explained in [11], if perturbations with large gradients are present and the extra assumption

$$v_{M,\psi} - v_{M,\psi}^{(0)} \ll v_{M,\psi}^{(0)} \tag{12}$$

is made, where $v_{M,\psi}^{(0)}$ is the radial magnetic drift corresponding to $\mathbf{B}_0$, then a scaling less favorable than α^2 but more favorable than α^0 is obtained. Hence, when designing a stellarator close to quasisymmetry large helicity perturbations should be avoided; but if this is not possible, and it probably is not, condition (12) should, in principle[4], be a design goal.

Denote by L_0 the typical variation length of B_0, $L_0^{-1} \sim |\nabla \ln B_0|$. In particular, the wells of B_0 along a magnetic field line have size L_0. It is easy to realize that when the perturbation $\alpha \mathbf{B}_1$ has strong gradients, $\mathbf{B} = \mathbf{B}_0 + \alpha\, \mathbf{B}_1$ can have, in addition, ripple wells of size $L_1 \sim |\nabla \ln B_1|^{-1}$ (see figure 1). Sometimes [15], it has been mistakenly thought that the original results of [16] apply to stellarators close to quasisymmetry. This has led to the assumption that ripple wells are the main cause of transport, and that in the so-called $1/\nu$ regime particle and energy fluxes scale with $\alpha^{3/2}$. In [11], it was shown that this is not

[4] In reference [14] it is proven that the extra assumption (12) can actually be relaxed in the $1/\nu$ regime. However, we do not have a similar proof for other collisionality regimes.

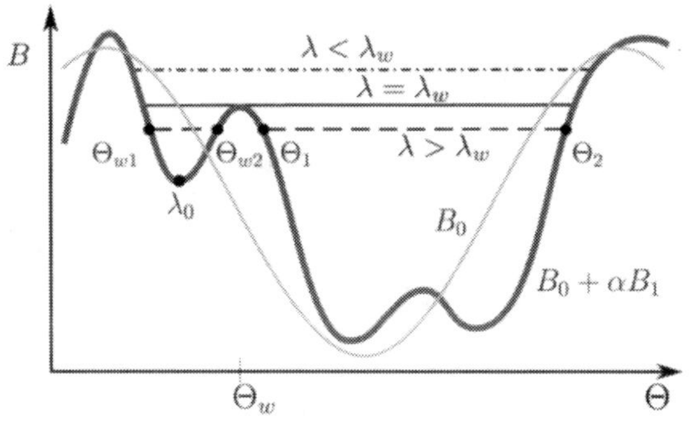

Figure 1. Magnitude of a quasisymmetric magnetic field along a field line segment (thin curve), and magnitude of the perturbed field (thick curve).

correct. A rigorous calculation was carried out assuming (9) and (12) and the result was found to be, in the $1/\nu$ regime,

$$\langle \mathbf{J}_n \cdot \nabla \psi \rangle_\psi \sim \frac{\alpha \epsilon_i^2 v_{ti}}{L_0 \nu_{ii}} e n_i v_{ti} |\nabla \psi|_0. \tag{13}$$

Furthermore, the calculation shows that ripple wells do not dominate transport. Even if they contribute to the fluxes with the scaling given on the right side of (13), the same scaling is produced by orbits trapped in the wells of size L_0; more specifically, by the part of the orbit near the bounce points. Here, v_{ti} is the ion thermal speed, n_i is the ion equilibrium density, ν_{ii} is the ion-ion collision frequency, $|\nabla\psi|_0$ is a characteristic value of $|\nabla\psi|$ on the flux surface, and $\epsilon_i := \rho_i/L_0$, where ρ_i is the ion Larmor radius. Throughout the paper the ion and electron temperatures are assumed to be comparable, and an expansion in the square root of the ratio of the electron and ion masses, $\sqrt{m_e/m_i}$, is employed.

The rest of the paper gives a brief summary of the calculations leading to the above results and presents the criteria for closeness to quasisymmetry implied by them. We also derive the scaling of the fluxes with α, and the corresponding criterion, in a new collisionality regime. In [11] passing particles, particles trapped in wells of size L_0, and particles trapped in wells of size L_1 were assumed collisionless (this is what is understood by $1/\nu$ regime). Here, we extend the computation to the case when particles in ripple wells are collisional and the rest are collisionless. Finally, we provide a treatment of the collisional boundary layers more detailed than in our previous works. The results of [10 and 11], and those presented for the first time here, also apply to tokamaks with ripple.

Before finishing this Introduction, it is pertinent to comment on the size of the plasma flows that we are dealing with. In this paper we always assume that the ion flow V_i is subsonic, i.e. $V_i \ll v_{ti}$. The presence of sonic flows, $V_i \sim v_{ti}$, immediately implies quasisymmetry [17]. Hence, the absence of damping of subsonic flows is a necessary condition to achieve sonic flows. However, it is not sufficient. It has recently been proven [18] that even in quasisymmetric stellarators there are global obstructions that forbid strictly sonic flows, but they do not exclude regimes with $\epsilon_i v_{ti} \ll V_i \ll v_{ti}$. The actual flow size that can be reached in a stellarator that is not perfectly quasisymmetric will be the subject of further research.

2. Formulation of the problem

We use the subindex σ to denote different species and define $\epsilon_\sigma := \rho_\sigma/L_0$, the Larmor radius over the macroscopic scale. We also need to define the collisionality $\nu_{*\sigma\sigma'} = \nu_{\sigma\sigma'}L_0/v_{t\sigma}$, where $\nu_{\sigma\sigma'}$ is the collision frequency of species σ with σ'. Here, $v_{t\sigma} = \sqrt{T_\sigma/m_\sigma}$, T_σ and m_σ are the thermal speed, temperature and mass of species σ. Denote by $\nu_{*\sigma}$ the largest of all collisionalities $\nu_{*\sigma\sigma'}$ when σ' runs over species. If $\epsilon_\sigma \ll \nu_{*\sigma}$ for all species, one can eliminate the degree of motion corresponding to the gyration of particles around the magnetic field by expanding the fields and the kinetic equations in $\epsilon_\sigma \ll 1$ and averaging order by order in the gyrophase.

We employ phase-space coordinates $(\psi, \Theta, \zeta, v, \lambda, s)$, where v is the magnitude of the velocity, $\lambda = B^{-1}v_\perp^2/v^2$ is the pitch-angle coordinate, $v_\perp$ is the magnitude of the velocity component perpendicular to the magnetic field, and $s = \pm 1$ is the sign of the parallel velocity $v_\parallel$, that can be expressed as

$$v_\parallel = sv\sqrt{1-\lambda B}\,. \tag{14}$$

Let us use the notation $F_\sigma = F_{\sigma 0} + F_{\sigma 1} + \ldots$, $\varphi = \varphi_0 + \varphi_1 + \ldots$ for the expansions of the distribution function and the electrostatic potential, where $F_{\sigma l}/F_{\sigma 0} = O(\epsilon_\sigma)$, $\varphi_1/\varphi_0 = O(\epsilon_i)$, $e\varphi_0/T_i = O(1)$ and e is the proton charge. To lowest order, one obtains that the distribution function is Maxwellian

$$F_{\sigma 0}(\mathbf{R}, u, \mu) = n_\sigma \left(\frac{m_\sigma}{2\pi T_\sigma}\right)^{3/2} \exp\left(-\frac{m_\sigma(u^2/2 + \mu B)}{T_\sigma}\right) \tag{15}$$

and that the density n_σ, the temperature T_σ and φ_0 are flux functions. The next order pieces $F_{\sigma l}$ and φ_1 have both neoclassical and turbulent components, but only the former, that varies in macroscopic length scales, matters for our calculation. Thus, $F_{\sigma l}$ and φ_1 stand for the neoclassical components of the corrections to $F_{\sigma 0}$ and φ_0. Denote by $G_{\sigma l} = F_{\sigma l} + (Z_\sigma e\, \varphi_1/T_\sigma)\, F_{\sigma 0}$ the non-adiabatic component of the correction to the distribution function. The equation that determines $G_{\sigma l}$ is called *the drift-kinetic equation* [19] and reads

$$v_\parallel \hat{\mathbf{b}} \cdot \nabla G_{\sigma 1} + \Upsilon_\sigma v_{\psi,\sigma} F_{\sigma 0} = C_\sigma^\ell [G_1]\,. \tag{16}$$

Here,

$$\Upsilon_\sigma := \frac{Z_\sigma e}{T_\sigma}\partial_\psi \varphi_0 + \frac{1}{n_\sigma}\partial_\psi n_\sigma + \left(\frac{m_\sigma v^2}{2T_\sigma} - \frac{3}{2}\right)\frac{1}{T_\sigma}\partial_\psi T_\sigma, \tag{17}$$

$v_{\psi,\sigma} = \mathbf{v}_{M,\sigma} \cdot \nabla \psi$ is the radial magnetic drift,

$$\mathbf{v}_{M,\sigma} = \frac{v^2(1-\lambda B)}{\Omega_\sigma}\hat{\mathbf{b}} \times (\hat{\mathbf{b}} \cdot \nabla \hat{\mathbf{b}}) + \frac{v^2 \lambda}{2\Omega_\sigma}\hat{\mathbf{b}} \times \nabla B, \tag{18}$$

$\Omega_\sigma = Z_\sigma e\, B/(m_\sigma c)$ is the gyrofrequency of species σ, $Z_\sigma e$ is the electric charge, c is the speed of light and $C_\sigma^\ell[G_1]$ is the linearized collision operator, whose explicit expression is not needed (see, for example, [20]).

The well-known neoclassical expression for the flux-surface averaged radial electric current is

$$\langle \mathbf{J}_n \cdot \nabla \psi \rangle_\psi$$

$$= \left\langle \sum_\sigma \sum_{s=-1}^{1} Z_\sigma e \int_0^\infty \mathrm{d}\, v \int_0^{B^{-1}} \mathrm{d}\lambda \frac{\pi v^2 B}{\sqrt{1-\lambda B}} v_{\psi,\sigma} G_{\sigma 1} \right\rangle_\psi . \quad (19)$$

From here on, and for the sake of simplicity, we only deal with the ion-drift kinetic equation. We note that if the ion and electron temperatures are comparable and just lowest-order terms in an expansion in $\sqrt{m_e / m_i}$ are kept, then only the ion particle flux needs to be taken into account in (19) and ion-electron collisions can be neglected.

Consider an expansion in ϵ_i of the total flux-surface averaged radial electric current, $\langle \mathbf{J} \cdot \nabla \psi \rangle_\psi$ (which, of course, vanishes due to quasineutrality). In a generic stellarator the right-hand side of (19) dominates because it scales with ϵ_i^2, whereas turbulent and higher-order neoclassical contributions to the total radial current are $O(\epsilon_i^3)$, as explained in detail in reference [10]. These $O(\epsilon_i^3)$ contributions include higher-order flow damping terms and the polarization current. Schematically,

$$\langle \mathbf{J} \cdot \nabla \psi \rangle_\psi = (\epsilon_i^2 A + \epsilon_i^3 C)\, e n_i v_{ti} |\nabla \psi|_0 + \dots \quad (20)$$

For perfectly quasisymmetric stellarators, A is identically zero. Then, for stellarators close to quasisymmetry, it is expected that $A \sim \tilde{A} \alpha^q \nu_{*i}{}^r$, with $q > 0$ and $\tilde{A} = O(1)$. Then,

$$\langle \mathbf{J} \cdot \nabla \psi \rangle_\psi = \left(\epsilon_i \frac{V_i}{v_{ti}} \tilde{A}\, \alpha^q \nu_{*i}^r + \epsilon_i^3 C \right) e n_i v_{ti} |\nabla \psi|_0 . \quad (21)$$

Note that we have replaced one instance of ϵ_i by V_i/v_{ti} in the first term of (20). We have done this just to point out that, in general, this is the actual scaling of that term; this comes from the fact that $G_{\sigma 1}$ depends linearly on φ'_0, and the latter sets the magnitude of the flow. The two terms in equation (21) are comparable when $V_i \sim \epsilon_i^2 \alpha^{-q} \nu_{*i}^{-r} v_{ti}$. The stellarator exhibits a quasisymmetric behavior if flows of size $V_i > \epsilon_i v_{ti}$ are reached, which gives the criterion

$$\alpha < \left(\epsilon_i \nu_{*i}^{-r} \right)^{1/q} . \quad (22)$$

Without further information, we have assumed $C = O(1)$ above. Our task consists of finding the powers q and r, that depend on the geometry of the deviation from quasisymmetry and on the collisionality regime.

3. Criteria for closeness to quasisymmetry

In this section we compute the scaling with α (and with ν_{*i} when the result depends on the collisionality regime) of the right side of (19). As advanced in the Introduction, this depends strongly on the size of the gradient of B_1, the magnitude of the perturbation to the quasisymmetric configuration $\mathbf{B}_0$.

3.1. Perturbations with small gradients

The calculation when (7) is satisfied was carried out in [10] and reviewed in [11]. Hence, we do not repeat it here. We simply recall that it consists of Taylor expanding the right side of (19) written in Boozer coordinates. The result is given in equation (8) and therefore the stellarator can be considered quasisymmetric if

$$\alpha < \epsilon_i^{1/2}, \quad (23)$$

for a generic value $\nu_{*i} \sim 1$. In the regime $1/\nu$ the function k appearing in (8) scales with ν_{*i}^{-1} and the criterion (23) can be more precisely written as

$$\alpha < \sqrt{\nu_{*i} \epsilon_i} . \quad (24)$$

In the next subsection we show that when B_1 has large gradients the calculation is more complicated and that the scaling with α is not quadratic, but more unfavorable.

3.2. Perturbations with large gradients

The straightforward approach of Taylor expanding the drift-kinetic equation (16) fails when the perturbation has large gradients in the sense of (9). For example, it is clear that the parallel streaming operator $s v \sqrt{1-\lambda B}\, \hat{\mathbf{b}} \cdot \nabla$ cannot be expanded at points where $\nabla B_1 \sim \alpha^{-1} L_0^{-1} B_0$.

As mentioned in the Introduction, if (9) is satisfied and no additional condition is imposed, the perturbation to the source term of the drift-kinetic equation (16), $v_{M,\psi} - v_{M,\psi}^{(0)}$, is $O(1)$ and the stellarator cannot be viewed as a perturbation of a quasisymmetric one. Hence, we assume $v_{M,\psi} - v_{M,\psi}^{(0)} \ll v_{M,\psi}^{(0)}$.

In [11] a detailed calculation of the scaling has been given in the $1/\nu$ regime, i.e. when passing particles, particles trapped in larges wells, and particles trapped in ripple wells are collisionless. Here, we will extend those results to the case when passing particles and particles trapped in larges wells are collisionless, but particles trapped in ripple wells are collisional.

It will be useful to employ $\{\psi, \chi, \Theta\}$ as spatial coordinates, with $\chi = \Theta - \iota\zeta$. The coordinate χ acts as a magnetic field line label and Θ as the coordinate that gives the position along the magnetic field line. We make $\epsilon_i \ll \nu_{*i} \ll 1$, that defines what is usually understood by *low collisionality*. The drift-kinetic equation and its solutions can then be expanded in ν_{*i}. Taking

$$G_i = G_i^{[-1]} + G_i^{[0]} + O(\nu_{*i} \epsilon_i F_{i0}) , \quad (25)$$

with $G_i^{[n]} \sim \nu_{*i}^n \epsilon_i F_{i0}$, one finds that

$$\hat{\mathbf{b}} \cdot \nabla G_i^{[-1]} = 0. \quad (26)$$

For passing particles this means that $G_i^{[-1]}$ is a flux function, whereas for trapped particles it implies that $G_i^{[-1]}$ does not depend on Θ and on the sign of the parallel velocity, s. Hence, one can write

$$G_i^{[-1]}(\psi, \chi, v, \lambda, s) = g_i(\psi, v, \lambda, s) + \partial_\chi h_i(\psi, \chi, v, \lambda) , \quad (27)$$

with $h_i \equiv 0$ in the passing region. The function $G_i^{[-1]}$ is determined by going to next order in the ν_{*i} expansion,

$$v_\| \hat{\mathbf{b}} \cdot \nabla G_i^{[0]} + \Upsilon_i v_{\psi,i} F_{i0} = C_{ii}^\ell \left[G_i^{[-1]} \right] . \quad (28)$$

For trapped trajectories, we multiply (28) by $(v_\| \hat{\mathbf{b}} \cdot \nabla \Theta)^{-1}$ and integrate over the orbit, finding the constraint

$$\oint \Upsilon_i \frac{v_{\psi,i}}{v_\| \hat{\mathbf{b}} \cdot \nabla \Theta} F_{i0} \mathrm{d}\Theta = \oint \frac{C_{ii}^\ell \left[G_i^{[-1]} \right]}{v_\| \hat{\mathbf{b}} \cdot \nabla \Theta} \mathrm{d}\Theta . \quad (29)$$

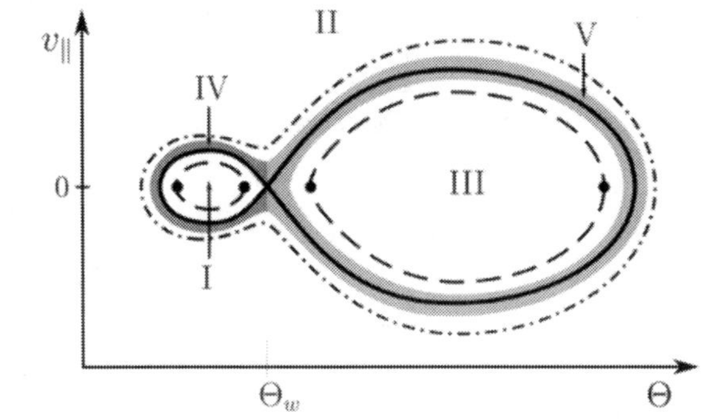

Figure 2. The different regions into which phase space is divided to solve the drift-kinetic equation.

An entropy production argument shows (see reference [11]) that $g_\sigma \equiv 0$ up to terms $O(\alpha)$. We will see below that terms $O(\alpha)$ are negligible, so that for our purposes $G_i^{[-1]}$ is zero in the passing region of velocity space and

$$\int_0^{2\pi} G_i^{[-1]} \mathrm{d}\chi = 0 \tag{30}$$

in the trapped region. Then, we focus on (29) to find the scaling of $G_i^{[-1]}$ for trapped particles. As discussed in the Introduction, when B_1 has large gradients, small wells of size $L_1 \sim \alpha\, L_0$ are typically created. Wells of size L_0 and ripple wells of size L_1 have to be treated separately. More specifically, the calculation is arranged by dividing the trapped part of phase-space into the regions depicted in figure 2. Region I corresponds to a ripple well of size L_1, and Regions II and III correspond to a well of size L_0 already present in B_0. We will give the pertinent form of the drift-kinetic equation in each region, derive the scaling of the distribution function, and then will investigate the matching conditions. We will see that Regions IV and V are collisional boundary layers that develop to heal discontinuities at the interfaces among the first three regions.

We start with particles trapped in wells of size L_0. The key is the computation of the scaling of the radial magnetic drift integrated over the trajectory. The rigorous calculation is rather complicated and is presented in detail in [11]. Here, we only state the result,

$$\oint \frac{v_{\psi,i}}{v_\parallel \hat{\mathbf{b}}\cdot\nabla\,\Theta}\mathrm{d}\Theta \sim \alpha^{1/2}\psi. \tag{31}$$

Employing (31), and the fact that in Regions II and III $\partial_\lambda \sim B_0$, it is easy to see that (29) implies

$$G_i^{\rm II} \sim \alpha^{1/2}\nu_{*i}^{-1}\epsilon_i F_{i0}, \tag{32}$$

$$G_i^{\rm III} \sim \alpha^{1/2}\nu_{*i}^{-1}\epsilon_i F_{i0} \tag{33}$$

and also

$$\partial_\lambda G_i^{\rm II} \sim \alpha^{1/2}\nu_{*i}^{-1}\epsilon_i B_0 F_{i0}, \tag{34}$$

$$\partial_\lambda G_i^{\rm III} \sim \alpha^{1/2}\nu_{*i}^{-1}\epsilon_i B_0 F_{i0}. \tag{35}$$

Observe that the size of these pieces of the distribution function justifies neglecting terms $O(\alpha)$ of the distribution function in the passing region, as advanced above.

Regions II and III give contributions to the neoclassical radial electric current

$$\langle \mathbf{J}_n\cdot\nabla\,\psi\rangle_\psi^{\rm II} \sim \frac{\alpha}{\nu_{*i}}\epsilon_i^2 e n_i v_{ti}|\nabla\,\psi|_0 \tag{36}$$

and

$$\langle \mathbf{J}_n\cdot\nabla\,\psi\rangle_\psi^{\rm III} \sim \frac{\alpha}{\nu_{*i}}\epsilon_i^2 e n_i v_{ti}|\nabla\,\psi|_0. \tag{37}$$

We turn to find the distribution function in the ripple well, Region I. We want to solve the regimes

$$\alpha^{-1/2}\nu_{*i} \ll 1, \tag{38}$$

corresponding to the case in which particles trapped in ripple wells are collisionless (the case solved in [11]), and

$$\alpha^{-1/2}\nu_{*i} \gg 1, \tag{39}$$

when those particles are collisional. For the moment, we take the maximal ordering $\alpha^{-1/2}\nu_{*i} \sim 1$. Since particles in the ripple wells are not necessarily collisionless, we cannot use (29), but we have to resort to (16). However, the latter equation can be simplified to

$$v_\parallel \hat{\mathbf{b}}\cdot\nabla\,\Theta\partial_\Theta G_i^{\rm I} - \frac{v_\parallel \nu_\lambda}{v^2 B_{\rm wmax}^2}\partial_\lambda\left(v_\parallel \partial_\lambda G_i^{\rm I}\right) = -\Upsilon_i v_{\psi,i} F_{i0}, \tag{40}$$

where $\nu_\lambda(v)$ is the pitch-angle scattering frequency. Here, we have approximated $B \approx B_{\rm wmax}$ and $\lambda \approx \lambda_w = B_{\rm wmax}^{-1}$, with λ_w defined in figure 1 along with other quantities that will be employed below. In order to get the above simplified equation, one needs the relations

$$\begin{aligned} &B_0^{-1}\partial_\lambda G_i^{\rm I} \gg v_{ti}\partial_v G_i^{\rm I} \sim v_{ti}\partial_v G_i^{\rm II} \\ &\sim v_{ti}\partial_v G_i^{\rm III} \sim B_0^{-1}\partial_\lambda G_i^{\rm II} \sim B_0^{-1}\partial_\lambda G_i^{\rm III}, \end{aligned} \tag{41}$$

obtained by using $v_\parallel \sim \alpha^{1/2} v_{ti}$, $\hat{\mathbf{b}}\cdot\nabla \sim (\alpha L_0)^{-1}$ and $\partial_\lambda \sim B_0/\alpha$.

If $\alpha^{-1/2}\nu_{*i} \ll 1$, then to lowest order $G_i^{\rm I}$ does not depend on Θ and it is determined by integrating (40) over the orbit,

$$\int_{\Theta_{w1}}^{\Theta_{w2}} \frac{\mathrm{d}\Theta}{v_\parallel \hat{\mathbf{b}}\cdot\nabla\,\Theta}\frac{v_\parallel \nu_\lambda}{v^2 B_{\rm wmax}^2}\partial_\lambda\left(v_\parallel\,\partial_\lambda G_i^{\rm I}\right) = \int_{\Theta_{w1}}^{\Theta_{w2}} \frac{\mathrm{d}\Theta}{v_\parallel \hat{\mathbf{b}}\cdot\nabla\,\Theta}\Upsilon_i v_{\psi,i} F_{i0}. \tag{42}$$

Imposing a regularity condition at the bottom of the well λ_0, we can find an explicit expression for $\partial_\lambda G_i^{\rm I}$ in the well. Using again that the size of the well in λ is $O(\alpha)$, we get

$$\partial_\lambda G_i^{\rm I} \sim \nu_{*i}^{-1}\epsilon_i F_{i0}. \tag{43}$$

If $\alpha^{-1/2}\nu_{*i} \gg 1$, one can drop the parallel streaming term in (40) to obtain

$$\frac{v_\parallel \nu_\lambda}{v^2 B_{\rm wmax}^2}\partial_\lambda\left(v_\parallel\,\partial_\lambda G_i^{\rm I}\right) = \Upsilon_i v_{\psi,i} F_{i0}. \tag{44}$$

This also yields $\partial_\lambda G_i^{\rm I} \sim \nu_{*i}^{-1}\epsilon_i F_{i0}$. In this case $\partial_\lambda G_i^{\rm I}$ can depend on Θ.

Hence, the size of $\partial_\lambda G_i^{\rm I}$ is the same for both regimes, $\alpha^{-1/2}\nu_{*i} \ll 1$ and $\alpha^{-1/2}\nu_{*i} \gg 1$. Whereas we have been able to determine the size of $\partial_\lambda G_i^{\rm I}$ by using only the drift-kinetic equation in the ripple well, the size of $G_i^{\rm I}$ is still unknown. It is determined by matching with $G_i^{\rm II}$ and $G_i^{\rm III}$, and found to be, therefore,

$$G_i^{\rm I} \sim \alpha^{1/2}\nu_{*i}^{-1}\epsilon_i F_{i0}. \quad (45)$$

Assuming a number of ripple wells per magnetic field line $O(\alpha^{-1})$, and a number of magnetic field lines with ripple wells $O(\alpha^{-1})$, we find the contribution of Region I to the flux-surface averaged radial current,

$$\langle \mathbf{J}_n\cdot\nabla\psi\rangle_\psi^{\rm I} \sim \frac{\alpha}{\nu_{*i}}\epsilon_i^2 e n_i v_{ti}|\nabla\psi|_0. \quad (46)$$

In deriving the scalings (36), (37) and (46), we have skipped the proof that the matching among the three regions can be done consistently. Actually, it cannot be done without Regions IV and V. One can show that the values of $\partial_\lambda G_i^{\rm I}$ and $\partial_\lambda G_i^{\rm II}$ do not match exactly, and the same happens with $\partial_\lambda G_i^{\rm II}$ and $\partial_\lambda G_i^{\rm III}$. The correct matching is provided by the emergence of collisional layers at the interfaces between I and II (denoted by Region IV) and II and III (denoted by Region V). A careful treatment of the collisional layers, Regions IV and V, gives

$$\langle \mathbf{J}_n\cdot\nabla\psi\rangle_\psi^{\rm IV} \sim \alpha\epsilon_i^2 e n_i v_{ti}|\nabla\psi|_0 \quad (47)$$

and

$$\langle \mathbf{J}_n\cdot\nabla\psi\rangle_\psi^{\rm V} \sim \alpha^{1/2}\epsilon_i^2 e n_i v_{ti}|\nabla\psi|_0, \quad (48)$$

where the assumption that the number of ripple wells is $O(\alpha^{-2})$ has been employed again. We postpone the details on the boundary layers to section 4. In particular, the estimations of section 4 show that the discontinuities in the absence of the collisional boundary layers are small enough that the estimates (36), (37) and (46) are correct.

It is clear that the contribution of Region IV is always negligible. When $\alpha^{-1/2}\nu_{*i} \ll 1$, Regions I, II and III contribute the same and the contribution of Region V is negligible, giving

$$\langle \mathbf{J}_n\cdot\nabla\psi\rangle_\psi \sim \frac{\alpha}{\nu_{*i}}\epsilon_i^2 e n_i v_{ti}|\nabla\psi|_0. \quad (49)$$

However, when $\alpha^{-1/2}\nu_{*i} \gg 1$, the contribution of Region V dominates and one obtains

$$\langle \mathbf{J}_n\cdot\nabla\psi\rangle_\psi \sim \frac{\alpha^{1/2}\epsilon_i^2 v_{ti}}{L_0} e n_i v_{ti}|\nabla\psi|_0. \quad (50)$$

Finally, we write the criteria to assess closeness to quasisymmetry inferred from the above results. Namely, if the perturbation has large gradients and $\alpha^{-1/2}\nu_{*i} \ll 1$, then the criterion is

$$\alpha < \nu_{*i}\epsilon_i. \quad (51)$$

If the perturbation has large helicities and $\alpha^{-1/2}\nu_{*i} \gg 1$, then the criterion is

$$\alpha < \epsilon_i^2. \quad (52)$$

One can compare these criteria with (24) and confirm that, for the same value of α, large helicity perturbations degrade more efficiently the quasisymmetric properties of the stellarator than small helicity ones, as expected.

4. The collisional boundary layers

Recall that the different regions into which we split phase space to compute the distribution function are shown in figure 2. We have explained how to determine the distribution function in Regions I, II, and III, and here we focus on the collisional boundary layers, Regions IV and V. In fact, as already pointed out, the correct treatment of the collisional boundary layers is also needed to ensure that the results on Regions I, II, and III are consistent. Below we show how to find two functions $\delta G_i^{\rm IV}$ and $\delta G_i^{\rm V}$ such that $G_i^{\rm II} + \delta G_i^{\rm IV}$ smoothly matches $G_i^{\rm I}$, and $G_i^{\rm II} + \delta G_i^{\rm V}$ smoothly matches $G_i^{\rm III} + \delta G_i^{\rm V}$. This section is based on the rigorous discussion given in [14]. In this reference, the boundary layers are calculated for a stellarator close to omnigeneity, proving that their effect on transport is negligible. As advanced in subsection 3.2, the boundary layers are more relevant in stellarators close to quasisymmetry because they can dominate flow damping in the regime in which particles trapped in ripple wells are collisional and the rest are collisionless.

Let us start by Region IV and recall the notation used in figure 1. We define a function $\delta G_i^{\rm IV}$ supported in $\lambda \in [\lambda_w - K\delta\lambda_{\rm IV}, \lambda_w]$ and $\Theta \in [\Theta_1(\lambda), \Theta_w]$. Here, $\delta\lambda_{\rm IV}$ is the characteristic width of the layer and $K \gg 1$. The equation for $\delta G_i^{\rm IV}$ is

$$v_{\|}\hat{\mathbf{b}}\cdot\nabla\Theta\partial_\Theta\delta G_i^{\rm IV} - \frac{v_{\|}\nu_\lambda}{v^2 B_{\rm wmax}^2}\partial_\lambda\left(v_{\|}\partial_\lambda\delta G_i^{\rm IV}\right) = 0, \quad (53)$$

which has to be solved with the boundary conditions $\partial_\lambda\delta G_i^{\rm IV} = \partial_\lambda G_i^{\rm I} - \partial_\lambda G_i^{\rm II}$ at $\lambda = \lambda_w$, $\delta G_i^{\rm IV} = 0$ at $\lambda = \lambda_w - K\delta\lambda_{\rm IV}$, $\delta G_i^{\rm IV} = \delta G_i^{\rm V}$ at $\Theta = \Theta_w$, the interface between the two collisional boundary layers, and $\delta G_i^{\rm IV}(s = 1) = \delta G_i^{\rm IV}(s = -1)$ at Θ_1.

If both terms in (53) have to be of the same order, and in the layer $v_{\|} \sim \alpha^{1/2} v_{ti}$ and $\hat{\mathbf{b}}\cdot\nabla \sim \alpha^{-1}L_0^{-1}$, one infers

$$\delta\lambda_{\rm IV} \sim \alpha^{3/4}\nu_{*i}^{1/2}B_0^{-1}. \quad (54)$$

The boundary condition on $\partial_\lambda\delta G_i^{\rm IV}$ at $\lambda = \lambda_w$ gives

$$\partial_\lambda\delta G_i^{\rm IV} \sim B_0\nu_{*i}^{-1}\epsilon_i F_{i0}, \quad (55)$$

and therefore

$$\delta G_i^{\rm IV} \sim \alpha^{3/4}\nu_{*i}^{-1/2}\epsilon_i F_{i0}. \quad (56)$$

We proceed to deal with the collisional layer denoted by Region V. In this case, $\delta G_i^{\rm V}$ satisfies

$$v_{\|}\hat{\mathbf{b}}\cdot\nabla\Theta\partial_\Theta\delta G_i^{\rm V} - \frac{v_{\|}\nu_\lambda}{v^2 B_{\rm wmax}^2}\partial_\lambda\left(v_{\|}\partial_\lambda\delta G_i^{\rm V}\right) = 0 \quad (57)$$

and is defined on $\lambda \in [\lambda_w - K\delta\lambda_{\rm V}, \lambda_w + K\delta\lambda_{\rm V}]$, $\Theta \in [\Theta_w, \Theta_2(\lambda)]$ for $\lambda < \lambda_w$, and $\Theta \in [\Theta_1(\lambda), \Theta_2(\lambda)]$ for $\lambda > \lambda_w$. The boundary conditions are $\partial_\lambda\delta G_i^{\rm V}(\lambda_w^+) - \partial_\lambda\delta G_i^{\rm V}(\lambda_w^-) = -\left(\partial_\lambda G_i^{\rm III}(\lambda_w) - \partial_\lambda G_i^{\rm II}(\lambda_w)\right)$ at $\lambda = \lambda_w$, $\delta G_i^{\rm V} = 0$ at $\lambda_w - K\delta\lambda_{\rm V}$ and $\lambda_w + K\delta\lambda_{\rm V}$, $\delta G_i^{\rm IV} = \delta G_i^{\rm V}$, for $\lambda < \lambda_w$ at $\Theta = \Theta_w$, $\delta G_i^{\rm V}(s = 1) = \delta G_i^{\rm V}(s = -1)$ for $\lambda > \lambda_w$ at Θ_1, and $\delta G_i^{\rm V}(s = 1) = \delta G_i^{\rm V}(s = -1)$, for any λ, at Θ_2.

Balancing the two terms in (57), and employing $v_{|} \sim v_{ti}$ and $\hat{\mathbf{b}}\cdot\nabla \sim L_0^{-1}$, one gets

$$\delta\lambda_{\rm V} \sim \nu_{*i}^{1/2}B_0^{-1}. \quad (58)$$

Estimating the size of $\delta G_i^{\rm V}$ is slightly more involved. It can be shown [14] that phase space continuity between Regions IV and V implies

$$\delta G_i^{\mathrm{V}} \sim \alpha^{3/2} \nu_{*i}^{-1/2} \epsilon_i F_{i0} \,. \tag{59}$$

Finally, we emphasize that the size of both δG_i^{IV} and δG_i^{V} is very small compared to G_i^{II} and G_i^{III}, given in (32). This implies, in particular, that the estimation (45) is correct.

5. Conclusions

In this paper we have first reviewed the results of references [10 and 11]. In those references, we started a systematic approach to understand how much one can deviate from perfect quasisymmetry without spoiling some of the properties that make quasisymmetric stellarators interesting; in particular, the absence of flow damping in the symmetry direction. We show that the answer depends on the size of the helicity of the deviations from quasisymmetry and, in general, on the collisionality regime. Formal criteria are derived in a variety of situations, to assess whether a stellarator can be considered quasisymmetric in practice. The survey of [10 and 11] is completed by extending those results to a new collisionality regime. All these results apply to tokamaks with ripple as well.

Acknowledgments

This work has been carried out within the framework of the EUROfusion Consortium and has received funding from the European Union's Horizon 2020 research and innovation programme under grant agreement number 633053. The views and opinions expressed herein do not necessarily reflect those of the European Commission. This research was supported in part by grant ENE2012-30832, Ministerio de Economía y Competitividad, Spain.

References

[1] Helander P *et al* 2012 *Plasma Phys. Control. Fusion* **54** 124009
[2] Connor J W, Fukuda T, Garbet X, Gormezano C, Mukhovatov V, Wakatani M, the I T, the I T and Internal Barrier Physics 2004 *Nucl. Fusion* **44** R1
[3] Nührenberg J *et al* 1995 *Trans. Fusion Technol.* **27** 71
[4] Helander P and Simakov A N 2008 *Phys. Rev. Lett.* **101** 145003
[5] Boozer A H 1983 *Phys. Fluids* **26** 496
[6] Nührenberg J and Zille R 1988 *Phys. Lett.* A **129** 113
[7] Garren D A and Boozer A H 1991 *Phys. Fluids* B **3** 2822
[8] Anderson F S B, Almagri A F, Anderson D T, Mathews P G, Talmadge J N and Shohet J L 1995 *Fusion Technol.* **27** 273
[9] Gerhardt S P, Talmadge J N, Canik J M and Anderson D T 2005 *Phys. Plasmas* **12** 056116
[10] Calvo I, Parra F I, Velasco J L and Alonso J A 2013 *Plasma Phys. Control. Fusion* **55** 125014
[11] Calvo I, Parra F I, Alonso J A and Velasco J L 2014 *Plasma Phys. Control. Fusion* **56** 094003
[12] Boozer A H 1981 *Phys. Fluids* **24** 1999
[13] Simakov A N and Helander P 2009 *Phys. Plasmas* **16** 042503
[14] Parra F I, Calvo I, Velasco J L and Alonso J A 2014 Neoclassical fluxes in optimized stellarators, in preparation
[15] Beidler C D *et al* 2011 *Nucl. Fusion* **51** 076001
[16] Ho D D M and Kulsrud R M 1987 *Phys. Fluids* **30** 442
[17] Helander P 2007 *Phys. Plasmas* **14** 104501
[18] Sugama H, Watanabe T H, Nunami M and Nishimura S 2011 *Phys. Plasmas* **18** 082505
[19] Hazeltine R D 1973 *Plasma Phys.* **15** 77
[20] Helander P and Sigmar D J 2002 *Collisional Transport in Magnetized Plasmas (Cambridge Monographs on Plasma Physics)* ed Haines M G *et al* (Cambridge: Cambridge University Press)

IOP Publishing

Plasma Physics and Controlled Fusion

Plasma Phys. Control. Fusion **57** (2015) 014015 (10pp)

doi:10.1088/0741-3335/57/1/014015

The physics and detection of nanodust in the solar system

N Meyer-Vernet[1], I Mann[2,3], G Le Chat[1], P Schippers[1], S Belheouane[1], K Issautier[1], A Lecacheux[1], M Maksimovic[1], F Pantellini[1] and A Zaslavsky[1]

[1] LESIA, Observatoire de Paris, CNRS, UPMC, Université Paris Diderot, Meudon, France

[2] EISCAT Scientific Association, Rymdcampus 1, 98192, Kiruna, Sweden

[3] Physics Department, Förvaltningshuset, Umea University 16, 90187, Sweden

E-mail: nicole.meyer@obspm.fr

Received 12 June 2014, revised 12 September 2014

Accepted for publication 19 September 2014

Published 28 November 2014

Abstract

The mass distribution of small bodies in the solar system extends over more than 35 orders of magnitude, from asteroids to nanodust, which bridge the gap between molecules and macroscopic submicron grains. The small size of nanograins compared to the relevant basic scales gives them peculiar properties. Some of these properties affect their electric charging and their large charge-to-mass ratio drives their acceleration to very high speeds in moving magnetised plasmas, as the solar wind and rotating planetary magnetospheres. The electric charge and/or high speed of nanograins have enabled them to be detected serendipitously in various parts of the solar system by several instruments designed to study larger dust, plasma particles, or waves, on a number of spacecraft. These discoveries have opened an emerging field of research, in which many open questions remain, in particular concerning the lower size limit of the particles.

Keywords: nanodust, solar system, interplanetary physics, space research instruments, solar wind, magnetospheres, nanograins

(Some figures may appear in colour only in the online journal)

1. Introduction

According to the International Council of Chemical Associations, a nano-object is defined as having at least one external dimension between 1 nm and 100 nm (norm ISO TS 27687). For compact spheres, this corresponds typically to a few tens to a few 10^7 atoms or small molecules (figure 1), so that nanodust lies at the frontier between macromolecules and bulk matter. This emerging field of research on nano-objects includes a various fauna, which is much more actively studied in the contexts of biology and nanotechnology than in astronomy and geophysics.

Nanodust has been inferred a long time ago in the interstellar medium [1]. It is revealed there by a variety of observed phenomena: optical luminescence, near- and mid-infrared continuum emission and spectral bands, microwave emission and far ultra-violet extinction [2]. The presence of nanodust outside the solar system is also inferred from the identification of presolar nanograins embedded in primitive meteorites [3] and in collected micro dust [4].

However, these grains cannot enter the solar system, as explained in section 4. Furthermore, the phenomena revealing nanodust in interstellar space are not likely to be observed in the solar system [2], mainly due to the small integrated depth involved, so that remote nanodust observations [5] are rare and nanodust detection is generally performed *in situ*. In the absence of dedicated instruments in operation, nanodust has thus been observed only recently in the solar system. It has been discovered in media as diverse as the solar wind [6] and cometary [7] and planetary environments (e.g. [8]). These discoveries were serendipitous, with some of them using instruments that were not designed to measure dust, but plasma particles or waves (e.g. [9]). Most of these detections were made possible by the high speed of nanoparticles, produced by acceleration by ambient moving magnetized plasmas: the solar wind [10] and/or rotating planetary magnetospheres [11].

Section 2 addresses small size effects, some of which affect the electric charge and increase its relevance. Section 3 studies the electric charge itself, which drives the high speed

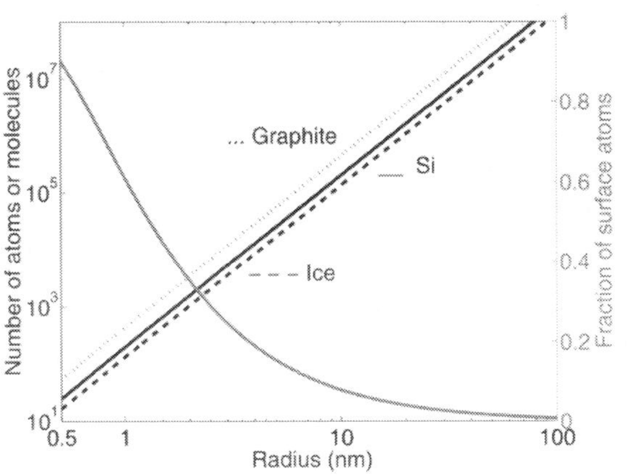

Figure 1. Number of atoms or molecules (black, left axis) and fraction of atoms or molecules located at the grain's surface (red, right axis) as a function of particle radius for compact spheres.

addressed in section 4. Finally, section 5 reviews nanodust detection, which was made possible by these latter effects.

2. What makes nanodust different?

First of all, nanodust has a large surface-to-volume ratio. Other differences arise when the particles' size is smaller than the basic scales determining their behaviour in relation to radiation and/or the ambient plasma. Further differences arise because of the strong increase of electric charging effects as size decreases.

2.1. Large surface-to-volume ratio

The surface-to-volume ratio varies in the inverse proportion of size for 3D compact particles and still faster for lower dimensional ones, including those having a fractal structure (e.g. [13]). Thus a large proportion of atoms are lying at the grain's surface (see e.g. [12] and figure 1). Surface atoms have too few bonding partners, yielding free radicals that produce chemical activity at the surface, surface reconstruction and coagulation (e.g. [14]), as well as changes in optical properties. Furthermore, the mean-square displacements of surface atoms are greater than within the grain, making the melting point and latent heat decrease and the diffusion coefficient in the grain increase [12].

2.2. Size smaller than de Broglie wavelengh: quantum confinement

From the Heisenberg uncertainty principle, confinement within a nanograin introduces an uncertainty in momentum, which plays an important role when the confinement energy becomes comparable to the thermal energy. Equivalently, the de Broglie wavelength $\hbar/mv$ of an elementary particle of mass m and speed v sets the scale below which quantum effects become conspicuous. Since electrons (of mass m_e) at temperature T have a thermal speed $\sim(k_BT/m_e)^{1/2}$ (k_B is Boltzmann's constant), one expects changes in radiating properties of semiconductors at sizes $\lesssim\hbar/(m_ek_BT)^{1/2}$. Since this is below the nanodust range for $T > 300$ K, quantum confinement effects are expected to be minor for solar system nanodust, although they are conspicuous in the cold interstellar medium [15].

2.3. Size smaller than mean electron free paths

The mean free path of electrons l_e in most materials increases with energy above a few 100 eV and has a minimum $\lesssim 1$ nm near tens of eV. Still below this energy, it increases as energy decreases up to $l_e \gtrsim 1$ nm near 1 eV [16].

This has important consequences for the electric charging of nanograins of size $\lesssim l_e$ in plasmas: the sticking coefficient of the plasma electrons impacting the grain decreases below unity [17–19], and the electron secondary emission induced by impacting electrons exceeds the value for bulk matter [20, 21].

2.4. Size smaller than scales relevant for ambient radiation

The attenuation length of photons producing photoelectron emission in most materials ($\sim$ 10–100 nm) largely exceeds the photoelectron escape length ($l_e \sim 0.5$–5 nm) [22]. Therefore, for nanograins of radius $a \lesssim l_e$, the photoelectrons have a better chance to escape than from bulk matter, which increases the photoelectron emission yield—i.e. the number of ejected photoelectrons per absorbed photon [22, 23]. This effect can be counterbalanced by two further effects. First, the removal energy of an electron increases for very small grains since the image charge contribution $\sim (3/8)(e^2/4\pi\epsilon_0 a)$ adds to the work function [24]; this increases the photon energy required for photoelectron emission. Second, the wavelength of radiation mainly responsible for ionization $\lambda \sim hc/E_{\rm Bohr} \sim 0.1\ \mu$m ($E_{\rm Bohr} \simeq 10$ eV is the Bohr energy) exceeds the nanodust size; hence the photon absorption cross-section is expected to decrease with grain size in proportion of a^3 (Rayleigh *oblige*), i.e. faster than the a^2 variation of the grain's cross-sectional area. The final result depends on the grain size and material and on the ambient radiation.

Finally, the heat capacity of a nanograin can become so small that the absorption of a single photon changes significantly its temperature. This stochastic heating affects the radiating properties of interstellar grains (see e.g. [25]), but this is not so in the solar system because the intense solar radiation field produces a photon impact rate exceeding the grain's heating rate (except below the nanometer size) [2].

2.5. Size smaller than the plasma Landau radius

In a plasma of temperature T, the so-called plasma Landau radius

$$r_L = e^2/(4\pi\epsilon_0 k_B T) \qquad (1)$$

is the distance below which the mutual electrostatic energy of two approaching particles of charge $\pm\ e$ exceeds the kinetic energy of their relative motion, so that they significantly perturb each other's trajectories. This scale determines the plasma particle cross-section for Coulomb collisions producing large

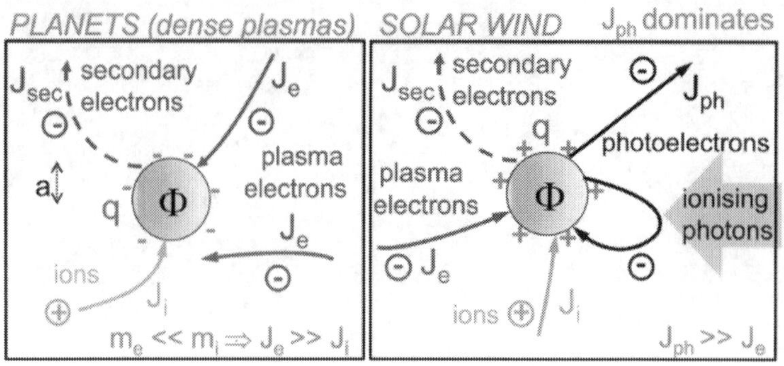

Figure 2. Basics of grain charging.

perturbations. It is also of major importance for dust grains since the plasma particles approaching a grain of radius $a \lesssim r_L$ induce polarization charges whose Coulomb attraction increases significantly the collected fluxes and thus modifies the grain charging (section 3.2). Furthermore, at this scale the charging becomes discretized [26] (section 3.2).

Since (1) can be written

$$r_{L(nm)} \simeq 1.44 / T_{eV} \quad (2)$$

these effects hold in cold plasmas, as the Earth's low ionosphere (see e.g. [27, 28]), the interstellar medium (see e.g. [29]) and the environment of outer planet satellites [26].

2.6. Nucleation on charged nanograins

The grain's electric charge (section 3) can affect its size via growth and/or disruption. Consider the growth via nucleation of water molecules on ice grains [30]. It is governed by the tendency of the system to minimize the Gibbs free energy, given for an uncharged grain by $G = 4\pi a^2 \sigma - N k_B T \ln S$ in the framework of liquid drop theory; here the first term is the surface energy ($\sigma \sim 0.1$ J m^{-2}); the second term represents the energetic preference for condensation, $N \propto a^3$ being the number of grain's molecules and S the supersaturation ratio (vapour pressure divided by saturation pressure). The first term produces an energy barrier which prevents nucleation on uncharged grains of size smaller than about 1 nm at $T \simeq 130$ K in the Earth thermosphere. The grains' electric charge changes this situation since it (Coulomb) attracts the incident dipolar water molecules, which decreases the Gibbs energy and suppresses the energy barrier for $T \lesssim 130$ K [31]. This enables the numerous subnanometer charged 'smoke' particles present in the thermosphere to act as condensation nuclei, therefore increasing by a large amount the formation of nanodust in this region [18] (see section 5.1).

2.7. Electrostatic and centrifugal disruption

The electrostatic field E at the surface of a nanograin of radius a and charge q increases in proportion of q/a^2; since q varies in proportion of a or less (section 3.2), E can become very high for nanograins. When the electrostatic stress ($\propto E^2$) exceeds the maximum grain's tensile strength against fracture S, it makes the grain explode (e.g. [20, 32]), so that the minimum size of a grain carrying Z elementary charges is [34]

$$a_{nm} \simeq 0.4 Z^{1/2} (S / 10^9 \mathrm{Nm}^{-2})^{-1/4} \quad (3)$$

where we have normalized the maximum nano grain tensile strength S to the typical value for polycrystalline matter or tektites. This limit should be considered together with field emission; for example, for compact ice grains carrying one electron in Enceladus plume (not less because $a \lesssim r_L$ and not more because of field emission, see section 3.2), (3) yields a minimum diameter of 1.4 nm [26].

Another size limitation can be produced by the centrifugal stress due to the grain's spin induced by impacts of ambient particles [20, 33]. Contrary to the electrostatic limit, it concerns also uncharged grains. When the grain is charged and the time scales for the grain to be charged and to spin up are both smaller than the other relevant time scales, both effects can act simultaneously. In that case the grain minimum size is determined by the largest of both minima, which is set by centrifugal disruption if the temperature of the species making the grain spin satisfies $T_{eV} > (S/10^9 \mathrm{~Nm}^{-2})^{1/4}$. This is so for fragile grains, or warm media as for example the innermost interplanetary medium [35]. The minimum radius is then given by $a_{nm} \simeq 0.5 \times T_{eV}^{1/3} (S / 10^9 \mathrm{Nm}^{-2})^{-1/3}$ (e.g. [34]) instead of (3).

3. How are nanograins charged in solar system plasmas?

3.1. Basics of electric charging

Dust grains charge by collecting and emitting charged particles, which changes their charge and electrostatic potential, which in turn changes the fluxes of incoming and outgoing charged particles, until an equilibrium is reached when the electric currents balance (see e.g. [36]).

In dense plasmas (figure 2, left), as in planetary ionospheres and the inner magnetospheres of outer planets, the photoelectron current is generally negligible. The plasma electron current largely exceeds the ion current because of the smaller electron mass (since both species generally have temperatures of the same order of magnitude), so that the grain charges negatively. This negative charge repels the incoming electrons and an equilibrium is reached when the grain's charge repels sufficiently the electrons so that their collected current balances that of the positive ions. For doing so, the grain's electrostatic potential Φ with respect to the ambient plasma must ensure that the potential energy $-e\Phi$ equals a few times the kinetic energy of the plasma electrons. This yields $-e\Phi \simeq \eta\, k_BT$, where η equals a few units. A roughly spherical grain of radius a much smaller than both the plasma Debye length L_D and the grains' separation carries the electric charge $q \simeq 4\pi\,\epsilon_0 a\,\Phi$. Substituting the above value of Φ and using (1) yields the normalized grain charge q/e at equilibrium

$$Z = -\eta a\,/\,r_L \qquad (4)$$

where η—of order of magnitude unity, can be calculated straightforwardly since the small size of the grain ensures that the particles are subjected to its Coulomb potential without intervening barriers of potential (the so-called orbit-limited condition, see e.g. [37]). In a Maxwellian plasma with one singly charged ion species of the same concentration and temperature as electrons and a sticking coefficient $\simeq 1$, we have $\eta \simeq 2.5 - 3.9$ for ion mass $m_i \simeq m_p - 30\, m_p$, m_p being the proton mass [36].

In contrast, in the interplanetary medium and in dilute regions of planetary magnetospheres, the charge flux is generally dominated by photoemission produced by solar photons of energy exceeding the work function of the grain material (figure 2, right). The grain thus charges positively until its positive charge binds sufficiently the photoelectrons to reduce their escaping flux so that it balances the flux of incoming plasma electrons. For doing so, the grain's electric potential must provide the photoelectrons with a potential energy $-e\Phi$ equal to a few times their typical kinetic energy. Since for most materials exposed to the solar flux, the photoelectron velocity distribution can be approximated by a Maxwellian (of temperature T_{ph}) [38], one has $e\Phi = +\ \eta\, k_BT_{ph}$ with η equal to a few units. Therefore, a sunlit grain carries a number of charge units

$$Z = +\eta a\,/\,r_L \qquad (5)$$

where r_L is given by substituting in (1) the effective photoelectron temperature of a few 10^4 K (in practice 1–4 eV).

These basic results can be modified by two effects. First, incident electrons of energy exceeding the typical atomic binding energy $E_{Bohr} \simeq 10$ eV can produce secondary electron emission. In addition to shifting the grain electric potential to more positive values, this effect can produce two stable equilibrium charges of opposite signs for a given grain in a given environment when the ambient electron velocity distribution is not Maxwellian [39]. Second, when the dust concentration is so high that the grains' Debye spheres overlap, the electrons are depleted with respect to the ions since many of them rest on the grains' surface, which in turn reduces the grains' charge [26, 36, 40–42].

3.2. Electric charging of nanodust

How does nanodust charging differ from this picture? First, as noted in section 2.5, when the grain radius is of order the Landau radius (1) or smaller, the polarization induced on a grain by approaching charges produces an electric potential that perturbs significantly their trajectories and modifies the charging currents. This polarization increases significantly the flux of particles whatever the grain charge (see e.g. [29]). Since the flux of repelled particles increases more than the other ones, this effect tends to make the negative charge more negative [26]. Such phenomena affect noctilucent cloud particles and meteoritic dust (see e.g. [28]) as well as nanograins in Enceladus plume [26] (section 5.1).

Second, one sees from (4)–(5) that $|Z|$ is not large when $a \lesssim r_L$, so that a statistical treatment is needed. In that case, one must calculate the probability $f(Z)$ that a grain carry the electric charge Ze. In the simple case when the charging currents come from the ambient electrons (J_e) and ions (J_i) in stationary conditions, the probability can be deduced from the recurrence relation $f(Z)/f(Z+1) = J_e(Z+1)/J_i(Z)$ [29]. However, because of the small surface area, the charging time scale is large, so that nanograins may not carry their equilibrium charge; in that case, the problem must be solved numerically (see e.g. [43]).

Finally, for nanograins we have $e^2/4\pi\,\epsilon_0 a \simeq 1.4$ eV, so that the surface Coulomb electric field deforms significantly the potential barrier at the surface, which enables electrons inside to tunnel efficiently. In practice, this process becomes efficient when the (inward) electric field exceeds 10^9 V m^{-1}. An ejected electron near the surface of a grain of charge $Ze < 0$ will be subjected to the field amplitude $|E| \simeq (Z+1)e/4\pi\,\epsilon_0 a^2$. Hence electron field emission limits the grain charge to $|Z| < (1 + 0.7a_{nm}^2)$ (e.g. [29, 44]). Note that ion field emission, which limits the positive charging, is generally negligible because it requires a much higher electric field.

In summary, when the grain's size is of order r_L or smaller, the equilibrium charge no longer decreases in proportion of the size, but remains comparable to one electron charge in a wide range of sizes, mainly because the probability that an uncharged grain collects an electron far exceeds the probability that a neutral or negatively charged grain collects an ion. This charging is crucial for their detection by plasma analyzers (section 5.1) and for their dynamics, which is addressed below.

4. Dynamics of nanodust in solar system plasmas

With the electric charge decreasing in proportion of the grain's radius a or less, the charge-to-mass ratio q/m of nanograins becomes fairly high and so does the acceleration by the Lorentz force, with major consequences on the dynamics in magnetized plasmas. Let us estimate q/m. In basic units, we have in order of magnitude $m/m_p \sim (a/r_B)^3$ for compact grains and $|q/e| \sim 1 + \eta\, a/r_L$ [26], where η equals a few units (section 3.1), $r_B \simeq 0.05$ nm is the Bohr radius and r_L the Landau radius of the charge species dominating the grain's charging. Hence, in terms of the grain size and plasma parameters

$$\begin{aligned}|q/m| &\gtrsim 3.5 \times 10^{-4} \times (e/m_p)/(r_{L(\mathrm{nm})}\, a_{nm}^2) \\ &\gtrsim 2.5 \times 10^{-4} \times (e/m_p) T_{eV}/a_{nm}^2\end{aligned} \qquad (6)$$

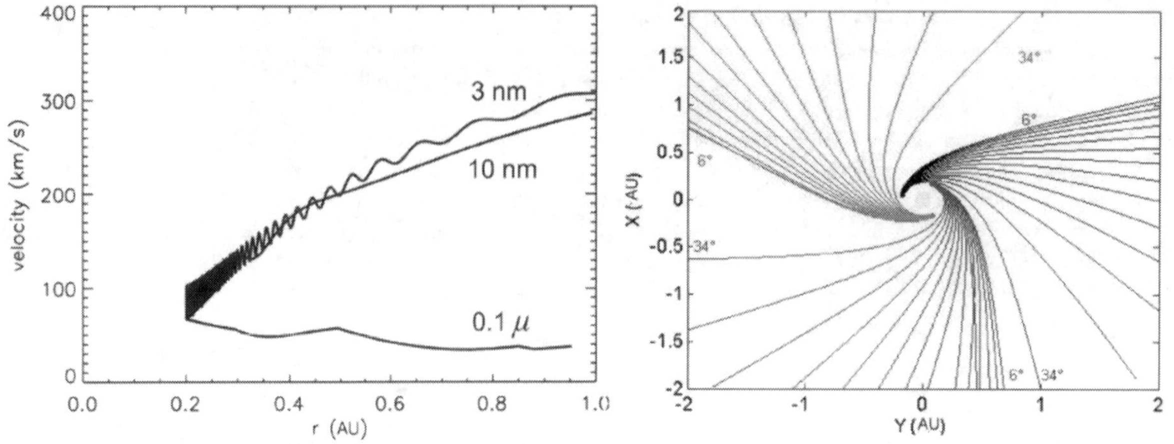

Figure 3. Speed of nanodust and 0.1 μm particles as a function of heliocentric distance, showing the nanodust accelerated at about 300 km s^{-1} at 1 AU (left, adapted from [49]). Trajectories of 3 nm radius particles projected on the solar magnetic equatorial plane for three initial longitudes and different initial latitudes (indicated on the figure), showing the particles tending to follow the solar wind spiral magnetic field lines (right, from [50]).

Consider first nanograins in the interstellar medium close to the heliosphere, of approximate radius 100 AU (1 AU $\simeq 1.5 \times 10^{11}$ m is the Sun–Earth distance). With a relative velocity $V \sim 25$ km s^{-1}, a plasma temperature $T \sim 0.5$ eV and magnetic field $B \sim 3 \times 10^{-10}$ T (e.g. [45]), (6) yields a gyroradius $r_{gyr} \sim mV/qB \sim 0.05a_{nm}^2$ AU. Thus $r_{gyr} < 100$ AU for $a < 50$ nm. Hence grains of this size are deflected and interstellar nanograins, for which $r_{gyr} \ll 100$ AU, follow the interstellar field lines and cannot enter the heliosphere (see e.g. [46]).

Now, consider the solar wind and the inner magnetospheres of outer planets, where the charging is dominated by respectively photoelectrons and plasma electrons, of energy around a few eV . From (2), r_L is in the nanometer range, so that (6) yields

$$|q/m| \sim (e/m_p)\, 10^{-3}/a_{nm}^2 \quad (7)$$

and the Lorentz force generally dominates the other forces in magnetized plasmas.

4.1. Solar wind

To a first approximation, the solar wind plasma expands radially at a speed $V \simeq 400$ km s^{-1} and carries a magnetic field, which has a spiral shape due to the solar rotation at $\Omega \simeq 2.7 \times 10^{-6}$ rad s^{-1}. The ratio of azimuthal to radial magnetic components at heliocentric distance r is thus $\Omega r/V$ in the ecliptic plane where lie most solar system objects and the magnetic field radial component varies with distance as r^{-2}. The magnetic field modulus is $B \simeq 4.5 \times 10^{-9}$ T at $r \simeq 1$ AU where $\Omega r/V \simeq 1$ (see e.g. [47]).

We deduce from (7) that the angular gyrofrequency (normalized to Ω) of a nanograin of radius a and velocity v is $\omega_{gyr}/\Omega = (qB/m)/\Omega \simeq 150/a_{nm}^2$ at 1 AU, whereas its gyroradius (normalized to r) $r_{gyr}/r \lesssim |(\mathbf{v}-\mathbf{V})_\perp|/(\omega_{gyr} r) \simeq 0.5 \times 10^{-2} a_{nm}^2$ at 1 AU (the subscript $\perp$ denotes the component $\perp\mathbf{B}$). Both quantities increase with distance inward of 1 AU, but the normalized gyroradius remains roughly constant farther out since the magnetic field there varies roughly as $1/r$.

Hence at any heliocentric distance a nanoparticle of radius <10 nm has a gyroradius smaller than the distance, so that it is quickly picked-up by the solar wind and accelerated roughly to the drift velocity, equal to the projection of the solar wind velocity normal to the magnetic field, which amounts to about 300 km s^{-1} at 1 AU [10] (figure 3). This holds provided that the nanoparticle is released farther than about 0.2 AU from the Sun; at closer distances, nanograins can be trapped by the combined actions of the solar gravitational attraction, the magnetic mirror force and the centrifugal force [48].

4.2. Rotating magnetospheres

Most solar system planets are rotating about an axis making a small angle with the normal to their orbit and have an intrinsic magnetic field. For the Earth and the giant outer planets Jupiter and Saturn, this field is roughly dipolar, with the magnetic dipole axis lying close to the spin axis Ω (see e.g. [47]).

Consider a planet of mass M, radius R and magnetic field amplitude B_R at equator. Assume a nanograin of charge q and velocity $\mathbf{v}$ at distance r, close to the equatorial plane where lie the satellites and other material, so that most grains are likely to be produced there. The plasma rotates with the azimuthal velocity $\mathbf{V_{rot}} = \Omega \times \mathbf{r}$ and the Lorentz force $F_L = q(\mathbf{v} - \mathbf{V_{rot}}) \times \mathbf{B}$ is directed outwards for $q > 0$, farther than the corotation distance. It therefore tends to eject nanograins small enough for the corotation potential energy $\Phi_{rot} \simeq q\Omega\, B_R R^3/r$ to exceed half the gravitational energy $\Phi_G/2 \simeq mMG/2r$; the grains released at radius r_0 are ejected with a speed v_{ej} approximately given by [11]

$$v_{ej}^2 \simeq (2MG/r_0)\,(\Phi_{rot}/\Phi_G - 1/2) \quad (8)$$

Note that grains of gyroradius smaller than the scale of magnetic field variation remain confined along the closed magnetic field lines, so that (8) does not hold for too small grains.

Nanograins produced by volcanoes of Jupiter's satellite Io are ejected in this way from Jupiter's magnetosphere at speeds of nearly 300 km s^{-1} if their size lies between a few nanometers to a few tens nanometers. Similarly, nanograins emerging from Saturn's icy moon Enceladus are ejected from Saturn's environment, albeit at somewhat smaller speeds due to the smaller rotation rate and magnetic field of Saturn (see e.g. [8]).

5. *In situ* detection of nanodust in solar system plasmas

5.1. Nanodust detection with dust and plasma analyzers

The presence of nanodust in the Earth lower ionosphere was inferred long ago (see e.g. [51] and references therein) from the observation of noctilucent ('night-luminous') clouds (NLC), which revealed the occasional presence of aerosol particles in the polar low ionosphere. And the ubiquitous presence of so-called meteor smoke particles produced by ablation and subsequent recondensation of meteoric matter was suggested a long time ago.

These inferences have been confirmed by *in situ* measurements (see e.g. [52]) and nanodust has been shown to play a major role in the physics of the Earth's polar mesosphere in summer at 80–90 km altitude, which is the coldest place on Earth. With a temperature <150 K, below the water frost point, large quantities of icy nanograins can be produced there by condensation of water vapour on meteor smoke (section 2.6). These nanograins are at the origin of three conspicuous phenomena. First, nanograins large enough to scatter visible light produce the above-mentioned NLC's (also known as Polar Mesospheric Clouds). Second, most nanograins are charged negatively (section 3.1) and since the Landau radius $r_L > 0.1$ μm at $T < 150$ K, most of them carry one electron (section 3.2) and act as sinks for ambient electrons, producing large decreases in plasma electron density. Third, the variations in plasma electron density strongly backscatter radio waves, producing Polar Mesosphere Summer Echoes [53].

At higher altitudes in the Earth environment, a flux of nanodust of radius $\lesssim 5$ nm (mass $\lesssim 10^{-21}$ kg) compatible with the interplanetary dust model at 1 AU (figure 4) has been measured by a penetration dust detector on the International Space Station in low Earth orbit; these particles may be either debris or interplanetary dust [54].

Farther out, the ion mass spectrometers on the spacecraft Vega's [55] and Giotto [7], designed to study the composition of cometary dust, detected unexpected signals at 10^6 km from the nucleus of comet Halley. These signals were first thought to be spurious noise because the instruments were not designed to operate with such small particles, but a detailed analysis has attributed these observations to nanodust of mass $\simeq 10^{-21}$ kg [7].

Nanodust has also been detected serendipitously in the environments of outer planets by instruments not designed to do so. A subsystem of the plasma spectrometer on Cassini, designed to measure electrons at Saturn, detected negatively charged particles of energy per charge corresponding to masses up to about 10^4 m_p and kinetic energy consistent

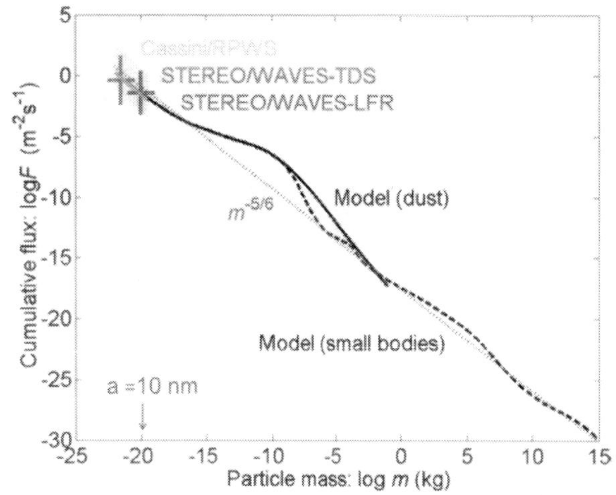

Figure 4. Cumulative flux of interplanetary dust and bodies near 1 AU from the Sun. The superposition of models for dust ([89], continuous line), small bodies ([90], dashed) and collisional equilibrium $\propto m^{-5/6}$ (dotted), is reproduced from [47], with measurements of nano dust by [6, 85, 86 and 87] superimposed.

with the spacecraft velocity, revealing large quantities of nanograins in Titan's atmosphere [56]. The mainly negative charge agrees with expectations (section 3.2), as well as do the grains' concentrations [57].

Still in Saturn's environment, the same instrument also discovered a large concentration of negatively charged nanograins in the plume ejected by the geologically active icy moon Enceladus [58, 59]. The mainly negative charge and the minimum size of about 1 nm [59] are in agreement with expectations [26] (sections 2.7 and 3.2).

Nanodust was also detected serendipitously by instruments designed to measure dust grains of much larger size. In that case, the detection was made possible by the high speed of these particles (section 4.2). The dust analyzer onboard the spacecraft Ulysses detected 'small events' which were initially attributed to streams of ~0.2 μm grains coming from Jupiter at about 50 km s^{-1} [60]. Shortly after, these streams were recognized, from dynamics calculations, as made instead of nanodust ~10^3 less massive and moving ~5–10 times faster [61], well outside the calibration range of the instrument. These results have been confirmed and completed by observations from dust analyzers onboard the spacecraft Galileo and Cassini (see e.g. [8]) and by a wave instrument [62].

Similarly, the dust analyzer on Cassini has detected nanodust streams ejected from the Saturn environment and further accelerated by the solar wind [63]. As for Jupiter nanodust streams, these measurements were mainly based on dynamics calculations [8] (section 4) since the nanodust size and speed are outside the calibration range of the dust detector.

5.2. Nanodust detection with wave instruments

Wave instruments are routinely used in space for *in situ* plasma measurements via quasi-thermal noise spectroscopy [64]. This technique is based on the voltage fluctuations

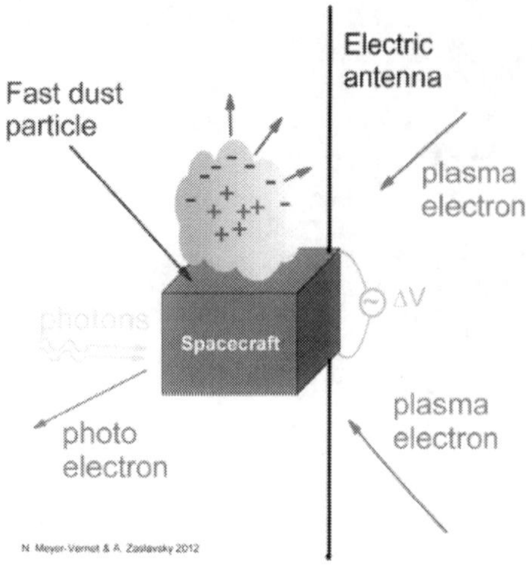

Figure 5. Principle of *in situ* measurements of plasma and dust via waves. Plasma particles passing-by the antennas (or impacting or being ejected) produce a quasi-thermal electrostatic noise whose power spectrum reveals plasma properties. Dust impacts at high speed produce partial ionisation of the dust and target, yielding an expanding plasma cloud. The cloud's charges produce voltage pulses whose analysis reveals some dust properties. Adapted from [9] with the kind permission of Springer Science and Business Media.

induced on electric antennas by the quasi-thermal motion of the ambient electrons and ions (figure 5) [65]. These fluctuations produce a peak in spectral density at the local plasma frequency, whose shape reveals the electron temperature and further plasma properties [66].

Charged dust grains passing closer than the Debye length (~ 10 m in the solar wind at 1 AU) from the electric antennas also produce electric voltage fluctuations, whose detection, however, requires very large grain sizes and/or number densities [67, 68]. In contrast, a dust grain impacting the spacecraft (or antennas) at a large speed produces a strong shock compression which vaporises and ionises the dust as well as a part of the impact crater. This produces an expanding plasma cloudlet whose residual ionisation exceeds by several orders of magnitude the initial grain charge and can be used to detect the grain (figure 5). This technique was pioneered when the spacecraft Voyager crossed the dusty rings of Saturn; at this occasion, the inboard radio [69] and plasma wave instruments [70] detected dust grains, opening the way to microdust measurements with wave instruments in various environments (see [68, 71, 72] and references therein).

The detection is based on the residual charge Q carried by the electrons and/or the ions of the impact cloudlet, which can be recollected by a biased target. A spacecraft in the solar wind is positively charged (section 3.1) and thus recollects the cloudlet's electrons [9, 73], making its floating potential change by $\simeq -Q/C_{SC}$ (C_{SC} being the spacecraft capacitance), so that a monopole antenna (which measures the difference of potential between an antenna arm and the spacecraft) measures a voltage pulse

$$\delta V \simeq +\Gamma Q / C_{SC} \qquad (9)$$

where Γ is the wave receiver gain. Modern wave instruments include two independent parts: a time domain sampler measuring the voltage as a function of time at a high rate and a frequency receiver measuring the time-integrated power spectrum. The individual voltage pulses produced by dust impacts can be detected by the time domain sampler, whereas the corresponding voltage power spectrum due to many impacts during the integration time can be detected by the frequency receiver of the instrument [74].

The impact charge Q depends on the grain mass m and speed v and this dependence is used by the classical impact ionisation dust detectors [75], with various relationships of the form $Q \propto m^{\alpha} v^{\beta}$ where $\alpha \simeq 1$ and $\beta \simeq 3 - 4.5$. The coefficients depend on mass, speed, angle of incidence, as well as grain and target composition [76, 77], and have not been measured for either nanodust or $v \gtrsim 70$ km s^{-1} [75]. Furthermore, the huge strain rate $v/a \simeq 3 \times 10^{13}$ s^{-1} for particles of radius $a \simeq 10$ nm impacting at $v \simeq 300$ km s^{-1} lies in a newly explored range of modelling (e.g. [78]), whereas existing N-body simulations [79] of the plasma cloud expansion corresponding to a fast nanodust impact only consider uncharged targets.

Nanodust detection was made possible by the high speed of nanodust together with the much faster increase of the impact charge with speed than with mass. Why does the impact charge increase much faster than the grain's kinetic energy? This is so because Q depends on vaporization and ionization, in a complex process involving plasma expansion and recombination. The grain speed determines not only the kinetic energy but also the time scale of the process, with incomplete recombination requiring short time scales. Indeed, for a grain of radius a and mass density ρ, the incident kinetic energy $\rho\,(4\pi a^3/3)v^2/2$ encounters a surface πa^2 of the target during the time $\sim a/v$, yielding the power flux $P \sim \rho\, v^3$. With $\rho \simeq 2.5 \times 10^3$ kg m^{-3} (typical for silicates) and $v \simeq 300$ km s^{-1}, this yields the huge power flux $W \sim 10^{20}$ W m^{-2}.

In the absence of laboratory calibrations or numerical simulations for impact ionization of high-speed nanodust, we have used the empirical relation [80, 81]

$$Q_{(\mathrm{Cb})} \simeq 0.7 m_{(\mathrm{kg})}^{1.02} v_{(\mathrm{kms^{-1}})}^{3.48} \qquad (10)$$

whose application to fast nanodust is consistent with independent results as explained below. According to (10), a 10 nm grain impacting at 300 km s^{-1} should produce a similar impact ionisation charge as a grain more massive by 3 orders of magnitude but less fast by a factor ~7. Even though this relationship was not calibrated for nanodust, it is remarkable that the initial (incorrect) identification of Jovian dust streams by traditional dust detectors based on calibrations [60] and the final (correct) identification based on dynamics (grains of mass smaller by 3 orders of magnitude but moving 5–10 times faster [61]), have similar values of the quantity $mv^{3.5}$. This suggests that this calibration holds also approximately for fast nanodust. Note also that impact ionisation yields of materials relevant for the STEREO and Cassini spacecraft have been reported recently for microdust at speeds below 40 km s^{-1} [82]. Extrapolating these results for nanodust impacting at 300 km s^{-1} produces charges of the same order of magnitude as those given by (10). We estimate that using (10) for fast nanodust introduces an uncertainty of about a factor of ten.

The measurement of the power spectrum due to the pulses of amplitude (9) with the calibration (10) has enabled us to detect Jovian nanodust streams with the Cassini/RPWS low-frequency receiver [62], simultaneously to the detection of these particles by the Cosmic Dust Analyser onboard the spacecraft Cassini and Galileo [83].

The serendipitous discovery of fast nanodust in the solar wind at 1 AU [6] by the frequency receiver of the instrument STEREO/WAVES, designed to study solar radio emissions, came as a surprise because the detection was not made by traditional dust detectors. This detection should not, however, have been surprising since (i) the production of such particles in the inner heliosphere by collisional fragmentation of larger dust and their acceleration to high speeds was suggested a few years before [84], (ii) the flux detected was compatible with the extrapolated interplanetary dust model [89] and (iii) the detection via impacts on the spacecraft surface is much more sensitive than traditional dust detectors since the collecting area and solid angle are larger by several orders of magnitude.

The STEREO discovery was confirmed by detailed studies of the data from the two independent parts of the WAVES instrument: the low frequency receiver [86] and the time domain sampler [85], as well as onboard another spacecraft: Cassini [87]. Indeed, during the one month period when the Cassini/RPWS instrument was turned on near 1 AU before reaching Jupiter and Saturn, the low-frequency receiver (using the antennas in monopole mode) was able to measure the power spectrum produced by the pulses given by (9) with a flux compatible with the STEREO measurements [87] (figure 4).

It is important to note that the dust detection via recollection of the impact charges by the spacecraft requires the antennas to be used in monopole mode in order for them to detect the pulses in spacecraft potential [72]. Unfortunately, many wave instruments use instead the antennas as dipoles, which measure the voltage between two antenna arms, because this provides a greater sensitivity and a better calibration.

In that case, the antennas can still detect voltage pulses produced by dust impacts—albeit by other mechanisms. First, if the antennas have a large surface area, they can collect some of the charges produced by impacts on the spacecraft or on their own surface. Second, the ions of the impact plasma cloud (remaining after recollection of the electrons by the positively charged spacecraft) can produce a voltage perturbation on an antenna close to the impact site, which is generally much smaller than (9) and in practice only detectable for large grains [72], except when these impact charges close to an antenna can perturb the photoelectron sheath surrounding it, in which case they can produce large voltage pulses, as was observed on STEREO [73, 85, 88].

Indeed, because of the intense solar radiation, a body in the solar wind tends to eject much more photoelectrons than it collects ambient electrons (section 3); equilibrium therefore requires that most photoelectrons remain trapped by the (positive) body's potential, producing a photoelectron sheath around it, of radius roughly the photoelectron Debye length ($L_{\rm Dph} \sim 0.5$ m at 1 AU). As photoelectrons move outwards from the surface of an antenna of radius $r_a \ll L_{\rm Dph}$ along their balistic orbits in the photoelectron sheath, their transverse energy decreases rapidly (conservation of angular momentum *oblige*). Hence the electric field perturbation produced by the dust impact charges is sufficient to eject the photoelectrons from their trapped trajectories. This interrupts momentarily the photoelectron return current on a fraction l/L of the antenna, producing a voltage pulse [73, 88]

$$\delta V_{\rm ph} \simeq \Gamma \left(k_{\rm B} T_{\rm ph} / e\right) (l / L) \tag{11}$$

This latter mechanism, which affects only one antenna arm and thus produces a pulse detected both in monopole and dipole mode, requires two conditions to operate: first, an adequate antenna geometry, as is the case for the STEREO antennas lying close to the spacecraft (which has plane faces), in contrast to other interplanetary probes carrying wave instruments; second, the antenna radius must be large enough for the interruption of the photoelectron current to produce a detectable voltage pulse [88].

The interplanetary nanodust fluxes detected near 1 AU on STEREO/WAVES [6, 85, 86] and Cassini/RPWS [87] by these mechanisms are shown in figure 4, superimposed to the flux models for interplanetary dust [89] and small bodies [90]. Both STEREO and Cassini measured large variations in flux—reaching two orders of magnitude, at several time scales. This is not surprising, given the expected variations in nanodust production [92] and the ubiquitous fluctuations in the solar wind properties that determine the Lorentz force governing nanodust dynamics [10, 36, 48, 91].

6. What next?

Most nanodust detections in the solar system were performed by instruments not designed to do so; in particular, neither the cosmic dust analyzers nor the wave instruments onboard solar system space probes in operation are calibrated for nanodust detection. Hence several major questions about these particles are still pending. Among them are the composition, physical state and lower size limit of interplanetary nanodust. The composition of nanodust streams coming from the environments of Jupiter [93] and Saturn [63] have been studied in detail (see e.g. [8]); the size distribution of the smaller nanodust in Titan's ionosphere (see e.g. [57]) and in the Enceladus plume have also been measured [59] and found to agree with the size limitations discussed in section 2.7 [26]. But this is not so for the interplanetary nanodust, for which there is presently no dedicated detector in operation, although such an instrument is being developed [94]. In particular, the smallest size is presently unknown, even though we can derive an educated guess for electrostatic disruption of nanograins charged by photoelectron emission near 1 AU, using (3), (5) and (2): for an interplanetary grain of tensile strength S, the minimum radius is expected to be $a_{\rm min} \simeq (0.5-1)(S/10^9\ {\rm N\ m^{-2}})^{-1/2}$ nm.

Acknowledgments

NM-V thanks the ISSI International team on 'Nanodust in the Solar System' for useful discussions on nanodust and the International Space Science Institute for having fostered them.

References

[1] Platt J R 1956 *Astrophys. J.* **123** 486–90
[2] Li A and Mann I 2012 *Nanodust in the Solar System: Discoveries and Interpretations* (*Astrophysics and Space Science Library*) vol 385, ed Mann I *et al* (Heidelberg: Springer) pp 5–30
[3] Lodders K 2005 *Chem. Erde: Geochem.* **65** 93
[4] Bernatowicz T J *et al* 1991 *Astrophys. J.* **373** L73
[5] Hervig M E *et al* 2009 *Geophys. Res. Lett.* **36** L18805
[6] Meyer-Vernet N *et al* 2009 *Solar Phys.* **256** 463–74
[7] Utterback N G and Kissel J 1990 *Astron. J.* **100** 1315–22
[8] Hsu H-W, Krüger H and Postberg F 2012 *Nanodust in the Solar System: Discoveries and Interpretations* (*Astrophysics and Space Science Library*) vol 385, ed Mann I *et al* (Heidelberg: Springer) pp 77–117
[9] Meyer-Vernet N and Zaslavsky A 2012 *Nanodust in the Solar System: Discoveries and Interpretations* (*Astrophysics and Space Science Library*) vol 385, ed Mann I *et al* (Heidelberg: Springer) pp 133–60
[10] Czechowski A and Mann I 2010 *Astrophys. J.* **714** 89–99
[11] Burns J A, Hamilton D P and Showalter M R 2001 *Interplanetary Dust* (*Astrophysics and Space Science Library*) vol 385, ed Grün E *et al* (Heidelberg: Springer) pp 641–725
[12] Kimura Y 2012 *Nanodust in the Solar System: Discoveries and Interpretations* (*Astrophysics and Space Science Library*) vol 385, ed Mann I *et al* (Heidelberg: Springer) pp 31–46
[13] Weibel E R 1991 *Am. J. Physiol.* **261** 361–9
[14] Lavvas P 2011 *Astrophys. J.* **728** 80
[15] Li A 2004 *Astrophysics of Dust* (*ASP Conference Series*) vol 309, ed Witt A N *et al* (Heidelberg: Springer) pp 417–52
[16] Fitting H-J *et al* 2001 *J. Electron Spectrosc. Relat. Phenom.* **119** 35–47
[17] Vostrikov A A and Dubov D Y 2006 *Tech. Phys.* **51** 1537–52
[18] Megner L and Gumbel J 2009 *J. Atmos. Solar Terr. Phys.* **71** 1236–44
[19] Michaud M and Sanche L 1987 *Phys. Rev.* A **36** 4672–83
[20] Draine B T and Salpeter E E 1979 *Astrophys. J.* **231** 77–94
[21] Pavlů J *et al* 2009 *Contrib. Plasma Phys.* **49** 169–86
[22] Watson W D 1972 *Astrophys. J.* **176** 103–10
[23] Weingartner J C and Draine B T 2001 *Astron. Astrophys. Suppl.* **134** 263–381
[24] Wong K, Vongehr S and Kresin V V 2001 *Phys. Rev.* B **67** 035406
[25] Draine B T and Li A 2001 *Astrophys. J.* **551** 807–24
[26] Meyer-Vernet N 2013 *Icarus* **226** 583–90
[27] Rapp M and Lübken F-J 2001 *J. Atmos. Solar Terr. Phys.* **63** 759–70
[28] Robertson S and Sternovsky Z 2008 *Phys. Plasmas* **15** 040702
[29] Draine B T and Sutin B 1987 *Astrophys. J.* **320** 803–17
[30] Rapp M and Thomas G E 2006 *J. Atmos. Solar Terr. Phys.* **68** 715–44
[31] Gumbel J and Megner L 2009 *J. Atmos. Solar Terr. Phys.* **68** 1225–35
[32] Hill J R and Mendis D A 1979 *Moon Planets* **21** 3–16
[33] Spitzer L Jr 1978 *Physical Processes in the Interstellar Medium* (New-York: Wiley) p 185
[34] Meyer-Vernet N 1984 *Icarus* **57** 422–31
[35] Misconi N Y 1993 *J. Geophys. Res.* **98** 18951–61
[36] Mann I, Meyer-Vernet N and Czechowski A 2014 *Phys. Rep.* **536** 1–39
[37] Whipple E C 1981 *Rep. Prog. Phys.* **44** 1197–250
[38] Grard R J L 1973 *J. Geophys. Res.* **78** 2885–906
[39] Meyer-Vernet N 1982 *Astron. Astrophys.* **105** 98–106
[40] Havnes O, Morfill G E and Goertz C K 1984 *J. Geophys. Res.* **89** 10999–1003
[41] Whipple E C, Northrop T G and Mendis D A 1985 *J. Geophys. Res.* **90** 7405–13
[42] Mendis D A and Rosenberg M 1994 *Annu. Rev. Astron. Astrophys.* **32** 419–63
[43] Hsu H *et al* 2011 *J. Geophys. Res.* **116** A09215
[44] Mendis D A and Axford W I 1974 *Ann. Rev. Earth Planet. Sci.* **2** 419–74
[45] Izmodenov V V 2009 *From the Outer Heliosphere to the Local Bubble* (Space Science Series of ISSI) ed Linsky J L *et al* (New York: Springer) pp 139–50
[46] Slavin J D et al 2010 *Solar Wind 12* (*AIP Conference Proceedings*) vol 1216, ed Maksimovic M *et al* (New York: American Institute of Physics) pp 497–501
[47] Meyer-Vernet N 2007 *Basics of the Solar Wind* (Cambridge: Cambridge University Press) pp 291–9, 337, 347, 339
[48] Czechowski A and Mann I 2012 *Nanodust in the Solar System: Discoveries and Interpretations* (*Astrophysics and Space Science Library*) vol 385, ed Mann I *et al* (Heidelberg: Springer) pp 47–75
[49] Mann I *et al* 2010 *Plasma Phys. Control. Fusion* **52** 124012
[50] Belheouane S 2014 Nanoparticules dans le milieu interplanétaire: observations spatiales et théorie *PhD Thesis* Université Pierre et Marie Curie
[51] Friedrich M and Rapp M 2009 *Surveys Geophys.* **30** 525–59
[52] Havnes O *et al* 1996 *J. Geophys. Res.* **101** 10839–47
[53] Rapp M and Lübken F-J 2004 *Atmos. Chem. Phys.* **4** 2601–33
[54] Carpenter J D *et al* 2007 *J. Geophys. Res.* **112** E08008
[55] Sagdeev R Z *et al* 1989 *Adv. Space Res.* **9** 263–7
[56] Coates A J *et al* 2007 *Geophys. Res. Lett.* **34** L22103
[57] Lavvas P *et al* 2013 *Proc. Natl Acad. Sci.* **110** 2729–34
[58] Jones G H *et al* 2009 *Geophys. Res. Lett.* **36** L16204
[59] Hill T W *et al* 2012 *J. Geophys. Res.* **117** A05209
[60] Grün E *et al* 1992 *Science* **257** 1550–2
[61] Zook H A *et al* 1996 *Science* **274** 1501–3
[62] Meyer-Vernet N *et al* 2009b *Geophys. Res. Lett.* **36** L03103
[63] Kempf S *et al* 2005 *Science* **307** 1274–6
[64] Meyer-Vernet N *et al* 1998 *Measurement Techniques in Space Plasmas: Fields Geophysical Monograph Series*) vol 103, ed Pfaff R *et al* (Washington, DC American Geophysical Union pp 205–10
[65] Meyer-Vernet N 1979 *J. Geophys. Res.* **84** 5373–7
[66] Meyer-Vernet N and Perche C 1989 *J. Geophys. Res.* **94** 2405–15
[67] Meuris P, Meyer-Vernet N and Lemaire J F 1996 *J. Geophys. Res.* **101** 24471–7
[68] Meyer-Vernet N 2001 *Proc. 7th Spacecraft Charging Technology Conf. (The Netherlands, 23–27 April 2001) ESA SP*) vol 476, ed R A Harris (Nordwijk: European Space Agency) pp 635–40
[69] Aubier M G, Meyer-Vernet N and Pedersen B M 1983 *Geophys. Res. Lett.* **10** 5–8
[70] Gurnett D A *et al* 1983 *Icarus* **53** 236–54
[71] Oberc P 1996 *Adv. Space Res.* **17** 105–10
[72] Meyer-Vernet N *et al* 2014 *Geophys. Res. Lett.* **41** 2716–20
[73] Pantellini F *et al* 2012 *Astrophys. Space Sci.* **341** 309–14
[74] Meyer-Vernet N 1985 *Adv. Space Res.* **5** 37–46
[75] Auer S 2001 *Interplanetary Dust* ed Grün E *et al* (Heidelberg: Springer) pp 385–444
[76] Göller J R and Grün E 1989 *Planet. Space Sci.* **37** 1197–206
[77] Burchell M J *et al* 1999 *Meas. Sci. Technol.* **10** 41–50
[78] Price M C *et al* 2012 *Meteorit. Planet. Sci.* **47** 684–95
[79] Pantellini F *et al* 2012 *Plasma Phys. Control. Fusion* **54** 045005
[80] McBride N and McDonnell J A M 1999 *Planet. Space Sci.* **47** 1005–13
[81] Lai S T, Murad E and McNeil W J 2002 *J. Spacecr. Rockets* **39** 106–14

[82] Collette *et al* 2014 *J. Geophys. Res.* **119** 1883–902
[83] Graps A L *et al* 2001 *Proc. Meteoroids Conf. (Kiruna, Sweden, 6–10 August 2001)* (Nordwijk: European Space Agency) pp 601–8
[84] Mann I, Murad E and Czechowski A 2007 *Planet. Space Sci.* **55** 1000–9
[85] Zaslavsky A *et al* 2012 *J. Geophys. Res.* **117** A05102
[86] Le Chat G *et al* 2013 *Solar Phys.* **286** 549–59
[87] Schippers P *et al* 2014 *Geophys. Res. Lett.* **41** 5382–88
[88] Pantellini F *et al* 2013 *Solar Wind 13 (AIP Conf. Proc.)* vol 1539, ed Zank G P *et al* (New York: American Institute of Physics) pp 414–7
[89] Grün E, Zook H A Fechtig H and Giese R H 1985 *Icarus* **62** 244–72
[90] Ceplecha Z *et al* 1998 *Space Sci. Rev.* **84** 327–471
[91] Juhasz A and Horanyi M 2013 *Geophys. Res. Lett.* **40** 2500–4
[92] Mann I and Czechowski A 2012 *Nanodust in the Solar System: Discoveries and Interpretations* (*Astrophysics and Space Science Library*) vol 385, ed Mann I *et al* (Heidelberg: Springer) pp 195–219
[93] Postberg F 2006 *Icarus* **183** 122–34
[94] O'Brien L *et al* 2014 *Rev. Sci. Instrum.* **85** 035113

Plasma Phys. Control. Fusion **57** (2015) 014016 (12pp) doi:10.1088/0741-3335/57/1/014016

Influence of the external electrical circuit on the regimes of a nanosecond repetitively pulsed discharge in air at atmospheric pressure

Fabien Tholin[1,2,3] **and Anne Bourdon**[1,2,4]

[1] CNRS, UPR 288 'Laboratoire d'Energétique Moléculaire et Macroscopique, Combustion', Grande voie des vignes, 92295 Châtenay-Malabry, France
[2] Ecole Centrale Paris, Grande voie des vignes, 92295 Châtenay-Malabry, France

E-mail: fabien.tholin@onera.fr and anne.bourdon@lpp.polytechnique.fr

Received 30 June 2014, revised 10 September 2014
Accepted for publication 19 September 2014
Published 28 November 2014

Abstract

This paper presents 2D simulations of nanosecond repetitively pulsed discharges in air at atmospheric pressure coupled with a model of the external electrical circuit used in experiments. Then, during the pulsed discharge, the voltage applied to the electrodes varies in time as a function of the time dependent value of the plasma channel conductivity. In this work, we have simulated several consecutive nanosecond pulsed discharges between two point electrodes in air initially at 1000 K at a frequency of 10 kHz. First, we have simulated three consecutive nanosecond spark discharges. We have shown that the air temperature increases significantly pulse after pulse in the discharge channel. As a consequence, for the three consecutive simulated nanosecond spark discharges, we have put forward a decrease in the discharge radius, pulse after pulse. Then, to further limit the discharge current, a ballast resistance R has been added into the electrical circuit and the results are presented for seven consecutive nanosecond discharges. For a value of $R = 1000\,\Omega$ in the conditions studied in this work, we have shown that the first nanosecond discharges are in the glow regime, with a small gas heating per pulse. However, as the number of pulses increases due to the gas heating by each pulse, the discharge may transit to a multipulse nanosecond spark regime. For a higher value of $R = 10\,000\,\Omega$, we have put forward that the gas heating by each nanosecond discharge becomes negligible and then the multipulse nanosecond discharge remains in this case in a stable 'quasi-periodic' multipulse glow regime.

Keywords: non-thermal discharges at atmospheric pressure, air discharges, spark discharges, glow discharges, nanosecond repetitively pulsed discharges, electrical circuit

(Some figures may appear in colour only in the online journal)

1. Introduction

For the last few years, nanosecond repetitively pulsed (NRP) discharges have been studied for various applications (Becker *et al* 2004) as plasma assisted combustion (Pilla *et al* 2006, Starikovskaia 2006) or nanomaterial synthesis (Pai 2011). In air at atmospheric pressure between two point electrodes, Pai *et al* (2010) studied the different discharge regimes of NRP discharges. It is interesting to note that even though NRP discharges are transient, the cumulative effect of repeated pulsing achieves a stable 'quasi-periodic' behavior. Pai *et al* (2010) identified three discharge regimes of NRP discharges as the applied voltage increases: Nanosecond corona, glow and spark regimes. Nanosecond corona and glow regimes are characterized by a negligible or small gas heating, contrary to

[3] Present address: ONERA, DMPH Department, 29 avenue de la Division Leclerc, 92322 Châtillon cedex, France
[4] Present address: Laboratory of Plasma Physics (CNRS, Ecole Polytechnique, Sorbonne Universities, University of Pierre and Marie Curie-Paris 6,University Paris-Sud), Ecole Polytechnique, route de Saclay, 91128 Palaiseau, France

nanosecond spark regimes. It is interesting to note that fast gas heating by nanosecond repetitively pulsed spark discharges has been shown to play a key role in plasma assisted combustion applications (Starikovskiy and Aleksandrov 2013). However, for other applications this gas heating may be problematic. Therefore, it is very important to be able to control and optimize the conditions to obtain the different regimes of NRP discharges.

In Pai *et al* (2010) and Tholin and Bourdon (2013a) it was shown experimentally and in numerical simulations that the nanosecond glow-to-spark transition depends strongly on the reduced Laplacian electric-field. During a multipulse glow discharge, as the gas heating is negligible, the applied reduced electric-field remains the same, pulse after pulse, as both the applied voltage and the neutral gas density remain the same. With no significant gas heating, the effects of previous pulsed discharges on the following one is mostly to change the pre-ionization level at the end of the interpulse period, as studied in Tholin and Bourdon (2013a).

In Xu *et al* (2011) and Tholin and Bourdon (2013b), it was shown experimentally and in numerical simulations that after a single nanosecond spark discharge, the fluid dynamics during the interpulse lead to the fast formation of a hot low density channel in the inter-electrode gap at the location of the nanosecond spark discharge. In Tholin and Bourdon (2013b) we only simulated one nanosecond spark discharge. Indeed, in the simulations, if one assumes that the voltage applied to the electrodes remains the same, the following pulsed discharges will then ignite and propagate in a lower density channel with a higher applied reduced electric field. As a consequence, the conductive current will increase, pulse after pulse. In experiments (Pai *et al* 2010, Rusterholtz *et al* 2013), the peak values of the conductive current remained less than or equal to about 40 A, pulse after pulse. Indeed, experimentally, the voltage applied to the electrodes decreases sharply as the current increases during the spark phase due to the external electrical circuit. Then the external electrical circuit limits the nanosecond spark discharge current, pulse after pulse.

So far, only a few studies have been devoted to the simulation of multi-pulse nanosecond spark discharges in air at atmospheric pressure (Naidis 2008, Naidis 2009). Naidis (2008) developed a 1D model to simulate the dynamics of multi-pulse nanosecond spark discharges and the neutral gas temperature evolution pulse after pulse. In particular, the establishment of a 'quasi-periodic' spark regime due to the balance between the gas heating by each discharge pulse and the different loss processes in the gas was shown. In Naidis (2009), the significant effect of the power-supply on the limitation of the current of nanosecond repetitively spark discharges in air at atmospheric pressure was simulated in 1D.

In this work, we propose to take into account the external electrical circuit used in the experiments (Pai *et al* 2010) and to couple it with a 2D model for nanosecond repetitively pulsed discharges. The objective is to study the influence of the external electrical circuit on the dynamics of successive nanosecond discharges and the transition between nanosecond repetitively pulsed glow and spark regimes. In section 2.1, we present the model used to simulate nanosecond repetitively pulsed discharges in air at atmospheric pressure in a point-to-point geometry. In section 2.2, we set out the model for the external electrical circuit. In section 3.1, we simulate several consecutive nanosecond spark discharges with an external electrical circuit. Finally, in sections 3.2 and 3.3, we study the influence of an additional ballast resistance in the electrical circuit on the multipulse discharge dynamics and on the transition between multipulse nanosecond glow and spark regimes.

2. Model formulation

2.1. Discharge model

We have shown in our previous numerical studies (Bourdon *et al* 2010, Tholin and Bourdon 2011, 2013a, 2013b), validated by a comparison with experiments in Tholin *et al* (2011), that during a nanosecond voltage pulse applied to two point electrodes in air at atmospheric pressure, first, a positive streamer propagates from the anode to the cathode and a negative streamer propagates in the opposite direction until they form a plasma channel. To simulate the streamer phase, we have used a classical 2D fluid model (Kulikovsky 1997), coupling drift-diffusion equations for electrons and positive and negative ions with Poisson's equation. In the charged species continuity equations, all the transport coefficients and rate coefficients are assumed to be functions of the local reduced electric field E/N, where E is the electric field magnitude and N is the air neutral density. In this work, we study axisymmetric streamers propagating between two hyperboloid point electrodes with a radius of curvature of R_p = 300 μm and separated by 2.5 mm. Then cylindrical coordinates (x, r), are used with the x-axis as the axis of the discharge. For all discharge simulations, a computational domain with a radius of R_{max} = 1.0 cm and a length L_{max} = 1.0 cm is used with a fixed non-uniform rectilinear grid, as defined in Tholin and Bourdon (2011).

In this work, we study the discharge dynamics at atmospheric pressure between two point electrodes in air initially at T_g = 1000 K. As discussed in our previous works, to take into account the fact that T_g = 1000 K in discharge simulations, we have only scaled the value of the total air density N_{1000} at 1000 K as $N_{1000} = 0.3 \times N_{300}$, where $N_{300} = 2.45 \times 10^{19}$ cm^{-3} is the density of air at 300 K. For transport parameters, reaction rates and the photoionization source term, the same models as in Tholin and Bourdon (2013a, 2013b) are used. The numerical methods and the boundary conditions are the same as in Tholin and Bourdon (2013a, 2013b). In particular, for the time integration during the voltage pulse, a simple first-order Euler time integration is used for fluxes and the photoionization source term and an explicit fourth order Runge Kutta method (Ferziger and Peric 2002) is used for other source terms. For stability and accuracy, the time step in our explicit simulations is limited by several considerations. The time scales of relevance are the Courant δt_c time scale, defined as in Vitello *et al* (1994), the chemistry time scale δt_S due to the ionization, attachment, recombination and detachment processes and the dielectric relaxation time scale $\delta t_D = \epsilon_0/(q_e\Sigma_i|n_i\,\mu_i|)$. The model time step is calculated as $\delta t = \min(A_c \delta t_c, A_S \delta t_S, A_D\, \delta t_D, 10^{-10})$ with $A_c = 0.5$, $A_S = 0.05$ and $A_D = 0.5$. The upper limit value of 10^{-10} s is used as a characteristic time step to follow accurately the time evolution of the nanosecond applied voltage. As an initial condition, we consider a low uniform preionization of the ambient air with 10^4 cm^{-3} electrons and positive ions, as discussed in Pancheshnyi (2005). The conductive current and the discharge energy are outputs of the discharge simulations and are calculated as in Tholin and Bourdon (2013b).

In Tholin and Bourdon (2011), it was shown that after the connection of both streamer discharges, the electric field in the plasma channel between the point electrodes rapidly becomes rather uniform and equal to the average electric field (i.e. the applied voltage divided by the gap length). If the value of the average electric field is higher than the breakdown field, the electron density goes on increasing while the electric field remains constant and the discharge transits to a spark phase. To model this spark phase, we use the same approach as in Tholin and Bourdon (2013b): The temperature during the discharge pulse is computed assuming that 30% of the discharge energy instantaneously heats the ambient air. In Tholin and Bourdon (2013b), we have shown that after a nanosecond spark discharge, due to the fast gas heating by the nanosecond spark, a shock wave propagates radially starting from the heated discharge channel. After the shock wave propagation, the pressure in the inter-electrode gap becomes uniform and equal to the atmospheric pressure. The characteristic time of the shock wave propagation is close to the gap size divided by the speed of sound ($\simeq 1\ \mu$s). In this work, we simulate nanosecond repetitively pulsed discharges at 10 kHz and then the inter-pulse duration is 100 μs, much longer than the characteristic time of the shock wave propagation. Then, as a first approximation, we assume that the neutral gas density is scaled according to the ideal gas law instantaneously just after the pulse, assuming a constant pressure. It was shown in Tholin and Bourdon (2013b) that the generated shock waves remove a significant fraction of the energy deposited into the inter-electrode gap ($\simeq 50\%$). Then the constant pressure approach overestimates the density decrease in the post-discharge. However, this constant pressure hypothesis allows us to avoid the resolution of the entire set of Euler equations during the inter-pulse and gives qualitative results on the influence of the air density variations pulse after pulse. Finally, during the interpulses, we consider that the gas temperature remains constant and we simply take into account the chemistry and diffusion of the charged species as in Tholin and Bourdon (2013a). This approximation is valid when the gas heating by each discharge pulse is small or for a few consecutive nanosecond spark discharges at 10 kHz. On longer timescales, different loss processes as thermal conduction in the gas (Naidis 2008, Naidis 2009) and heat transfer to the electrodes have to be considered.

2.2. External circuit model

In the experiments (Pai *et al* 2010, Xu *et al* 2011, Rusterholtz *et al* 2013), FID Technologies generators (Model FPG 1030MS and FPG 30100MC4K) are used to produce repetitive nanosecond voltage pulses and are connected with BNC coaxial cables to the two point electrodes. For 10 ns voltage pulses, the propagation distance of a voltage pulse through electric cables is around 3 m. It is then reasonable to assume that the applied voltage on the electrodes is independent of the power supply during the pulsed discharge. In this work, we model the 5 ns-long applied voltage pulse delivered by the power supply with a sigmoid shape as in Tholin and Bourdon (2013b) with rise and decrease times of 2 ns and a voltage plateau. When the voltage pulse reaches the anode, one part of the voltage and the current will be transmitted and one part will be reflected depending on the impedances of the plasma and of the coaxial cable. Detailed calculations about the propagation of current and voltage waves in BNC cables can be found in many textbooks on the transmission line theory as in Ramo *et al* (1993). Basically, a BNC cable is characterized by a linear inductance Λ (H m^{-1}) and a linear capacitance Γ (F m^{-1}). If we neglect the linear resistance of the cable, it can be modeled by the simple circuit as shown in figure 1 and the voltage $u(x, t)$ and the intensity $i(x, t)$ at the axial position x of the cable and at time t obey the following system of differential equations, often referred to as the telegraph equations:

$$\begin{cases} \partial_x i(x,t) &= -\Gamma \partial_t u(x,t) \\ \partial_x u(x,t) &= -\Lambda \partial_t i(x,t) \end{cases} \tag{1}$$

Combining both equations results in a wave equation for both $u(x, t)$ and $i(x, t)$ in the form:

$$\partial_{tt} u(x,t) - \frac{1}{c^2}\partial_{xx} u(x,t) = 0 \tag{2}$$

where $c = 1/\sqrt{\Gamma\Lambda}$ is the speed of the voltage wave in the cable. If we consider a voltage wave propagating in a BNC cable from the power supply to the anode of a point-to-point electrode set-up, the voltage $u(x, t)$ follows the wave equation (equation (2)) which means that it can be written in the form $u(x - ct)$. Then, using this expression in equation (1) one can derive:

$$u(x,t) = \frac{1}{\Gamma c} i(x,t) \tag{3}$$

For a reflected voltage wave propagating in the opposite direction, from the anode to the power supply, $u_{\mathrm{ref}}(x, t)$ can be written in the form $u_{\mathrm{ref}}(x + ct)$. In this case the relation between $u_{\mathrm{ref}}(x, t)$ and $i_{\mathrm{ref}}(x, t)$ is given by:

$$u_{\mathrm{ref}}(x,t) = -\frac{1}{\Gamma c} i_{\mathrm{ref}}(x,t) \tag{4}$$

Finally, the total voltage in the BNC cable is given by:

$$u_{\mathrm{tot}}(x,t) = u(x,t) + u_{\mathrm{ref}}(x,t) = Z_{\mathrm{cable}}\,(i(x,t) - i_{\mathrm{ref}}(x,t)\,) \tag{5}$$

where $Z_{\mathrm{cable}} = \sqrt{\Lambda/\Gamma}$ is the real impedance of the BNC cable. As shown in figure 1, the plasma and the electrode set-up can be modelled by an output impedance Z_o. The main difficulty of the model is to estimate accurately the value of Z_o. In experiments, as the BNC cable is connected to the electrodes, the plasma and the electrode set-up forms a loop, as shown on figure 1. Assuming that the plasma resistance is much higher than the resistance of the wires of the electrode set-up, such

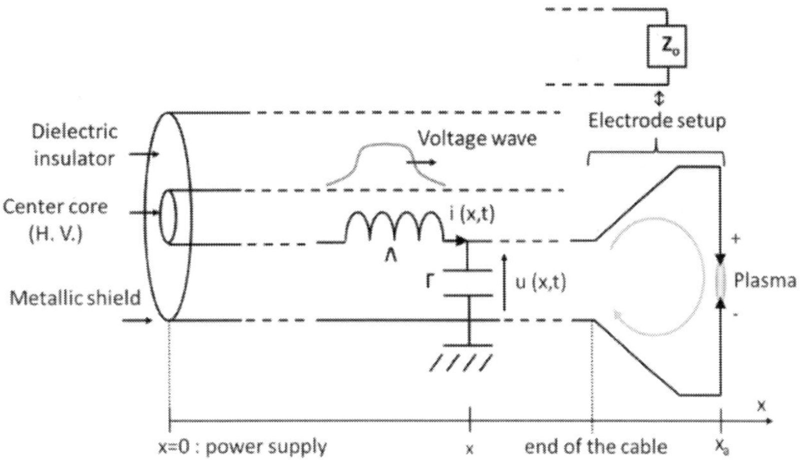

Figure 1. Schematic view of a BNC coaxial cable and the electrode set-up and its equivalent electrical circuit model.

a loop can be modeled by a circuit in series with a resistance R_{plasma}, corresponding to the resistance of the plasma and an inductance $\mathcal{L}$, corresponding to the self-inductance of the loop. The value of the self-inductance of a circular loop may be estimated by Elliott (1993):

$$\mathcal{L}_{\text{circle}} \approx r\mu_0 \ln\left(\frac{8r}{a} - 2\right) \tag{6}$$

where r is the radius of the loop, a is the radius of the wire and μ_0 is the magnetic constant. Assuming that $r \simeq 10$ mm and $a \simeq 1$ mm, we have estimated that $\mathcal{L}_{\text{circle}} \simeq 10^{-8}$ H. For a rectangular loop, or any other geometry of the same scale, the self-inductance remains quite similar and depends only slightly on the value of a. Then it is reasonable to assume that even in a real experimental electrode set-up with a loop of typical size of around $r \simeq 10$ mm, the inductance will remain of the same order. In those conditions, the inductance of the electrode set-up and the plasma is much smaller than the inductance of a BNC cable a few meters in length with a typical linear inductance of $\Lambda \simeq 0.25\,\mu\text{H m}^{-1}$ and a linear capacitance of $\Gamma \simeq 100$ pF m^{-1}. Then, as a first approximation, we have neglected the inductance of the electrode set-up and the plasma and we have assumed that $Z_0 \simeq R_{\text{plasma}}(t)$. Then, applying Ohm's law to the plasma one gets:

$$u_{\text{tot}}(x_a, t) = R_{\text{plasma}}(t)\,(i(x_a, t) + i_{\text{ref}}(x_a, t)\,) \tag{7}$$

where x_a is the axial position of the anode along the cable. Combining equation (5) with $x = x_a$ and equation (7) gives the relation between the transmitted and the reflected components of the voltage and the current at the anode:

$$\frac{i_{\text{ref}}}{i} = -\frac{u_{\text{ref}}}{u} = \frac{Z_{\text{cable}} - R_{\text{plasma}}}{Z_{\text{cable}} + R_{\text{plasma}}} = r_{\text{cable}} \tag{8}$$

where r_{cable} is called the reflection coefficient. Based on equation (8) several situations can be considered. First, with no plasma in the inter-electrode gap, we can consider that $Z_{\text{plasma}} \to \infty$. Then, according to equation (8) $u_{\text{ref}} = u$ and the total applied voltage u_{tot} is two times the voltage of the incident voltage wave u. On the other hand in this case, the transmitted and reflected currents are opposite, such that the total conductive current is exactly zero. Conversely, if we consider a plasma of infinite conductivity (i.e. $R_{\text{plasma}} = 0\,\Omega$), then $i_{\text{ref}} = i$ and the reflected voltage is exactly the opposite of the incident voltage. In this case, the total applied voltage in the inter-electrode gap is zero. Finally, if the impedance of the plasma is exactly the same as the impedance of the cable, no reflection is observed and all the incident voltage and current are transmitted into the plasma. In the discharge code, it is possible to calculate explicitly the resistance of the plasma as a function of time by considering the total voltage and the transmitted current at the previous time-step:

$$u_{\text{tot}}^{t+\text{d}t} = u^{t+\text{d}t}(1 - r_{\text{cable}}{}^{t}) \tag{9}$$

with:

$$r_{\text{cable}}{}^{t} = \frac{Z_{\text{cable}} - R_{\text{plasma}}^{t}}{Z_{\text{cable}} + R_{\text{plasma}}^{t}} \quad \text{and} \quad R_{\text{plasma}}^{t} = \frac{u_{\text{tot}}^{t}}{i_{\text{tot}}^{t}}$$

where u^t is the voltage delivered by the power supply. This approach is quite accurate provided that the variation of R_{plasma} remains small during one time-step dt. According to the Drude–Lorentz formula for conductivity, this hypothesis is verified as long as the variations of the averages (over the whole plasma volume) of the electron density and of the collision frequency remain small during one time-step. As explained in section 2.1, the chemical time-step during the streamer discharge ensures that the variation of the electron density remains less than 10% during each iteration. During the streamer propagation phase, the electric-field may change rapidly but in a very narrow region, such that the averaged electric-field in the whole discharge volume is almost proportional to the applied voltage and does not change very rapidly. Moreover during this phase, the conductive current is very small and the reflection coefficient r_{cable} is almost always equal to −1. During the spark phase, the chemistry is computed implicitly but the electron density varies slowly due to the relatively low homogeneous electric-field and the applied voltage is updated frequently (Tholin and Bourdon 2013b). It is the same for the collision

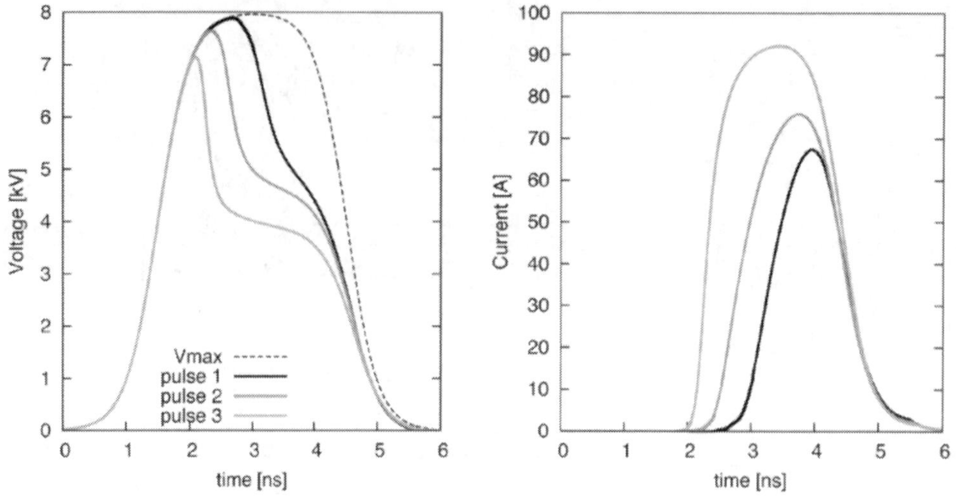

Figure 2. Time evolutions of the applied voltage to the electrodes and the discharge conductive current for three consecutive nanosecond spark discharges at 10 kHz with the circuit model in section 2.2 without ballast resistance. The applied voltage is 8 kV and the gap size is 2.5 mm. The initial air temperature is $T_g = 1000$ K. For each pulse, the start time is the beginning of each voltage pulse. The dashed line represents the ideal applied voltage.

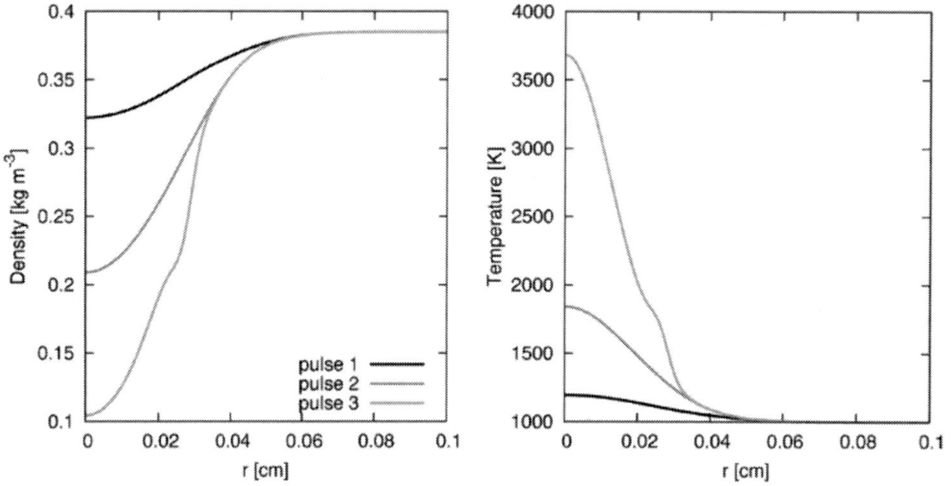

Figure 3. Radial profiles in the middle of the gap of the air density and temperature at the end of each voltage pulse (i.e. 6 ns after the start of each voltage pulse) for the same conditions as shown in figure 2. As explained in the text, these radial evolutions are identical at the beginning of the next voltage pulse.

frequency as the electric-field during the spark phase evolves proportionally to the applied voltage.

Then, with the circuit model presented in this section, the voltage applied to the electrodes varies in time as a function of the time dependent value of the plasma channel conductivity. As already mentioned, without plasma, the applied voltage is two times the voltage delivered by the power supply. In practice, the impedance of the plasma is much higher than the impedance of the cable. When the conductivity of the plasma increases, its impedance decreases and slowly converges to a value close to the cable impedance. In this case, no reflection is observed and the applied voltage is exactly the one delivered by the power supply. The value of the real impedance of the cable Z_{cable} is therefore a key parameter of the coupling between the discharge and the external electrical circuit. As in the experiments, in this work, we have used $Z_{\text{cable}} = 50\Omega$. When Z_{cable} increases, all other conditions being unchanged, the impedance of the plasma converges more rapidly to this value during the spark phase and a faster voltage decrease and a more significant current limitation can be obtained. The cable impedance can be increased either by changing the BNC cable properties (e.g. its linear inductance), or by adding a ballast resistance in series with the anode. Adding a ballast resistance is a very well-known experimental method to limit the current of a DC discharge (Packan 2003, Becker *et al* 2004). In section 3.1, we consider as a reference, an external electrical circuit without ballast resistance. Then, in sections 3.2 and 3.3, the influence of an additional ballast resistance in the electrical circuit on the multipulse discharge dynamics is studied.

A limitation of the circuit model presented in this section is that the propagation of the applied potential wave inside the plasma volume is considered as instantaneous.

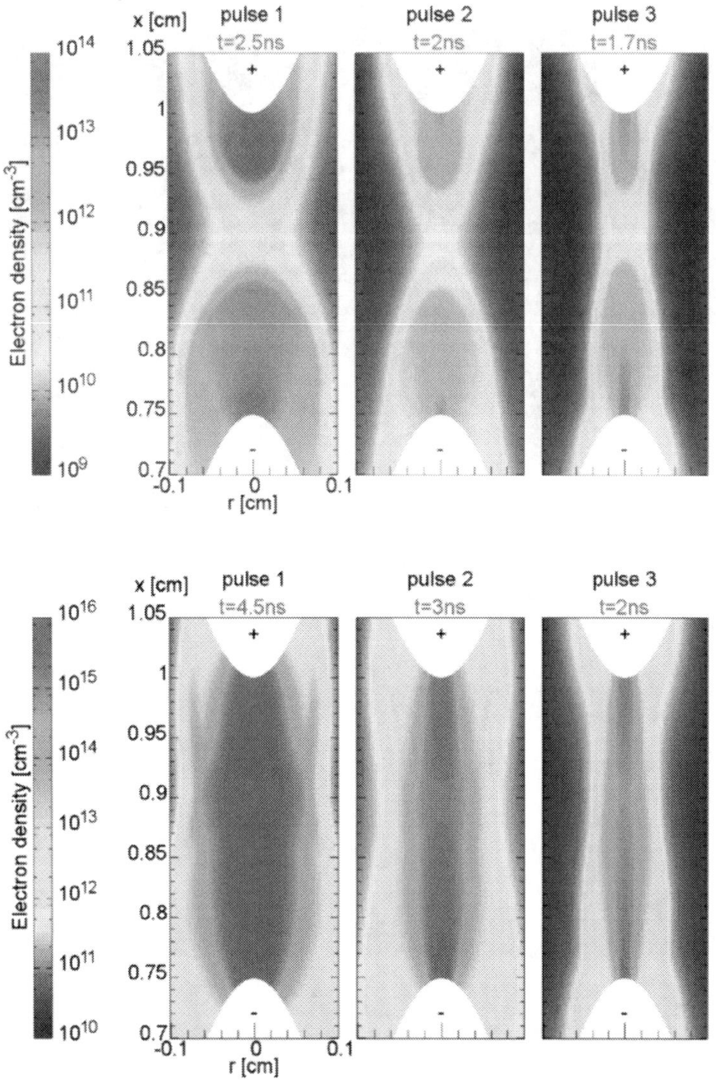

Figure 4. Cross-sectional views of the electron density for the three consecutive discharges as shown in figure 2. For each pulse, the origin of time is the beginning of the considered voltage pulse. For the top figures, the times are chosen to correspond to the same axial position of the positive discharge front in the inter-electrode gap. For the bottom figures, the times correspond to the arrival of the positive discharge at the cathode and the beginning of the spark regime.

A better description would require computing the potential and the electric-field with Maxwell's equations instead of simply Poisson's equation. However, the propagation time of light inside an inter-electrode gap of 2.5 mm is around 10^{-12} s and in the conditions studied in this work, u_{tot} at the anode is varying on much longer time-scales ($\simeq 10^{-10}$ s) and therefore the electrostatic approximation remains reasonable.

3. Results and discussion

3.1. Simulation of multipulse nanosecond spark discharges with an external electrical circuit without ballast resistance

As a reference test-case for multipulse nanosecond spark air discharges between two point electrodes separated by 2.5 mm, we consider a nanosecond repetitively pulsed voltage at 10 kHz with an applied voltage of 8 kV and an external circuit without ballast resistance ($R = 0\ \Omega$). The initial air temperature is $T_g = 1000$ K. Figure 2 shows the time evolutions of the applied voltage to the electrodes and the discharge conductive current during the first three voltage pulses. The dashed line represents the voltage applied to the electrodes without an external electrical circuit as in Tholin and Bourdon (2013b). It is referred to as the ideal applied voltage in the following. It is clearly apparent that during the three consecutive voltage pulses, the applied voltage is equal to the ideal applied voltage during the first phase of each voltage pulse. This phase corresponds to the ignition and propagation of the two streamer discharges in the inter-electrode gap up to their connection. During this phase, the discharge conductive current remains very small ($r_{cable} \simeq -1$). After the connection and during the nanosecond spark phase, the discharge current increases significantly as the electron density and the plasma conductivity increase. For this reason, the applied voltage to the point

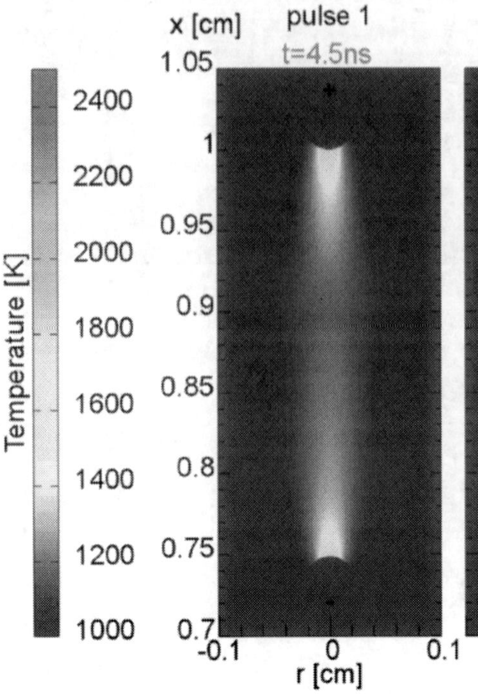

Figure 5. Cross-sectional views of the air temperature for the three consecutive discharges as shown in figure 2. The times are the same as those of the bottom figures as shown in figure 4.

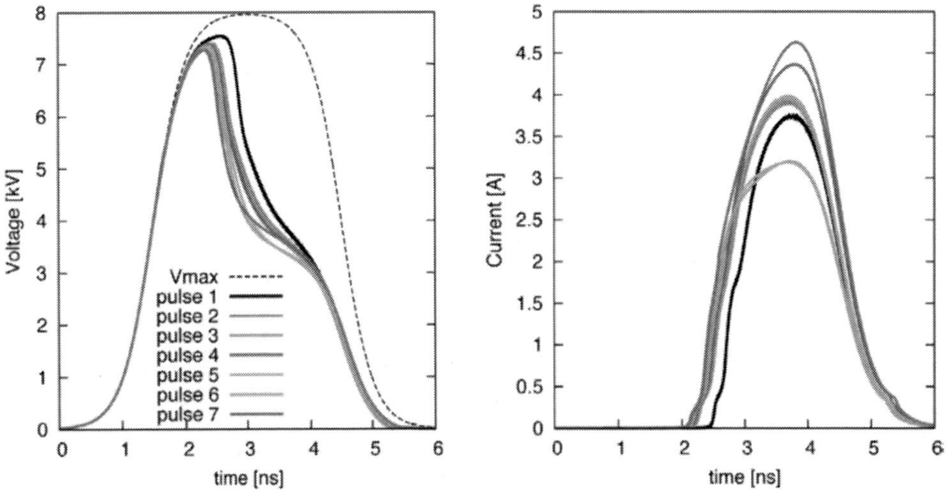

Figure 6. Time evolutions of the applied voltage to the electrodes and the discharge conductive current for seven consecutive nanosecond discharges at 10 kHz with the circuit model in section 2.2 with an additional ballast resistance $R = 1000\ \Omega$. The applied voltage is 8 kV and the gap size is 2.5 mm. The initial air temperature is $T_g = 1000$ K. For each pulse, the start time is the beginning of the considered voltage pulse. The dashed line represents the ideal applied voltage.

electrodes decreases rapidly compared to the ideal applied voltage. Then, it forms a plateau around a voltage value which is slightly higher than one half of the ideal applied voltage. As explained in section 2.2 this phase corresponds to a situation in which the plasma resistance becomes nearly equal to the cable's real impedance. Between the first voltage pulse and the third one, we note that the voltage plateau becomes longer and at a lower voltage value.

For the same conditions as in figures 2 and 3 shows the 1D radial evolutions of the air density and temperature in the middle of the inter-electrode gap, 6 ns after the start of each voltage pulse (i.e. at the end of each voltage pulse). It is important to note that in this work, the gas temperature and air density are assumed to be constant during the inter-pulse and then these radial evolutions are identical at the beginning of the next voltage pulse. While the air temperature increases significantly pulse after pulse, the air density decreases by almost a factor of three between the first pulsed discharge and the third one. For this reason the spark regime occurs earlier pulse after pulse and is more and more energetic pulse after pulse. We

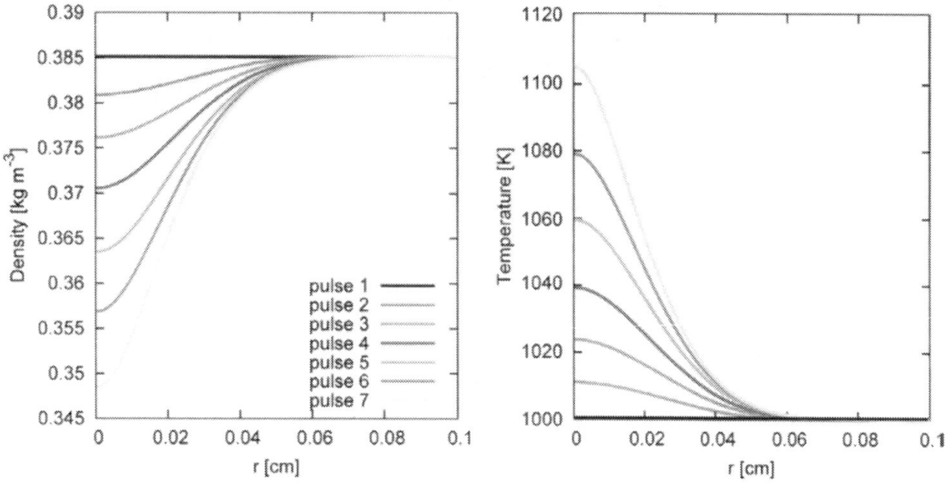

Figure 7. Radial profiles in the middle of the gap of the air density and temperature just at the beginning of each voltage pulse for the seven consecutive discharges as shown in figure 6.

note that the maximum of the discharge conductive current increases pulse after pulse and reaches a high value of 90 A for the third pulsed discharge. However, it is important to note that the electrical circuit model used in this work allows to significantly reduce the current in the spark discharge. Indeed, we have checked that for the same conditions as shown in figure 2, without the electrical circuit model, the discharge conductive current reaches in a few pulses values of about 2000 A. The fact that the current increases pulse after pulse, as shown in figure 2, indicates that the electrical circuit model is not able to compensate completely for the effect of the increasing reduced electric-field due to the fast gas-heating.

To show that the air density decrease has a major influence on the discharge structure, figure 4 shows the 2D distribution of the electron density for the three consecutive pulses as shown in figures 2 and 3 during the propagation phase of the two streamers in the inter-electrode gap. The times presented refer to the beginning of each voltage pulse. For the top figures, the times are chosen to correspond to the same axial position of the positive discharge front in the inter-electrode gap for the three consecutive voltage pulses. For the bottom figures, the times correspond to the arrival of the positive discharge front at the cathode and then to the beginning of the spark regime as discussed in Tholin and Bourdon (2011). We observe that both positive and negative discharges are influenced by the air density variation pulse after pulse: Both discharges tend to ignite and to propagate faster when they are ignited in air with a lower density. We also note that the discharge filament radius decreases pulse after pulse. It is interesting to note that the rapid decrease in the radius of an air nanosecond discharge pulse after pulse due to gas-heating was been obtained in the simulations carried out by Nikipelov *et al* (2011). Figure 5 shows the 2D distribution of the air temperature due to the fast gas-heating by the three consecutive nanosecond spark discharges for the same times as for the bottom figures in figure 4. For each pulse, two hot spots are observed close to the electrode tips and as expected the gas heating becomes more intense pulse after pulse.

In this work, we have simulated only the three first consecutive nanosecond spark discharges at a frequency of 10 kHz. In the experiments and 1D simulations carried out in Naidis (2009), for a larger number of nanosecond spark discharge pulses, the establishment of a 'quasi-periodic' spark regime was observed. The simulation of the 'quasi-periodic' spark regime requires taking into account the different loss processes occurring on longer timescales as the thermal conduction in the gas, the heat transfer to the electrodes and the influence of the air flow used in Pai *et al* (2010). These loss processes have a negligible influence on the first multipulse spark discharges at 10 kHz and then on the results presented in this work. In a future work, it would be interesting to take into account the different loss processes and to simulate the dynamics of formation of the 'quasi-periodic' spark regime.

3.2. Simulation of multipulse nanosecond spark discharges with an external electrical circuit with a ballast resistance

To increase the current limitation by the electrical circuit, in this section we add a ballast resistance R to the cable impedance Z_{cable} in the electrical circuit model presented in section 2.2. Figure 6 shows the time evolutions of the applied voltage and the discharge conductive current for the same conditions as in the previous section, but with an additional ballast resistance of $R = 1000\ \Omega$. In this case, we simulate seven consecutive nanosecond discharges at a repetition frequency of 10 kHz. It is interesting to note that the maximum current during the seventh voltage pulse is around 4.5 A, while it is about 90 A for the third voltage pulse as shown in figure 2 for $R = 0\ \Omega$. As expected, these results clearly show that the ballast resistance has a major influence on the multipulse discharge dynamics and structure. However, it is interesting to note that the evolution of the maximum of the discharge conductive current shows a complex non-monotonous evolution pulse after pulse since it increases between pulses 1 and 4, then decreases for pulses 5 and 6 and finally increases again for pulse 7. Such a complex evolution is due to the strong non-linear coupling between the electrical circuit model, the reduced electric-field, the evolution of the discharge structure and the energy released in the discharge. Figure 7 shows the

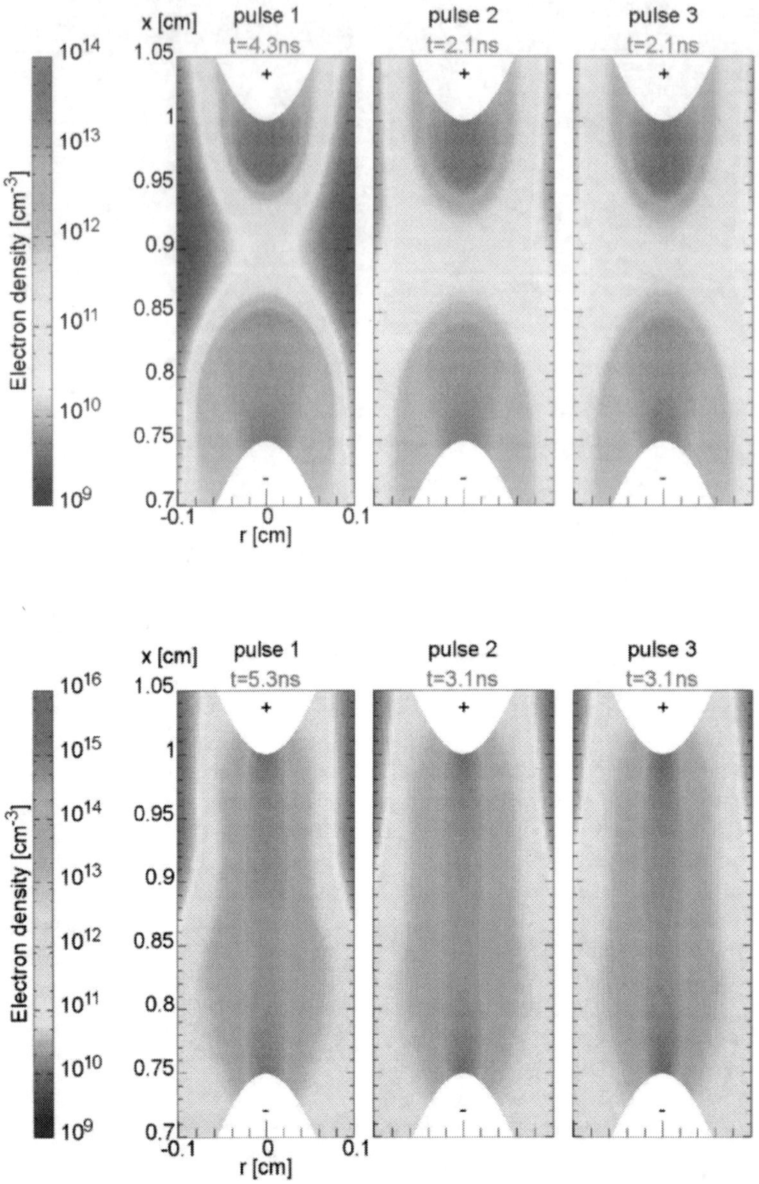

Figure 8. Cross-sectional views of the electron density for the three first consecutive discharges as shown in figure 6. For each pulse, the start time is the beginning of the considered voltage pulse. For the top figures, the times are chosen to correspond to the same axial position of the positive discharge front in the inter-electrode gap. For the bottom figures, the times correspond to the arrival of the positive discharge at the cathode and the beginning of the spark regime.

radial profiles in the middle of the gap of the air density and the temperature just at the beginning of each voltage pulse for the seven consecutive discharges as shown in figure 6. It is interesting to note that after three voltage pulses, the air temperature increase as shown in figure 7 is at maximum of 40 K instead of 2500 K as shown in figure 3. Correspondingly, the air density decrease after three voltage pulses is of about 2%, as shown in figure 7 and of 70% in figure 3.

Figure 8 shows the 2D distribution of the electron density for the three first consecutive discharges at two different times during each voltage pulse for the same conditions as shown in figure 6. The times presented refer to the beginning of each voltage pulse. For the top figures, the times are chosen to correspond to the same axial position of the positive discharge front in the inter-electrode gap for the three consecutive pulses. For the bottom figures, the times correspond to the arrival of the positive discharge front at the cathode and then to the beginning of the spark regime as discussed in Tholin and Bourdon (2011). We note that the discharge structure is very close for the three voltage pulses, even if during the first voltage pulse the positive discharge ignites slightly later than for the two next voltage pulses due to the low initial preionization level. Figure 9 shows the 2D distribution of the air temperature during these three first discharges for the same times as for the bottom figures in figure 8. For the two first voltage pulses, we observe that a slightly heated channel is obtained in the discharge path, with a non-uniform axial temperature profile with a hot spot close to the cathode tip and a more

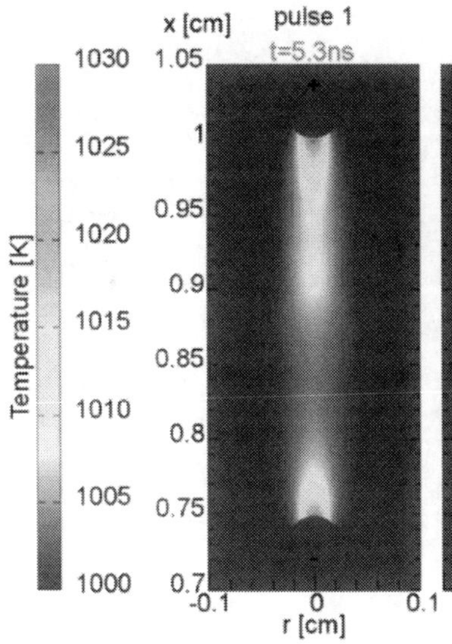

Figure 9. Cross-sectional views of the air temperature for the three first consecutive discharges as shown in figure 6. The times are the same as those for the bottom figures in figure 8.

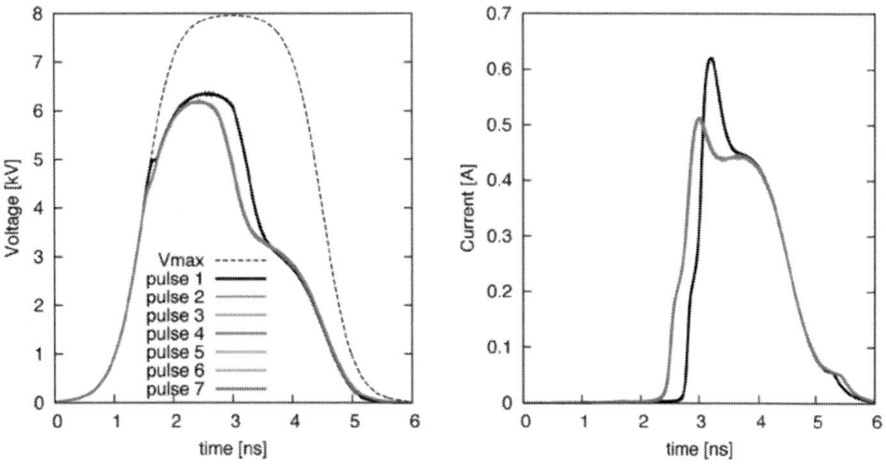

Figure 10. Time evolutions of the applied voltage to the electrodes and the discharge conductive current for seven consecutive nanosecond discharges at 10 kHz with the circuit model in section 2.2 with an additional ballast resistance $R = 10000\ \Omega$. The applied voltage is 8 kV and the gap size is 2.5 mm. The initial air temperature is $T_g = 1000$ K. For each pulse, the start time is the beginning of the considered voltage pulse. The dashed line represents the ideal applied voltage.

extended heated region close to the anode tip. For the third pulse, the axial temperature evolution is smoother. Then, due to the current limitation with $R = 1000\ \Omega$, the intense multipulse spark discharge obtained in the previous section is inhibited during the first voltage pulses. However, after seven consecutive pulses, figure 7 shows a gas temperature increase of 160 K. Then, on longer timescales, with a larger number of pulses, this multipulse discharge may evolve towards a more energetic multipulse nanosecond spark discharge.

3.3. Use of a ballast resistance to obtain a stable 'quasi-periodic' multipulse glow regime

In Tholin and Bourdon (2013a), we simulated the stable 'quasi-periodic' multipulse glow discharge observed in experiments in Pai *et al* (2010). In these simulations, no external electric circuit was considered and we neglected the gas heating by each pulsed discharge. In this section, we propose to check that with a sufficiently high value of the ballast resistance in the external electrical circuit, it is possible to obtain the stable 'quasi-periodic' multipulse glow discharge regime and to avoid the glow to spark transition after a few discharge pulses. For example, for a ballast resistance of $10000\ \Omega$, figure 10 shows that the discharge current and the applied voltage for seven consecutive discharges are almost identical. The maximum current remains very small for all the pulses and is around 0.5 A. Figure 11 shows the 2D distribution of the electron density for the three first discharges during each voltage pulse. For each pulse, the start time is the beginning of the considered voltage pulse. The results are presented 3 ns after the beginning of each voltage pulse, which corresponds to the arrival of the positive discharge front at the cathode.

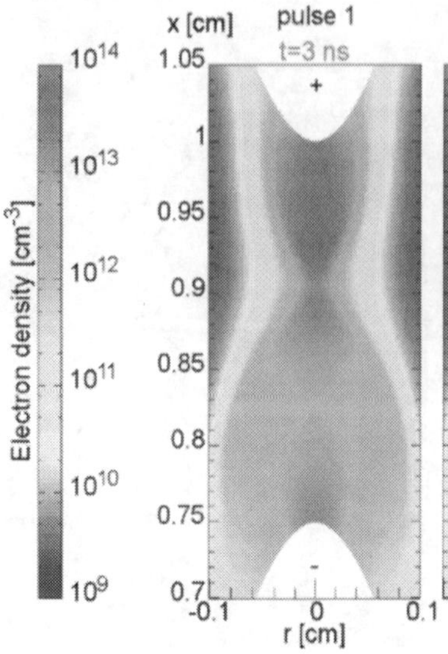

Figure 11. Cross-sectional views of the electron density for the three first consecutive discharges as shown in figure 10. For each pulse, the start time is the beginning of the considered voltage pulse. The results are presented 3 ns after the beginning of each voltage pulse, which corresponds to the arrival of the positive discharge front at the cathode.

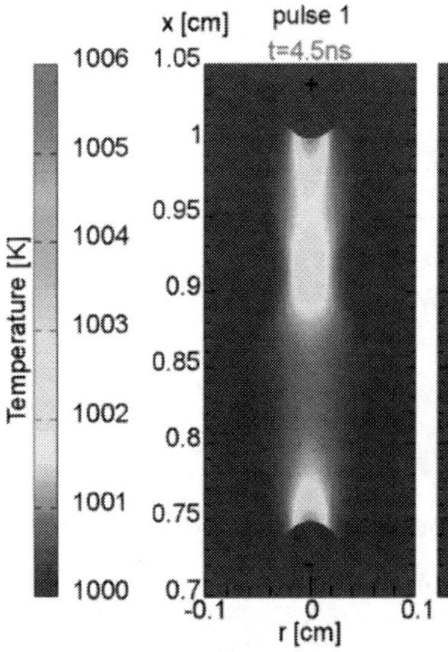

Figure 12. Cross-sectional views of the air temperature for the three first consecutive discharges as shown in figure 10. For each pulse, the start time is the beginning of the considered voltage pulse. The results are presented 4.5 ns after the beginning of each voltage pulse, which corresponds to the time for which the maximum of air temperature is reached.

We note that the discharge structure looks exactly the same between the first three voltage pulses with a constriction in the middle of the gap and a large variation in the discharge radius as a function of the axial position in the gap. This discharge structure is very different from the one obtained in figure 4 with the external electrical circuit model but no ballast resistance. Figure 12 shows the 2D distribution of the air temperature during these three first discharges with $R = 10\,000\,\Omega$. The results are presented 4.5 ns after the beginning of each voltage pulse, which corresponds to the time for which the maximum of air temperature is reached. We observe that the current limitation is so strong with $R = 10\,000\,\Omega$ that the nanosecond discharge remains in the glow regime at 1000 K with a very small temperature increase of about 6 K at maximum, located in three hot spot regions (close to electrode tips and in the gap). Then, significantly increasing the ballast resistance is an efficient way to obtain a stable 'quasi-periodic' multipulse glow regime.

4. Conclusions

The principal results and contributions, which follow from the studies presented in this paper on the influence of the external electrical circuit on the regimes of a nanosecond repetitively pulsed discharge in air at atmospheric pressure, can be summarized as follows:

- In taking into account a model for the external circuit used in experiments, we have shown that the voltage applied to the electrodes decreases dynamically when the conductivity of the plasma increases. Therefore, we have been able to simulate several consecutive nanosecond discharges in air at atmospheric pressure at a frequency of 10 kHz.
- First, we simulated three consecutive nanosecond spark discharges. We showed that the air temperature increases, whereas the air density correspondingly decreases significantly pulse after pulse in the discharge channel. As a consequence, for the three simulated consecutive nanosecond spark discharges, we put forward a decrease of the discharge radius, pulse after pulse.
- Then, we showed that an additional ballast resistance in the electrical circuit could be used to further limit the discharge current. For a value of $R = 1000\ \Omega$ in the conditions studied in this work, we simulated seven consecutive nanosecond discharges. We showed that the first nanosecond discharges are in the glow regime, with a small gas heating per pulse. However, as the number of pulses increases, due to the small gas heating by each pulse, the discharge may transit to a multipulse nanosecond spark regime. For a higher value of the ballast resistance ($R = 10\,000\ \Omega$ in the conditions studied in this work), we put forward that the gas heating by each nanosecond discharge becomes negligible and then the multipulse nanosecond discharge remains in this case in a stable 'quasi-periodic' multipulse glow regime.

Acknowledgments

The authors thank the Agence Nationale de la Recherche for its support of the PLASMAFLAME (Grant No. ANR-11-BS09-025-01) and the DRACO (Grant No. ANR-13-IS09-0004) projects and acknowledge the computational resources of the Mesocentre of Ecole Centrale Paris. The authors thank Professor J-H Paillol for their very helpful and stimulating discussions on this work and Drs D Xu, D Rusterholtz-Duval, D Pai, D Lacoste and Professor C Laux for helpful discussions on their experimental results on NRP discharges.

References

Becker K H, Kogelschatz U, Schoenbach K H and Barker R J 2004 *Non-Equilibrium Air Plasmas at Atmospheric Pressure* (Boca Raton, FL: CRC Press)

Bourdon A, Bonaventura Z and Celestin S 2010 Influence of the pre-ionization background and simulation of the optical emission of a streamer discharge in preheated air at atmospheric pressure between two point electrodes *Plasma Sources Sci. Technol.* **19** 034012

Elliott R S 1993 *Electromagnetics: History, Theory and Applications (IEEE Press Series on Electromagnetic Waves)* (Piscataway, NJ: IEEE Press)

Ferziger J and Peric M 2002 Computational Methods for Fluid Dynamics 3rd edn (New York: Springer)

Kulikovsky A A 1997 Positive streamer between parallel plate electrodes in atmospheric pressure air *J. Phys. D: Appl. Phys.* **30** 441–50

Naidis G V 2008 Simulation of spark discharges in high-pressure air sustained by repetitive high-voltage nanosecond pulses *J. Phys. D: Appl. Phys.* **41** 234017

Naidis G V 2009 Simulation of streamer-induced pulsed discharges in atmospheric pressure air *Eur. Phys. J. Appl. Phys.* **47** 22803

Nikipelov A, Popov I B, Pancheshnyi S and Starikovskii A Y 2011 Localized and distributed behavior of nanosecond pulse-periodic discharge in air *Proc. of the 30th Int. Conf. on Phenomena in Ionized Gases (Belfast, Northern Ireland, 28 August–2 September 2011)* pp 28–31

Packan D 2003 Repetitive nanosecond glow discharge in atmospheric pressure air *PhD Thesis* Stanford University

Pai D Z 2011 Nanomaterials synthesis at atmospheric pressure using nanosecond discharges *J. Phys. D: Appl. Phys.* **44** 174024

Pai D Z, Lacoste D A and Laux C O 2010 Transitions between corona, glow and spark regimes of nanosecond repetitively pulsed discharges in air at atmospheric pressure *J. Appl. Phys.* **107** 093303

Pancheshnyi S 2005 Role of electronegative gas admixtures in streamer start, propagation and branching phenomena *Plasma Sources Sci. Technol.* **14** 645–53

Pilla G, Galley D, Lacoste D A, Lacas F, Veynante D and Laux C O 2006 Stabilization of a turbulent premixed flame using a nanosecond repetitively pulsed plasma *IEEE Trans. Plasma Sci.* **34** 2471–7

Ramo S, Whinnery J R and Duzer T V 1993 *Fields and Waves in Communication Electronics* (New York: Wiley)

Rusterholtz D L, Lacoste D A, Stancu G D, Pai D Z and Laux C O 2013 Ultrafast heating and oxygen dissociation in atmospheric pressure air by nanosecond repetitively pulsed discharges *J. Phys. D: Appl. Phys.* **46** 464010

Starikovskaia S M 2006 Plasma assisted ignition and combustion *J. Phys. D: Appl. Phys.* **39** R265

Starikovskiy A and Aleksandrov N 2013 Plasma-assisted ignition and combustion *Prog. Energy Combust. Sci.* **39** 61–110

Tholin F and Bourdon A 2011 Influence of temperature on the glow regime of a discharge in air at atmospheric pressure between two point electrodes *J. Phys. D: Appl. Phys.* **44** 385203

Tholin F and Bourdon A 2013a Simulation of the stable 'quasi-periodic' glow regime of a nanosecond repetitively pulsed discharge in air at atmospheric pressure *Plasma Sources Sci. Technol.* **22** 045014

Tholin F and Bourdon A 2013b Simulation of the hydrodynamic expansion following a nanosecond pulsed spark discharge in air at atmospheric pressure *J. Phys. D: Appl. Phys.* **46** 365205

Tholin F, Rusterholtz D L, Lacoste D A, Pai D Z, Celestin S, Jarrige J, Stancu G D, Bourdon A and Laux C O 2011 Images of a nanosecond repetitively pulsed glow discharge between two point electrodes in air at 300 K and at atmospheric pressure *IEEE Trans. Plasma Sci.* **39** 2254–5

Vitello P A, Penetrante B M and Bardsley J N 1994 Simulation of negative-streamer dynamics in nitrogen *Phys. Rev.* E **49** 5574–98

Xu D A, Lacoste D A, Rusterholtz D L, Elias P-Q, Stancu G D and Laux C O 2011 Experimental study of the hydrodynamic expansion following a nanosecond repetitively pulsed discharge in air *Appl. Phys. Lett.* **99** 121502

IOP Publishing

Plasma Physics and Controlled Fusion

Plasma Phys. Control. Fusion **57** (2015) 014017 (6pp)

doi:10.1088/0741-3335/57/1/014017

Fast sawtooth reconnection at realistic Lundquist numbers

S Günter[1,3], Q Yu[1,3], K Lackner[1,3], A Bhattacharjee[2,3] and Y-M Huang[2,3]

[1] Max-Planck-Institut für Plasmaphysik, Garching, Germany
[2] Princeton Plasma Physics Laboratory, Princeton, New Jersey 08543, USA
[3] Max-Planck/Princeton Center for Plasma Physics

E-mail: Sibylle.Guenter@ipp.mpg.de

Received 26 June 2014, revised 12 September 2014
Accepted for publication 23 September 2014
Published 28 November 2014

CrossMark

Abstract

Magnetic reconnection, a ubiquitous phenomenon in astrophysics, space science and magnetic confinement research, frequently proceeds much faster than predicted by simple resistive MHD theory. Acceleration can result from the break-up of the thin Sweet–Parker current sheet into plasmoids, or from two-fluid effects decoupling mass and magnetic flux transport over the ion inertial length v_A / ω_{ci} or the drift scale $\sqrt{T_e / m_i} / \omega_{ci}$, depending on the absence or presence of a strong magnetic guide field. We describe new results on the modelling of sawtooth reconnection in a simple tokamak geometry (circular cylindrical equilibrium) pushed to realistic Lundquist numbers for present day tokamaks. For the resistive MHD case, the onset criteria and the influence of plasmoids on the reconnection process agree well with earlier results found in the case of vanishing magnetic guide fields. While plasmoids are also observed in two-fluid calculations, they do not dominate the reconnection process for the range of plasma parameters considered in this study. In the two-fluid case they form as a transient phenomenon only. The reconnection times become weakly dependent on the S-value and for the most complete model—including two-fluid effects and equilibrium temperature and density gradients—agree well with those experimentally found on ASDEX Upgrade ($\leq 100\ \mu s$).

Keywords: magnetic reconnection, sawteeth, plasmoids, two-fluid effects

(Some figures may appear in colour only in the online journal)

1. Introduction

Magnetic reconnection is a central phenomenon in astrophysics, space science and magnetic confinement research. The study of its dynamics is essential to the understanding of the conversion of magnetic into thermal and kinetic energy in explosive events such as solar flares, coronal mass ejection and sawtooth crashes in magnetic fusion plasmas. For the low-collisionality plasmas to be considered, the observed reconnection rates are generally much higher than predicted by simple resistive MHD theory, such as the classical Sweet–Parker model.

In Sweet-Parker theory [1, 2] reconnection occurs in a thin current layer of length L of the order of the system size and width $\delta_{SP} \sim L / \sqrt{S}$ with the Lundquist number $S = \mu_0 L v_A / \eta$, the Alfvén velocity v_A and plasma resistivity η. As the maximum outflow velocity in MHD is the Alfvén velocity, the inflow velocity and thus the reconnection rate according to the Sweet–Parker theory decreases with resistivity $\sim v_A / \sqrt{S}$.

However it has already been found that the thin current layer becomes tearing unstable [3, 4] once its aspect ratio exceeds a critical value $L/\delta_{SP} \sim 100$, corresponding to a critical Lundquist number S_C of about 10^4 [5]. This break-up of the current layer and the formation of plasmoids have been observed, e.g. in solar flares and in the magnetotail [6–11].

Analytically, it has been shown that the plasmoid instability of a thin current layer shows super-Alfvénic growth, with the linear growth rate increasing with Lundquist number ($S^{1/4}$) and the number of plasmoids scaling as $S^{3/8}$ [12, 13]. These scalings have been verified in several numerical simulations [14–16], which also find a critical S value for the onset of the instability $S_C \sim 10^4$ [5, 13, 14, 18]. Even more interesting for the reconnection process are however the non-linear properties that have been studied in several numerical simulations, both in fluid [5, 18] as well as in fully kinetic models [19]. The plasmoids have been found to significantly accelerate the reconnection process, leading to a reconnection rate $\frac{1}{v_A B}\frac{d\psi}{dt}$ of about 0.01…0.02 nearly independent of S for high Lundquist numbers [13, 14, 17] in the resistive MHD model.

However an MHD description becomes questionable when the current layer width decreases to the scale at which two-fluid and kinetic effects become important. Beginning in the 90s, it was shown by several authors [20–27] how two-fluid effects can significantly accelerate reconnection, by mitigating the constraints on the plasma flow in the proximity of the reconnection line. Thus, plasmoids in particular should play a role in accelerating the reconnection process as long as the current layer width is larger than the ion inertial length $d_i = v_A/\omega_{ci}$ in cases without a strong guide field (due to the Hall effect) or the ion sound Lamor radius $\rho_s = \sqrt{T_e / m_i} / \omega_{ci}$ in cases with a strong guide field. Several numerical studies using either fully kinetic [19] or Hall MHD models [28] support this hypothesis for cases without a magnetic guide field.

In this paper we investigate possible similarities between these observations in 2D reconnection without magnetic guide fields and reconnection events in magnetic fusion plasmas. We concentrate on the most ubiquitous fast reconnection events in tokamaks, the quasi-periodic sawtooth crashes. In such events, reconnection times of the order of ~100μs are observed in present-day devices. On the other hand, resistive MHD theory for the Sweet–Parker type reconnection (the Kadomtsev model) [29]) would predict reconnection times several orders of magnitude longer for medium to large-size tokamaks, increasing with the Lundquist number as ~$S^{1/2}$.

2. How the formation of plasmoids accelerates reconnection

In [17] and [30] a scaling argument was proposed as to why the generation of plasmoids leads to a reconnection rate independent of S, consistent with numerical simulations. The proposed fractal model implies that for each generation of plasmoids the current sheet between them (figure 1 of [30]) becomes further fragmented when it reaches the critical inverse aspect ratio that gives a characteristic length between two neighbouring plasmoids, $l \sim LS_c / S$. This implies a continuous thinning of the current sheet, and hence—at given resistivity—an increase of the electric field (at each generation of plasmoids) and thus in the reconnection rate. This explanation however, does not resolve one major bottleneck: the plasma mass flow out of the current layer needs to be increased compared to the Sweet–Parker theory as well.

In the Sweet–Parker theory, the transport of mass (or in our case also of toroidal flux) occurs at a rate $v_A\delta_{SP} \sim v_A L / \sqrt{S}$. If plasmoids are formed, the transfer of mass into each individual plasmoid, by the same process, occurs at a rate $v_A\delta$ with the current sheet width between plasmoids, given by [30] as $\delta \sim l / \sqrt{S_c} \sim L\sqrt{S_c} / S$. Assuming that the plasmoids are expelled out of the current layer with Alfvénic speed, they will grow during the transit to an area $v_A\delta L / v_A \sim L\delta \sim L^2\sqrt{S_c} / S$. The total number of plasmoids exiting the macroscopic Sweet–Parker sheet per unit time then becomes proportional to $v_A / l \sim v_A S / (LS_c)$, with the plasmoid density being the inverse of their distance. The total mass ejected from the sheet is thus proportional to $L^2\sqrt{S_c} / S \cdot v_A S / (LS_c) \sim Lv_A / \sqrt{S_c}$. This rate is independent of S and larger by the ratio of $\sqrt{S / S_c}$ than the simple Sweet-Parker result. This illustrates how plasmoids offer a way to escape the limitation on the reconnection rate within the Sweet–Parker model, which identifies the outflow scale with the resistive layer. If plasmoids are formed, the resistive layer becomes much thinner than the effective outflow layer, whose mass transport proceeds via the plasma blobs. Although they are only intermittent, the above argument highlights how they can indeed lead to a large enhancement of the mass flow.

3. Results of resistive MHD simulations

Non-linear simulations of sawtooth reconnection using the resistive MHD model have been performed recently at high Lundquist numbers [31]. To simplify the simulations, cylindrical geometry with a circular plasma cross section is assumed. The magnetic field is thus given by $B = B_{0t}\, \mathbf{e}_t - (kr/m)B_{0t}\mathbf{e}_\theta + \nabla\psi \times \mathbf{e}_t$ with ψ being the helical flux function, m/r and k = n/R the wave numbers in $\mathbf{e}_\theta$ (poloidal) and $\mathbf{e}_t$ (toroidal) directions, respectively [32].

The resistive MHD simulations in [26] were pushed to Lundquist numbers as realistic in present day tokamaks ($S^* = 10^7 \ldots 10^9$, defined in terms of the toroidal magnetic field). To compare this to the simulation results without a magnetic guide field as discussed above, a proper definition of Lundquist number and reconnected magnetic flux must be used. As the field component with the helicity of the equilibrium magnetic field at the resonant surface enters the single fluid model only by ensuring incompressibility of the flow, the component perpendicular to it determines the actual reconnection dynamics. Hence the proper field to be used in the Sweet–Parker model for S is the helical magnetic field $B_{\rm hel} = (1 - q)\, B_{\rm pol}$ with the safety factor $q = \frac{r}{R}\frac{B_{\rm tor}}{B_{\rm pol}}$ in cylindrical geometry. In the simulations [31] an equilibrium current profile of the form $j(r) = j_0(1 - r^2 / a^2)^2$ was used with j_0 chosen such that the q value on the axis is 0.9. For that current profile, the maximum helical field (at radius $r/a = 0.17$) is $B_{\rm hel} = 0.0135\, a/R\, B_{\rm tor} = 0.004 B_{\rm tor}$, for an aspect ratio of 3.3. To compare Lundquist numbers to the simulations without a guide field, one should therefore use the normalization $S = 0.004\, S^*$. This brings the critical S number for the occurrence of plasmoids in the case of strong magnetic guide fields very close to the values found in the simulations without a magnetic guide field: plasmoids do appear for $S \geq 10^4$. Furthermore the critical aspect ratio for the formation of plasmoids was found to be between 60 and 80 in our simulations, close to the value of 100 quoted above for the case with no guide field.

The formation of plasmoids during the reconnection process is shown in figure 1: The initial growth of a (1,1) internal kink mode results in a thin current layer at the q = 1 surface

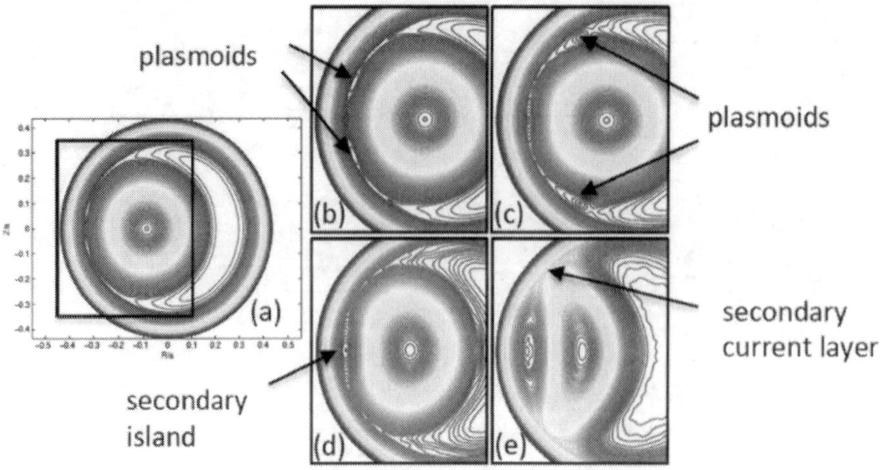

Figure 1. Time sequence of contours of constant helical flux during sawtooth reconnection for $S = 10^6$. Figures 1(*b*)–(*e*) show zooms into the region surrounded by the rectangle in figure 1(*a*). Figure 1(*b*) shows the occurrence of plasmoids that eventually merge with the (1,1) island. In figure 1(*c*), the secondary island appears. While it grows to larger amplitude (figure 1(*d*)) it eventually gives rise to a secondary current layer (figure 1(*e*)) that might become tearing unstable.

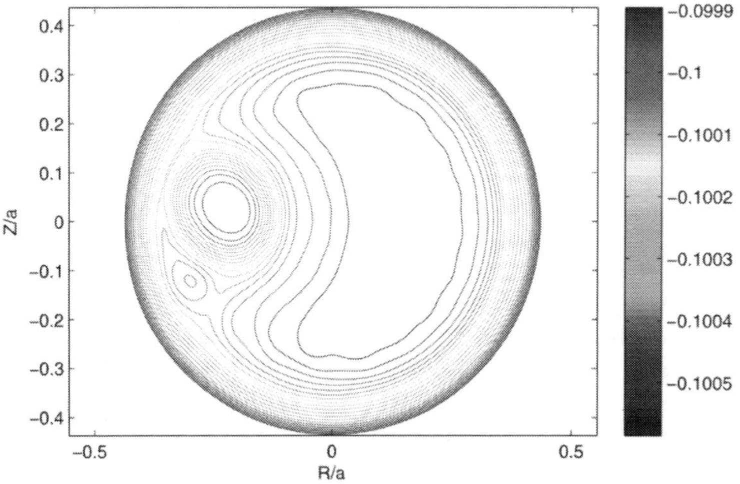

Figure 2. Contour plot of constant helical flux surfaces for $S = 10^5$, demonstrating the merging of the secondary island with the (1,1) island.

and the formation of a (1,1) magnetic island. The thin current layer becomes unstable against tearing modes at sufficiently high S numbers (large aspect ratio of the current layer), as shown in figure 1(*a*) for $S = 10^6$. The resulting plasmoids move towards the (1,1) island (figure 1(*b*)) and finally merge with it (figure 1(*c*)), thereby increasing the reconnection rate. After some time, a magnetic island grows to significant size and is referred to as the 'secondary island' (figure 1(*d*)). When this island reaches a sufficient size, thin current layers form between the original (1,1) and the secondary island (figure 1(*e*)). These current layers have a significantly smaller aspect ratio than the original current layer and thus should give rise to a much shorter Sweet–Parker reconnection time.

The number of plasmoids generated in the beginning of the reconnection process increases with the S number, concomitant with a narrowing of the current layer (see figure 2 of [31]). As discussed in [31], the growth of the secondary island shown in figure 1(*e*) further influences the reconnection process. In [31] ultimately, a stable helical state with two magnetic islands was reported, which changes only on a global current diffusion time. However the simulations in [31] have been initialized by a small up-down symmetric perturbation of (1,1) helicity only. Thus, only up-down symmetric solutions were possible. In the simulations reported here, we added high harmonic perturbations of different phases with respect to the 1/1 component, seeding also up-down asymmetric perturbations. This does not change the fast dynamics of the sawtooth crash itself, but the helical equilibrium with two magnetic islands is not stable against an up-down asymmetric perturbation. Therefore, the secondary island eventually merges with the (1,1) island (see figure 2) and full reconnection occurs.

4. Two fluid simulations

To extend our calculations to the two-fluid case, we use the following set of equations [33], normalizing the length to the minor radius a, the time t to τ_R, ψ to aB_{0t}, **v** to a/τ_R and T_e and n_e to their values at the magnetic axis (a being the minor radius and $\tau_R = a^2/\eta$ the resistive time):

$$\frac{dn_e}{dt} = d_1 \nabla_{\|} j - \nabla_{\|}(n_e v_{\|}) + \nabla \cdot (D_{\perp} \nabla n_e) + S_n, \tag{1}$$

$$\frac{d\psi}{dt} = E_0 - \eta j - \frac{\eta}{\nu_{ei}}\frac{dj}{dt} - \eta\frac{\mu_e}{\nu_{ei}}\nabla_{\perp}^2 j + \Omega\left(\nabla_{\|} n_e + \nabla_{\|} T_e\right), \tag{2}$$

$$\frac{dv_{\|}}{dt} = -C_s^2 \nabla_{\|} p / n_e + \mu \nabla_{\perp}^2 v_{\|}, \tag{3}$$

$$\frac{dU}{dt} = S^{*2} \Delta_{\|} j + \mu_{\perp}^2 U, \tag{4}$$

$$\frac{3}{2} n_e \frac{dT_e}{dt} = d_1 T_e \nabla_{\|} j - T_e n_e \nabla_{\|} v_{\|} + n_e \nabla \cdot \left(\chi_{\|} \nabla_{\|} T_e\right) + n_e \nabla \cdot \left(\chi_{\perp} \nabla_{\perp} T_e\right) + S_p \tag{5}$$

In the above equations the total time derivative includes the perpendicular plasma flow $d/dt = \partial/\partial t + \nu_{\perp} \cdot \nabla$ with the ion velocity $\boldsymbol{\nu} = \nu_{\|}\boldsymbol{e}_{\|} + \boldsymbol{\nu}_{\perp}$, where $v_{\|}$ and $\boldsymbol{\nu}_{\perp} = \nabla \phi \times \boldsymbol{e}_t$ are the parallel (to the magnetic field) and the perpendicular velocity, respectively. The current density in the toroidal direction is given by

$$j = -\nabla_{\perp}{}^2 \psi - 2nB_{0t} / \mathrm{mR} \tag{6}$$

and the plasma vorticity by

$$U = -\nabla_{\perp}{}^2 \phi. \tag{7}$$

μ is the ion viscosity, χ the heat conductivity and D the particle diffusivity. Cold ions are assumed, so that $p = p_e = n_e T_e$. S_n and S_p are a particle and a heat source, respectively. The equilibrium electric field $E_0 = \eta j_0$ is needed to maintain the original equilibrium plasma current density j_0 against resistive decay. The parameters in equation (1)–(5) are given by $d_1 = \omega_{ce} / \nu_{ei}$, $\Omega = \beta_e d_1$, $C_s = [T_e/m_i]^{1/2}/(a/\tau_R)$ and $S^* = \tau_R/\tau_{A^*}$, with $\beta_e = 4\pi n_e T_e/B_{0t}{}^2$ and the electron cyclotron frequency ω_{ce}. μ_e is a perpendicular electron viscosity (introduced for numerical reasons), υ_{ei} the electron-ion collisional frequency, $\tau_A{}^* = a/v_{A^*}$ the toroidal Alfven time, with v_{A^*} being defined in terms of the toroidal field.

The simulations based on two-fluid equations use — unless specified otherwise —typical ASDEX Upgrade experimental parameters as input: $B_{0t} = 2$ T, a = 0.5 m, R = 1.7 m, $T_e = 2$ keV and $n_e = 3 \times 10^{19}\,\mathrm{m}^{-3}$ at the q = 1 surface. These parameters result in $S^* = 2.6 \times 10^8$ (corresponding to $S = 10^6$), $\Omega = 9.4 \times 10^4$, $C_s = 2.0 \times 10^7(a/\tau_R)$, $d_1 = 3.1 \times 10^7$ and $\nu_{ei} = 2.2 \times 10^4\,\mathrm{s}^{-1}$. Anomalous transport parameters are chosen: $\mu_e/\upsilon_{ei} = 10^{-4}a^2$, $\chi_{\|}/\chi_{\perp} = 8.0 \times 10^8$, $\chi_{\perp} = \mu_{\perp} = 19\,(a^2/\tau_R)$ and $D_{\perp} = \mu/5$. Thus the Prandtl number corresponds to realistic experimental values if one assumes anomalous values for plasma viscosity.

The results of our simulations are summarized in figure 3. While in resistive MHD simulations, the formation of plasmoids strongly accelerates the reconnection rate, in the case of two-fluid simulations, plasmoids do not affect the sawtooth reconnection rate as much, in particular if the equilibrium pressure gradients are included and diamagnetic rotation exerts an additional stabilizing effect on them.

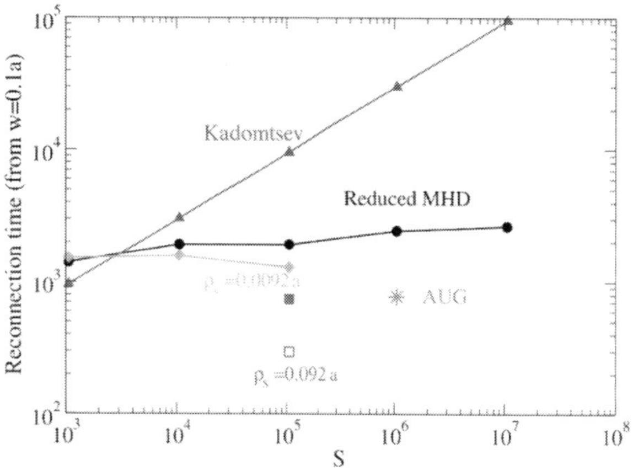

Figure 3. Comparison of the reconnection time (defined as the interval between the attainment of an island width of 0.1 a to the final reconnected state, normalized to the resistive time) in reduced MHD and two-fluid simulations, compared to the Kadomtsev scaling. The (green) diamonds correspond to a constant value of $\rho_s = 0.0092$ a. The (violet) squares demonstrate the influence of diamagnetic rotation. The squares correspond to $\rho_s = 0.092$ a with the full square including diamagnetic equilibrium rotation, the open does not. The star corresponds to realistic parameters for the ASDEX Upgrade tokamak.

In two-fluid simulations, the critical S value above which plasmoids are observed increases significantly. This is to be expected, as the inclusion of two-fluid effects decreases the aspect ratio of the current layer by forming an x-point geometry. While we found a critical S value of 10^4 for resistive MHD, it increases to 10^6 for two-fluid simulations. For S values above 10^6, plasmoids are always present independent of the ρ_s value, at least at the start of the reconnection process. Early in the reconnection process, the x-point geometry has not yet been fully established and thus a thin current layer forms. Due to the increased reconnection rate, the width of this current layer can become much smaller than the width of the Sweet–Parker current layer: The current layer width during the reconnection process is given by the skin depth, $\delta_{\mathrm{res}} = \sqrt{2\eta / \mu_0 \gamma}$. Here, instead of the Sweet-Parker reconnection rate, the realistic reconnection rate γ resulting from two-fluid simulations is used. As the reconnection rate in two-fluid theory often significantly exceeds the Sweet–Parker reconnection rate, the current layer width can become much smaller than δ_{SP}. Thus, the critical aspect ratio for the formation of plasmoids can still be reached. This is the case for all ρ_s values, at least during the initial phase before the x-point geometry has fully developed. (The latter happens only on a longer time scale as shown, e.g. in [34].) To give an example, for $S = 10^6$ and $\rho_s = 0.01$ a ($> \delta_{\mathrm{SP}} = 10^{-3}$ a), we find an aspect ratio of 80 ($\delta_{res} = 7 \cdot 10^{-4}\,a$) at the beginning of the reconnection process.

Despite its stabilizing influence on the plasmoids, the inclusion of two-fluid effects leads to a further acceleration

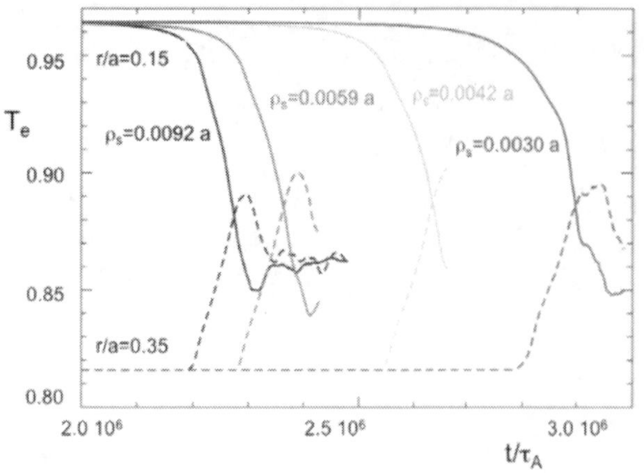

Figure 4. Temporal evolution of the electron temperature inside (r/a = 0.15) and just outside (r/a = 0.35) the q = 1 surface ($r_{q=1}$ = 0.3 a) for various ρ_s values. S = 10^6.

of the reconnection process as compared to resistive MHD. The green diamonds in figure 3 demonstrate how two-fluid effects become dominant when the ion sound Lamor radius ρ_s becomes larger than the Sweet-Parker current layer width δ_{SP}. With increasing Lundquist number at constant ρ_s value, δ_{SP} becomes smaller. For S < 10^4 two-fluid effects are not important since δ_{SP} is larger than ρ_s. For S = 10^4 the Sweet-Parker current layer width is approximately equal to the ion sound Lamor radius ρ_s. Therefore, for higher S values, the reconnection time in two-fluid simulations decreases below that of the MHD case. Figure 3 also contains a scan of ρ_s at constant S (= 10^5) showing that reconnection indeed becomes faster with the increasing influence of two-fluid effects (increasing ρ_s value): the one-fluid result (full black circle) corresponds to ρ_s = 0, the full green diamond to ρ_s = 0.0092 a and the open violet square to ρ_s = 0.092 a. For sufficiently large ρ_s values, an x-point rather than a Sweet–Parker current sheet forms, allowing for larger plasma outflow velocities as shown in [34].

A comparison between the open and full violet square demonstrates the stabilizing influence of the diamagnetic rotation. While for all results discussed above, no equilibrium gradients of density and temperature were included, for the case shown by the full square, a realistic temperature profile ($T_e = T_{e0}[1 - (r/a)^2]^2$) was used. Finally the red star corresponds to the reconnection time for realistic parameters of a mid-size tokamak like ASDEX Upgrade. Indeed we find reconnection rates as measured in present-day tokamaks (of the order of 100μs).

In our simulations we also include the time evolution of the electron temperature (equation (5)). In that way, we are able to follow the evolution of the equilibrium gradients in a more realistic way, as the time evolution of the temperature is much faster than that of the density which evolves with the ion sound speed. Our numerical method allows us to use realistic values for the ratio of parallel to perpendicular heat conductivity. In fact, most experimental observations of sawteeth rely on measurements of the electron temperature. Simulated time traces for the electron temperature inside and outside the $q = 1$ radius are given in figure 4 for several values of ρ_s, demonstrating the heat flux associated with the sawtooth crashes. While the crash time is very similar for the various ρ_s values, the growth time of the (1,1) magnetic island, and thus the time before the crash happens, increases significantly for lower ρ_s values.

5. Summary and discussion

We have compared the influence of plasmoids on sawtooth reconnection with earlier findings for a vanishing magnetic guide field and have found many similarities. Plasmoid formation occurs for the same critical S values and consequently the same critical aspect ratio of the current layer. The plasmoid formation strongly accelerates the reconnection process, with a dependence of the rate on the Lundquist number much weaker than predicted by the Kadomtsev model.

As expected, in two-fluid simulations plasmoids become less relevant as soon as the width of the current layer δ_{SP} becomes smaller than the ion sound Lamor radius ρ_s. While in our two-fluid simulations, plasmoids have hardly any effect on the sawtooth reconnection rate, in [35] there is a discussion that at large values of the system size (with respect to the ion inertial length in the absence of a guide field), there is a regime of plasmoid dominated reconnection. We nevertheless find plasmoid formation for S numbers above 10^6, in particular at the onset of the reconnection process when the x-point geometry is not yet fully developed and thin current layers form. For plasma parameters being realistic in magnetic fusion devices however, plasmoid formation does not significantly influence the sawtooth reconnection time. Nevertheless, as already discussed in [20–27], two-fluid effects lead to very fast sawtooth reconnection, and the reconnection time becomes independent of the Lundquist number. Including all relevant two-fluid effects (electron inertia, parallel electron pressure gradient, diamagnetic drift due to finite radial density and temperature gradients) and using parameters realistic for present-day experiments in our simulations, we find sawtooth crash times in agreement with experimental observations.

In contrast to several experimental observations [36–39], however, our simulations always result in full reconnection: after the sawtooth, the q value on the axis is always above 1. Incomplete reconnection might require simulations in toroidal geometry where the drive of the (1,1) kink mode is much smaller than in cylindrical geometry and the influence of several helicities allow for x-point ergodization and hence pressure flattening. In that case, the reduced drive might be sufficient to prevent full reconnection, in spite of the reduced effect of diamagnetic stabilization.

References

[1] Sweet P A 1958 in *Electromagnetic Phenomena in Cosmical Physics* ed B Lehnert (Cambridge University Press: Cambridge)

[2] Parker E N 1957 *J. Geophys. Res.* **62** 509

[3] Bulanov S V, Sakai J and Syrovatskii S I 1979 *Sov. J. Plasma Phys.* **5** 157

[4] Lee L C and Fu Z F 1986 *J. Geophys. Res.* **91** 6807
[5] Biskamp D 1986 *Phys. Fluids* **29** 1520
[6] Kliem B, Karlicky M and Benz A O 2000 *Astron. Astrophys.* **360** 715
[7] Ko Y K *et al* 2003 *J. Astrophys.* **594** 1068
[8] Asai A *et al* 2004 *J. Astrophys.* **605** L77
[9] Zong Q G *et al* 2004 *Geophys. Res. Lett.* **31** 18803
[10] Lin J *et al* 2005 *J. Astrophys.* **622** 1251
[11] Nishizuka N *et al* 2010 *J. Astrophys.* **711** 1062
[12] Loureiro N F, Schekochihin A A and Cowley S C 2007 *Phys. Plasmas* **14** 100703
[13] Bhattacharjee A *et al* 2009 *Phys. Plasmas* **16** 112102
[14] Huang Y M and Bhattacharjee A 2010 *Phys. Plasmas* **17** 062104
[15] Samtaney R *et al* 2009 *Phys. Rev. Lett.* **103** 105004
[16] Ni L *et al* 2010 *Phys. Plasmas* **17** 052109
[17] Loureiro N F *et al* 2012 *Phys. Plasmas* **19** 042303
[18] Loureiro N F *et al* 2005 *Phys. Rev. Lett.* **23** 235003
[19] Daughton W *et al* 2009 *Phys. Rev. Lett.* **103** 065004
[20] Wesson J 1990 *Nucl. Fusion* **30** 2545
[21] Ottaviani M and Porcelli F 1995 *Phys. Plasmas* **2** 4104
[22] Aydemir A Y 1992 *Phys. Fluids B* **4** 3469
[23] Wang X and Bhattacharjee A 1993 *Phys. Rev. Lett.* **70** 1627
[24] Birn J *et al* 2001 *J. Geophys. Res.* **106** 3715
[25] Fitzpatrick R 2004 *Phys. Plasmas* **11** 3961
[26] Shay M A, Drake J F, Swisdak M and Rogers B N 2004 *Phys. Plasmas* **11** 2199
[27] Bhattacharjee A, Germaschewski K and Ng C S 2005 *Phys. Plasmas* **12** 042305
[28] Shepherd L S and Cassak P A 2010 *Phys. Rev. Lett.* **105** 015004
[29] Kadomtsev B B 1975 *Sov. J. Plasma Phys.* **1** 389
[30] Huang Y M and Bhattacharjee A 2013 *Phys. Plasmas* **20** 055702
[31] Yu Q, Günter S and Lackner K 2014 *Nucl. Fusion* **54** 072005
[32] Yu Q, Günter S and Lackner K 2004 *Phys. Plasmas* **11** 140
[33] Yu Q *et al* 2010 *Nucl. Fusion* **50** 025014
[34] Schmidt S, Günter S and Lackner K 2009 *Phys. Plasmas* **16** 072302
[35] Huang Y M, Bhattacharjee A A and Sullivan B 2011 *Phys. Plasmas* **18** 072109
[36] Soltwisch H 1992 *Plasma Phys. Control. Fusion* **34** 1669
[37] Soltwisch H 1988 *Rev. Sci. Instrum.* **59** 1599
[38] Yamada M M *et al* 1994 *Phys. Plasmas* **1** 3269
[39] Letsch A *et al* 2002 *Nucl. Fusion* **42** 1055

Plasma Phys. Control. Fusion **57** (2015) 014018 (15pp) doi:10.1088/0741-3335/57/1/014018

Quantification of the impact of large and small-scale instabilities on the fast-ion confinement in ASDEX Upgrade

B Geiger[1], M Weiland[1], A Mlynek[1], M Reich[1], A Bock[1], M Dunne[1], R Dux[1], E Fable[1], R Fischer[1], M Garcia-Munoz[2], J Hobirk[1], C Hopf[1], S Nielsen[3], T Odstrcil[1], C Rapson[1], D Rittich[1], F Ryter[1], M Salewski[3], P A Schneider[1], G Tardini[1], M Willensdorfer[1] and the ASDEX Upgrade Team[1]

[1] Max Planck Institute for Plasma Physics, Boltzmannstr. 2, 85748 Garching, Germany
[2] Faculty of Physics, University of Seville, Calle San Fernando 4, 41004 Sevilla, Spain
[3] Technical University of Denmark, Department of Physics, Dk-2800 Kgs. Lyngby, Denmark

E-mail: benedikt.geiger@ipp.mpg.de

Received 27 June 2014, revised 17 September 2014
Accepted for publication 19 September 2014
Published 28 November 2014

Abstract

The confinement fast ions, generated by neutral beam injection (NBI), has been investigated at the ASDEX Upgrade tokamak. In plasmas that exhibit strong sawtooth crashes, a significant sawtooth-induced internal redistribution of mainly passing fast ions is observed, which is in very good agreement with the theoretical predictions based on the Kadomtsev model. Between the sawtooth crashes, the fishbone modes are excited which, however, do not cause measurable changes in the global fast-ion population. During experiments with on- and off-axis NBI and without strong magnetohydrodynamic (MHD) modes, the fast-ion measurements agree very well with the neo-classical predictions. This shows that the MHD-induced (large-scale), as well as a possible turbulence-induced (small-scale) fast-ion transport is negligible under these conditions. However, in discharges performed to study the off-axis NBI current drive efficiency with up to 10 MW of heating power, the fast-ion measurements agree best with the theoretical predictions that assume a weak level anomalous fast-ion transport. This is also in agreement with measurements of the internal inductance, a Motional Stark Effect diagnostic and a novel polarimetry diagnostic: the fast-ion driven current profile is clearly modified when changing the NBI injection geometry and the measurements agree best with the predictions that assume weak anomalous fast-ion diffusion.

Keywords: current drive, sawtooth, fast ion, tokamak, neutral beam, FIDA

(Some figures may appear in colour only in the online journal)

1. Introduction

The confinement of fast ions is an important topic in fusion research because these suprathermal particles are responsible for plasma heating [1] and current drive [2]. In present day fusion devices, fast ions are generated by neutral beam injection (NBI) or ion cyclotron heating (ICRH) and their slowing down distribution in unperturbed plasmas can be described by collisions with electrons and ions (neo-classical). In the presence of magnetohydrodynamic (MHD) modes and turbulence, the fast-particle confinement can be affected by additional anomalous fast-ion redistribution. The physics background of this anomalous transport must be investigated in view of future fusion devices that will exhibit large fast-ion populations: strong MHD modes and turbulence may not only limit the heating and current drive performance but could also damage the device due to the redistribution of the fast ions onto the first wall [3].

A significant anomalous transport of fast ions has been observed due to sawtooth crashes and fishbones. Both are

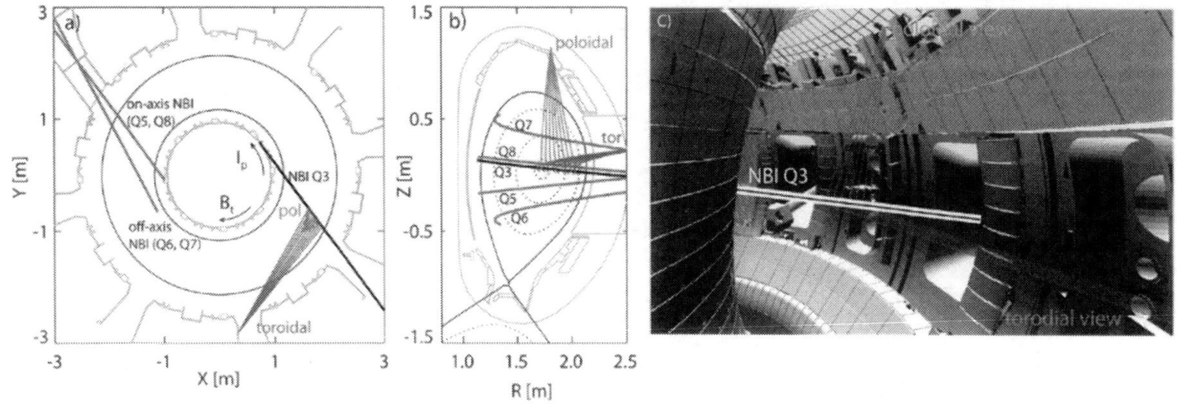

Figure 1. Top–down (*a*) poloidal (*b*) and 3D (*c*) view on the AUG tokamak. The neutral beams with 93 keV maximum injection energy are plotted in green and NBI Q3 (60 keV), which is used as a diagnostic beam for the FIDA diagnostic, is given in black. The toroidal LOS of the FIDA diagnostic are indicated in blue and the poloidal ones are displayed in red.

two prominent MHD instabilities that often occur in tokamak plasmas. They are linked to the growth of a $n = 1$, $m = 1$ kink mode [4] at the $q = 1$ surface (here, n and m are the toroidal and poloidal mode numbers, respectively and q is the safety factor) and appear periodically. While the fishbone instability mainly interacts with trapped fast ions (the fishbone instability is resonant with the precession motion of the banana-tips of trapped particles), sawtooth crashes cause a fast magnetic reconnection in the plasma center and, hence, redistribute heat, particles and momentum. Already 20 years ago, a clear reduction in measured neutron rates was observed at the JET [5] and TFTR [6] tokamaks during sawtooth crashes, indicating a significant fast-ion transport. In recent years, measurements at DIII-D [7] and TEXTOR [8] showed that sawteeth redistribute passing fast ions more strongly than trapped ions (as predicted by theory [9] above a given critical fast-ion energy) and measurements at MAST [10] and JET [11] evidenced a reduction in the density of confined fast ions in the presence of fishbones.

The effects of small-scale turbulence on fast ions are, to a first approximation, expected to be negligible because the turbulent fields are supposed to average out over the fast ions' large gyro orbits and trajectories that deviate from the flux surfaces [12]. However, gyro-kinetic simulations show that the turbulence-induced variations of the magnetic field can hamper this orbit averaging mechanism [13] and hence also yield anomalous diffusion for the suprathermal particles. Experimentally, differences between measurements and neo-classical expectations were observed at DIII-D [14]. Furthermore, the measurements from a Motional Stark Effect (MSE) diagnostic [15] at the ASDEX Upgrade (AUG) were almost identical for on-axis and off-axis NBI heating, which cannot be explained by the neoclassical fast-ion current drive [16]. The turbulence-induced anomalous fast-ion transport was therefore attributed to these observations as the analyzed discharges were MHD-quiescent, i.e. no significant mode activity was observed using soft x-ray and magnetic measurements. This is, however, in contradiction to more recent measurements of the fast-ion slowing down distribution which do not show a clear impact of turbulence [17–19].

Over the last few years, AUG has been equipped with a set of new fast-ion diagnostics. In particular, a fast-ion D-alpha (FIDA [20]) spectroscopy diagnostic has been put into operation [21], which is absolutely calibrated and, hence, allows us to compare predicted fast-ion densities with the experimental data. Moreover, a revised MSE diagnostic and a novel DCN polarimeter [22] permit us to investigate the effects of the fast ions on the current profile. The effects of large and small-scale instabilities on the fast-ion confinement and on the fast-ion current drive efficiency can, consequently, be studied and are presented here.

This paper is organized as follows: first, the experimental setup at AUG is presented, focusing on the FIDA diagnostic. Then, experiments using NBI heating are presented that show strong sawtooth crashes and fishbone activity. A comparison is given between the experimental data and theoretical predictions from TRANSP [23] that include the Kadomtsev model [24]. In the fourth section, investigations of the fast-ion confinement in MHD-quiescent discharges with 5 MW of on- and off-axis NBI are presented. Section five is devoted to the analysis of a discharge with up to 10 MW of heating power, which was performed to study the off-axis NBI current drive efficiency. The measured radial fast-ion profiles are compared with simulations and changes in the current profile due to the application of off-axis NBI are studied. Finally, a short conclusion and summary are given.

2. Experimental setup

ASDEX Upgrade [25] is a medium size tokamak with minor and major radii of 0.5 m and 1.65 m, respectively, resulting in an aspect ratio similar to ITER [26]. The experiments presented here were performed in deuterium plasmas with a central toroidal magnetic field strength of about −2.5 T and a feedback controlled plasma current at either 800 kA or 1 MA. Five different neutral beams, delivering each 2.5 MW of heating power were used for this investigation. The geometry of these NBIs, injecting fast deuterium neutrals in the plasma current direction, is described in figures 1(*a*) and (*b*) on a top-down and poloidal view. As can be seen in the poloidal view,

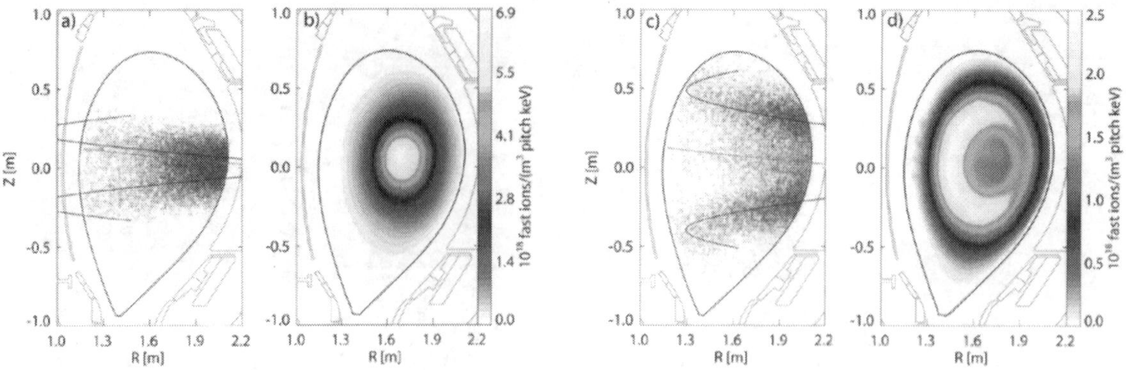

Figure 2. Simulated fast-ion deposition positions (of the Monte Carlo markers in TRANSP) and the corresponding fast-ion densities (integrated over all pitch angles and energies above 30 keV) on a poloidal view for discharge #30 840. Figures (*a*) and (*b*) represent the fast ions from NBI sources Q3, Q5 and Q8 (#30 840 at 3.9 s). Figures (*c*) and (*d*) represent the fast ions from NBI sources Q3, Q6 and Q7 (#30 840 at 5.0 s). The deposition positions calculated for the 93 keV beams are given in blue. The simulated deposition positions of NBI Q3 (60 keV) are plotted in red.

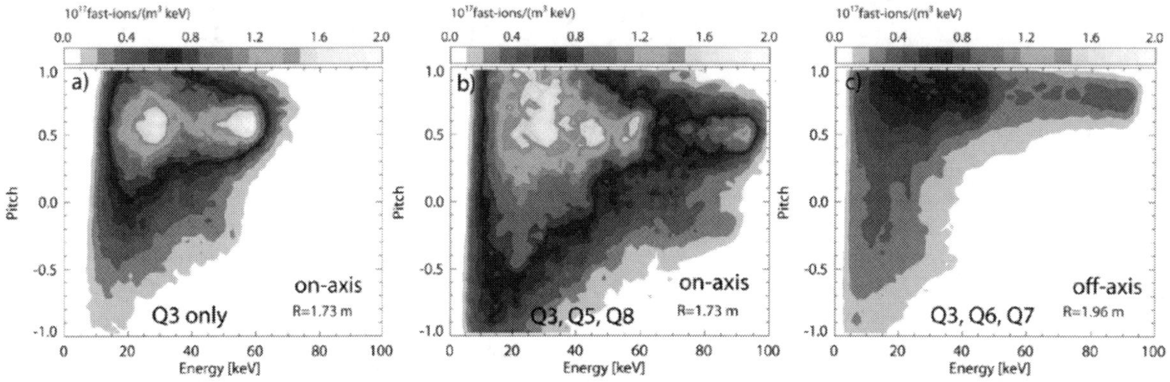

Figure 3. Velocity-space distribution of the fast ions as predicted by TRANSP. (*a*) Central fast-ion velocity-space distribution during the operation of NBI Q3 in discharge #30 809 at 2.5 s. (*b*) Central velocity-space distribution when sources Q3, Q5 and Q8 are on in #30 840 at 3.9 s. (*c*) Off-axis velocity-space distribution when sources Q3, Q6 and Q7 are on in #30 840 at 5.0 s

three of the beam sources inject radially into the plasma center, whereas two beams, i.e. Q6 and Q7, have off-axis and tangential geometry and are used for off-axis current drive experiments. The simulated (TRANSP) deposition positions of the injected neutrals from these NBIs, as well as the corresponding fast-ion densities, are plotted in figure 2 for two representative cases. Figures 2(*a*) and (*b*) represent the fast-ion deposition and density profiles during the operation of the three on-axis NBI sources Q3, Q5 and Q8. Figures 2(*c*) and (*d*) show the deposition and density profiles for the operation of Q3 together with the two off-axis sources (Q6 and Q7). The representative velocity-space distributions for the two cases are depicted in figures 3(*b*) and (*c*) as a function of energy and pitch for two radial positions (pitch = $v_{||}/v$, where v is the absolute value of the fast-ion velocity and $v_{||}$ is its component anti-parallel to the magnetic field). In addition, figure 3(*a*) shows the central velocity-space distribution of the fast ions from just NBI Q3. Most of the fast ions have positive pitch values, corresponding to the co-current neutral beam injection geometry. In figure 3(*a*), two peaks can be observed in the fast-ion velocity space distribution at pitch values close to 0.6. These can be attributed to the full injection energy of NBI Q3 of 60 keV and to the injection of half (30 keV) and one third (20 keV) energetic particles, which is characteristic for positive ion sources in NBI systems. The maximum energies in figures 3(*b*) and (*c*) correspond to the full injection energy of 93 keV of sources Q5–Q8. The pitch values of the fast ions from sources Q5 and Q8 are similar to those of the fast ions injected by Q3 as the geometry of the sources is comparable. In contrast, the paths of the off-axis sources are close to tangential to the magnetic field lines. Therefore, these neutral beams generate fast ions with pitch values close to 1 (see figure 3(*c*)).

The main diagnostic used for this investigation is a fast-ion D-alpha (FIDA) spectroscopy diagnostic. This diagnostic utilizes the principle of charge exchange recombination spectroscopy (CXRS [27]) and is well suited for measuring the fast ions present in the velocity-space region populated by the individual NBIs. The strongly Doppler shifted Balmer alpha radiation ($n = 3 \rightarrow n = 2$, 656.1 nm) is analyzed, which is emitted by the fast ions after receiving an electron along the path of the neutral beam Q3. By using several lines of sight (LOS) that intersect this NBI path, good radial coverage of the plasma is obtained. The LOS geometry of the FIDA diagnostic is plotted in figure 1 with the toroidal views illustrated in blue and the poloidal views shown in red. Example spectra from a central toroidal LOS and a central poloidal LOS are shown

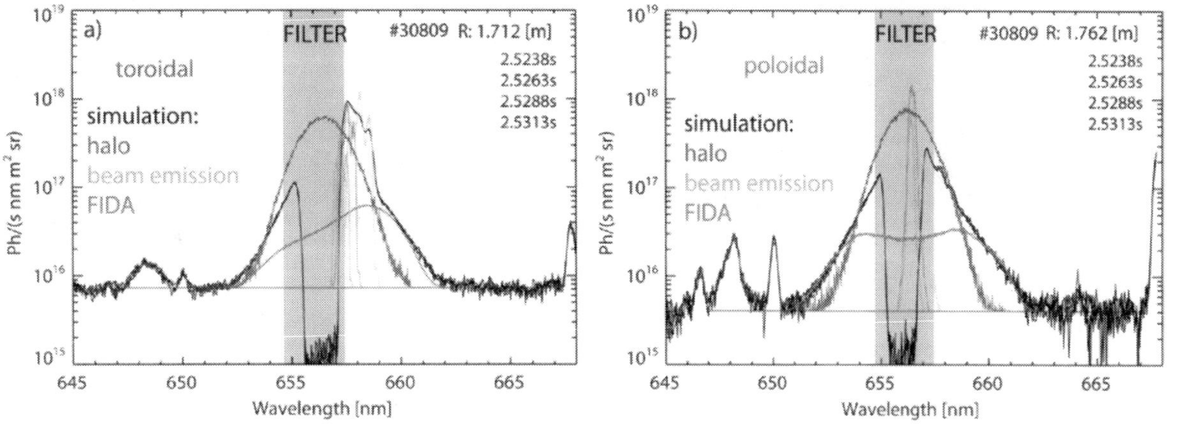

Figure 4. Spectra from a poloidal and a toroidal line of sight of the FIDA diagnostic measured before a sawtooth crash in discharge #30 809 (note the semi-logarithmic scale). In addition, the simulated spectra from FIDASIM are plotted. The simulated halo radiation is displayed in green, the beam emission in yellow/orange and the simulated FIDA radiation, based on a theoretical fast-ion distribution function from TRANSP, is plotted in red.

in figure 4 for four successive time points during the operation of NBI Q3 only (actually measured in discharge #30 809, which will be discussed in detail in section 3). Between 655 nm and 657 nm, the very intense D-alpha radiation from the plasma edge is blocked by a filter (actually a wire in the FIDA spectrometer close to the camera of the diagnostic) because it would cause the saturation of the diagnostic's CCD camera. The experimental data in figure 4 are compared with simulated spectra from FIDASIM [28]. FIDASIM is a Monte Carlo code on a 3D simulation grid that calculates synthetic FIDA spectra based on a given theoretical fast-ion distribution function. The simulation plotted in red in figure 4 shows the FIDA radiation that corresponds to a neo-classical fast-ion distribution function from TRANSP [23]. The FIDA radiation can only be analyzed at strongly red and blue-shifted wavelengths, because at lower wavelength shifts it is obscured by other spectral components. The strongest contributions, in addition to the FIDA radiation, are the beam emission from the injected neutrals (the corresponding prediction from FIDASIM is plotted in yellow/orange) and the halo emission (the simulation shown in green), which is charge exchange radiation from thermal deuterium ions. Moreover, the passive components (also present without NBI Q3) limit the observable wavelength range. In particular, the impurity line radiation of oxygen (OV at 650.02 nm) limits the observation of blue-shifted FIDA radiation.

For the investigation presented here, the FIDA radiation is analyzed between 651 nm and 652 nm and between 660 nm and 661 nm. The Bremsstrahlung radiation, additionally present in these wavelength intervals, is modeled as a constant that is independent of the wavelength. The region in the velocity space that is accessed when analyzing the FIDA radiation in a given wavelength range depends on the LOS geometry, i.e. on the projection angle between the magnetic field and the LOS. Hence, different regions in velocity space are monitored by the toroidal ($\sim 10°$) and poloidal LOS ($\sim 70°$). The velocity-space regions, addressed when analyzing red and blue-shifted FIDA radiation from the two viewing geometries, are illustrated in figures 5(*a*)–(*d*) in terms of weight functions [29, 30]. As can be seen in figure 5(*a*), the toroidal LOS on the blue-shifted side of the spectra can only detect counter rotating passing fast ions above 30 keV with pitches close to −1. In contrast, the toroidal LOS on the red-shifted side (figure 5(*b*)) are sensitive to co-rotating fast ions. The poloidal LOS cover a broad range in velocity space. The blue-shifted part of the spectra is more sensitive to negative pitch values (see figure 5(*c*)) and the red-shifted part measures the fast ions with pitch values biased towards the positive side (figure 5(*d*)). It should be noted that, based on these weight functions, a tomographic reconstruction of the fast-ion velocity space distribution can be performed [31].

Due to the injection geometry of the NBI sources at AUG, the fast-ion distribution functions have mainly positive pitch angles (see figure 3). Fast ions with negative pitch values are only produced by pitch angle scattering (which becomes significant at fast-ion energies below $18.6 \cdot T_e$ [1]) and by trapped orbits. Therefore, the overlap of the weight functions with the distribution functions is small for blue-shifted wavelengths and large for red-shifted wavelengths. Figures 5(*e*)–(*h*) show the product of the weight functions with the representative velocity-space distribution of the fast ions generated by NBI Q3, as shown in figure 3(*a*). The significant overlap of the toroidal weight function for red-shifted wavelengths with the fast-ion distribution function (see figure 5(*f*)) explains the asymmetric shape of the FIDA radiation in the toroidal spectrum (figure 4(*a*)): the strongest FIDA contribution is visible at red-shifted wavelengths. Interestingly, the poloidal LOS also allows us to measure the FIDA radiation between 651 nm and 652 nm (see figure 4(*b*)). This radiation corresponds to fast particles with pitch values in the vicinity of zero. When analyzing this radiation we thus obtain information on the mainly trapped fast particles.

3. Sawtooth and fishbone experiments

Discharges with 2.5 MW of heating power from the NBI source Q3 were conducted with 1 MA of current and −2.4 T of magnetic field (see table 1). Figures 6(*a*)–(*c*) display the

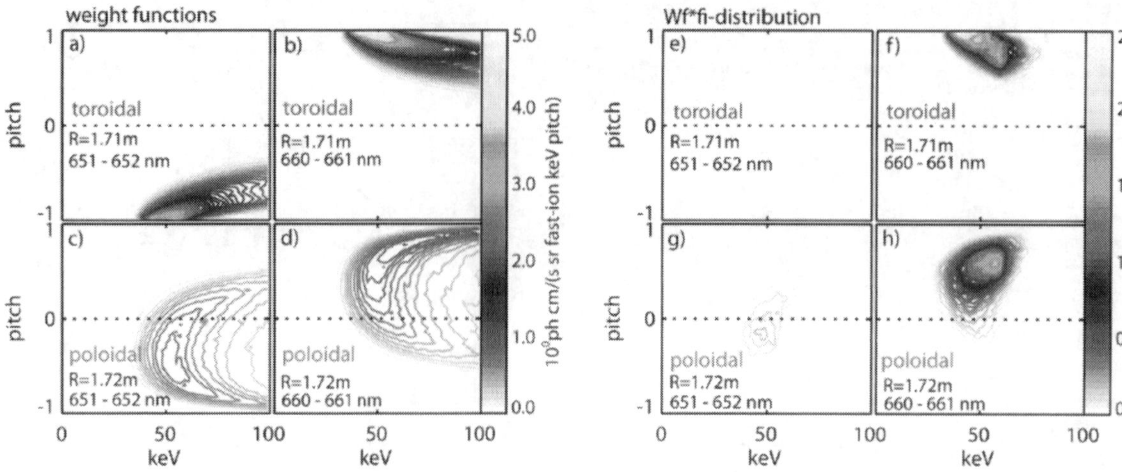

Figure 5. (*a*)–(*d*) Weight functions for two LOS of the FIDA diagnostic for wavelengths between 651 nm and 652 nm as well as for wavelengths between 660 nm and 661 nm. The viewing angle of the toroidal view in the upper panels is ~10° and that of the poloidal view in the lower panel is ~70° relative to the magnetic field. (*e*)–(*h*) Product between the velocity space distribution shown in figure 3(*a*) and of the weight functions plotted in figures 5(*a*)–(*d*).

Table 1. Main discharge parameters of #30 809 consisting of the toroidal magnetic field, B_t, the plasma current, I_p, the average electron density, n_e, the safety factor at 95% of magnetic flux, $q95$, the position of the $q = 1$ surface, $\rho_t(q = 1)$ and the approximate maximum magnetic perturbation amplitude induced by $n = 1$, $m = 1$ modes, $\delta B_{\max}$.

B_t	I_p	n_e	$q95$	$\rho_t(q = 1)$	$\delta B_{\max}$
−2.4 T	1.0 MA	3.9×10^{19} m^{-3}	4.09	0.30	3 mT

characteristic time traces, a spectrogram and the approximate magnetic perturbation amplitude induced by $n = 1, m = 1$ modes for a representative discharge. The approximate magnetic perturbation amplitude was calculated based on measurements of a magnetic pick-up coil at $R = 2.25$ m by assuming that the amplitude decays with $1 / r^{m+1}$. The sawtooth crashes become visible as spikes in the magnetic perturbation amplitude and by drops in the central electron density and temperature, the central ion temperature and the plasma rotation (see figure 6(*a*)). This behavior is well known and expected as the sawtooth instability is linked to the magnetic reconnection of the field lines in the plasma center and redistributes heat, particles and momentum.

The relatively low electron densities and high temperatures in discharge #30 809 yield low collision frequencies ($\sim n_e/T_e^{3/2}$), long slowing down times and consequently a significant fast-ion fraction in the plasma. Hence, the FIDA diagnostic measures with good signal to noise ratio and a clear fast-ion redistribution due to the sawtooth crashes can be observed. As an example, figures 7(*a*) and (*b*) show the spectra from a central toroidal LOS and a central poloidal LOS, respectively. The data plotted in black correspond to four time points before the sawtooth crash at about 2.5 s in discharge #30 809. The spectra in red represent the measurements after that sawtooth crash. As can be seen, the FIDA radiation observed with the toroidal LOS shows a very clear change at red-shifted wavelengths, while the FIDA radiation measured with the poloidal LOS shows a weaker effect of the sawtooth crashes. The radial profiles of the FIDA radiation, measured with the toroidal and poloidal LOS, are shown in figures 8(*a*) and (*b*) as a function of the normalized toroidal flux coordinate, ρ_t. The error-bars of the measurement correspond to the readout and photon noise of the FIDA diagnostic. The profiles were calculated by integrating the FIDA radiation per channel between 660 nm and 661 nm and by subtracting the level of Bremsstrahlung. In addition, constant offsets were subtracted from the toroidal ($0.2 \cdot 10^{16}$ ph/(s sr m^2)) and poloidal ($0.1 \cdot 10^{16}$ ph/(s sr m^2)) profiles in order to match the simulation. These offsets can be explained by the passive FIDA radiation, which is caused by charge exchange reactions between fast ions and neutrals at the plasma edge and which can be roughly described by a constant in the radial profiles (see [21]).

The simulated profiles from FIDASIM represent a fast-ion distribution function from TRANSP. This transport code has been supplied with time-dependent information on the plasma current and separatrix position, on the geometry, power and timing of NBI and on the kinetic profiles. TRANSP (the module NUBEAM) first determines the deposition position of NBI and then follows the corresponding fast-ion orbits in the poloidal plane. The orbits are simulated in the gyro center approximation and the interaction with the electrons and ions is considered using a Monte Carlo collision operator. The evolution of the fast-ion density is simulated by starting new fast-ion markers in the orbit simulation every 10 ms and by removing the thermalized fast ions as soon as their energy is below $3/2 \cdot T_i$. The sawtooth redistribution in TRANSP is simulated by applying the Kadomtsev model [24] for predefined time points. The Kadomtsev model assumes the complete reconnection of the helical magnetic field component, $\vec{B}_*$, which is defined relative to the helically twisted field lines in the $q = 1$ surface. $\vec{B}_*$ changes its sign at the $q = 1$ surface since the field lines are more twisted inside $q = 1$ than outside. The helical flux, i.e. the integral over $\vec{B}_*$, has consequently

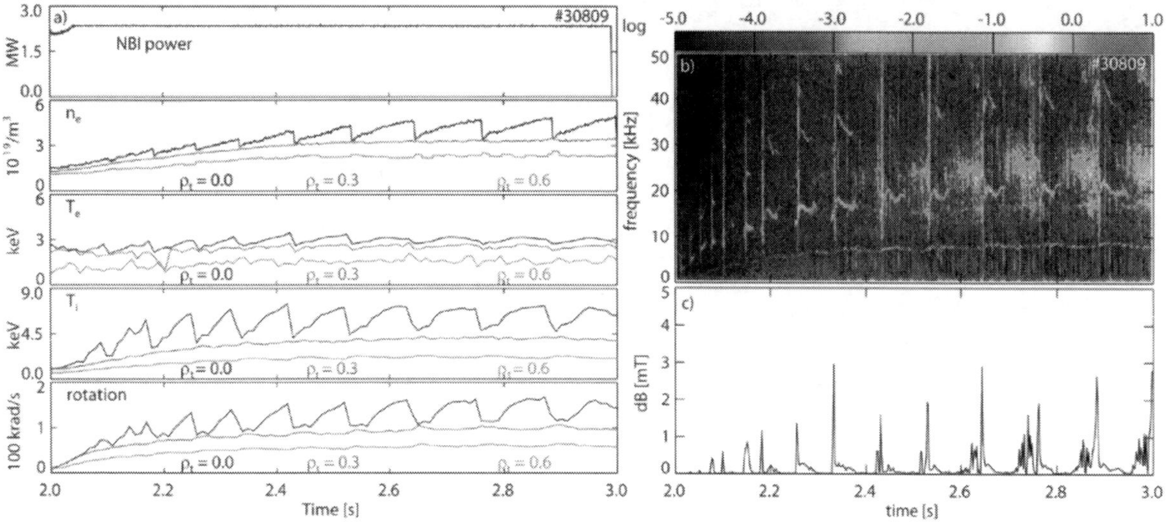

Figure 6. (*a*) Characteristic time traces of discharge #30 809 showing the NBI heating power, the electron density and temperature, the ion temperature and the toroidal plasma rotation frequency. (*b*) Magnetic spectrogram of #30 809 showing the sawtooth crashes that appear at all frequencies and the fishbone instabilities (before the sawteeth from 2.5 s on). (*c*) Approximate magnetic perturbation amplitude of $n = 1$, $m = 1$ modes based on measurements of a magnetic pick-up coil.

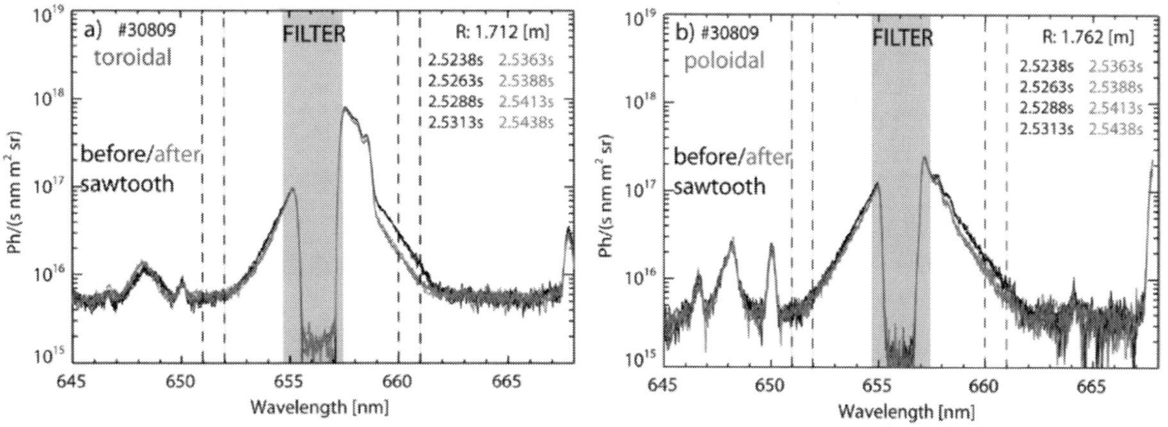

Figure 7. FIDA spectra from a central toroidal LOS (*a*) and a central poloidal LOS (*b*). In black, four successive spectra are shown, measured before a sawtooth crash. The data in red correspond to four time points after a sawtooth crash. The dashed lines illustrate the two wavelength-regions that are used when analyzing the FIDA radiation (blue and red-shifted FIDA radiation).

an extremum at $q = 1$. The model assumes that the field lines inside $q = 1$ reconnect with those field lines outside $q = 1$ that are located at the same helical flux. The new positions of the reconnected field lines after a sawtooth crash are calculated based on the conservation of the toroidal flux. Finally, the sawtooth-induced fast-ion redistribution is determined assuming that fast ions are bound to the reconnecting field lines and that their quantity is conserved.

As shown in figures 9(*a*) and (*b*), a drop in the central fast-ion density of more than 50% is simulated when applying the Kadomtsev model. In addition, an increase in the fast-ion density is predicted at $\rho_t \sim 0.4$, according to the radial redistribution of the fast ions. The profiles from FIDASIM in figures 8(*a*) and (*b*) represent this TRANSP simulation. Their good agreement with the measurements gives evidence that the sawtooth crashes internally redistribute a large part of the confined fast ions.

In figure 10(*e*), the temporal evolution of the measured neutron rates (black) is compared with the neutron flux predicted by the TRANSP code (red). The drops in the predicted neutron rates after every sawtooth crash are in good agreement with the measurement. The relatively weak variation by only about 10% (compared to a fast-ion redistribution of about 50%) agrees with the observation from the FIDA spectroscopy: the sawtooth crashes mainly cause an internal fast-ion redistribution but no significant losses. Indeed, if the sawteeth ejected fast ions from the plasma, the neutron rates would drop more significantly. Here, it should be noted that the absolute value of the neutron rate is not perfectly matched by the simulation, which may be explained by uncertainties in the measured effective charge. The latter leads to uncertainty in the deuterium density, which yields uncertainty in the calculation of the neutron rates: the fusion reaction between the fast ions and the thermal deuterium ions is typically the dominant source of neutrons.

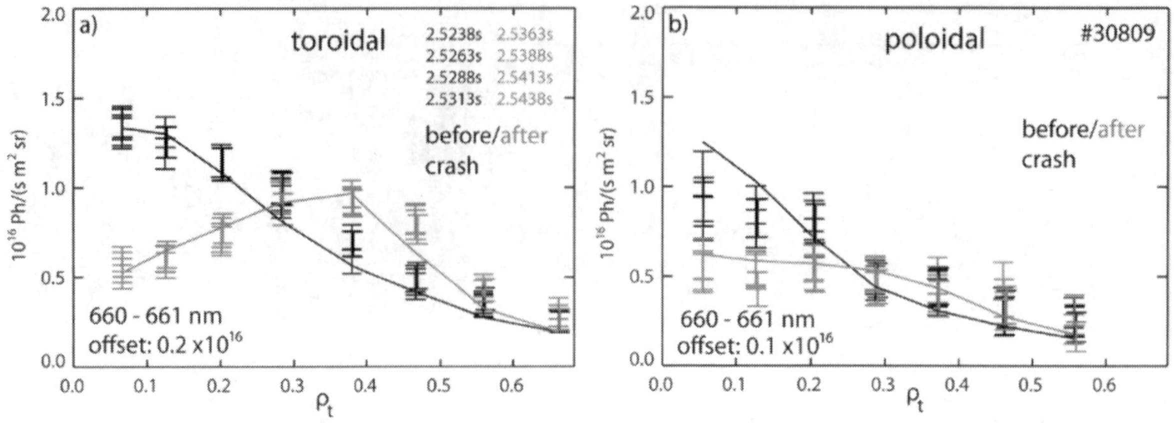

Figure 8. Red-shifted radial FIDA intensity profiles acquired by the toroidal (*a*) and poloidal (*b*) LOS. The profiles were measured before (black) and after (red) a sawtooth crash in discharge #30 809 at 2.532 s. The error bars correspond to the readout and photon noise contained in the FIDA spectra. The solid lines show synthetic FIDA intensity profiles from FIDASIM that are based on the TRANSP predictions, as shown in figure 9(*b*).

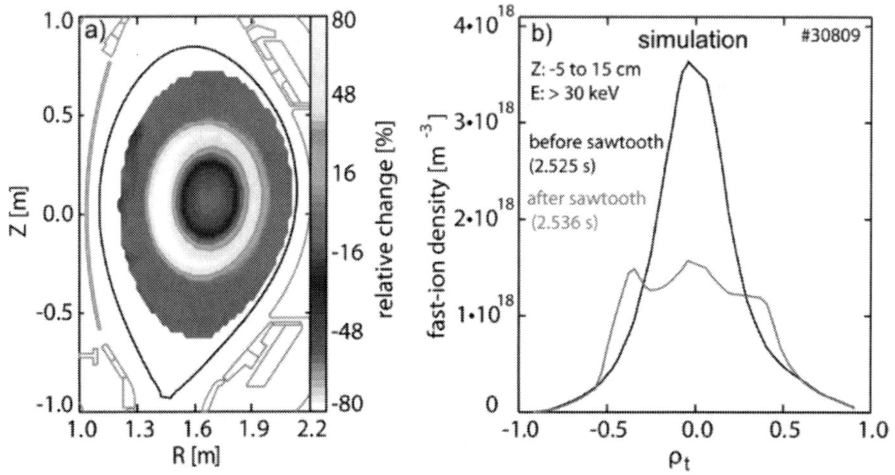

Figure 9. (*a*) Poloidal view on the relative change of the density of fast ions above 30 keV due to a sawtooth crash as simulated by TRANSP using the Kadomtsev model. (*b*) TRANSP predicted radial profiles of fast ions with energies above 30 keV that are present between $Z = -0.05$ m and $Z = 0.15$ m before (black) and after (red) a sawtooth crash.

In addition to the neutron rates, figure 10 compares the temporal evolution of the FIDA data from the toroidal and poloidal LOS with the measurement of a magnetic pick-up coil. The FIDA data represents the integrated FIDA radiation (in wavelength), divided by the level of the beam emission which is measured in the spectra of the toroidal LOS (see figure 4). The beam emission depends on the attenuation of the injected neutrals and, hence, allows us to account for changes in the density of the neutrals present along the NBI path. The latter could modify the FIDA signal even if the fast-ion density was constant. The ratio between the FIDA radiation I_{FIDA} and the beam emission I_{BES} does account for this effect and is a more appropriate signal to study the temporal evolution of the fast-ion density. The $I_{\mathrm{FIDA}}/I_{\mathrm{BES}}$ signals from a central toroidal (blue) and poloidal (red) LOS are shown in figure 10(*b*) for red-shifted wavelengths (between 660 nm and 661 nm). Clearly, the measured signals drop simultaneously with the appearance of the sawtooth crashes in the magnetic signal and then recover slowly. Interestingly the $I_{\mathrm{FIDA}}/I_{\mathrm{BES}}$ signals from the two viewing geometries first show a similar trend, which then becomes different from about 2.5 s on: while the signal from the toroidal LOS increases monotonically between the crashes, the $I_{\mathrm{FIDA}}/I_{\mathrm{BES}}$ signal from the poloidal view saturates. This different behavior might be explained by the rise in the electron and impurity density during the observed time interval and, hence, by the increase in the collision frequency. Increased collision frequencies cause stronger pitch angle scattering and consequently a more efficient redistribution of energetic particles to passing orbits, to which the toroidal views are very sensitive (see figure 5). In contrast, the poloidal views measure a saturated density as these views are sensitive to fast ions at pitch values in the broad vicinity of 0.5, which corresponds to the injection geometry of NBI Q3. For these ions to exist, no pitch angle scattering is necessary. Pitch angle scattering in fact rather reduces the fast-ion density in the observed region of velocity space. The data shown

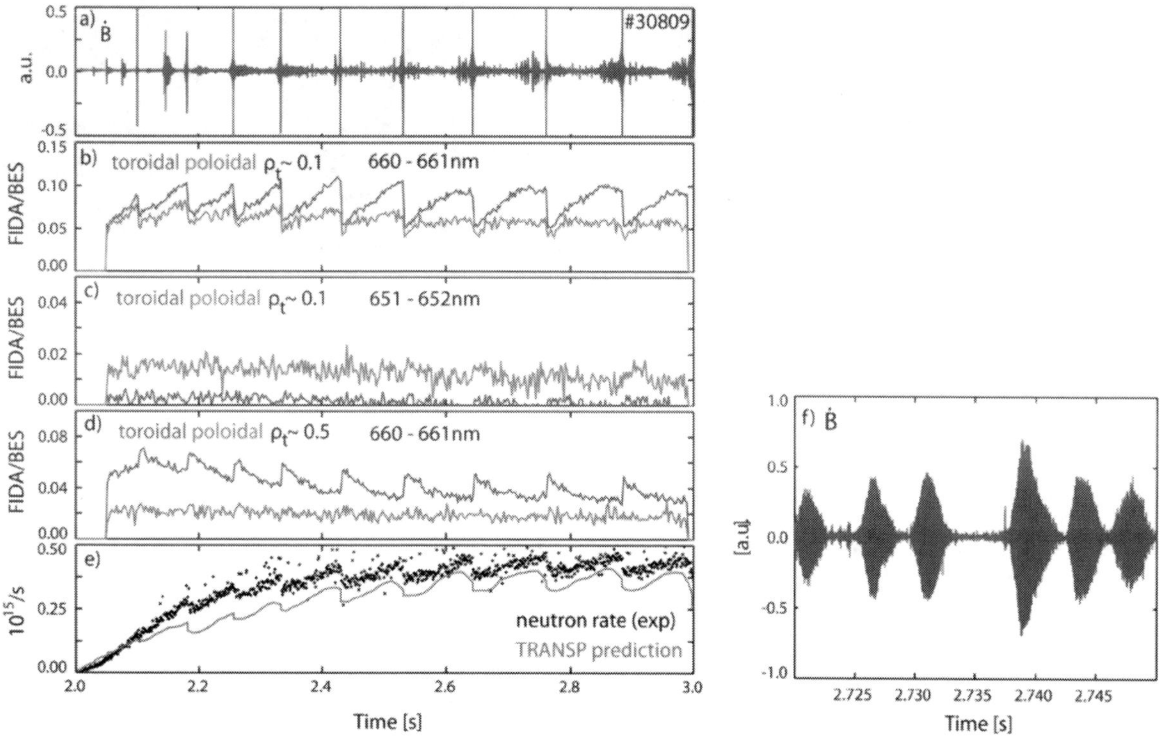

Figure 10. (*a*) Magnetic signal observed by a pick-up coil. (*b*) Temporal evolution of the measured FIDA radiation from a central toroidal LOS (blue) and a central poloidal LOS (red) between 660 nm and 661 nm (I_{FIDA}), divided by the measured beam emission (I_{BES}). (*c*) I_{FIDA}/I_{BES} signal of the toroidal and poloidal LOS for wavelengths between 651 nm and 652 nm. (*d*) I_{FIDA}/I_{BES} signal between 660 nm and 661 nm measured outside the sawtooth inversion radius. (*e*) Measured and TRANSP predicted neutron rates. (*f*) Expansion of figure 10(*a*) between 2.72 s and 2.75 s.

Table 2. Main discharge parameters of #27 237.

B_t	I_p	n_e	$q95$	$\rho_t(q=1)$	δB_{max}
−2.5 T	0.8 MA	6.9×10^{19} m^{-3}	5.27	0.15	3 mT

in figure 10(*c*) correspond to the blue shifted wavelengths, i.e. the mainly counter-rotating fast ions. The I_{FIDA}/I_{BES} signal from the poloidal LOS (red) covers the velocity space of the trapped fast ions. Interestingly, the measurement shows no correlation with the appearance of the sawtooth crashes, indicating that the sawtooth-induced fast-ion redistribution is weak in the observed part of velocity space. It should be noted that the toroidal measurement (blue) exhibits almost no FIDA signal. The density of the purely counter-rotating fast ions, probed by this view at blue shifted wavelengths, is negligible because the NBI geometry is co-current.

The I_{FIDA}/I_{BES} signals measured outside the $q = 1$ surface (at $\rho_t = 0.5$) are plotted in figure 10(*d*) for the red-shifted wavelengths. Clearly, the toroidal LOS measures an increase in the I_{FIDA}/I_{BES} signal directly after the sawtooth crashes, followed by a slow decay within about 50 ms. This shows that passing fast ions are radially redistributed by the crashes and that these ions are subsequently slowed down. In contrast to this observation, the poloidal data outside $q = 1$ does not exhibit an increase after the crashes. This is in agreement with the observation of a weaker redistribution of the fast ions in the broad vicinity of pitch values of 0.5. Moreover, the absence of an increased signal can be explained since the redistributed fast ions change their pitch angle. The fast particles that are ejected to the low-field side keep their magnetic moment and, hence, get onto more passing orbits. These more passing orbits are rarely detected by the poloidal views, while the toroidal LOS are very sensitive to such a modification of the pitch angle distribution.

The clear observation of fast-ion redistribution due to sawtooth crashes demonstrates the high sensitivity and good signal to noise ratio of the FIDA diagnostic. Therefore, it is also possible to study the fast-ion transport induced by other instabilities such as fishbones. Fishbone instabilities are clearly visible in discharge #30 809 before the sawtooth crashes from about 2.5 s on. Figure 10(*f*) shows a zoom of the magnetic data plotted in figure 10(*a*) between 2.72 s and 2.75 s. The increasing and then decreasing oscillations in the magnetic measurement are a clear footprint of fishbone activity. The modes are firstly excited by the resonant interaction with the trapped fast ions. The subsequent decay of the mode amplitude is then explained by anomalous fast-ion transport: the instability redistributes the resonant particles in phase space, loses its drive and, consequently, decays. This anomalous fast-ion transport is, however, relatively small or localized in a very specific part of the phase space. Neither the measured neutron rates, nor the FIDA data exhibit a variation that is correlated with the presence of the individual fishbone modes. Therefore, we can state that, under the investigated conditions,

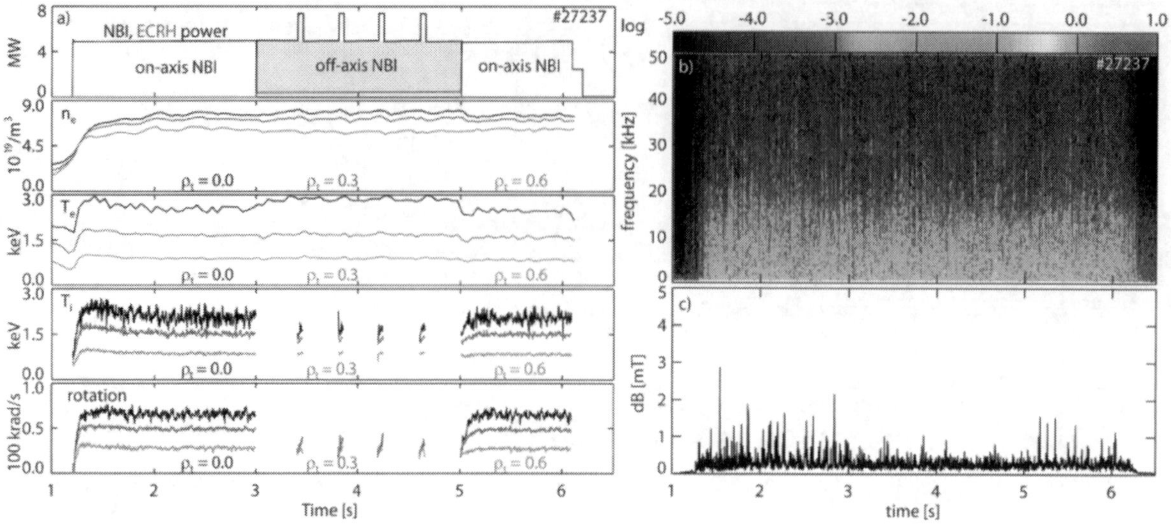

Figure 11. (*a*) Characteristic time traces of discharge #27 237 showing the NBI and ECRH heating power, the electron density and temperature, the ion temperature and the toroidal plasma rotation frequency. (*b*) Magnetic spectrogram of discharge #27 237. (*c*) Approximate magnetic perturbation amplitude of $n = 1$, $m = 1$ mode activity based on the measurements of a magnetic pick-up coil.

the fishbone instability is a small perturbation of the fast-ion distribution function. In particular, a strong anomalous fast-ion transport of the passing fast ions can be excluded because these particles are very well probed. In order to obtain a better signal to noise ratio in the measurement of the trapped fast-ion population, new experiments are planned using more radial NBI geometry. This produces more trapped fast particles and increases the drive of the fishbone modes.

The experimental evidence of a very strong sawtooth-induced redistribution of co-rotating fast ions, as presented here, is in good agreement with recent observations at ASDEX Upgrade [19]. There also, a sawtooth-induced increase in the fast-ion density outside the sawtooth inversion radius was observed, which was followed by a relatively slow decay. Good agreement between the corresponding decay times and the fast-ion slowing down times therefore suggests that the fast-ion transport is neo-classical in the phase between the crashes.

4. MHD-quiescent experiments using on- and off-axis NBI

As already mentioned in the introduction, previous experiments at the ASDEX Upgrade suggested that small-scale turbulence is responsible for the observation of a weaker off-axis NBI current drive than predicted by neo-classical theory. Motivated by our new diagnostic capabilities, MHD-quiescent discharges, similar to these prior experiments, were performed with a toroidal magnetic field of −2.5 T and a plasma current of 800 kA (see table 2). Characteristic time traces of a representative discharge (#27 237) are plotted in figure 11(*a*). 5 MW of on-axis NBI heating power was replaced by off-axis NBI between 3 s and 5 s simultaneously, with the application of electron cyclotron resonance heating (ECRH) to maintain a constant central electron temperature. In addition, short pulses from NBI Q3 were made as the FIDA and CXRS measurements rely on that beam line. Thus, ion temperature and rotation measurements can be performed between 3 s and 5 s. A significant drop in the central ion temperature and rotation is observed during this off-axis NBI phase, indicating that the heat and momentum deposition profile of the fast ions has changed. This is in agreement with the temporal evolution of the radial FIDA intensity profiles from the toroidal LOS, as shown in figure 12. The shape of the profiles clearly changes when switching from on-axis NBI to off-axis NBI heating. Between 1.2 s and 3 s and between 5 s and 6 s, the profiles peak at $\rho_t \approx 0.3$, which corresponds to the on-axis NBI injection scheme. During the application of the off-axis NBI (between 3 s and 5 s) the active FIDA radiation, only visible when NBI Q3 is operational, peaks at $\rho_t \approx 0.5$. Here, it should be noted that the intensity of the FIDA radiation not only depends on the fast-ion density, but also on the density of the neutrals along NBI Q3. The latter is attenuated towards the plasma center and, thus, the FIDA profiles do not peak in the very center during on-axis NBI heating, but slightly off-axis. Forward modeling by FIDASIM is consequently needed to interpret the absolute value and shape of the FIDA profiles. Figure 13(*a*) shows the radial FIDA intensity profiles that were measured during on-axis NBI at about 2.98 s and figure 13(*b*) shows the profiles measured during off-axis NBI at about 4.25 s. Passive radiation has been subtracted from the data, based on measurements when NBI Q3 was off. In addition to the experimental data, the FIDASIM results are shown for two different fast-ion slowing down distributions from TRANSP. The simulated radial FIDA intensity profiles in blue represent a neo-classical fast-ion behavior, while the profiles in red are calculated with the assumption of a constant (in time and space) anomalous fast-ion diffusion of 0.5 $m^2 s^{-1}$. The radial FIDA intensity profiles measured during on-axis NBI, but also those measured during off-axis NBI, are in good agreement with the neo-classical simulation. The assumption

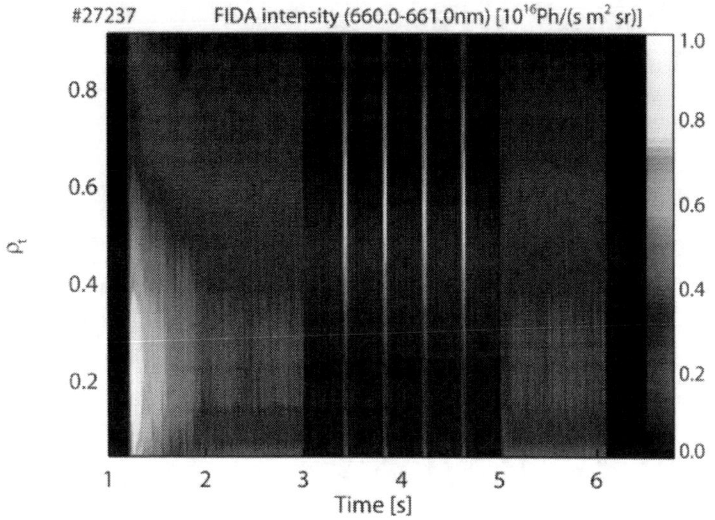

Figure 12. Temporal evolution of radial FIDA intensity profiles as a function of the normalized toroidal flux. The profiles correspond to the FIDA radiation, observed by the toroidal LOS that has been integrated between 660 nm and 661 nm.

of an additional fast-ion transport does, on the contrary, not fit most parts of the experimental data. This indicates that, in the presence of weak MHD activity (see figures 11(*b*) and (*c*)), the anomalous fast-ion diffusion is well below 0.5 $m^2 s^{-1}$. The effects of small-scale turbulence on the radial fast-ion transport are consequently small.

It should be noted that the detailed investigation of the current driven by the application of on- and off-axis NBI in discharge #27 237 is not possible. The most important diagnostic to study the current distribution at AUG, i.e. the MSE diagnostic, relies on the continuous operation of NBI Q3. Since this beam source was only operated in short blips during off-axis NBI, the evolution of the current profile cannot be analyzed. For this reason, new NBI current drive experiments have been conducted in the 2014 experimental campaign. The results of these experiments are presented in the next section.

5. Off-axis NBI current drive experiments using 10 MW of heating power

The analysis of the NBI current-drive (NBCD) efficiency is closely linked to the study of the transport of beam-generated fast ions: if fast ions were redistributed by an anomalous diffusion, the current driven by these ions would also change. Consequently, the combined analysis of the fast-ion density and the fast-ion driven current can provide a more detailed view of fast-ion transport processes. Moreover, the analysis of NBCD is a very important topic for future fusion devices since the continuous operation of tokamak discharges, as well as the capability to tailor the q-profile, relies in part on an efficient NBCD. In the 2014 experimental campaign, discharges with −2.6 T of toroidal magnetic field and a plasma current of 800 kA were performed to study the effects of the off-axis NBI sources on the fast-ion and current profiles (see table 3). The characteristic time traces of a representative discharge (#30 840) are shown in figure 14(*a*). As can be seen, 7.5 MW of NBI heating power were applied from 2 s until 9 s. To provide continuous FIDA, CXRS and MSE measurements, NBI Q3 was turned on during that whole phase. In addition, three 3 s long phases of 5 MW of on-axis NBI (Q5 and Q8) and of off-axis NBI (Q6 and Q7) were applied. These relatively long phases are needed to study the evolution of the current profile: when turning on a given current drive source, the injected net current creates a magnetic field that generates a second mirror current carried by electrons. Both currents first compensate each other until the mirror current decays on the relatively long resistive time scale of a few seconds.

In addition to the application of NBI heating, electron cyclotron current drive (co-ECCD) was used in the plasma center throughout the discharge in order to avoid sawtooth crashes. The latter could interfere with our measurements because sawteeth broaden the fast-ion and current profiles. Moreover, 0.6 MW of modulated (30 Hz) preemptive ECCD was successfully applied at the q = 1.5 surface to prevent the possible destabilization of a tearing mode that could also affect our measurements. When replacing two on-axis NBI sources by the off-axis sources between 4 s and 7 s, additional central ECRH was applied to keep the temperature profile constant. The ohmic current profile strongly depends on the plasma resistivity and, hence, on the electron temperature profile. If the central electron temperature dropped, the current profile would broaden, even without the influence of fast particles. To maintain constant electron temperature profiles, a new real-time algorithm was developed and applied for the first time in this experiment. The algorithm received the electron temperature measured by the ECE diagnostic at ρ_t = 0.2 and accordingly controled the ECRH power. As can be seen in figure 15(*a*), the feedback-controlled application of the central ECRH results in comparable electron temperature profiles during on- and off-axis NBI from ρ_t = 0.2 outwards. In the very center, slightly too high electron temperatures are measured which is, however, not crucial to the analysis of the NBI current drive efficiency: the plasma

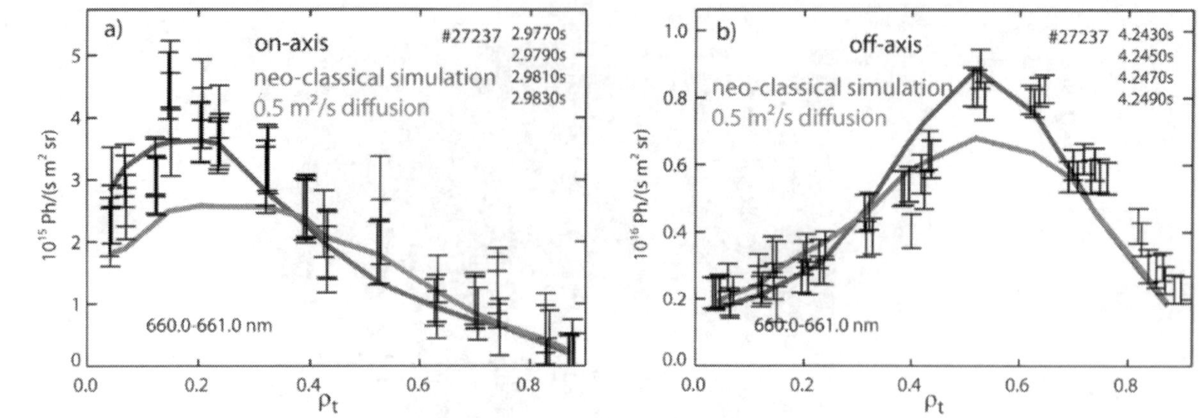

Figure 13. Radial FIDA intensity profiles (between 660 nm and 661 nm) measured during on-axis NBI (*a*) and during off-axis NBI (*b*) in discharge #27 237 for four successive time intervals. The passive radiation, measured during the phase without the operation of NBI Q3, has been subtracted. The error-bars correspond to the statistical noise in the spectra. In blue, the simulated profiles from FIDASIM are shown which correspond to a neo-classical fast-ion distribution function from TRANSP. In red, a simulation based on an anomalous fast-ion transport of 0.5 $m^2\ s^{-1}$ is plotted.

Table 3. Main discharge parameters of #30 840.

B_t	I_p	n_e	$q95$	$\rho_t(q=1)$	δB_{max}
−2.6 T	0.8 MA	$5.3\times10^{19}\ m^{-3}$	5.38	0.15	10 mT

volume is small for $\rho_t < 0.2$ and too high temperatures yield more peaked current profiles, which is opposite to what we expect from off-axis NBI current drive. The electron density, as well as the effective charge (determined from the level of Bremsstrahlung in the spectra of the CXRS diagnostics [32]) remains unchanged between 2 s and 9 s. In contrast, the central ion temperature and rotation (neither of which impact the current profile) clearly drop when switching to off-axis NBI, explained by the different torque and heat deposition profiles of off-axis NBI.

The temporal evolution of the radial FIDA intensity profiles is plotted in figure 16. The modification of the FIDA profiles is clearly observed when switching from on-axis NBI to off-axis NBI: during on-axis NBI, the radial intensity profiles peak at $\rho_t \approx 0.2$ (the FIDA intensity profiles do not peak in the plasma center as they also depend on the attenuation of NBI Q3) and during off-axis NBI, the maximum of the radial FIDA intensity profiles is located at $\rho_t \approx 0.5$.

In figures 17(*a*) and (*b*) the experimental profiles are plotted in black for four successive time points during on-axis NBI (at 3.88 s) and during off-axis NBI (at 4.98 s). The profiles correspond to the spectra from the toroidal LOS, which were integrated between 660 nm and 661 nm and from which the passive radiation was subtracted. In addition, the simulated profiles (FIDASIM + TRANSP) are shown in color. The profiles plotted in blue represent the FIDA radiation expected from the neo-classical fast-ion distribution functions from TRANSP (already presented in section 2, figures 2 and 3). In addition, figures 17(*a*) and (*b*) show profiles plotted in red and green that represent simulations assuming a constant anomalous transport of 0.2 $m^2\ s^{-1}$ and 0.5 $m^2\ s^{-1}$, respectively. This very basic assumption of anomalous fast-ion diffusion is a 0-order approach to assess the effects of instabilities on the fast-particle confinement. However, a relatively good agreement between the measured profiles and the simulation assuming an anomalous fast-ion diffusion of 0.2 $m^2\ s^{-1}$ is found. This indicates a finite, but small anomalous transport and we can define an approximate upper boundary of the anomalous fast-ion transport in discharge #30 840 : 0.5 $m^2\ s^{-1}$.

It should be noted that, despite the anomalous fast-ion diffusion of up to 0.5 $m^2\ s^{-1}$, the radial fast-ion density profiles change when switching from on-axis NBI to off-axis NBI. This consequently changes the current driven by the on- and off-axis fast-ion populations. Figures 15(*b*) and (*c*) illustrate the TRANSP predicted contributions to the current profile (except the ohmic current). The solid lines correspond to the on-axis NBI phase, while the dashed lines represent the off-axis NBI heating case. The contribution from the bootstrap current is plotted in black in figure 15(*b*). It is nearly unaffected when comparing the on-axis phase with the off-axis phase because the electron density and temperature profiles remain roughly constant during the discharge. Also, the current generated by the application of ECCD, shown in green, does not vary appreciably. The total current profile (blue) changes when replacing on-axis NBI with off-axis NBI because the fast-ion current density changes. The NBI driven current, as calculated by TRANSP using [33], is shown in figure 15(*c*) in blue. It varies significantly when switching from on-axis NBI to off-axis NBI. In addition to this neo-classical prediction, the simulated current density profiles are given in red which represent an anomalous fast-ion diffusion of 0.5 $m^2\ s^{-1}$ (the upper boundary of the fast-ion transport observed by FIDA spectroscopy). As can be seen, the anomalous fast-ion diffusion results in a reduction in the fast-ion driven current density by up to 50%.

In the following, this prediction is compared with the experimental data from the DCN polarimetry, the magnetic signals (the internal inductance) and with data from the MSE diagnostic. The MSE diagnostic analyzes a part of the beam

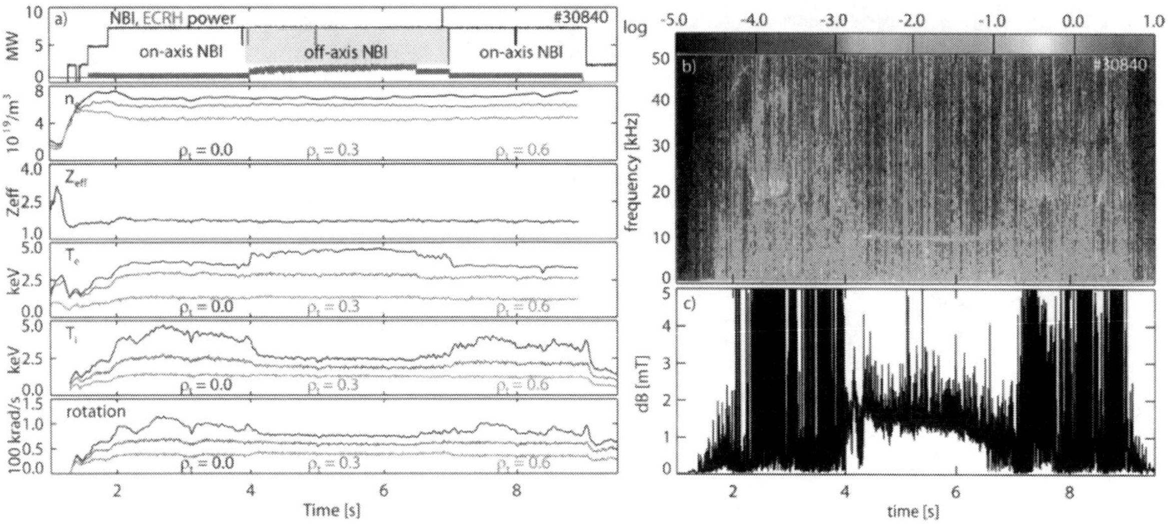

Figure 14. (*a*) Characteristic time traces of discharge #30 840 showing the NBI and ECRH heating power, the electron density, the effective charge, the electron and ion temperatures and the toroidal plasma rotation frequency. The kinetic data is plotted for three different radial positions. (*b*) Spectrogram of a magnetic pick-up coil showing the footprints of fishbone instabilities (i.e. the frequency chirping between 15 kHz and 25 kHz) and that of a continuous mode at the $q = 1$ surface between 4 s and 7 s. (*c*) Approximate magnetic perturbation induced by these modes based on measurements of a magnetic pick-up coil.

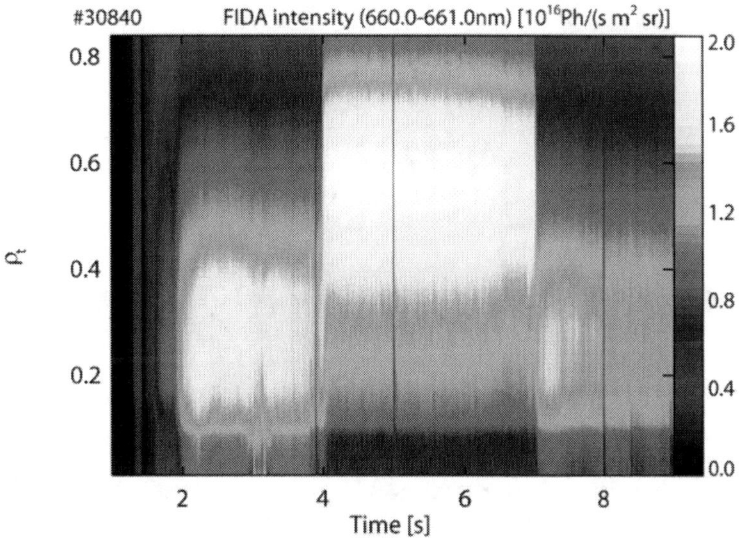

Figure 15. Temporal evolution of radial FIDA intensity profiles as a function of the normalized toroidal flux. The profiles correspond to the FIDA radiation, observed by the toroidal LOS that has been integrated between 660 nm and 661 nm.

emission spectrum emitted by the fully energetic neutrals from NBI Q3 (the sigma component of D-alpha light split by the Stark effect). The polarization direction/angle of this radiation depends on the projection of the v_{NBI} x B electric field on the plane perpendicular to the MSE's LOS and is sensitive to changes in the poloidal magnetic field. It thus allows us to investigate the plasma current distribution. In figure 18, the measured projection angles from five LOS of the MSE diagnostic are plotted in gray. The measurement positions of these individual LOS are indicated by the values of the normalized toroidal flux. A clear variation of the MSE measurement is visible at 3 s. This change is caused by a radial sweep of the plasma column by about 2.5 cm, which was performed to validate the measurement. By sweeping the plasma, different field line angles are scanned over the LOS. The clear effect of this sweep proves that the diagnostic measures changes in the field line angles accurately. When replacing the on-axis NBI sources by the off-axis sources in discharge #30 840 at 4 s, a fast change in the field line angles is observed in the outer LOS. This change is related to a change in the pressure profile: when switching from on-axis NBI heating to off-axis NBI, the ion and fast-ion pressure profiles change on the time

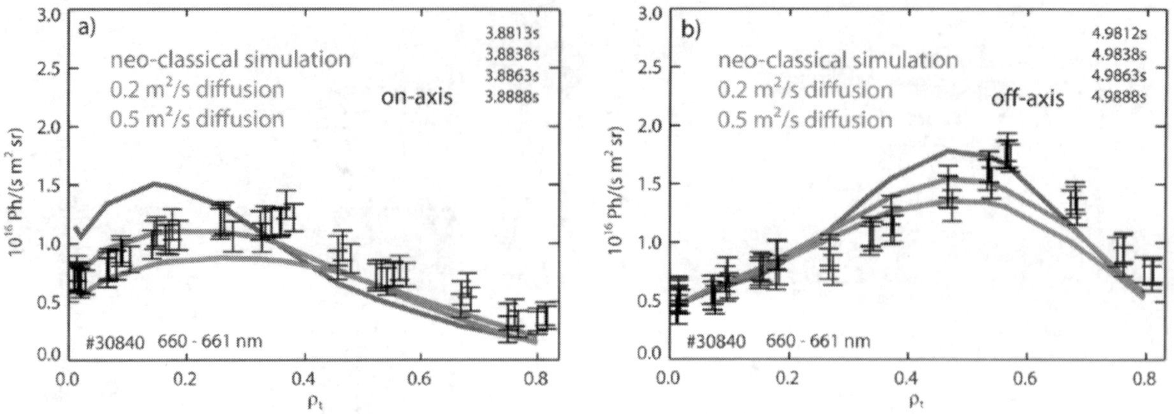

Figure 16. Measured and simulated radial FIDA profiles during on-axis NBI (*a*) and during off-axis NBI (*b*). In blue, a neo-classical simulation is shown. A simulation based on an anomalous fast-ion transport of 0.2 $m^2 s^{-1}$ is plotted in green. In red, a simulation is given based on an anomalous fast-ion transport of 0.5 $m^2 s^{-1}$.

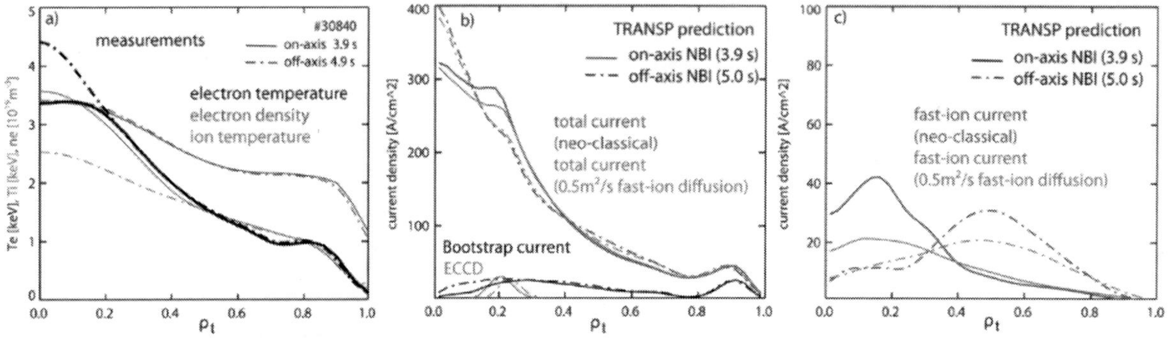

Figure 17. (*a*) Measured electron temperature (black), ion temperature (red), and electron density (blue) profiles during on-axis NBI (solid lines) and during off-axis NBI (dashed lines). (*b*) TRANSP predicted current density profiles during on-axis NBI (solid lines) and during off-axis NBI (dashed lines). In blue, the total current density is shown. The bootstrap current density is given in black and the current driven generated by the application of central ECCD is illustrated in green. In red, the predicted total current density is plotted when assuming anomalous fast-ion transport in the simulation. (*c*) The predicted fast-ion current density from NBI for the on-axis and off-axis cases. In blue, the neo-classical simulation is plotted. In red, the simulation assuming an anomalous fast-ion diffusion of 0.5 $m^2 s^{-1}$ is shown.

scale of the fast-ion slowing-down time, which is in the range of 40 ms in discharge #30 840. This change results in a different Shafranov shift of the plasma center (about 1.5 cm as predicted by TRANSP) and, thus, in a relatively fast change in the angles observed by the MSE. After this step, a slow and continuous evolution is visible in the outer channels of the MSE diagnostic, which can be attributed to the off-axis NBI current drive.

The simulated MSE angles from TRANSP are plotted in color, which are normalized to the measured data before the off-axis NBI heating phase. The variation in the measured MSE angles during the radial sweep is well matched by the simulation. The neo-classical simulation is shown in blue, which predicts much stronger changes in the MSE angles than actually observed. The data in red represent a TRANSP simulation assuming an anomalous fast-ion diffusion of 0.5 $m^2 s^{-1}$. This simulation agrees well with the data. The assumption that the fast ions do not drive current at all (in orange) does not fit.

This result is in agreement with investigations of the internal inductance (ℓ_i). The internal inductance depends on the volume

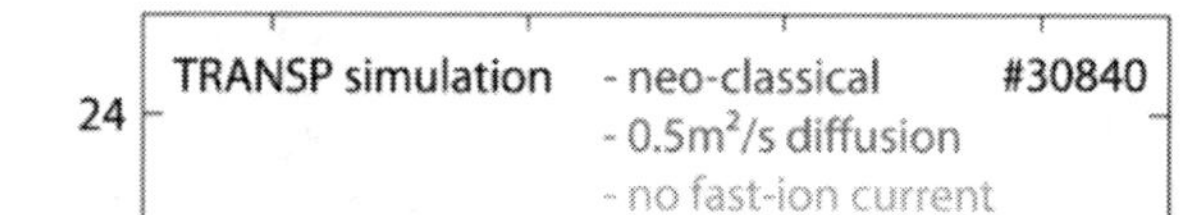

Figure 18. Time traces of the measured MSE angles during discharge #30 840. The experimental data are compared with simulations from TRANSP assuming a neo-classical fast-ion behavior (blue), an anomalous fast-ion diffusivity of 0.5 $m^2 s^{-1}$ (red) and no current drive from the fast ions (orange).

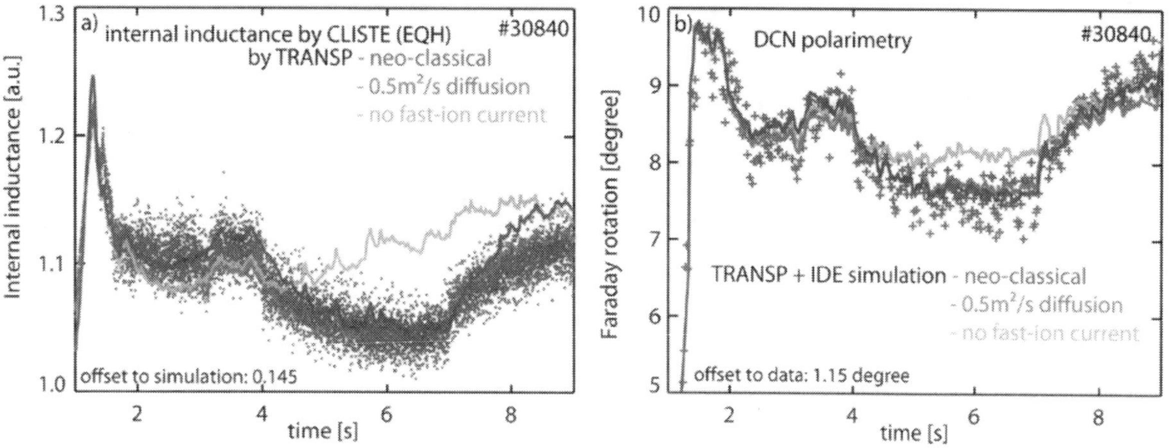

Figure 19. Time traces of the internal inductance as reconstructed by CLISTE (*a*) and the measured polarization degree of a DCN laser that propagates horizontally through the plasma (*b*). The experimental data are compared with simulations from TRANSP assuming a neo-classical fast-ion behavior (blue), an anomalous fast-ion diffusivity of 0.5 $m^2 s^{-1}$ (red) and no current drive from the fast ions (orange).

integral of the square of the poloidal magnetic field inside the separatrix [34] and is high when the current profile peaks in the plasma center and small when it is broad. It can be reconstructed by equilibrium codes that are based on magnetic coil measurements which yield $\ell_i/2 + \beta_{pol}$ [35], where β_{pol} is the ratio of the plasma pressure and the poloidal magnetic field energy. The internal inductance based on magnetic measurements has been reconstructed by the CLISTE equilibrium code [36] and is shown in gray in figure 19. It clearly shows the effects of the off-axis NBI sources between 4 s and 7 s, which shift the current distribution outwards, causing the reduction in ℓ_i. The slow decrease during the off-axis NBI heating and the subsequent increase after switching back to on-axis NBI at 7 s can be explained by the long current diffusion times ($\sim$ 1 s). Shown in blue, the time trace of the simulated ℓ_i (TRANSP) is illustrated, which is calculated from the predicted current distribution assuming a neo-classical fast-ion behavior (a constant offset of 0.145 has been added to the simulations in order to match the measurement, explained by slightly different definitions of ℓ_i in the codes). The temporal evolution of the simulation matches the measurement well. Also, the simulated ℓ_i that corresponds to an anomalous fast-ion transport of 0.5 $m^2 s^{-1}$, shown in red, is consistent with the ℓ_i from CLISTE. Assuming no fast-ion driven current, however, clearly does not match. Also, the measurements yielded by a novel polarimetry diagnostic [22] show the clear effect of the off-axis neutral beam current drive. The polarimetry diagnostic measures the Faraday rotation of the polarization direction of a DCN laser beam, which propagates radially through the plasma at $z = 0.3$ m. The Faraday rotation is sensitive to the magnetic field component parallel to the laser path and, hence, allows us to investigate the plasma current distribution. Figure 19(*b*) shows in gray the measurement of the polarimetry diagnostic compared with a forward model (called IDE) that uses TRANSP predicted equilibria. The simulated polarimetry data, which correspond to a neo-classical TRANSP simulation/equilibrium, are plotted in blue. Shown in red, a simulation is plotted, which assumes an anomalous fast-ion transport of 0.5 $m^2 s^{-1}$. The simulation illustrated in orange corresponds to a prediction assuming no fast-ion driven current. It should be noted that the experimental data have been corrected by a constant offset (1.45 degrees, likely explained by uncertainties in the offset determination of the DCN diagnostic). Moreover, the present polarimeter setup shows increased noise levels during density ramps. In the time window of interest of discharge #30 840, however, the density was constant and the uncertainty in the polarimetry signal therefore small. We can consequently conclude that the simulation plotted in orange, i.e. the assumption that no current is driven by the fast ions, does not match the experiment.

In conclusion, the overall agreement of four independent measurements shows that the application of off-axis NBI produces a clear change in the radial fast-ion profiles. Not only is the FIDA radiation shifted outwards during off-axis NBI, the current profile is also modified when switching from on-axis NBI to off-axis NBI. The predictions from TRANSP that assume an anomalous fast-ion diffusion of up to 0.5 $m^2 s^{-1}$ are in agreement with all the measurements and, hence, show that under the investigated conditions a weak fast-ion transport is present. It can be speculated whether this anomalous transport is caused by turbulence. However, MHD instabilities might also be responsible for the mismatch between the experimental data and the neo-classical simulations. Despite the suppression of the large tearing modes in discharge #30 840, the magnetic spectrogram, as shown in figure 14(*b*), indicates fishbone activity during the on-axis phase and a continuous $n = 1$, $m = 1$ mode during the off-axis phase. An effect of these modes on the fast-ion confinement cannot be excluded, which makes a clear conclusion on the origin of the anomalous transport impossible. However, it should be noted that these results imply that in future fusion devices, the application of off-axis NBI current drive is likely to be effective for tailoring and modifying the current profiles.

6. Conclusion

The impact of sawtooth crashes on the distribution function of fast ions with 60 keV maximum energy has been studied. A sawtooth-induced redistribution of more than 50% of the central fast-ion population has been observed by a novel fast-ion D-alpha (FIDA) diagnostic. The diagnostic uses two arrays of lines of sight and monitors different parts of the fast-ion velocity space. The theoretical predictions by FIDASIM and TRANSP that use the Kadomtsev model to account for the sawtooth-induced fast-ion redistribution are in good agreement with the measured spectra and radial profiles. Moreover, the purely co-rotating fast ions are found to be mostly affected by the crashes, while a strong redistribution of fast ions in the broad vicinity of the zero pitch is not detected. In contrast to the significant fast-ion redistribution due to the sawtooth crashes, the fishbone modes, observed between the subsequent sawtooth crashes, do not exhibit a measurable effect on the fast-ion distribution function. This can possibly be explained as fishbone modes only redistributing fast ions in a small region of the phase space.

In dedicated experiments with 5 MW of NBI heating power, the FIDA measurements evidenced a good confinement of fast ions. The radial profiles of the FIDA radiation, measured during on-axis NBI and during off-axis NBI, are in very good agreement with the neo-classical predictions from TRNASP. The MHD-induced fast-ion transport, but also a possible fast-ion transport due to turbulence, is consequently below the measurement uncertainties.

In discharges with up to 10 MW of total heating power, the FIDA profiles are slightly flatter than predicted by purely collisional transport and the additional fast-ion diffusion coefficient is in the range of 0.5 $m^2\ s^{-1}$. An investigation of the NBI current-drive efficiency based on data from an MSE diagnostic, a DCN polarimetry diagnostic and the internal inductance agrees with this transport level, but also shows that the fast-ion current profiles are modified when varying the NBI injection scheme. A clear broadening of the current profile is measured when replacing on-axis NBI with off-axis NBI.

Acknowledgments

This work was carried out within the framework of the EUROfusion Consortium and has received funding from the European Unions Horizon 2020 research and innovation programme under grant agreement number 633053. The views and opinions expressed herein do not necessarily reflect those of the European Commission.

References

[1] Stix T H 1972 *Plasma Phys.* **14** 367
[2] Fisch N J 1987 *Rev. Mod. Phys.* **59** 175
[3] Fasoli A *et al* 2007 *Nucl. Fusion* **47** S264
[4] Chen L *et al* 1984 *Phys. Rev. Lett.* **52** 1122
[5] Marcus F B *et al* 1991 *Plasma Phys. Control. Fusion* **33** 277
[6] Lovberg J A *et al* 1989 *Phys. Fluids B: Plasma Phys.* **1** 874
[7] Muscatello C M *et al* 2012 *Plasma Phys. Control. Fusion* **54** 025006
[8] Nielsen S K *et al* 2010 *Plasma Phys. Control. Fusion* **52** 092001
[9] Kolesnichenko Y *et al* 2000 *Nucl. Fusion* **40** 1325
[10] Jones O M *et al* 2013 *Plasma Phys. Control. Fusion* **55** 085009
[11] Von Thun C P *et al* 2012 *Nucl. Fusion* **52** 094010
[12] Angioni C *et al* 2008 *Phys. Plasmas* **15** 052307
[13] Hauff T *et al* 2009 *Phys. Rev. Lett.* **102** 075004
[14] Heidbrink W W *et al* 2009 *Phys. Rev. Lett.* **103** 175001
[15] Reich M 2007 *34th EPS Conf. on Plasma Physics (Warsaw, Poland, 2–6 July 2007)* vol 31F
[16] Guenter S *et al* 2007 *Nucl. Fusion* **47** 920
[17] Pace D C *et al* 2013 *Phys. Plasmas* **20** 056108
[18] Geiger B *et al* 2011 *Plasma Phys. Control. Fusion* **53** 065010
[19] Geiger B *et al* 2014 *Nucl. Fusion* **54** 022005
[20] Heidbrink W W *et al* 2004 *Plasma Phys. Control. Fusion* **46** 1855
[21] Geiger B *et al* 2013 *Rev. Sci. Instrum.* **84** 113502
[22] Mlynek A *et al* 2014 *Rev. Sci. Instrum.* **85** 11D833
[23] Pankin A *et al* 2004 *Comput. Phys. Commun.* **159** 157
[24] Porcelli F *et al* 1996 *Plasma Phys. Control. Fusion* **38** 2163
[25] Herrmann A and Gruber O 2003 *Fusion Sci. Technol.* **44** 569–77
[26] ITER Physics Basis Editors *et al* 1999 *Nucl. Fusion* **39** 2137
[27] Isler R C 1994 *Plasma Phys. Control. Fusion* **36** 171
[28] Heidbrink W *et al* 2011 *Commun. Comput. Phys.* **10** 716–41
[29] Heidbrink W W *et al* 2007 *Plasma Phys. Control. Fusion* **49** 1457
[30] Salewski M *et al* 2014 *Plasma Phys. Control. Fusion* **56** 105005
[31] Salewski M *et al* 2014 *Nucl. Fusion* **54** 023005
[32] Rathgeber S K *et al* 2010 *Plasma Phys. Control. Fusion* **52** 095008
[33] Hirshman S P 1978 *Phys. Fluids* **21** 145
[34] Giannone L *et al* 2013 *Fusion Eng. Des.* **88** 3299
[35] Lao L *et al* 1985 *Nucl. Fusion* **25** 1421
[36] Carthy P J M *et al* 2012 *Plasma Phys. Control. Fusion* **54** 015010

IOP Publishing

Plasma Physics and Controlled Fusion

Plasma Phys. Control. Fusion **57** (2015) 014019 (8pp)

doi:10.1088/0741-3335/57/1/014019

Charging of multiple grains in subsonic and supersonic plasma flows

D Block[1] **and W J Miloch**[2]

[1] IEAP, Christian-Albrechts-Universität, D-24098 Kiel, Germany
[2] Department of Physics, University of Oslo, Box 1048 Blindern, N-0316 Oslo, Norway

E-mail: w.j.miloch@fys.uio.no

Received 27 June 2014, revised 9 September 2014
Accepted for publication 19 September 2014
Published 28 November 2014

Abstract

The role of wake effects in the charging of dust grains by plasmas with subsonic and supersonic ion flows is studied with numerical simulations. Significant ion focusing which is common for supersonic flows is also observed for subsonic regimes. In both regimes, the charge on a downstream grain aligned with the flow depends linearly on the intergrain distance. For subsonic flows and systems with several grains, the complex ion dynamics can lead to significant modifications of the charge on grains located close to the boundary of a dust lattice and the charge distribution on the grains depends on the detailed grain arrangement. The studies are carried out with DiP3D, a self-consistent particle-in-cell code (Miloch 2010 *Plasma Phys. Control. Fusion* **52** 124004; Miloch and Block 2012 *Phys. Plasmas* **19** 123703).

Keywords: plasma, dust, wake, charge

(Some figures may appear in colour only in the online journal)

1. Introduction

Complex plasmas i.e. plasmas containing finite-sized dust grains, have attracted much attention in the last two decades [1]. Most studies investigated the strong coupling regime and used the possibility to access the individual grain dynamics experimentally [2, 3]. This approach has offered fascinating views on various dynamical processes at a kinetic rather than the well-known statistical level [1, 4, 5]. For this purpose complex plasmas are most often described by a one-component approach [6, 7]. The underlying idea is that the ambient plasma is assumed to provide enough electrons and ions to charge all dust grains identically. Combined with the fact that all dust dynamics are several orders of magnitude slower than the ion and electron dynamics, the only remaining plasma property in a one-component description of a complex plasma is the Debye shielding of the dust grains.

Nevertheless, dust charging and shielding are fundamental problems in complex plasmas. At usual laboratory conditions, dust grains are levitating above an electrode or are found in other regions with notable electric fields. Thus they are usually exposed to a directed flow of ions. From the sheath region in rf-discharges it is well known that the flowing plasma gives rise to wakes behind dust grains and that the resulting wakefield strongly affects the dynamics of other charged dust grains [3, 5, 8–12]. A good understanding of the interaction of a single grain with flowing plasma is already important to aim towards an appropriate description of the collective interactions in complex plasmas. Thus, much effort has been directed towards analytical [13–18] and numerical [19–24] studies of dust grain charging in plasma flows.

In most laboratory conditions, dust grains experience sonic or supersonic ion flows. At such conditions, the dust charge and potential agree in general with the shifted orbit-motion-limited theory and capacitance models [25–27]. The wakefield of a single grain can be successfully studied with the linear response theory as well as with numerical simulations using particle-in-cell codes [17, 19, 22, 23, 27–29]. With such simulations, in which the plasma particle trajectories are followed in self-consistent fields, it has been confirmed that a negatively charged grain causes an electrostatic lensing of ions into the wake region. This process is often termed *ion focusing* [3, 20, 30]. Good agreement between analytical solutions and numerical simulations of the wake field [16, 31] proves that its size and shape are well understood.

Unfortunately, this changes as soon as pairs or groups of dust grains are investigated. In supersonic flows, dust grains tend to form chains in the direction of the flow. The plasma response in such a wake allows for non-reciprocal forces between dust grains where the motion of the downstream grain is controlled by the upstream one but not vice versa [8, 10, 32]. While the preference for dust chains has been explained with the positive space charge downstream of a grain very early [33], the decharging of the downstream grain in the wake and the role of this process on the grain dynamics and their alignment with the flow have only recently been investigated [31, 34]. Fortunately, different simulation approaches and experiments on dust pairs now confirm that the downstream grain is significantly decharged by the enhanced ion flow which is produced by the electrostatic lensing by the upstream grain. Thus these results nicely extend the single grain picture. However, they raise a question about how larger groups of dust grains will charge and whether their charges can be predicted.

In many experimental situations dust grains are exposed to subsonic ion flows. These include dust grains that are located far in the pre-sheath from electrodes, naturally occurring complex plasmas and experiments under microgravity conditions [1, 2]. Although such conditions are ubiquitous, to date very few studies have been conducted to address the dust charging in subsonic flows [20, 23, 35, 36]. In recent studies the role of drag force in the dust alignment in subsonic flows has been addressed [37, 38]. However, there are still many open questions regarding the subsonic regime. For example, in supersonic flows, the decharging of the downstream grain in the wakefield is important for the system dynamics and stability and it has been shown that this effect is nearly linear with respect to the flow speed and with respect to the intergrain distance [27]. If this linear behaviour can be applied to the subsonic regime, it would also facilitate the modelling of dust dynamics in subsonic regimes [36]. Finally, in subsonic flows, the plasma particle trajectories may be non-trivial due to strong electric fields in the vicinity of grains. Thus, they may affect the dust charging in various multi-grain systems.

The modelling of dust grains in subsonic plasma flows poses some challenges: with an analytical approach one should account for nonlinear phenomena and self-consistent charging of many dust grains, while numerical simulations with methods that include a finite spatial grid can be difficult due to the stability of a simulated system. Systems with subsonic flows can become unstable due to physical instabilities, such as Pierce instability [39, 40], as well as numerical instabilities due to finite grid and particle weighting, such as the ringing instability for subsonic flows [41]. This could explain why so few numerical studies of dust charging in subsonic flows have been conducted in the past.

In this paper we push the limits of our numerical code DiP3D to study the charging of dust grains and wake effects in the range of accessible, numerically stable subsonic flow velocities. We consider a single grain, a pair of aligned grains, as well as multi-grain systems.

2. Numerical simulation

To study the charging of dust in subsonic plasma flows we use DiP3D, a 3D Particle-In-Cell (PIC) numerical code. The DiP3D code has been designed for studies of object-plasma interactions in various environments [22]. The details of the DiP3D code are available in previous works [27, 31] and here we provide only its main features in its basic version.

DiP3D is an electrostatic code that operates in a 3D Cartesian coordinate system. It simulates dynamics of electrons and ions in self-consistent force fields and can also account for plasma flows. The boundaries of the simulated system are open for plasma particles. Thus the particles can leave the simulation box, while new particles are introduced through the boundaries into the box according to the pre-computed flux. We use the Dirichlet boundary conditions at the edges of the simulation box for solving the Poisson equation. We find these boundary conditions to be applicable for the present study, provided that the simulation domain is large in terms of the Debye length and that the dust grains are far from the boundaries, i.e. when the plasma at the boundaries is affected very little by the charged dust.

Dust grains of spherical shape are introduced into the system, far away from the boundaries. The grains are charged self-consistently by plasma currents throughout the simulation. Since forces acting on plasma particles are smoothed due to the spatial grid, close to the grain surface we correct them by carrying out the direct force calculation with the P^3M-technique [42, 43]. This approach allows us to simulate dust grains with shapes that are independent of the grid structure, calculate proper dynamics of the plasma particles close to the dust surface and avoid self-forces due to the grid. We assume the dust grains to be rigid and immobile.

In the present study we simulate electrons and singly charged positive ions, with a mass ratio of numerical particles $m_i/m_e = 120$. Using reduced mass ratios is common in numerical plasma simulations as it allows speeding up the evolution of the system. We find that mass ratios $m_i/m_e > 100$ give credible results that are comparable in normalized units to the results with real mass ratios [44]. At the start of the simulation and at the boundaries of the simulation box, the velocities of plasma particles are assigned according to shifted Maxwellian velocity distributions. We use plasma density of $n = 10^{13}\ \mathrm{m}^{-3}$, electron temperature of $T_e = 3$ eV and electron to ion temperature ratio $T_e/T_i = 100$. The electron Debye length is $\lambda_{\mathrm{De}} = 4.03 \cdot 10^{-3}$ m.

We follow the dynamics of approximately 10^7 particles of each species with time resolution of $\Delta t < 0.01\omega_{\mathrm{pe}}^{-1}$, where ω_{pe} is the electron plasma frequency and run the code up to ten ion plasma periods τ_i. This is much longer than the time needed to reach stationary conditions for the charge on the dust (for supersonic flows this time is typically four τ_i). The simulated box is 5×10^{-2}m in each direction, with the grid spacing of $0.097\lambda_{\mathrm{De}}$. The radius of dust grains is $r \approx 0.1\lambda_{\mathrm{De}}$. Thus, we are in the same parameter regime that was used in our previous numerical experiment that considered supersonic flows [27].

The size of the grain is comparable to the grid spacing, but since we use the P^3M-technique, we can well resolve its shape and plasma dynamics in its vicinity. The size of the grain is large as compared to those in real experiments, which is a compromise between accuracy and computational effort. For smaller grains we find that the numerical noise in the charging characteristic of the grain becomes dominant. However, since the grain is still smaller than the Debye length, we see that the processes associated with the wake formation and the wake topology remain unchanged. For instance, the strength of the ion focus is determined by the plasma density and the charge of the dust grains, and we compensate higher dust charge by slightly lower plasma density. Note that experimental findings agree rather well with simulations of relatively large grains [34].

We concentrate on subsonic plasma flow velocities, where the speed of sound is defined as $C_s = \sqrt{kT_e/m_i}$. Subsonic flows are much slower than electron thermal velocity and they can also be considered in terms of flowing ions. For velocities $v_d \in [0.3, 0.5]C_s$, we find that the results for a single grain are not numerically stable, which we attribute to the ringing instability. Since we use the first order linear weighting on the grid, it seems consistent that the ringing instability sets in when $v_d \leqslant 0.5C_s$ [41]. Thus, the single grain results for $v_d \in [0.3, 0.5]C_s$ are not shown, as they are burden with large inaccuracy and possibly systematic errors. However for multigrain arrangements the simulations are stable for all flow velocities. For the completeness of discussion we also consider supersonic plasma flows, $v_d \leqslant 1.5C_s$.

3. Results

To study the effects of subsonic plasma flows on charging of multiple grain arrangements, we first consider a single grain and then continue with two grains aligned with the flow before addressing arrangements of 10 and 15 grains.

3.1. Single grain

In figure 1 we show the equilibrium charge q_s on a single grain from simulations with different flow velocities. The results are normalized to the absolute value of the charge on the grain in stationary plasma $q_0 = q(v_d{=}0)$ and the velocities to the sound speed C_s. The charge value on a grain gradually increases with the flow velocity within the considered velocity range. This charge increase is substantial already for subsonic flows, indicating that even a small break in the symmetry in the charging process can lead to a notable reduction of the ion collection current to the grain surface. This charge increase (i.e. the grain becomes more negatively charged) is due to reduction of the effective collecting area for ions. Note, that in agreement with standard theories [1, 26], the charge will reach a maximum at supersonic velocities and then decrease again, with the decrease at supersonic velocities that is due to increased ion flux to the grain surface at a fixed collection cross section [27]. In figure 1, the charge for subsonic velocities reaches up to double of the grain charge in stationary plasma. It does not change much with further increase in the velocity, indicating that the maximum charge is at $v_d > 1.5C_s$.

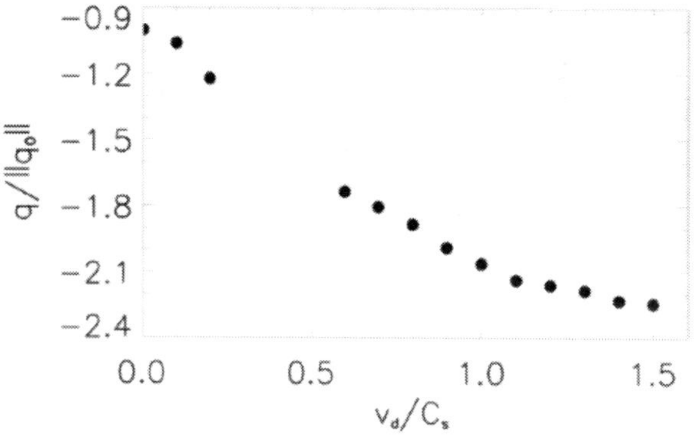

Figure 1. Charge q_s on a single grain, as a function of the flow velocity v_d. Results are for the data that is time averaged over three ion plasma periods for time-stationary conditions at the end of simulation. Charge q is normalized to the absolute value of the grain charge q_0 with vanishing flow velocity. Data points for $v_d \in [0.3, 0.5]C_s$ are omitted because they are affected by numerical instabilities.

3.2. Two grains

If the charge on a single grain depends that much on the flow velocity, one can expect, that the charge enhancement will have an effect on other grains, especially on downstream grains and their tendency to align in the direction of the flow. We have thus run a set of simulations with two grains aligned with the flow, where the flow velocity v_d and the distance $d \in (0.7, 1.3)\lambda_{De}$ between the grains have been varied systematically.

When looking at the influence of a downstream grain on the charge on the upstream one, we see deviations from the charge on a single grain which are less than 3% in the supersonic regime and less than 5% in the subsonic regime. Thus for the considered intergrain distances d and plasma parameters, the influence of the downstream grain on the charging of upstream grain in the two-grain arrangement can be neglected in most of the studies.

However, the situation is completely different for the downstream grain. Already for supersonic flows the charge on the downstream grain was found to be reduced significantly [31]. For subsonic flow we observe the same trend. The results in figure 2(*a*) show the relative charge on the downstream grain q_2 with respect to the charge on the upstream grain q_1 as a function of relative distance d and flow velocity v_d. For a fixed v_d, the charge scales linearly with the intergrain distance, that is clearly seen in figure 2(*b*) where the relative charge is plotted as a function of d for selected v_d. Note that q_1 used for normalization changes with v_d. Thus each curve has been normalized to a different value of q_1. Note as well, that the previously shown linear scaling for supersonic flows [27], obviously holds even in the subsonic flow regime. Finally, figure 2 shows that the charge reduction increases with the flow velocity.

Thus, the charge on the downstream grain is clearly influenced by the wake effects. These wake effects are not only important for supersonic flows, but they are already significant for subsonic flows, where they cause notable charge reductions on the downstream grain only.

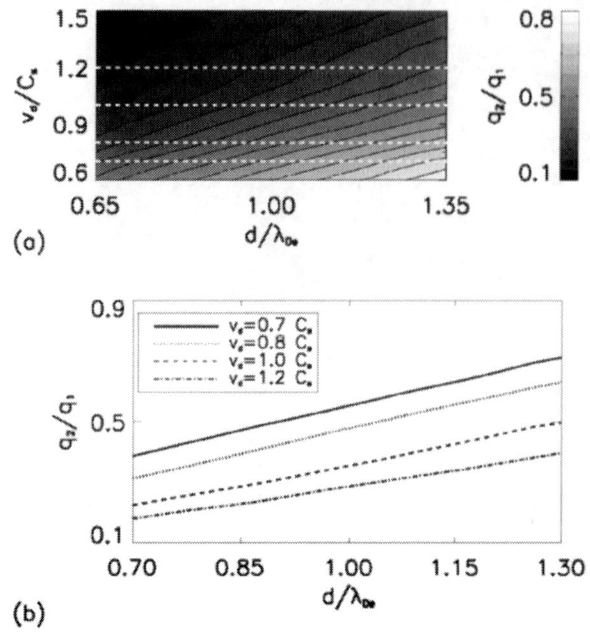

Figure 2. The charge q_2 on a downstream grain from the simulations of two grains aligned with the flow for different intergrain distances d and flow velocities v_d, normalized to the the charge on the upstream grain $q_1(a)$. In (b) the charge q_2 on the downstream grain is shown as a function of d for selected flow velocities v_d. The velocities are normalized to the speed of sound C_s and intergrain distances d to the electron Debye length λ_{De}. For completeness, results for supersonic flow velocities are also shown.

Modification of the charge on the downstream grain will have implications on the equilibrium distance between two dust grains aligned with the flow. The two charged grains will interact with each other and with the surrounding plasma. The downstream grain will in particular experience electrostatic force from the upstream grain and from the net positive charge due to the ion focus in the wake. Eventually, the grains will find the equilibrium position d_{eq}, at which the forces will balance each other. To find the equilibrium distance between the grains, we employ the model for finding the force f_2 acting on the downstream grain [27]:

$$f_2 = \frac{q_2}{4\pi\epsilon_0}\left(\frac{q_1 \overrightarrow{r}_{1,2}}{r_{1,2}^3} + \int_0^d \int_A \frac{\rho(\overrightarrow{r})\,\overrightarrow{r}_{\rho,2}}{r_{\rho,2}^3} \mathrm{d}\overrightarrow{r}\right). \quad (1)$$

In this model, we account for the electrostatic force contribution from the charge q_1 on the upstream grain, as well as from the net positive charge, which is volume integrated in the wake region between the two grains. This model does not account for some other forces that can be present in the experiments, such as the gravity, ion drag, or forces due to confining potential. Nevertheless our force calculation is a reasonable approximation as gravity and confinement forces basically cancel, drag forces will only shift the grains further downstream, while the difference between the drag on different grains will increase the equilibrium distance. Thus, this simplified model provides a first estimate for the typical intergrain distance for grains aligned with the flow.

Table 1. Equilibrium distance d_{eq} between two grains aligned with the flow for different flow velocities v_d.

$v_d(C_s)$	0.6	0.7	0.8	0.9	1.0	1.1	1.2	1.3
d_{eq} (λ_{De})	0.78	0.78	0.87	0.97	0.97	1.0	1.07	1.07

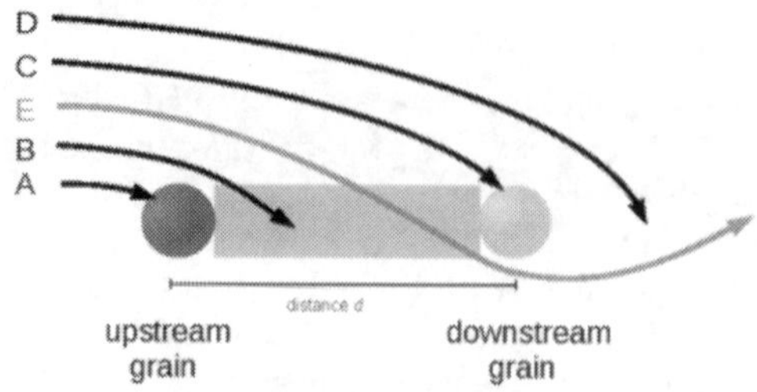

Figure 3. Schematic drawing of particle arrangement and different types of ion trajectories. Trajectories of type A and C end either on the upstream or on the downstream grain. Trajectories of type B only pass the wake region of the upstream grain (blue box). Trajectories of type D are those passing only the wake of the downstream grain (yellow box). Type E trajectories are special as they pass both wake regions.

The results are summarized in table 1. While for supersonic flows, d_{eq} is larger than the electron Debye length, for subsonic flows, the equilibrium distance is reduced to less than λ_{De}. The equilibrium distances all lie well in the simulated parameter regime (see figure 2).

To understand the charging process it is instructive to study the ion trajectories in the wake. For this purpose we distinguish five types of ion trajectories (see figure 3). Trajectories of type A and C are those which are responsible for the grain charging as they end either on the upstream or on the downstream grain. Trajectories of type B and D are those which only cross one of the wake regions. They contribute to the ion focus observed behind each grain. The fifth trajectory type (E) is special as it crosses both wake regions. We will refer to these types of trajectories and their color-coding in the following figures.

In figure 4 we show trajectories of sample cold ions for subsonic $v_d = 0.7C_s$ (a) and supersonic $v_d = 1.2C_s$ (b) flows. The results are for time-stationary conditions at the end of simulations and they are overplotted on the corresponding charge density plot in the x-y plane crossing the center of the grain. Only a small, central part of the simulation domain is shown, so that the wake is clearly demonstrated. The charge density is characterized by a strong maximum in the wake. This net positive charge density reflects the ion focusing by the grain, which is due to electrostatic lensing. The wake regions of the upstream (light blue box) and downstream (yellow box) are marked according to the color-code introduced in figure 3.

First, it is seen that the focus region in the wake of the upstream grain is due to ions deflected in the vicinity of that grain for both flow regimes. However, for the subsonic flow the ions passing the focus region are so strongly deflected that they can even travel against the flow after leaving the focus region. Second, figures 4(a)–(b) demonstrate that the focus region is less pronounced and less elongated for the subsonic case. In both cases the focus region behind the downstream grain

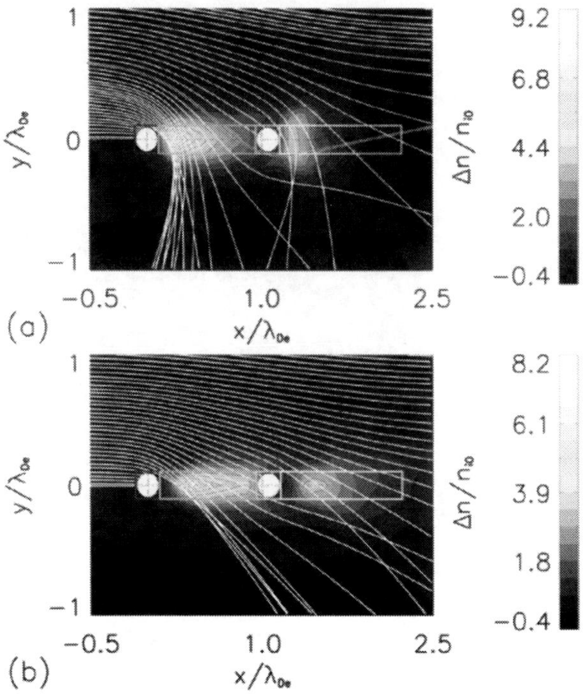

Figure 4. Charge density in the x-y plane through the centre of dust grains for two aligned grains with $d = 1.1\lambda_{De}$ with overplotted trajectories of sample cold ions for $v_d = 0.7C_s$ (a) and $v_d = 1.2C_s$ (b). Different regions of the wake are marked by rectangles, see text for details. Only a small, central part of the simulation domain is shown. The flow is in the positive x-direction.

is weaker and more diffuse. Finally, for both flow regimes, ions from the focus region of the upstream grain do not contribute much to the charging of the downstream grain. The main charging is due to ions being electrostatically lensed by the upstream grain directly to the downstream one and having a significant velocity component perpendicular to the average flow direction, see figure 4. While in the supersonic regime (figure 4(b)) only trajectories of types A–D are observed, we find type E trajectories for the subsonic case as well. The red trajectory in figure 4(a) proves that some of the ions entering the ions focus region of the upstream grain are passing the wake region behind the the downstream grain. Although, this *slalom* trajectory is not of importance for the charging, it is important for the momentum transfer to the grains and thus for the stability of the chain as discussed in [37].

Figure 5 visualizes the mapping of ions to the different regions in more detail. Here, the injection plane (y-z) for flowing cold ions is plotted. The colors indicate the region and thus the type of trajectory (see figure 3) where the particle injected at this y-z-position is mapped to. From these plots, we see that the respective regions upstream from the grains are well separated (their boundaries will slightly overlap for finite ion temperature). The exception is for subsonic flows, where the a thin red ring corresponds to ions that contribute to both ion density enhancements. In addition, the effective collection area for the ions decreases with increasing flow velocity, which will have effect on the grain charge (the grain charge becoming more negative, see figure 1). Further, note that for

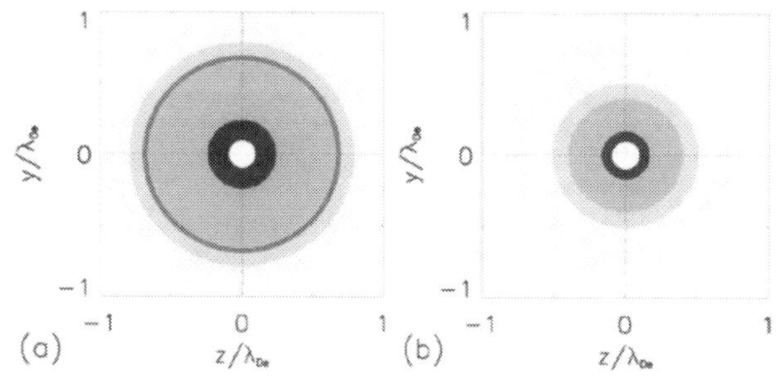

Figure 5. Regions in plasma in the y-z plane far upstream from the grains that correspond to the cold, flowing ions, whose trajectories would enter the regions marked in the charge density plot for $v_d = 0.7C_s$ (a) and $v_d = 1.2C_s$ (b). The particles in a yellow ring would enter a yellow box, etc. White circle in the center denotes the size of the grain. The red circle in (a) corresponds to ions that in the wake have *slalom* trajectories.

subsonic flows the light blue region, i.e. those ions are mapped to the focus region of the upstream grain, ends at about $0.7\lambda_{De}$. This is considerably larger than for supersonic flows (see figure 5(b)) and has implications for multi-grain systems.

3.3. Multi-grain systems

For the two grain-system we observe significant electrostatic lensing at both flow regimes, but in addition for subsonic flows we see that a significant fraction of ions is subject to large angle scattering giving rise to an upstream ion flow. Thus, subsonic flows should have important implications on the charging of grains in larger dust clusters. In this section we focus on the systems of 9 grains arranged in the layer, with one grain downstream in the wake and a system of 5 rods, each made of three grains.

Figure 6(a) shows the grain arrangement for the 9 + 1-system. The charge on the individual grains in this arrangement as a function of ion flow velocity is plotted in figure 6(b). The intergrain distance is set to $b = \lambda_{De}$. Thus, the Debye-spheres of individual particles overlap and the screening parameter $\kappa = b/\lambda_{De} = 1$ is typical for dust clouds. As a consequence the charges are not identical at $v_d = 0$. The charge is slightly more negative for particles with many nearest neighbors. Therefore, the charge of grain 1 is used as reference and all charges are normalized to $q_{1,0}$, i.e. its charge at $v_d = 0$.

Figure 6(b) clearly shows that the charge reduction of the downstream grain (grain 2) increases with v_d. At the same time the charge on all upstream grains increases. For supersonic flows all grains in the layer nearly have a similar charge, but as soon as we are in the subsonic regime, the charge on the grains differs. While the charges on grain 5 and 6 increase first, the same is observed later for grain 3 and 4 and last for the central grain (grain 1). Obviously, the number of nearest neighbors is of importance. Recalling figure 5, it is clear that for all flow conditions the collection radius is smaller than $0.5\lambda_{De}$. Thus, the collection areas (dark blue area) do not overlap and all grains could collect the same ion current. However, each grain will give rise to back-scattered ions at substantially low velocities (see figure 4(a)). These provide an additional ion current

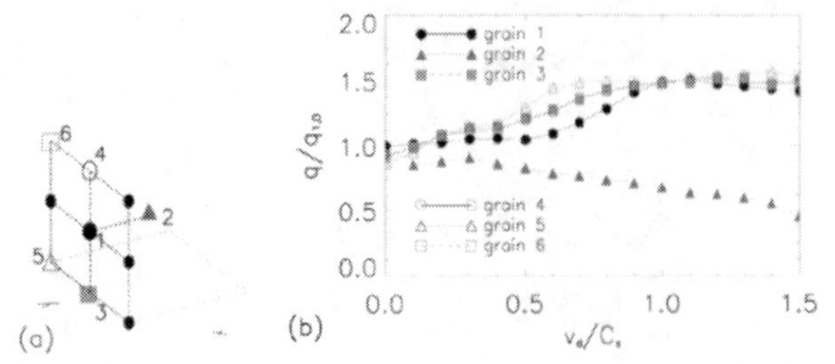

Figure 6. (*a*) Schematics of grain arrangements in the layer of nine grains and one downstream grain. (*b*) Charge on particular grains as a function of the flow velocity. The charge is normalized to the charge of grain 1 at conditions without the flow.

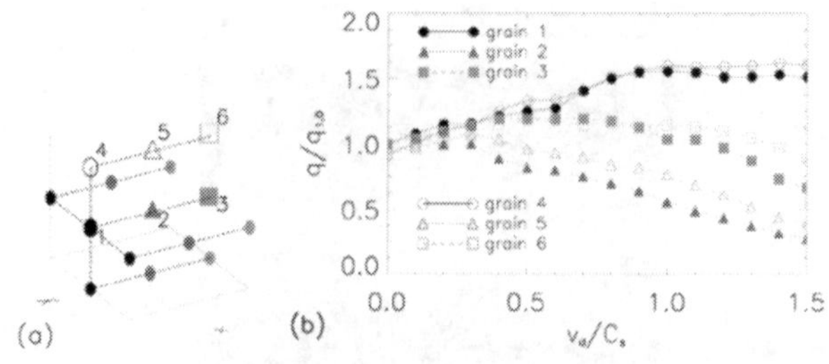

Figure 7. (*a*) Schematics of 15 grain arranged in five chains. (*b*) Charge on particular grains as a function of flow velocity, normalized to the charge on grain 4 at time-stationary conditions with no flow.

which will scale with the number of nearest neighbors. Note, that for $v_d = 0.7C_s$ the large angle scattered ions are mapped to the nearest neighbor position at $y = \lambda_{De}$. Thus, in flowing plasma the central grain should have the lowest charge and the grains 5 and 6 at the corners the highest charge. However, the charging of grain 1 is typical for a grain in an infinite layer. Interestingly, its charge is constant for $v_d < 0.5C_s$ and $v_d > C_s$. Only in the regime $0.5C_s < v_d < C_s$, the charge increases by about 50 percent.

If we now compare the charge of grain 1 (upstream grain) and grain 2 (downstream grain), we observe that the initial charge difference (at $v_d = 0$), which is caused by overlapping Debye-spheres, remains almost unchanged up to $v_d = 0.3C_s$. For $v_d > 0.4C_s$ first indications of wake effects and additional ion currents due to lensing towards the downstream grain are observed. Instead of a further increase of charge (see upstream grains), its charge starts to decrease till a significant charge reduction is observed at supersonic flow conditions. Although, it should be noted that the charge reduction differs with respect to two isolated grains. For significant subsonic ion flows we observe a weaker ion focusing in the wake of grain 1, because the collection radius (see figure 5(*a*), light blue area) of an isolated grain is larger than $0.7\lambda_{De}$ and thus the nearest neighbors in the layer truncate this to less than approximately $0.5\lambda_{De}$.

A very similar charging characteristics is observed for 15 grains arranged in chains (see figure 7). The charge of the grains in the upstream layer gradually increases with flow velocity until it saturates at sonic velocities. The second layer is highly decharged by the upstream layer, due to strong ion focusing. Focus effects are observed for $v_d > 0.4C_s$. This matches the results of the 9 + 1-particle system well. It is interesting, that the charge reduction of the third layer is less than on the second. Obviously the low charge on the grains of the second layer results in a weak ion focus behind grains of the second layer. This characteristics is significant for both supersonic and subsonic flows, down to $v_d \approx 0.4C_s$.

4. Discussion

The significant change in the charge on the grain for subsonic flows, which is due to reduced ion collection cross section, has implications on the charging of downstream grains in this regime. The electrostatic lensing associated with ion focusing gives rise to enhanced ion density region in the wake and results in the reduction of the charge on the downstream grain. As shown in figure 2, for the two grains aligned with the flow, this charge scales linearly with the distance between grains in subsonic flows, within the considered velocity regime. The linear dependence on the charge on the downstream grain on the intergrain distance has been earlier reported for supersonic flows [27] and the charge reduction has been confirmed in experiments [34]. The fact that decharging of the downstream grain follows the linear scaling with the distance also for subsonic flows regime can largely facilitate modeling of larger dusty plasma systems in subsonic flows [36]. The same approach and analysis can be therefore used in both flow regimes. Furthermore, this indicates that both the ion focusing and wake effects are already substantial for subsonic flows and there seem not to be a very clear boundary for the charging characteristics and mechanisms between the two flow regimes in collisionless plasmas. Note that with further decrease of the flow velocity, the relative charge on the grains goes towards unity for larger distances. This is to be expected, as in the limit of very slow velocities and large distances the grains should charge independently and acquire equal charge.

For both subsonic and supersonic flows, decharging of the downstream grains is primarily due to ions being electrostatically lensed by the upstream grain, as it is also indicated in figure 4. The ion focus region will have an effect on the equilibrium position of the downstream grain, but unless the intergrain distance is small enough, these ions will not contribute significantly to the charging of the downstream grain. The effective collection area for the downstream grain will change with the flow velocity (see again figure 5) and for subsonic flows its radius will be in general larger than $\lambda_{De}/2$. Thus, in systems with extended dust layers perpendicular to the flow, it is expected that charging of the downstream grain can be significantly different for sub- and supersonic ion flows.

A significant feature of the self-consistent charging in subsonic flows is the presence of *slalom* trajectories. Some ions that enter the ion focus region of the upstream grain are electrostatically lensed into the focus region of the downstream grain, see again figure 4. These slalom trajectories of ions will contribute to the perpendicular drag force and will have implications on the stability and alignment of the system of two grains. This mechanism has also been recently discussed in [37] and the role of perpendicular component of the drag force has been addressed in PIC simulations [38], although the latter did not account for self-consistent charging of the dust grains. While for supersonic flows the momentum transfer from an ion to the downstream grain will be larger, for subsonic flows the larger collection will imply a significant net momentum transfer to the grain. Notably, in subsonic flows, the slalom trajectories will have an impact on the stability of the system.

The charging of a downstream grain (grain 2) in the system of 9 + 1 grains, can be considered as an example of the grain charging behind an extended layer of dust. In this case the collection cross section for that grain will be determined by a Wigner cell within the layer [1]. The geometric effect due to the presence of the neigbors is important, in particular in the subsonic flow regime. Here the effective collecting area that would otherwise be larger than λ_{De}, is limited to the interparticle distance in the layer. The charge on the grain remains approximately constant for the wide range of subsonic flow velocities and it starts to decrease for $v_d > 0.5C_s$. Thus, the geometric effect limiting the effective collection area influences the charge on the downstream grain.

On the other hand, the charge on grain 1 in the system of 9 + 1 is approximately constant for subsonic velocities $v_d \leqslant 0.5C_s$. This is attributed to a strong back-scattering of ions in that regime. The collective response of the cluster can be considered as a large charged object and it is characterized by a strong potential enhancement in the wake. Both the potential enhancement and the neighboring charged grains will contribute to the scattering of ions. In the subsonic flow regime there is a significant number of ions that compensate for the reduced collection cross section on grain 1, while the number of back-scattered ions is less for other grains, for which the charge value slowly increases. For $v_d > 0.5C_s$ the population of backscattered ions is gradually reduced and the charge on all the grains increases. The geometry of the cluster implies that the charge increase is fastest for grains with the least number of neighbors (i.e. grains 5 and 6) and slowest for grain 1. For supersonic flows the charge on all upstream grains is approximately the same and the charging characteristics flattens already at $v_d = 1C_s$. The flow regime of $v_d \in (0.5, 1.0)C_s$ is thus an interesting one, where the most inhomogeneities in the charge distribution can be expected.

The charging mechanisms for dust strings is different. Here, the fraction of back-scattered ions is reduced due to the extension of the dust cluster in the flow direction and the corresponding shift of the potential enhancement further downstream. Thus, already for subsonic flows the grains in the upstream layer (grain 1 and 4) follow a similar characteristic and the charge difference for $v_d \in (0.5, 1.0)C_s$ is significantly reduced, also due to the fact that the number of nearest neighbors is only four for grain 1 in this cluster. Consistent with the picture for two grain aligned with the flow, already for $v_d > 0.5C_s$ the charge on the middle layer is significantly lower than that on the upstream grain. However, an important effect is that the charge on the third layer is almost constant for the subsonic flows and remains significantly less decharged in the supersonic regime. Seemingly, in the subsonic regime, the two first layers significantly disturb the flow and the last layer is effectively charged in more homogeneous plasma conditions.

Thus, our simulations clearly show that the charging process of grains significantly depends on the grain arrangement. However, we would like to recall that our simulations are based on some simplifications. First, we studied flowing ions which is not the same as sheath conditions. Second, we focused on collisionless plasmas. Third, the grains are rather large and at fixed positions. Thus, our results are not directly applicable to experimental situation. However, the observed charging processes are expected to play a role for small grains. As a consequence future investigations should carefully check whether the frequently used one-component approach for dust clouds is really valid and whether OML-theory or other isotropic charging models yield resonable charge approximations.

5. Conclusions

To conclude, our simulations of several dust grains embedded in a collisionless plasma with fowing ions show that the grain charge depends significantly on the detailed grain arrangement. We find that the grain charge of downstream grains is altered by the wake effects for supersonic as well as for subsonic ion flows. For specific flow conditions at subsonic velocities we even find that the charge of neighboring grains in a layer is influenced by backscattered ions. Thus, our results show that an identical charge on all grains in many-grains arrangements should not be expected as soon as ion flows are present.

Acknowledgments

We gratefully acknowledge financial support by Deutsche Forschungsgemeinschaft: SFB-TR24 project A3b.

References

[1] Fortov V E, Ivlev A V, Khrapak S A, Khrapak A G and Morfill G E 2005 *Phys. Rep.* **421** 1
[2] Morfill G E and Ivlev A V 2009 *Rev. Mod. Phys.* **81** 1353
[3] Shukla P K and Eliasson B 2009 *Rev. Mod. Phys.* **81** 25
[4] Piel A and Melzer A 2002 *Plasma Phys. Control. Fusion* **44** R1
[5] Ishihara O 2007 *J. Phys. D: Appl. Phys.* **40** R121
[6] Shukla P K and Mamun A A 2002 *Introduction to Dusty Plasmas* (Bristol: Institute of Physics Publishing)
[7] Vladimirov S V, Ostrikov K and Samarian A A 2005 *Physics and Applications of Complex Plasmas* (London: Imperial College Press)
[8] Melzer A, Schweigert V A and Piel A 1999 *Phys. Rev. Lett.* **83** 3194
[9] Hebner G A and Riley M E 2004 *Phys. Rev.* E **69** 026405
[10] Samarian A A, Vladimirov S V and James B W 2005 *JETP Lett.* **82** 758
[11] Kroll M, Schablinski J, Block D and Piel A 2010 *Phys. Plasmas* **17** 013702
[12] Block D, Carstensen J, Ludwig P, Miloch W J, Greiner F, Piel A, Bonitz M and Melzer A 2012 *Contrib. Plasma Phys.* **52** 804
[13] Ivlev A V, Morfill G and Fortov V E 1999 *Phys. Plasmas* **6** 1415

[14] Khrapak S A, Ivlev A V, Zhdanov S K and Morfill G E 2005 *Phys. Plasmas* **12** 042308
[15] Lampe M, Joyce G, Ganguli G and Gavrishchaka V 2000 *Phys. Plasmas* **7** 3851
[16] Lampe M, Joyce G and Ganguli G 2005 *IEEE Trans. Plasma Sci.* **33** 57
[17] Ludwig P, Miloch W J, Kählert H and Bonitz M 2012 *New J. Phys.* **14** 053016
[18] Willis C T N, Coppins M, Bacharis M and Allen J E 2012 *Phys. Rev.* **83** 036403
[19] Lapenta G 1999 *Phys. Plasmas* **6** 1442
[20] Maiorov S A, Vladimirov S V and Cramer N F 2000 *Phys. Rev.* E **63** 017401
[21] Miloch W J, Vladimirov S V, Pécseli H L and Trulsen J 2008 *Phys. Rev.* E **78** 036411
[22] Miloch W J 2010 *Plasma Phys. Control. Fusion* **52** 124004
[23] Hutchinson I H 2005 *Plasma Phys. Control. Fusion* **47** 71
[24] Hutchinson I H 2011 *Phys. Plasmas* **18** 032111
[25] Allen J 1992 *Phys. Scr.* **45** 497
[26] Whipple E C 1981 *Rep. Prog. Phys.* **44** 1197
[27] Miloch W J and Block D 2012 *Phys. Plasmas* **44** 123703
[28] Vladimirov S V and Nambu M 1995 *Phys. Rev.* E **52** R2172
[29] Ishihara O and Vladimirov S V 1997 *Phys. Plasmas* **4** 69
[30] Miloch W J, Trulsen J and Pécseli H L 2008 *Phys. Rev.* E **77** 056408
[31] Miloch W J, Kroll M and Block D 2010 *Phys. Plasmas* **17** 103703
[32] Steinberg V, Sutterlin R, Ivlev A V and Morfill G 2001 *Phys. Rev. Lett.* **86** 4540
[33] Melzer A, Schweigert V A, Schweigert I V, Homann A, Peters S and Piel A 1996 *Phys. Rev.* E **54** R46
[34] Carstensen J, Greiner F, Block D, Schablinski J, Miloch W J and Piel A 2012 *Phys. Plasmas* **19** 033702
[35] Arp O, Goree J and Piel A 2012 *Phys. Rev.* E **85** 046409
[36] Patrick Ludwig H K and Bonitz M 2012 *Plasma Phys. Control. Fusion* **54** 045011
[37] Piel A 2011 *Phys. Plasmas* **18** 073704
[38] Hutchinson I H 2011 *Phys. Rev. Lett.* **107** 095001
[39] Iizuka S, Saeki K, Sato N and Hatta Y 1979 *Phys. Rev. Lett.* **43** 1404
[40] Hur M S, Lee H J and Lee J K 1999 *IEEE Trans. Plasma Sci.* **27** 1449
[41] Brackbill J U 1988 *J. Comput. Phys.* **75** 469
[42] Hockney R W and Eastwood J W 1988 *Computer Simulation Using Particles* (Bristol: IOP Publishing)
[43] Matyash K, Schneider R, Ikkurthi R, Lewerentz L and Melzer A 2010 *Plasma Phys. Control. Fusion* **52** 124016
[44] Miloch W J, Pécseli H L and Trulsen J 2007 *Nonlinear Process. Geophys.* **14** 575

IOP Publishing | Plasma Physics and Controlled Fusion

Plasma Phys. Control. Fusion **57** (2015) 014020 (10pp) doi:10.1088/0741-3335/57/1/014020

Non-linear MHD modeling of edge localized mode cycles and mitigation by resonant magnetic perturbations

François Orain[1], M Bécoulet[1], J Morales[1], G T A Huijsmans[2], G Dif-Pradalier[1], M Hoelzl[3], X Garbet[1], S Pamela[4], E Nardon[1], C Passeron[1], G Latu[1], A Fil[1] and P Cahyna[5]

[1] CEA-IRFM, Cadarache, 13108 Saint-Paul-Lez-Durance, France
[2] ITER Organization, Route de Vinon-sur-Verdon, CS 90 046, 13067 Saint-Paul-Lez-Durance, France
[3] Max Planck Institute for Plasma Physics, Boltzmannstr. 2, 85748 Garching, Germany
[4] CCFE, Culham Science Centre, Oxon, Abingdon, Oxfordshire OX14 3DB, UK
[5] Institute of Plasma Physics ASCR, Za Slovankou 1782/3, 182 00 Prague 8, Czech Republic

E-mail: francois.orain@cea.fr

Received 27 June 2014, revised 7 September 2014
Accepted for publication 22 September 2014
Published 28 November 2014

Abstract

The dynamics of a multi-edge localized mode (ELM) cycle as well as the ELM mitigation by resonant magnetic perturbations (RMPs) are modeled in realistic tokamak X-point geometry with the non-linear reduced MHD code JOREK. The diamagnetic rotation is found to be a key parameter enabling us to reproduce the cyclical dynamics of the plasma relaxations and to model the near-symmetric ELM power deposition on the inner and outer divertor target plates consistently with experimental measurements. Moreover, the non-linear coupling of the RMPs with unstable modes are found to modify the edge magnetic topology and induce a continuous MHD activity in place of a large ELM crash, resulting in the mitigation of the ELMs. At larger diamagnetic rotation, a bifurcation from unmitigated ELMs—at low RMP current—towards fully suppressed ELMs—at large RMP current—is obtained.

Keywords: ELM, RMP, MHD, cycles, mitigation, tokamaks, plasmas

(Some figures may appear in colour only in the online journal)

1. Introduction

Edge localized modes (ELMs) are magnetohydrodynamic (MHD) events occurring in high-confinement regime (H-mode) of tokamak plasmas. They are initiated with the growth of instabilities triggered by the large edge pressure gradient (so-called ballooning modes) or by the large edge current (so-called peeling modes) [1]. These instabilities generate a quasi-periodic relaxation of the edge plasma (also called pedestal), and represent a cyclical dynamics of the plasma. Experimentally, ELM precursors (possibly related to peeling-ballooning instabilities) are observed to grow and rotate in the pedestal in the electron diamagnetic direction, followed by the expulsion of filaments through the edge transport barrier [2, 3]. A possible understanding is that the ELM precursors are triggered when the peeling-ballooning stability boundary is crossed, and grow until the plasma non-linearly reorganizes; as a result, the pressure profile relaxes and filaments are expelled from the plasma. The relaxation brings back the plasma to a stable state, yet as the pedestal is rebuilt, the profiles steepen and the plasma is re-destabilized, inducing another ELM crash [4]. The heat flux reaching the divertor due to an ELM is foreseen to be likely to erode the divertor in ITER if the ELMs are not mitigated, thus the ELMs have to be controlled in ITER [5]. A promising control method is the application of resonant magnetic perturbations (RMPs) by dedicated coils. This method proved to be successful in either suppressing the ELMs in the DIII-D [6, 7] and KSTAR tokamaks [8] or in mitigating the ELM power in ASDEX Upgrade [9], JET [10], MAST [11] and NSTX [12], which validated its use for ITER operation.

So far, only a single ELM crash could be reproduced in modeling in realistic geometry with non-linear MHD codes such as M3D [13], BOUT + + [14, 15], NIMROD [16] and JOREK [17–19]; however multi-ELM cycling regimes were not reproduced in modeling up to now. As for the ELM mitigation by RMPs, its observation in modeling is a new result. The reduced MHD code JOREK was recently extended to introduce the bi-fluid diamagnetic effects, the neoclassical friction and a source of toroidal rotation, in order to self-consistently describe the plasma flows. Also, RMPs were added in the code, the vacuum RMP field being set as boundary conditions for the perturbation of the magnetic flux. The full model is described in [20]. These improvements now enable to simulate both multi-ELM cycles—we show in this paper that the diamagnetic stabilization is a key parameter that allows for obtaining them—and the ELM mitigation or suppression by RMPs. Section 2 presents the modeling of ELM cycles in JET-like configuration. We show that a competition between the diamagnetic stabilization and the plasma destabilization by the growth of the pressure gradient due to the heat source generates the cyclical dynamics of the plasma relaxation and pedestal reconstruction. The near-symmetric repartition of the ELM power on the inner and outer divertor target plates due to the convection of the filaments by the diamagnetic drifts is also presented. In section 3, the mechanism of the ELM mitigation or suppression by RMPs is described. Under a certain RMP current threshold, the ELM mitigation is due to the continuous MHD activity produced by the non-linear interaction between RMPs and the unstable plasma modes. When the RMP current is sufficiently increased, the plasma is stabilized under the peeling-ballooning instability threshold and the ELMs are fully suppressed.

2. ELM cycle studies

ELM simulations are performed for JET-like plasma parameters and geometry, at low triangularity shape, similar to [20]: major radius $R_0 = 3$ m, minor radius $a = 1$ m, toroidal magnetic field $B_t = 2.9$ T and safety factor $q_{95} \sim 3$. Established H-mode experimental profiles are taken initially, with central electron density $n_{e,0} = 6 \times 10^{19}$ m^{-3} and central temperature $T_{e,0} = 5$ keV. The pedestal density and temperature are $n_{e,\text{ped}} = 3.8 \times 10^{19}$ m^{-3} and $T_{e,\text{ped}} = 2.5$ keV. The edge transport barrier is maintained by reducing the perpendicular diffusive coefficients of heat and particles in the pedestal. *Ad hoc* heat and particle sources are chosen to mimic e.g. the heating power due to neutral beam injection: the heat source follows the pedestal profile and the density source is constant over the plasma. These sources are chosen to ensure that the pedestal can reconstruct fast enough such that several ELM relaxations can be observed in a simulation, thus these simulations allow to give a qualitative comparison of the ELM frequency evolution with respect to parameters such as the heating power, but cannot give a quantitative comparison against experiment. Because of computational restrictions, the central resistivity is taken $\eta_0 = 2.5 \times 10^{-7}$ Ω.m, which is 2 orders of magnitude larger than the Spitzer resistivity $\eta_{\text{Spitzer}} = 2.5 \times 10^{-9}$ Ω.m. The resistivity profile follows a $T^{-3/2}$ radial dependence. A JET realistic diamagnetic velocity is taken for both species s (ions and electrons): $\vec{V}_s^* = Z_s(R/R_0)^2\,\tau_{IC}/\rho{\cdot}\vec{b} \times \nabla P_s$. Z_s and P_s are respectively the charge number and the scalar pressure of the species s, R is the horizontal coordinate, ρ is the mass density and $\vec{b}$ is the magnetic field normalized to B_t. The realistic value of the diamagnetic parameter (inverse of the ion cyclotron frequency) $\tau_{IC} = m_i/e\,B_t \sim 7 \times 10^{-9}$ s is taken, and ion and electron temperatures are assumed to be equal. The model is the MHD model extended with bi-fluid diamagnetic flows, as described on [20]. Note that the neoclassical friction and the source of toroidal rotation are not included in this section in order to point out the role of diamagnetic stabilization.

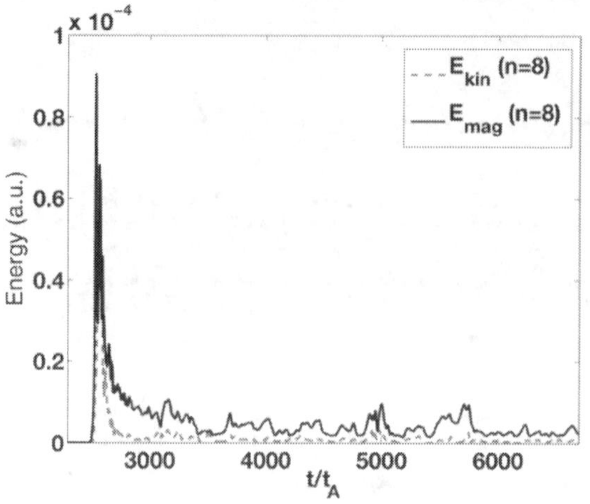

Figure 1. Time variation of the magnetic and kinetic energies of the mode $n = 8$ without diamagnetic effects. Time is normalized to the Alfvén time t_A.

The ELM crash is simulated first without diamagnetic rotation (section 2.1), then the same simulation is performed with diamagnetic drifts (section 2.2).

2.1. simulations without diamagnetic drifts: single ELM crash

In the modeling without diamagnetic drifts, the ELM crash is generated by the chosen initially unstable pressure profile. After the crash, the unstable modes remain unstable, and the residual magnetic activity expels the plasma outside the separatrix, which prevents the pedestal from building-up again. The example is given for the case of an $n = 8$ unstable toroidal mode. The ELM crash corresponds to a large peak of magnetic and kinetic energy (figure 1). At the crash, the large magnetic activity leads to a strong ergodization of the edge (the magnetic field is ergodic for a normalized flux $\psi > 0.85$, as shown in figure 2(*a*)). After the crash, magnetic islands remain for $\psi > 0.85$ and an ergodic layer subsists at the very edge ($\psi > 0.95$, see figure 2(*b*)). This magnetic activity and particularly the edge stochastization increase both the heat parallel diffusivity and the $E \times B$ convection of particles. The enhanced transport prevents the reconstruction of the pedestal profiles and keeps the plasma below the peeling-ballooning stability limit, thus a second ELM cannot be obtained in the simulations without diamagnetic effects.

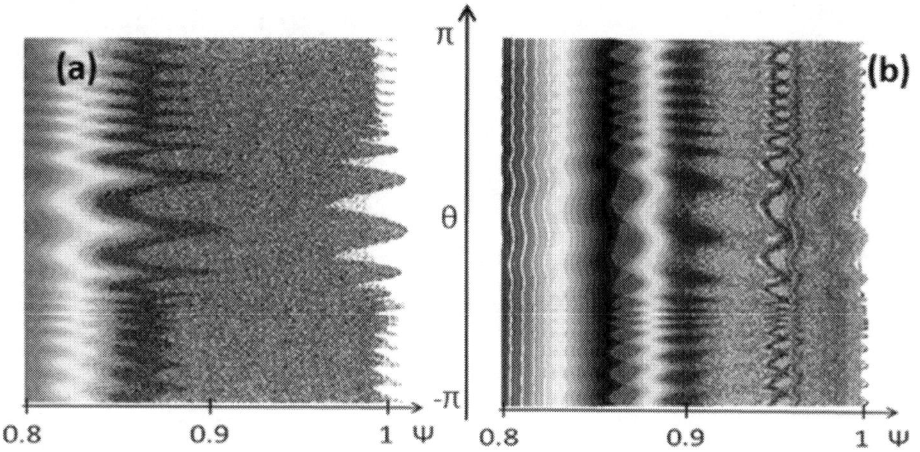

Figure 2. Poincaré plot of the magnetic topology at the plasma edge (normalized flux $\psi \geqslant 0.8$) in the $n = 8$ ELM simulation without diamagnetic effect, given at the peak of the ELM crash (*a*) and after the ELM crash at $t \sim 7 \times 10^3\ t_A$(*b*). Note the full edge reconnection at the peak of the crash, and the remaining partial reconnection after the crash.

2.2. Simulations with diamagnetic drifts: ELM cycling dynamics

The diamagnetic effects are known to prevent the magnetic field from reconnecting and to reduce the growth rate of ideal and resistive instabilities [21, 22] and thus have a stabilizing effect on plasma instabilities. In particular they are evidenced to be a key parameter for simulating cycles of sawtooth crashes [23]. We show here that the diamagnetic drifts also enable to simulate cycles of ELM crashes. The diamagnetic stabilization has two major effects on ELMs. First, it reduces the amplitude of the ELM perturbation and crash. Indeed, compared to the simulation without diamagnetism (figure 1), the magnetic and kinetic energies of the same $n = 8$ ELM with diamagnetic drifts included (figure 3) are 1–2 orders of magnitude lower. In figure 3, the multi-harmonic simulation of the modes $n = 0$, 2, 4, 6 and 8 with diamagnetic drifts is presented. The most unstable $n = 8$ mode is hardly coupled with the other modes during the first ELM crash and the energy of the other modes remain several orders of magnitude lower than the energy of the $n = 8$ mode. So the dynamics of the first ELM is almost only governed by the $n = 8$ mode. Note that the modes $n > 9$ are stabilized by the diamagnetic stabilization; for this reason and in order to reduce the time-consumption of the simulation, the highest $n > 9$ modes were not kept in simulation.

After the first ELM crash, we observe the second major role of the diamagnetic stabilization: instead of remaining unstable after the crash, the plasma is stabilized by the diamagnetic rotation (figure 3). Only when the pressure profile is built up by the applied heating power, the plasma is destabilized again. The ballooning modes are growing again until a threshold in pressure gradient is reached, triggering the second ELM relaxation. Therefore the ELM dynamics results from a competition between the diamagnetic stabilization and the destabilization of the plasma by the heat and particle sources filling the pedestal and steepening the pressure profile. The ELM cycles obtained in simulation (figure 3) can be decomposed into two periods. The first three transient ELMs are largely dominated by the most unstable modes ($n = 8$ during the first two ELMs then $n = 6$), and their triggering threshold is dependent on the initial chosen state. However, after 3–4 ELMs, the memory of the initial state is lost, and a quasi-periodic regime of ELM crashes is obtained. These ELMs are characterized by a strong non-linear coupling between the $n = 2$–8 modes. The $n = 6$ mode is growing first, directly followed by the other modes. After a crash in this quasi-periodic regime, the plasma reorganizes into a self-consistent state. Thus, the ELM frequency does not depend any more on initial conditions, but depends on intrinsic parameters: the diamagnetic rotation (which have a stabilizing effect on ELMs) and the applied heating power (destabilizing), as described in the next section 2.3. The competition between these two parameters governs the cyclical dynamics of the ELMs. Each ELM in the quasi-periodic regime can be decomposed into three steps (figure 4): first, the instability grows as the pressure gradient (and in a smaller extent the edge current)

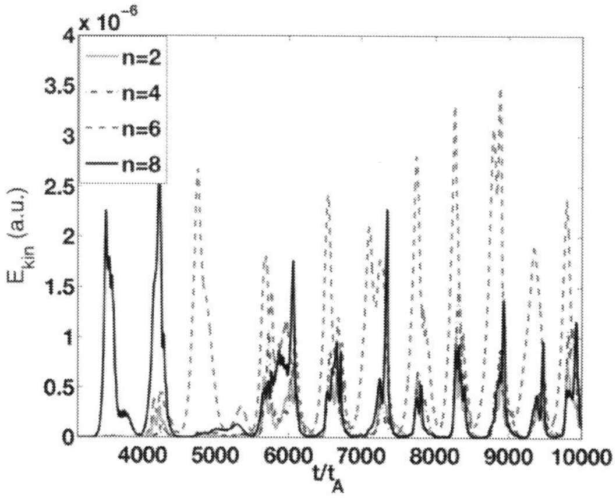

Figure 3. Kinetic energy of the modes $n = 2$–8 in the multi-harmonic $n = 0:2:8$ simulation with diamagnetic effects.

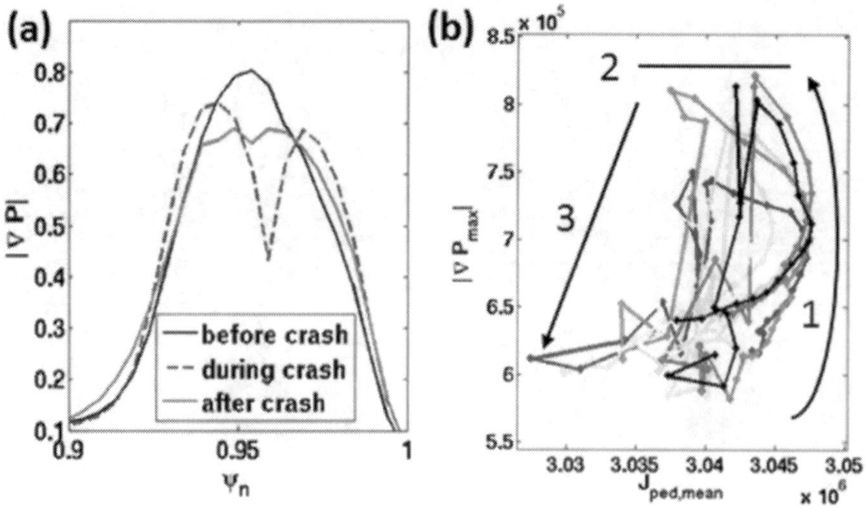

Figure 4. In quasi-periodic regime: (*a*) edge pressure gradient before, during and after an ELM. (*b*) Peeling-ballooning diagram of the ELMs.

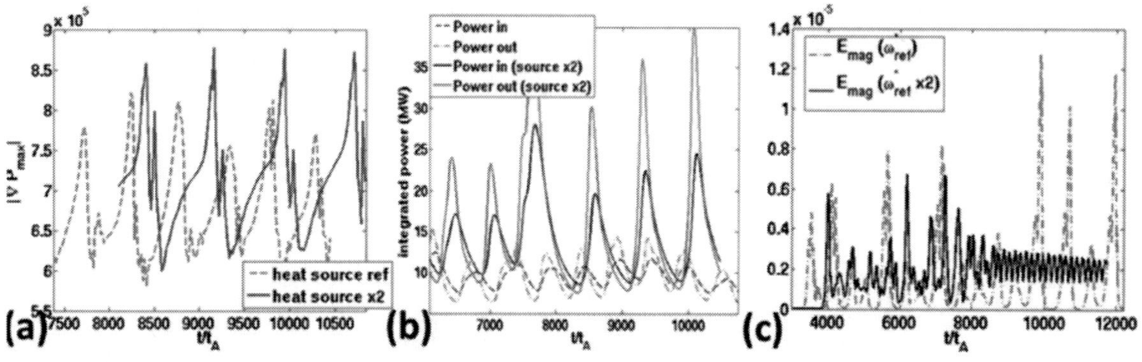

Figure 5. (*a*)–(*b*) Time variation of the maximal edge pressure gradient (*a*) and integrated power on the inner and outer divertor plates (*b*) in quasi-periodic $n = 8$ ELMy regime in the reference case (dash) and with a double injected power. (*c*) Magnetic energy of the $n = 8$ ELM depending on the diamagnetic parameter: realistic value (dash) and doubled (full).

increase. Second, the ELM crash occurs when a limit of pressure gradient is reached. And third, the pedestal profiles relax: particles and energy are expelled out of the plasma, and the pressure gradient collapses, as presented in figure 4(*a*). Note that the bootstrap current is not included in the model (the current is forced to tend to the initial realistic current profile via a current source, but does not depend on the pressure gradient evolution, as it is in experiments), so the dynamics of the instabilities simulated here shows only a small dependence on the pedestal current (figure 4(*b*)).

2.3. Characterization of the ELM frequency

The ELM frequency obtained in figure 3 is about 3 kHz. This frequency cannot be compared quantitatively with experiments, due to the *ad hoc* sources taken in simulations; however a qualitative study on the parameters impacting the ELM frequency can be done. The ELM frequency depends on both the stabilizing effect of the diamagnetic rotation and the destabilization by the steepening of the pressure profile due to the heat source. To assess these effects, simulations are performed first with a twice larger heat source and second with a twice larger diamagnetic parameter τ_{IC}. First, we notice that the enhanced heat source has the effect of raising the maximal pressure gradient reached at the ELM crash, which is about 5% larger with the enhanced source (figure 5(*a*)). This can be explained by two combined effects. First, whilst the peeling-ballooning modes are growing, the enhanced heating power makes that the pressure gradient rise more rapidly during the delay needed by the non-linear effects to induce the pedestal relaxation. Second, due to the larger heating power, the mean equilibrium (inter-ELM) temperature T_0 is raised after several ELMs. Consequently, the resistivity (proportional to $T_0^{-3/2}$) is reduced, so the boundary stability is increased. For both reasons, a larger pressure gradient develops in the pedestal before the ELM crash, which induces a larger amount of particles and energy expelled by the ELM. Subsequently, the power reaching the divertor (figure 5(*b*)) is increased. This means that the pedestal has emptied more consequently and that more time is needed to reconstruct the pedestal. Therefore the ELM frequency is decreased as the heat power is increased. This behavior is closer to the type-III ELMs obtained in the

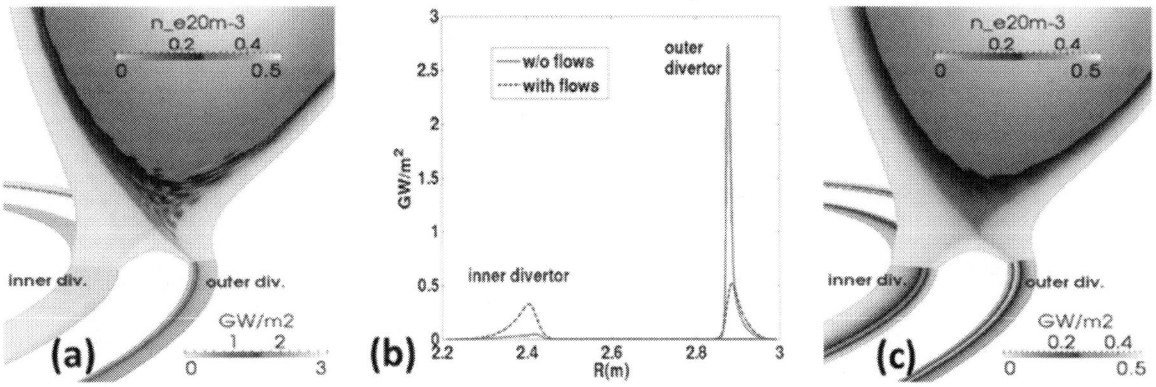

Figure 6. Heat flux (in MW/m^2) reaching the inner and outer divertor target plates after an ELM: without flows (*a*) or with diamagnetic, neoclassical and toroidal flows (*c*). A radial section along the divertor is plotted in (*b*): almost all flux reaches the outer divertor (full line) without flows whereas the deposition is near-symmetric with flows (dash line).

experiments [24]. Yet the type-III ELM instabilities are so far not clearly understood theoretically, and since both the type-I and type-III ELMs are linked with the same instabilities (peeling-ballooning), it is still not clear what fundamentally differentiates these two types. More modeling with realistic parameters (allowing for a close comparison to experiments) is planned to better characterize the difference between type-I and type-III ELMs.

Second, simulations at larger diamagnetic rotation present a larger ELM frequency. Actually, due to the diamagnetic stabilization, smaller instabilities develop at the edge plasma and a smaller energy is expelled from the pedestal. So the plasma after the crash remains just below the stability limit, such that less time is needed to cross again the stability threshold, which explains a larger ELM frequency. In the case with a diamagnetic parameter twice larger than the realistic one (figure 5(*c*)), not only the ELM frequency (obtained after the transient phase) is increased, but the large diamagnetic rotation gradually decreases the amplitude of the ELMs, so that in terms of peeling-ballooning cycles, the plasma moves from the initially unstable state towards a 'fix point' in the P-B diagram by doing smaller and smaller cycles. As it represents a bifurcation from a cycling dynamics towards a stable state, it may present similarities with the so-called dithering cycles [4].

2.4. Near-symmetric power deposition on divertor target plates

The diamagnetic rotation also affects the dynamics of the ELM precursors and filaments. Without diamagnetic rotation, the instabilities at the onset of the ELM have a static growth, whereas the diamagnetic rotation makes the ELM precursors grow and rotate in the electron diamagnetic direction at the speed $V_{\text{mode}} \approx V_{E\times B} + V_i^* / 2$ [25]. In the non-linear stage, the plasma inside the separatrix rotates in the electron diamagnetic direction, and the ELM filaments are sheared out of the plasma in the ion diamagnetic direction. The temperature is conducted by the heat diffusivity from the reconnected edge plasma to the divertor plates, whereas the density is convected by the $E \times B$ and diamagnetic rotation. It is important to notice that this diamagnetic advection of the density makes that more density reaches the inner divertor plate than the outer plate, hence the heat flux reaching the divertor at the sound speed is deposited near-symmetrically in the inner and outer divertor plates: even though the temperature is larger on the outer than on the inner side, the density is larger in the inner region so a similar power, proportional to both the plasma temperature and density, reaches the inner and outer divertor target plates due to an ELM. As an example, figures 6(*b*) and (*c*) present the near-symmetric power deposition of the heat flux in the simulation of a realistic JET shot (#77329, described in [26]) where the diamagnetic, neoclassical and toroidal flows have been taken into account. In comparison, in simulations made without flows (figures 6(*a*) and (*b*)), the outer divertor received almost all the heat power. This was in contradiction with the experimental observations where the deposit is either symmetric on inner and outer divertor plates or two times larger in the inner divertor [27, 28]. Thus simulations with diamagnetic drifts allow for a more realistic reproduction of the ELM dynamics up to the deposition on the divertor.

3. ELM control by RMPs

The plasma rotation and in particular the diamagnetic rotation is also known to affect the penetration of the RMPs in the plasma [20, 26, 29, 30]. Previous studies demonstrated that the plasma perpendicular ($E \times B$ + diamagnetic) rotation is likely to prevent the RMPs from penetrating into the plasma. Thus RMPs can just penetrate on particular rational surfaces where the perpendicular rotation is zero and at the very edge where the resistivity is the highest. In [20], we found that even though RMPs are screened in the bulk plasma, they still affect the magnetic topology: small islands are generated on the rational surfaces $q = m/n$ and an ergodic layer is formed at the very edge, which induces an enhanced transport at the edge. In this section, we find that the perpendicular diamagnetic rotation also plays an important role in the interaction between ELMs and RMPs in the plasma.

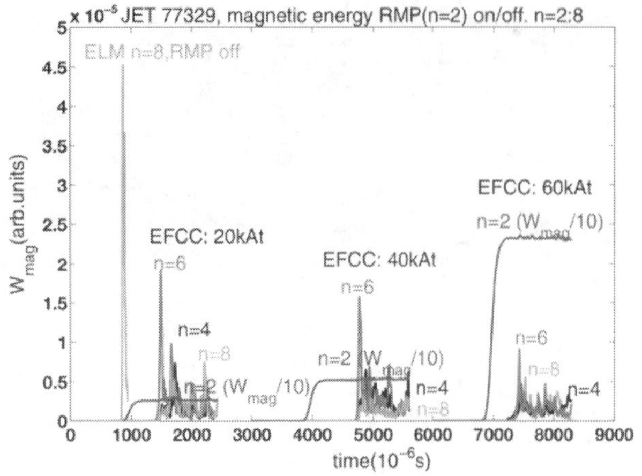

Figure 7. Magnetic energy of the modes $n = 2$–8; from left to right: without RMP, with RMP current $I_{coil} = 20kAt$, $40kAt$ and $60kAt$.

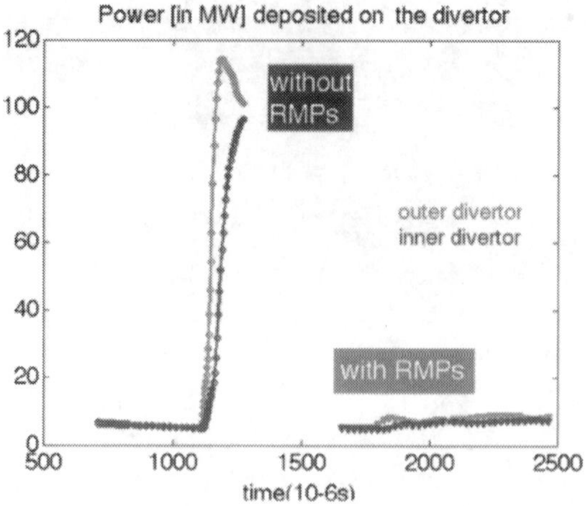

Figure 8. Power on inner/outer divertor targets without (left) and with RMPs (right) for $I_{coil} = 40kAt$.

3.1. ELM power deposition divided by ten with RMPs

In this section, experimental JET parameters corresponding to the shot #77329 [26] are used. Typical JET values are taken for the diamagnetic parameter and the neoclassical coefficients as in [20, 26]. The ($n = 2$) RMP spectrum due to the error field correction coils (EFCC) is first calculated in the vacuum with the ERGOS code ([31]) and applied as boundary conditions for the magnetic flux perturbation in JOREK. Once the plasma response to ($n = 2$) RMPs has stabilized to an equilibrium (affected by RMPs), the other toroidal modes are added to the simulation. The effect of RMPs on ELMs is tested first in a simulation with $n = 0, 2, 4, 6$ and 8, the RMP coil current being scanned between 20 and $60kAt$. In the simulation run without RMPs, the most unstable mode is the $n = 8$, which leads to a large ELM crash (figure 7, left). As compared to the ELM cycle considered in section 2, where the ELMs induced rather small pedestal relaxations (the integrated peak power reaching the inner and outer divertor targets was about 10–30 MW depending on the applied heating power, as shown in figure 5(b)), the ELM considered in this part induces a much larger relaxation (around 100 MW of integrated power reach each divertor target, as plotted in figure 8) when unmitigated. This large ELM is thus rather related to large type-I ELMs observed in experiments, yet further modeling is foreseen to compare more quantitatively this ELM with experimental observations. When RMPs are applied, instead of this large ELM crash, a more continuous activity of the modes $n = 2, 4, 6$ and 8 coupled altogether is observed (figure 7). This activity starts with a peak of energy of the mode $n = 6$, which is the most unstable with RMPs. This peak is smaller compared to the peak of the $n = 8$ mode without RMPs thus it leads to a mitigated ELM with a smaller peak energy released. Moreover, as the RMP current is increased from 20 to $60kAt$, the $n = 6$ peak size is reduced, which means that a stronger ELM mitigation is obtained when the RMP current is larger. The power reaching the divertor in the $40kAt$ case is plotted in figure 8. The deposited power is divided by ten when RMPs are applied compared to the 'natural' ELM (without RMPs).

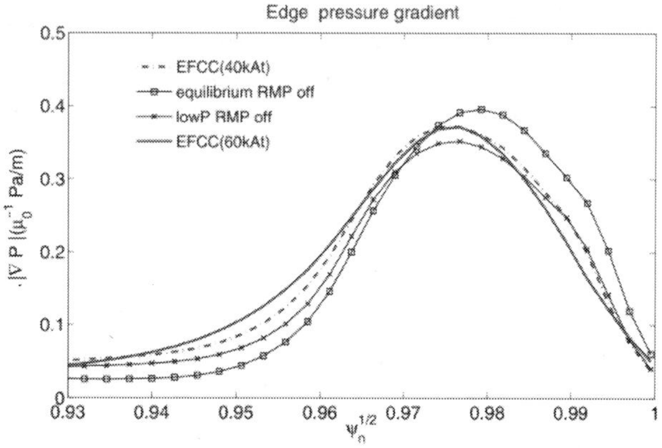

Figure 9. Edge pressure gradient: cases without RMP (squares), with RMP at $I_{coil} = 40kAt$ (dash–dot) and $60kAt$ (full line) and without RMP at reduced pressure gradient (crosses).

3.2. Mechanism of the ELM mitigation by RMPs

We try to understand the mechanism at stake in the ELM mitigation. The first hypothesis is that the edge ergodization due to RMPs which results in a lower edge pressure gradient (as shown in figure 9) may generate a (partial) stabilization of the ELMs. To test this hypothesis, a simulation is run without RMPs with a reduced pressure gradient similar to the pressure gradient obtained with RMPs (figure 9, line with crosses). In fact, the reduction of the edge pressure gradient delays the ELM crash (full line in figure 10) compared to the reference ELM (dashed line in figure 10), but still leads to a large ELM crash. So the only reduction of the pressure gradient by RMPs does not explain the observed ELM mitigation. The second tested hypothesis in that the modification of the magnetic topology due to RMPs induces the mitigation. A linear run of an ELM with RMPs is thus performed, where the magnetic topology is affected by both the ELM and the RMPs but where the RMPs and the unstable modes are not coupled. This linear run also leads to a large ELM crash.

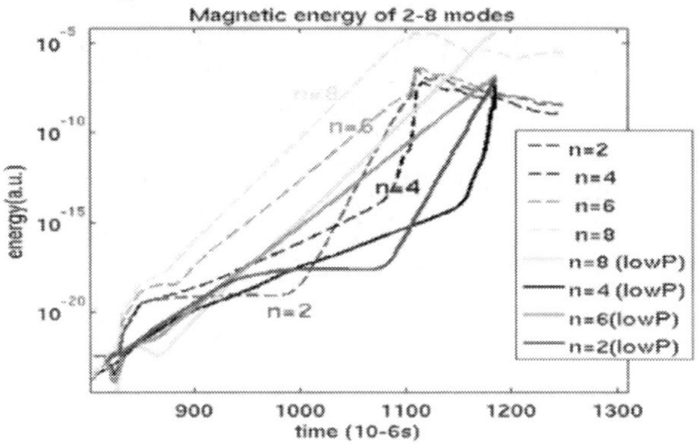

Figure 10. Magnetic energy of the modes $n = 2:8$ without RMP: in the simulation run at reduced pressure gradient (full line), the ELM crash is delayed compared to the reference case, but remains large.

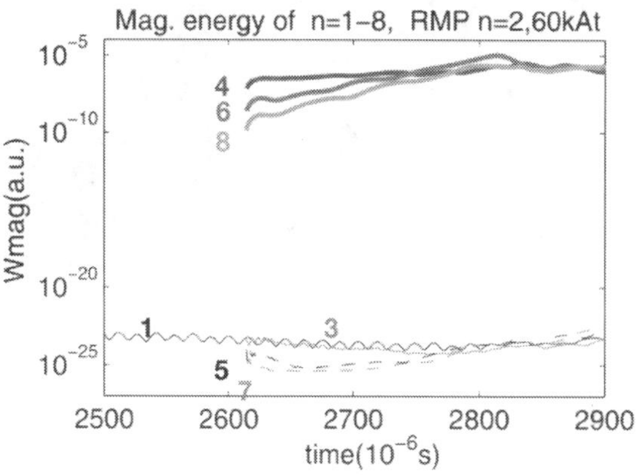

Figure 11. Magnetic energy of the modes $n = 1$–8 with RMPs ($I_{coil} = 60kAt$). The magnetic activity due to the even modes driven by RMPs totally damps the odd modes.

This proves that it is really the coupling between the ($n = 2$) RMPs and its harmonics (even modes) that generate the ELM mitigation. Figure 11 presents the simulation at $60kAt$ with $n = 0 - 8$. We notice that the initial growth rate of the even modes $n = 4$, 6 and 8 is large due to the non-linear drive by ($n = 2$) RMPs. The magnetic energy between these even modes is equally redistributed after the first relaxation of the $n = 6$ modes and the magnetic energy non-linearly cascades from the naturally unstable $n = 8$ mode towards the lower n even modes [32]. As for the odd modes (figure 11), they are totally damped due to the activity of the even modes driven by RMPs, and hence they remain at the noise level. The reduction of the ELM toroidal mode number when mitigated by RMPs was observed in KSTAR [33], but not in the general case [34]. Actually, in another simulation performed at 15% larger diamagnetic rotation (not presented here), the $n = 6$ mode is the most unstable for the natural ELM due to the diamagnetic stabilization of the larger n modes, and remains the most unstable mode while non-linearly coupled with ($n = 2$) RMPs. Thus the energy cascade implies a redistribution between the non-linearly coupled modes but does not necessarily mean a reduction of the main toroidal mode number.

The coupling of ELM and RMPs induce a change in the edge magnetic structure, as plotted in figure 13. The magnetic topology of the natural ELM (figure 13(a)) is dominated by the $n = 8$ ballooning perturbation that induces a large reconnection at the edge (for $\psi_{norm} > 0.85$). In the case of RMPs without ELMs (figure 13(b)), $n = 2$ magnetic islands are formed on the rational surfaces $q = m/n$ due to the forced magnetic reconnection, and present a tearing-like parity ($\psi_{mn} \neq 0$ on the rational surfaces). It is important to notice that the perpendicular ($E \times B$ + diamagnetic) electron rotation is zero on the rational surface $q = 5/2$ (at $\psi_{norm} \approx 0.85$) which leads to the RMP penetration and the formation of large islands on this surface, as demonstrated in [20, 26, 29]. As for the magnetic topology of the ELMs with RMPs (figure 13(c)), it is dominated by the modes $n = 4$ and $n = 6$, with magnetic islands observed on the rational surfaces $q = 9/4$, 14/6 and 15/6. The overlapping of the surfaces $q = 16/6$, 17/6 and 18/6 generates an ergodic zone deeper in the pedestal than in the

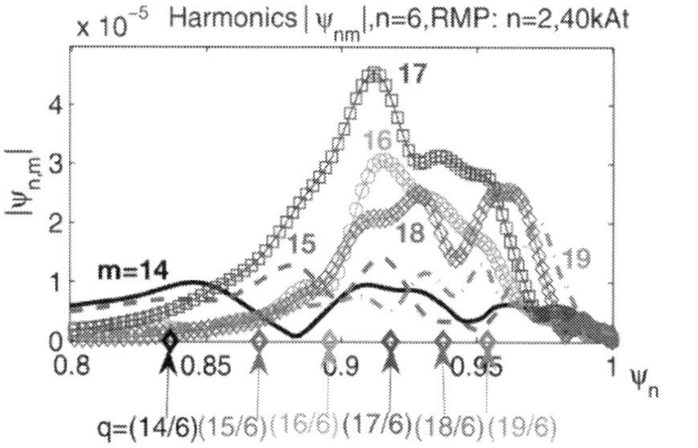

Figure 12. Radial profiles of the Fourier harmonics of the magnetic flux perturbation $|\psi_{mn}|$, for $n = 6$ and $14 < m < 19$. The rational surfaces $q = m/n$ are indicated by the diamonds.

case of RMPs alone. The $n = 4$ and 6 modes have a tearing-parity ($\psi_{mn} \neq 0$) on these rational surfaces that is imposed by the driving by RMPs (as plotted in figure 12), in place of the ballooning parity in the natural ELM case ($\psi_{mn} = 0$). The magnetic islands and above all the ergodic zone at the edge induce an enhanced transport at the edge that expels continuously the heat and particles out of the plasma. Note that no clear density pump-out is observed in our simulations of ELMs mitigated by RMPs: the density is only affected in the pedestal and less than 1% of the total plasma density is lost due to the RMP application. Actually two groups of RMP experiments can be distinguished. At high collisionality, the RMP application usually leads to the replacement of large ELMs by small frequent ELMs or MHD turbulence, and the heat flux on the divertor is reduced by RMPs. Small changes in the edge pressure gradient and MHD stability are reported [8–12, 35]. On the other hand, at low collisionality resulting from the strong density pump-out induced by RMPs, ELMs have been totally suppressed in DIII-D and the plasma edge appears to be stable to peeling/ballooning modes [7]. These simulations thus reproduce quite accurately the main features of the ELM mitigation at high collisionality. A study is under way to be able to also model the pumpout observed at low

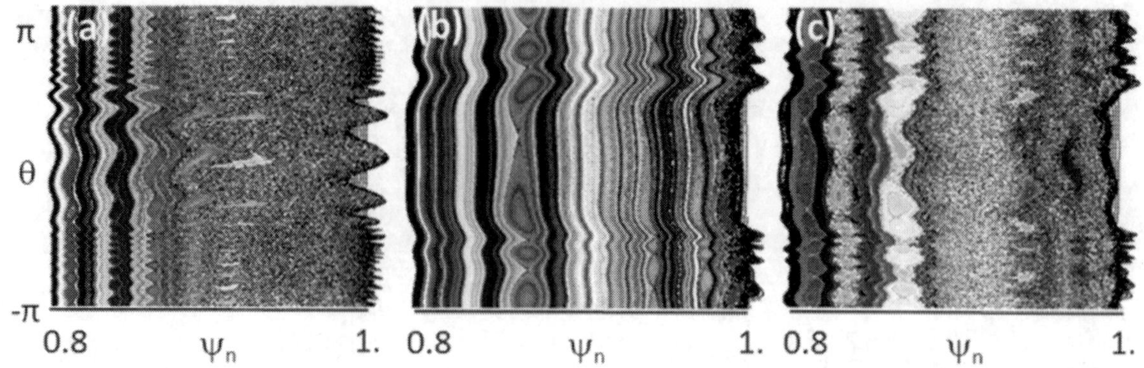

Figure 13. Poincaré plot of the magnetic topology at the edge ($0.8 < \psi_{cnorm} < 1$): in case of an ELM without RMP (*a*), in case of RMPs without ELM (*b*) and in case of an ELM with RMPs (*c*).

collisionality. Reference [36] demonstrated that the parallel current flowing along perturbed magnetic field lines (phenomenon called 'magnetic flutter') generates an enhanced radial transport of the electron density, which may be responsible for the pump-out of density observed in RMP experiments at low collisionality. In this respect, the parallel current flow has been added into the electron density equation in the JOREK model so as to test the effect of the magnetic flutter on the density transport, and thus observe the density pump-out in modeling. This aspect will be raised in future works.

It is also important to note that depending on the q profile taken (in experiments as well as in simulations), the penetration of the RMPs and thus their effect on ELMs may be different. As an example, the ELM suppression was obtained for a narrow q_{95} 'resonant window' in DIII-D [7]. This narrow resonant window is believed to be linked with the presence of a rational surface at the top of the pedestal where the perpendicular electron velocity vanishes [26, 29], inducing the penetration of the RMPs (i.e. the formation of an island chain) on this surface. When the q_{95} parameter is varied, the resonant surface is shifted towards a position where the perpendicular electron flow is larger and thus screens the RMPs. This might explain the presence of a q_{95} window for which RMPs penetrate and thus mitigate or suppress the ELMs. However the ELM mitigation was observed for a wide q_{95} window in JET [37] and Asdex Upgrade [38], so the importance of this 'resonance effect' is not clear. In our simulations, the presence of the island chain induced by RMPs at the position where the perpendicular electron velocity is close to zero (at $\psi_{norm} \approx 0.85$) might play an important role on the ELM mitigation, yet the region of importance where the ergodic layer is formed is located closer to the edge (around $\psi_{norm} = 0.9$). More modeling for different q_{95} values is planned to check if a resonance effect is found.

3.3. ELM cycle with RMPs

A multi-harmonic ($n = 2$, 4, 6 and 8) simulation of an ELM cycle with ($n = 2$) RMPs is performed for a larger diamagnetic rotation (this is done both by increasing the diamagnetic parameter τ_{IC} and by taking a steeper pressure profile). In this case, the most unstable mode without RMPs is the $n = 6$ mode (in red in figure 14) due to the diamagnetic stabilization of the larger mode numbers. At rather low applied RMP current (40*kAt*, left of figure 14), the ELMs are not mitigated by RMPs. Neither the ELM amplitude or frequency is affected by the RMPs. This may be explained by the large screening of RMPs at large diamagnetic (perpendicular) rotation, which does not allow for a large coupling between unstable modes and RMPs. However at twice larger applied RMP current (80*kAt*, right of figure 14), the modes $n = 4$, 6 and 8 are more strongly coupled to RMPs, thus they are now fully driven by $n = 2$ RMPs and present a large initial amplitude due to the coupling with RMPs. Contrary to the case presented in section 3.1, these coupled modes do not generate a turbulent MHD activity. Instead, static islands driven by RMPs are formed. The presence of these islands—reconnected at the edge—generates a permanently enhanced heat and particle transport, which reduces the edge pressure gradient and thus keeps the plasma under the stability threshold. Therefore, by increasing the RMP current from 40 to 80*kAt*, a bifurcation from a largely screened state toward a reconnected state takes place [39], which leads to the full ELM suppression. Subsequently we have observed three different regimes: ELM mitigation by RMPs, no effect of RMPs on ELMs and ELM suppression by RMPs. These regimes depend on both the RMP amplitude and the diamagnetic rotation. Future work will focus on quantifying more precisely the access to these different regimes.

4. Discussion and conclusion

The ELM cyclical dynamics as well as the ELM mitigation by RMPs was observed in non-linear modeling performed with the extended MHD code JOREK. In both cases, the diamagnetic rotation was found to be a key parameter. The main limitations of our model is the large resistivity which is two orders of magnitude larger than the experimental value due to computational limitations and the *ad hoc* heat and particle sources taken in simulation that affect the ELM frequency. In the simulation of ELM cycles, the diamagnetic stabilization allows for suppressing the MHD activity after the crash, and

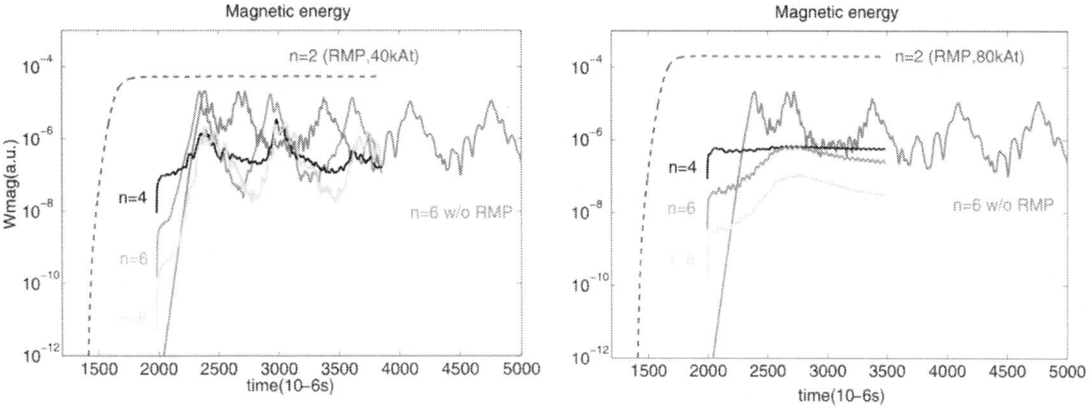

Figure 14. Comparison of the magnetic energy of the $n = 6$ mode without and with RMPs (in log scale). At 'low' RMP current ($40kAt$, left), RMPs have no effect on the ELM amplitude and frequency, whereas at larger RMP current ($80kAt$, right), the ELM is fully stabilized by RMPs.

a competition between the diamagnetic stabilization and the destabilization by the steepening of the pressure profile—due to the applied heating power—results in a cyclical dynamics. If the diamagnetic rotation is largely increased, the ELM frequency is increased and the ELM size is progressively reduced, so that the plasma gravitates towards an attracting point in the peeling-ballooning diagram. In our particular case of resistive (type-III like) ELMs, an enhanced heat source leads to a larger maximal pressure gradient reached at the ELM crash, resulting in a larger ELM frequency and a larger heat flux reaching the divertor. Concerning the deposition on the divertor, the diamagnetic drifts are found to convect the density towards the inner plate, which generates a near-symmetric deposition on the inner and outer divertor plates: this is closer to the experimental results compared to previous simulations without diamagnetic rotation.

As for the ELM control by RMPs, three regimes have been found in the JET case with $n = 2$ RMPs. First, at rather low diamagnetic rotation, the ELM mitigation is obtained, where the large ELM crash is replaced by a continuous (in time) magnetic activity or turbulence, also called type-II ELMy activity [7–9]. The size of the small initial relaxation is reduced as the RMP coil current is increased. For $I_{\text{coil}} = 40kAt$, the integrated peak power on the divertor is found to be divided by a factor of ten. The ELM mitigation is not due to the reduction of the edge pressure gradient but is rather due to the non-linear coupling of the even modes which are driven by the $n = 2$ RMPs. These RMP-driven modes generate plasma reconnection—characterized by magnetic island chains in the pedestal and an edge ergodic zone—and thus continuously expel heat and particles out of the plasma. Then, at larger diamagnetic rotation, two other regimes are observed in simulation. At rather low RMP current ($I_{\text{coil}} = 40kAt$), the RMPs have no effect on the ELM dynamics: the ELM amplitude and frequency are not affected by the RMPs since the RMPs are too 'shielded' by the plasma rotation to be coupled with unstable modes. However, at larger RMP current ($I_{\text{coil}} = 80kAt$), RMPs penetrate and are strongly coupled with the even modes, such that these modes are 'locked' to the RMP driving: hence the large permanent transport induced by these RMP-driven modes stabilizes the plasma under the stability threshold and the ELMs are fully suppressed. Current and future works are dedicated to a more precise understanding of the parameters that demarcate these three regimes.

Acknowledgments

This work has been carried out within the framework of the EUROfusion Consortium. It has received funding from the European Union Horizon 2020 research and innovation programme under grant agreement number 633053 and from the National French Research Program (ANR): ANEMOS (2011) and E2T2 (2010). This work was granted access to the HPC resources of Aix-Marseille Université financed by the project Equip@Meso (ANR-10-EQPX-29-01). A part of this work was carried out using the HELIOS supercomputer system (IFERC-CSC), Aomori, Japan, under the Broader Approach collaboration, implemented by Fusion for Energy and JAEA, and using the CURIE supercomputer, operated into the TGCC by CEA, France, in the framework of GENCI and PRACE projects. The views and opinions expressed herein do not necessarily reflect those of the European Commission or the ITER Organization.

References

[1] Snyder P B, Wilson H R, Osborne T H and Leonard A W 2004 Characterization of peeling-ballooning stability limits on the pedestal *Plasma Phys. Control. Fusion* **46** A131

[2] Kirk A *et al* 2014 Recent progress in understanding of the processes underlying the triggering of and energy loss associated with type I ELMs *Nucl. Fusion* accepted

[3] Dunai D *et al* 2014 Measurements of type I ELM triggering mechanism in MAST and KSTAR *Proc. of the 41st EPS Conf. (Berlin, 23–27 June 2014)*

[4] Zohm H 1996 Edge localized modes (ELMS) *Plasma Phys. Control. Fusion* **38** 105

[5] Hawryluk R J *et al* 2009 Principal physics developments evaluated in the ITER design review *Nucl. Fusion* **49** 065012

[6] Fenstermacher M E *et al* 2008 Effect of island overlap on edge localized mode suppression by resonant magnetic perturbations in DIII-D *Phys. Plasmas* **15** 056122
[7] Evans T E *et al* 2008 RMP ELM suppression in DIII-D plasmas with ITER similar shapes and collisionalities *Nucl. Fusion* **48** 024002
[8] Jeon Y M *et al* 2012 Suppression of edge localized modes in high-confinement KSTAR plasmas by nonaxisymmetric magnetic perturbations *Phys. Rev. Lett.* **109** 035004
[9] Suttrop W *et al* 2011 First observation of edge localized modes mitigation with resonant and nonresonant magnetic perturbations in ASDEX Upgrade *Phys. Rev. Lett.* **106** 225004
[10] Liang Y *et al* 2007 Active control of type-I edge-localized modes with $n = 1$ perturbation fields in the JET tokamak *Phys. Rev. Lett.* **98** 265004
[11] Kirk A *et al* 2010 Resonant magnetic perturbation experiments on MAST using external and internal coils for ELM control *Nucl. Fusion* **50** 034008
[12] Canik J M *et al* 2010 ELM destabilization by externally applied non-axisymmetric magnetic perturbations in nstx *Nucl. Fusion* **50** 034012
[13] Sugiyama L E and Strauss H R 2010 Magnetic x-points, edge localized modes, and stochasticitya *Phys. Plasmas* **17** 062505
[14] Xu X Q *et al* 2011 Nonlinear ELM simulations based on a nonideal peeling-ballooning model using the bout + + code *Nucl. Fusion* **51** 103040
[15] Xi W, Xu P Q and Diamond P H 2014 Phase dynamics criterion for fast relaxation of high-confinement-mode plasmas *Phys. Rev. Lett.* **112** 085001
[16] Sovinec C R *et al* 2007 Two-fluid studies of edge relaxation events in tokamaks *J. Phys.: Conf. Ser.* **78** 012070
[17] Huysmans G T A, Pamela S, van der Plas E and Ramet P 2009 Non-linear MHD simulations of edge localized modes (ELMS) *Plasma Phys. Control. Fusion* **51** 124012
[18] Krebs I, Hoelzl M, Lackner K and Guenter S 2013 Nonlinear excitation of low-n harmonics in reduced magnetohydrodynamic simulations of edge-localized modes *Phys. Plasmas* **20** 082506
[19] Pamela S J P *et al* 2013 Resistive mhd simulation of edge-localized-modes for double-null discharges in the mast device *Plasma Phys. Control. Fusion* **55** 095001
[20] Orain F *et al* 2013 Non-linear magnetohydrodynamic modeling of plasma response to resonant magnetic perturbations *Phys. Plasmas* **20** 102510
[21] Drake J F, Antonsen T M, Hassam A B and Gladd N T 1983 Stabilization of the tearing mode in high-temperature plasma *Phys. Fluids* **26** 2509
[22] Huysmans G T A, Sharapov S E, Mikhailovskii A B and Kerner W 2001 Modeling of diamagnetic stabilization of ideal magnetohydrodynamic instabilities associated with the transport barrier *Phys. Plasmas* **8** 4292–305
[23] Halpern F D, Ltjens H and Luciani J-F 2011 Diamagnetic thresholds for sawtooth cycling in tokamak plasmas *Phys. Plasmas* **18** 102501
[24] Sartori R *et al* 2004 Study of type-III elms in JET *Plasma Phys. Control. Fusion* **46** 723
[25] Morales J A *et al* 2014 Rotation of edge localized modes and Their filaments *Proc. of the 41st EPS Conf.* (*Berlin, 23–27 June 2014*)
[26] Bécoulet M *et al* 2012 Screening of resonant magnetic perturbations by flows in tokamaks *Nuc. Fus.* **52** 054003 (*5th Int. Workshop on Stochasticity in Fusion Plasmas* (*Julich, Germany, 11–14 April 2011*))
[27] Pitts R A *et al* 2007 ELM transport in the JET scrape-off layer *Nucl. Fusion* **47** 1437
[28] Eich T, Herrmann A and Neuhauser J 2003 Nonaxisymmetric energy deposition pattern on ASDEX Upgrade divertor target plates during type-I edge-localized modes *Phys. Rev. Lett.* **91** 195003
[29] Nardon E, Tamain P, Bécoulet M, Huysmans G and Waelbroeck F L 2010 Quasi-linear MHD modelling of H-mode plasma response to resonant magnetic perturbations *Nucl. Fusion* **50** 034002
[30] Ferraro N M 2012 Calculations of two-fluid linear response to non-axisymmetric fields in tokamaks *Phys. Plasmas* **19** 056105
[31] Bécoulet M *et al* 2008 Numerical study of the resonant magnetic perturbations for type-I edge localized modes control in iter *Nucl. Fusion* **48** 024003
[32] Biskamp D 1997 *Nonlinear Magnetohydrodynamics* vol 1 (Cambridge: Cambridge University Press)
[33] Yun G S *et al* 2011 2D visualization of growth and burst of the edge-localized filaments in KSTAR H-mode plasmas *Phys. Rev. Lett.* **107** 045004
[34] Kirk A *et al* 2013 Effect of resonant magnetic perturbations with toroidal mode numbers of 4 and 6 on edge-localized modes in single null H-mode plasmas in MAST *Plasma Phys. Control. Fusion* **55** 015006
[35] Evans T E *et al* 2004 Suppression of large edge-localized modes in high-confinement DIII-D plasmas with a stochastic magnetic boundary *Phys. Rev. Lett.* **92** 235003
[36] Waelbroeck F L, Joseph I, Nardon E, Bécoulet M and Fitzpatrick R 2012 Role of singular layers in the plasma response to resonant magnetic perturbations *Nucl. Fusion* **52** 074004
[37] Liang Y *et al* 2007 Active control of type-I edge localized modes on JET *Plasma Phys. Control. Fusion* **49** B581
[38] Suttrop W *et al* 2011 Studies of edge localized mode mitigation with new active in-vessel saddle coils in asdex upgrade *Plasma Phys. Control. Fusion* **53** 124014
[39] Fitzpatrick R 1998 Bifurcated states of a rotating tokamak plasma in the presence of a static error-field *Phys. Plasmas* **5** 3325–41

IOP Publishing Plasma Physics and Controlled Fusion

Plasma Phys. Control. Fusion **57** (2015) 014021 (10pp) doi:10.1088/0741-3335/57/1/014021

Doppler tomography in fusion plasmas and astrophysics

M Salewski[1], B Geiger[2], W W Heidbrink[3], A S Jacobsen[1], S B Korsholm[1], F Leipold[1], J Madsen[1], D Moseev[2,4], S K Nielsen[1], J Rasmussen[1], L Stagner[3], D Steeghs[5], M Stejner[1], G Tardini[2], M Weiland[2] and the ASDEX Upgrade Team[2]

[1] Technical University of Denmark, Department of Physics, 2800 Kgs. Lyngby, Denmark
[2] Max Planck Institute for Plasma Physics, Boltzmannstr. 2, D-85748 Garching, Germany
[3] University of California, Irvine, CA 92697, USA
[4] FOM Institute DIFFER, PO Box 1207, NL-3430 BE Nieuwegein, The Netherlands
[5] Department of Physics, University of Warwick, Coventry, CV4 7AL, UK

E-mail: msal@fysik.dtu.dk

Received 26 June 2014, revised 3 September 2014
Accepted for publication 22 September 2014
Published 28 November 2014

Abstract

Doppler tomography is a well-known method in astrophysics to image the accretion flow, often in the shape of thin discs, in compact binary stars. As accretion discs rotate, all emitted line radiation is Doppler-shifted. In fast-ion D_α (FIDA) spectroscopy measurements in magnetically confined plasma, the D_α-photons are likewise Doppler-shifted ultimately due to gyration of the fast ions. In either case, spectra of Doppler-shifted line emission are sensitive to the velocity distribution of the emitters. Astrophysical Doppler tomography has lead to images of accretion discs of binaries revealing bright spots, spiral structures and flow patterns. Fusion plasma Doppler tomography has led to an image of the fast-ion velocity distribution function in the tokamak ASDEX Upgrade. This image matched numerical simulations very well. Here we discuss achievements of the Doppler tomography approach, its promise and limits, analogies and differences in astrophysical and fusion plasma Doppler tomography and what can be learned by comparison of these applications.

Keywords: tomography, fast ions, accretion discs, spectroscopy, fast ion D-alpha spectroscopy, tokamak

(Some figures may appear in colour only in the online journal)

1. Introduction

Doppler tomography has been used to image a fast-ion velocity distribution function in a fusion plasma [1]. While this application of Doppler tomography is in its infancy, it has been used to study astrophysical accretion discs for more than 25 years [2–4]. Readily observable accretion discs form in pairs of stars, called interacting binaries, in which matter flows from one star to its companion. Angular momentum tends to confine these discs within the orbital plane of the binary with the gas orbiting around the more massive, compact component in the system, often a stellar remnant. They form when this compact object pulls matter towards it. Angular momentum in the accretion disc is transported outwards and hence matter spirals inwards and eventually reaches the accretor. Astrophysical Doppler tomography has provided images of accretion discs for several classes of binaries [3–9].

Magnetically confined laboratory plasmas are heated to ~10 keV mostly by fast ions generated by injected energetic neutrals (~ 30 keV–1 MeV), by electromagnetic wave acceleration (up to MeVs), or finally in a fusion power plant by the DT fusion reaction (3.5 MeV). Fast ions are magnetically forced on twisted trajectories within the donut-shaped plasma until they become part of the thermal ions. At the

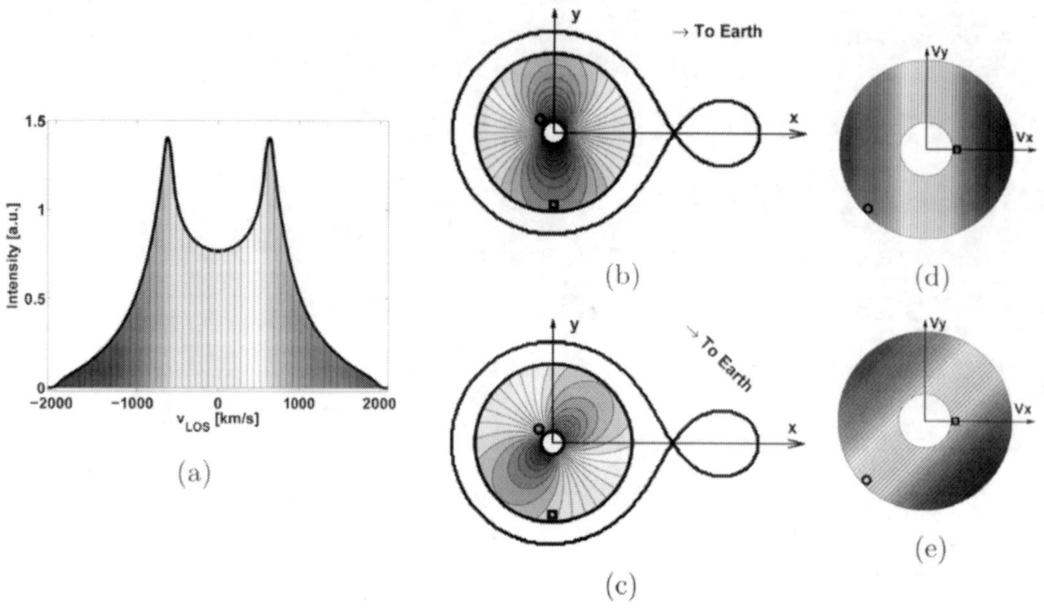

Figure 1. (*a*) Model spectrum from an accretion disc in LOS velocity space. Negative v_{LOS} corresponds to blue-shift and positive v_{LOS} to red-shift. (*b*)–(*d*) show where these v_{LOS} appear on the disc in position space and in velocity space in the same colours as in (*a*). (*b*) and (*c*) show the disc and the Roche lobe of a binary in position space at different phases. The Roche lobe is an equipotential surface of the rotating two-body system. Matter within the Roche lobe of a star is gravitationally bound to this star. (*d*) and (*e*) show the same accretion disc in velocity space at the phases of (*b*) and (*c*), respectively. The circles and the squares illustrate the mapping from position space to velocity space.

tokamak ASDEX Upgrade we can generate a variety of fast-ion populations by using neutral beams or electromagnetic waves at the ion cyclotron frequency [10–12].

Doppler tomography is analogous to standard tomography, but the images are constructed in velocity space rather than in position space. This is possible due to the Doppler shift of line radiation from emitters on trajectories with a near-circular component. Such trajectories are typical for rotating accretion discs of binary stars and gyrating ions in magnetized plasma that have locally helical trajectories. The velocity vector of the emitter and its projected velocity v_{LOS} onto the line-of-sight (LOS) of a detector depend on the angular position of the emitter in its orbit with respect to the line-of-sight. We refer to this angle as phase angle $[0, 2\pi]$ or alternatively as phase $[0, 1]$ following the astrophysical literature. The wavelength shift of emitted photons is proportional to v_{LOS} according to the Doppler shift $\lambda - \lambda_0 = v_{LOS}\,\lambda_0/c$ where c is the speed of light, λ_0 is the unshifted wavelength of the line emission and λ is its Doppler-shifted wavelength. The goal of Doppler tomography is to infer 2D velocity distributions of the emitters from spectroscopic measurements. In fusion plasmas such spectra are measured by fast-ion D_α (FIDA) spectroscopy [1, 13–19] where fast ions are neutralized to become excited neutrals emitting D_α-photons. The Doppler shift is determined by the phase angle of the fast ion at the time of the charge-exchange reaction.

The available measurements in astrophysical and fusion plasma Doppler tomography lead to different flavours of Doppler tomography. Since the two stars in an interacting binary orbit each other, we can view the binary at any phase angle in their orbit. The observations are made mostly using ground-based telescopes but sometimes also satellite-based telescopes such as the Hubble space telescope. Provided observations are obtained across a substantial fraction of the period of the binary system, spectra for various phase angles can be adequately sampled using time-series observations. These spectra change with the phase since several prominent features in the accretion disk are phase-locked to the binary. Astrophysical Doppler images are then inferred for two velocity coordinates in the orbital plane of the binary. The out-of-plane velocity component of matter in the disc is negligible as the disc is thin compared with its diameter. On the contrary, line radiation from fusion plasmas comes from many emitters at all phases and hence spectra are not phase-resolved but phase-averaged. But since the ion velocity distribution function is to a good approximation rotationally symmetric about the strong and slowly varying local magnetic field, resolution of different phases is not necessary. Fusion plasma Doppler tomograms are imaged in velocity components parallel and perpendicular to the magnetic field. Astrophysical Doppler tomograms are 2D by assuming zero out-of-plane velocity whereas fusion plasma Doppler tomograms are 2D by assuming rotational symmetry.

The incentives for imaging in velocity space are different in the two fields. In astrophysical Doppler tomography one is actually interested in the spatial structure of accretion discs. Each point of an accretion disc can be mapped onto velocity space by for example assuming flow velocities obeying Kepler's law which may, however, be a crude assumption. Such Keplerian mapping is illustrated in figure 1. Here we observe that typical rotational speeds in the disc are much larger than the thermal speed of the atoms, so that the line broadening is mostly caused by the rotation of the accretion disc. Tomographic reconstructions in velocity space are preferable as they do not require any mapping assumptions and make the method applicable to a wide range of flow geometries including emission sources not originating within discs. In fusion plasma Doppler tomography, on the contrary, knowledge of the fast-ion phase-space distribution function $f(\mathbf{u}, \mathbf{x})$ itself is essential for the successful operation of a fusion power plant.

In this paper we discuss Doppler tomography in astrophysical and nuclear fusion applications. In section 2 we discuss general principles appearing in both applications as well as their differences. We compare the equations describing the projection onto the LOS, the forward models and the most wide-spread inversion methods. In section 3 we highlight achievements of astrophysical Doppler tomography and in section 4 we present fusion plasma Doppler images. In section 5 we discuss the two applications of the Doppler tomography method in light of each other and what can be learned by comparison. In section 6 we draw conclusions.

2. Principles of velocity-space tomography

2.1. Line-of-sight velocity

Here we derive how v_{LOS} relates to the astrophysical 2D velocity space of the orbital plane (v_x, v_y) and to the fusion plasma 2D velocity space $(u_\parallel, u_\perp)$. Consider a coordinate system with unit vectors $(\hat{\mathbf{u}}_x, \hat{\mathbf{u}}_y, \hat{\mathbf{u}}_z)$ and components (u_x, u_y, u_z) in velocity space as illustrated in figure 1 where the u_z-axis is aligned with the rotation axis. The orientation of u_x and u_y-axes is arbitrary for fusion plasma due to rotational symmetry. Let the LOS be at an inclination angle i to the rotation axis and have an azimuthal angle ϕ from the $(\hat{\mathbf{u}}_x, \hat{\mathbf{u}}_z)$ -plane and let γ be a systemic or drift velocity along the LOS. Then the unit vector along the LOS $\hat{\mathbf{v}}$ and the emitter velocity $\mathbf{u}$ are

$$\hat{\mathbf{v}} = -\cos\phi\sin i\hat{\mathbf{u}}_x + \sin\phi\sin i\hat{\mathbf{u}}_y - \cos i\hat{\mathbf{u}}_z, \tag{1}$$

$$\mathbf{u} = \gamma\hat{\mathbf{v}} + u_x\hat{\mathbf{u}}_x + u_y\hat{\mathbf{u}}_y + u_z\hat{\mathbf{u}}_z. \tag{2}$$

The projected velocity v_{LOS} along the LOS from 3D velocity space is then [2]

$$v_{\mathrm{LOS}} = \gamma - u_x\sin i\cos\phi + u_y\sin i\sin\phi - u_z\cos i. \tag{3}$$

In the following we make further assumptions to deduce simplified projection equations for astrophysical and fusion plasma Doppler tomography. In astrophysics the inclination i of the accretion disc is often unknown and so one substitutes [2]

$$v_x = u_x\sin i, \quad v_y = u_y\sin i, \quad v_z = u_z\cos i \tag{4}$$

to get a projection equation not containing the inclination:

$$v_{\mathrm{LOS}} = \gamma - v_x\cos\phi + v_y\sin\phi - v_z. \tag{5}$$

Further, the out-of-plane flow v_z is assumed to be zero and we arrive at the projection equation used in many astrophysical applications [2]:

$$v_{\mathrm{LOS}} = \gamma - v_x\cos\phi + v_y\sin\phi. \tag{6}$$

In fusion plasma Doppler tomography, the magnetic field and the LOS vectors and hence the inclination i are known, making transformation to (v_x, v_y, v_z)-coordinates unnecessary. To exploit rotational symmetry, we transform to cylindrical coordinates:

$$u_x = u_\perp\cos\overline{\phi}, \quad u_y = u_\perp\sin\overline{\phi}, \quad u_z = u_\parallel \tag{7}$$

so that the projection equation becomes

$$\begin{aligned} v_{\mathrm{LOS}} &= \gamma - u_\perp\cos\phi\sin i\cos\overline{\phi} + u_\perp\sin\phi\sin i\sin\overline{\phi} - u_\parallel\cos i \\ &= \gamma - u_\perp\sin i\cos(\phi + \overline{\phi}) - u_\parallel\cos i. \end{aligned} \tag{8}$$

As the distribution is rotationally symmetric, we can choose $\phi = 0$. Further, it is assumed that there is no systemic or drift velocity γ. Hence we obtain the usual projection equation used for Doppler tomography in fusion plasma

$$v_{\mathrm{LOS}} = -\left(u_\parallel\cos i + u_\perp\sin i\cos\overline{\phi}\right) \tag{9}$$

where normally the vector along the LOS is defined in the opposite direction so that the minus disappears [19, 20]. Thus the two projection equations are consistently derived but describe projections from different image planes and rely on different assumptions.

2.2. Forward models

The projection of an arbitrary 3D function f_u^{3D} onto the LOS is [2, 20]

$$f_{v,\mathrm{LOS}}(v,\phi) = \int_{-\infty}^{\infty}\int_{-\infty}^{\infty}\int_{-\infty}^{\infty} f_u^{3D}(u_x, u_y, u_z) \times\delta(v - v_{\mathrm{LOS}})\,\mathrm{d}u_x\mathrm{d}u_y\mathrm{d}u_z. \tag{10}$$

We now reduce equation (10) by making the same assumptions as in the previous section. In astrophysics we transform the velocity coordinates to (v_x, v_y, v_z) using equation (4) and substitute for v_{LOS} using equation (5):

$$f_{v,\mathrm{LOS}}(v,\phi) = \int_{-\infty}^{\infty}\int_{-\infty}^{\infty}\int_{-\infty}^{\infty} f_v^{3D}(v_x, v_y, v_z) \times\delta\left(v - \gamma + v_x\cos\phi - v_y\sin\phi + v_z\right)\mathrm{d}v_x\mathrm{d}v_y\mathrm{d}v_z. \tag{11}$$

Assuming the out-of-plane velocity to be negligible, we write

$$f_v^{3D}(v_x, v_y, v_z) = f_v^{2D}(v_x, v_y)\,\delta(v_z) \tag{12}$$

and integrate over v_z to find the common astrophysical 2D projection equation [3]

$$f_{v,\mathrm{LOS}}(v,\phi) = \int_{-\infty}^{\infty}\int_{-\infty}^{\infty} f_v^{2D}(v_x, v_y) \times\delta\left(v - \gamma + v_x\cos\phi - v_y\sin\phi\right)\mathrm{d}v_x\mathrm{d}v_y. \tag{13}$$

In fusion plasma physics, u_z is allowed to be arbitrary, but we assume the $f_u^{3D}(u_x, u_y, u_z)$ to be rotationally symmetric so that it can be described by two coordinates $(u_\parallel, u_\perp)$. We define a 2D velocity distribution function

$$f_u^{2D}(u_\parallel, u_\perp) = \int_0^{2\pi} f_u^{3D}(u_\parallel, u_\perp)\,u_\perp d\phi = 2\pi u_\perp f_u^{3D}(u_\parallel, u_\perp), \tag{14}$$

transform equation (10) to cylindrical coordinates and substitute equation (14) [20]

$$f_{v,\mathrm{LOS}}(v,i) = \int_{-\infty}^{\infty}\int_{0}^{\infty}\frac{1}{2\pi}\int_{0}^{2\pi}\delta\left(u_{\parallel}\cos i + u_{\perp}\sin i \times \cos\phi - v\right)\mathrm{d}\phi f_u^{2D}(u_{\parallel},u_{\perp})\,\mathrm{d}u_{\perp}\mathrm{d}u_{\parallel}. \tag{15}$$

As one actually measures the number of photons in a small wavelength range rather than at one wavelength, the measurable quantity is the integral of f_v over a small velocity range [19]. Noting that $f_u^{2D}(u_{\parallel},u_{\perp})$ does not depend on v, we arrive at the forward model used in fusion plasma Doppler tomography:

$$\begin{aligned} f_{\mathrm{LOS}}(v_1,v_2,i) &= \int_{v_1}^{v_2} f_{v,\mathrm{LOS}}(v,i)\,\mathrm{d}v \\ &= \int_{0}^{\infty}\int_{-\infty}^{\infty} w\left(v_1,v_2,i,u_{\parallel},u_{\perp}\right) f_u^{2D}(u_{\parallel},u_{\perp})\,\mathrm{d}u_{\parallel}\mathrm{d}u_{\perp} \end{aligned} \tag{16}$$

where we have introduced a weight function w in analogy with position-space tomography. The weight function can be explicitly calculated [19, 20]:

$$\begin{aligned} &w\left(v_1,v_2,i,u_{\parallel},u_{\perp}\right) \\ &= \int_{v_1}^{v_2}\frac{1}{2\pi}\int_{0}^{2\pi}\delta\left(u_{\parallel}\cos i + u_{\perp}\sin i\cos\phi - v\right)\mathrm{d}\phi\mathrm{d}v \\ &= \int_{v_1}^{v_2}\frac{1}{\pi u_{\perp}\sin i\sqrt{1-\left(\frac{v-u_{\parallel}\cos i}{u_{\perp}\sin i}\right)^2}}\mathrm{d}v \\ &= \frac{1}{\pi}\left(\arccos\frac{v_1-u_{\parallel}\cos i}{u_{\perp}\sin i} - \arccos\frac{v_2-u_{\parallel}\cos i}{u_{\perp}\sin i}\right). \end{aligned} \tag{17}$$

Weight functions in this form enable us construct a transfer matrix W with which we can rapidly calculate the implied function f_{LOS} from an arbitrary image in 2D velocity space f_u^{2D}. The forward model can be written as a matrix equation of the form

$$F_{\mathrm{LOS}} = WF_u^{2D} \tag{18}$$

where F_{LOS} is a vector holding the measurements and F_u^{2D} is a vector holding the image pixel values [1, 20–22]. Refined forward models accounting for Stark splitting, charge-exchange reaction probabilities and electron transition probabilities are discussed in reference [19]. In this paper we focus on the Doppler shift to emphasize the analogy between astrophysical and fusion plasma Doppler tomography.

2.3. A rotationally symmetric accretion disc with no out-of-plane flow

Here we derive explicit formulas for the observable spectrum of a rotationally symmetric accretion disc with velocities from $v_{\perp 1}$ to $v_{\perp 2}$ and no out-of-plane flow so that $f_v^{2D}(v_{\perp},v_{\parallel}) = f_v^{1D}(v_{\perp})\,\delta(v_{\parallel})$. Exploiting the analogy with fusion plasma Doppler tomography, we integrate equation (16) over $v_{\parallel}$ using equation (17):

$$f_{\mathrm{LOS}}(v) = \int_{0}^{\infty}\frac{1}{\pi}\left(\arccos\frac{v_1}{v_{\perp}} - \arccos\frac{v_2}{v_{\perp}}\right) f_v^{1D}(v_{\perp})\,\mathrm{d}v_{\perp}. \tag{19}$$

We can evaluate the integral over $v_{\perp}$ by assuming a functional form of f_v^{1D}. Similar models used position coordinates [23–26] whereas we use velocity coordinates. In these earlier treatments power laws were assumed and then matched to experimental data. As for Keplerian flow power laws in position space map to power laws in velocity space, we also take the emitted intensity to follow a power law of the form $f_v^{1D}(v_{\perp}) = f_0 / v_{\perp}^a$ between $v_{\perp 1}$ and $v_{\perp 2}$. Hence we find theoretical spectra for rotationally symmetric discs with no out-of-plane flow for $a = (0, 2, 4)$:

$$\begin{aligned} &f_0 \rightarrow f_{\mathrm{LOS}}(v_1,v_2) \\ &= \frac{f_0}{\pi}\left(v\,\mathrm{arctanh}\frac{1}{\sqrt{1-\left(\frac{v}{v_{\perp}}\right)^2}} - v_{\perp}\arccos\frac{v}{v_{\perp}}\right)\Big|_{v_{\perp}=v_{\perp 1}}^{v_{\perp}=v_{\perp 2}}\Big|_{v=v_1}^{v=v_2} \end{aligned} \tag{20}$$

$$\begin{aligned} &\frac{f_0}{v_{\perp}^2} \rightarrow f_{\mathrm{LOS}}(v_1,v_2) \\ &= \frac{f_0}{\pi}\left(\frac{1}{v_{\perp}}\arccos\frac{v}{v_{\perp}} - \frac{1}{v}\sqrt{1-\left(\frac{v}{v_{\perp}}\right)^2}\right)\Big|_{v_{\perp}=v_{\perp 1}}^{v_{\perp}=v_{\perp 2}}\Big|_{v=v_1}^{v=v_2} \end{aligned} \tag{21}$$

$$\begin{aligned} &\frac{f_0}{v_{\perp}^4} \rightarrow f_{\mathrm{LOS}}(v_1,v_2) \\ &= \frac{f_0}{\pi}\left(\frac{1}{3v_{\perp}^3}\arccos\frac{v}{v_{\perp}} - \frac{1}{9v}\left(\frac{1}{v_{\perp}^2}-\frac{2}{v^2}\right)\sqrt{1-\left(\frac{v}{v_{\perp}}\right)^2}\right)\Big|_{v_{\perp}=v_{\perp 1}}^{v_{\perp}=v_{\perp 2}}\Big|_{v=v_1}^{v=v_2} \end{aligned} \tag{22}$$

These spectra are shown in figure 2. The value of a is unknown and could be found by matching experimental data. For $a = 0$ the 1D velocity distribution is uniform corresponding to a $1/v$ curve in 2D velocity space. For $a = 4$ very little emission comes from the rapidly rotating regions of the accretion disc which cover only a small area close to the accretor in position space. The models reproduce the characteristic often observed double-peak. This illustrates the analogy between astrophysical and fusion plasma Doppler tomography as we derived the astrophysical observation from the fusion plasma formula. Further, the model gives direct insight in the relation between 2D velocity space of the accretion disc and the line-of-sight velocity space of the measurement and it can be used to validate inversion algorithms. The detailed shape of the double-peak is also influenced by the optical depth [25] and magnetohydrodynamic turbulence [26].

2.4. Inversion methods

Several inversion methods have been applied to solve the velocity-space tomography problem. Astrophysical applications usually apply the maximum entropy method [2] or filtered back-projection method [3]. Nuclear fusion applications of velocity space tomography have relied on the singular value decomposition [1, 21, 22], an iterative technique akin to the back-projection method [20] and a maximum entropy method [27]. In all methods velocity space

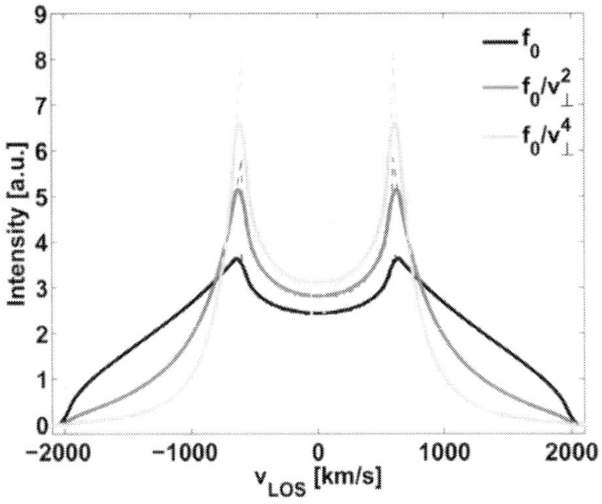

Figure 2. Theoretical spectra of rotationally symmetric accretion discs with no out-of-plane flow and intensities following power laws of the form $f \sim f_0 / v_\perp^a$. Solid lines show the analytic formulas and dashed lines the numeric integration of the velocity image not accounting for finite resolution.

is divided into many elements or pixels. In the maximum entropy method we calculate synthetic data for possible images and quantify the difference between the synthetic data with the measured data by a goodness-of-fit parameter χ^2. χ^2 could be decreased to very low values by changing the image, but this tends to amplify noise in the images. Instead we set a target χ^2 such that we judge the synthetic and the measured data to be consistent. The reduced χ^2 should then be of order one but the precise value is open for discussion. Out of the many tomograms that achieve this target χ^2 one selects the one with maximum entropy which can be found using Lagrange multipliers in an iterative procedure [28]. The standard definition of entropy

$$S = -\sum p_i \ln p_i, \quad p_i = f_i \Big/ \sum_j f_j \tag{23}$$

selects the most uniform image [2]. In astrophysical Doppler tomography one frequently instead uses a modified entropy

$$S = -\sum p_i \ln \frac{p_i}{q_i}, \quad p_i = f_i \Big/ \sum_j f_j, \quad q_i = D_i \Big/ \sum_j D_j \tag{24}$$

where D is an default image [2]. Prior information can be encoded in the default image which can be set to be for example axisymmetric or a heavily blurred version of the image. Such adaptive defaults are better than axisymmetric defaults as the Doppler images have well-defined spots of emission and discs can be strongly asymmetric. An advantage of the maximum entropy method is the enforced positivity that reduces high frequency jitter in the image which is often found in linear methods.

The filtered back-projection method is a linear method in which the inversion is computed in two steps. First the spectra are filtered to damp high frequency components which would otherwise lead to jitter in the tomogram. This step also blurs the tomogram. The filtered profiles $\tilde{f}\ (v, \phi)$ are

$$\tilde{f}\ (v, \phi) = \int F(\hat{v}, v) f(v - \hat{v}, \phi)\, \mathrm{d}\hat{v}\,. \tag{25}$$

where F is the filter function. The second step is the so-called back-projection which is

$$f(v_x, v_y) = \int_0^{0.5} \tilde{f}\ \left(\gamma - v_x \cos 2\pi\phi' + v_y \sin 2\pi\phi', \phi'\right) \mathrm{d}\phi'\,. \tag{26}$$

In this method each image value is found by integrating the 2D function $\tilde{f}\ (v, \phi)$ over the sinusoidal path which would be traced out by a bright light source with the velocity coordinates of the image point. One may also regard this as smearing the filtered profile measured at the angle ϕ across the image at the same angle ϕ.

The singular value decomposition method is another linear method, in which we formulate a forward model based on weight functions as a matrix equation. The tomographic inversion is then given by

$$F^+ = \widehat{W}^+ \widehat{F}_{\mathrm{LOS}}\,. \tag{27}$$

$\widehat{W}^+$ is the truncated Moore–Penrose pseudoinverse found by singular value decomposition of the transfer matrix $\widehat{W}$ from equation (18).

3. Doppler tomography of accreting binary stars

Astrophysical Doppler tomography is a standard technique to image accretions discs of binary star systems such as cataclysmic variables [6], Algols [8] and x-ray binaries [9] including neutron stars [29, 30] and black holes [31, 32]. It is also very useful to map the magnetically controlled accretion stream in polars where the strong magnetic fields prevent formation of an accretion disc [7] or in intermediate polars [33]. Often spectra of the strong emission lines from *H* and *He* are measured, but recently spectra of the Ca*II* line even revealed the presence of the faint donor star [34]. Here we highlight two particularly instructive achievements of astrophysical Doppler tomography. Figure 3 shows the observed time-series of emission line profiles, so-called trailed spectra, and the corresponding Doppler tomogram of the interacting binary CE315. The maximum entropy method was used for the inversion of the 34 measured spectra. The strong emission near 0 km s^{-1} comes from the compact, massive white dwarf. The trailed spectra in figure 3(*a*) have a half-width of about 1000 km s^{-1}. They show the characteristic S-curve of a bright source of emission that is phase-locked to the binary. Figure 3(*b*) shows the Doppler image constructed from the trailed spectra. The accretion disk appears as a ring with velocities between 400 km s^{-1} and 1000 km s^{-1}. The highest speeds show emission from the inner edge of the disc in position space that is close to the white dwarf. The lowest speeds show emission from the outer edge of the disc in position space. The accreting white dwarf sits at the center of the Doppler image. The mass donor sits at v_y = 400 km s^{-1} and by definition of the coordinate system at $v_x = 0$. The Doppler image reveals a bright spot causing the S-curve in the trailed spectra. At this location the gas flow from the donor star to the white dwarf

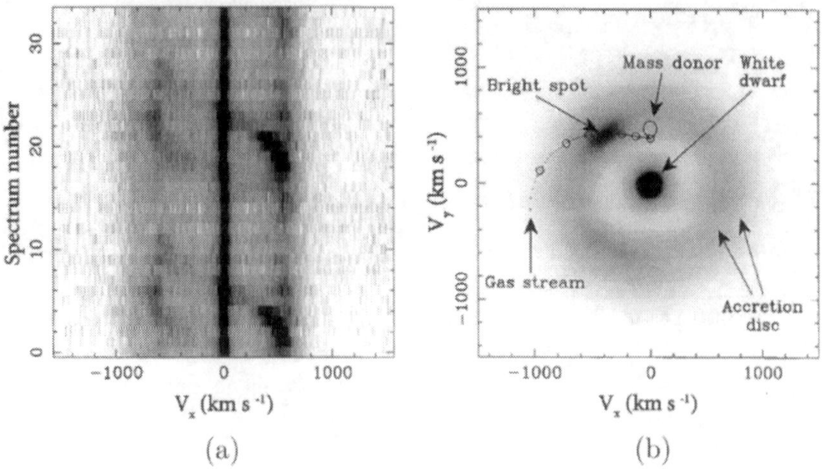

Figure 3. (*a*) The observed line emission from CE315 as a function of the binary orientation and projected velocity which is here called V_x [4]. Each of the 34 spectra is measured for a new phase angle of the binary, corresponding to a new line-of-sight. (*b*) The corresponding Doppler image reveals a bright spot where the gas stream from the donor to the accretor hits the accretion disc. The dotted line shows a calculated ballistic trajectory of the gas stream where the circles show 10% steps of the distance which the gas stream covers [4]. The V_x-coordinate in (*a*) corresponds to the V_x-coordinate in (*b*) for one particular phase.

hits the disc. In the Doppler image this gas flows from the L1 Lagrange point between the donor and the accretor to the high-speed outer edge of the disc which, in position space, corresponds to the inner edge of the disc close to the accretor.

In figure 4 we show an example of the very variable spectra observed over the binary phase of the binary IP Peg [35–37]. The Doppler images reveal that these variable spectra, which would otherwise be very hard if not impossible to interpret, are due to spiral arms in the accretion disc [35]. A numerical simulation and derived synthetic observations match respectively the Doppler image and observations well. These spiral arms are phase-locked to the binary suggesting they are some form of tidal wave [35], perhaps tidally induced shock waves [37]. However, the nature of the spiral arms is still controversial. Spiral arms have been observed in many other binaries, e.g. [33]. Other asymmetric features in the discs have also been observed, e.g. eccentricity [38] or alternating radial flow velocities [39]. These examples illustrate how the complex emissivity profiles observed from binaries can be conveniently mapped into images that are much more straightforward to interpret and at the same time offer quantitative tests against models.

4. Doppler tomography of fast-ion velocity distribution functions

Tomography in position space is a standard analysis method in nuclear fusion research [40, 41] just as in other fields throughout physical and medical sciences [42, 43]. Fusion plasma Doppler tomography has been studied theoretically for some years [20–22, 44]. The method has been theoretically demonstrated and made tractable by formulating the forward model in terms of weight functions [20] previously used to estimate the velocity-space sensitivity of FIDA measurements [14, 19, 45].

Figure 5 shows Doppler images of the fast-ion velocity distribution function in ASDEX Upgrade discharge 29578 on 17 × 8 grid points. Here we study the number of singular values and the effect of filtering the spectra as in the filtered back-projection method. From the uppermost row to the bottom row the number of singular values in the tomogram decreases. In the left column we use synthetic measurements, in the center column actual measurements and in the right column filtered actual measurements.

We measured in three FIDA views and used the singular value decomposition method to invert the spectra [1]. We used experimentally accessible parts of the spectrum with red- and blue shifts with a wavelength resolution of 0.1 nm over 16 nm. Of the resulting 3 × 160 measurement points, 217 were not obscured by other features in the FIDA spectra and covered the velocity-space region of interest. The inclination angles i of the three FIDA LOS are 12°, 69° and 156°. The left column in figure 5 shows reconstructions from synthetic FIDA measurements computed from a simulated fast-ion velocity distribution function using the FIDASIM code [46]. Figure 5(*a*) closely matches simulations which we show in reference [1].

This distribution function is typical for fast ions generated by neutral beam injection and its form can be explained by classical slowing down due to collisions. Neutral deuterium atoms at $E = 60$ keV are injected and ionized to D ions, forming a peak at about $(v_\parallel, v_\perp) \approx (-1, 2) \times 10^6$ m s^{-1}. D_2 and D_3 in the neutral beam lead to further injection peaks at $E/2 = 30$ keV and $E/3 = 20$ keV which are merged to form the second, larger peak at about $(v_\parallel, v_\perp) \approx (-1, 1) \times 10^6$ m s^{-1}. These injection peaks are the sources of fast ions which then slow down due to collisions. In collisions with electrons the ions lose energy while their pitch $p = -V_\parallel / \sqrt{V_\parallel^2 + V_\perp^2}$ does not change significantly. In collisions with ions the pitch also changes.

The Doppler images of the synthetic measurements show as expected that the larger the number of singular values, the finer features of the functions can be reconstructed but the more the noise is amplified. 50 singular values are not enough to recover the two peaks in the functions whereas for

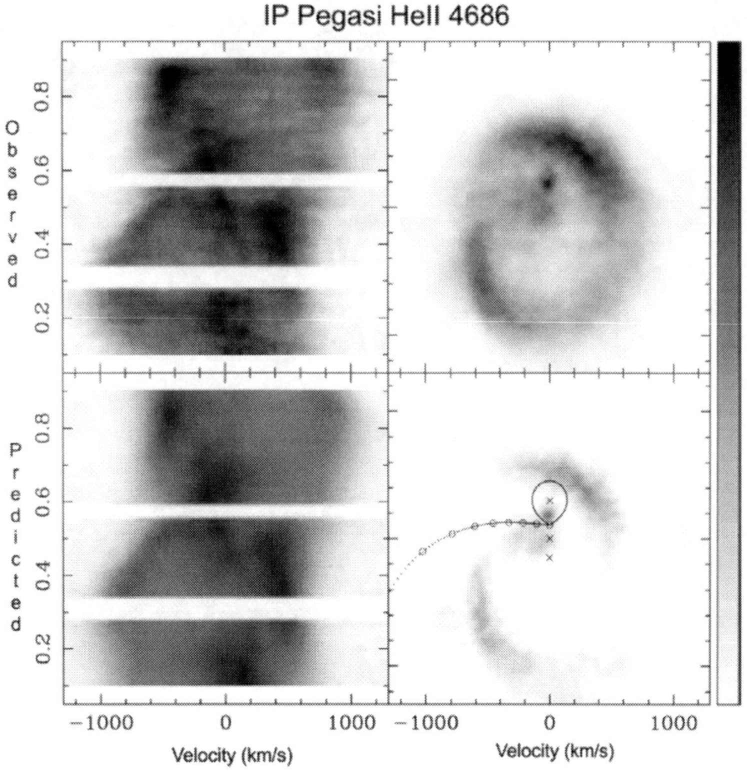

Figure 4. In the upper panels we show the observed line emission from IP Peg as a function of the binary orientation and projected velocity (left) and corresponding Doppler image revealing spiral arms in the accretion disc (right) [37]. The binary phase in the left panel runs from 0 to 1 during a period of the binary. Both axes of the Doppler image in the right panel are in identical units. The velocities are defined analogous to those in figure 3. The lower panels show a numerical simulation of the disc (right) and the implied observable line emission (left). The gas stream is marked as in figure 3.

60–70 singular values the two peaks appear. For more than 70 singular values the Doppler images become hard to interpret due to noise. There is no obvious objective rationale how to choose the number of singular values and this is a weakness of the method. A possible remedy could be the L-curve technique [47]. The center column in figure 5 shows Doppler images from actual FIDA measurements at ASDEX Upgrade. The large-scale shape of the function including the location of the beam injection peaks are reproduced well if 60–70 singular values are used as predicted. The right column shows Doppler images calculated from filtered spectra as in the filtered back-projection method. We use a box filter with a stencil of three points. The filter decreases the amplitude of the jitter as expected.

In figure 6 we present a maximum entropy reconstruction of a fast-ion velocity distribution typical for the tokamak DIII-D on a 25 × 25 element energy-pitch grid from noisy synthetic measurements. Here we used four views with viewing angles of 8°, 19°, 45° and 95°, a mean signal-to-noise ratio of 20 : 1 and a wavelength resolution of 0.14 nm over 14 nm, which corresponds to a total of 4 × 100 measurement points in the spectra. This simulated distribution assumes a 60 keV NBI in co-current direction (positive pitch) and an 80 keV NBI in counter-current direction (negative pitch) which generate the dominant beam injection peaks. Further peaks appear at half and third energies for both beams. The overall shape of the reconstruction matches the original function used to compute the synthetic measurements well, in particular for positive pitch. The peak at 80 keV at negative pitch in the reconstruction is barely visible, blurred and much smaller than in the original function. In the actual experiment the spectrum will be obscured by other emissions at small Doppler shifts. An experimental demonstration of the Doppler tomogaphy method on DIII-D with these or similar parameters is in preparation.

5. Discussion

Doppler tomography in nuclear fusion research and astrophysics rely on the same techniques, but substantially different measurement data lead to different implementations of the method. The imaging plane is the orbital plane in astrophysics. The reduction from 3D is achieved by assuming that the out-of-plane velocity components are much smaller than the flow within the binary plane. Several researchers have attempted 3D imaging using all three velocity coordinates, though this is a much less well-constrained inversion problem from the observable time-series of 1D spectra. The assumption of rotational symmetry in fusion plasmas is rather good, so it would likely not lead to new insight to do 3D imaging in velocity space. Nevertheless, the inclusion of a spatial

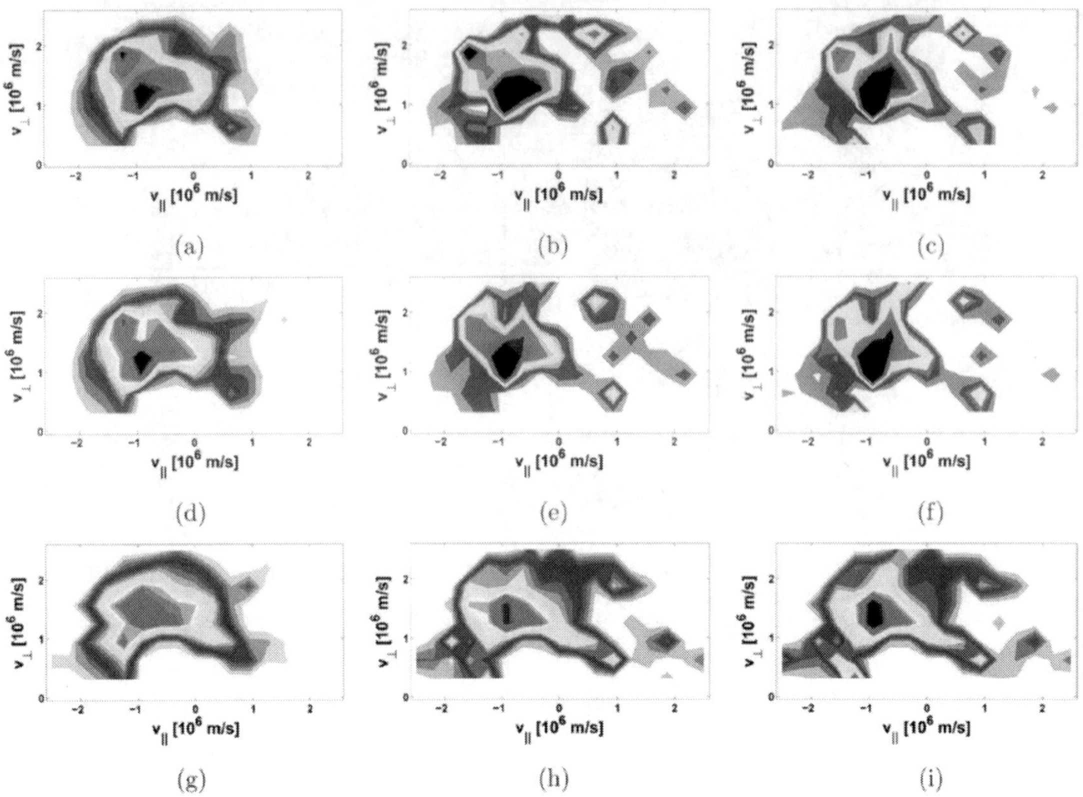

Figure 5. Doppler images from synthetic FIDA spectra (left column), from measured FIDA spectra (center column) and from filtered, measured FIDA spectra (right column). The number of singular values is 70 in the uppermost row, 60 in the center row and 50 in the bottom row. (*a*) Synthetic, 70 SV, (*b*) measured, 70 SV, (*c*) filtered, 70 SV, (*d*) synthetic, 60 SV, (*e*) measured, 60 SV, (*f*) filtered, 60 SV, (*g*) synthetic, 50 SV, (*h*) measured, 50 SV, (*i*) filtered, 50 SV.

dimension in fusion plasma Doppler tomography would likely improve the inference.

Astrophysical and fusion plasma Doppler tomography are photon-starved compared with many position-space tomography applications: The data is often scarce and the signal-to-noise ratio unfavourable. A spectrum in astrophysical Doppler tomography is analogous to a set of measurements with parallel or fanned beams along a LOS in position-space tomography. In astrophysical Doppler tomography the number of spectra or LOS's is limited by the signal-to-noise ratio, so that typically much fewer LOS's are used in astrophysical Doppler tomography (tens) than in position-space tomography (hundreds). In fusion plasmas actually only one LOS would be necessary for a measurement without noise since the fast-ion velocity distribution function is rotationally symmetric to a good approximation. However, due to noise in practise we need to use all available fast-ion measurements. So far three LOS's have been used and this may be increased to seven or eight LOS's in the future. In fusion plasma Doppler tomography at ASDEX Upgrade, we can combine the FIDA measurements with other measurements [22] such as collective Thomson scattering [20, 48–51], neutron emission spectroscopy or neutron yield measurements [52–56] or gamma-ray spectroscopy [57]. Similar combinations are possible at the tokamak DIII-D [27, 45], the stellarator LHD [58–60] or the spherical tokamak MAST [61–63] as well as the next step fusion experiment ITER [64–69]. In astrophysical Doppler tomography, one can use several emission lines from various elements such as hydrogen, helium, or calcium [34].

The astrophysical Doppler tomography allows for a systemic velocity γ along the LOS. This velocity is analogous to perpendicular drift in tokamak plasmas, such as the often dominant $\mathbf{E} \times \mathbf{B}$-drift or the often smaller grad-B, curvature or polarization drifts. Since parallel velocities are allowed in fusion plasma Doppler tomography, any drift velocities parallel to the magnetic field can already be handled, but not perpendicular drift velocities. It should be possible and beneficial to introduce a perpendicular drift velocity in the fusion plasma Doppler tomography approach as well, in particular when applied to the thermal ion population. This would possibly allow us to infer the perpendicular drift velocity and would probably also improve the Doppler image itself.

If there is a significant magnetic field, the line emission has finer structure. A moving D-atom in a magnetic field experiences an electric field in its own rest frame which causes the Balmer alpha line to split into 15 lines. This is referred to as Stark splitting. Stark shifts are usually substantially larger than Zeeman shifts which occur due to the magnetic field. Stark shifts are routinely accounted for in fusion plasma Doppler tomography by calculating the emission from the 15 lines and summing over them whereas Zeeman

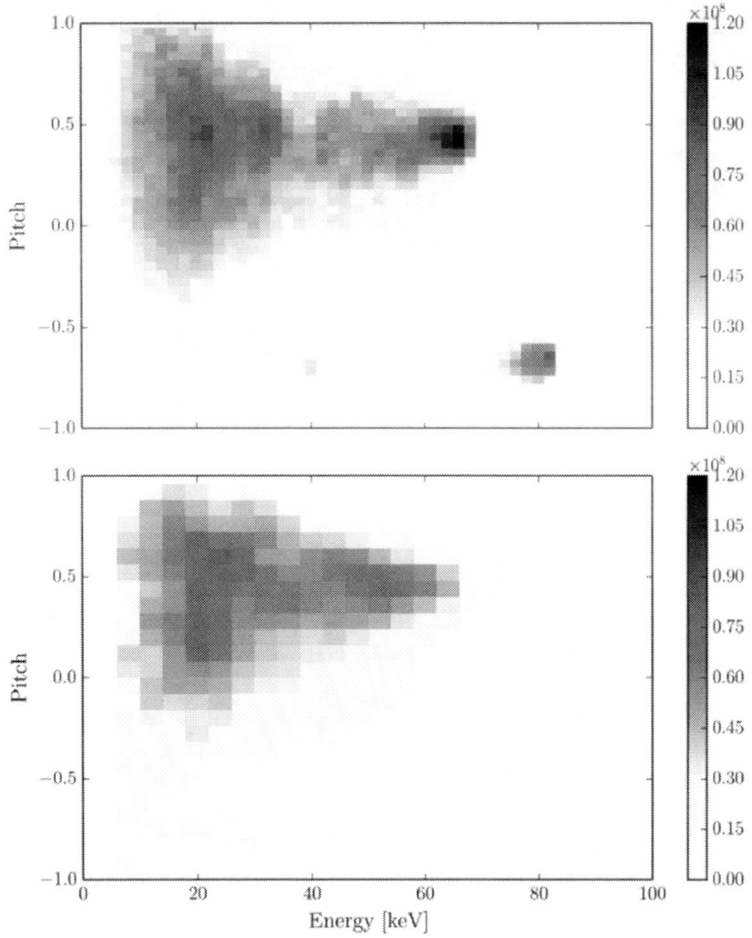

Figure 6. Simulation (up) and Doppler image (below) from synthetic measurements of a fast-ion velocity distribution function at the tokamak DIII-D. Here the function is presented in the widespread (energy, pitch)-coordinates with $E = 1/2\ \mathrm{m}v^2$ and $p = -v_{\|}/v$.

shifts are neglected. The effect of Stark splitting is significant for fusion plasmas [19]. In astrophysical Doppler tomography, Stark and Zeeman shifts have so far been neglected in Doppler tomography in binaries due to the large bulk velocities of the gas in the disc though Zeeman Doppler imaging has been successfully applied to resolve the stellar surfaces of magnetic stars. If the magnetic fields are strong, Stark and Zeeman shifts might also be a nuisance for Doppler tomography in binaries. They could be taken into a account in astrophysical Doppler tomography in binaries analogous to fusion plasma Doppler tomography.

Finally, the inversion algorithms are readily transferable between astrophysical and nuclear fusion Doppler tomography. The formulation as a matrix problem is possible in astrophysics. Maximum entropy inversion algorithms are already used in both fields. The filtered back-projection method is widely used in astrophysics. A closely related method also using back-projection has been used in nuclear fusion Doppler tomography [20]. In this paper we borrowed the idea to filter the spectra and found that improvements may be possible with this technique. Filtering the measurement data is also a common technique in position-space tomography. Improvement of inversion algorithms will clearly benefit both fields.

6. Conclusions

We outline basic principles of astrophysical and fusion plasma Doppler tomography by deriving projection equations and forward models from their common 3D framework. This enables us to derive the shape of observed spectra of light coming from accretion discs from the forward model of fusion plasma Doppler tomography. We present inversions of filtered measured spectra from the tokamak ASDEX Upgrade with the singular value decomposition method and of synthetic spectra with the maximum entropy method in preparation of Doppler tomography on the tokamak DIII-D. Prominent astrophysical Doppler images are discussed and compared with simulations. We already highlighted an example where an idea from one discipline was applied to the other. We further find that an inclusion of a perpendicular drift velocity in the fusion plasma forward model analogous to the systemic velocity of the binary in astrophysics will be valuable. Further, Stark and Zeemann splitting have so far been neglected in astrophysical Doppler tomography of binaries whereas Stark splitting is routinely accounted for in fusion plasma Doppler tomography. One could introduce similar models for the line splitting in astrophysical Doppler tomography even though additional models of the magnetic field would be required. Similar approaches have been successfully applied to image stellar surfaces of magnetic stars. Ideas in the inversion algorithms are readily transferable, for example the formulation as matrix equation used for fusion plasmas, filtering in linear methods to reduce noise, or different formulations of the entropy. In conclusion, using Doppler tomography we can conveniently map measured spectra into images that are much more straightforward to interpret and at the same time offer quantitative tests against models.

Acknowledgments

This project has received funding from the Euratom research and training programme 2014–2018 and from the European Union's Horizon 2020 research and innovation programme under grant agreement number 633053. The views and opinions expressed herein do not necessarily reflect those of the European Commission.

References

[1] Salewski M *et al* 2014 *Nucl. Fusion* **54** 023005
[2] Marsh T and Horne K 1988 *Mon. Not. R. Astron. Soc.* **235** 269–86
[3] Marsh T 2001 *Astrotomography, Indirect Imaging Methods in Observational Astronomy, Doppler tomography Lecture Notes in Physics*) vol 573 (Berlin: Springer) pp 1–26
[4] Marsh T R 2005 *Astrophys. Space Sci.* **296** 403–15
[5] Steeghs D 2004 *Astronomische Nachr.* **325** 185–8
[6] Morales-Rueda L 2004 *Astronomische Nachr.* **325** 193–6
[7] Schwope A D *et al* 2004 *Astronomische Nachr.* **325** 197–200
[8] Richards M T 2004 *Astronomische Nachr.* **325** 229–32
[9] VrtíLek S *et al* 2004 *Astronomische Nachr.* **325** 209–12
[10] Gruber O *et al* 2007 *Nucl. Fusion* **47** S622–34
[11] Kallenbach A *et al* 2012 *IEEE Trans. Plasma Sci.* **40** 605–13
[12] Stroth U *et al* 2013 *Nucl. Fusion* **53** 104003
[13] Heidbrink W W *et al* 2004 *Plasma Phys. Control. Fusion* **46** 1855–75
[14] Heidbrink W W 2010 *Rev. Sci. Instrum.* **81** 10D727
[15] Geiger B *et al* 2011 *Plasma Phys. Control. Fusion* **53** 065010
[16] Geiger B *et al* 2013 *Rev. Sci. Instrum.* **84** 113502
[17] Geiger B *et al* 2014 *Nucl. Fusion* **54** 022005
[18] Geiger B *et al* 2014 Quantification of the impact of large and small scale instabilities on the fast-ion confinement in ASDEX Upgrade, submitted
[19] Salewski M *et al* 2014 *Plasma Phys. Control. Fusion* **56** 105005
[20] Salewski M *et al* 2011 *Nucl. Fusion* **51** 083014
[21] Salewski M *et al* 2012 *Nucl. Fusion* **52** 103008
[22] Salewski M *et al* 2013 *Nucl. Fusion* **53** 063019
[23] Smak J 1969 *Acta Astron.* **19** 155–64
[24] Smak J 1981 *Acta Astron.* **31** 395–408
[25] Horne K and Marsh T R 1986 *Mon. Not. R. Astron. Soc.* **218** 761–73
[26] Horne K 1995 *Astron. Astrophys.* **297** 273–84
[27] Stagner L *et al* 2013 *Bull. Am. Phys. Soc.* **58** BP8.00104
[28] Skilling J and Bryan R K 1984 *Mon. Not. R. Astron. Soc.* **211** 111
[29] Steeghs D and Casares J 2002 *Astrophys. J.* **568** 273–8
[30] Bassa C G *et al* 2009 *Mon. Not. R. Astron. Soc.* **399** 2055–62
[31] Marsh T, Robinson E and Wood J 1994 *Mon. Not. R. Astron. Soc.* **266** 137–54
[32] Neilsen J, Steeghs D and Vrtilek S D 2008 *Mon. Not. R. Astron. Soc.* **384** 849–62
[33] Bloemen S *et al* 2010 *Mon. Not. R. Astron. Soc.* **407** 1903–12
[34] van Spaandonk L *et al* 2010 *Mon. Not. R. Astron. Soc.* **401** 1857–68
[35] Steeghs D, Harlaftis E T and Horne K 1997 *Mon. Not. R. Astron. Soc.* **290** L28 –32 (arXiv: astro-ph/9708005)
[36] Steeghs D and Stehle R 1999 *Mon. Not. R. Astron. Soc.* **307** 99–110
[37] Harlaftis E T *et al* 1999 *Mon. Not. R. Astron. Soc.* **306** 348–52
[38] Roelofs G H A *et al* 2006 *Mon. Not. R. Astron. Soc.* **371** 1231–42
[39] Copperwheat C M *et al* 2012 *Mon. Not. R. Astron. Soc.* **421** 149–58
[40] Ertl K *et al* 1996 *Nucl. Fusion* **36** 1477–88
[41] Anton M *et al* 1996 *Plasma Phys. Control. Fusion* **38** 1849–78
[42] Kak A and Slaney M 1988 *Principles of Computerized Tomographic Imaging* (New York: IEEE Press)
[43] Herman G 2009 *Fundamentals of Computerized Tomography* 2nd edn (New York: Springer)
[44] Egedal J and Bindslev H 2004 *Phys. Plasmas* **11** 2191
[45] Heidbrink W W *et al* 2007 *Plasma Phys. Control. Fusion* **49** 1457–75
[46] Heidbrink W W *et al* 2011 *Commun. Comput. Phys.* **10** 716–41
[47] Hansen P C 1992 *SIAM Rev.* **34** 561–80
[48] Meo F *et al* 2008 *Rev. Sci. Instrum.* **79** 10E501
[49] Salewski M *et al* 2010 *Nucl. Fusion* **50** 035012
[50] Meo F *et al* 2010 *J. Phys.: Conf. Ser.* **227** 012010
[51] Furtula V *et al* 2012 *Rev. Sci. Instrum.* **83** 013507
[52] Giacomelli L *et al* 2011 *Rev. Sci. Instrum.* **82** 123504
[53] Tardini G *et al* 2012 *J. Instrum.* **7** C03004
[54] Tardini G *et al* 2013 *Nucl. Fusion* **53** 063027
[55] Jacobsen A S *et al* 2014 *Rev. Sci. Instrum.* **85** 11E103
[56] Jacobsen A S 2014 Velocity-space sensitivity of neutron emission spectroscopy, submitted
[57] Nocente M *et al* 2012 *Nucl. Fusion* **52** 094021
[58] Kubo S *et al* 2010 *Rev. Sci. Instrum.* **81** 10D535
[59] Nishiura M *et al* 2014 *Nucl. Fusion* **54** 023006
[60] Ito T *et al* 2010 *Rev. Sci. Instrum.* **81** 10D327
[61] Michael C A *et al* 2013 *Plasma Phys. Control. Fusion* **55** 095007
[62] Jones O M *et al* 2013 *Plasma Phys. Control. Fusion* **55** 085009
[63] Jones O M *et al* 2014 arXiv: 1401.6864
[64] Salewski M *et al* 2008 *Rev. Sci. Instrum.* **79** 10E729
[65] Salewski M *et al* 2009 *Nucl. Fusion* **49** 025006
[66] Salewski M *et al* 2009 *Plasma Phys. Control. Fusion* **51** 035006
[67] Kappatou A *et al* 2012 *Nucl. Fusion* **52** 043007
[68] Bertalot L *et al* 2012 *J. Instrum.* **7** C04012
[69] Chugunov I N *et al* 2011 *Nucl. Fusion* **51** 083010

IOP Publishing | Plasma Physics and Controlled Fusion

Plasma Phys. Control. Fusion **57** (2015) 014022 (10pp) | doi:10.1088/0741-3335/57/1/014022

Shock ignition: a brief overview and progress in the design of robust targets

S Atzeni, A Marocchino and A Schiavi

Dipartimento SBAI, Università di Roma 'La Sapienza' and CNISM, Via A. Scarpa 14, 00161 Roma, Italy

E-mail: stefano.atzeni@uniroma1.it

Received 8 July 2014, revised 9 September 2014
Accepted for publication 23 September 2014
Published 28 November 2014

Abstract

Shock ignition is a laser direct-drive inertial confinement fusion (ICF) scheme in which the stages of compression and hot spot formation are partly separated. The fuel is first imploded at a lower velocity than in conventional ICF, reducing the threats due to Rayleigh–Taylor instability. Close to stagnation, an intense laser spike drives a strong converging shock, which contributes to hot spot formation. This paper starts with a brief overview of the theoretical studies, target design and experimental results on shock ignition. The second part of the paper illustrates original work aiming at the design of robust targets and computation of the relevant gain curves. Following Chang *et al* (2010 *Phys. Rev. Lett.* **104** 135002) a safety factor for high gain, ITF* (analogous to the ignition threshold factor ITF introduced by Clark *et al* (2008 *Phys. Plasmas* **15** 056305)), is evaluated by means of parametric 1D simulations with artificially reduced reactivity. SI designs scaled as in Atzeni *et al* (2013 *New J. Phys.* **15** 045004) are found to have nearly the same ITF*. For a given target, such ITF* increases with implosion velocity and laser spike power. A gain curve with a prescribed ITF* can then be simply generated by upscaling a reference target with that value of ITF*. An interesting option is scaling in size by reducing the implosion velocity to keep the ratio of implosion velocity to self-ignition velocity constant. At a given total laser energy, targets with higher ITF* are driven to higher implosion velocity and achieve a somewhat lower gain. However, a 1D gain higher than 100 is achieved at an (incident) energy below 1 MJ, an implosion velocity below 300 km s^{-1} and a peak incident power below 400 TW. 2D simulations of mispositioned targets show that targets with a higher ITF* indeed tolerate larger displacements.

Keywords: inertial confinement fusion, laser fusion, shock ignition

(Some figures may appear in colour only in the online journal)

1. Introduction

The necessary conditions for net energy production by laser-driven inertial confinement fusion (ICF) are compression of the fuel (a deuterium-tritium, DT, mixture) to very high density (at least one thousand times the DT solid density, $\rho_{DT} = 0.25$ g cm^{-3}) and hot spot ignition [1, 2]. Compression is achieved by the laser-driven implosion of a hollow spherical target. In the standard approaches to ICF, either by laser direct-drive, or by laser indirect-drive, a hot spot with a temperature of 5–10 keV is created at the center of the compressed fuel at the end of the implosion. Achievement of ignition requires that the shell fuel layer is imploded at a sufficiently high velocity u_{imp} and at the same time is kept on an adequately small *in-flight isentrope* or *adiabat* α_{if}. (The adiabat α_{if} is the ratio of the pressure of the imploding fuel to the pressure of the fully degenerate matter at the same density). For instance, an implosion velocity of 370 km s^{-1} and adiabat $\alpha_{if} < 1.5$ are deemed necessary to achieve ignition by indirect-drive with a 1.5 MJ laser energy [3, 4] on the National Ignition Facility (NIF [5]). Theory and simulations [6] also predict that the minimum laser energy required for ignition decreases very strongly with the implosion velocity, being approximately proportional to $u_{imp}^{-6}\alpha_{if}^{2}$.

On the other hand, target sensitivity to hydrodynamic Rayleigh–Taylor instabilities (RTI) grows with implosion velocity (at fixed α_{if}) [2]. RTI can lead to material mixing or cold–hot fuel mixing, hindering hot spot formation. This has been evidenced by recent NIF experiments [4, 7]. Also, once ignition is achieved, the target energy gain $G = E_{fus}/E_{laser}$, the ratio of fusion energy to laser energy, decreases with increasing implosion velocity.

Reducing the implosion velocity below the above *self-ignition* threshold u_{ig} might have significant advantages. In this case, the ignition hot spot must be generated by an additional mechanism. In shock ignition [8], laser-driven implosion at velocity $u_{imp} < u_{ig}$ creates a hot spot with a temperature and pressure below the values required for ignition. Ignition conditions are attained with the help of a converging strong shock wave, generated by an additional intense laser pulse. The pulse intensity (up to 10^{16} W cm^{-2}) is such that strongly non-linear laser-plasma interactions (LPI's [9]) are expected. This is an interaction regime not adequately investigated so far. Its study is the first priority for shock ignition. Two major assets of shock ignition are instead that hot spot generation is a purely hydrodynamic process and that the whole scheme can be investigated using existing facilities. According to present studies (see below) a full-scale demonstration could require a laser power and energy already available on the NIF and, in the near future on the Laser MegaJoule (LMJ [10]).

In this paper, we first describe the basic principle of shock ignition and briefly mention the potential advantages and issues (section 2). We then present in section 3 a brief overview of the achievements, including both the target design and experimental activities. The second part of the paper (sections 4–6) reports the original studies on target design by our group. In particular, we discuss how the robustness of shock ignition targets can be measured by an analogue of the ignition threshold factor ITF [11]. This factor, which we call ITF*, can be obtained by means of 1D numerical simulations, following a process described in [12]. By upscaling reference cases (concerning the HiPER baseline target [13, 14]) we generate gain curves with different values of ITF* and show how important quantities (such as the implosion velocity and peak laser intensity) vary along these curves. By 2D simulations, we also show that cases with a higher ITF* can tolerate larger deviations from implosion symmetry.

2. Principle and issues

2.1. Principle: shock ignition as shock-assisted central ignition

Shock ignition [8] is in fact laser direct-drive *shock-assisted* central ignition. The target, a hollow shell with an outer ablator layer surrounding the cryogenic DT layer, is uniformly irradiated by a conventional time-shaped laser pulse, which generates the ablation pressure driving the implosion. The pulse power is, however, lower than the one required for conventional central ignition (see figure 1) and a lower implosion velocity is attained. Notice that the compression pulse is preceded by an adiabat-shaping (A-s) picket [16], which helps in reducing RTI growth [14, 16, 17]. Towards the end of the implosion an intense laser pulse (*spike*) generates an ablation pressure of 200–400 Mbar and drives a strong converging shock wave. Such a shock is first amplified by spherical convergence. In proximity to the shell's inner surface it is further amplified by the collision with the shock bouncing from the center of the target and drives the hot spot to ignition. In typical cases, the intensity of the compression pulse is in the range 0.3–1 × 10^{15} W cm^{-2} (for UV laser light with wavelength λ = 0.351 μm). The spike intensity is instead one order of magnitude higher, and its duration ranges from a few hundred ps to 1–2 ns. A particularly attractive feature is that high gain targets can be driven by pulses with a total energy of below 1 MJ and a peak power of below 400–500 TW, within the operational limits of NIF or LMJ.

An important feature of shock ignition is that the hot spot generation can in principle be controlled by two main parameters, namely the implosion velocity u_{imp} and the spike power P_s. This leads to some flexibility in target design [19]. This flexibility increases with target (and driver) size, since the self-ignition velocity decreases with fuel mass $u_{ig} \propto m_{DT}^{\delta}$, with the exponent δ in the range 0.105–0.14 (see [19]). Here, and throughout this paper, we define the implosion velocity of shock ignition targets as the mass-averaged implosion velocity at the time of the peak kinetic energy of the imploding shell, computed in simulations without an ignition spike.

In previous papers [19] we have shown that shock ignition targets can be effectively upscaled using analytically derived expressions in two important cases: (*a*) Scaling at a constant implosion velocity; (*b*) scaling at a constant ratio u_{imp}/u_{ig} of the implosion velocity to the self-ignition velocity. In both cases the shell aspect ratio and the adiabat are kept constant. The expressions of the scaling (*b*), which will be used in section 5, are listed for convenience in table 1; the variable s is the target linear scale (radius); $s = 1$ for the reference case. It is immediately seen that the implosion velocity, compression pulse intensity and spike intensity all decrease with target size.

2.2. Issues

Shock ignition is not free of issues [20]. As already mentioned, the spike intensity is such that laser plasma interaction occurs in a regime so-far not well investigated. Experiments have to demonstrate that the required pressure can be achieved efficiently; and that LPIs do not cause excessive back-scattering, nor do they generate too many and too energetic hot electrons.

An emerging issue concerns cross-beam energy transfer [21], clearly evidenced in NIF indirect-drive experiments [4]. This energy transfer can also affect direct-drive, particularly if many overlapping beams are used to achieve uniform irradiation at high intensity.

High gain shock ignition requires a low adiabat implosion. Experiments at the NIF have shown that a low-adiabat (and then a high in-flight-aspect-ratio) increases subsceptibility to RTI-induced mixing [22]. However, we observe that shock ignition foresees an implosion velocity smaller than conventional central ignition and direct-drive allows for effective adiabat-shaping techniques [16] that reduce RTI growth.

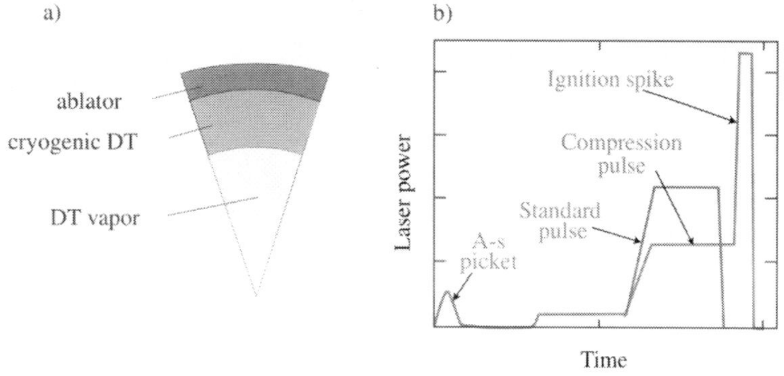

Figure 1. (*a*) Target for shock ignition, consisting of an ablator layer, a cryogenic DT fuel layer, and inner DT vapor. (*b*) Laser pulse for shock ignition, compared with a pulse for conventional central ignition. Adapted from [15].

Table 1. Scaling relations at fixed wavelength and constant ratio u_{imp}/u_{ig}.

Shell inner radius	s
Implosion velocity	$s^{-0.32}$
Time scale	$s^{1.32}$
Compression pulse energy	$s^{2.36}$
Compression pulse power	$s^{1.04}$
Compression pulse intensity	$s^{-0.96}$
Payload mass	s^{3}
Ablator mass	$s^{2.52}$
Spike power	$s^{0.5}$
Spike energy	$s^{1.82}$
Spike intensity	$s^{-3/2}$

In any case, uncertainty about LPIs, the limited predictability of cross-beam-energy-transfer, and the possible effects of mixing all call for designs allowing for large safety margins.

3. A brief overview of the target design and experimental results

Since its proposal in 2007 [8], shock ignition has been the subject of intense theoretical, computational, and experimental research. Extensive critical reviews, with reference to the original research papers, have been recently published [15, 20]. Here, we only briefly present a selection of the target design studies and experimental results on a few key issues for shock ignition.

3.1. Target design

Targets for shock ignition have been studied by several groups [8, 13, 14, 18, 19, 24–29]. We can consider such studies as scoping investigations, aiming at assessing concept potential and identifying issues requiring experimental and theoretical research. As shown in figure 2, target concepts differ for ablator material and aspect ratios. The aspect ratio $R_0/(R_0 - R_2)$ varies in the range 2.5–6; the aspect ratio $R_1/(R_1 - R_2)$ in the range 3–9; the radii R_0, R_1, and R_2 are defined in figure 2. Thicker targets require a higher laser compression power (and therefore could be more sensitive to LPIs), but they should be less sensitive to hydrodynamic instabilities. In all cases, the implosion velocity and peak intensity at compression are smaller than in conventional central ignition targets. All concepts foresee a low-adiabat compression; RTI growth is reduced by adiabat shaping and beam smoothing. The intensity of the shock-driving spike is instead in the range 10^{15}–10^{16} W cm^{-2}, above the threshold for parametric instabilities, in a regime that has not been widely investigated so far (see below). Target simulations indeed take into account refraction and collisional absorption, but neglect any LPI. Such simulations typically show 70–80% absorption of the compression pulse and 40–50% absorption of the spike pulse. Whether this latter efficiency is optimistic (neglecting any backscattering) or conservative (neglecting absorption by other mechanisms than by inverse bremsstrahlung), has to be assessed experimentally. The effects of hot electrons have been considered in a few studies [27, 30]. It is found that electrons with an energy below 100 keV do not compromise compression by preheating the fuel. Instead, they strengthen the spike-driven converging shock wave [30]. (Note that at the time of spike irradiation the shell areal density ρR significantly exceeds the range of the above hot electrons.)

Gain estimates by different groups substantially agree. A summary of gain curves can be found in [15]. A gain well in excess of 100 is obtained at a laser energy below 1 MJ for a laser light with wavelength $\lambda = 0.351$ μm. Higher gains are predicted for $\lambda = 0.25$ μm. Systematic 1D studies show that the required power pulse accuracy and timing accuracy are compatible with current laser specifications. 1D ignition windows in the absorbed power—spike timing plane are considered large enough to allow safe operation margins [13, 30].

Several aspects of target robustness to instabilities and implosion asymmetries have been investigated by detailed 2D simulations. For example, simulations by the NRL group show that targets with a DT-filled CH foam ablator, driven by KrF laser, can tolerate surface perturbations larger than NIF specifications [23, 24]. Simulations by the LLE group addressed the sensitivity of high gain SI targets (with a solid CH ablator) to inner and outer target surface roughness, imprint, power imbalance, mistiming, and mispositioning [27].

Shock ignition targets are characterized by values of the hot spot convergence ratio of 35–45, comparable to those of

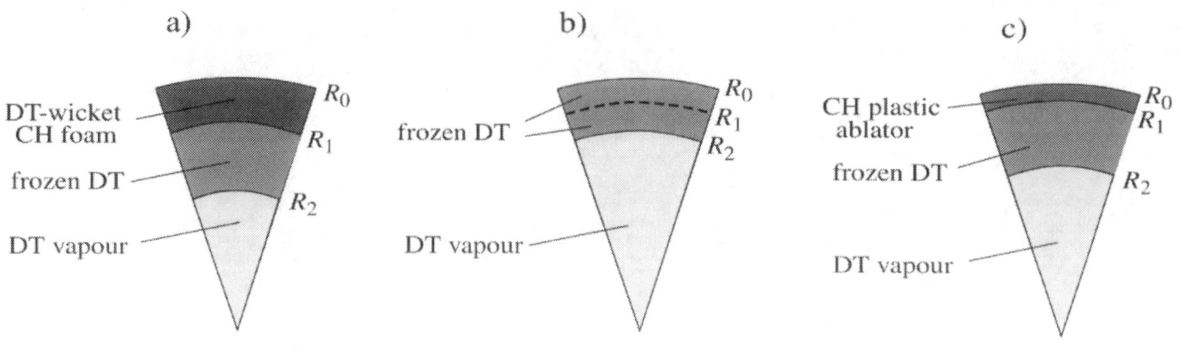

Figure 2. Targets for shock ignition, differing for the ablator layer and the initial aspect ratio. (*a*) Target with a wetted DT—plastic foam ablator; (*b*) all DT-target; (*c*) target with plastic ablator. In these targets, the cryogenic DT density is 0.25 g cm^{-3}; the vapor density is in the range 0.2–0.6 mg cm^{-3}; the total density of the wetted foam is 0.33 g cm^{-3}; the CH plastic density is about 1.05 g cm^{-3}. In target (*b*) R_1 is the initial radius of the imploded portion of the cryogenic DT. Reprinted from [15].

conventional central ignition targets. This makes them sensitive to long-scale asymmetries, resulting, e.g. from target mispositioning, as was shown in [14, 31]. Sensitivity to mispositioning (and presumably to other sources of long-scale perturbations) depend on target safety factors, as will be discussed in section 6. Instead, fluid simulations (either with flux-limited electron thermal conductivity [13, 14, 31] or with non-local electron transport [50]) indicate that large spike asymmetries are tolerated, probably due to thermal smoothing in the rather thick thermal conduction layer between the critical radius and ablation front at the time of spike irradiation.

According to the above studies, SI can achieve high gain at a laser energy and power achievable on the operating NIF facility [5] and on the LMJ facility [10], currently under construction. However, both the NIF and LMJ have been desighed for indirect-drive and do not allow for spherical irradiation. Several studies have, however, suggested that the NIF and LMJ could drive SI targets in the so-called polar direct-drive scheme [32]. A first proposal was put foward soon after the first paper on shock ignition [18, 33]. Other studies were published more recently [26, 27]. The most detailed one [27] concerns the design of a target (and corresponding laser pulse) for a shock ignition demonstration on the NIF. The target parameters are chosen to achieve ignition with relatively large safety margins, with the power and energy of each beam below the maximum that can be delivered by the present NIF laser. The rise times are also compatible with the NIF specifications. The target consists of a shell, with an outer radius of about 1.08 mm, a 30 μm thick CH ablator layer and a 163 μm thick frozen DT fuel layer. The target implosion is driven by a 400 kJ, 130 TW pulse delivered by half of the 192 NIF beams. The final shock is driven by a 290 kJ, 420 TW pulse delivered by all the NIF beams. According to 1D simulations, the target achieves a (1D) gain of 58. The polar drive scheme requires tuning the power of the different groups of beams of the NIF, as well as repointing the beams with respect to the original indirect-drive configuration, changes in focusing, and beam intensity distribution. 2D simulations with 3D raytracing confirmed the robustness of the design. However, one should be aware that these simulations neglected laser-plasma instabilties (and the related cross-beam-energy-transfer [21]).

3.2. Experimental results

The SI proposal was immediately followed by a proof-of-principle integrated experiment [34], performed at the OMEGA laser facility. A deuterium-filled CH-plastic shell (with a radius of 430 μm) was imploded by a standard time-shaped pulse (picket, foot, main pulse, with peak power of about 7 TW and duration of 3 ns), followed by a 15 TW spike, delivered with a variable delay. It was reported that a properly timed spike increased the neutron yield by a factor of 3–4, and the areal density by a factor of about 2. Subsequent experiments [35] have confirmed these findings.

SI requires the effective generation of an imploding spherical shock, driven by ablation pressure in the range 200–400 Mbar. The achievement of this pressure, which requires laser irradiation at an intensity well above 10^{15} W cm^{-2} has to be demonstrated. A few preliminary experiments have been conducted in relatively simple planar geometry [36–38]. The most detailed [37], performed at OMEGA, refers to intensities in the range 0.6–1.2 × 10^{15} W cm^{-2} and wavelength $\lambda = 0.351$ μm. It shows very modest levels of backscattering (<3%) and the generation of pressure up to 75 Mbar, approximately scaling with intensity as $I^{0.7}$, in agreement with the theory and previous experiments at lower intensities [40]. Somewhat higher pressures (90 Mbar) have been inferred in another experiment [38], where the laser beam parameters were not accurately determined. At higher intensities, however, planar experiments are dominated by 2D effects and shock-ignition relevant conditions cannot be reached with medium-size facilities. Very recently, to overcome the above limitations, experiments have been performed at OMEGA on small spherical targets [39]. Peak ablation pressures exceeding 300 Mbar were inferred at absorbed intensities of 3 × 10^{15} W cm^{-2} (and wavelength $\lambda = 0.35$ μm). Up to about 8% of the incident energy is converted into hot electrons with an energy below 100 keV, which contribute to strengthening the shock wave.

Concerning LPIs, a few experiments (reviewed in [20]) report a total level of backscattering below 15% in the whole

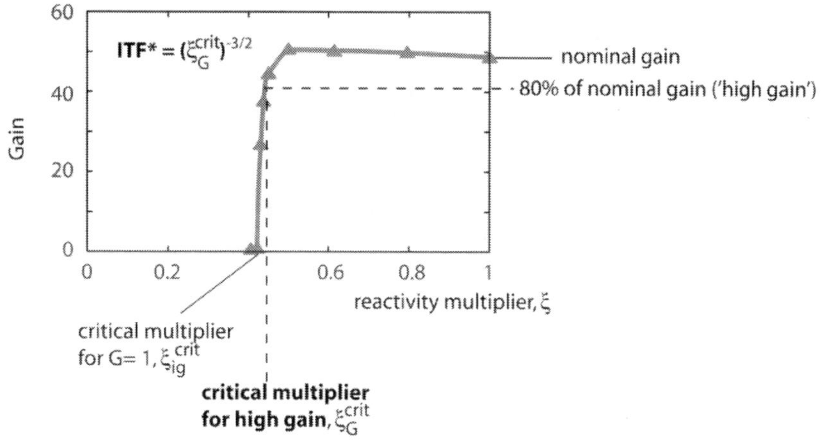

Figure 3. Definition of critical reactivity multipliers and of ITF*.

range of the relevant intensities. Spherical geometry experiments at OMEGA, however, show the backscattering rising from 10% at $I = 10^{15}$ W cm^{-2} to about 30% at $I = 10^{16}$ W cm^{-2}.

Detailed experiments on backscattering in preformed plasmas have been performed at LULI, with green light 1.5 ns pulses and an intensity up to 4×10^{15} W cm^{-2} [41]. Brillouin backscattering was found to peak at the beginning of the pulse and subsequently decreased substantially. For example, at 3×10^{15} W cm^{-2}, the initial reflectivity was 20%, while the pulse-averaged reflectivity was below 10% . The levels of Raman reflectivity have also been measured; see the review by Batani *et al* [20] and the original papers referred to therein. It is interesting to note that Particle-in-Cell simulations referring to shock ignition relevant conditions also show a high reflectivity during the early stages of irradiation and a much lower reflectivity after the first 60–80 ps (see [42] for 1D simulations and [43] for 2D simulations).

4. Measuring and increasing target robustness

The potential and favorable prospects of shock ignition outlined above motivate further design studies. However, the issues and uncertainties still remaining call for robust designs, i.e. designs that can tolerate significant departures from the predicted behavior. It is therefore important to define a quantitative measure of robustness, or a safety factor of a given target/pulse design. In this section, we first introduce a safety factor for shock-ignited targets. We then study its dependence on implosion velocity and spike power and discuss how our previous designs [14, 19] can be improved to increase their robustness.

4.1. ITF, a safety factor for high-gain shock ignition*

Several safety factors for indirect-drive ignition have been proposed and shown to be nearly equivalent. In particular, we refer to the ignition threshold factor ITF [11, 44], the experimental ignition threshold factor ITFX [11] and the GLC (a factor based on a generalized Lawson criterion) [45]. ITF is based on simulated data, ITFX on experimentally measured parameters, GLC is expressed in terms of a combination of measured data and simulated ones. While ITF and ITFX are approximately numerically equivalent, ITF $\approx$ GLC^m, with m in the range 2.2–2.6 [4, 46]. Such factors are equal to 1 when ignition is marginally achieved (i.e. when an NIF target releases 1 MJ). Larger values indicate increased robustness.

The use of the above factors for shock ignition is not straightfoward. However, Anderson *et al* [12], extending the previous work by Chang *et al* [46], showed how a threshold factor can be computed for any target by simply using 1D simulations. The process consists of running 1D simulations with an artificially decreased DT fusion reactivity, e.g.

$$< \sigma v >_{\mathrm{DT}}^{\mathrm{eff}} = < \sigma v >_{\mathrm{DT}} f_{\sigma v} \tag{1}$$

with

$$f_{\sigma v} \begin{cases} \xi, & T \leqslant T_{\mathrm{ig}}, \\ \xi + (1-\xi)\,(T - T_{\mathrm{ig}})\,/\,(T_{\mathrm{b}} - T_{\mathrm{ig}}), & T_{\mathrm{ig}} \leqslant T \leqslant T_{\mathrm{b}}, \\ 1, & T \geqslant T_{\mathrm{b}}, \end{cases} \tag{2}$$

that is, with the reactivity multiplied by a factor $\xi \leqslant 1$ when the central temperature T is below some ignition threshold T_{ig} and is again set to a nominal value once ignition is attained. As in [46], we use $T_{\mathrm{ig}} = 10$ keV and $T_{\mathrm{b}} = 15$ keV, while in the intermediate temperature range $f_{\sigma v}$ varies linearly with temperature. Critical values of ξ can then be found, in correspondence to which the target achieves ignition ($\xi_{\mathrm{ig}}^{\mathrm{crit}}$), or, e.g. 80% of the nominal 1D clean gain ($\xi_{\mathrm{G}}^{\mathrm{crit}}$): see figure 3. According to [12] the ITF defined for NIF targets is approximately ITF $\simeq (\xi_{\mathrm{ig}}^{\mathrm{crit}})^{-3/2}$.

Since we are more interested in a safety factor for high gain, we use:

$$\mathrm{ITF}^* = (\xi_{\mathrm{G}}^{\mathrm{crit}})^{-3/2}\,. \tag{3}$$

We use the name ITF* to distinguish it from the ITF, referring to ignition rather than to high gain and referring to conventional central ignition. Of course, ITF* is smaller than ITF. However, for the shock ignition targets discussed in this

Table 2. Main parameters of the selected cases.

Case	HiPER baseline (H)	Reference (R)	$s = 1.53$ (scaled from R)
Inner target radius	0.833 mm	0.833 mm	1.27 mm
Total target mass	0.586 mg	0.586 mg	1.90 mg
Main pulse			
Spot radius HM	0.45 mm	0.45 mm	0.68 mm
Flat-top power P_c	46 TW	65 TW	101 TW
Energy E_c	180 kJ	243 kJ	569 kJ
Abs. energy $E_{c\text{-abs}}$	133 kJ	168 kJ	437 kJ
SI Pulse			
Spot radius (HM)	0.28 mm	0.28 mm	0.38 mm
Peak power P_s	240 TW	240 TW	297 TW
Δt (FWHM)	500 ps	500 ps	870 ps
Energy E_s	120 kJ	120 kJ	257 kJ
Abs. energy $E_{s\text{-abs}}$	45 kJ	44 kJ	132 kJ
1D Fusion performance			
Fusion energy	19.4 MJ	17.6 MJ	85.1 MJ
Gain	65	48.4	103
ITF*	1.48	3.4	3.1

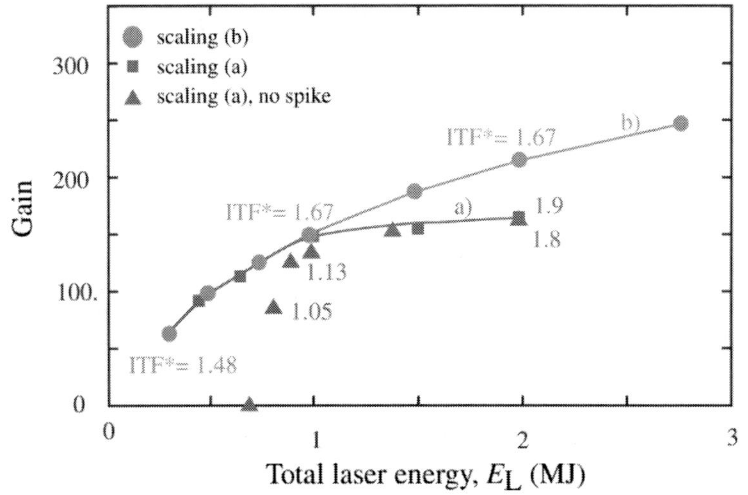

Figure 4. Gain curves for targets scaled from the HiPER baseline (see [19]), with an indication of ITF* for selected points. Scaling (*a*): at a constant implosion velocity; scaling (*b*) at a constant ratio of implosion velocity to self-ignition velocity.

paper, the relative difference is not large and decreases as ITF increases. For example, for a few cases discussed in the next subsection, ITF = 1.35 when ITF* = 1.1; ITF = 2.59 when ITF* = 2.32; ITF = 5.8 when ITF* = 5.5. Notice that the artificial reduction of the hot spot reactivity prior to full ignition mimics the reduction of volume-integrated fusion power due to hot spot mixing or hot spot cooling by hot spot deformation.

4.2. Increasing ITF of SI designs*

In a previous publication [19] we generated gain curves by upscaling the so-called HiPER baseline design for shock ignition [13, 19]. The target is a simple all-DT shell, with an inner radius of 833 μm and an outer radius of 1044 μm, filled with DT vapor at a density of 0.3 mg cm^{-3} [17]. The main pulse parameters of a baseline case (H in the following), as well as of another reference case (R), and of a scaled target discussed later, are listed in table 2.

All the 1D and 2D simulations presented in this paper have been performed with the code DUED [47, 48], with the same physical model as outlined in [19].

In [19] we obtained two different curves for the targets and pulses scaled from case H at (a) constant implosion velocity and (b) constant u_{imp}/u_{ig} ratio. We also computed the gain achieved by imploding the targets at the same implosion velocity, but without a final spike (i.e. in the conventional direct-drive mode).

We computed the ignition factor ITF* for many points along these curves (see figure 4). We found that all the scaled shock-ignited targets have approximately the same ITF*. This confirms that the targets, as well compression and spike pulses, are properly scaled. However, all the scaled targets

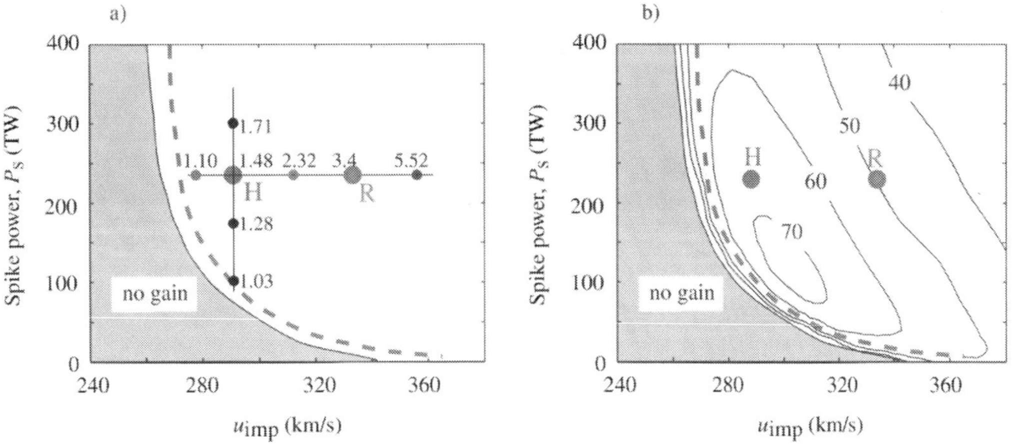

Figure 5. (*a*) ITF*, for selected points in the implosion velocity-spike power plane, u_{imp}, P_s plane. (*b*) Gain contours in the same u_{imp}, P_s plane (from 1D simulations with 2D ray-tracing) of the HiPER baseline target. The HiPER SI baseline case and the reference case considered in this paper are indicated by labels H and R, respectively.

have rather small safety factors, i.e. ITF* $\simeq$ 1.57 ± 0.1 when scaling (b) is used. Of course, the cases without the final spike have an even smaller ITF*.

This led us to analyze how the ITF* depends on the main implosion parameters in order to proceed to the design of more robust targets. We refer again to the HiPER target. Although very simple and probably unrealistic, this has been analyzed in-depth [13, 14, 19, 29, 31, 49] and serves well as a propotype to understand trends. In figure 5 we show how the ITF* varies with the implosion velocity and spike power (The implosion velocity is varied by changing the peak power of the compression pulse; the duration of the peak flat-top is also adjusted to match the duration of the implosion. Consequently, the compression laser energy varies as the velocity changes. The spike energy is proportional to the spike power, since the spike duration is kept fixed when the spike power is varied.). For completeness, we also show the gain contours in the same plane. It is apparent that the ITF* increases as we move away from the ignition boundary. The target of our previous study, indicated in the figure by the label *H*, had a gain close to the maximum achievable, with the in-flight-adiabat α if $\simeq$ 1.3, but had quite a small ITF*. (We define the in-flight adiabat as the mass-averaged adiabat of the imploding shell at the time of peak kinetic energy, computed in a simulation without a spike.) The figure shows that by increasing the spike power and/or the implosion velocity we can increase the ITF* significantly, although at the expense of the decreased gain. Of course, both options are in principle possible. As expected, in both cases an additional price has to be paid for the increased ITF*: Either a higher intensity (increasing the risks due to laser-plasma instabilities) or a higher velocity (increasing the RTI-related risks). (Note that in this study, we do not consider the dependence of ITF* on the fuel adiabat. All the designs have a mass-averaged in-flight adiabat α if $\simeq$ 1.3–1.4.)

We chose to increase the implosion velocity, keeping the spike pulse at the same value as in the previous study, i.e. 240 TW. Instead, the compression power is increased from 46 TW to 65 TW, to increase the implosion velocity (measured without a spike) from 292 km s^{-1} to 336 km s^{-1}. The corresponding new reference case is indicated by point *R* in figure 5, indicating a more robust design. Note that, due to slighly reduced imploding mass and the (very small) adiabat increase, the yield of the reference target *R* is 9% smaller than that of target *H*. The implosion velocity of these targets may appear large; however, it is still 10–15% smaller than the value needed for ignition on the NIF [3]. In addition, if this target is subsequentely upscaled following the above strategy (*b*), the implosion velocity decreases with target size, so that one can design a somewhat larger target with a high ITF* and, at the same time, a moderate implosion velocity (see below).

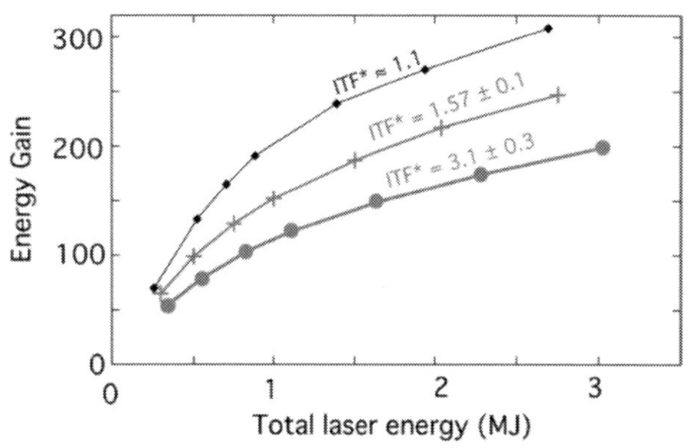

Figure 6. Gain curves for the target scaled from the HiPER target, with different values of ITF*.

5. Gain curves and ITF*

We then generated a gain curve by upscaling the reference target R at constant u_{imp}/u_{ig} according to the prescriptions of [19], which, for convenience, are listed in table 1. The target dimensions and pulse power are simply scaled. No adjustment has been made to the computed values. The timing of the adiabat-shaping pulse and of the onset of the foot have been adjusted instead (they do not scale as the implosion time).

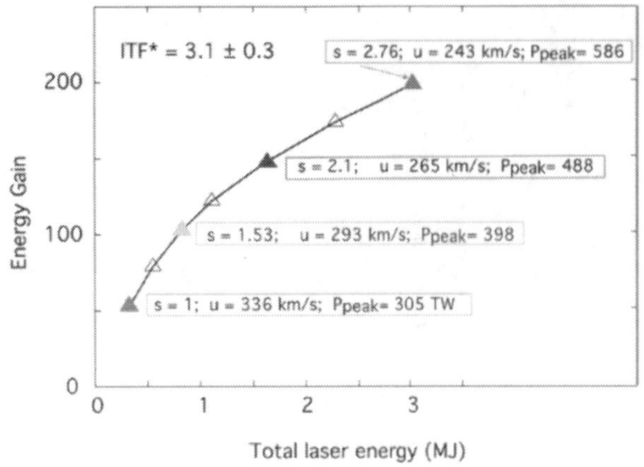

Figure 7. Gain curve for the targets scaled from the reference scale $s = 1$ case (point R in figure 5). Target scale s, the implosion velocity and the peak laser power are indicated for selected cases.

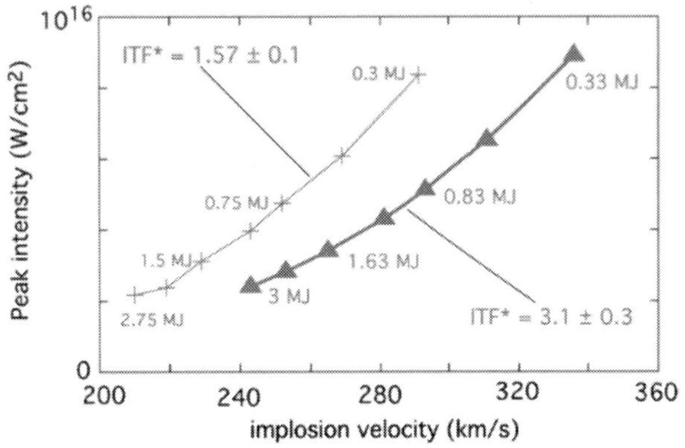

Figure 8. Peak intensity versus implosion velocity for the target scaled, respectively, from the cases H (red curve and crosses) and R (green curve and triangles) as shown in figure 5. The total laser energy is also indicated for selected cases.

For each case, we have also determined the ITF*, finding that the scaled targets have similar (although not exactly the same) values of ITF*. The resulting gain curve is shown in figure 6, where it is compared with the curves for the cases scaled from the HiPER baseline case (H) and for the marginal cases (with an ITF* close to 1). It can be seen that the (ideal 1D) gain decreases with target robustness. More details on the points on the curve with a higher ITF* are given in figure 7, where the values of the linear scale s, the implosion velocity, and the peak total (spike plus compression) laser power are shown. In figure 8 the points on the two above gain curves are compared in the implosion velocity-peak intensity plane. One can see that larger energy (but lower peak intensity) is required by the more robust case, at a given implosion velocity.

One of the purposes of shock ignition is just the reduction of the implosion velocity below those (350 km s^{-1} or higher) required by conventional central ignition, without increasing too much the risks related to the high irradiation intensity. We see from figure 8 that it is possible to obtain ITF* = 3.1 with a scale $s = 1.53$ target, imploded at 293 km s^{-1}, irradiated at a peak intensity of 5×10^{15} W cm^{-2} by a spherically symmetric

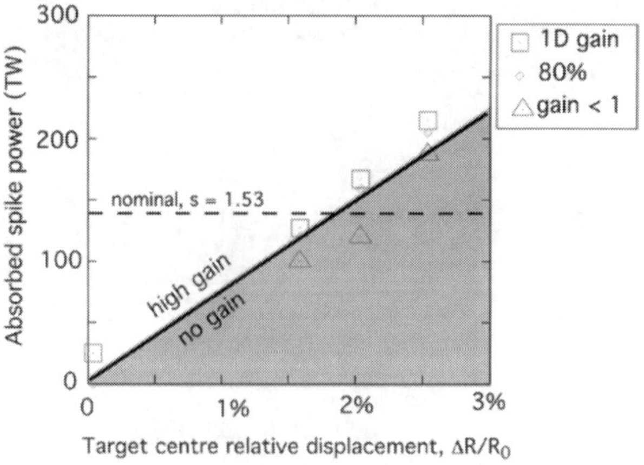

Figure 9. 2D simulations of cases scaled from the case R in figure 5: The absorbed spike power required for high gain versus relative displacement of the target center. The target irradiated with the nominal pulse tolerates 1.8% displacement. Slightly larger displacements require increasing the spike power.

pulse of total energy of 826 kJ. (Here the peak intensity is measured by considering the peak power at a radius equal to one half of the initial target radius). The parameters of this scale $s = 1.53$ target are summarized in table 2. Of course, a larger pulse energy is required for the polar direct drive.

6. 2D simulations: sensitivity to large-scale asymmetries versus ITF*

The previous analysis is based on 1D simulations (including, however, 2D laser ray-tracing and the appropriate laser beam focal spots and intensity profiles). However, ITF* was introduced to measure the robustness to deviations from a nominal and spherically symmetric performance, including, e.g. the effects of asymmetries and mixing. Testing the relevance and predictivity of ITF* would therefore require 2D (as well as 3D) simulations accounting for the most important deviations from nominal.

Here, as a first approach to the problem, we are studying a single effect, namely the displacement of the target center with respect to the center of the laser beams. We consider the simplified set up used in previous studies performed in the frame of the HiPER collaboration [13, 14, 31, 50]. We assume radial laser rays and also neglect refraction. We choose the beam power in such a way as to accurately reproduce the evolution computed by the 1D simulations with 2D ray-tracing. We have verified that this does not change the ITF* of the targets. We assume that the radial power distribution is characterized by a time-independent perturbation described by a linear combination of Legendre polynomials. The chosen spectrum corresponds to the initial illumination spectrum generated by the reference HiPER irradiation scheme [53], as applied to a target displaced by a given amount from the center of the beams. The adopted treatment of laser irradiation is of course a rough approximation of the actual 3D and time-dependent irradiation, but can nevertheless serve to gain a physical insight. Of course, it is not suited to get design specifications.

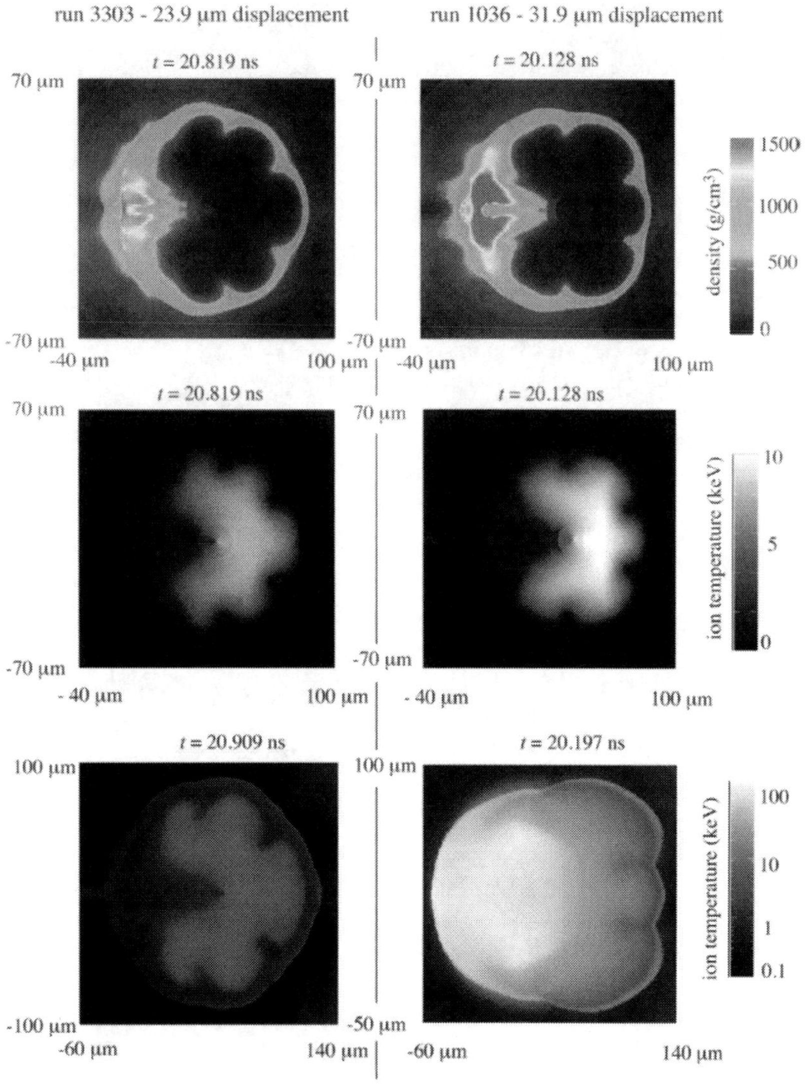

Figure 10. 2D Simulation of the scale $s = 1.53$ targets. Left hand side: case 3303, scaled from point H in figure 5, displaced by 23.9 μm; right-hand side: case 1036, scaled from point R in figure 5 and displaced by 31.9 μm. In both cases the peak absorbed power (compression pulse plus spike) is 230 TW. Case 1036 achieves a full yield (87 MJ), while case 3303 does not ignite. Upper row: density maps at stagnation; middle row: ion temperature maps at the same times as in the upper row, respectively; lower row: ion temperature maps at the respective times of peak temperature. The coordinate system is centered at the target center at $t = 0$.

The preliminary DUED simulations, with 2D hydrodynamics and full 3D ray-tracing, show that the targets considered here tolerate somewhat larger irradiation non-uniformity, since the laser energy is absorbed further out in the corona and thermal smoothing is active on a longer distance with respect to the 2D hydrodynamic-1D ray-tracing model used here.

We have run a series of simulations for the scale $s = 1.53$ target described in the previous section (case scaled from the reference case R in figure 5; the main parameters are listed in the rhs column in table 2). For a few values of target displacement, we have varied the spike power in order to determine the minimun absorbed power required to obtain a) approximately the same gain as in 1D simulations and b) 80% of the 1D gain. The results are presented in figure 9, where cases with gain $G < 1$ are also shown. The (hand-drawn) line in the figure indicates an approximate boundary between high gain and no gain. As expected, the power required for high gain increases with displacement. Notice that the nominal scaled target tolerates displacements of up to about 1.8% . Still larger displacements can be tolerated if the spike power is increased.

We have then run a few simulations for the ITF* $\simeq$ 1.57 scale $s = 1.53$ case (the same target, but with the pulse scaled from the reference H case in figure 5). In this case, even for displacements as small as 1%, the power has to be increased above the nominal value.

The different robustness of the two cases is illustrated in figure 10: With the same absorbed peak total power of 230 TW, the more robust case achieves a full yield (87 MJ) when the target is displaced by about 32 μm, while the less robust one does not ignite when it is displaced by 24 μm.

The above simulations were performed using flux-limited thermal conductivity (with a constant flux-limiter $f = 7.5\%$).

We also ran a few cases with the non-local transport model proposed in [51] and recently introduced in our DUED code [50, 52]. We found results very close to the above; the ITFs* are only slightly reduced; the achievement of the same 2D performances of the flux-limited case typically requires increasing the absorbed spike power by 5–12%.

7. Conclusions

The brief overview of research on shock ignition presented in section 3 shows that considerable efforts are underway to address the crucial issues for the scheme's feasibility. The availability of large-scale facilities where shock ignition could be tested at full scale allows us to predict a roadmap, beginning with proof-of-principle experiments and experiments on specific, isolated physics issues, followed by scaled-down integrated experiments and then full-scale tests (which, however, require some modification to the laser beam configuration).

Modelling uncertainties motivate the design of targets characterized by large values of safety factors (or threshold factors). In sections 4–6 we first introduced the safety factor ITF* and then studied the dependence of the ITF* on two crucial design parameters, i.e. the implosion velocity and spike power. We then generated gain curves for different values of ITF*. Finally we showed that 2D simulations indeed confirm that target robustness to large-scale implosion asymmetries, like those due to target mispositioning, increases with IFT*. In the near future we plan to significantly extend the present work. We will run simulations with a combination of perturbations (e.g. target displacement, larger irradiation non-uniformity, smaller scale perturbations, artificial reactivity reduction to mimic small-scale mixing). We will also replace the simple radial beams with a realistic 3D model (already included in the DUED code).

Acknowledgments

This work was partially supported by the Italian projects PRIN 2012AY5LEL and Sapienza 2012 C26A12CZH2. We thank G Schurtz for collaborating on the target modelling and scaling, K Anderson for the information on the computation of ITF, and R Betti, R Nora, and W Theobald for kindly communicating to us their very recent results prior to publication.

References

[1] Atzeni S and Meyer-ter-Vehn J 2004 *The Physics of Inertial Fusion* (Oxford: Clarendon)
[2] Lindl J D 1995 *Phys. Plasmas* **2** 3933
[3] Haan S W *et al* 2011 *Phys. Plasmas* **18** 051001
[4] Lindl J D *et al* 2014 *Phys. Plasmas* **21** 020501
[5] Miller G H, Moses E I and Wuest C R 2004 *Nucl. Fusion* **44** S228
[6] Herrmann M C, Tabak M and Lindl J D 2001 *Nucl. Fusion* **41** 99
[7] Ma T *et al* 2013 *Phys. Rev. Lett.* **111** 085004
[8] Betti R, Zhou C D, erson K S, Perkins L J, Theobald W and Solodov A A 2007 *Phys. Rev. Lett.* **98** 155001
[9] Kruer W L 1988 *The Physics of Laser Plasma Interactions* (Redwood City: Addison-Wesley)
[10] Ebrardt J and Chaput J M 2010 *J. Phys.: Conf. Ser.* **244** 032017
[11] Spears B K *et al* 2012 *Phys. Plasmas* **19** 056316
[12] Anderson K S *et al* 2013 *LLE Rev.* **133** 1
[13] Ribeyre X, Schurtz G, Lafon M, Galera S and Weber S 2009 *Plasma Phys. Control. Fusion* **51** 015013
[14] Atzeni S, Schiavi A and Marocchino A 2011 *Plasma Phys. Control. Fusion* **53** 035010
[15] Atzeni S, Ribeyre X, Schurtz G, Schmitt A J, Canaud B, Betti R and Perkins L J 2014 *Nucl. Fusion* **54** 054008
[16] Anderson K S and Betti R 2004 *Phys. Plasmas* **11** 5
[17] Atzeni S, Schiavi A and Bellei C 2007 *Phys. Plasmas* **14** 052702
[18] Perkins L J, Betti R, LaFortune K N and Williams W H 2009 *Phys. Rev. Lett.* **103** 045004
[19] Atzeni S, Marocchino A, Schiavi A and Schurtz G 2013 *New J. Phys.* **15** 045004
[20] Batani D *et al* 2014 *Nucl. Fusion* **54** 054009
[21] Igumenshchev I V *et al* 2012 *Phys. Plasmas* **19** 056314
[22] Hurricane O A *et al* 2014 *Nature* **506** 343
[23] Bates J W, Schmitt A J, Fyfe D E, Obenschain S P and Zalesak S T 2010 *High Energy Density Phys.* **6** 128
[24] Schmitt A J, Bates J W, Obenschain S P, Zalesak S T and Fyfe D E 2010 *Phys. Plasmas* **17** 042701
[25] Canaud B and Temporal M 2010 *New J. Phys.* **12** 043037
[26] Terry M R, Perkins L J and Sepke S M 2012 *Phys. Plasmas* **19** 112705
[27] Anderson K S *et al* 2013 *Phys. Plasmas* **20** 056312
[28] Atzeni S, Marocchino A and Schiavi A 2012 *Phys. Plasmas* **19** 090702
[29] Lafon M, Ribeyre X and Schurtz G 2010 *Phys. Plasmas* **17** 052704
[30] Betti R *et al* 2008 *J. Phys.: Conf. Ser.* **112** 022024
[31] Atzeni S *et al* 2013 *EPJ Web Conf.* **59** 01005
[32] Skupsky S *et al* 2004 *Phys. Plasmas* **11** 2763
[33] Perkins L J *et al* 2010 *Proc. 23rd Int. Fusion Energy Conf. (Daejon, 10–16 October 2010)* (Vienna: IAEA) CD-ROM file [IFE/P6-13] and www-naweb.iaea.org/napc/physics/FEC/FEC2010/html/index.htm
[34] Theobald W *et al* 2008 *Phys. Plasmas* **15** 056306
[35] Theobald W *et al* 2009 *Plasma Phys. Control. Fusion* **15** 124052
[36] Baton S *et al* 2012 *Phys. Rev. Lett.* **108** 195002
[37] Hohenberger M *et al* 2014 *Phys. Plasmas* **21** 022702
[38] Batani D *et al* 2014 *Phys. Plasmas* **21** 032710
[39] Nora R*etal* 2014 *Phys. Rev. Lett.* submitted
[40] Fratanduono D E *et al* 2011 *J. Appl. Phys.* **110** 073110
[41] Depierreux S *et al* 2012 *Phys. Plasmas* **19** 012705
[42] Klimo O *et al* 2011 *Phys. Plasmas* **18** 082709
[43] Riconda C *et al* 2011 *Phys. Plasmas* **18** 092701
[44] Clark D S, Haan S and Salmonson J D 2008 *Phys. Plasmas* **15** 056305
[45] Zhou C D and Betti R 2008 *Phys. Plasmas* **15** 102707
[46] Chang P Y *et al* 2010 *Phys. Rev. Lett.* **104** 135002
[47] Atzeni S 1986 *Comput. Phys. Commun.* **43** 107
[48] Atzeni S *et al* 2005 *Comput. Phys. Commun.* **169** 153
[49] Ribeyre X, Lafon M, Schurtz G, Olazabal-Loumé M, Breil J, Galera S and Weber S 2009 *Plasma Phys. Control. Fusion* **51** 124030
[50] Marocchino A, Atzeni S and Schiavi A 2014 *Phys. Plasmas* **21** 012701
[51] Schurtz G P, Nicolaï P D and Busquet M 2000 *Phys. Plasmas* **7** 4238
[52] Marocchino A *et al* 2013 *Phys. Plasmas* **20** 022702
[53] Hallo L *et al* 2009 *Plasma Phys. Control. Fusion* **51** 014001

IOP Publishing | Plasma Physics and Controlled Fusion
Plasma Phys. Control. Fusion **57** (2015) 014023 (7pp) | doi:10.1088/0741-3335/57/1/014023

Implosion dynamics in direct-drive experiments

D T Michel, R S Craxton, A K Davis, R Epstein, V Yu Glebov, V N Goncharov, S X Hu, I V Igumenshchev, D D Meyerhofer, P B Radha, T C Sangster, W Seka, C Stoeckl and D H Froula

Laboratory for Laser Energetics, University of Rochester, Rochester, 14636 NY, USA

E-mail: tmic@lle.rochester.edu

Received 27 June 2014, revised 15 September 2014
Accepted for publication 22 September 2014
Published 28 November 2014

Abstract

Increasing the ablation pressure is a path to achieving cryogenic implosion performance on the OMEGA laser that will hydrodynamically scale to ignition on the National Ignition Facility. An increased ablation pressure will allow a more-massive shell (i.e. thicker and more hydrodynamically stable) and a higher adiabat to achieve ignition-relevant velocities ($>3.5 \times 10^7$ cm s^{-1}), areal densities (>300 mg cm^{-2}) and hot-spot pressures (>100 Gbar). Two approaches have demonstrated increased ablation pressure: (1) a target design is shown that uses a Be ablator to increase the hydrodynamic efficiency, resulting in a ~10% increase in the ablation pressure, in comparison to a CH ablator; (2) reducing the beam size is shown to recover all of the ablation pressure lost to cross-beam energy transfer (CBET), i.e. the ablation pressure calculated without CBET, but the degraded illumination uniformity reduces the integrated target performance. The hydrodynamic efficiency is measured for the current cryogenic design, multiple ablator material design and CH capsule designs with various beam focal-spot sizes. In each case, an excellent agreement is observed with 1D hydrodynamic simulations that include CBET and nonlocal heat-transport models.

Keywords: fusion, hydrodynamic efficiency, direct drive, cryogenic, self-emission, ablator, zooming

(Some figures may appear in colour only in the online journal)

1. Introduction

Direct-drive inertial confinement fusion uses laser beams to implode a spherical shell [1]. The laser energy is absorbed near the critical surface of the target, transferred through the conduction zone to the ablation region and converted into the kinetic energy of the shell through the rocket effect. Near peak compression, a fraction of the kinetic energy of the imploding shell is converted to the internal energy of the fuel. When the ion temperature of the central region (hot spot) and the areal density of the compressed fuel are sufficiently large, a burn wave originating from the alpha particles produced by the fusion of deuterium (D) and tritium (T) will propagate through the confined fuel in the shell (ignition).

The Omega Laser System [2] is used to study the physics of direct-drive fusion and could demonstrate a hydrodynamically equivalent implosion that when scaled to the available energy on the National Ignition Facility (NIF, 1.5 MJ) would produce ignition [3, 4]. Implosions are defined as hydro-equivalent when the laser intensity, the shell adiabat and the implosion velocity are conserved. This theory is limited due to not scalable parameters (surface roughness, laser-plasma instabilities, laser-beam geometry, etc).

One dimensional (1D) hydrodynamic simulations performed using the code *LILAC* [5], including nonlocal thermal conduction [6] and cross-beam energy transfer (CBET) models [7, 8], show that a hydrodynamically equivalent ignition design on OMEGA requires a final implosion velocity $V_{imp} > 3.5 \times 10^7$ cm s^{-1}, an areal density $\rho R > 300$ mg cm^{-2} and a final hot-spot pressure $P_{hs} > 100$ Gbar [9]. To achieve these conditions with the current ablation pressure on OMEGA, a target design with

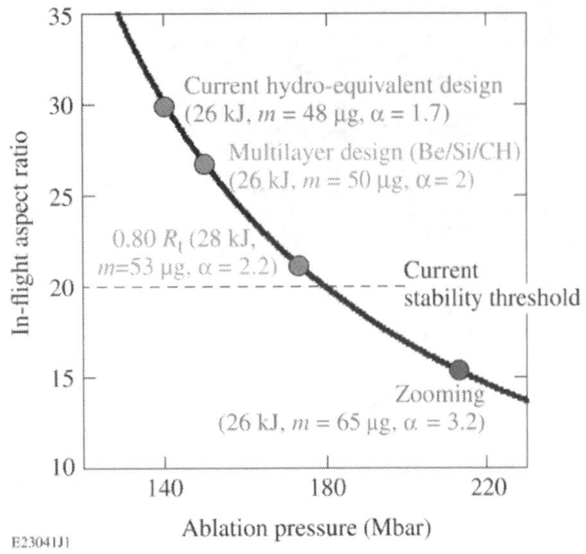

Figure 1. In-flight aspect ratio calculated at a convergence ratio of 1.5 as a function of ablation pressure (black curve) calculated at the end of the laser pulse for OMEGA direct-drive designs where the areal density ($\rho R = 300$ mg cm^{-2}), implosion velocity ($V_{imp} = 3.7 \times 10^7$ cm s^{-1}) and hot-spot pressure ($P_{hs} = 180$ Gbar) are held constant. The designs used triple-picket laser pulses with a maximum intensity of 9×10^{14} W cm^{-2}. Simulations for the current cryogenic configuration (blue circle), a multilayer configuration (green circle), a configuration with a 20% -reduced beam size (orange circle) and a zooming configuration (purple circle) are shown. The shell masses, adiabats and laser energies used for each simulation are indicated in the figure. The current experimental stability threshold is indicated (dashed line).

an in-flight aspect ratio (IFAR) of 30 is required, where the IFAR is a measure of the hydrodynamic stability of the implosion [10] given by the ratio of the radius to the thickness of the shell at a convergence ratio (defined as the ratio of the ablation front of the imploding shell over the initial inner shell radius) of 1.5 (figure 1).

An extensive set of experiments on OMEGA has been used to study the implosion performance in targets made of an outer layer of deuterated plastic (CD) and an inner layer of cryogenic DT ice surrounding DT gas [9, 11]. In these experiments, a threshold was characterized in the shell adiabat and IFAR space, where below the IFAR threshold, the areal density calculated in 1D simulations is recovered in the experiments. Simulations showed that a hydro-equivalent ignition design on OMEGA is above the threshold (figure 1).

Figure 1 shows a 1D design curve generated by increasing the ablation pressure while maintaining a constant areal density ($\rho R = 300$ mg cm^{-2}), implosion velocity ($V_{imp} = 3.7 \times 10^7$ cm s^{-1}) and hot-spot pressure ($P_{hs} = 180$ Gbar). Increasing the ablation pressure requires thicker shells and larger adiabats to be used to maintain a constant implosion velocity, areal density and hot-spot pressure. The higher adiabat and thicker shells leads to more stable implosions (lower IFAR). Figure 1 illustrates that increasing the ablation pressure from 140 MBar (current design) to 180 MBar is likely to achieve ignition hydro-equivalent implosions on OMEGA, assuming the current level of target nonuniformity seeds.

In this paper, the transfer of laser energy to the kinetic energy of the shell (hydrodynamic efficiency) is studied for three different direct-drive implosion experiments: (1) cryogenic experiments [11], (2) ablator material experiments [12] and (3) reduced-beam-size experiments [13]. Experimental measurements of both the laser absorption and the implosion velocity are compared with 1D hydrodynamic simulations. In each case, excellent agreement is observed, indicating that the simulations accurately reproduce the hydrodynamic efficiency. Simulations showed that an ~10% increase in the ablation pressure was obtained for a Be ablator compared with a CH ablator. When the radius of the laser beams was reduced by 50%, nearly all of the ablation pressure lost to CBET was recovered and the ablation pressure was increased by ~60%.

2. Results

The experiments discussed here were conducted on the Omega Laser Facility, where sixty 351 nm laser beams illuminated spherical shells. The hydrodynamic coupling was studied in cryogenic experiments, ablator material experiments and reduced-beam-size experiments. In each case, the scattered-light power was measured to determine the absorption, while the shell trajectory, velocity and neutron bang time were measured to quantify the transfer of the absorbed energy into the shell's kinetic energy. The experimental observables were compared with hydrodynamic simulations performed using the code *LILAC* [5], which includes nonlocal thermal transport [6] and CBET [8] models. The simulated trajectories were obtained by post-processing the 1D hydrodynamic parameters with *Spect3D* [14] to determine synthetic self-emission images. The simulations were used to determine the ablation pressure for the various cases and to assess the potential improvements in ignition hydro-equivalent OMEGA designs.

2.1. Cryogenic experiments

The total laser energy in the cryogenic experiment was 26.6 kJ. The laser beams were smoothed by polarization smoothing (PS) [15], smoothing by spectral dispersion (SSD) [16] and distributed phase plates (SG4-DPP, fourth-order super-Gaussian with a 650 μm full width at half maximum (FWHM)) [17]. A laser pulse with three short pickets was used to set the implosion target on a moderate adiabat ($\alpha = 3.7$) [18], followed by a 1.2 ns square pulse that accelerated a shell with an initial radius of a 435 μm. The shell was made with a 7.4 μm CD ablator on top of a 52.2 μm-thick cryogenic DT ice layer.

The total unabsorbed laser energy was measured by five calorimeters located around the target chamber with an uncertainty of 5% . The scattered-light power was measured at four locations by multiplexing the signal into a 1.5 m spectrometer with a high-dynamic-range streak camera. The system had a 100 ps (FWHM) temporal resolution.

Figure 2 compares the measured and simulated time-resolved unabsorbed laser light. The excellent agreement between them indicates that the CBET model accurately

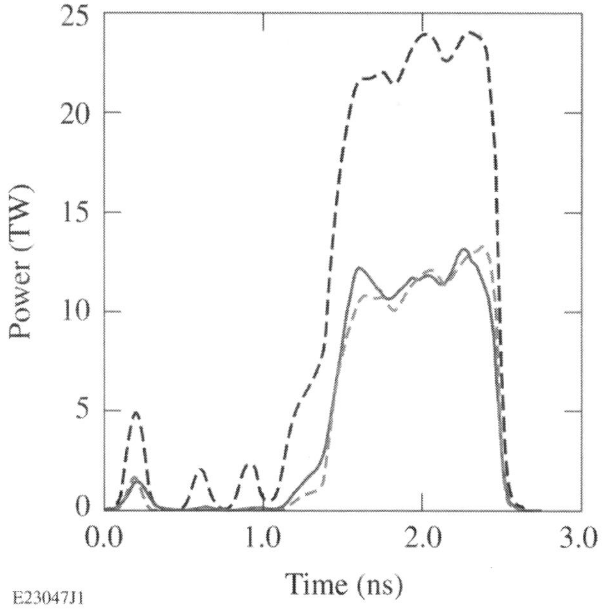

Figure 2. Comparison of the calculated (dashed red line) and measured (solid line) scattered-light power. The laser pulse is plotted (dashed black line).

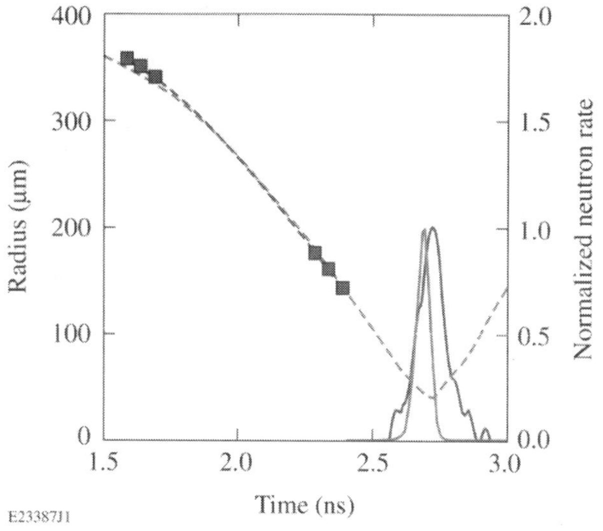

Figure 3. Comparison of the calculated (dashed red line) and measured (square) shell trajectory and normalized neutron production obtained for the current cryogenic design. The third-order polynomial fit to the experimental data is plotted (dashed blue line).

reproduces the laser power absorbed in the plasma; the simulation also shows that nearly half of the scattered light is a result of CBET. The small discrepancy observed around 1.6 ns could be due to an error in the CBET calculation during the rise of the laser pulse, where the hydrodynamic conditions evolve rapidly.

Figure 3 shows excellent agreement between the measured and simulated shell trajectories. These results suggest that the nonlocal transport model accurately calculates this coupling since the transfer of the absorbed laser energy to the shell motion depends primarily on the thermal transport. A small discrepancy was obtained in the ablation-front position at early times when the simulated radius was $\sim$10 μm smaller than the measured radius. The early discrepancy may be explained by the small difference in laser absorption measured during the rise of the main pulse.

The agreement observed between the simulated and measured shell trajectory is consistent with the good agreement observed in the time of neutron production (figure 3). The total neutron production was 5 × lower in the experiment than in simulation, likely related to shell perturbations that compromise the implosion. The self-emission shadowgraphy technique [19] was used to determine the shell trajectories. The soft x rays emitted by the imploding target were integrated over 40 ps and imaged with an array of pinholes onto a four-strip fast x-ray framing camera (XRFC) [20]. The distinct peak in the images emission profile resulted from the high density and temperature at the ablation surface and the large integration distance for the emission that reaches the XRFC (limb effect). In the cryogenic target experiments [9], the primarily DT shell is optically thin to the CD emission at early times (figures 4(*a*) and (*b*)) and the ablation front is determined from the peak of the emission [21]. At late times (figures 4(*e*) and (*f*)), the shell is optically thick to the DT emission and the ablation front is determined from the position of the inner gradient. During the transition time when the CD is being ablated (figures 4(*c*) and (*d*)), the CD emission peak separates from the ablation surface and cannot be used to determine the ablation-front position. During this time, the ablation-front trajectory is inferred by fitting the ablation-front position measured at early and late times with a third-order polynomial. A standard deviation of the peak (inner gradient) location around the image of $\sigma_{\text{abl}} = 2\ \mu$m ($\sigma_{\text{abl}} = 5\ \mu$m) is obtained. These positions are averaged as $N = 2\pi / \theta_{\text{av}} \sim 60$ independent measurements (where each lineout is averaged over $\theta_{\text{av}} = 5°$) to give a resulting accuracy for the averaged radius of $\sigma_{\text{abl}}/\sqrt{N} < 1\ \mu$m [21].

2.2. Ablator material experiments

In the ablator experiments, the overlapped intensity was varied by changing the total energy on target from 18.5 kJ (low intensity, $I = 4.5 \times 10^{14}$ W cm^{-2}) to 23 kJ (high intensity, $I = 7.2 \times 10^{14}$ W cm^{-2}). The laser beams were smoothed by PS, SSD and SG4-DPP. Three 100 ps-long pickets were used to set the target implosion onto a low adiabat followed by a 1.2 ns (high intensity) and 1.6 ns (low intensity) square pulse that drove the target to its final velocity. The three ablators had mass densities of 1.03 g cm^{-3} (CH), 3.35 g cm^{-3} (C) and 1.83 g cm^{-3} (Be). Their thicknesses were varied to maintain the initial total ablator mass to be equivalent to a 27 μm-thick CH shell. The outer radius at low energy and at high energy was 440 μm and 445 μm, respectively and the total mass was 62 ± 1 μg and 64 ± 2 μg, respectively.

Figure 5 shows the simulated and measured shell trajectories and scattered powers for three different ablator materials: beryllium (Be, $\langle A\rangle/\langle Z\rangle = 2.25$), high-density carbon (C, $\langle A\rangle/\langle Z\rangle = 2$) and glow-discharge polymer (CH, $\langle A\rangle/\langle Z\rangle = 1.85$). In C and Be ablators, multiple experiments

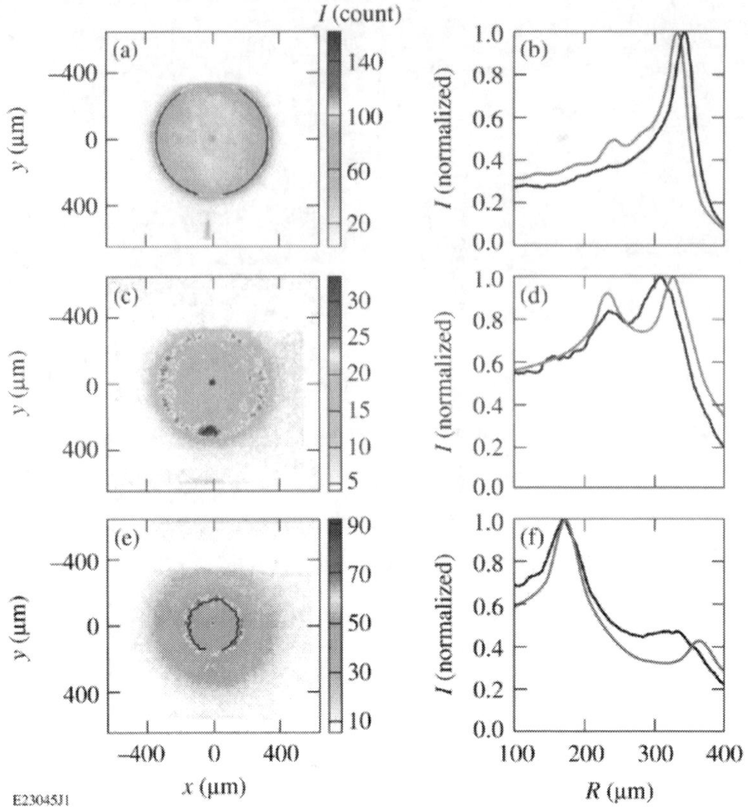

Figure 4. (*a, c, e*) Self-emission x-ray images from a cryogenic experiment; (*b, d, f*) comparison of the lineouts averaged over the entire experimental image (black curve) with lineouts calculated from the simulations (red curve) at t = 1680 ps (*a, b*), t = 2145 ps (*c, d*) and t = 2325 ps (*e, f*).

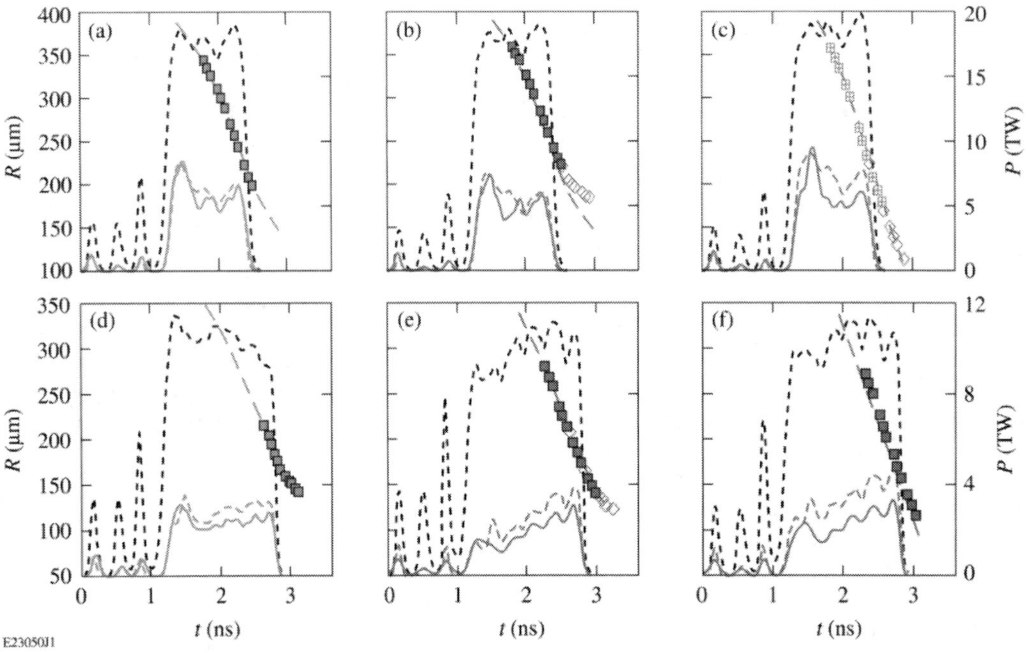

Figure 5. Comparison of the calculated (dashed curves) and measured (symbols) shell trajectories (the different symbols represent different shots) and of the calculated (dashed curves) and measured (solid curve) scattered-light power. The results are presented for (*a, d*) CH, (*b, e*) C and (*c, f*) Be ablators at high and low laser power (dashed black curves, right axis).

were performed to demonstrate reproducibility of these results. The ability of the simulations to accurately reproduce both the measured trajectories and scattered powers suggests that the hydrodynamic efficiency is well modeled with *LILAC* for all three ablators at both high and low intensities.

The shell radii measured on two consecutive XRFC strips, 200 ps apart, were used to calculate time-averaged shell velocities. The relative timing between the XRFC strips was known to within 5 ps and the absolute timing to the laser pulse to within 30 ps [12]. The accuracy in the velocity (dV/V) was ~4% for a shell velocity of 200 km s^{-1}, given by $dV/V = ((d(\Delta R)/\Delta R)^2 + (d(\Delta t)/\Delta t)^2)^{1/2}$, where the error in the radius was $d(\Delta R)/\Delta R = 0.7/40$ μm = 1.8% and the error in the timing was $d(\Delta t)/\Delta t = 7/200$ ps = 3.5%.

Figures 6(*a*) and (*b*) show that the velocity of the shell at the end of the laser pulse is increased by 20% in the Be ablators at both low and high intensities compared to the CH ablators. As the initial target mass was held constant, a higher velocity suggests a larger kinetic energy of the shell, but this is slightly reduced by a smaller final mass due to a higher mass ablation rate. For the three ablators, the total laser absorption was similar at low intensity (~ 80%) and high intensity (~ 70%). This indicates that the increased velocity in Be ablators results from the transfer of the absorbed laser energy to the kinetic energy of the shell. Increasing $\langle A\rangle/\langle Z\rangle$ improves the coupling of the absorbed energy to the kinetic energy of the shell by increasing the mass density near the critical density (i.e. where the laser energy is absorbed) [22].

Figure 6(*c*) compares the ablation pressure calculated during the implosion at a similar convergence ratio of 1.5 for the three ablators at low and high laser intensities. An ~10% increase is observed for the Be ablator compared to the C and CH ablators as a result of the increased $\langle A\rangle/\langle Z\rangle$. Simulations suggest that hydro-equivalent ignition designs for OMEGA that use a Be layer can increase the ablation pressure by ~7% compared with the standard CD/DT design, allowing the IFAR to be reduced to 27 (figure 1) [9].

2.3. Reduced-beam-size experiments

The reduced-beam-size experiments used laser beams smoothed by PS and DPP. Fifty-seven phase plates designed to produce elliptical spots were oriented with their minor axes aligned in the direction of the wedge dispersion of the PS crystal, producing a nearly round $R_b = 215$ μm (95% encircled energy) laser spot at best focus. The ellipticities of these laser spots were measured to be less than 15% . Three round DPP's with a best-focus radius of $R_b = 210$ μm were used to complete the set. The radii of the laser spots were varied by defocusing the laser beams. A triple-picket laser pulse shape with a 1.6 ns square drive pulse and 18 kJ of total energy was used to set the shell on a low adiabat ($\alpha \approx 3$). The outer radius of the CH capsule was $R_t = 430$ μm, with a wall thickness of 27 μm.

Figure 7(*a*) shows the measured shell trajectories when the radius of each laser-beam spot was scaled. The smaller beam spots resulted in a higher ablation pressure, which accelerated the target significantly faster. Even for a moderate reduction in beam size ($R_b/R_t = 1$ to $R_b/R_t = 0.9$), a significant increase in velocity was observed: 186–194 km s^{-1} (~ 5% increase). Good agreement between the simulated and measured shell trajectories and scattered-light powers was observed for all focal-spot radii. The improved coupling was a result of both more-normal rays that deposit their energy closer to the ablation surface and the reduction of CBET [13]. Figure 7(*b*) shows the ablation pressure calculated from the simulations at a convergence ratio of 1.5. A 50% increase in the ablation pressure was calculated when R_b/R_t was varied from 1.1 to 0.5.

Figure 8(*a*) shows the standard deviation in the variation of the position of the inner gradient around the image for various laser-beam diameters. When the laser-beam size is reduced, the standard deviation of the low-mode nonuniformities

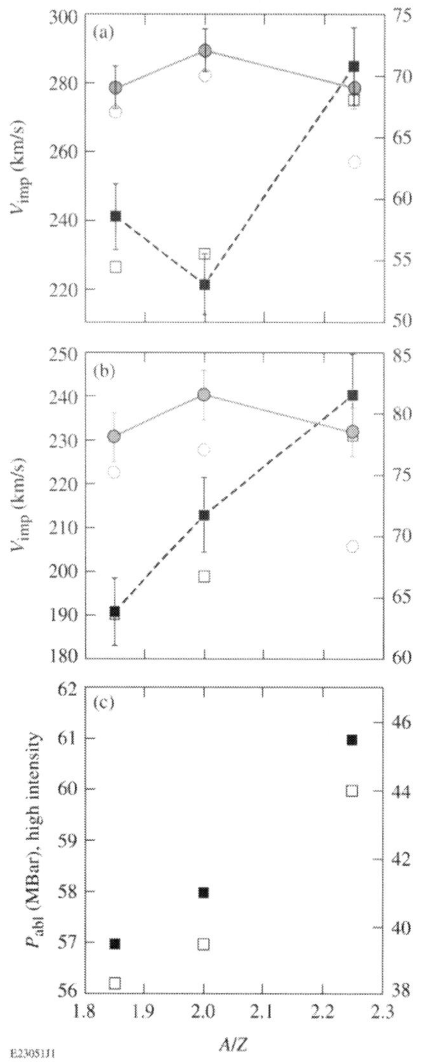

Figure 6. Comparison of the shell velocities measured at the end of the laser pulse (solid squares) and the total laser absorption determined from the measured scattered light (solid circles) for CH ($\langle A\rangle/\langle Z\rangle = 1.85$), C ($\langle A\rangle/\langle Z\rangle = 2$) and Be ($\langle A\rangle/\langle Z\rangle = 2.25$) ablators at high (*a*) and low (*b*) laser intensities; the corresponding calculated values are shown as open symbols. (*c*) Comparison of the simulated ablation pressure at a convergence ratio of 1.5 for CH ($\langle A\rangle/\langle Z\rangle = 1.85$), C ($\langle A\rangle/\langle Z\rangle = 2$) and Be ($\langle A\rangle/\langle Z\rangle = 2.25$) ablators at low (open square) and high (solid square) laser intensities.

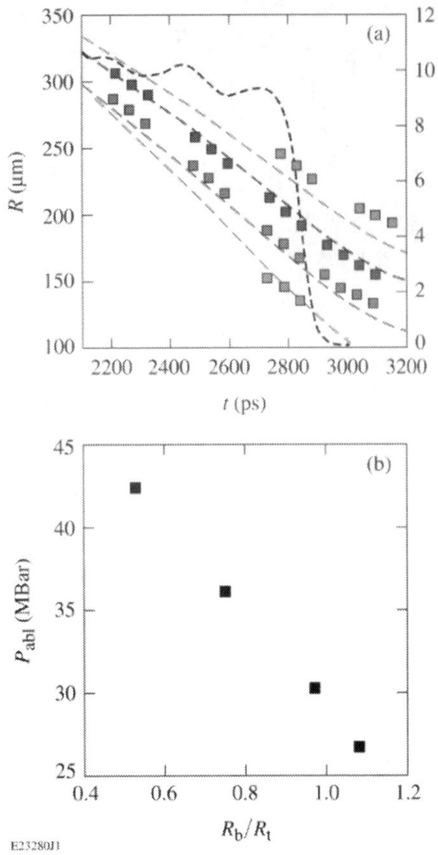

Figure 7. (*a*) Comparison of the measured (squares) and simulated (dashed curves) shell trajectories for $R_b/R_t = 0.5$ (green), $R_b/R_t = 0.9$ (red), $R_b/R_t = 1$ (blue) and $R_b/R_t = 1.1$ (light blue). A similar laser pulse was measured for each shot (black curve, right axis). (*b*) The ablation pressure calculated at a convergence ratio of 1.5.

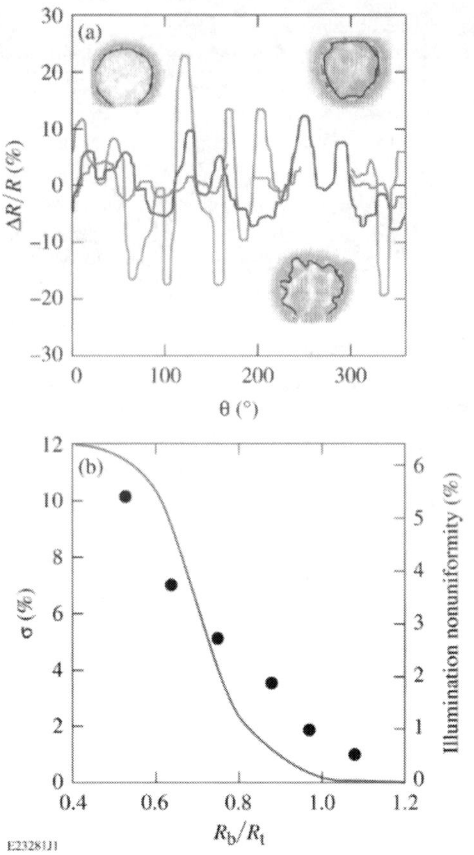

Figure 8. (*a*) Comparison of the low-mode shell perturbations measured for $R_b/R_t = 0.5$ (light grey line, bottom insert), $R_b/R_t = 0.75$ (grey line, top right insert) and $R_b/R_t = 1$ (black line, top left insert) at an averaged radius of 175 μm. (*b*) Comparison of the standard deviation of the shell perturbation obtained in experiments (left axis) with the standard deviation of the simulated nonuniformity in the absorbed laser power averaged over the entire laser pulse (red curve, right axis).

measured at the ablation surface increases from 1% to more than 10% . The structure of the perturbations grows at positions consistent with regions where the illumination is lower. The standard deviation of the low-mode perturbations measured from the XRFC images is compared to the standard deviation of the low-mode nonuniformity of the absorbed laser power averaged over the entire laser pulse for different R_b/R_t (figure 8(*b*)). When the size of the focal spot is reduced, the illumination nonuniformity increases and the shell perturbations compromise the target compression. Simulations suggest that the laser-beam diameters can be reduced to 80% of the target diameter before significant degradation in yield is observed [8]. In this case, simulations indicate that the ablation pressure was increased by 24%, allowing for a hydro-equivalent ignition design on OMEGA with an IFAR of 22—close to the current stability threshold (figure 1).

To further reduce the beam size, a two-stage zooming scheme has been proposed where large beams are used during the pickets and small beams are used during the main drive when the conduction zone is large enough to smooth low-mode laser nonuniformity [23, 24]. By using $R_b/R_t = 0.6$ during the main drive, nearly all of the ablation pressure lost to CBET is recovered and a hydro-equivalent ignition design is well below the current stability threshold (figure 1).

3. Conclusions

Demonstrating hydro-equivalent ignition on the Omega Laser Facility is a first step toward direct-drive ignition on the NIF. Achieving hydro-equivalent ignition on OMEGA requires an implosion velocity > 3.5×10^7 cm s^{-1}, an areal density > 300 mg cm^{-2} and a hot-spot pressure above 100 Gbar. Currently, the best-performing implosions, which are driven to ignition-relevant velocities, do not reach hydro-equivalent areal densities, limiting the peak hot-spot pressure to ~40 Gbar [9]. One approach to recovering the hydro-equivalent areal density is to decrease the IFAR by using thicker shells. To drive the thicker shells to the relevant velocities, the hydrodynamic efficiency must be improved.

In this manuscript, the hydrodynamic efficiency has been studied in three different direct-drive experiments. The coupling of the laser energy to the plasma (absorption) was determined by measuring the scattered-light power. The plasma energy transferred to the kinetic energy of the shell was studied by measuring the shell trajectory, the shell velocity and the neutron burn history. These experimental observables were compared with 1D hydrodynamic simulations (conducted by the code *LILAC*) that include CBET and nonlocal thermal-transport models. In each case, excellent agreement was observed, suggesting that the code was able to reproduce the hydrodynamic parameters of the imploding shell. For Be targets, the ablation pressure of the laser was shown to increase by ~10% compared with the C and standard CH ablators. When the spot size of the laser was reduced by ~50%, the ablation pressure was shown to increase by a factor of 1.6, but the increased illumination nonuniformity compromised the integrated target performance. In future experiments, three new designs will be tested: (1) multilayer ablators [9], (2) beam diameters that are 80% of the target diameter and (3) two-stage zooming. Simulations indicate that the ablation pressure in hydro-equivalent designs will be increased by 7% for the first case, 24% for the second case and 50% for the third case.

Acknowledgments

This material is based upon work supported by the Department of Energy National Nuclear Security Administration under Award Number DE-NA0001944, the University of Rochester and the New York State Energy Research and Development Authority. The support of DOE does not constitute an endorsement by DOE of the views expressed in this article.

References

[1] Nuckolls J, Wood L, Thiessen A and Zimmerman G 1972 *Nature* **239** 139
[2] Boehly T R *et al* 1997 *Opt. Commun.* **133** 495
[3] Nora R *et al* 2014 *Phys. Plasmas* **21** 056316
[4] Paisner J *et al* 1994 *Laser Focus World* **30** 75
[5] Delettrez J 1986 *Can. J. Phys.* **64** 932
[6] Goncharov V N *et al* 2008 *Phys. Plasmas* **15** 056310
[7] Michel P *et al* 2009 *Phys. Rev. Lett.* **102** 025004
[8] Igumenshchev I V *et al* 2012 *Phys. Plasmas* **19** 056314
[9] Goncharov V N *et al* 2014 *Phys. Plasmas* **21** 056315
[10] Radha P B *et al* 2011 *Phys. Plasmas* **18** 012705
[11] Sangster T C *et al* 2013 *Phys. Plasmas* **20** 056317
[12] Michel D T *et al* 2013 *Phys. Rev. Lett.* **111** 245005
[13] Froula D H *et al* 2012 *Phys. Rev. Lett.* **108** 125003
[14] MacFarlane J J *et al* 2007 *High Energy Density Phys.* **3** 181
[15] Boehly T R *et al* 1999 *J. Appl. Phys.* **85** 3444
[16] Skupsky S *et al* 1989 *J. Appl. Phys.* **66** 3456
[17] Kessler T J, Lin Y, Armstrong J J and Velazquez B 1993 *Laser Coherence Control: Technology and Applications* **1870**, Powell H T and Kessler T J (Bellingham, WA: SPIE) p 95
[18] Goncharov V N *et al* 2010 *Phys. Rev. Lett.* **104** 165001
[19] Michel D T *et al* 2012 *Rev. Sci. Instrum.* **83** 10E530
[20] Bradley D K *et al* 1995 *Rev. Sci. Instrum.* **66** 716
[21] Davis A K *et al* 2014 *Rev. Sci. Instrum.* **85** 11D616
[22] Manheimer W M, Colombant D G and Gardner J H 1982 *Phys. Fluids* **25** 1644
[23] Froula D H *et al* 2013 *Phys. Plasmas* **20** 082704
[24] Igumenshchev I V *et al* 2013 *Phys. Rev. Lett.* **110** 145001

IOP Publishing Plasma Physics and Controlled Fusion
Plasma Phys. Control. Fusion **57** (2015) 014024 (10pp) doi:10.1088/0741-3335/57/1/014024

Energetic particles and multi-scale dynamics in fusion plasmas

F Zonca[1,2], L Chen[2,3], S Briguglio[1], G Fogaccia[1], A V Milovanov[1], Z Qiu[2], G Vlad[1] and X Wang[4]

[1] ENEA C.R. Frascati, Via Enrico Fermi 45, CP 65-00044 Frascati, Italy
[2] Institute for Fusion Theory and Simulation and Department of Physics, Zhejiang University, Hangzhou 310027, People's Republic of China
[3] Department of Physics and Astronomy, University of California, Irvine, CA 92697-4575, USA
[4] Max-Planck-Institut für Plasmaphysik, Boltzmannstraße 2, Garching D-85748, Germany

E-mail: fulvio.zonca@enea.it

Received 28 July 2014, revised 22 September 2014
Accepted for publication 23 September 2014
Published 28 November 2014

CrossMark

Abstract

The role of energetic particles (EPs) in fusion plasmas is unique as they could act as mediators of cross-scale couplings. More specifically, EPs can drive instabilities on the macro- and meso-scales and intermediate between the microscopic thermal ion Larmor radius and the macroscopic plasma equilibrium scale lengths. On one hand, EP driven shear Alfvén waves (SAWs) could provide a nonlinear feedback onto the macro-scale system via the interplay of plasma equilibrium and fusion reactivity profiles. On the other hand, EP-driven instabilities could also excite singular radial mode structures at SAW continuum resonances, which, by mode conversion, yield microscopic fluctuations that may propagate and be absorbed elsewhere, inducing nonlocal behaviors. The above observations thus suggest that a theoretical approach based on advanced kinetic treatment of both EPs and thermal plasma is more appropriate for burning fusion plasmas. Energetic particles, furthermore, may linearly and nonlinearly (via SAWs) excite zonal structures, acting, thereby, as generators of nonlinear equilibria that generally evolve on the same time scale of the underlying fluctuations. These issues are presented within a general theoretical framework, discussing evidence from both numerical simulation results and experimental observations. Analogies of fusion plasmas dynamics with problems in condensed matter physics, nonlinear dynamics, and accelerator physics are also emphasized.

Keywords: Alfvén waves, fast particle effects, fusion plasmas, BGK modes, nonlinear phenomena, phase locking, solitons

(Some figures may appear in colour only in the online journal)

1. Introduction

There is increasing interest in fusion plasmas as complex systems [1–4], where many interacting degrees of freedom yield a variety of interesting nonlinear behaviors characterized by a broad range of spatiotemporal scales [5–8].

Adopting the conceptual framework of [2, 3], where magnetized fusion plasmas are assumed to be close to marginal stability, thereby supporting the Ansatz that their dynamics is governed by self-organized criticality (SOC) [9], single transport events (avalanches) may exhibit characteristic aspects of sandpile physics involving SOC [4]. The application of this approach, aimed at characterizing and interpreting strong nonlinear phenomena in fusion as well as astrophysical plasmas, was reviewed in [10, 11] and is beyond the scope of the present work. Here, we emphasize that fusion plasmas may be (and generally are) strongly driven systems, where an ideal SOC state can be destabilized above a certain critical level of external forcing, and that the dynamics may become dominated by bursting periodic relaxation events, for which the name 'fishbone-like instability of SOC' has been suggested [12, 13] by analogy with the nonlinear dynamics of 'fishbone' fluctuations in fusion plasmas [14].

Multi-scale dynamics suggests the necessity to go beyond theoretical formulations of fluctuation induced transport based on local closures [15–19] and to replace the corresponding description of turbulent fluxes connected to thermodynamic forces through nonlocal modifications of Fick's law [20–22]. In particular, meso-scale dynamics plays a crucial role in turbulent transport for two fundamental reasons: (i) drift-wave turbulence (DWT) itself can spontaneously generate patterns [zonal flows and currents/fields, generally zonal structures (ZS)] [23–26] on scales that are typically larger than the perpendicular (w.r.t. the equilibrium magnetic field $\boldsymbol{B}_0$) wavelength of the underlying fluctuations, $\lambda_\perp$, but still shorter than the equilibrium nonuniformity scale length; and (ii) the interplay of DWT and ZS may result in turbulence spreading [27–29] and the self-consistent evolution of equilibrium radial profiles in the form of avalanches [30]. These phenomena have been observed in recent global gyrokinetic simulations [21, 30–32].

Less attention has been devoted to multi-scale dynamics and complex behaviors connected with energetic particles and fusion products (EPs) [33–37], despite the fact that EPs are expected to dominate the power balance in burning fusion plasmas. This is the focus of the present work, with the aim of emphasizing general aspects of nonlinear behaviors with their similarities and differences with respect to those discussed above, and of highlighting issues of common interest of fusion science with neighboring fields of physics. In section 2, we first introduce and discuss the spatiotemporal structures of shear Alfvén waves (SAWs) and, more generally, of DWT. Section 3 then addresses the excitation of SAW fluctuations by resonant wave-EP interactions in nonuniform plasmas and the role of the SAW continuous spectrum [38] by which EPs could act as mediators of cross-scale couplings. For these analyses, we adopt the unified theoretical framework of the general fishbone like dispersion relation (GFLDR), recently presented, discussed, and reviewed in [39, 40]. Furthermore, we discuss the importance of the so-called 'Alfvénic state', where SAWs can exist in uniform, incompressible MHD plasmas independently of their amplitude due to the cancellation of Reynolds and Maxwell stresses and the incompressible plasma motion produced by SAWs [41–43]. Section 4 introduces nonlinear dynamics of ZS excited by SAWs [44–46] and/or DWT [47, 48] in the presence of EPs. In this case, interesting analogies may be drawn with the competition of nonlinearity versus randomness in the discrete Anderson nonlinear Schrödinger equation (DANSE) (see, e.g. [49–52]). In section 5, we illustrate nonlinear dynamics of energetic particle modes (EPMs) [53], excited within the SAW continuous spectrum as discrete fluctuations at the frequency maximizing wave-EP power exchange above the threshold condition set by continuum damping. EPMs are an example of autoresonance [54] and superradiance [55] in fusion plasmas [38, 56] and suggest similarities with the amplification of a short optical pulse in a free electron laser (FEL) [57]. Finally, section 6 provides a summary and conclusions.

2. Spatiotemporal structures of SAW and DWT

We adopt straight magnetic field line toroidal flux coordinates (r, θ, ζ), with r as the radial 'magnetic flux' variable and θ and ζ as periodic angular coordinates along poloidal and toroidal directions, respectively. Thus, denoting $\boldsymbol{b} = \boldsymbol{B}_0/B_0$, the safety factor $q(r)$

$$q(r) = \boldsymbol{b}\cdot\nabla\zeta / \boldsymbol{b}\cdot\nabla\theta, \tag{1}$$

describing the pitch of $\boldsymbol{B}_0$ field lines, is a flux function. SAWs as well as DWT, represented by $\delta\phi(r, \theta, \zeta) = \Sigma_{m,n}\exp(\mathrm{i}n\zeta - \mathrm{i}m\theta)\,\delta\phi_{m,n}(r)$, can be decomposed as [58]

$$\begin{aligned}\delta\phi(r,\theta,\zeta) &= 2\pi \sum_{\ell,n\in\mathbb{Z}} \mathrm{e}^{\mathrm{i}n\zeta-\mathrm{i}nq(\theta-2\pi\ell)}\delta\hat{\phi}_n(r,\theta-2\pi\ell)\\ &= \sum_{m,n\in\mathbb{Z}} \mathrm{e}^{\mathrm{i}n\zeta-\mathrm{i}m\theta}\int \mathrm{e}^{\mathrm{i}(m-nq)\vartheta}\delta\hat{\phi}_n(r,\vartheta)\,\mathrm{d}\vartheta\\ &= \sum_{m,n\in\mathbb{Z}} \mathrm{e}^{\mathrm{i}n\zeta-\mathrm{i}m\theta}\int \mathrm{e}^{\mathrm{i}(m-nq)\vartheta}\mathcal{P}_{Bn}(r,\vartheta)\,[\delta\phi]\,\mathrm{d}\vartheta,\end{aligned} \tag{2}$$

where the mapping $\mathcal{P}_{Bn}(r,\vartheta)$: $\delta\phi(r,\theta,\zeta)\mapsto\delta\hat{\phi}_n(r,\vartheta)$ [38–40] of the scalar potential $\delta\phi(r, \theta, \zeta)$ to $\delta\hat{\phi}_n(r,\vartheta)$ holds for arbitrary mode numbers and follows from the general properties of the Poisson summation formula [58]. Here, time dependence is suppressed to simplify notation, and the same mapping can be adopted for the parallel vector potential $\delta A_\parallel$ and any other fluctuating field, including the perturbed particle distribution δf, whose value at equilibrium can be denoted by f_0. In equation (2), ϑ plays the role of an 'extended poloidal angle' following $\boldsymbol{B}_0$ field lines. When the radial length scale of $\delta\hat{\phi}_n(r,\vartheta)$ is longer than $|nq'|^{-1}$, equation (2) reduces to the well-known ballooning representation (see e.g. [59, 60]).

Collective SAW fluctuations excited by EPs generally have a dense spectrum of modes with characteristic frequencies and spatial locations [34, 61]. Thus, for SAWs as well as DWT, the mode structure decomposition of equation (2) can be simplified, introducing

$$\delta\hat{\phi}_n(r,\vartheta) = A_n(r)\,\delta\hat{\phi}_{0n}(r,\vartheta) \simeq A_n(r)\,\delta\hat{\phi}_{0n}(\vartheta)\,. \tag{3}$$

Equations (2) and (3) show that mode structures can be represented by three degrees of freedom: the toroidal mode number n, the radial envelope $A_n(r)$, and the parallel (to $\boldsymbol{b}$) mode structure $\delta\hat{\phi}_{0n}(r,\vartheta)$, with only a slow radial variation on the equilibrium scale length L. Correspondingly, nonlinear interactions can take the following three forms: mode coupling between two ns, modulation of the radial envelope, and distortion of the parallel mode structure [62, 63]. These three degrees of freedom also have their own characteristic spatiotemporal scales. The toroidal mode number n of most unstable modes is determined by the condition that $\lambda_\perp$ is of the order of the characteristic orbit width ($\propto\rho_L$, the Larmor radius) of destabilizing particle species (EPs for SAWs and thermal plasma particles for DWT/SAWs). The corresponding time scale is set by three-wave coupling cross sections, i.e. it tends to infinity in the linear limit, where n is fixed for each separate fluctuation because of the toroidal equilibrium symmetry. The radial envelope $A_n(r)$, meanwhile, varies on 'meso-scales'

L_A ($\rho_L \ll L_A \ll L$) that are different for SAWs excited by EPs and DWT, and its formation time scale, τ_A, is typically $\tau_A \sim \gamma_L^{-1}$, with γ_L as the fluctuation linear growth rate. Finally, the parallel mode structure $\delta\widehat{\phi}_{0n}(r,\vartheta)$ has a typical parallel scale length $\sim qR_0$ (the connection length of a torus of major radius R_0), while its radial scale is $\sim L$ and its characteristic time is $\sim |\omega_0|^{-1}$ (the inverse mode frequency).

For the sake of simplicity, we assume that the nonlinear time, τ_{NL}, is such that $\tau_{NL} \sim \tau_A \sim \gamma_L^{-1} \gg |\omega_0|^{-1}$, which is still a rather general condition of practical interest [38–40, 64–66]. In this way, the parallel mode structure $\delta\widehat{\phi}_{0n}(r,\vartheta)$ remains essentially unmodified on the shortest time scale (longer time scale dynamics is not considered here), while nonlinear dynamics can be understood as coupling of the remaining relevant degrees of freedom [n and $A_n(r)$] on time scale $\tau_{NL} \sim \tau_A \sim \gamma_L^{-1}$. A possible way to visualize this 'reduced' nonlinear dynamics is conceiving the plasma as made of particles and quasi-particles. Unlike particles, which are conserved in number, quasi-particles are 'rigid' field-aligned elongated structures carrying energy and momentum that can be emitted and reabsorbed due to the nonuniform particle distribution. Their nonlinear interaction takes place via spectral transfers in n-space and change in number density [$\propto A_n(r)$], generally due to nonlocal processes.

3. SAWs and EPs as mediators of cross-scale couplings

Further to the spatiotemporal structures of SAWs and DWT, discussed in section 2, SAWs in nonuniform toroidal plasmas are characterized by a continuous spectrum with gaps, due to periodic poloidal variation of the Alfvén speed, v_A, along $\boldsymbol{B}_0$ field lines and/or finite plasma compressibility. In fact, SAW group velocity is directed along $\boldsymbol{B}_0$ and thus SAWs behave as (electron) wave-packets in a one-dimensional (1D) periodic lattice. In general, frequency gaps are formed when the Bragg reflection condition is met, i.e. when two degenerate and counter-propagating SAWs can form a standing wave for $\omega^2 = \ell^2 v_A^2/(2qR_0)^2, \ell \in \mathbb{N}$ [34]. Continuing this analogy, nonuniform plasma equilibrium plays the role of 'defects', causing discrete modes to exist inside the frequency gaps. Different plasma equilibria and related profiles correspond to various types of Alfvén Eigenmodes (AEs) inside different SAW continuum frequency gaps [34] (the Alfvén 'zoology' [67]), the first example of which are Toroidal AEs (TAEs) [68].

The importance of AEs lies in the fact that they are readily excited by resonant interactions with EPs since they are essentially unaffected by SAW continuum damping [68–71] and may have strong impact on EP transports and confinement [72, 73]. However, when resonant EP drive exceeds the threshold condition set by continuum damping, EPMs [53] are excited within the SAW continuous spectrum as discrete fluctuations. For EPMs, EPs play the role of a free energy source as well as 'defects', i.e. EPMs are excited at the resonant EP characteristic frequency [53] and are radially localized so as to maximize wave-EP power exchange [74, 75]. This property of EPMs has important consequences on nonlinear mode dynamics and EP particle transport [76–80] (see section 5). Linear and nonlinear dynamics of AEs and EPMs excited by EPs in fusion plasmas can be described within the unified theoretical framework of the general fishbone like dispersion relation (GFLDR) [38–40], which reads

$$\left[i\Lambda_n - \left(\delta\overline{W}_f + \delta\overline{W}_k\right)_n \right] A_n(r) = D_n(r,\theta_k,\omega) A_n(r) = 0. \quad (4)$$

Here, we have assumed the eikonal form $A_n(r) \sim \exp i \int nq' \theta_k(r)\, dr$ [81, 82] and $D_n(r, \theta_k, \omega)$ plays the role of a local WKB dispersion function, obtained by projection of nonlinear gyrokinetic quasineutrality condition and vorticity equation on the parallel mode structures $\delta\widehat{\phi}_{0n}(r,\vartheta)$. In fact, equation (4) is written for a single n, but, in general, it should be intended as the n-th row of a matrix equation for the complex amplitudes $(\ldots, A_n, \ldots)$ involving nonlinear mode couplings [33, 38]. It is a kinetic energy principle, which reduces to well-known forms of kinetic MHD energy principle in the appropriate limits [83–87]. Here, it is specialized to localized mode structures, but it can also be written for global fluctuations. The Λ_n term represents a generalized 'inertia' due to the plasma response in the kinetic/singular layer at the radial location(s) where the SAW continuous spectrum is resonantly excited [70, 71]; and the Alfvén fluctuation spectrum is characterized by $|k_r| \gg |k_\theta|$. Meanwhile, $\delta\overline{W}_{fn}$ and $\delta\overline{W}_{kn}$ represent the contributions from the regular regions with $|k_r| \sim |k_\theta|$ and can be interpreted as fluid and kinetic 'potential energies' [14]. General expressions of Λ_n, $\delta\overline{W}_{fn}$ and $\delta\overline{W}_{kn}$ are given in [39, 40]. Here, we only note that the various types of AEs are obtained for $\mathbb{R}\mathrm{e}\Lambda_n^2<0$, while EPMs are described by $\mathbb{R}\mathrm{e}\Lambda_n^2>0$. Consistent with this, the general (kinetic) structure of the SAW continuous spectrum, for $q^2R_0^2|\omega|^2 \ll v_A^2$, is given by

$$\Lambda_n^2 = k_\parallel^2 q^2 R_0^2 . \quad (5)$$

Adopting the time scale ordering $\tau_{NL} \sim \tau_A \sim \gamma_L^{-1} \gg |\omega_0|^{-1}$, discussed in section 2, we can examine the spatiotemporal evolution of SAW wave packets expanding the solutions of equation (4) about the linear characteristics

$$D_n^L(r, \theta_{k0}(r), \omega_0) = 0, \quad (6)$$

where D_n^L is the linearized dispersion function introduced in equation (4). Thus, letting $A_n(r) = \exp(-i\,\omega_0 t)\, A_{n0}(r, t)$, the initial value problem for $A_{n0}(r, t)$ becomes [39, 58]

$$\frac{\partial D_n^L}{\partial \omega_0}\left(i\frac{\partial}{\partial t}\right)A_{n0} + \frac{\partial D_n^L}{\partial \theta_{k0}}\left(-\frac{i}{nq'}\frac{\partial}{\partial r} - \theta_{k0}\right)A_{n0} + \frac{1}{2}\frac{\partial^2 D_n^L}{\partial \theta_{k0}^2} \times\left[\left(-\frac{i}{nq'}\frac{\partial}{\partial r} - \theta_{k0}\right)^2 A_{n0} - \frac{i}{nq'}\frac{\partial \theta_{k0}}{\partial r}A_{n0}\right] = S_n(r,t). \quad (7)$$

The $S_n(r, t)$ term on the right hand side represents a generic source term, including external forcing, $S_n^{\mathrm{ext}}(r,t)$, and/or nonlinear interactions [38, 39, 58]

$$S_n(r,t) = S_n^{\mathrm{ext}}(r,t) - e^{i\omega_0 t}\left[i\Lambda_n^{NL} - \left(\delta\overline{W}_f^{NL} + \delta\overline{W}_k^{NL}\right)_n\right]e^{-i\omega_0 t}A_{n0}. \quad (8)$$

Thus, equation (7) has the general form of a nonlinear Schrödinger equation [33, 38] with integro-differential

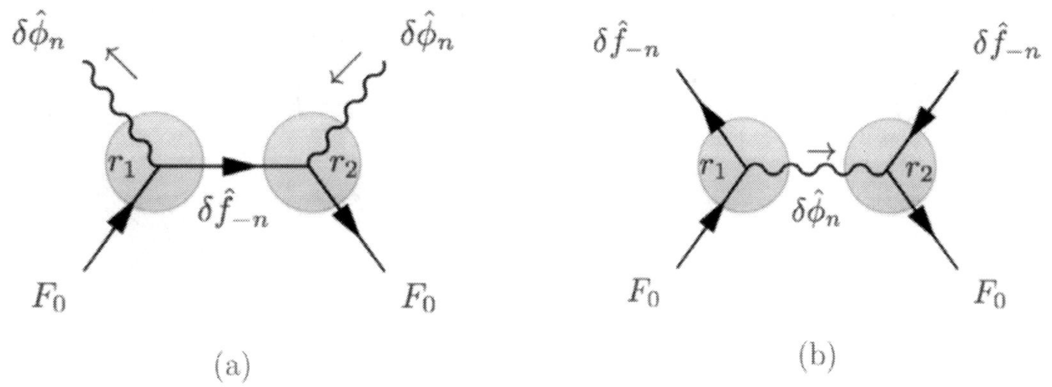

Figure 1. Processes underlying EP transport due to emission and reabsorption of two similar SAW modes (*a*) and the same quasi-particle (*b*).

nonlinear terms [39]. For global modes, for which $\partial_r D_n = \partial_{\theta_k} D_n = \partial^2_{\theta_k} D_n = 0$, equation (7) still holds as an evolution equation, provided that A_{n0} is interpreted as mode amplitude at a fixed radial position and D_n is suitably redefined as global dispersion function [39].

The spatiotemporal structures for SAWs excited by EPs described equation (7) are those generally discussed in section 2 and demonstrate the peculiar role of EPs as mediators between different spatiotemporal scales. In fact, as emphasized in section 1, EPs as fusion products provide the dominant power density source in burning plasmas and, when produced by external heating and current drive systems, they are used to tailor plasma profiles [35]. Thus, they provide a stirring of the system on the macro-scales (L) and are responsible for collective excitation of macroscopic fluctuations with $n = \mathcal{O}(1)$. Furthermore, EPs eventually determine long time scale plasma evolution due to the mutual feedback of fusion reactivity and thermal plasma profiles (see section 6). Energetic particles also provide a free energy source for SAW and drift-Alfvén wave (DAW) [88–92] excitations on the micro- and meso-scales (see section 2), where $1 \ll |n|q^2 \sim (r/\rho_{LE}) \ll (r/\rho_{Li})$ (ρ_{LE} and ρ_{Li} are energetic and thermal ion Larmor radii, respectively). It is worthwhile noting that the micro-scale for SAWs/DAWs excited by EPs ($\lambda_\perp \sim \rho_{LE}$) is typically the same as that belonging to meso-scale nonlinear structures generated by DWT, e.g. avalanches (see sections 1 and 4) and that, for the same reason, meso-scale nonlinear structures generated by EPs (L_A) may extend into the macro-scales (L). However, the existence of the SAW continuous spectrum, whose linear and nonlinear structures are accounted for by Λ_n in equation (7), naturally brings an even shorter spatial scale into the problem under consideration, due to mode conversion of SAW/DAW to short wavelength kinetic Alfvén waves (KAWs) with $\lambda_\perp \sim \rho_i$ at the resonances with the SAW continuum [89]. The importance of short spatial scales is further emphasized by the cancellation of Reynolds and Maxwell stresses in the 'Alfvénic state' [41–43], where long wavelength SAWs can exist in uniform, incompressible MHD plasmas independently of their amplitude. Nonlinear behaviors of SAWs/DAWs are thus intimately connected with the mechanisms that 'break the Alfvénic state' [45], i.e. plasma compressibility, the violation of ideal MHD constraint ($\delta E_\parallel = 0$), and nonuniformity due to plasma equilibrium profiles and magnetic field geometry (see section 4). Because of finite compressibility, a SAW can decay into a sideband SAW and an ion sound wave (ISW) [93]. At sufficiently short wavelength, this MHD process is taken over by the parametric decay of a KAW into a sideband KAW and ISW [89]. The branching ratio of kinetic to MHD decay cross sections is $\sim (\Omega_i/\omega_0)^2\ (k_\perp\ \rho_{Li})^4$, with Ω_i as the ion cyclotron frequency. Thus, the kinetic process is dominant for $|k_\perp \rho_{Li}|^2 > |\omega_0|/|\Omega_i| \sim \mathcal{O}(10^{-2})$ [89, 94]. However, most importantly, the transition from MHD to kinetic regimes corresponds to a crucial change in the consequence of SAW/KAW parametric decay on transport [94]. In fact, in the MHD regime [93], the perpendicular sideband wave vector is parallel to that of the original SAW, $\boldsymbol{k}_{-\perp} \| \boldsymbol{k}_{0\perp}$, whereas $\boldsymbol{k}_{-\perp} \perp \boldsymbol{k}_{0\perp}$ in the kinetic regime [89, 94]. Thus, assuming an initial spectrum of EP driven 'mode converted KAWs' ($\boldsymbol{k}_{0\perp} \simeq k_{0r}\ \nabla\ r$), the MHD parametric decay does not break poloidal angular momentum, P_θ, conservation, causing negligible cross-field transport, while the kinetic parametric decay tends to isotropize the perpendicular KAW spectrum, thereby causing P_θ breaking and significant transport, as observed in numerical simulations [95]. All these elements confirm the unique roles played by EPs as mediators between different spatiotemporal scales in fusion plasmas, also noted in [96] when discussing the properties of the geodesic acoustic mode (GAM) [97] spectrum in the presence of a spatially broad EP beam (see section 4).

The processes discussed so far can be schematically illustrated as in figure 1 (see section 2). Figure 1(*a*) shows fluctuation induced EP transport due to the emission and reabsorption of two similar SAW modes, labeled by $\delta\widehat{\phi}_n$ and omitting $\delta\widehat{A}_{\|n}$ for brevity. The EP distribution function averaged over (θ, ζ), $F_0 = f_0 + \delta f_0$, consists of an equilibrium part f_0 and a fluctuation induced part δf_0. It is unstable and, therefore, emits $\delta\widehat{\phi}_n$ at the radial position r_1, generating a perturbation $\delta\widehat{f}_{-n}$ of the particle distribution function. When the quasi-particle (not necessarily the same as they are continuously emitted and reabsorbed) is reabsorbed at the radial position r_2, the resulting distribution function F_0 is modified by transport processes due to emission and reabsorption of toroidal symmetry breaking quasi-particles. If $|r_2 - r_1| \ll |nq'|^{-1}$, the transport process is local and can generally be described by a diffusive process. If $|r_2 - r_1| \gtrsim |nq'|^{-1}$, transport becomes nonlocal and the diffusive transport paradigm is generally (although not necessarily) broken (see

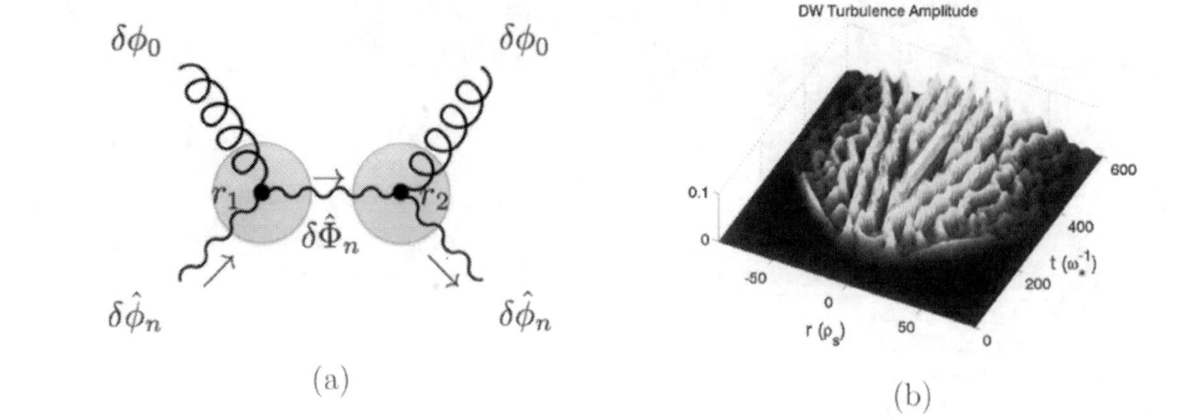

Figure 2. (*a*) Generation of the quasi-particle 'multiplet' $\delta\widehat{\Phi}_n$ due to the spontaneous emission of the ZS $\delta\,\phi_0$. (*b*) DWT spreading via soliton formation due to DWT-ZS interactions. Reprinted with permission from Guo *et al* [112]. Copyright 2009, by the American Physical Society.

section 5). The EP transport process of figure 1(*a*) is what is usually observed in experimental conditions, e.g. during 'TAE avalanches' in the National Spherical Torus Experiment (NSTX) [98, 99], where rapid EP losses occur in bursts of non-adiabatic frequency sweeping modes [100, 101], which are consistent with the general features of EPMs, following the activity of quasi-periodic TAE fluctuations with limited frequency chirping. The same processes, in a two-stage combination, underlie the idea of 'alpha channeling' [102], i.e. 'the diversion of energy from energetic alpha particles to waves' [103]. In the first stage, radio-frequency waves are used to extract energy from EPs with negligible radial displacement ($r_1 \simeq r_2$) [104] and selectively heat the thermal plasma, while in the second stage, after EPs are sufficiently 'cooled', low-frequency macroscopic waves are adopted (stimulated emission) to remove the particles from the plasma ($|r_2 - r_1| \sim L$) without further appreciable energy exchange [104].

Figure 1(*b*) illustrates EP transport processes due to emission and reabsorption of the same quasi-particle $\delta\widehat{\phi}_n$. After reabsorption, energy and momentum of the quasi-particle is transported from r_1 to r_2 and thus occurs on the typical time scale associated with quasi-particle propagation rather than particle transport. Moreover, while propagating between r_1 and r_2, $\delta\widehat{\phi}_n$ can exchange energy and momentum with the EP distribution (by change in the number density, A_n) and, more generally, with the plasma equilibrium (corresponding more articulated diagrams are omitted here). Transport processes of figure 1(*b*) may be connected with KAW that are generally not absorbed near the mode conversion layer in high temperature plasmas [105–107]. These properties further show that mode structures and stability of SAWs are truly kinetic and global in character and that, consequently, transport in fusion plasmas can be significantly affected. Evidence of such KAW 'action at a distance' is given by bursting Alfvén instabilities in Wendelstein 7-AS [108, 109], where frequency chirping is attributed to radial KAW propagation and time varying Doppler shift. At the same time, modulation of plasma energy by the Alfvén wave bursts is explained as heat convection due to wave excited at r_1 and absorbed at r_2 (see figure 4 of [109], reproduced from the original of [108]). This mechanism of 'spatial channeling', proposed in [110], is different from that of 'alpha channeling' [102] discussed above and has also been invoked for explaining similar evidence in NSTX [109].

4. Nonlinear dynamics of zonal structures

Zonal structures [23–26] are patterns spontaneously generated by DWT as well as SAWs/DAWs on meso-scales (L_A; see section 2), which are typically larger than $\lambda_\perp$ of the underlying fluctuations, but still shorter than the equilibrium length scale L (see section 1). Due to finite magnetic shear, which suppresses convective cells in toroidal fusion plasmas, the prevalent ZS have $n = m = 0$ and can be interpreted as radial corrugations of nonuniform plasma equilibrium [111]. Spontaneously generated ZS scatter original fluctuations into shorter radial wavelength stable domain, thus acting as self-regulation mechanism of plasma instabilities. Continuing the schematic picture proposed in section 2 for DWT and SAWs/DAWs, ZS are generated by 'decay' of the $\delta\widehat{\phi}_n$ quasi-particle via modulational instability [64], by which ZS 'modulate' the number density $A_n(r)$ in the radial direction. The spontaneous emission of ZS, denoted for brevity with $\delta\,\phi_0$ (having omitted the parallel vector potential $\delta A_\parallel\,|_{n\,=\,m\,=\,0}$), is depicted in figure 2(*a*), where it generates a 'multiplet' $\delta\widehat{\Phi}_n$ of the original quasi-particle made of $\delta\widehat{\phi}_n$ and its radial 'modulations' due to the ZS action on $A_n(r)$.

Due to the properties of the Alfvénic State [41–43] discussed in section 3, the SAW continuous spectrum makes negligible contributions to nonlinear ZS generation. However, finite amplitude waves of the discrete AE spectrum and EPMs may spontaneously excite ZS via modulational instability of the radial envelope $A_n(r)$, due to the breaking of the Alfvénic State via geometry and/or finite plasma/EP compressibility effects [44–46, 113], in addition to the short wavelength kinetic effects discussed in section 3 [45, 89, 94]. Thus, a theoretical approach based on advanced kinetic treatment of both EPs and thermal plasma, accounting for realistic equilibrium geometries and nonuniformities, is necessary for a proper analysis of cross-scale couplings in burning fusion plasmas. Meanwhile, ZS generated by EP driven instabilities reflect the different spatiotemporal structures characterizing the SAW/DAW fluctuation spectrum with respect to DWT, as discussed in section 3. A further distinctive element of ZS generated in the presence of EPs is that ZS may not be linearly stable and, hence, the SAWs/DAWs interaction with ZS and DWT should be formulated taking that into account. This is the case in, for example, EP driven GAMs (EGAMs) [114, 115], where the picture of parametric GAM excitation by DWT [116–118] acquires additional twists due to the active/reactive EP response [119]. In fact, recent global nonlinear gyrokinetic simulations of the interplay between EGAMs and DWT [47, 48] have demonstrated that this interaction may be more complicated than expected and that EGAMs could enhance turbulent transport and couple to meso-scale avalanches, with obvious impact on nonlocal properties of the system. The self-consistent nonlinear interplay of SAWs/DAWs, DWT, and ZS in the presence of a non-perturbative EP population is one of the most challenging and open problems in magnetic fusion research.

Figure 2(*a*) also suggests that interactions of ZS with DWT and SAWs/DAWs have two components: a coherent part [64, 116, 120, 121], due to the interaction with the self-generated ZS, and a random contribution [25, 26], due to interaction with ZS produced by other incoherent components of the fluctuation spectrum. Assuming non-dispersive waves and the local character of the associated nonlinear interactions in n-space, equation (7) can be cast in the mathematical form of a discrete Anderson nonlinear Schrödinger equation (DANSE) with randomness, e.g. [50, 51, 122],

$$i\hbar\frac{\partial}{\partial t}\psi_n = \widehat{H}_L\psi_n + \beta|\psi_n|^2\psi_n, \quad (9)$$

as shown in [38]. Localization of longitudinal and transverse electron plasma waves by disorder in uniform plasmas was investigated first in [123]. Here, the focus is on the competition between nonlinearity and randomness. In equation (9), $\widehat{H}_L$ is linear Hamiltonian in the tight binding approximation [124] and accounts for random transitions between nearest-neighbor states through a hopping matrix with disorder. In magnetized fusion plasmas, it corresponds to random shearing by ZS incoherent coupling. Meanwhile, β is the strength of nonlinearity and describes self-generated ZS interaction. The competition of nonlinearity versus randomness in equation (9) yields different behaviors depending on the sign and the value of β. For repulsive nonlinearity ($\beta > 0$), localization by disorder is destroyed and unlimited spreading is observed for $\beta > \beta_{\max} = 1/\ln 2 \simeq 1.4427$. This result, which is mathematically exact, is obtained through a topological approximation of the nonlinear Anderson problem in equation (9) using a mapping procedure on a Cayley tree (Bethe lattice) and percolation theory in Hilbert space [122, 125]. For attractive nonlinearity ($\beta < 0$), instead, soliton solutions are generally obtained [126] as in the typical case of ZS interacting with DW turbulence. Figure 2(*b*), taking plasma nonuniformity into account, illustrates turbulence spreading via soliton formation due to self-consistent DWT-ZS interaction [112]. There, r is expressed in units of ρ_s, i.e. ρ_{Li} computed at the electron temperature, while t is in units of the inverse diamagnetic frequency ω_*^{-1}.

5. Phase space zonal structures, phase locking and convective amplification

The nonlinear evolution of the EP distribution function, F_0, illustrated in figure 1(*a*) (see section 3), is characterized by the formation of 'phase space zonal structures' [38, 127–129], i.e. δf_0, which are the EP phase space counterpart of the ZS discussed in section 4 [111]. When the fluctuation spectrum consists of a predominant quasi-particle $\delta\widehat{\phi}_n$ (nearly periodic fluctuation), the evolution of F_0 is given by all possible emission and reabsorption processes, obtained by closing the quasi-particle paths in figure 1(*a*) into a loop and accounting for all possible loops as in the Dyson series shown in the upper part of figure 3(*a*), which correspond to the solution of the Dyson equation [38, 127, 128, 130]. This is the case, for example, of an externally forced-driven mode or of a single (most) unstable mode, spontaneously 'emitted' by F_0; thus, the nonlinear interaction of $\delta\widehat{\phi}_n$ with F_0 can be schematically depicted as in the lower part of figure 3(*a*). Focusing on one isolated resonance, the EP response is obtained from the universal description of a non-autonomous system with one degree of freedom [131]. Sufficiently close to marginal stability such that the fluctuation induced (nonlinear) EP displacement, Δr_{NL}, satisfies $\Delta r_{NL} \ll \lambda_\perp$, the system can be described by the 'bump-on-tail' paradigm, widely used by Berk, Breizman, and coworkers (see the review [132]). This paradigm adopts a 1D uniform beam-plasma system with sources and collisions for analyzing nonlinear behaviors of AEs near marginal stability [133] and has been used for interpretation of various experimental evidences of SAW collective instabilities excited by EPs in tokamak plasmas [132–136]. As $\Delta r_{NL} \lesssim \lambda_\perp$, however, self-consistent SAW/DAW nonlinear interactions with EPs, in general, are significantly influenced by nonuniformities in realistic plasmas [38, 80, 127, 128]. In fact, the EP response fully reflects the 3D geometries although it is still described as a non-autonomous system with one degree of freedom, as demonstrated also by numerical simulations [129, 137].

Near marginal stability and with perturbative EP dynamics, AEs evolve adiabatically and, when their frequency 'chirps', $|\dot{\omega}| \ll \omega_B^2$ and $|\dot{\omega}_B| \ll \omega_B^2$ [138, 139], with ω_B the wave-particle trapping frequency. Thus, there exists a conserved phase space action connected with wave-particle trapping and hole-clump pairs are formed [134–136, 138, 139] whose frequency slowly evolves by balancing the rate of energy extraction of the moving holes/clumps in the phase space with the fixed background dissipation. These processes are similar to the EP transport in 'buckets' introduced in [140]. They are also similar to 'autoresonance' [54, 141], by which a nonlinear pendulum can be driven to large amplitude, evolving in time to instantaneously match its frequency with that of an external drive with sufficiently slow downward frequency sweeping. Here, it is worthwhile noting that original theories on adiabatic frequency sweeping of holes/clumps [134–136, 138] have been significantly extended in [139, 142], taking into account the local TAE radial mode structure near one

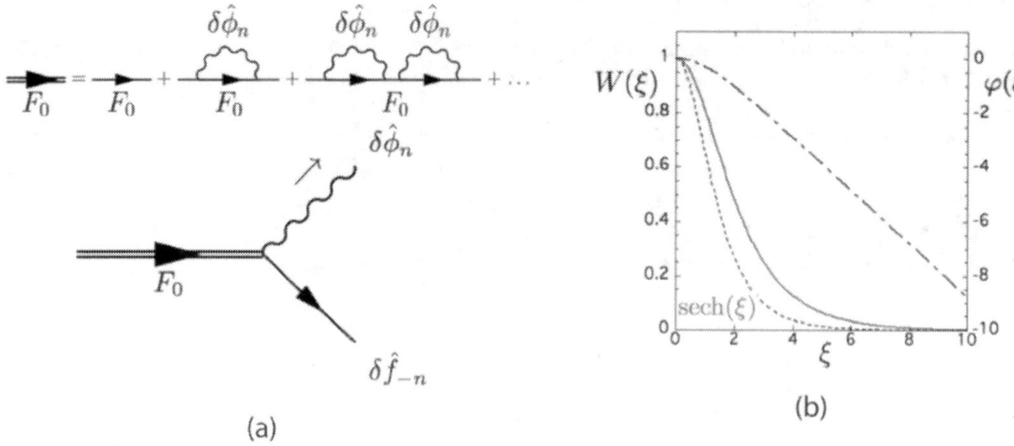

Figure 3. (*a*) Coherent nonlinear interaction of the EP distribution function F_0 with the quasi-particle $\delta\hat{\varphi}_n$. (*b*) Amplitude $W(\xi)$ (solid line) and phase $\varphi(\xi)$ (dash–dotted line) of the EPM soliton; sech(ξ) (dashed line) is shown as a reference for $W(\xi)$.

(radially) isolated gap in the SAW continuous frequency spectrum[5], but preserving the Ansatz of proximity to marginal stability and perturbative EP dynamics [142]. In particular, time evolution of the local TAE mode structure is demonstrated to be crucial for describing chirping events with nonlinear frequency shifts comparable with the distance of linear mode frequency from the SAW continuum accumulation point [142], consistent with the results of prior analyses reported in [80, 143] and of hybrid MHD-gyrokinetic simulations [76–79, 137]. In this way, it has been shown that the predicted chirping may be fast, i.e. non-adiabatic, $|\dot{\omega}| \lesssim \omega_B^2$; thereby challenging the self-consistency of assumptions made for the derivation of model equations in [139, 142]. These works, nonetheless, suggest that non-adiabatic chirping is naturally developed in nonlinear dynamics of phase space holes and clumps, as anticipated in other works [75, 80, 144][6]. Furthermore, [142] extended model equations predict the possible penetration of downward frequency sweeping clumps connected with chirping TAEs into the lower SAW continuum. There, as the mode structure evolves into that of an EPM [39, 40, 53, 74, 145] (see section 3 and further discussions hereafter), a non-perturbative treatment of EP nonlinear dynamics is, however, necessary in general. This can be understood, since the condition $\Delta r_{NL} \lesssim \lambda_\perp$ can be readily reached for EPs, unless the mode is close to marginal stability, while, at the same time, continuum damping is further increasing [38, 127, 128].

For strong drive and non-perturbative EPs, meanwhile, mode structure change and nonlinear frequency shift/chirping reflect EP transport, consistent with nonlinear change in D_n (GFLDR; see section 3) on the right hand side of equation (7). For increasing EP source strength, non-perturbative EP effects are visible on both AEs and EPMs [74, 76, 145, 146] and have been recently documented in a comparison of experimental measurements of TAE mode structures in DIII-D with numerical simulation results [147]. The effects of non-perturbative EP dynamics are particularly important for EPMs, which (see section 3) are excited at the resonant EP characteristic frequency [53] and are radially localized near the maximum of the resonant EP drive [74, 75]. Hence, they maximize wave-EP power exchange by preserving the resonance condition throughout the nonlinear evolution. That is, by 'phase locking' [38, 127–129, 137, 148, 149] EPM frequency adapts to local resonance condition, which can be shown to imply non-adiabatic frequency chirping, i.e.

$$\dot{\omega}_0 \simeq \delta\dot{\boldsymbol{X}}_\perp \cdot \nabla\, \omega_{\rm res} \sim \omega_B^2, \tag{10}$$

where $\delta\dot{\boldsymbol{X}}_\perp$ is the fluctuation induced EP velocity and $\omega_{\rm res}$ is the space-dependent resonance frequency. Equation (10) may be viewed as an expression of the 'intrinsic autoresonant character' of nonlinear EPM dynamics, unlike the usual 'autoresonance' case, where the drive frequency is controlled externally [54, 141]. Note that 'phase locking' is generally used in the description of 'autoresonant' nonlinear evolution of phase space holes/clumps to describe the instantaneous matching of the nonlinear oscillator frequency with that of the external drive [54, 141]. 'Phase locking' is also used in [142] to describe the slow (adiabatic) evolution of phase space holes/clumps that, by moving to a lower energy state, compensate the energy dissipation due to background damping. This confirms the analogy of theories on adiabatic frequency sweeping of holes/clumps [134–136, 138, 139, 142] with 'autoresonance' [54, 141], discussed above in this section. In both cases, the shortest nonlinear time scale is set by wave-particle trapping, with the conservation of the corresponding phase space action on the longer time scale. Equation (10), meanwhile, illuminates the qualitatively different nature of 'phase locking' for sufficiently strong instability drive and non-perturbative EP response. In this case, the wave-particle phase evolves slowly on the shortest nonlinear time scale and wave-particle trapping is *de facto* suppressed [38, 80, 128, 129, 137, 149].

Because of phase locking, the finite interaction length of EP with EPM is extended; thus, $\Delta r_{NL} \gtrsim \lambda_\perp$ and the EPM radial envelope evolves as a convectively amplified soliton in an active medium [38, 79, 80, 127]. Assuming the EPM mode structure is radially localized about $r_0(t)$ with

[5] Please see [39, 40] for a recent in depth discussion and review of the multiple radial scales characterizing Alfvénic fluctuations.

[6] Please see [38] for a recent in-depth review of this and other related issues.

$|\theta_{k0}| \ll 1$, the EPM frequency and growth rate at lowest order are obtained from equation (6) as $D_n^L(r_0, \omega_0) = 0$, i.e. the mode locks onto the local resonance frequency [53, 74, 75], consistent with equation (10). Then, by letting $A_n(\xi, t) = U(\xi)\exp[-i\int^t \Delta\omega(t')\,dt'] \equiv W(\xi)\exp[i\varphi(\xi) - i\int^t \Delta\omega(t')\,dt']$, with $\Delta\omega(t)$ the complex nonlinear frequency shift, $\xi \equiv k_{n0}(r - r_0(t)) \equiv k_{n0}\left(r - r_0(0) - \int_0^t v_g(t')\,dt'\right)$, k_{n0} denoting the nonlinear wave-vector, and v_g the nonlinear group velocity, equation (7) can be reduced to the following nonlinear Schrödinger equation [38, 127, 128]

$$\partial_\xi^2 U = \lambda_0 U - 2iU|U|^2. \tag{11}$$

Here, phase locking has been assumed, i.e. $|\dot{\omega}_0 - \delta\dot{X}_\perp \cdot \nabla\, \omega_{\rm res}|\tau_{NL}^2 < 1$. Furthermore, given $v_{E\times B,r}$, the EP radial $\boldsymbol{E}\times\boldsymbol{B}$ drift at the peak EPM amplitude, we have $v_g = \lambda_g v_{E\times B,r}$, with λ_g a parameter to be determined and $k_{n0} = \kappa\lambda_g^{-1}k_\theta$, with $\kappa = \mathcal{O}(1)$ a constant depending on local plasma equilibrium parameters [38, 127]. Thus, the product $k_{n0}v_g$ is fixed for a given EPM amplitude and local plasma equilibrium, but equation (7) allows us to compute a one-parameter (λ_g) family of EPM wave-packet solitons. These are shown in figure 3(*b*), illustrating the solution of equation (11) for $\lambda_0 \simeq -0.47 + i\,1.33$, corresponding to the ground state of the complex nonlinear oscillator. Meanwhile, the EPM frequency, as a solution of equations (6) and (7), is obtained from the condition

$$D_n^L(r_0, \omega_0) + P_n(r_0, \omega_0; \ldots)\,\lambda_0 / \lambda_g^2 = 0, \tag{12}$$

with $P_n(r_0, \omega_0; \ldots)$ as a known function also depending on local plasma equilibrium parameters [38, 127, 128]. Thus, $\omega_0 = \Omega\left(r_0, \lambda_0/\lambda_g^2; \ldots\right)$, with $\mathbb{R}e\lambda_0$ and $\mathbb{I}m\lambda_0$ accounting for, respectively, the (downward) EPM nonlinear frequency shift and the strengthening of mode drive, due to self-consistent EPM-EP interactions. Furthermore, the fastest growing EPM wave packet is obtained for λ_g satisfying $\partial\mathbb{I}m\Omega/\partial\lambda_g = 0$. The convective EPM wave packet amplification is locked onto the radial EP gradient steepening and, hence, it is an avalanche phenomenon [33, 34, 38, 80]. However, unlike SOC avalanches, which can be considered as chain reactions of local events with no characteristic scale, the EPM avalanche scale length is set by phase locking. Convective amplification continues until it is quenched by plasma nonuniformities [38, 127, 128], accounted for by the dependence on r_0, e.g. increased continuum damping, followed by radial EP gradient relaxation [79, 80]. Because of phase locking, EPs coherently emit EPM quasi-particles as in figure 3(*a*) over a finite cooperation length and then lose the resonance condition before they begin gaining energy back. This suggest similarities with the amplification of a short optical pulse in a free electron laser (FEL) [57] as, similar to a superradiant FEL pulse, the intensity of EPM wave packet grows as $[r_0(t) - r_0(0)]^2$ [80].

6. Conclusions and discussion

In the recent years, significant progress has been made in understanding the individual nonlinear processes that are important in fusion plasmas; the corresponding spatiotemporal structures are reasonably well understood. One of the most significant open problems is how the interplay of these individual processes determine long time scale behaviors. Mutual positive feedback between theory, simulation, and experiment are necessary for further progress.

In this work, we have presented and discussed the multi-scale dynamics and complex behaviors connected with EPs in burning plasmas. Wave–wave and wave-particle nonlinear interactions with EPs are the two routes that yield saturation of SAW/DAW instabilities and cause a variety of fluctuation induced transport processes. Here, we have introduced a conceptual framework that treats such processes as general interactions of quasi-particle and particles and sheds light onto the mechanisms that may cause non-local behaviors. In particular, due to the existence of the SAW continuous spectrum and the properties of the Alfvénic state, the role of EPs in fusion plasmas is unique as they could act as mediators of cross-scale couplings, ranging from the system size macro-scales to the micro-scales of the thermal ion Larmor radius. Important examples of these behaviors are the impact on plasma transport of parametric decays of EP driven 'mode converted KAWs'; 'alpha channeling' as 'diversion of energy from energetic alpha particles to waves'; and the more recently proposed 'spatial channeling', by which an 'action at a distance' is possible, with KAW transferring energy and/or momentum between two spatially separated regions. Furthermore, similar to DWT and Alfvén waves, EPs may generate zonal flows and fields and their resonant structure counterpart in the EP phase space. It is suggestive that these meso-scale dynamics may influence the long time scale behaviors of fusion plasmas and the cross-scale couplings between DWT and EP driven SAW/DAW. However, addressing these issues requires accounting for additional physics not discussed in this work, e.g. collisions and finite magnetic compression effects. Here, we emphasize again that effects of plasma nonuniformities and realistic equilibrium magnetic field geometries are crucially important for predicting SAW/DAW nonlinear dynamics and ensuing EP transport in conditions of practical interest. This occurs because the nonlinear displacement of resonant EPs becomes comparable with the perpendicular fluctuation wavelength for sufficiently strong drive. The interplay of SAW/DAW instability mode structures with non-perturbative EP dynamic response also underlies rapid (non-adiabatic) frequency sweeping of phase space zonal structures. Because of phase locking, i.e. the preservation of wave-particle resonance through the nonlinear evolution phase, fast frequency chirping is characterized and accompanied by secular radial motion of a 'steepened' pressure profile, consistent with the changing SAW/DAW radial structures. Such a meso-scale 'avalanche' could, depending on the plasma nonuniformity and equilibrium magnetic field geometry, redistribute EPs up to the system macro-scales.

The general problems posed by investigations of burning plasma physics often have broader applications than just fusion science. Here, we have given examples of how fusion plasmas may yield a wide class of nonlinear problems, which may be readily extended to neighboring fields, e.g. condensed matter and accelerator physics, nonlinear dynamics, and plasma astrophysics.

Acknowledgments

This work was supported by the European Union's Horizon 2020 Research and Innovation Program under grant agreement number 633053 as Enabling Research Project CfP-WP14-ER-01/ENEA_Frascati-01 and by the US DoE, ITER-CN No. 2013GB104004, and NSFC No. 11235009 grants. Useful discussions with R Dumont, X Garbet, and D Zarzoso are kindly acknowledged. We are also grateful to Z Guo for granting his permission of reproducing one figure from his original work [112].

References

[1] Kadomtsev B B 1993 *Tokamak Plasma: a Complex Physical System* (Boca Raton, FL: CRC Press)
[2] Newman D E, Carreras B A, Diamond P H and Hahm T S 1996 *Phys. Plasmas* **3** 1858
[3] Carreras B A, Newman D, Lynch V E and Diamond P H 1996 *Phys. Plasmas* **3** 2903
[4] Dendy R O and Helander P 1997 *Plasma Phys. Control. Fusion* **39** 1947
[5] Bak P 1996 *How Nature Works: the Science of Self-Organised Criticality* (New York: Copernicus Press)
[6] Badii R and Politi A 1999 *Complexity: Hierarchical Structures and Scaling in Physics* (Cambridge: Cambridge University Press)
[7] Sornette D 2000 *Critical Phenomena in Natural Sciences* (Heidelberg: Springer)
[8] Johnson N 2009 *Simply Complexity: a Clear Guide to Complexity Theory* (Oxford: Oneworld)
[9] Bak P, Tang C and Wiesenfeld K 1987 *Phys. Rev. Lett.* **59** 381
[10] Dendy R O and Chapman S C 2006 *Plasma Phys. Control. Fusion* **48** B313
[11] Dendy R O, Chapman S C and Paczuski M 2007 *Plasma Phys. Control. Fusion* **49** A95
[12] Milovanov A V 2010 *Europhys. Lett.* **89** 60004
[13] Milovanov A V 2011 *New J. Phys.* **13** 043034
[14] Chen L, White R B and Rosenbluth M N 1984 *Phys. Rev. Lett.* **52** 1122
[15] Hazeltine R D and Meiss J D 1992 *Plasma Confinement* (Redwood City, CA: Addison-Wesley)
[16] Itoh K, Itoh S-I and Fukuyama A 1999 *Transport and Structural Formation in Plasmas* (Bristol: Institute of Physics Publishing)
[17] Balescu R 2005 *Aspects of Anomalous Transport in Plasmas* (Bristol: Institute of Physics Publishing)
[18] Spatschek K H 2008 *Plasma Phys. Control. Fusion* **50** 124027
[19] Diamond P H, Itoh S-I and Itoh K 2010 *Modern Plasma Physics. Volume 1. Physical Kinetics of Turbulent Plasmas* (Cambridge, UK: Cambridge University Press)
[20] Sánchez R, van Milligen B and Carreras B A 2005 *Phys. Plasmas* **12** 056105
[21] Dif-Pradalier G, Diamond P H, Grandgirard V, Sarazin Y, Abiteboul J, Garbet X, Ghendrih Ph, Strugarek A, Ku S and Chang C S 2010 *Phys. Rev.* E **82** 025401
[22] Milovanov A V and Rasmussen J J 2014 *Phys. Lett.* A **378** 1492
[23] Hasegawa A, Maclennan C G and Kodama Y 1979 *Phys. Fluids* **22** 2122
[24] Lin Z, Hahm T S, Lee W W, Tang W M and White R B 1998 *Science* **281** 1835
[25] Diamond P H, Itoh S-I, Itoh K and Hahm T S 2005 *Plasma Phys. Control. Fusion* **47** R35
[26] Itoh K, Itoh S-I, Diamond P H, Hahm T S, Fujisawa A, Tynan G R, Yagi M and Nagashima Y 2006 *Phys. Plasmas* **13** 055502
[27] Garbet X, Laurent L, Samain A and Chinardet J 1994 *Nucl. Fusion* **34** 963
[28] Diamond P H, Lebedev V B, Newman D E and Carreras B A 1995 *Phys. Plasmas* **2** 3685
[29] Diamond P H and Hahm T S 1995 *Phys. Plasmas* **2** 3640
[30] Gürcan Ö D *et al* 2013 *Nucl. Fusion* **53** 073029
[31] Wang W X, Hahm T S, Lee W W, Rewoldt G, Manickam J and Tang W M 2007 *Phys. Plasmas* **14** 072306
[32] Jolliet S and Idomura Y 2012 *Nucl. Fusion* **52** 023026
[33] Zonca F, Briguglio S, Chen L, Fogaccia G, Hahm T S, Milovanov A V and Vlad G 2006 *Plasma Phys. Control. Fusion* **48** B15
[34] Chen L and Zonca F 2007 *Nucl. Fusion* **47** S727
[35] Fasoli A *et al* 2007 *Nucl. Fusion* **47** S264
[36] Zonca F and Chen L 2008 Nonlinear dynamics and complex behaviors in magnetized plasmas of fusion interest *Frontiers in Modern Plasma Physics* vol CP1061, Shukla P K *et al* (Melville, NY: American Institute of Physics) p 34
[37] Zonca F 2008 *Int. J. Mod. Phys.* A **23** 1165
[38] Chen L and Zonca F 2014 Physics of Alfvén waves and energetic particles in burning plasmas *Rev. Mod. Phys.* submitted
[39] Zonca F and Chen L 2014 *Phys. Plasmas* **21** 072120
[40] Zonca F and Chen L 2014 *Phys. Plasmas* **21** 072121
[41] Alfvén H 1942 *Nature* **150** 405
[42] Walén C 1944 *Ark. Mat. Astron. Fys.* **30A** 1
[43] Alfvén H 1950 *Cosmical Electrodynamics* (Oxford, UK: Clarendon)
[44] Chen L and Zonca F 2012 *Phys. Rev. Lett.* **109** 145002
[45] Chen L and Zonca F 2013 *Phys. Plasmas* **20** 055402
[46] Qiu Z, Chen L and Zonca F 2013 *Europhys. Lett.* **101** 35001
[47] Zarzoso D *et al* 2013 *Phys. Rev. Lett.* **110** 125002
[48] Dumont R J *et al* 2013 *Plasma Phys. Control. Fusion* **55** 124012
[49] Shepelyansky D L 1993 *Phys. Rev. Lett.* **70** 1787
[50] Pikovsky A S and Shepelyansky D L 2008 *Phys. Rev. Lett.* **100** 094101
[51] Iomin A 2010 *Phys. Rev.* E **81** 017601
[52] Krivolapov Y, Fishman S and Soffer A 2010 *New J. Phys.* **12** 063035
[53] Chen L 1994 *Phys. Plasmas* **1** 1519
[54] Fajans J and Friedland L 2001 *Am. J. Phys.* **69** 1096
[55] Dicke R H 1954 *Phys. Rev.* **93** 99
[56] Zonca F and Chen L 2013 *Bull. Am. Phys. Soc.* **58** 75
[57] Bonifacio R, De Salvo L, Pierini P and Piovella N 1990 *Nucl. Instrum. Methods Phys. Res.* A **296** 358
[58] Lu Z X, Zonca F and Cardinali A 2012 *Phys. Plasmas* **19** 042104
[59] Connor J W, Hastie R J and Taylor J B 1978 *Phys. Rev. Lett.* **40** 396
[60] Pegoraro F and Schep T J 1978 *Plasma Physics and Controlled Nuclear Fusion Research* vol 1 (Vienna: International Atomic Energy Agency) p 507
[61] Chen L 2008 *Plasma Phys. Control. Fusion* **50** 124001
[62] Chen L, Zonca F and Lin Z 2005 *Plasma Phys. Control. Fusion* **47** B71
[63] Lin Z, Chen L and Zonca F 2005 *Phys. Plasmas* **12** 056125
[64] Chen L, Lin Z and White R B 2000 *Phys. Plasmas* **7** 3129
[65] Chen L, White R B and Zonca F 2004 *Phys. Rev. Lett.* **92** 075004
[66] Zonca F, White R B and Chen L 2004 *Phys. Plasmas* **11** 2488
[67] Heidbrink W W 2002 *Phys. Plasmas* **9** 2113
[68] Cheng C Z, Chen L and Chance M S 1985 *Ann. Phys.* **161** 21
[69] Grad H 1969 *Phys. Today* **32** 34

[70] Hasegawa H and Chen L 1974 *Phys. Rev. Lett.* **32** 454
[71] Chen L and Hasegawa A 1974 *Phys. Fluids* **17** 1399
[72] Wong K L *et al* 1991 *Phys. Rev. Lett.* **66** 1874
[73] Heidbrink W W, Strait E J, Doyle E and Snider R 1991 *Nucl. Fusion* **31** 1635
[74] Zonca F and Chen L 1996 *Phys. Plasmas* **3** 323
[75] Zonca F and Chen L 2000 *Phys. Plasmas* **7** 4600
[76] Briguglio S, Zonca F and Vlad G 1998 *Phys. Plasmas* **5** 3287
[77] Briguglio S, Vlad G, Zonca F and Fogaccia G 2002 *Phys. Lett.* A **302** 308
[78] Zonca F, Briguglio S, Chen L, Dettrick S, Fogaccia G, Testa D and Vlad G 2002 *Phys. Plasmas* **9** 4939
[79] Vlad G, Briguglio S, Fogaccia G and Zonca F 2004 *Plasma Phys. Control. Fusion* **46** S81
[80] Zonca F, Briguglio S, Chen L, Fogaccia G and Vlad G 2005 *Nucl. Fusion* **45** 477
[81] Dewar R L, Manickam J, Grimm R C and Chance M S 1981 *Nucl. Fusion* **21** 493
[82] Dewar R L, Manickam J, Grimm R C and Chance M S 1982 *Nucl. Fusion* **22** 307
[83] Kruskal M D and Oberman C R 1958 *Phys. Fluids* **1** 275
[84] Rosenbluth M N and Rostoker N 1959 *Phys. Fluids* **2** 23
[85] Antonsen T M, Lane B and Ramos J J 1981 *Phys. Fluids* **24** 1465
[86] Antonsen T M and Lee Y C 1982 *Phys. Fluids* **25** 132
[87] Van Dam J W, Rosenbluth M N and Lee Y C 1982 *Phys. Fluids* **25** 1349
[88] Mikhailovskii A B and Rudakov L I 1963 *Sov. Phys.—JETP* **17** 621
[89] Hasegawa H and Chen L 1976 *Phys. Fluids* **19** 1924
[90] Tang J T and Luhmann N C Jr 1976 *Phys. Fluids* **19** 1935
[91] Chen L, Hsu J Y, Kaw P K and Rutherford P H 1978 *Nucl. Fusion* **18** 1371
[92] Tang W M, Connor J W and Hastie R J 1980 *Nucl. Fusion* **20** 1439
[93] Sagdeev R Z and Galeev A A 1969 *Nonlinear Plasma Theory* (New York: W A Benjamin)
[94] Chen L and Zonca F 2011 *Europhys. Lett.* **96** 35001
[95] Lin Y, Johnson J R and Wang X 2012 *Phys. Rev. Lett.* **109** 125003
[96] Qiu Z, Zonca F and Chen L 2012 *Phys. Plasmas* **19** 082507
[97] Winsor N, Johnson J L and Dawson J M 1968 *Phys. Fluids* **11** 2448
[98] Fredrickson E D *et al* 2009 *Phys. Plasmas* **16** 122505
[99] Podestà M *et al* 2009 *Phys. Plasmas* **16** 056104
[100] Podestà M, Bell R E, Crocker N A, Fredrickson E D, Gorelenkov N N, Heidbrink W W, Kubota S, LeBlanc B P and Yuh H 2011 *Nucl. Fusion* **51** 063035
[101] Podestà M *et al* 2012 *Nucl. Fusion* **52** 094001
[102] Fisch N J and Rax J M 1992 *Phys. Rev. Lett.* **69** 612
[103] Fisch N 2000 *Nucl. Fusion* **40** 1095
[104] Chen L, Vaclavik J and Hammett G W 1988 *Nucl. Fusion* **28** 389
[105] Jaun A, Fasoli A and Heidbrink W W 1998 *Phys. Plasmas* **5** 2952
[106] Jaun A, Fasoli A, Vaclavik J and Villard L 2000 *Nucl. Fusion* **40** 1343
[107] Kolesnichenko Ya I, Yakovenko Yu V, Weller A, Werner A, Geiger J, Lutsenko V V and Zegenhagen S 2005 *Phys. Rev. Lett.* **94** 165004
[108] Weller A, Anton M, Geiger J, Hirsch M, Jaenicke R, Werner A, W7-AS Team, Nührenberg C, Sallander E and Spong D A 2001 *Phys. Plasmas* **8** 931
[109] Kolesnichenko Ya I, Yakovenko Yu V, Lutsenko V V, White R B and Weller A 2010 *Nucl. Fusion* **50** 084011
[110] Kolesnichenko Ya I, Yakovenko Yu V and Lutsenko V V 2010 *Phys. Rev. Lett.* **104** 075001
[111] Chen L and Zonca F 2007 *Nucl. Fusion* **47** 886
[112] Guo Z, Chen L and Zonca F 2009 *Phys. Rev. Lett.* **103** 055002
[113] Qiu Z, Chen L and Zonca F 2014 *JPS Conf. Proc.* **1** 015007
[114] Berk H L, Boswell C J, Borba D, Figueiredo A C A, Johnson T, Nave M F F, Pinches S D, Sharapov S E and JET EFDA Contributors 2006 *Nucl. Fusion* **46** S888
[115] Fu G Y 2008 *Phys. Rev. Lett.* **101** 185002
[116] Zonca F and Chen L 2008 *Europhys. Lett.* **83** 35001
[117] Qiu Z, Chen L and Zonca F 2014 *Phys. Plasmas* **21** 022304
[118] Qiu Z, Chen L and Zonca F 2014 *Nucl. Fusion* **54** 033010
[119] Qiu Z, Chen L and Zonca F 2014 *Proc. of the 41.st EPS Conf. on Plasma Physics (Berlin, Germany, 23–27 June 2014)* vol 38F, Paper No. P4.004
[120] Chen L, Lin Z, White R B and Zonca F 2001 *Nucl. Fusion* **41** 747
[121] Guzdar P N, Kleva R G and Chen L 2001 *Phys. Plasmas* **8** 459
[122] Milovanov A V and Iomin A 2014 *Phys. Rev.* E **89** 062921
[123] Escande D F and Souillard B 1984 *Phys. Rev. Lett.* **52** 1296
[124] Anderson P W 1958 *Phys. Rev.* **109** 1492
[125] Milovanov A V and Iomin A 2012 *Europhys. Lett.* **100** 10006
[126] Zelenyi L M and Milovanov A V 2004 *Phys.—Usp.* **47** 749
[127] Zonca F, Briguglio S, Chen L, Fogaccia G, Vlad G and Wang X 2013 Nonlinear dynamics of phase-space zonal structures and energetic particle physics *Proc. of the 6th IAEA Technical Meeting on 'Theory of Plasmas Instabilities' (Vienna, Austria, 27–29 May 2013)*
[128] Zonca F, Chen L, Briguglio S, Fogaccia G, Vlad G and Wang X 2014 Nonlinear dynamics of phase-space zonal structures and energetic particle physics in fusion plasmas *New J. Phys.* submitted
[129] Briguglio S, Wang X, Zonca F, Vlad G, Fogaccia G, Di Troia C and Fusco V 2014 *Phys. Plasmas* **21** 112301
[130] Al'tshul' L M and Karpman V I 1966 *Sov. Phys.–JETP* **22** 361
[131] Chirikov B V 1979 *Phys. Rep.* **52** 263
[132] Breizman B N and Sharapov S E 2011 *Plasma Phys. Control. Fusion* **53** 054001
[133] Berk H L and Breizman B N 1990 *Phys. Fluids* B **2** 2246
[134] Berk H L, Breizman B N and Pekker M 1996 *Phys. Rev. Lett.* **76** 1256
[135] Berk H L, Breizman B N and Petiashvili N V 1997 *Phys. Lett.* A **234** 213
[136] Breizman B N, Berk H L, Pekker M, Porcelli F, Stupakov G V and Wong K L 1997 *Phys. Plasmas* **4** 1559
[137] Wang X, Briguglio S, Chen L, Di Troia C, Fogaccia G, Vlad G and Zonca F 2012 *Phys. Rev.* E **86** 045401
[138] Berk H L, Breizman B N, Candy J, Pekker M and Petiashvili N V 1999 *Phys. Plasmas* **6** 3102
[139] Wang G and Berk H L 2012 *Nucl. Fusion* **52** 094003
[140] Mynick H E and Pomphrey N 1994 *Nucl. Fusion* **34** 1277
[141] Meerson B and Friedland L 1990 *Phys. Rev.* A **41** 5233
[142] Wang G 2013 *PhD Thesis* University of Texas, Austin, TX
[143] Zonca F, Briguglio S, Chen L, Fogaccia G and Vlad G 2000 Theoretical aspects of collective mode excitations by energetic ions in tokamaks *Theory of Fusion Plasmas* Connor J W *et al* (Bologna: SIF) p 17
[144] Gorelenkov N N *et al* 2000 *Nucl. Fusion* **40** 1311
[145] Chen L and Zonca F 1995 *Phys. Scr.* T **60** 81
[146] Briguglio S, Vlad G, Zonca F and Kar C 1995 *Phys. Plasmas* **2** 3711
[147] Wang Z, Lin Z, Holod I, Heidbrink W W, Tobias B, Van Zeeland M and Austin M E 2013 *Phys. Rev. Lett.* **111** 145003
[148] White R B, Goldston R J, McGuire K, Boozer A H, Monticello D A and Park W 1983 *Phys. Fluids* **26** 2958
[149] Vlad G, Briguglio S, Fogaccia G, Zonca F, Fusco V and Wang X 2013 *Nucl. Fusion* **53** 083008

IOP Publishing
Plasma Physics and Controlled Fusion
Plasma Phys. Control. Fusion **57** (2015) 014025 (8pp)
doi:10.1088/0741-3335/57/1/014025

H-mode transitions and limit cycle oscillations from mean field transport equations

Gary M Staebler and R J Groebner

General Atomics, PO Box 85608, San Diego, CA 92186–5608, USA

E-mail: staebler@fusion.gat.com

Received 27 June 2014, revised 24 September 2014
Accepted for publication 1 October 2014
Published 28 November 2014

Abstract

The mean field toroidal and parallel momentum transport equations will be shown to admit both one-step transitions to suppressed transport (L/H) and limit cycle oscillations (LCO). Both types of transitions are driven by the suppression of turbulence by the mean field ExB velocity shear. Using experimental data to evaluate the coefficients of a reduced transport model, the observed frequency of the LCO can be matched. The increase in the H-mode power threshold above and below a minimum density agrees with the trends in the model. Both leading and lagging phase relations between the turbulent density fluctuation amplitude and the ExB velocity shear can occur depending on the evolution of the linear growth rate of the turbulence. The transport solutions match the initial phase of the L/H transition where the poloidal and ExB velocities are observed to change, and the density fluctuations drop, faster than the diamagnetic velocity.

Keywords: mean field transport, limit cycle oscillations, H-mode transitions

(Some figures may appear in colour only in the online journal)

1. Introduction

The focus of this paper is to show that the mean field momentum transport equations have both one-step L/H transitions and limit cycle oscillations (LCO) or dithering transitions. A simplified model, based on the properties of gyrokinetic turbulence, is used to demonstrate the two types of solutions. The data from the L-mode phase of a DIII-D [1] tokamak discharge 140426 at 1 265 ms at the major radius of $R = 2.265$ m (toroidal flux surface $\rho = 0.98$) prior to the LCO will be used to evaluate the model parameters as much as is possible. This will be referred to as the reference data. A summary of the data used is given in table 1. Extensive turbulence measurements during the LCO phase have been made for this discharge and similar ones and published in [2]. The calculated velocities and density fluctuation amplitudes in this paper can be compared with the measurements in [2] by the reader. The reasonable agreement with the data achieved with such a simple model verifies that the mean field transport equations contain the right physics to simulate the LCO and L/H transitions. The simple model also shows trends in qualitative agreement with the power threshold scaling with density. Oscillating solutions of the toroidal and parallel (or poloidal) momentum transport solutions have been observed in previous phenomenological modeling of internal transport barrier dynamics [3, 4]. Analysis of the LCO solutions in this paper identify the cause of the oscillations. It is well established that ExB velocity shear supresses turbulence and improves transport in all channels: energy, particle, momentum [5]. One outstanding puzzle has been the observation that the velocity of the plasma changes first and faster than the pressure or density profiles [6]. This is known as the L/H 'trigger'. In this paper it will be shown that the mean field momentum transport provides this fast change in velocity that triggers the suppressed turbulence. Particle and energy transport would also be improved but will not be included in the modeling in order to clearly demonstrate that the oscillation is due to momentum transport physics. This work establishes that the mean field transport equations can, in principle be a predictive theory of the L/H transition, given an accurate model of the turbulence driven fluxes in all channels. There are many physical effects that can be important in momentum transport that will not be considered in this work. In particular, the ion orbit loss cone at the separatrix [7, 8] and the reduction of the poloidal damping

Table 1. Local data for DIII-D discharge 140426 at 1 265 ms and $\rho = 0.98$.

n_e	T_e	T_i	L_{ne}	L_{Te}	L_{Ti}	χ_{tot}	c_s
$1.7 \times 10^{19}\,m^{-3}$	64 eV	150 eV	9.26 cm	1.37 cm	4.16 cm	$10.3\,m^2 s^{-1}$	$5.5 \times 10^4\,m\,s^{-1}$
R	r	R_{ref}	B_{ref}	q	B_z	B_{unit}	ρ_s
2.265 m	58.0 cm	1.68 m	−2.01 T	4.5	−0.34 T	−2.93 T	0.039 cm
c_{par}	c_{per}	c_{tor}	ν_{exch}	ν_i	ν_{pol}	Ω_s	ν_i^*
0.502	0.663	1.063	$684\,s^{-1}$	$4790\,s^{-1}$	$632\,s^{-1}$	$6.6 \times 10^7\,s^{-1}$	4.5

term at high mach number [7, 8] are neglected. The turbulent momentum transport and poloidal damping by Coulomb collisions are assumed to be the dominant contributions.

2. Mean fields and fluctuations

It is important to verify that there is sufficient time scale separation to define the mean fields and fluctuations. The mean fields are defined as an ensemble average of the total fields involving both space and time averages [9]. For example the ensemble average $\langle \langle A \rangle \rangle$ of some field A (i.e. electric, magnetic or particle distribution) could be a magnetic flux surface average $\langle A \rangle$ followed by a moving boxcar time and radius average with averaging time τ and length L. The length and time scales of the averaging are chosen to give separation $A = A_0 + \widetilde{A}$ between the fluctuating $\widetilde{A}$ and mean fields A_0. The fluctuations are annihilated by the ensemble average $\langle \langle \widetilde{A} \rangle \rangle = 0$. In general, the time average has to be longer than the decorrelation time of the turbulence. For the experimental conditions of the reference L-mode plasma prior to the onset of the LCO (table 1) the measured decorrelation time of the density fluctuations was $2\,\mu s$ [2]. The LCO period for the reference data was $625\,\mu s$ and the fastest timescale (fluctuation drop) of the L/H transition is about $100\,\mu s$ [10]. The solutions to the momentum transport equations will be shown to reproduce both the LCO and L/H timescales so, for the purpose of separating fluctuations and mean fields, a $100\,\mu s$ time average is optimal. This is 50 times the decorrelation time so there is enough time separation to average out the fluctuations. The Doppler shift velocity of the turbulence measured with Doppler backscatter (DBS) is averaged over $100\,\mu s$ [2]. Thus, the Doppler shift velocity measured by DBS should be considered the mean field ExB velocity contribution that can be compared with the transport solution for the ExB velocity. The term zonal flows has often been used in the literature to refer to the ExB velocity due to an axisymmetric radial electric field without separation of timescales. Using the ensemble average separates *zonal fluctuations* from the mean field ExB velocity. The zonal fluctuations, are governed by the gyrokinetic equation on the decorrelation timescale that is far faster than the observed LCO oscillations and are filtered out of the time average DBS data.

The separation of faster fluctuations from slower mean fields is the basis for traditional ordering expansions. Systematic derivations of the gyrokinetic equation (9) for the fluctuations and transport equations for the mean fields, including both turbulence driven fluxes and Coulomb collision terms, have been published [11–13]. Because of the conservation of total momentum (summed over all species a = ions, electrons) by the coulomb collision operator there are no Coulomb force terms in either the total toroidal or parallel momentum transport equations [14]. Hence, the parallel momentum balance equation will be the most susceptible to modification by turbulence from the neoclassical result [13]. The total toroidal and parallel to the magnetic field ($\vec{B}$) momentum transport equations in the right handed toroidal coordinates (R, φ, Z) averaged over a magnetic flux surface are [13]

$$\sum_a \left[\frac{\partial}{\partial t} \langle m_a n_a R^2 \vec{\nabla}\varphi \cdot \vec{u}_a \rangle + \left\langle R^2 \vec{\nabla}\varphi \cdot \left(\vec{\nabla} \cdot \vec{\Pi}_a \right) \right\rangle \right] = \sum_a \langle R^2 \vec{\nabla}\varphi \cdot \vec{S}_a \rangle, \quad (1)$$

$$\sum_a \left[\frac{\partial}{\partial t} \langle m_a n_a \vec{B} \cdot \vec{u}_a \rangle + \left\langle \vec{B} \cdot \left(\vec{\nabla} \cdot \vec{\Pi}_a \right) \right\rangle \right] = \sum_a \langle \vec{B} \cdot \vec{S}_a \rangle . \quad (2)$$

The mean velocity is $\vec{u}_a$, the mean density is n_a and the mass is m_a. The external momentum sources are $\langle \vec{S}_a \rangle$. There are, in general, turbulent and collisional contributions to the stress tensor ($\vec{\Pi}_a = \vec{\Pi}_a^{turb} + \vec{\Pi}_a^{col}$). The importance of the parallel momentum transport equation for H-mode was first emphasized by Rozhanski and Tendler [15].

The 1st order in ρ_s / L mean field species velocities are within flux surfaces and incompressible [13]. For low mach number they can be written in terms of two flux functions (K_a, Ω_a): $\vec{u}_a = \vec{B} K_a + R^2 \vec{\nabla}\varphi \Omega_a$. The flux surface average velocities are

$$\langle R^2 \vec{\nabla}\varphi \cdot \vec{u}_a \rangle / R_{ref} = c_{per} u_{a,pol} + c_{tor} (u_{ExB} + u_{a,dia}), \quad (3)$$

$$\langle \vec{B} \cdot \vec{u}_a \rangle / B_{ref} = c_{par} u_{a,pol} + c_{per} (u_{ExB} + u_{a,dia}), \quad (4)$$

where $c_{per} = \langle R^2 \vec{\nabla}\varphi \cdot \vec{B} \rangle / (R_{ref} B_{ref})$, $c_{par} = \langle B^2 \rangle / B_{ref}^2$, $c_{tor} = \langle R^2 \rangle / R_{ref}^2$, $u_{a,pol} = B_{ref} K_a$, $u_{ExB} = R_{ref} \vec{\nabla}\varphi \cdot \vec{E} \times \vec{B} / B^2$, $u_{a,dia} = -(R_{ref} / e_a n_a) \vec{\nabla}\varphi \cdot \vec{\nabla} P_a \times \vec{B} / B^2$, $u_{ExB} + u_{a,dia} = R_{ref} \Omega_a$. The reference magnetic field B_{ref} is the vacuum field at the geometric center of the vacuum chamber at major radius R_{ref} (table 1). There are three mean field velocities for the main ions $(u_{i,pol}, u_{ExB}, u_{i,dia})$. The diamagnetic velocity is determined by the gradient of the pressure Pa and hence particle and energy transport. The ExB velocity is the same for all species. The total parallel and toroidal momentum transport equations [equations (1) and (2)] solve for the ExB velocity and a poloidal velocity common to all species [14]. The neoclassical equations are used to compute all of the poloidal velocity *differences* that contribute to the collision operator. For a pure plasma, this is just the poloidal (bootstrap) current. This is the minimal modification of neoclassical theory due to turbulence [13]. Note that $u_{i,\text{pol}}$ is related to the vertical velocity $v_{i,z}$ that is usually the measured 'poloidal' velocity by $v_{i,z} = \vec{u}_i \cdot \vec{\nabla} Z = u_{i,\text{pol}} B_z / B_{\text{ref}}$. Charge exchange recombination spectroscopy [6] can measure the mean poloidal velocity. The turbulence based methods [16–18] measure the time averaged Doppler shifted phase velocity of the turbulence. Only the mean ExB toroidal velocity contributes to these measurements even if the measurement is of the time delay in the poloidal direction [17, 18]. The mean poloidal velocity is parallel to the magnetic field and hence contributes very little to the Doppler shift due to the small parallel wavenumbers of the gyrokinetic turbulence compared to the perpendicular wavenumbers. The ExB velocity that is measured by the Doppler shifted phase velocity is $v_{\text{ExB}} = u_{\text{ExB}}\, r / (R_{\text{ref}} q)$. Hence the '$u$' velocities are approximately toroidal projections and the 'v' velocities are approximately perpendicular projections. The poloidal velocity shear only contributes to the parallel velocity shear in the gyrokinetic equations and this is a destabilizing term [19]. Only the shear in the ExB mean velocity Doppler shift is stabilizing to the turbulence apart from 'profile shear' effects that cause radial variations in the eigenmode frequencies [20]. Since the mean ExB velocity is purely toroidal, it also contributes to the destabilizing parallel velocity shear. The mean diamagnetic torodial velocity contributes to the phase velocity of the turbulence through its frequency. It also contributes to the parallel velocity.

3. L/H and LCO solutions

In order to make analysis of the solutions easier the following simplifications are made. A pure plasma is assumed with a single ion species (suppressed species index). The electron inertia and stress are neglected due to the small electron mass. The ion density (n) and the geometry coefficients are taken to be constant. Radial derivatives of only the velocities are retained. The momentum transport equations [equations (1) and (2)] are reduced to:

$$\frac{\partial}{\partial t} mn \left[c_{\text{per}} u_{\text{pol}} + c_{\text{tor}} \left(u_{\text{ExB}} + u_{\text{dia}} \right) \right] + \frac{\partial \Pi_{\text{tor}}}{\partial r} = S_{\text{tor}}, \tag{5}$$

$$\frac{\partial}{\partial t} mn \left[c_{\text{par}} u_{\text{pol}} + c_{\text{per}} \left(u_{\text{ExB}} + u_{\text{dia}} \right) \right] + \frac{\partial \Pi_{\text{par}}}{\partial r} + mn\nu_{\text{pol}} \left(u_{\text{pol}} - u_{\text{pol}}^{\text{neo}} \right) = S_{\text{par}}. \tag{6}$$

The neoclassical parallel collisional stress yields the poloidal damping terms in equation (6). The poloidal damping rate $\left(\nu_{\text{pol}} = \mu_{11} / m_i n_i\right)$ is determined from the NCLASS [21] code coefficient (μ_{11}). The poloidal damping rate ν_{pol} has a complex dependence on the ion-ion collision frequency ν_i [14]. At low collision frequency, in the banana regime, the poloidal damping rate is proportional to collision frequency. At high ν_i, in the Pfirsch-Schluter regime, the poloidal damping rate decreases with collision frequency $\nu_{\text{pol}} \propto \omega_t^2 / \nu_i$ where $\omega_t = \sqrt{T_i / m_i} / (Rq)$ is the ion transit frequency. The L-mode reference data (table 1) is in the Pfirsch-Schluter regime ($\nu^* = \nu_i / (\omega_t (r/R)^{3/4}) = 4.5$) and the poloidal damping rate is decreasing as the separatrix is approached (increasing ν_i).

The most natural two independent velocity variables for the turbulent stresses are parallel and ExB toroidal ($u_{\text{par}}, u_{\text{ExB}}$), where $u_{\text{par}} = \langle \vec{B} \cdot \vec{u} \rangle / c_{\text{per}} B_{\text{ref}}$. The poloidal velocity is eliminated using $u_{\text{pol}} = (u_{\text{par}} - u_{\text{ExB}} - u_{\text{dia}})\, c_{\text{per}} / c_{\text{par}}$ and $a_{\text{per}} = c_{\text{tor}} c_{\text{par}} / c_{\text{per}}^2 - 1$. Motivated by gyrokinetic turbulence results [20] the stresses are taken to have the form:

$$\begin{aligned} \Pi_{\text{tor}}^{\text{turb}} &= \frac{c_{\text{per}}^2}{c_{\text{par}}} mn \left\{ -\Phi^2 d_{\text{tor}} \frac{\partial}{\partial r} \left(u_{\text{par}} - \beta_{\text{tor}} u_{\text{ExB}} \right) \right\}, \\ \Pi_{\text{par}}^{\text{turb}} &= c_{\text{per}} mn \left\{ -\Phi^2 d_{\text{par}} \frac{\partial}{\partial r} \left(u_{\text{par}} - \beta_{\text{par}} u_{\text{ExB}} \right) \right\} \end{aligned} \tag{7}$$

The momentum diffusion coeffients $\left(d_{\text{par}}, d_{\text{tor}}\right)$ are taken to be constants as are the two coefficients of the stress due to the shear in the ExB velocity Doppler shift $\left(\beta_{\text{par}}, \beta_{\text{tor}}\right)$. The contribution due to parallel velocity and up/down symmetry breaking of the flux surfaces and other effects [20] are neglected. The geometry coefficients in equation (7) were chosen to reduce the number of coefficients that appear in the final form of the equations. Gyrokinetic turbulence simulations have shown that the ExB velocity shear causes a shift in the radial wavenumber spectrum of electric potential fluctuations. The normalized peak of the shifted potential spectrum that appears in equation (7) is well modeled by the 'spectral shift' model [22] which for a single poloidal wavenumber k_y is

$$\Phi = \frac{\gamma / \gamma_0}{\left[1 + 0.56 \langle k_x / k_y \rangle^2 \right] \left[1 + \left(1.15 \langle k_x / k_y \rangle \right)^4 \right]}, \tag{8a}$$

$$\langle k_x / k_y \rangle \cong -0.36 \gamma_{\text{ExB}} / \gamma, \quad \gamma_{\text{ExB}} = -\frac{r}{R_{\text{ref}} q} \frac{\partial u_{\text{ExB}}}{\partial r}. \tag{8b}$$

The linear growthrate γ is here a representative value for the k_y-spectrum. The linear growthrate at the peak of the potential spectrum is a good choice. This is typically 1/3 smaller than the maximum linear growthrate of the spectrum. The potential is normalized by γ_0. The spectral average radial wavenumber shift $\langle k_x / k_y \rangle$, induced by the ExB velocity shear has been simplified from its more general form [23]. The coefficients in equation (8) are not adjustable parameters to fit experiment but were determined from the properties of the potential spectrum in gyrokinetic turbulence simulations [22].

The collisional contributions to the perpendicular stresses are:

$$\begin{aligned} \Pi_{\text{tor}}^{\text{col}} &= \frac{c_{\text{per}}^2}{c_{\text{par}}} mn \left\{ -d_{\text{neo}} \frac{\partial}{\partial r} \left[u_{\text{par}} + a_{\text{per}} \left(u_{\text{ExB}} + u_{\text{dia}} \right) \right] \right\}, \\ \Pi_{\text{par}}^{\text{col}} &= c_{\text{per}} mn \left\{ -d_{\text{neo}} \frac{\partial u_{\text{par}}}{\partial r} \right\}. \end{aligned} \tag{9}$$

The turbulent coefficients $\left(d_{\text{tor}}, d_{\text{par}}\right)$ are typically two orders of magnitude larger than the Coulomb one d_{neo} [24].

The equations are further reduced to a local flux surface r_s by expanding the three mean velocites: $u_{\text{ExB}} = \hat{u}_{\text{ExB}}(t)\, h(r)$,

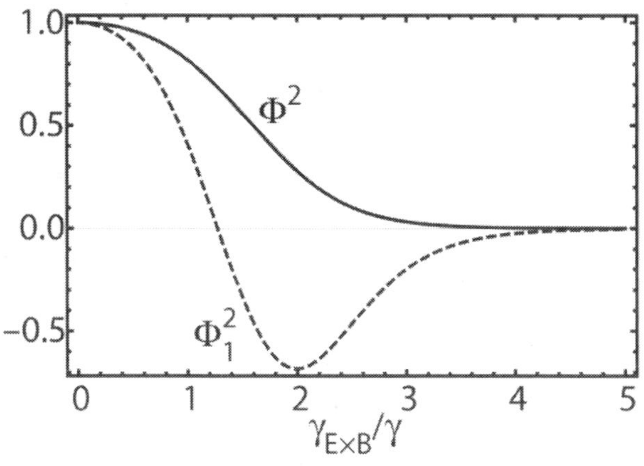

Figure 1. Spectral shift model [equation (8)] for the peak of the spectrum of electric potential fluctuations Φ^2 (solid) and $\Phi_\mathrm{I}^2 = \Phi^2 + \gamma_\mathrm{ExB}\partial\Phi^2 / \partial\gamma_\mathrm{ExB}$ (dashed) as a function of $\gamma_\mathrm{ExB} / \gamma$ for a constant linear growth rate γ.

$u_\mathrm{par} = \hat{u}_\mathrm{par}(t)\, h(r)$, $u_\mathrm{dia} = \hat{u}_\mathrm{dia}(t)\, h(r)$. A common radial profile function is assumed $h = 1 - (r - r_s) / \Delta_r - 0.5(r - r_s)^2 / \Delta_r^2$ to reduce the number of free fitting parameters to just the one length Δ_r. The diamagnetic velocity is treated as an external driver and will be ramped linearly in time $\hat{u}_\mathrm{dia} = \hat{u}_0(1 + c_t t)$. The neoclassical ion poloidal velocity is simplified to be proportional to the diamagnetic velocity $u_\mathrm{pol}^\mathrm{neo} = c_\mathrm{neo} u_\mathrm{dia}$. Substituting the local radial expansion into the momentum transport equations and evaluating the result at the flux surface r_s gives the final local equations that will be solved.

$$\frac{\partial}{\partial t}\left[\hat{u}_\mathrm{par} + a_\mathrm{per}(\hat{u}_\mathrm{ExB} + \hat{u}_\mathrm{dia})\right] - \Phi_\mathrm{I}^2 \frac{d_\mathrm{tor}}{\Delta_r^2}\left[\hat{u}_\mathrm{par} - \beta_\mathrm{tor}\hat{u}_\mathrm{ExB}\right] - \frac{d_\mathrm{neo}}{\Delta_r^2}\left[\hat{u}_\mathrm{par} + a_\mathrm{per}(\hat{u}_\mathrm{ExB} + \hat{u}_\mathrm{dia})\right] = \hat{S}_\mathrm{tor}, \tag{10}$$

$$\frac{\partial}{\partial t}\hat{u}_\mathrm{par} - \Phi_\mathrm{I}^2 \frac{d_\mathrm{par}}{\Delta_r^2}\left[\hat{u}_\mathrm{par} - \beta_\mathrm{par}\hat{u}_\mathrm{ExB}\right] - \frac{d_\mathrm{neo}}{\Delta_r^2}\hat{u}_\mathrm{par} + \nu_\mathrm{pol}\left(\hat{u}_\mathrm{par} - \hat{u}_\mathrm{ExB} - (1 + c_\mathrm{neo})\,\hat{u}_\mathrm{dia}\right) = \hat{S}_\mathrm{par}, \tag{11}$$

where $m\, n\, \hat{S}_\mathrm{tor} = S_\mathrm{tor} c_\mathrm{par} / c_\mathrm{per}^2$, $m\, n\, \hat{S}_\mathrm{par} = S_\mathrm{par} / c_\mathrm{per}$. The common intensity factor has been modified to $\Phi_\mathrm{I}^2 = \Phi^2 + \Delta_r \partial\Phi^2 / \partial r$. For the special case of a constant linear growthrate $\gamma = \gamma_0$ this reduces to $\Phi_\mathrm{I}^2 = \Phi^2 + \gamma_\mathrm{ExB}\partial\Phi^2 / \partial\gamma_\mathrm{ExB}$. This 'intensity slope' function Φ_I^2 and the intensity Φ^2 are plotted in figure 1. The reduction of the intensity with increasing γ_ExB causes Φ_I^2 to become negative above a critical value of γ_ExB. A negative Φ_I^2 causes the normally dissipative turbulence terms in equations (10) and (11) to become drive terms which is the sole cause of both the L/H and the LCO transitions. It is important that Φ_I^2 is positive for low γ_ExB so that there is a stable L-mode equilibrium.

Solving the local momentum transport system, without sources, using Mathematica™ gives the particular case with limit cycle oscillations of the 2D momentum transport system shown in figure 2 for the ExB (Doppler phase) velocity (a) The density fluctuation amplitude $\tilde{n} / n = 0.25\Phi$ normalized to the arbitrary unit level of the L-mode measurement in [2] (b) and the vertical velocity (c). The parameters of the model were taken from the data as much as possible: $a_\mathrm{per} = 0.215$, $c_\mathrm{neo} = 0.2$, $d_\mathrm{par} = d_\mathrm{tor} = 10\,\mathrm{m^2\,s^{-1}}$, $d_\mathrm{neo} = 0.01 d_\mathrm{par}$, $\nu_\mathrm{pol} = 632\,\mathrm{s^{-1}}$. The two coefficients of the Doppler shear pinch, the gradient scale length and the reference linear growth rate were chosen to approximately fit the LCO measurements in figure 4(*a*, *d*) of [2] at $R = 2.27\,\mathrm{m}$ $\left(\beta_\mathrm{tor} = 0.1, \beta_\mathrm{par} = 0.17, \Delta_r = 0.018\,\mathrm{m}, \gamma = \gamma_0 = 0.766 \times 10^5\,\mathrm{rad\,s^{-1}}\right)$. The dashed line in figure 2(*a*) is minus the diamagnetic velocity. During the LCO, the ExB velocity is not equal to the diamagnetic velocity but it becomes close to it after the oscillations cease in the H-mode. The LCO frequency given by the model is 1.5 kHz which is a good fit to the observed value of 1.6 kHz. The magnitude of the ExB velocity is well matched. The oscillating poloidal and ExB velocities tend to cancel in the parallel and toroidal velocity but there is a net parallel velocity oscillation during the LCO. The ExB velocity shear rate is just the ExB velocity [figure 2(*a*)] divided by the constant gradient scale length Δ_r and is about a factor of 2 lower than the measured value. The fitting of the model to the data is quite constrained but not unique. The gradient scale length can be reduced to better fit the ExB velocity shear but then the LCO frequency is reduced. The LCO frequency can still be matched by reducing the coefficient β_par. However, the maximum vertical velocity during the LCO then increases and it is already too large. The case in figure 2 is a compromise.

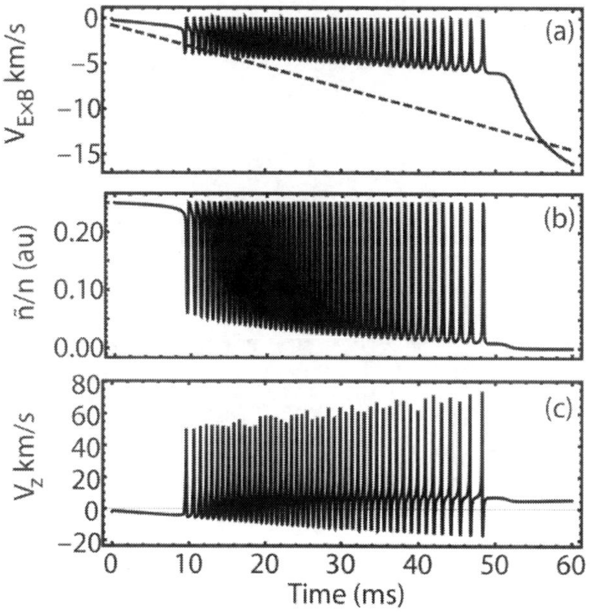

Figure 2. Time history of the ExB velocity (*a*) relative density fluctuation amplitude (*b*) and vertical velocity (*c*) for an LCO case with constant growth rate.

Scanning only the poloidal damping rate shows that the frequency of the LCO goes up linearly with increasing poloidal damping rate as shown in figure 3. This qualitatively agrees with the trend in the DIII-D data that shows the

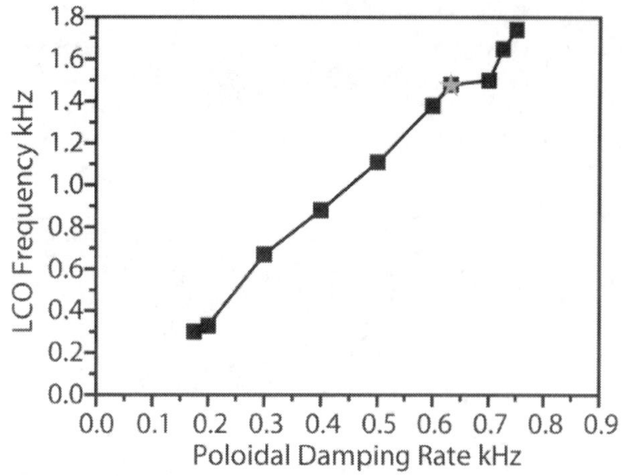

Figure 3. Variation of the LCO frequency with poloidal damping rate about the reference data point (grey star).

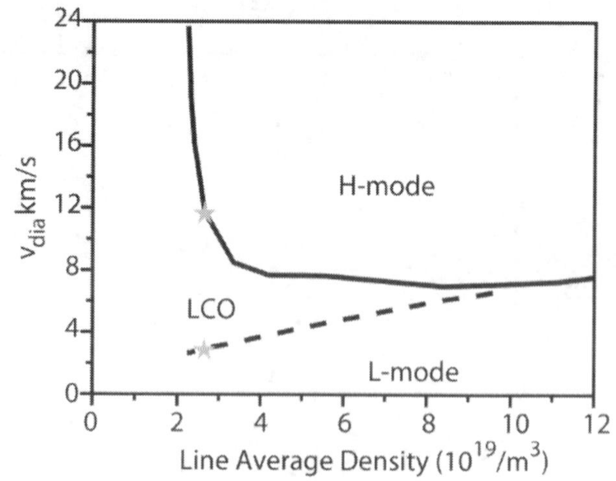

Figure 4. Critical value of the diamagnetic velocity to trigger and LCO phase (dashed) or an L/H transition (solid) vs line average density scaled from reference data point (grey stars).

LCO frequency increases with decreasing density (recall the poloidal damping rate scales inversely with the ion-ion collision rate which increases with density). The line average density of the reference discharge was $\bar{n}_{\text{ref}} = 2.65 \times 10^{19}\,\text{m}^{-3}$. Using the poloidal damping rate to scale the line average density gives $\bar{n} = \bar{n}_{\text{ref}}632 / \nu_{\text{pol}}$. The value of the diamagnetic velocity is a proxy for the heating power. The critical value of the diamagnetic velocity at the onset of the LCO and the value required to lock in the H-mode is shown as a function of the scaled line average density in figure 4. This is qualitatively similar to the dependence of the H-mode power threshold (diamagnetic velocity at H-mode) on the line average density with both a linear rise in the power threshold with density above the minimum and a strong increase in the power threshold at low density [16]. All of the other parameters were held fixed in the poloidal damping rate scan. The LCO frequency also depends on the diffusion rate $d_{\text{par}} / \Delta_r^2$ and the difference of the Doppler shear coefficients $\left(\beta_{\text{par}} - \beta_{\text{tor}}\right)$ in a linear way. This gives a measure of the sensitivity of the results to errors in the date used to constrain the model. The high-density increase of the L/H threshold in figure 4 is caused by the decoupling of the L-mode ExB velocity from the diamagnetic velocity at low poloidal damping rate. The reduction of the L/H threshold with decreasing density carriers onto the LCO threshold curve in figure 4 but the L/H threshold rises sharply below a minimum. This sharp rise in the L/H threshold is due to the ExB velocity being prevented from growing enough to reach the H-mode by the LCO as explained in the next section.

The large vertical velocity excursion during the LCO [figure 2(*c*)] can be reduced by allowing the parallel and ExB velocity gradients lengths to differ. It can also be reduced by including the linear growth rate dependence upon the parallel velocity shear: $\gamma = \gamma_0 + \alpha_{\text{KH}}0.08\left|\partial u_{\text{par}} / \partial r\right|$. This is a physical effect in gyrokinetic theory [19] (Kelvin Helmoltz type mode). The coefficient 0.08 is determined from the GA standard case [20] but a fitting parameter α_{KH} is included. Using the more physical growthrate in the spectral shift model gives a rapid LCO (2.2 kHz) that decays to a quiet phase followed by an L/H transition. This three-stage transition is not seen in the data. In order to get back the two-stage transition adjustments were made $\left(\beta_{\text{par}} = 0.24,\ \gamma_0 = 0.38 \times 10^5\,\text{rad s}^{-1},\ \alpha_{\text{KH}} = 0.11\right)$. The ramp rate of the diamagnetic velocity was also reduced. The resulting waveforms are shown in figure 5. The parallel velocity shear KH drive suppresses the large excursion of the vertical velocity in figure 5(*c*) so it is now in the observed range for the main ions [25]. The LCO in figure 5 has a lower frequency than the data (0.5 kHz). This model demonstrates that the vertical velocity can be brought into the experimental range by additional physics. This model also introduces a phase shift between the density fluctuation amplitude and the ExB velocity shear as shown in figure 6. Comparing figure 6 to [2] figure 3(*a*, *b*) it is clear that the density fluctuation amplitude is rising at the time of maximum ExB velocity shear so the counter clockwise rotation of this LCO cycle agrees with the direction measured at this location. It can be shown in general, for the spectral shift model, that if the linear growth rate γ is rising at the peak in the magnitude of the ExB shear, there is a counter clockwise orbit, but if the linear growth rate is falling at the peak in the ExB shear magnitude, the orbit is clockwise. A reduction of the linear growth rate could occur due to a greater steepening of the density gradient than the temperature gradient causing a reduction in the ion temperature gradient mode growthrate. Both directions of the phase orbit are observed in experiments [16].

A one-step L/H transition occurs in the model with KH drive for a smaller gradient scale length ($\Delta_r = 0.0154$ m) as shown in figure 7. The density fluctuations [figure 7(*a*)] drop within 0.1 ms and the ExB (black) and vertical (grey) velocities [figure 7(*b*)] re-arrange within 0.5 ms at the transition. These are the same timescales that have been observed for the 'trigger' event at the start of the L/H transition [6]. This fast change in velocity has been called a trigger because it precedes the slower change in the diamagnetic velocity. This is also the case in figure 7(*b*) since the diamagnetic velocity (dashed) is being ramped at a slow rate.

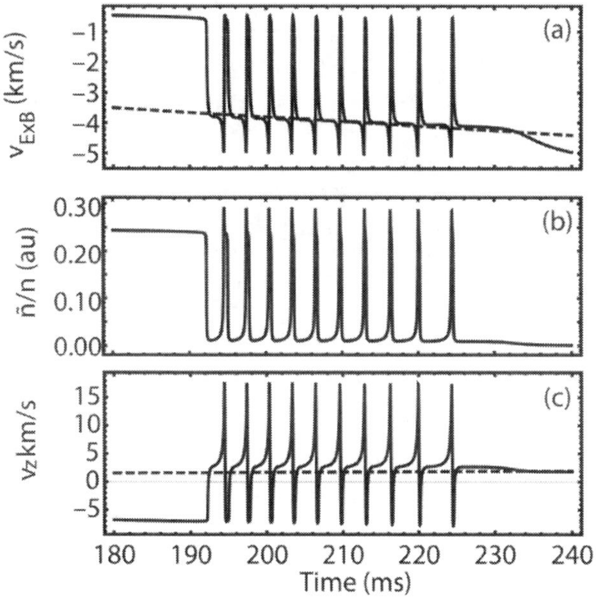

Figure 5. Time history of ExB velocity (solid) and minus the diamagnetic velocity (dashed) (*a*) relative density fluctuation amplitude (*b*) and vertical velocity (solid) and neoclassical vertical velocity (dashed) (*c*) for the LCO case with KH drive.

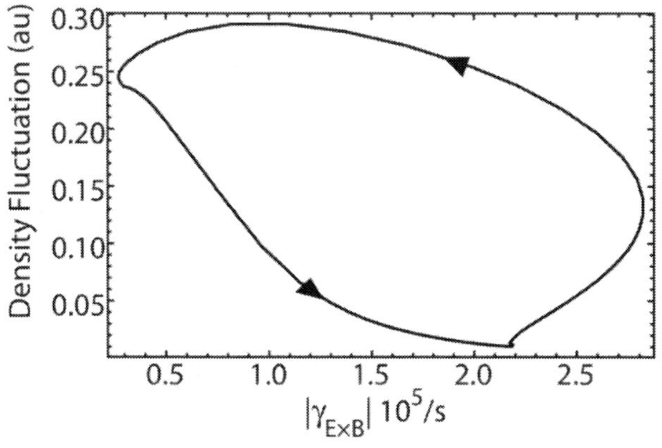

Figure 6. Relative density fluctuation amplitude versus $|\gamma_{ExB}|$ for the LCO orbit from 202–205 ms in figure 5.

4. Origin of the momentum transport LCO

The origin of the LCO type transition can be understood by examination of the acceleration of the parallel and ExB velocities. The local momentum transport equations [equations (10) and (11)] can be re-written in the following way by diagonalizing the time derivatives and re-grouping terms.

$$a_{\text{per}}\frac{\partial \hat{u}_{\text{ExB}}}{\partial t} = A_{11}\hat{u}_{\text{ExB}} + A_{12}\hat{u}_{\text{par}} + s^{\text{ext}}_{\text{ExB}}, \tag{12a}$$

$$\frac{\partial \hat{u}_{\text{par}}}{\partial t} = A_{22}\hat{u}_{\text{par}} + A_{21}\hat{u}_{\text{ExB}} + s^{\text{ext}}_{\text{par}}. \tag{12b}$$

The acceleration matrix has the elements:

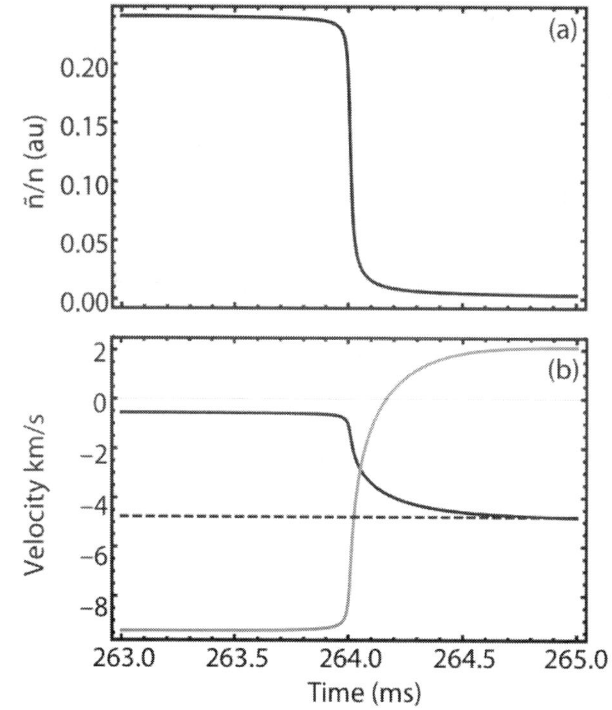

Figure 7. L/H transition relative density fluctuation amplitude (*a*) and ExB (black), vertical (grey) and minus the diamagnetic (dashed) velocities (*b*) vs time (ms).

$$A_{11} = -\Phi_{\text{I}}^2\frac{d_{\text{par}}\beta_{\text{par}} - d_{\text{tor}}\beta_{\text{tor}}}{\Delta_r^2} - a_{\text{per}}\frac{d_{\text{neo}}}{\Delta_r^2} - \nu_{\text{pol}},$$
$$A_{22} = -\Phi_{\text{I}}^2\frac{d_{\text{par}}}{\Delta_r^2} - \frac{d_{\text{neo}}}{\Delta_r^2} - \nu_{\text{pol}}, \tag{13a}$$

$$A_{12} = \Phi_{\text{I}}^2\frac{d_{\text{par}} - d_{\text{tor}}}{\Delta_r^2} + \nu_{\text{pol}},\ A_{21} = \Phi_{\text{I}}^2\frac{\beta_{\text{par}}d_{\text{par}}}{\Delta_r^2} + \nu_{\text{pol}}. \tag{13b}$$

And the external sources including the diamagnetic velocity terms are collected into

$$s^{\text{ext}}_{\text{ExB}} = \hat{S}_{\text{tor}} - \hat{S}_{\text{par}} - \left[a_{\text{per}}\left(\frac{\partial}{\partial t} + \frac{d_{\text{neo}}}{\Delta_r^2}\right) + \nu_{\text{pol}}(1 + c_{\text{neo}})\right]\hat{u}_{\text{dia}}, \tag{14a}$$

$$s^{\text{ext}}_{\text{par}} = \hat{S}_{\text{par}} + \nu_{\text{pol}}(1 + c_{\text{neo}})\,\hat{u}_{\text{dia}}. \tag{14b}$$

When the intensity slope is positive $\Phi_{\text{I}}^2 > 0$ the acceleration matrix has negative diagonal terms and causes a de-acceleration (decay) of the velocities. For sufficient ExB velocity shear, the intensity slope is negative $\Phi_{\text{I}}^2 < 0$ (figure 1). Assuming both diagonal elements become positive for some value of the ExB velocity shear, the parallel velocity will have a larger diagonal acceleration (A_{22}) than the ExB velocity (A_{11}) since the parallel momentum diffusion is larger than the ExB effective diffusion: $d_{\text{par}} > d_{\text{par}}\beta_{\text{par}} - \beta_{\text{tor}}d_{\text{tor}}$. There will also be a range of ExB velocity shear where only the parallel diagonal element A_{22} has changed sign causing a growth in parallel velocity and an increase in the turbulence.

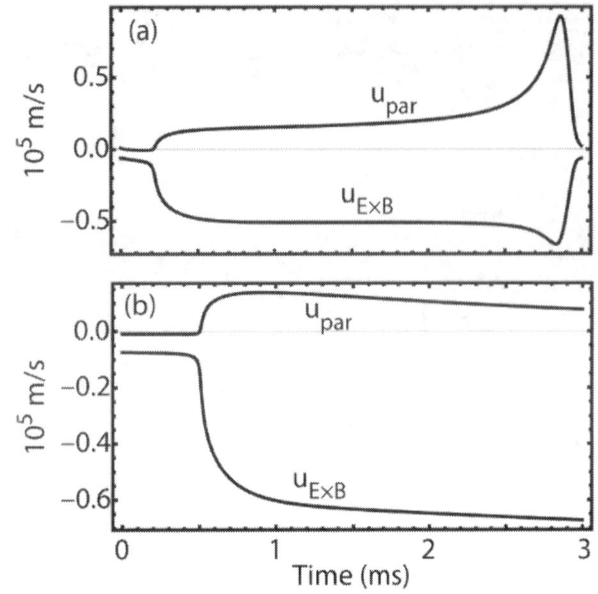

Figure 8. Parallel velocity u_{par} (positive), ExB velocity u_{ExB} (negative) for a single LCO cycle (*a*) and for a single step L/H transition (*b*) over a 3 ms time window.

The off diagonal elements play a critical role in the LCO. If the off-diagonal element in the ExB velocity equation A_{12} is positive, then when the parallel velocity has the opposite sign of the ExB velocity, the off-diagonal term causes a de-acceleration of the ExB velocity. The evolution of the parallel and ExB velocities are plotted in figure 8 for a single LCO cycle (a) and for a single step L/H transition (b) over a 3 ms time frame. Both cases start with a rapid growth of the velocities. The difference between the LCO and L/H cases is that for the L/H cases the ExB velocity reaches a larger amplitude and the parallel velocity decays due to the elimination of the turbulence drive. For the LCO, the parallel velocity continues to grow in amplitude until it becomes large enough to cause a decay of the ExB velocity and a return to the high turbulence L-mode state. The off-diagonal coupling between velocities is responsible for this. The strong rise in parallel velocity at the end of each LCO cycle has a short duration that makes it difficult to measure with CER but is a signature feature of the LCO predicted by this model. If the parallel and ExB velocities could be made to have the same sign, then the LCO would not occur because the off-diagonal coupling would enhance rather than retard the growth of the ExB velocity resulting in an L/H transition.

5. Summary

In this paper it has been shown that the mean field parallel and toroidal momentum equations can have both one-step transitions to suppressed turbulence (L/H) and limit cycle oscillations of the parallel and ExB velocities. Using data from a DIII-D L-mode discharge, it was shown that a simplified model, with the turbulent momentum transport constrained by the power balance diffusivities and the computed neoclassical poloidal damping rate, can fit the LCO frequency, the ExB velocity and the density fluctuation amplitude waveforms in reasonable agreement with the measurements. It was further shown that the frequency scales linearly with the neoclassical poloidal damping rate and that the density dependence of the L/H power threshold is well represented by the model. The counter clockwise phase shift between the density fluctuations and the ExB velocity shear observed in the data is explained by the model as being due to the parallel velocity shear drive of the linear growth rate. The LCO of the momentum equations occurs because there is a weaker growth of the ExB velocity shear than the parallel velocity shear amplitudes. For a high poloidal velocity damping rate, the ExB velocity shear cannot make a full transition to suppressed turbulence and the parallel velocity shear growth de-accelerates the ExB velocity forcing a return to L-mode at the end of the LCO cycle. For weaker poloidal velocity damping, the ExB shear can complete the transition to H mode before the parallel flow grows to a critical level for the LCO. The goal of this work was to establish that the mean field momentum equations can fit the measurements. The next goal is to show that a calibrated quasilinear model of the turbulence, like TGLF [26], can predict the L/H transition and LCO properties without adjustable parameters. There are still many obstacles along the path to a realistic theoretical prediction of the L/H transition including all of the transport channels. This paper has taken the step of verifying that the mean field momentum transport equations capture the non-linear dynamics between the turbulence and the evolution of the mean velocities well enough to reproduce the observed phenomenology of LCO and L/H transitions.

Acknowledgments

This material is based upon work supported by the US Department of Energy, Office of Science, Office of Fusion Energy Sciences, Theory Program, using the DIII-D National Fusion Facility, a DOE Office of Science user facility, under Awards DE-FG02-95ER54309 and DE-FC02-04ER54698.

References

[1] Luxon J L *et al* 1986 *Plasma Physics Controlled Nuclear Fusion Research* vol 1 (Vienna: International Atomic Energy Agency) p 159
[2] Schmitz L *et al* 2012 *Phys. Rev. Lett.* **108** 155002
[3] Staebler G M *et al* 2001 *Nucl. Fusion* **41** 891
[4] Newman D *et al* 1999 *Phys. Plasmas* **6** 854
[5] Burrell K H 1997 *Phys. Plasmas* **4** 1499
[6] Burrell K H *et al* 1996 *Plasma Phys. Control. Fusion* **38** 1313
[7] Shaing K C and Crume E C Jr 1989 *Phys. Rev. Lett.* **63** 2369
[8] Tendler M and Rozhansky V 1990 Controlled fusion and plasma heating *Proc. of 17th Euro. Conf. (Amsterdam, 1990)* vol 14B (Switzerland: European Physical Society, Petit-Lancy) p 744
[9] Frieman E A and Chen L 1982 *Phys. Fluids* **25** 502

[10] Doyle E J *et al* 1991 *Phys. Fluids* **B3** 2300
[11] Shaing K C 1988 *Phys. Fluids* **31** 2249
[12] Sugama H and Horton W 1997 *Phys. Plasmas* **4** 405
[13] Staebler G M 2004 *Phys. Plasmas* **11** 1064
[14] Hirshman S P 1978 *Phys. Fluids* **21** 224
[15] Rozhansky V and Tendler M 1992 *Phys. Fluids* **B4** 1877
[16] Conway G D *et al* 2011 *Phys. Rev. Lett.* **106** 065001
[17] Xu G S *et al* 2011 *Phys. Rev. Lett.* **107** 125001
[18] Yan Z *et al* 2014 *Phys. Rev. Lett.* **112** 125002
[19] Catto P, Rosenbluth M N and Liu C S 1973 *Phys. Fluids* **16** 1719
[20] Waltz R E *et al* 2007 *Phys. Plasmas* **14** 122507
[21] Houlberg W *et al* 1997 *Phys. Plasmas* **4** 3230
[22] Staebler G M *et al* 2013 *Phys. Rev. Lett.* **110** 055003
[23] Staebler G M *et al* 2013 *Nucl. Fusion* **53**, 113017
[24] Hinton F L and Wong S K 1985 *Phys. Fluids* **28** 3082
[25] Kim J C *et al* 1994 *Plasma Phys. Control. Fusion* **36** A183
[26] Staebler G M *et al* 2005 *Phys. Plasmas* **12** 102508

IOP Publishing

Plasma Physics and Controlled Fusion

Plasma Phys. Control. Fusion **57** (2015) 014026 (11pp)

doi:10.1088/0741-3335/57/1/014026

Thomson scattering on non-equilibrium low density plasmas: principles, practice and challenges

Emile Carbone and Sander Nijdam

Department of Applied Physics, Eindhoven University of Technology, PO Box 513, 5600 MB Eindhoven, The Netherlands

E-mail: emile@epgmod.phys.tue.nl

Received 14 June 2014, revised 8 September 2014
Accepted for publication 24 September 2014
Published 28 November 2014

Abstract

In this paper, we review the main challenges related to laser Thomson scattering on low temperature plasmas. The main features of the triple grating spectrometer used to discriminate Thomson and Raman scattering signals from Rayleigh scattering and stray light are presented. The main parameters influencing the detection limit of Thomson scattering are reviewed. Laser stray light and plasma emission are two limiting factors, but Raman scattering from molecules inside the plasma will further decrease it.

In the case of non-thermal plasmas at high pressure, Thomson scattering is the only technique which allows us to obtain the electron density without any prior knowledge of the plasma properties. Moreover, very high 3D spatial and temporal resolutions can easily be achieved. However, special care still needs to be taken to verify that Thomson scattering is non intrusive. The mechanisms that will lead to possible measurement errors are discussed. The wavelength-resolved scattering signal also allows us to get direct information about the electron energy distribution function in the case of incoherent light scattering.

Finally, we discuss some recent applications of Thomson scattering on atmospheric pressure plasma jets, but also in the field of electron collision kinetics. Thomson scattering can be applied on atomic but also molecular plasmas. In the latter case, one needs to take into account the possible contribution of rotational Raman scattering.

Keywords: Thomson scattering, Rayleigh scattering, rotational Raman scattering, low density plasmas, non-equilibrium, laser diagnostics, electron density and temperature

(Some figures may appear in colour only in the online journal)

1. Introduction

In low temperature non-equilibrium plasmas, internal plasma parameters such as species densities and temperatures become decoupled and their steady state values cannot be described as a function of external parameters such as input power, pressure and initial species densities as in local thermal equilibrium conditions [1]. Plasma kinetics depends to a very large extent on the local electron density n_e and temperature T_e. This is due to the much lighter electron mass which allows it to follow the electric field much faster compared to the ions and to absorb energy from the external electric field [2]. In these cases, the rotational T_{rot} and vibrational T_{vib} temperatures can also significantly differ from the translational gas temperature T_g and lead to extra degrees of departure from thermal equilibrium [3]. In some cases, rotational and vibrational states distribution functions can also deviate from the Boltzmann equilibrium [4]. The non-applicability of the definition of a temperature for a single population is also widely found for electronic states of atoms and molecules both for low and high pressure plasmas. This renders the determination of an electronic temperature from optical diagnostics very tedious and inaccurate. Active diagnostic methods based for instance on electrical probes and laser techniques were developed to solve this problem [5]. The electron energy distribution function can strongly deviate from the Maxwellian equilibrium as well [6].

In this paper, we will focus on Thomson scattering which has been developed as an accurate diagnostic technique to determine the electron density and temperature [7]. The technique allows us to obtain the electron density accurately without any prior assumption on the state of departure from the equilibrium of the plasma. The method has been proven to be particularly valuable in the case of high pressure plasmas where electrical probes data interpretation becomes very difficult [8]. Moreover, Bowden *et al* [9] compared Thomson scattering and Langmuir probe measurements and showed that the latter perturbs the plasma significantly. Other methods such as microwave interferometry and resonator probes can provide only line of sight measurements or spatial profiles with limited temporal and/or spatial resolution [10, 11].

In section 2, the general equations needed to describe and interpret Thomson, Raman and Rayleigh scattering are given. Then, we present in section 3 the triple-grating spectrometer (TGS) that is currently used at the Eindhoven University of Technology (TU/e) and was built by van de Sande *et al* [12]. The TGS allows for the physical removal of the Rayleigh scattering from atoms and molecules and selectively collects light scattered from free electrons as well as rotational Raman scattering from molecules. In addition to removing Rayleigh scattering, the TGS blocks the central laser wavelength scattered by any surface in the vicinity of the collection volume (i.e. stray light) [13]. The scattered photons are collected by an iCCD camera in a photon-counting mode which allows a high signal to noise ratio [14]. The various factors influencing the detection limit and possible perturbations of the plasma by the laser are discussed in sections 4 and 5 respectively. In section 6, the link between the measured spectral distribution of Thomson scattered photons and the electron energy distribution function is briefly reviewed in the case of high pressure, non-Maxwellian plasmas. Finally, in section 7, we discuss rotational Raman scattering from molecules inside the plasma. In conclusion, a brief overview is given of some recent applications that were developed at the TU/e.

2. Laser light scattering

When a laser beam is guided through a plasma, a fraction of the laser light is scattered by the species present inside the laser beam. The measured scattered power per unit of wavelength P^i_λ by a species i in the plasma is given by [15]

$$P^i_\lambda(\lambda) = P_0 n_i f\left(\frac{d\sigma_i}{d\Omega}\right)\Delta\Omega S^i_\lambda(\lambda)\, L \qquad (1)$$

where f is a constant factor that takes into account the efficiency of the optics and camera, L the length of the detection volume along the laser path, P_0 the incident laser power, $\Delta\Omega$ the solid angle of detection, n_i the density of the scattering particle i, and $d\sigma_i/d\Omega$ its differential scattering cross section. The factor $S^i_\lambda(\lambda)$ represents the scattering spectral distribution as a function of wavelength and is normalized such that $\int S_\lambda(\lambda)d\lambda = 1$. Note that in the following, the subscript λ denotes that the quantity is per unit of wavelength. The wavelength integrated value P is written without subscript, so that $P = \int P_\lambda d\lambda$.

The total Thomson scattering differential cross section $d\sigma^T/d\Omega$ is given by [5]

$$\frac{d\sigma^T}{d\Omega} = r_e^2(1 - \sin^2\theta\cos^2\varphi) \qquad (2)$$

where θ is the scattering angle (i.e. the angle between the incident and scattered wave vector), φ is the angle between the plane of scattering and the polarization of the incident beam and r_e is the classical electron radius ($r_e = e^2/(4\pi\,\epsilon_0 m_e c^2)$). In our setup, the scattering angle is $\theta = 90°$ while the laser beam is vertically polarized so that $\varphi = 90°$. For this configuration, we obtain a numerical value of $d\sigma^T/d\Omega = 7.94 \cdot 10^{-30}$ m^2 sr^{-1}.

Depending on the electron density and temperature, laser light can be either scattered in coherent or incoherent regimes which is defined by the coherence parameter α_{coh}

$$\alpha_{coh} = \frac{\lambda}{4\pi\sin\frac{\theta}{2}}\sqrt{\frac{4\pi n_e e^2}{k_B T_e}} \qquad (3)$$

where e is the electron charge and k_B the Boltzmann constant. For values of $\alpha_{coh} << 1$, the laser light is scattered by independent free electrons, whereas for $\alpha_{coh} >> 1$ the laser photons are scattered coherently by the electron gas [5].

In the case of incoherent Thomson scattering and a Maxwellian velocity distribution function, the spectral distribution $S^T_\lambda(\lambda)$ is a Gaussian function around the laser wavelength λ_i

$$S^T_\lambda(\lambda) = \frac{1}{\Delta\lambda\sqrt{\pi}}\cdot\exp\left(-\frac{\lambda-\lambda_i}{\Delta\lambda_{1/e}}\right)^2 \qquad (4)$$

where $\Delta\,\lambda_{1/e}$ is the $1/e$ half-width of the Gaussian fitting of the Thomson spectrum. Assuming a Maxwell distribution, a Gaussian fitting of the Doppler- broadened Thomson photons gives the electron temperature [15]

$$T_e = \left(\frac{m_e c^2}{8k_B\lambda_i^2\sin^2(\theta/2)}\right)\Delta\lambda^2_{1/e}. \qquad (5)$$

In figure 1, an example of a typical Thomson scattering spectrum is shown. On the vertical axis, the Thomson signal is resolved in one direction over a length of approximatively 11 mm [16]. In the case of non-Maxwellian distribution functions of the electrons, we refer to the paper of Huang and Hiefje [17] for a general derivation of the relation betweeen the $S^T_\lambda(\lambda)$ function and the EEDF shape.

The differential Rayleigh scattering cross section depends on the gas species and more particularly on their volume polarizability α^i_v through the following relation [18]

$$\frac{d\sigma^{Ry}}{d\Omega} = \frac{16\pi^4\alpha^i_v}{\lambda^4}(1 - \sin^2\theta\cos^2\varphi). \qquad (6)$$

Raman scattering is the inelastic light scattering on a molecule with a change in the rotational (and/or vibrational) structure of the molecule. Either, part of the incident photon energy is converted into rotational excitation and production of a photon

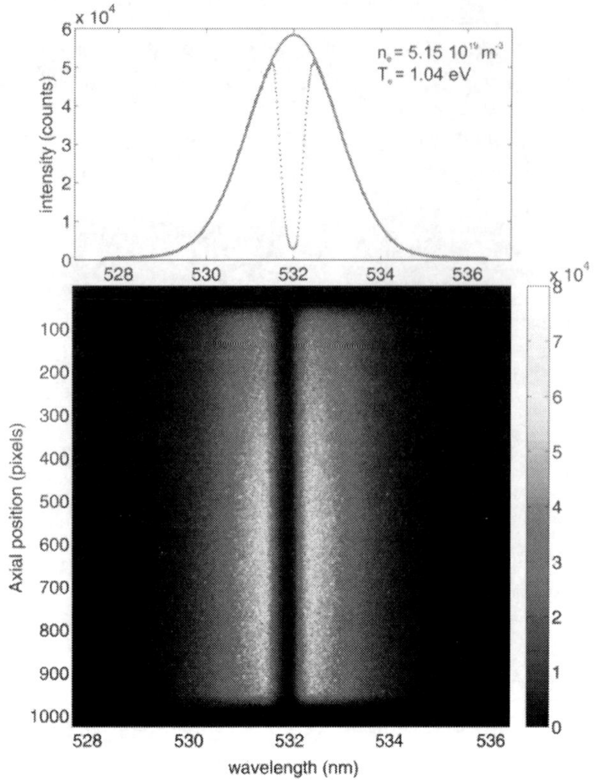

Figure 1. Bottom: iCCD image of a typical Thomson scattering spectrum for a homogeneous plasma at a gas pressure of 40 mbar [16]; the horizontal direction corresponds to the wavelength dispersion and the vertical direction to the position along the plasma axis. The inhomogeneity in vertical direction is mainly due to the TGS flat field sensitivity. Top: vertical binning of the spectrum with the corresponding Gaussian fit. Note that the central part is excluded from the fit and stems from the center laser wavelength removal by the notch filter.

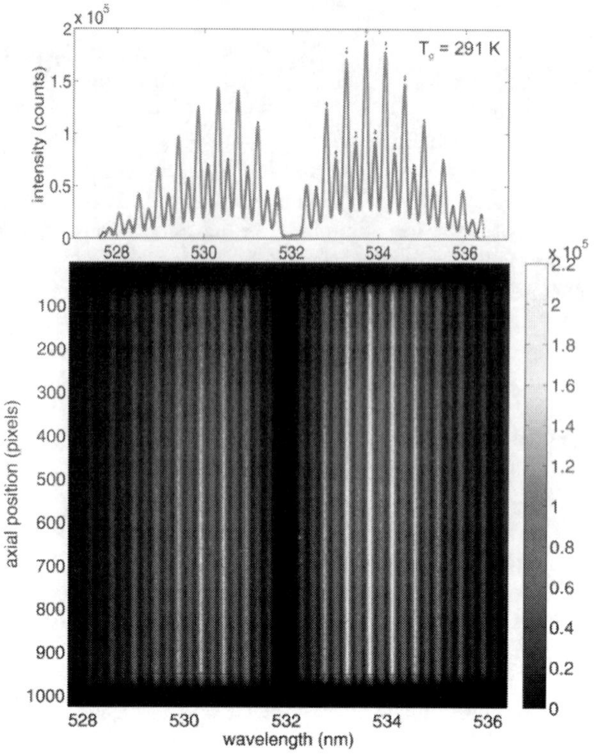

Figure 2. Bottom: 1D spatially resolved Raman scattering from a nitrogen gas filling at a pressure of 60 mbar. Top: the resulting vertical binned spectrum together with a fit composed of Gaussian profiles for the rotational lines. The shape of the combined fit gives T_g, which perfectly matches room temperature.

with lower energy or superelastic quenching of a rovibrational state occurs with the production of a photon with higher energy. This leads, respectively, to the Stokes (red shift) and anti-Stokes (blue shift) Raman structures. In figure 2, an example of a rotational Raman spectrum of nitrogen gas is shown[1]. The transitions from rotational quantum numbers J to J' leads to peaks in the spectrum at different wavelengths

$$\lambda_{J\rightarrow J'} = \lambda_i + \frac{\lambda_i^2}{hc} \cdot B\left[\,(J'^2 + J') - (J^2 + J)\,\right] \qquad (7)$$

where λ_i is the laser wavelength, h is the Planck's constant and B the molecule's rotational constant. The allowed transitions are those given by $J' = J + 2$ (Stokes) and $J' = J - 2$ (anti-Stokes).

The total Raman scattering power-density spectrum $P_\lambda^{\mathrm{Rm}}(\lambda)$ is given by

$$P_\lambda^{\mathrm{Rm}}(\lambda) = P_0 n_i f L \Delta\Omega \sum_{J'=J\pm 2} \left(\frac{\mathrm{d}\sigma_{J\rightarrow J'}^{\mathrm{Rm}}}{\mathrm{d}\Omega}\right) S_\lambda^{\mathrm{Rm}}(\lambda - \lambda_{J\rightarrow J'}) \,. \qquad (8)$$

[1] The slight asymmetry in the experimental Raman spectrum compared to the theoretical one is attributed to weak stimulated emission of the Stokes band [19].

The spectral distribution function $S_\lambda^{\mathrm{Rm}}(\lambda - \lambda_{J\rightarrow J'})$ of Raman photons is much narrower than the instrumental profile of the TGS and so the latter is taken for the fit of the Raman spectra [20]. If we take the ratio between the total Thomson and Raman photon counts, the parameters related to the particular instrumental configuration, which are given by the $fLP_0\Delta\Omega$ factor in equation (1) for $P_\lambda^T(\lambda)$ and $P_\lambda^{\mathrm{Rm}}(\lambda)$,simply cancel out. The absolute electron density n_e is then given by

$$n_e = n_i \frac{P^{\mathrm{T}}}{P_i^{\mathrm{Rm}}} \frac{\left(\mathrm{d}\sigma_i^{\mathrm{Rm}} / \mathrm{d}\Omega\right)}{\left(\mathrm{d}\sigma^T / \mathrm{d}\Omega\right)} \,. \qquad (9)$$

The Raman scattering function is well known for various diatomic species. In our experiments, fixed pressures of nitrogen gas are usually injected inside the plasma reactor to absolutely calibrate the Thomson scattered signal [16]. In the case of atmospheric pressure plasmas produced in an ambient environment, the Raman scattering signal from ambient air is directly fitted to calibrate the signal [20]. To take into account the variability of the laser pulse energy during a measurement campaign, a fast photodiode is installed to measure the drift in laser power. These values are then used to rescale the Raman absolute calibration.

The Rayleigh scattering intensity is proportional to the density of heavy particles n_i. In general, the pressure p is controlled experimentally. By applying the ideal gas law

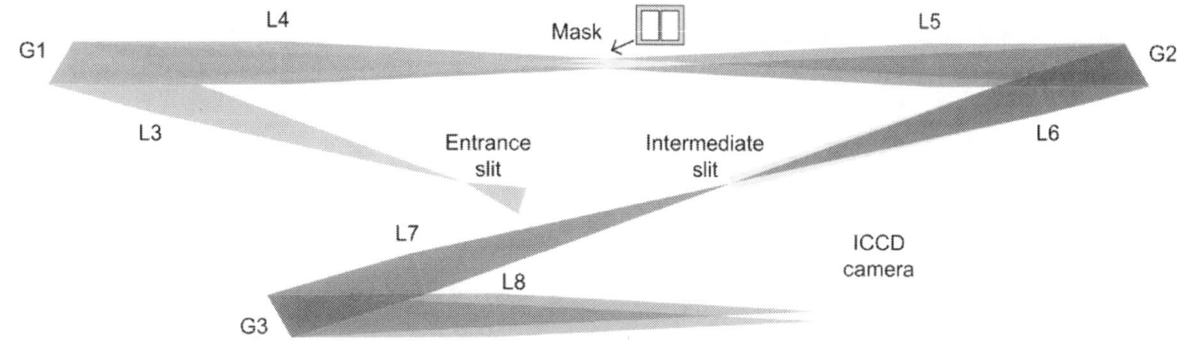

Figure 3. Schematic depiction of the working principle of the triple grating spectrometer. In this image all mirrors are omitted, including the image rotator (mirror tower). Therefore, the entrance slit is oriented vertically instead of horizontally. The diffraction angles have been reduced by about a factor two for compactness. The color separation has been greatly exaggerated with blue and red representing light with wavelengths slightly smaller and larger than 532 nm respectively. Stray light is indicated by the light green color up to the intermediate slit. Gratings and lenses are labeled identically to figure 4.

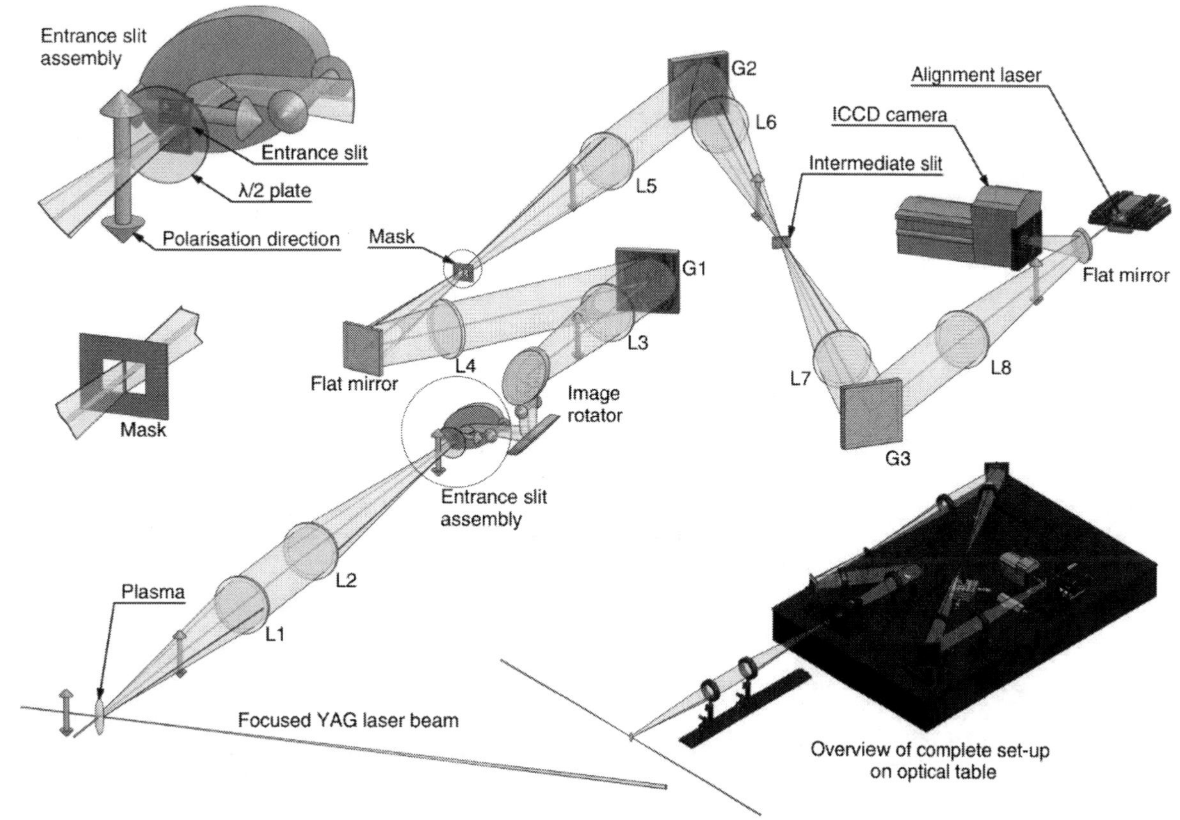

Figure 4. Layout of the triple grating spectrometer and representation of its different optical components: the arrows indicate the polarization of the photons. The small narrow beam indicates the central optical path defined by the alignment laser.

$p = \Sigma_i n_i k_B T_g$, Rayleigh scattering directly gives the plasma translational gas temperature T_g.

3. Experimental design of the triple grating spectrometer (TGS)

In figure 3, a schematic drawing of the physical principle of a triple grating spectrometer is shown. The scattered light from the plasma is collected along a 1D axis and dispersed in wavelength by the first grating (G1). A notch filter is then used to physically block the central wavelength of the laser. Only the spectral wings of the scattered signal can pass and are then re-collected by the second grating (G2). Its function is to re-create a one to one image of the scattered photons on the intermediate slit identical to the image on the entrance slit of the TGS. The third grating (G3) then acts as a normal spectrograph and disperses the photons in wavelength. These photons are then collected on the iCCD camera, where spatial resolution is retained along the length of the entrance slit.

The experimental design to collect the Thomson and Raman scattered photons is presented in figure 4. The plasma is aligned at the intersection between the focus of the laser beam and the focal point of lens L1. The second lens (L2) in front of the TGS focuses the scattered photons on the entrance slit of the TGS, and, in this case, a one to one image of the scattered photons along the laser beam in the plane parallel to the entrance slit is obtained. The collection system of light in front of the TGS can be changed depending on the particular applications, and, for instance, an optical fiber coupling system can be used instead of a pair of achromatic lenses.

After the entrance slit, an image rotator is used to rotate the sheet of scattered light by 90°. The 1D horizontal imaging of the plasma and laser beam intersection is thus transferred into the vertical plane of the TGS. In this way, the gratings can be used to disperse the photons in the horizontal plane. A $\lambda/2$ plate is placed in front of the image rotator in order to recover the vertical polarization of the scattered photons. With the grooves of the gratings being vertically oriented, the maximum diffraction efficiency of the gratings is for vertically-polarized photons [21].

To minimize the space inside the TGS black box, a flat mirror was inserted to reflect the light from grating G1 in the opposite direction and place the second grating (G2) behind it. The 600 mm focal length lenses L3 and L4 are aligned in such a way that the light from the entrance slit is focused on the mask. The mask removes the center laser wavelength from the scattering spectrum[2], which leads to the characteristic dip in the Thomson spectrum as seen in figure 1. This mask is placed on a mounting so that it can be removed and replaced easily to collect either Thomson (+ Raman) or Rayleigh photons. The physical suppression of the central wavelength allows us to remove very efficiently the signal from the laser wavelength by a factor 10^6 [12] and then works better than atomic notch filters [22].

The anti-dispersing grating G2 acts in the opposite way as the grating G1 and its diffracted light is then focused via lens L6 on the intermediate slit. The latter allows for the removal of most of the stray light coming from the first and second gratings. In the same way as in the imaging of L3 and L4 lenses, the lens L5 collects the light focused on the mask by lens L4 and refocuses it on the intermediate slit via lens L6.

The diverging beam coming from the intermediate slit is collected by lens L7 which makes it co-linear and directs it onto the third grating. The photons are then spectrally dispersed by the grating G3 and spatially refocused on the iCCD camera via the lens L8. The lenses L3–L8 all have the same focal length, therefore the spatial information collected along the entrance slit of the TGS is transferred onto the iCCD as a one to one image.

For the alignment of the TGS, a small 532 nm laser was installed behind the removable flat mirror placed in front of the iCCD camera. This laser defines the optical path of the TGS and its beam should be aligned first on the intermediate slit and then on the mask. The beam must then finally be aligned through the image rotator on the entrance slit of the TGS. This beam can then be used to find the intersection point between the pulsed laser beam in 3D space and the focal point of the TGS through lenses L1 and L2. The lenses inside the TGS are then placed and their position is optimized to obtain a focused laser scattering signal on each slit, the mask and the camera.

[2] In the standard TGS configuration, the width of the mask is $800\,\mu m$ while the entrance and intermediate slits have $250\,\mu m$ width.

4. Stray light rejection and detection limit

In addition to scattering with atoms, molecules and electrons (+ ions), the laser light is also scattered by any solid surface present near it. The resulting scattered light is called stray light [23] and is collected both at the laser center wavelength and around it as well due to the spectral shape of the laser linewidth. Various strategies need to be developed to minimize this stray light [24]. The stray light signal P^{St} is usually independent of plasma or gas-flow conditions and can be subtracted using a reference measurement which is taken for a pressure $p = 0$ for Rayleigh scattering or simply in the absence of plasma in the case of Thomson scattering. Stray light has roughly the shape of the Rayleigh scattering signal with some small far wings contributions (see for instance [25]).

In the case of Rayleigh scattering, the neutral gas density is then defined as follows

$$n_g = \frac{P^{p+\mathrm{St}} - P^{\mathrm{St}}}{P^{\mathrm{ref+St}} - P^{\mathrm{St}}} \cdot n_{\mathrm{ref}} \tag{10}$$

where n_{ref} is a known gas density injected into the plasma chamber determined by a fixed pressure and (room) temperature. In the case of atmospheric pressure plasmas in open environment, it is not possible to take a zero pressure measurement. To remedy this, one can use helium which has a much lower Rayleigh cross section and obtain by extrapolation the stray light intensity for zero pressure [26].

The detection limit of the Thomson setup can only be defined for a given plasma and laser geometry, but it also depends on the plasma emission $P^p\,(n_e, T_e)$ itself. The sensitivity of the iCCD camera is defined by the conversion efficiency of photons into electrical counts P and the thermal and electrical noise N. The latter is usually the limiting factor and substracting it leads to a negative noisy signal as seen in figure 5. It defines the minimum ratio between the Thomson signal P^T and the summation over the plasma light P_p and stray light P^{St} that can be accurately subtracted from each other. All the contributions to the total photon count on the iCCD camera P_{tot} can be defined as follows

$$P_{\mathrm{tot}} = P^T(n_e) + P^{\mathrm{St}} + P^p_{\mathrm{cont}}(n_e, T_e) + P^p_{\mathrm{line}}(n_e, T_e) + N \tag{11}$$

The Thomson scattering $P^T(n_e)$ and continuum radiation $P^p_{\mathrm{cont}}(n_e, T_e)$ contributions are directly proportional to the electron density. Depending on the conditions, plasma line emission $P^p_{\mathrm{line}}(n_e, T_e)$ can also be directly proportional to the electron density. The various contributions to the total iCCD signal, can then be schematically factorized as follows

$$P_{\mathrm{tot}} = n_e\left[\Pi^T + \Pi^p_{\mathrm{cont}}(T_e) + \gamma\Pi^p_{\mathrm{line}}(T_e)\right] + (1-\gamma)P^p_{\mathrm{line}}(n_e, T_e) + P^{\mathrm{St}} + N \tag{12}$$

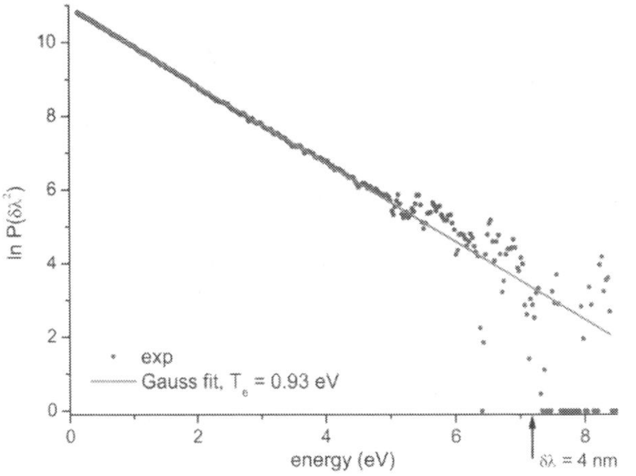

Figure 5. Logarithmic plot of the Thomson scattered signal $P(\delta\lambda)$ as a function of the electron energy that is proportional to the wavelength shift squared $\delta\lambda^2$. After baseline correction, a linear trend with $-1/k_BT_e$ for the central $\delta\lambda = 4$ nm is found. For higher $\delta\lambda$-values the noise obscures the signal. The negative counts are reset to zero for the logarithmic plotting.

where γ is a factor between 0 and 1. From this relation, it is clear that an absolute detection limit cannot be defined for a given Thomson scattering setup. Plasma line emission can always induce systematic errors in the determination of the electron temperature. The minimum detection limit for the electron density will depend on the strength of continuum radiation and atomic/molecular band emission. The Thomson count detection limit is at least 3 times larger than the thermal iCCD noise intensity as three different measurements are needed for the individual determination of the Thomson, plasma light and stray light contributions. Furthermore, the absolute accuracy is reduced by any (non averaged) fluctuation and especially by drift in the plasma or laser power (typically a few percents for the latter). On the surfatron plasma [16], the detection limit in the center of the 3 mm radius quartz tube is roughly $1 \cdot 10^{18}$ m^{-3}. On the other hand, for measurements 2 mm out of the center, this detection limit increases up to $1 \cdot 10^{19}$ m^{-3}. Still on the same setup, it was possible to determine the relative axial variation of the electron density down to the very end of the visible plasma column [27, 28]. The errors in electron temperature were however quite large due to the poor signal to noise ratio, but electron density profiles in the range of $5 \cdot 10^{17}$ m^{-3} could still be estimated. For Thomson scattering on atmospheric pressure plasma jets in open air (no nearby surfaces), a detection limit of $3 \cdot 10^{19}$ m^{-3} was given in reference [20] for Thomson scattering with 10% air entrainment inside the plasma jet. Using an Andor DH534 iCCD camera, electron densities $n_e < 1 \cdot 10^{19}$ m^{-3} could not be detected in this study. This limit was however recently significantly improved using a new Andor camera (iStar 734) with lower thermal noise and higher quantum efficiency. The signal to noise ratio was improved by almost a factor of ten [29] and electron densities as low as $1 \cdot 10^{18}$ m^{-3} could be measured on an RF argon plasma jet [30]. Furthermore, in a large, low pressure, coaxial discharge, electron densities down to a few 10^{17} m^{-3} could be detected [28, 31]. The latter measurements were however performed with an optical fiber coupling, which reduces the overall performance of the Thomson scattering detection system quite significantly.

On the other hand, the accuracy in the determination of the mean electron energy is very high and only at low electron densities some systematic errors can occur if the signal is not averaged over sufficiently long periods so that a sufficient signal to noise ratio is obtained. This is crucial for the substraction of all contributions to the iCCD camera signal that are not caused by Thomson scattered photons. The main sources of uncertainty in the electron mean energy determination will come from the stability of the plasma itself. Otherwise, a reproducibility of the measured Doppler width of the Thomson signal corresponding to variations less than 0.01 eV was observed in many cases. However, because the parameter of main interest is the electron temperature and not its mean energy, these errors are small compared to the actual deviations from a Maxwellian EEDF (see section 6).

5. Plasma perturbations during Thomson scattering

5.1. Laser heating

In a recent paper [32], the critical laser fluence, for which no electron gas heating occurs, was revisited in the case of non-thermal plasmas. In the limit of no losses of electron energy by elastic and inelastic collisions, the net absorption of energy from the laser by the electrons, and, consequently the increase of electron temperature can be expressed as

$$\frac{3}{2}k_B\Delta T_e V n_e = E_l k_\omega L \tag{13}$$

with E_l the laser pulse energy, L the absorption length, V the volume of interaction between the plasma and the laser and ΔT_e the fractional increase of electron temperature due to heating by the laser. The electrons gain energy through photon energy absorption via inverse bremsstrahlung. The absorption coefficient k_ω of energy by the electrons from the laser field can be written as [33]

$$k_\omega = \frac{\nu_{eh}}{\mu c}\frac{\omega_{pl}^2}{\omega_{\text{laser}}^2 + \nu_{eh}^2} \tag{14}$$

where μ is the plasma refractive index, $\omega_{pl} = [n_e e^2/(m_e\,\epsilon_0)]^{1/2}$ is the plasma circular frequency and $\omega_{\text{laser}} = 2\pi\, c/\lambda$ is the laser circular frequency [34]. The total electron collision frequency with heavy particles is given by

$$\nu_{eh} = \langle\sigma_{ei}^k v_e\rangle n_e + \sum_i \langle\sigma_i^{ea} v_e\rangle\ n_i, \tag{15}$$

where σ_{ei} and σ_i^{ea} are the electron impact momentum transfer cross sections with the ionic and neutral species i respectively and v_e is the electron velocity. The electron-ion collision frequency is independent of the nature of the ion [35].

The critical laser fluence F_{crit} can then be written as

$$F_{\text{crit}} = 6\pi^2\frac{m_e\epsilon_0 c^3}{e^2}\frac{k_B T_e}{\nu_{eh}\lambda^2}\xi \tag{16}$$

where ϵ_0 is the vacuum permittivity and $\xi = \Delta T_e/T_e$ is a threshold fixed by the experimentalist for the maximum allowed perturbation of the electron temperature by the laser. This value is given for a laser energy per pulse and should be not mistaken with the total laser power which also depends on the laser pulse repetition frequency. The latter will not play a role except at very high frequencies (i.e. in the same order as the relaxation time scales of the plasma) when successive laser pulses can contribute to the global perturbation of the plasma. It is noteworthy that the absolute laser perturbation of the electron temperature scales directly with the electron temperature itself. This criterion from [32, 34] gives a safe value for which no perturbation of the plasma by absorption of power from the laser will occur. However, to obtain correct values of the electron temperature, much larger laser fluences can be used. Indeed, the small gain of momentum of the electrons in the laser field is damped by inelastic collisions. Equation (16) gives an upper value of electron heating in the absence of any cooling mechanisms [36]. The experimental increase of electron temperature is always lower than the one expected in the case of collision-free plasma.

In [32], we showed experimentally that laser heating can occur in low ionization degree, but high pressure plasmas. This is because of the high electron-neutral collision frequency that is typically found in the case of non-thermal atmospheric pressure plasmas.

We stress here that small errors in the determination of the electron temperature (and EEDF shape) can generate large over or under-estimations of reaction rates. If one describes electron impact-induced reactions via effective Arrhénius rate coefficients ($k = k_0 \exp(-E_i/k_B T_e)$), the errors in electron temperature will propagate in the densities via the exponential dependence of k in the electron temperature.

Considering the extremely high electric field frequency of the laser compared to the one used to ignite the plasma, whenever an increase in the electron temperature is observed, an increase in the electron density could be expected on the grounds of the ω^2 dependence in equation (14). Although it was not directly reported in [32], we indeed observed an increase in the electron density in the order of a few percents. However, as the signal to noise ratio increases for lower laser fluences, an exact quantification of this relative increase in electron density could not be performed. This is something that could be done in the future.

5.2. Negative ions

In the case of electronegative plasmas, it is also necessary to take into account the photo-detachment of electrons from negative ions

$$A^- + h\nu \rightarrow A + e^- .$$

This effect increases the measured electron densities compared to the case without the Thomson laser pulse. These electrons are produced with an excess translational energy $\Delta E = E_{\text{bound}} - h\nu$ where E_{bound} is the electron affinity of the species A. In some conditions, it is thus possible to detect a distortion in the EEDF due to the production of these extra electrons [37, 38]. However, this can only be seen for low electron density plasmas where the electron thermalization time is sufficiently long compared to the acquisition time of Thomson scattered photons during the laser pulse.

For the typical laser fluences used for Thomson scattering, complete photo-detachment of all negative ions present in the plasma is expected, even for strongly electronegative plasmas [39]. The measured electron densities will therefore be the summation over electrons and all negative ions (i.e. the total positive ions density for quasi-electroneutral plasmas).

In addition to photo-detachment of electrons from negative ions, one also needs to consider the charge desorption from a dielectric surface by photons [40]. This can, in turn, drastically change the pulsed breakdown behavior of a discharge as it was shown for instance by Ambrico et al. for a dieletric barrier discharge [41].

5.3. Optical pumping

A common method to choose the laser wavelength to avoid perturbation of the plasma is to look at the plasma emission spectrum. However, optical transitions with small oscillator strengths can still be easily saturated due to the very high power of the laser needed to perform Thomson scattering. The laser produces energy pulses in the order of 100 mJ. This is about 3 orders of magnitude higher than typical dye laser pulse energies used to perform laser induced fluorescence measurements in the saturation regime [42]. The cross section for resonant absorption is given by [43]

$$\sigma(\nu_{ji}) = \frac{h\nu_{ji}}{c} B_{ij} S(\nu_{ji}) = \frac{1}{4} \lambda_{ji}^2 S(\nu_{ji}) \frac{g_j}{g_i} A_{ji} \tag{17}$$

where $\nu_{ji} = c/\lambda_{ji}$ is the frequency of the laser, $S(\nu_{ji})$ is the normalized line function, B_{ij} and A_{ji} are the Einstein absorption and spontaneous emission coefficient for the upper j and lower i levels.

The A_{ji} of the transitions used for LIF experiments usually lies in the range of 10^6 s^{-1}. Very weak transitions with A_{ij} values below 10^4 s^{-1} can be easily saturated by the high power laser used to perform Thomson scattering measurements in the case of resonant absorption. Wesseling and Kronast [44] found such an effect in a pure helium plasma. A LIF signal due to optical pumping of weak He_2 molecular bands can even affect Rayleigh scattering measurements [45]. These lines often have small Einstein coefficients and will not be detected in a standard optical emission spectrum. It is then recommended, whenever possible, to check the energy diagrams of the species expected to be present in the plasma before choosing the laser wavelength used to perform Thomson scattering [46].

In terms of electron density perturbation, the importance of optical pumping depends on how the excited states relax and whether a significant part of the optically pumped level j will be ionized or not. As an example, in the case of lines from the 4s–4p groups of argon, the maximum increase of electron density due to optical pumping depends on the ratio between the 4s and ion densities and the laser pulse length [42, 47]. Multi-photon ionization processes need also to be considered, but they usually require at least 3 or 4 532 nm photons. However, they can still play a role in some cases [7]. An *a posteriori* check can be performed by monitoring the electron density versus the laser fluence [48].

6. Deviations from Maxwellian equilibrium

Low temperature, non-thermal plasmas usually have electron energy distribution functions which deviate from Maxwell equilibrium [49, 50]. The definition of an 'electron temperature' from Thomson scattering needs to be revisited. The spectrometer that was presented in section 3, is designed to collect only photons in a range of ± 4 nm around the laser wavelength. This means only electrons within 0.1–7 eV. In the simple case of an argon plasma, there are no excited states below 11.5 eV which means that only electrons with energy of 12 eV or higher can excite argon atoms from the ground state. The low energy part of the EEDF thermalizes fast via electron–electron collisions as it can be seen experimentally in figure 5. This means that Thomson scattering detects only the bulk of the electron energy distribution function.

For an intermediate pressure argon microwave plasma discharge, we showed that Thomson scattering measures the temperature of the bulk electrons of the EEDF, and to a very good approximation, it corresponds to the mean electron energy $< E_e >$ [51]. Indeed, more than 95% of the electrons in the EEDF are embedded in the first 5 eV of kinetic energy for an electron 'temperature' in the range of a few eV. The 'electron temperature' measured by Thomson scattering corresponds to the mean electron energy $T_e = 2/3 \cdot < E_e >$. The first correction for non-Maxwellian EEDF is to apply a two-temperature formulation. For a characterization of the EEDF, we can define a bulk temperature T_e (bulk) and a tail temperature T_e (tail) (electrons with energies above 12 eV in the case of an argon plasma). The bulk electron temperature is directly measured by Thomson scattering and the tail of the EEDF can be probed with optical diagnostics. These results can then be compared with a Boltzmann solver to obtain full insight into the shape of the EEDF (for more detail, see for instance [51, 52]).

7. Raman scattering due to air entrainment and molecular admixtures

In the case of an atomic plasma with a small molecular admixture (or a pure molecular plasma), Thomson and rotational Raman scattering are both collected (section 2). The Raman signal overlapping with the Thomson signal comes from the molecules in their ground electronic (E^0) and vibrational state (v=0). The density of rovibrational states can be expressed as [53]

$$n_i(J, v) = n_i \frac{g_J(2J+1)}{Z} \exp\left(\frac{-B \cdot J(J+1)}{k_B T_{rot}}\right) \exp\left(\frac{-E_v}{k_B T_{vib}}\right) \tag{18}$$

where Z is the partition function, g_J is the nuclear spin degeneracy of the molecule, B is the rotational constant and E_v is the energy of the vibrational level. To obtain a correct estimation of the molecular density, it is then necessary to correct for its vibrational temperature. In the case of nitrogen with T_{vib} = 5000 K, almost 50% of the ground state is already lost in higher vibrational levels. One way to take into account the depletion of $n_i(J, 0)$ is to estimate T_{vib} via optical emission spectroscopy [3]. A more accurate method would be to perform vibrational Raman scattering [54] or CARS [15].

In reference [20], significant amounts of cold nitrogen molecules were found inside the plasma jet, mixed with the Thomson signal. A serious discrepancy between the Rayleigh and Raman temperatures was observed (see figure 13 in [20]) and most curiously, the rotational temperature was found to be lower than the translational temperature. The rotational temperature in atmospheric pressure plasmas is usually very close to the translational gas temperature [4] which is directly measured by Rayleigh scattering due to fast thermalization. However, lateral oscillations of the plasma were reported in this study as well. One should be aware that the nitrogen molecules transported in the plasma will be excited in vibrational states as well and will not be detected by ground state rotational Raman scattering. Consequently, the nitrogen gas heated up by electron collisions with a higher rotational temperature will not be measured due to the depletion of the $n_i(J, 0)$ state due to the simultaneous vibrational heating. This could explain why the Raman temperature was found to be lower than the Rayleigh temperature. Indeed, Raman scattering will detect only nitrogen molecules present around the plasma jet and only a fraction of the molecules which are heated up (rotationally and vibrationally) while diffusing into the plasma region. Last but not least, because of its limited spectral range, the TGS does not allow precise measurements of Raman spectra with rotational temperatures above ≈800 K because of the large spectral range of the rotational bands.

8. Some applications of Thomson scattering and outlook

A rather extensive review of the possible applications of Thomson scattering was done recently by Muraoka and Kono [7]. Therefore, we will only review some of the recent work done at the Elementary Processes in Gas Discharges (EPG) group in Eindhoven.

Over the last few years, Thomson scattering was successfully developed to measure the electron density in cold atmospheric pressure plasmas in ambient air [20, 29, 30, 55]. This was made possible by developing a new fitting routine to take into account Raman and Thomson scattered signals simultaneously. This allowed us to significantly improve the Thomson's scattering method accuracy for the determination of the electron density and mean electron energy (see sections 4 and 7).

Thomson scattering also allows us to obtain spatial profiles of the electron density with high resolution both in space [16, 55] and time domains [30, 56]. A temporal resolution of 5 ns is easily achievable [30]. Current iCCD camera gating times cannot be reduced much more than 200 ps [57]. The lowest time resolution available to perform Thomson scattering measurements will in the end be defined either by the gating time of the camera or the laser pulse length (and the jitter of both). It is interesting to note that Kritcher *et al* succeeded in obtaining an unprecedented time resolution of 10 ps using x-rays to perform Thomson scattering [58]. Although, in this paper, we discussed only the cases of plasmas for which the scattering parameter α is much smaller than 1 (i.e. Thomson scattering is in the incoherent regime), measurements were also done previously in our group on very high density plasmas with $n_e > 1 \cdot 10^{22}$ m^{-3} [59]. These are more simple measurements as the signal to noise ratio is usually very good and the plasmas are very close to thermal equilibrium [60]. For electron densities in that range, it is actually possible to perform single shot Thomson scattering experiments [61].

Thomson scattering has also many advantages in the field of analytical plasmas [62]. In [42], a laser pump-probe experiment was used to determine the electron impact transfer rate coefficients between argon 4s states inside a microwave plasma. Besides the inherent accuracy of Thomson scattering, it allows us to obtain local values of the electron density and temperature at the same point of space where the laser pump-probe experiments are performed. High electron density plasmas are needed to study electron collisional transfer kinetics. The ratio between electron and heavy particle total quenching rates for electronic states used in optical emission spectroscopy are typically in the order of 10^{-4}. An ionization degree in the order of 10^{-3} is thus preferable to perform sensitive measurements. This can be achieved only with relatively high pressure plasmas for which probe measurements cannot be used to determine the electron density. Thomson scattering is then an essential tool to obtain accurate measurements of the electron parameters for such studies. In this respect, microwave plasmas are candidates of choice as they are very stable plasma sources with comparatively much higher electron densities than other non-thermal plasma sources [63].

The Thomson scattering setup described in this paper is optimized for the measurements of EEDF in the low energy range (0.1–7 eV), This can be changed if one wants to gain insight into the tail of the EEDF [59] but it would require some serious re-design of the setup. In section 4, the importance of the iCCD camera sensitivity was emphasized. The overall performance of the TGS could however be improved significantly by investing in newer gratings with much higher efficiency than those currently used. Another current improvement is the installation of a second CCD camera after the grating G1 to detect Rayleigh scattering directly. It will allow us to measure the gas temperature at the exact same location as Thomson scattering, but with much higher sensitivity than in the current design. This is planned for future experiments.

Acknowledgments

The authors wish to thank the Dutch Technology Foundation STW for funding under the project number 10744. The authors would like also to acknowledge all their colleagues who participated in the experiments over the last few years, and especially L Baede, E van Veldhuizen, S Hübner and B van Gessel.

References

[1] Lieberman M A and Lichtenberg A J 2005 *Principles of Plasma Discharges and Materials Processing* (New York: Wiley)
[2] Muraoka K and Maeda M 2010 *Laser-Aided Diagnostics of Plasmas and Gases* (Boca Raton, FL: CRC Press)
[3] Dilecce G 2014 Optical spectroscopy diagnostics of discharges at atmospheric pressure *Plasma Sources Sci. Technol.* **23** 015011
[4] Bruggeman P J, Sadeghi N, Schram D C and Linss V 2014 Gas temperature determination from rotational lines in non-equilibrium plasmas: a review *Plasma Sources Sci. Technol.* **23** 023001
[5] Hutchinson I H 2005 *Principles of Plasma Diagnostics* (Cambridge: Cambridge university press)
[6] Hagelaar G J M and Pitchford L C 2005 Solving the Boltzmann equation to obtain electron transport coefficients and rate coefficients for fluid models *Plasma Sources Sci. Technol.* **14** 722
[7] Muraoka K and Kono A 2011 Laser thomson scattering for low-temperature plasmas *J. Phys. D: Appl. Phys.* **44** 043001
[8] Godyak V A and Demidov V I 2011 Probe measurements of electron-energy distributions in plasmas: what can we measure and how can we achieve reliable results? *J. Phys. D: Appl. Phys.* **44** 233001
[9] Bowden M D, Kogano M, Suetome Y, Hori T, Uchino K and Muraoka K 1999 Comparison of electron property measurements in an inductively coupled plasma made by langmuir probe and laser Thomson scattering techniques *J. Vac. Sci. Technol.* A **17** 493
[10] Overzet L J and Hopkins M B 1993 Comparison of electrondensity measurements made using a Langmuir probe and microwave interferometer in the gaseous electronics conference reference reactor *J. Appl. Phys.* **74** 4323
[11] Piejak R B, Al-Kuzee J and St J BraithwaiteN 2005 Hairpin resonator probe measurements in rf plasmas *Plasma Sources Sci. Technol.* **14** 734
[12] van de Sande M J 2002 Laser scattering on low temperature plasmas-high resolution and stray light rejection *PhD Thesis* Eindhoven University of Technology, The Netherlands
[13] Siemon R E 1974 Polychrometer with extreme rejection of stray light *Appl. opt.* **13** 697–9
[14] Kono A and Nakatani K 2000 Efficient multichannel Thomson scattering measurement system for diagnostics of low-temperature plasmas *Rev. Sci. Instrum.* **71** 2716
[15] Kempkens J and Uhlenbusch H 2000 Scattering diagnostics of low-temperature plasmas (Rayleigh scattering, Thomson scattering, CARS) *Plasma Sources Sci. Technol.* **9** 492–506
[16] Carbone E A D, Hübner S, Palomares J M and van der Mullen J J A M 2012 The radial contraction of argon microwave plasmas studied by Thomson scattering *J. Phys. D: Appl. Phys.* **45** 345203
[17] Huang M, Warner K, Lehn S and Hieftje G M 2000 A simple approach to deriving an electron energy distribution from an incoherent Thomson scattering spectrum *Spectrochim. Acta Part B: At. Spectrosc.* **55** 1397–410

[18] Sneep M and Ubachs W 2005 Direct measurement of the Rayleigh scattering cross section in various gases *J. Quant. Spect. Radiat. Transfer* **92** 293–310
[19] Mack M E, Carman R L, Reintjes J and Bloembergen N 1970 Transient stimulated rotational and vibrational Raman scattering in gases *Appl. Phys. Lett.* **16** 209
[20] van Gessel A F H, Carbone E A D, Bruggeman P J and van der Mullen J J A M 2012 Laser scattering on an atmospheric pressure plasma jet: disentangling Rayleigh, Raman and Thomson scattering *Plasma Sources Sci. Technol.* **21** 015003
[21] Palmer Christopher A, Loewen Erwin G and Thermo R G 2005 *Diffraction Grating Handbook* (New York: Newport Corporation Springfield)
[22] Bakker L P, Freriks J M, De Hoog F J and Kroesen G M W 2000 Thomson scattering using an atomic notch filter *Rev. Sci. Instrum.* **71** 2007–14
[23] Evans D E and Katzenstein J 1969 Laser light scattering in laboratory plasmas *Rep. Prog. Phys.* **32** 207
[24] Gerry E T and Rose D J 1966 Plasma diagnostics by thomson scattering of a laser beam *J. Appl. Phys.* **37** 2715
[25] Hassaballa S, Yakushiji M, Kim Y-K, Tomita K, Uchino K and Muraoka K 2004 Two-dimensional structure of PDP micro-discharge plasmas obtained using laser Thomson scattering *IEEE Trans. Plasma Sci.* **32** 127–34
[26] Huang M and Hieftje G M 1985 Thomson scattering from an ICP *Spectrochim. Acta Part B: At. Spectrosc.* **40** 1387–400
[27] Carbone E A D, Hübner S, Iordanova E, Palomares J M and van der Mullen J J A M 2011 Thomson scattering imaging of the very end of surfatron plasmas *IEEE Trans. Plasma Sci.* **39** 2558–9
[28] Huebner S 2013 Poly-diagnostic study of low pressure microwave plasmas *PhD Thesis* Technical University of Eindhoven
[29] van Gessel B, Brandenburg R and Bruggeman P 2013 Electron properties and air mixing in radio frequency driven argon plasma jets at atmospheric pressure *Appl. Phys. Lett.* **103** 064103
[30] Hübner S, Hofmann S, van Veldhuizen E M and Bruggeman P J 2013 Electron densities and energies of a guided argon streamer in argon and air environments *Plasma Sources Sci. Technol.* **22** 065011
[31] Hübner S, Palomares J M, Carbone E A D, van der Mullen J J A M 2012 Investigating a coaxial microwave plasma *Microwave Discharges: Fundamentals, Applications (MD-8)* p 40
[32] Carbone E A D, Palomares J M, Hübner S, Iordanova E and van der Mullen J J A M 2012 Revision of the criterion to avoid electron heating during laser aided plasma diagnostics (LAPD) *J. Instrum.* **7** C01016
[33] Hughes T 1975 *Plasmas and Laser Light* (Bristol: Adam Hilger)
[34] Carbone E A D, Palomares J M, Hübner S, Iordanova E and van der Mullen J J A M 2013 Erratum: revision of the criterion for avoiding electron heating during Laser Aided Plasma Diagnostics (LAPD) *J. Instrum.* **8** E05001
[35] Mitchner M and Kruger C H 1973 *Partially Ionized Gases* (New York: Wiley)
[36] Dzierzega K, Zawadzki W, Pokrzywka B and Pellerin S 2006 Experimental investigations of plasma perturbation in Thomson scattering applied to thermal plasma diagnostics *Phys. Rev.* E **74** 026404
[37] Noguchi M, Ariga K, Hirao T, Suanpoot P, Yamagata Y, Uchino K and Muraoka K 2002 Laser Thomson scattering measurements of negative ion density in a glow discharge plasma *Plasma Sources Sci. Technol.* **11** 57
[38] Noguchi M, Hirao T, Shindo M, Sakurauchi K, Yamagata Y, Uchino K, Kawai Y and Muraoka K 2003 Comparative studies of the laser Thomson scattering and Langmuir probe methods for measurements of negative ion density in a glow discharge plasma *Plasma Sources Sci. Technol.* **12** 403
[39] van de Wetering F M J H, Beckers J and Kroesen G M W 2012 Anion dynamics in the first 10 milliseconds of an argonacetylene radio-frequency plasma *J. Phys. D: Appl. Phys.* **45** 485205
[40] Guaitella O, Marinov I and Rousseau A 2011 Role of charge photodesorption in self-synchronized breakdown of surface streamers in air at atmospheric pressure *Appl. Phys. Lett.* **98** 071502
[41] Ambrico P F, Ambrico M, Šimek M, Colaianni A, Dilecce G and De Benedictis S 2009 Laser triggered single streamer in a pin-to-pin coplanar dielectric barrier discharge *Appl. Phys. Lett.* **94** 231501
[42] Carbone E A D, Hübner S, van der Mullen J J A M, Kroesen G M W and Sadeghi N 2013 Determination of electron-impact transfer rate coefficients between argon $1s_2$ and $1s_3$ states by laser pump-probe technique *J. Phys. D: Appl. Phys.* **46** 415202
[43] Hilborn R C 1982 Einstein coefficients, cross sections, *f* values, dipole moments, and all that *Am. J. Phys.* **50** 982–6
[44] Wesseling H-J and Kronast B 1996 Thomson light scattering measurements of electron temperature and density in the $\alpha - \gamma$ transition of a capacitive rf discharge in helium *J. Phys. D: Appl. Phys.* **29** 1035
[45] Soltwisch H and Kaczor M 1999 Potential and limitations of Thomson scattering in reactive plasmas *Proc. Front. Low Temp. Plasma Diag.* **3** 77
[46] Suanpoot P, Narishige S, Yamagata Y, Uchino K, Muraoka K and Noguchi M 2002 Applicabilities of laser Thomson scattering to various kinds of discharge plasmas *J. Plasma Fusion Res.* **78** 242–7
[47] Carbone E A D 2013 Laser diagnostics and modelling of microwave plasmas *PhD Thesis* Technical University of Eindhoven
[48] Muraoka K, Uchino K, Yamagata Y, Noguchi Y, Mansour M, Suanpoot P, Narishige S and Noguchi M 2002 Laser Thomson scattering studies of glow discharge plasmas *Plasma Sources Sci. Technol.* **11** A143
[49] Godyak V A 2006 Nonequilibrium EEDF in gas discharge plasmas *IEEE Trans. Plasma Sci.* **34** 755
[50] Kortshagen U 1993 On the influence of energy transfer efficiency on the electron energy distribution function in HF sustained rare gas plasmas: experimental and numerical study *J. Phys. D: Appl. Phys.* **26** 1230
[51] Carbone E A D, Hübner S, Jimenez-Diaz M, Palomares J M, Iordanova E, Graef W A A D, Gamero A and van der Mullen J J A M 2012 Experimental investigation of the electron energy distribution function (EEDF) by Thomson scattering and optical emission spectroscopy *J. Phys. D: Appl. Phys.* **45** 475202
[52] Crintea D L, Czarnetzki U, Iordanova S, Koleva I and Luggenhölscher D 2009 Plasma diagnostics by optical emission spectroscopy on argon and comparison with thomson scattering *J. Phys. D: Appl. Phys.* **42** 045208
[53] McQuarrie D A 2000 *Statistical Mechanics* (Sausalito, Calif: University Science Books)
[54] Lapp M, Penney C M and Goldman L M 1973 Vibrational Raman scattering temperature measurements *Opt. Commun.* **9** 195–200
[55] van Gessel A F H, Carbone E A D, Bruggeman P J and van der Mullen J J A M 2011 Simultaneous Thomson and Raman scattering on an atmospheric-pressure plasma jet *IEEE Trans. Plasma Sci.* **39** 2382–3

[56] Hübner S, Carbone E, Palomares J M and van der Mullen J 2014 Afterglow of argon plasmas with H_2, O_2, N_2, and CO_2 admixtures observed by Thomson scattering *Plasma Process. Polym.* **11** 482–8
[57] Kieft E R, Groothuis C H J M, van der Mullen J J A M and Banine V 2005 Subnanosecond Thomson scattering setup for space and time resolved measurements with reduced background signal *Rev. Sci. Instrum.* **76** 093503
[58] Kritcher A L *et al* 2008 Ultrafast x-ray Thomson scattering of shock-compressed matter *Science* **322** 69–71
[59] Kieft E R, van der Mullen J J A M, Kroesen G M W, Banine V and Koshelev K N 2004 Collective Thomson scattering experiments on a tin vapor discharge in the prepinch phase *Phys. Rev.* E **70** 056413
[60] Murphy A B 2002 Electron heating in the measurement of electron temperature by Thomson scattering: are thermal plasmas thermal? *Phys. Rev. Lett.* **89** 8–11
[61] van der Mullen J J A M, van de Sande M J, de Vries N, Broks B, Iordanova E, Gamero A, Torres J and Sola A 2007 Single-shot thomson scattering on argon plasmas created by the microwave plasma torch; evidence for a new plasma class *Spectrochim. Acta Part B: At. Spectrosc.* **62** 1135–46
[62] Warner K and Hieftje G M 2002 Thomson scattering from analytical plasmas *Spectrochim. Acta Part B: At. Spectrosc.* **57** 201–41
[63] Moisan M and Wertheimer M R 1993 Comparison of microwave and r.f. plasmas: fundamentals and applications *Surf. Coat. Technol.* **59** 1–3

IOP Publishing | Plasma Physics and Controlled Fusion

Plasma Phys. Control. Fusion **57** (2015) 014027 (9pp) | doi:10.1088/0741-3335/57/1/014027

Magnetic perturbations as a viable tool for edge turbulence modification

N Vianello[1], C Rea[1], M Agostini[1], R Cavazzana[1], G Ciaccio[1], G De Masi[1], E Martines[1], A Mazzi[2], B Momo[1], G Spizzo[1], P Scarin[1], M Spolaore[1], P Zanca[1], M Zuin[1], L Carraro[1], P Innocente[1], L Marrelli[1], M E Puiatti[1], D Terranova[1] and the RFX-mod Team

[1] Consorzio RFX (CNR, ENEA, INFN, Università di Padova, Acciaierie Venete SpA), Corso Stati Uniti 4, 35127 Padova, Italy

[2] Dipartimento di Fisica, Università degli Studi di Trento, I-38123 Povo, Trento, Italy

E-mail: nicola.vianello@igi.cnr.it

Received 28 July 2014, revised 12 September 2014
Accepted for publication 24 September 2014
Published 28 November 2014

Abstract

A complete description of the effects of magnetic perturbation on the edge region of RFX-mod is here reported. The flexibility of the RFX-mod device [1] allows for the operation of the machine both as a reversed field pinch (RFP, with maximum current 2 MA) and as a low-current, circular ohmic tokamak ($I_{p,max} = 0.15$ MA). The present paper summarizes the most recent results obtained in both configurations with either spontaneous or induced edge radial magnetic perturbation. Emphasis will be devoted to the experimental characterization of the edge flow, focusing on the phase relation between flow and perturbed magnetic field. These informations are provided for natural and stimulated helical discharges in RFPs, and for tokamak safely operated, thanks to the unique RFX-mod MHD control system, in a wide range of edge safety factor $1.9 \lesssim q(a) \leqslant 3.4$ with externally imposed helical boundary. For the first time a detailed comparison between this phenomenology in tokamaks and RFPs will be presented, providing experimental measurement of the streamline of $\mathbf{E} \times \mathbf{B}$ flow around the magnetic perturbation and of the density modulation which exhibits the same periodicity of the perturbation. Strong new indication of the modification of the small scale turbulence in presence of magnetic perturbation is reported: this modification is deeply connected to the variation of turbulence induced particle transport.

Keywords: electric field, RMP, turbulence

(Some figures may appear in colour only in the online journal)

Magnetic islands and magnetic perturbations in the edge of fusion experiments are almost ubiquitous in present tokamak, stellarator and reversed field pinch (RFP) devices. They are inherent to the configuration dynamics (possibly interacting with remnant magnetic field errors, see for example [2, 3]), or externally produced with resonant and non resonant magnetic perturbation (RMP) experiments [4, 5]. The interaction between magnetic islands and plasma is a subject rather more general than fusion-related topic, and it is observed for example in astrophysical system and magnetoplasma dynamics as well [6]. The interest for a fusion plasma is primarily related to the fact that the use of a slightly stochastic magnetic field is observed to mitigate the explosive events known as edge localized modes (ELMs): future devices like ITER strongly rely on RMPs for a safe operation in ELMy H-Mode, in order to reduce the power load to divertor plates, which are expected to be unacceptable during unmitigated large ELM events [7].

Although the large interest and a wide experimental effort in all the present devices, a large number of unanswered questions still remains: among them we can cite the transport mechanism responsible for the observed density pump-out [8] during RMP experiments, the evaluation of magnetic flutter transport in the pedestal region [9], the estimate of the predicted modification of zonal-flow [10] partially confirmed by changes observed in long range correlation (LRC) during DED experiment in TEXTOR [11], the different behaviour of the applied perturbation as a function of pedestal collisionality ν^* [4] and edge safety factor parameter dependence [12].

Within this general framework, the present paper reports the most recent experimental observations and theoretical interpretations of the effects of spontaneous and applied magnetic perturbation in the edge region of RFX-mod, operated both as a reversed field pinch and as an ohmic circular tokamak. We will provide a detailed description of the $\mathbf{E} \times \mathbf{B}$ flow and anomalous transport which are found to be tightly linked to magnetic topology. For the RFP case, data will be presented both in spontaneous high current helically shaped plasmas and in low current discharge with an applied helical magnetic perturbation. For the tokamak case, the flexibility of the RFX-mod MHD control system [13] allows for safely operating the discharges also with $q(a) \gtrsim 2$: we have thus used the MHD control to stimulate a helical perturbation also in this unusual condition where a large MHD $(m, n) = (2, 1)$ mode resonates close to the wall. In both configurations a large modification of the plasma flow is observed, with a large vortex-like pattern building up around the edge magnetic islands. This velocity pattern produces local density accumulation/convection with a consequent modification of local gradients. Variation of flow and gradients are found to profoundly affect the properties of small scale electromagnetic fluctuations: in particular a process of selection of intrinsic vorticity based on local variation of $\mathbf{E} \times \mathbf{B}$ flow pattern is observed and spatial distribution of vortices is found to follow local plasma boundary deformation. The variation of small scale fluctuations will have consequences on turbulence induced particle losses. The present paper will consequently address the issue of transport in presence of magnetic perturbation, focusing on electrostatic fluctuations and flow modification, an providing the information in a wide range of edge safety factor $q(a)$.

The paper is organized as follow: section 1 describes the RFX-mod device and the diagnostic used. In section 2 experimental results obtained in the helical boundary, RFP low-current plasma are presented. In section 3 the tokamak results are summarized whereas in section 4 a general comparison will be done between the two configurations. Finally, the conclusion will be drawn in the last section.

1. The RFX-mod device

RFX-mod [1] is the largest RFP device presently in operation with a major radius $R = 2$ m and a minor radius $a = 0.459$ m. It is equipped with an advanced system for MHD control, composed of 192 independently fed saddle coils [14]. Through the amelioration of the plasma boundary and reduction of edge radial magnetic field, high current operation with plasma current up to 2 MA has been possible. At high plasma current ($I_p \geqslant 1$ MA) the RFP exhibits a transition from a globally chaotic magnetic topology with a broad spectrum of kink-tearing modes, towards helical equilibrium set by a single mode (usually the innermost resonant one). The amplitude of the mode is sufficient to modify the flux surfaces, creating a helical core surrounded by a quasi-symmetric boundary. The appearance of these helically states is accompanied by an amelioration of plasma performance, and in some cases by the appearance of an internal electron transport barrier [15]. In these helical regimes the helical ripple at the edge is relatively small, of the order of 1% of the total field, but nevertheless it is sufficient to modulate edge kinetic properties and plasma wall interaction (PWI) [16, 17], consistently with what has been reported in [18] in the case of distorted plasma boundary in presence of saturated MHD modes. Some preliminary indication of the effects of helical boundary on small scale turbulence has been already reported [19] but in order to allow probe insertion and turbulence analysis the helical plasma has been forced at lower current ($I_p \lesssim 450$ kA) with an externally applied helical perturbation [20]. The insertion of probe head, extensively described elsewhere [21, 22], allows for a proper characterization of the perpendicular flow, density, temperature and fluxes, plus the determination of the electromagnetic properties of small scale blobs/filaments [21]. The probe head indeed consists of a 2D array of electrostatic pins, arranged in a five-pin triple balanced probe [23] in order to provide high frequency estimate of plasma potential, electron density and electron temperature plus. The arrangement of the pins allows also the estimate of the plasma vorticity as the $\nabla_\perp^2 V_f$ [24] and the local current filaments fluctuations from the direct measurement of the $\nabla \times \tilde{b}$ obtained from a 2D array of tiny pick-up coils. In particular data reported hereafter have been taken in discharges with $I_p \lesssim 450$ kA, normalized greenwald fraction $n/n_G \simeq 0.3$–0.4 and edge safety factor $q(a) \approx -0.004$. A helical perturbation of helicity $(m, n) = (1, -7)$, rotating with a frequency 10 Hz, has been applied throughout the entire flat-top.

In addition to RFP operation RFX-mod can now routinely operate as a low current ohmic circular tokamak with the first wall representing the limiter, as already reported (see for example [25]). The maximum plasma current achieved is $I_p = 0.15$ MA, with a maximum toroidal magnetic field $B_\phi = 0.55$ T, corresponding to an edge safety factor $q(a) = 1.7$. This low-q operation was made possible only by means of a feedback control on the $(m, n) = (2, 1)$ mode, which otherwise grows too large and disrupts the discharge. The discharges presented in the paper exhibit a controlled variation from $q(a) \simeq 2.1$ to $q(a) > 3$ obtained by decreasing the plasma current. The flexibility of the MHD control system has allowed to vary the helical boundary condition within the same discharge with a (2, 1) radial perturbation during the $q < 3$ phase, whereas in the $q(a) > 3$ a magnetic perturbation on modes with the same toroidal periodicity (both $n = 1$ and $n = 2$) and poloidal mode numbers $m = -1, 0, 1, 2$ has been applied. This produces a perturbation of the magnetic field with a bulging of the plasma towards the external equatorial plane, similar to the one obtained with the so-called C-Coils in DIII-D [26]. Data presented hereafter refer to the first part of the discharge with a single applied harmonic.

In figure 1, examples of the discharges analysed in the present paper are shown: they represent both High and Low current RFP discharges, respectively with a spontaneous helical perturbation and an imposed helical boundary, plus a tokamak discharges.

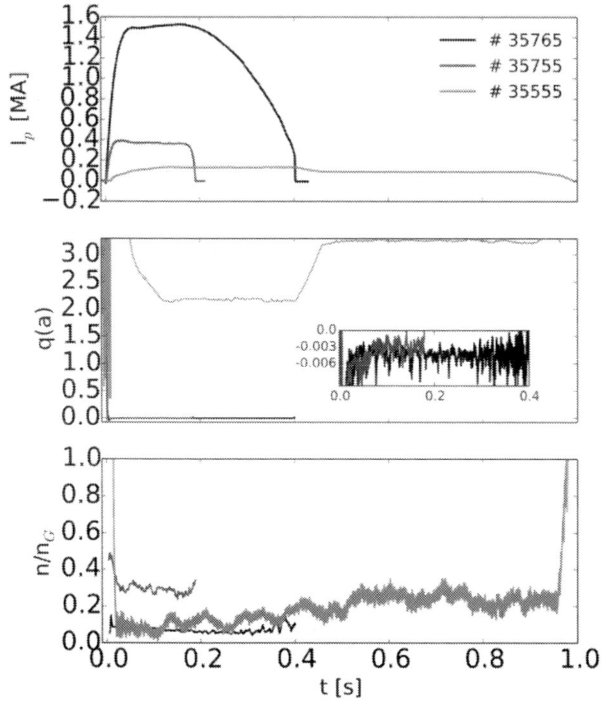

Figure 1. Top: plasma current waveform in high current QSH RFP (black, #35765), low current with helical boundary RFP (blue, #35755) and tokamak discharges (red, #35555). With the same color code are also reported and the vale of $q(a)$ (middle) and the normalized greenwald fraction (bottom). In the middle panel the inset reports the zoomed values of $q(a)$ for the two RFP cases.

It is worthwhile placing the RFX-mod database in the classic operational plane for describing ELM suppression, featuring normalized pedestal collisionality against Greenwald density fraction. Figure 2 shows this operational space as taken from [4]: three type of RFX-mod discharges are shown on top of the existing database. In the case of RFX-mod, lacking an appropriate definition of a pedestal region, we will consider values at $r/a \approx 0.95$. It is also worth to remember that RFX-mod tokamak operation is limited to ohmic L-mode where ELMs are not present. For the RFP case, due to the different magnetic topology, an adaptation of the standard formula for bounce frequency [27] has been derived, where the maximum bounce frequency can be written as

$$\omega_{\max} = \frac{v_{\rm th}}{\sqrt{2}R}\sqrt{\frac{R}{rf}}\sqrt{\frac{r}{r-\Delta'}} \qquad (1)$$

being $f = B_{\hat{\theta}} / B_0$. Corrections are needed in order to take into account the vanishing $q = 0$ surface and the fact that magnetic field is mainly poloidal towards the edge. The operational space covered by the present experiments enlarge the established one at higher collisionality, in particular at lower plasma current. On the other side the RFP is also known to generally operate at higher beta with respect to Tokamaks and for the present experiments a local electron β of about 1% may be estimate. This is important as a β dependence is expected to modify the behavior of magnetic field penetration [28], and in general to enhance the electromagnetic nature of

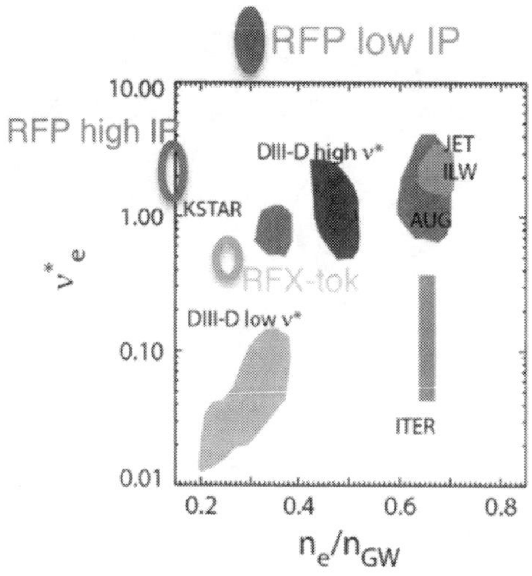

Figure 2. Operational parameter space in terms of normalized greenwald fraction and pedestal/edge collisionality. Figure adapted from [4].

plasma turbulence [29, 30], exploring ranges interesting for future devices but generally not covered by present machines. To complete the characterization of local plasma parameters relevant for transport studies, we can mention that the typical values of $\rho^* = \rho_s/a$ are in the range of 0.002 for the RFP plasmas up to 0.006 for the RFX-mod tokamak discharges, thus in the same range, in terms of a-dimensional parameters, as larger machines [31]. We can consequently conclude that information presented hereafter are in the correct framework for the comprehension of the interaction between magnetic perturbation and kinetic properties. The already reported commonalities between tokamaks and RFPs in terms of ambipolar electric field [32], anomalous transport [33], electromagnetic structure characteristics [21, 24], suggest that a comparison of the effects of the topology on flow and transport on these two configuration could allow to disentangle the inherent fundamental mechanisms which regulates this physical problem, allowing a deeper comprehension of the topic.

2. RFP operation

High current RFP operation reveals the tendency of the plasma to produce an helical equilibrium featuring ordered magnetic surfaces where a single $m = 1$ mode (generally the most internally resonating $n = -7$) becomes dominant. This RFP state, dubbed quasi single helicity state (QSH) is substantially different with respect to what is generally observed at lower current, where a wide spectrum of $m = 1$ modes produces magnetic chaos in the plasma core. The chaotic case is dubbed multiple helicity (MH) state. It has been widely reported [16, 17, 19] that although the residual edge helical ripple is of the order of 1% this is sufficient to modulate all the plasma kinetic quantities. In particular density, pressure and flow are all found to exhibit an oscillation correlated with the toroidally rotating dominant magnetic perturbation. In order to properly handle the phase relation with the magnetic

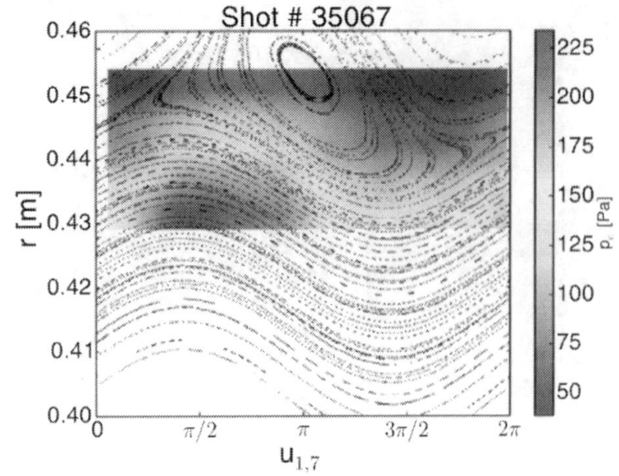

Figure 3. Colour map of the pressure as a function of the helical angle and radius superimposed to the Poincarè plot of the underlying magnetic structure.

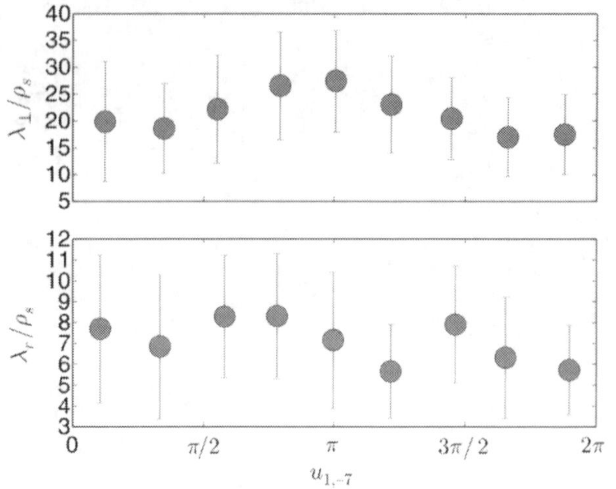

Figure 4. Toroidal (top) and radial (bottom) blob dimension, normalized to the local ion sound gyroradius ρ_s as a function of the helical angle. Measurements refer to $r/a = 0.97$.

perturbation, the helical angle $u_{m,\, n}$ has been introduced [17]. This is defined as:

$$u_{m,n}(t) = m\theta - n\phi + \varphi^{m,n}(t) \tag{2}$$

where $\varphi^{m,\, n}$ is the proper phase velocity of the rotating magnetic perturbation. The helical angle assumes, by construction, the value of $\pi/2$ whenever the O-point (OP) of the magnetic island passes in front of the observer, whereas assumes value of $3\pi/2$ at the X-point (XP). As an example of the effect of the helical boundary on edge plasma the plasma pressure as a function of the helical angle and radius is shown in figure 3. The pressure pattern is superimposed to the Poincaré map of the magnetic perturbation, where only the dominant mode $(m, n) = (1, 7)$ and its toroidally coupled $m = 0$ component $(m, n) = (0, 7)$ are retained. Poincaré plot has been obtained with the field line tracing code FLiT [34]. The eigenvalues given as input to FLiT are obtained by solving Newcomb's equations in toroidal geometry through the code NCT [35]: constraints are the edge magnetic field measurements of B_r, B_θ, B_ϕ. Although much richer, we have limited the spectrum included in the computation of the Poincaré map to highlight the space relation with respect to the dominant periodicity. Plasma pressure is observed to oscillate as a function of the helical angle, with increased pressure located at $u_{1,\,-7} \approx \pi/2$ at the maximum of the radial displacement induced by the dominant mode [17], mainly due to an increasing of the local plasma density confirmed also by microwave reflectometer measurements [36]. It is worth recalling [19] that the modulation is also observed on the entire edge profile, with a reduction of the pressure characteristic length $L_p = -P_e / \nabla P_e$ at the OP of the $(m, n) = (1, -7)$ island, $u \approx \pi/2$. Investigation obtained through the Gas Puff Imaging diagnostic [36] revealed that the modification of L_p and in general of the edge properties determine the modification also of local small scale fluctuations. The edge region of RFX-mod is indeed characterized by the presence of localized fluctuations, dubbed as blobs/coherent structures, which are filaments elongated in the parallel direction featuring a parallel current, parallel vorticity and local density accumulation [21, 24]. The perpendicular dimensions of these blobs are found to exhibit a modulation consistent with the helicity of the existing perturbation as shown for example in figure 4 where the dimensions have been normalized to the local sound gyroradius $\rho_s = c_s/\Omega_i$. A clear reduction of both dimensions is observed in the region $3/2\pi \lesssim u \lesssim 2\pi$, close to the location of the XP of the $(1, -7)$ islands, or correspondingly in the region of inward plasma shift.

It has also been reported that perpendicular flow is modulated by the helical angle, and this has been confirmed independently by the helicity of the perturbation [37]. Owing to the small contribution coming from the diamagnetic flow the observed velocity modulation reflects the modification of the radial electric field, which has been theoretically explained in terms of ambipolar constraints [37, 38]. Indeed species behave differently in presence of a magnetic ripple, or magnetic islands, with electrons strictly following magnetic field lines due to their smaller drift. In some of cases, in high density regimes, it has been shown that the modulation of the flow causes a reversal in a toroidal sector of the machine, which determines the appearance of a point of density accumulation causing a radiative collapse of the discharge [2]. This was observed as due to a macroscopic $(m, n) = (0, 1)$ magnetic island characterizing those high density regimes.

In order to properly investigate flow and turbulence as a function of the magnetic perturbation, QSH states have been forced at lower plasma current ($I_p \leqslant 450$ kA), imposing a rotating helical boundary with a $(1, -7)$ perturbation. The lower current allows insertion of probes, providing detailed information on both perpendicular components of the $\mathbf{E} \times \mathbf{B}$ flow, high frequency electromagnetic turbulence, plus a direct estimate of the parallel current and parallel vorticity fluctuations [24]. The rationale behind these experiments is the possibility to provide a detailed map of the perpendicular flow perturbation caused by the magnetic perturbation, and to investigate how the global changes affect the turbulence responsible for anomalous transport. A first important result

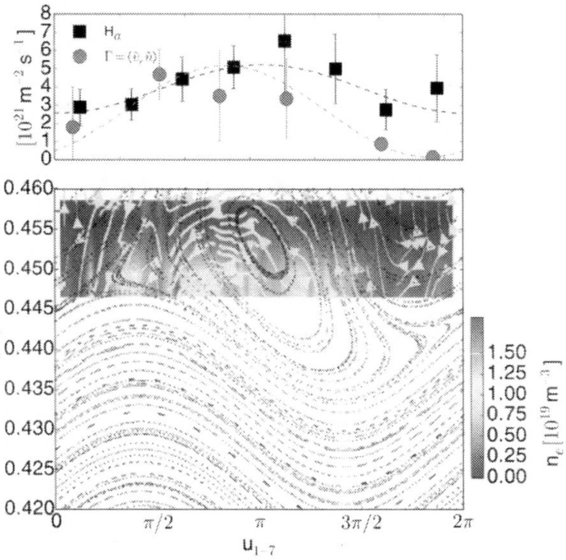

Figure 5. Top: particle influx from H_α emission as a function of the helical angle (black square) and particle flux (at $r = 0.446$) estimated from electrostatic fluctuations (red circle). Bottom: density map in colour code superposed to the **E** × **B** streamline of electric drift flow and Poincarè map. Thickness of the streamlines is proportional to the absolute value of the velocity.

is shown in figure 5: in this plot we report in color-code, as a function of radius and helical angle, the value of density as measured simultaneously by an array two five-pins balanced triple probes. Superimposed with arrows, the streamline plot of the **E** × **B** flow resulting from the combination of spatially resolved measurements of the two perpendicular components of the velocity is reported (light blue in the online color version). To properly compare with the underlying topology the Poincaré map of the magnetic field, reconstructed similarly to figure 2 is also shown. As observed in the high current regimes (see figure 2), density is increased around $u \approx \pi/2$, at the maximum of the helical radial displacement (OP). What is far more important is the observation that pattern of the underlying flow follows rather strictly the magnetic field line, creating sort of vortex-like pattern around the $m = 0$ island. On the same plot, in the top panel the particle influx estimated from H_α emission as a function of the helical angle is shown. One can clearly seen that the emission is enhanced at $u \approx \pi$ where the $m = 0$ island is touching the wall. On the other side following the pattern of the flow, particles injected in the region of higher emission are convected and tend to accumulate around $\pi/2$ where the X-point of the (0, 7) island is located. Thus the observed modulation of the density, which is not homogeneous along the helical angle seems motivated by a modulation of the perpendicular components of the velocity and the already observed trapping effect caused by the presence of a magnetic XP [38]. Still in figure 5 in red dots the particle flux resulting from electrostatic fluctuation, $\Gamma = \langle \tilde{v}_r \tilde{n} \rangle$ is shown, exhibiting an increase in the region $\pi/2 \lesssim u \lesssim \pi$. Particle flux has been derived from density and radial velocity fluctuations both of the quantities sampled at 5 MHz. Computation has been done using standard technique in the frequency domain [39] with an estimate every 10 ms. The data refer to the position $r = 0.446$ or $r/a = 0.97$. As a matter of fact, the modulation of the magnetic topology provides a modulation of the particle losses due to small scale electrostatic fluctuations, with the same periodicity. It is worth remembering that the particle fluxes due to electrostatic fluctuations, account for almost all the particle losses in the edge region of an RFP plasma, despite the high level of magnetic fluctuations. A large fraction of these losses can be imputed to the presence of blobs [40]. In order to verify the possible effect of the magnetic topology on blob properties, the detection frequency of the blob f_b, defined as the number of blobs detected for second, has been computed as a function of the helical angle and reported in figure 6(*a*). We observe that increasing of the particle fluxes coincides with an increase of the frequency of the observed blobs, as also clearly demonstrated in figure 6(*b*), thus strongly relating the enhanced transport to a substantial increase of blob-driven particle losses.

The presence of a magnetic perturbation modifies the entire profile of the **E** × **B** flow. This can be observed for example in figure 7 where the **E** × **B** shear at $r = 0.446$ is shown as a function of helical angle. Velocity profile is found profoundly modified, with an inversion of the value of the shear around $u \approx \pi/2$. On the other side shear is well known to act on turbulent eddies. In particular we are interested in the process of eddies shear selection which is a know feature in both fluids [41] and plasmas [42]. Indeed in plasmas there have been observed blobs with both positive and negative vorticity [42], where positive and negative are defined as parallel and antiparallel with respect to the guiding magnetic field, although with different abundance [42]. If we now compute the ratio, along the helical angle, between blobs with vorticity antiparallel to B_0 with respect to the total number of structures (shown in figure 7 with orange dots), we observe that it is not homogeneous and the fraction is peaked whenever the shear change sign, in accordance to the shear selection rule.

3. Tokamak operation

In order to establish the universality of the mechanism described so far, we have exploited the versatility of RFX-mod device to study the same phenomenology in the tokamak configuration. As shown in figure 1, the operational space of RFX-mod in terms of edge safety faction $q(a)$ is enlarged, thanks to the possibility to operate with $q(a) \gtrsim 2$ with MHD control system mitigating the $(m, n) = (2, 1)$ perturbation. In terms of collisionality, RFX-mod tokamak is roughly positioned at the same values of KSTAR or DIII-D at high ν^*. The experiments presented hereafter refer to the part of the discharge where $q(a) \approx 2.1$, during which an helical radial magnetic perturbation $(m, n) = (2, 1)$ rotating with a frequency of 10 Hz has been applied. As shown in the Poincaré map in figure 8, the perturbation creates a large $(m, n) = (2, 1)$ island whose external convolution touches directly the first wall shown in a thick blue line. The map is the result of the already described field line tracing code FLiT [34], where only the eigenfunctions for perturbation $m = 1$ with toroidal periodicity

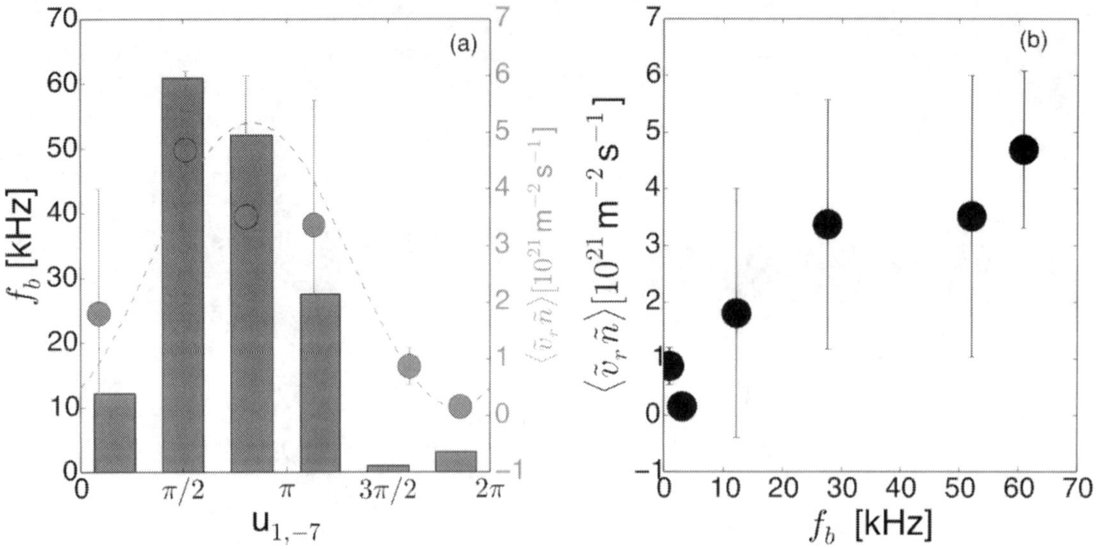

Figure 6. (*a*) Bar plot of blobs frequency f_b, estimated as the number of blobs detected per second, as a function of the helical angle. On the same plot, with right-hand scale the profile of the particle fluxes as a function of the helical angle. (*b*) Particle flux from electrostatic fluctuations as a function of blob detection frequency. Both the measurements refer to position $r = 0.446$.

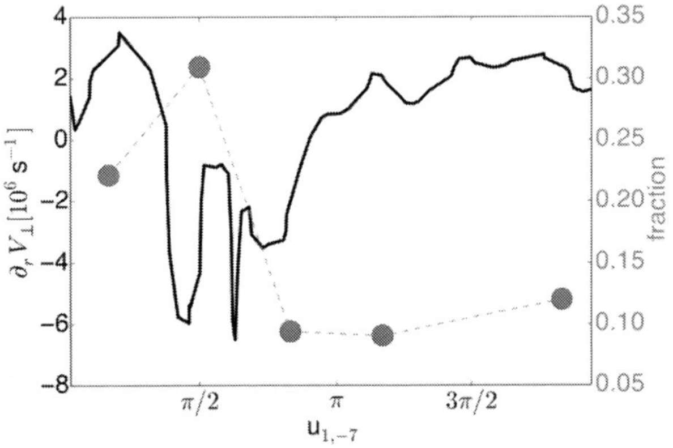

Figure 7. $\partial_r V_\perp$ shear as a function of the helical angle (black line). In red dots the fraction of blobs with vorticity antiparallel to guiding magnetic field with respect to total blob frequency. Both the measurements refer to $r = 0.446$.

$n = 2, 3$ are kept. The eigenfunctions are still calculated with code NCT [35] extended to the tokamak configuration. The measurements presented hereafter are obtained at the radial location indicated by the red dashed line. Assuming that the magnetic island is rigidly rotating with the applied perturbation, we can infer the poloidal map of the various kinetic quantities. The evolution along the perturbation of the radial electric field estimated by plasma potential measurements is shown in the same figure 8. The electric field exhibits an $m = 2$ periodicity and it is strongly reduced at the locations of the X-points of the (2, 1) island. It is worth noting that these are measurement that according to the reconstruction lies inside the magnetic island itself. A plausible interpretation of the profile observed is the following: the presence of the island touching directly the material wall creates a wide scrape off layer region (SOL), with the internal separatrix of (2, 1) mode representing the separatrix between confined plasma and SOL.

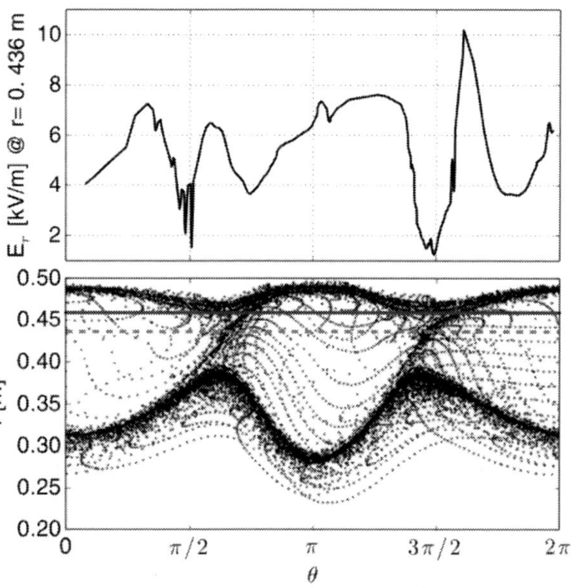

Figure 8. Top: radial electric field at $R = 0.436$, estimated from plasma potential gradient as a function of the poloidal angle, calculated assuming rigid body rotation of the $(m, n) = (2, 1)$ mode. Bottom: Poincaré map of the magnetic field as a function of radius and poloidal angle). The red dashed line indicates the location of the measurements whereas the blue thick line the location of the first wall acting as limiter.

The positive radial electric field is consistent with an higher electron mobility (generally observed in the far SOL, see for example [43]). On the other side at the X-point the separatrix position is moving towards the wall and the reduction of the E_r may be interpreted as a reduction of the difference in electron-ion mobilities and is equivalent to provide a measurement of the electric field in a region closer to the separatrix, where also in standard discharges the E_r approaches the value of 0 and

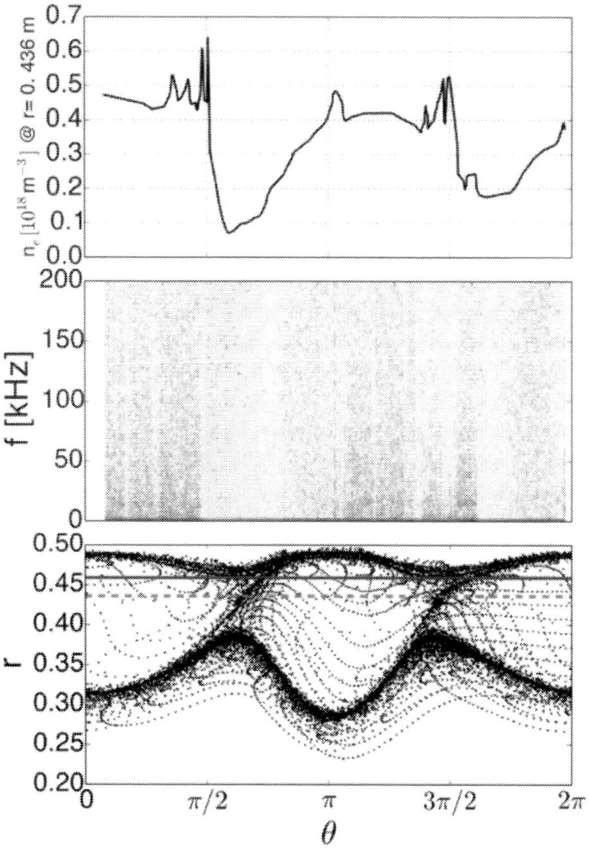

Figure 9. Top: electron density, low-pass filtered at 5 kHz, as a function of the poloidal angle, calculated assuming rigid body rotation of the $(m, n) = (2, 1)$ mode. Middle: electron density spectrogram along the magnetic perturbation. Bottom: Poincaré map of the magnetic field as a function of radius and poloidal angle).

than changes sign (e.g. [43]). This is consistent with measurements performed in LHD [44], where the change of sign of the E_r is suggested to provide a valid measurement of the effective plasma boundary in presence of a 3D magnetic field. Within the same assumption of rigid body rotation we can obtain the poloidal distribution of the electron density shown in figure 9 after a low pass filtering: a clear depletion at the location of the X-points and an enhancement in correspondence to the O-point of the $m = 2$ island can be distinguished. In the same figure we also show the spectrogram of the density fluctuations up to 200 kHz which suggest a strong damping in all the spanned frequency range at the location of the two X-point of the islands. Also in the tokamak configuration the turbulence modification has influence on the particle losses due to electrostatic turbulence. In figure 10 the poloidal profile of the radial particle flux due to electrostatic turbulence is shown. A clear $m = 2$ periodicity can be derived and reduction of the flux is observed in correspondence to the damping of the fluctuations shown in figure 9. To confirm the observation drawn in RFP configuration on the effects of the presence of magnetic island on small scale electrostatic turbulence figure 11 shows the poloidal modulation of the $\mathbf{E} \times \mathbf{B}$ shear as derived from the electric field profile obtained by radially resolved

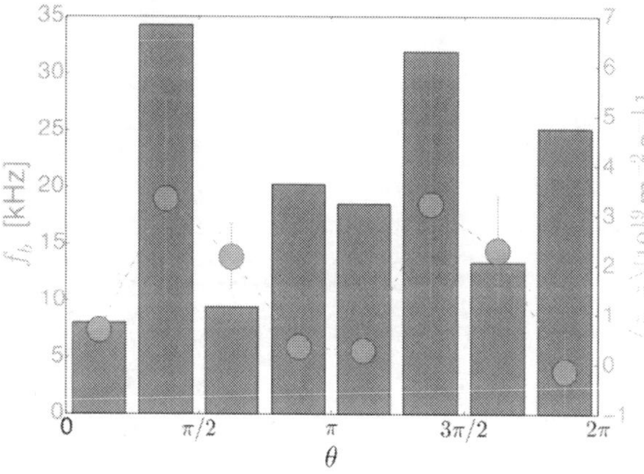

Figure 10. Bar plot with the poloidal distribution of blobs frequency. Superimposed the poloidal distribution of electrostatic turbulence driven particle flux.

plasma potential measurement. Consistently, $m = 2$ periodicity is observed also on the shear, with an abrupt reduction at the location of the O-points of the island, which confirms flat radial electric field profile at the O-point of the island as observed for example in [45]. On the same plot also the fraction of vortices with vorticity anti-parallel to B_0 is shown, and in analogy to the observation obtained in the RFP configuration, a strict correlation with the shear profile is confirmed.

4. Physical interpretation of particle losses

It has been postulated [46] that the presence of eddies with opposite vorticity may change the nature of turbulent transport. This is basically related to the change of probability for particles leaving a given eddies to encounter other eddies with the same or opposite axial vorticity. On the other side it is also known that internal blob spin change their dynamics [47], enhancing their coherence and that for spinning blobs the interaction of the blob with the ambient velocity field depends on the sign of the external shear with respect to the spin. As an indication that this mechanism is likely to take place we report in figure 12 particle transport as a function of the relative abundance of eddies with vorticity anti-parallel to the guiding magnetic field. The analysis seems to indicate an enhancement of losses whenever the fraction increasing suggesting that the presence of eddies with opposite vorticity with approximately the same probability increases the particle losses. This represents an indirect confirmation of what theoretically suggested [46] and represents a confirmation of preliminary indication given in [42]. The mechanism is confirmed by the observation of similar results independently on the configuration.

5. Conclusion

A detailed study of the effect of magnetic perturbation on edge flow and small scale turbulence has been done in RFX-mod. The unique versatility of the device offers the possibility to analyse the same phenomenology in two different

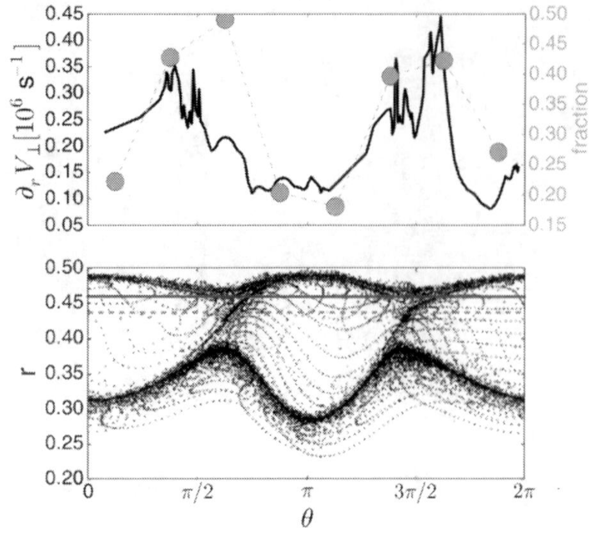

Figure 11. Top panel: in black the poloidal distribution of the $\mathbf{E} \times \mathbf{B}$ shear derived from plasma potential profile. In dots the fraction of vortices anti-parallel to B_0. Bottom panel: Poincaré plot of magnetic field line clearly exhibiting the $m = 2$ magnetic island. The red dashed line indicates the location of the measurements whereas the blue thick line the location of the first wall acting as limiter.

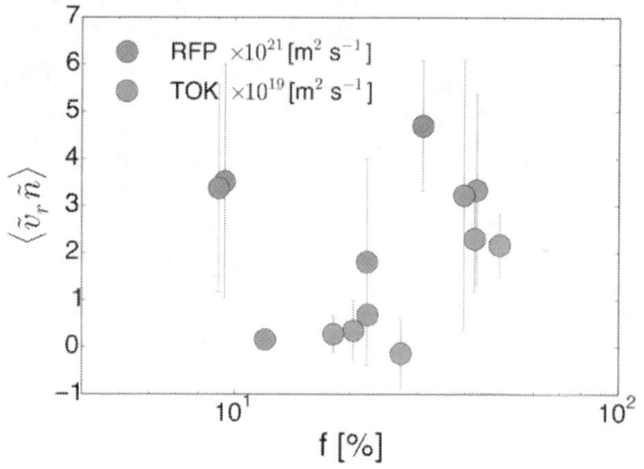

Figure 12. Electrostatic induced particle fluxes as a function of the relative fraction of vortices with axial vorticity anti-parallel to the guiding magnetic field.

configurations, Reversed Field Pinch and Tokamak, on the same machine and with the same diagnostic. Both the configurations reveal a modulation of the electron density which exhibits the same periodicity of the underlying topology. The radial electric field is modulated as well and it also exhibits the same symmetry as the magnetic topology. It has been demonstrated that the presence of magnetic island close to the wall modifies the pattern of $\mathbf{E} \times \mathbf{B}$ flow consistently with the geometry of the island, with the appearance of a convective-cell around the island. The observed flow acts as convective motion which contributes to the observed modulation of plasma density with local accumulation, for the RFP case at the X-point of the islands. Particle fluxes induced by electrostatic turbulence is found to be modulated as well, still exhibiting the same symmetry. The variation of transport is ascribed to a modification of small scale turbulence which is not homogeneous in presence of magnetic perturbation. Indeed both the appearence frequency of radially propagating blobs, and the intrinsic vorticity of these structures are modified, and particles losses increases whenever the frequency increase and the relative abundance of counter-rotating eddies are found to be approximately equivalent. The observations reported indicate that a complete understanding of the processes occurring during RMP experiments should take into account both the convective motion caused by modified electric field and enhanced particle losses due to small scale variation. Furthermore we provide indication that shear flow modification effects on transport is much richer than pure fluctuation amplitude suppression and a that a complete comprehension of transport events should take into account intrinsic properties of turbulent eddies. The unique versatility of RFX-mod, in term of operational space and diagnostic for turbulence analysis, has allowed for a configuration-independent physical description of detailed interaction between topology and transport.

Acknowledgments

This project has received funding from the European Unions Horizon 2020 research and innovation program under grant agreement number 633053. The views and opinions expressed herein do not necessarily reflect those of the European Commission.

References

[1] Sonato P *et al* 2003 *Fusion Eng. Des.* **66–8** 161–8
[2] Puiatti M E *et al* 2013 *Plasma Phys. Control. Fusion* **55** 124013
[3] Feng Y, Kobayashi M, Lunt T and Reiter D 2011 *Plasma Phys. Control. Fusion* **53** 024009
[4] Kirk A, Chapman I T, Evans T E, Ham C, Harrison J, Huijsmans G, Liang Y, Liu Y Q, Loarte A and Suttrop W 2013 *Plasma Phys. Control. Fusion* **55** 124003
[5] Cappello S *et al* 2012 Nonlinear modeling for helical configurations in toroidal pinch systems *Proc. of the 23rd IAEA Fusion Energy Conf. (San Diego, CA, 8–13 October 2012)* pp TH/P2-16
[6] Graham D B, Khotyaintsev Y V, Vaivads A, André M and Fazakerley A N 2014 *Phys. Rev. Lett.* **112** 215004
[7] Loarte A *et al* 2014 *Nucl. Fusion* **54** 033007
[8] Evans T E *et al* 2006 *Nat. Phys.* **2** 419–23
[9] Callen J D, Hegna C and Cole A J 2013 *Nucl. Fusion* **53** 113015
[10] Leconte M, Diamond P H and Xu Y 2014 *Nucl. Fusion* **54** 013004
[11] Xu Y *et al* 2011 *Nucl. Fusion* **51** 063020
[12] Liang Y F *et al* 2011 *Nucl. Fusion* **51** 073001
[13] Zanca P, Marchiori G, Marrelli L and Piron L 2012 *Plasma Phys. Control. Fusion* **54** 124018
[14] Marrelli L *et al* 2007 *Plasma Phys. Control. Fusion* **49** B359
[15] Lorenzini R *et al* 2009 *Nat. Phys.* **5** 570–754
[16] Scarin P *et al* 2011 *Nucl. Fusion* **51** 073002
[17] Vianello N *et al* 2013 *Nucl. Fusion* **53** 073025
[18] Chapman I T, Brunetti D, Buratti P, Cooper W A, Graves J P, Harrison J R, Holgate J, Jardin S, Sabbagh S A and Tritz K 2014 *Nucl. Fusion* **54** 083007

[19] Agostini M, Scaggion A, Scarin P, Spizzo G and Vianello N 2012 *Plasma Phys. Control. Fusion* **54** 065003
[20] De Masi G, Martines E, Spolaore M, Vianello N, Cavazzana R, Innocente P, Momo B, Spagnolo S and Zuin M 2013 *Nucl. Fusion* **53** 083026
[21] Vianello N, Spolaore M, Martines E, Cavazzana R, Serianni G, Zuin M, Spada E and Antoni V 2010 *Nucl. Fusion* **54** 042002
[22] De Masi G *et al* 2010 *Contrib. Plasma Phys.* **50** 824–9
[23] Tsui H, Bengtson R, Li G, Lin H, Meier M, Ritz C P and Wootton A 1992 *Rev. Sci. Instrum.* **63** 4608–10
[24] Spolaore M, Vianello N, Agostini M, Cavazzana R, Martines E, Scarin P, Serianni G, Spada E, Zuin M and Antoni V 2009 *Phys. Rev. Lett.* **102** 165001
[25] Piovesan P *et al* 2013 *Phys. Plasmas* **20** 056112
[26] Fenstermacher M E, Evans T E, Osborne T H, Schaffer M J, Degrassie J S, Gohil P and Moyer R A 2008 *Nucl. Fusion* **48** 122001
[27] White R B 2006 *The Theory of Toroidally Confined Plasmas* 2nd edn (London: Imperial College Press) pp 87–88 ISBN 1860946399 www.icpress.co.uk/physics/p440.html
[28] Sakakibara S *et al* 2013 *Plasma Phys. Control. Fusion* **55** 014014
[29] Scott B 1997 *Plasma Phys. Control. Fusion* **39** 1635–68
[30] Manz P, Carralero D, Birkenmeier G, Müller H W, Müller S H, Fuchert G, Scott B D and Stroth U 2013 *Phys. Plasmas* **20** 102307
[31] Beurskens M *et al* 2011 *Phys. Plasmas* **18** 056120
[32] Antoni V, Desideri D, Martines E, Serianni G and Tramontin L 1996 *Nucl. Fusion* **36** 1561–70
[33] Antoni V, Cavazzana R, Desideri D, Martines E, Serianni G and Tramontin L 1998 *Phys. Rev. Lett.* **80** 4185
[34] Innocente P and Terranova D 2014 *Comput. Phys. Commun.* in preparation
[35] Zanca P and Terranova D 2004 *Plasma Phys. Control. Fusion* **46** 1115–41
[36] Scarin P, Agostini M, Carraro L, Cavazzana R, Ciaccio G, De Masi G, Spizzo G, Spolaore M and Vianello N 2014 Edge plasma physics modificatins due to magnetic ripple in RFX-mod *Proc. of the 21 Int. Conf. on Plasma Surface Interaction (Kanazawa, Japan, 26–30 May 2014)* pp P2–042
[37] Spizzo G *et al* 2014 *Phys. Plasmas* **21** 056102
[38] Spizzo G, Agostini M, Scarin P, Vianello N, White R B, Cappello S, Puiatti M E, Valisa M and the RFX-Mod Team 2012 *Nucl. Fusion* **52** 054015
[39] Levinson S, Beall J, Powers E and Bengtson R D 1984 *Nucl. Fusion* **24** 527
[40] Spolaore M, Antoni V, Spada E, Bergsåker H, Cavazzana R, Drake J R, Martines E, Regnoli G, Serianni G and Vianello N 2004 *Phys. Rev. Lett.* **93** 215003
[41] Marcus P S, Kundu T and Lee C 2000 *Phys. Plasmas* **7** 1630–40
[42] Spolaore M, Serianni G, Antoni V, Cavazzana R, Regnoli G, Spada E, Vianello N, Bergsåker H and Drake J R 2002 *Phys. Plasmas* **9** 4110–3
[43] Müller H W *et al* 2011 *Nucl. Fusion* **51** 073023
[44] Suzuki Y, Ida K, Kamiya K, Yoshinuma M, Sakakibara S, Watanabe K Y, Yamada H and LHD Experiment Group 2013 *Nucl. Fusion* **53** 073045
[45] Ida K *et al* 2004 *Nucl. Fusion* **44** 290–5
[46] Sánchez R, Newman D E, Leboeuf J N and Decyk V K 2011 *Plasma Phys. Control. Fusion* **53** 074018
[47] Myra J R, D'ippolito D A, Krasheninnikov S I and Yu G Q 2004 *Phys. Plasmas* **11** 4267

IOP Publishing Plasma Physics and Controlled Fusion
Plasma Phys. Control. Fusion **57** (2015) 014028 (7pp) doi:10.1088/0741-3335/57/1/014028

Dynamics of low–intermediate–high-confinement transitions in the HL-2A tokamak

Y Xu[1], J Cheng[1], J Q Dong[1,2], Y B Dong[1], M Jiang[1], W L Zhong[1], L W Yan[1], Z B Shi[1], Z H Huang[1], Y G Li[1], L Nie[4], M Xu[1], K J Zhao[1,3], X Q Ji[1], D L Yu[1], Y Liu[1], Z Y Cui[1], W Chen[1], Q W Yang[1], X T Ding[1], X R Duan[1], Yong Liu[1] and the HL-2A team

[1] Southwestern Institute of Physics, Chengdu 610041, People's Republic of China
[2] Institute for Fusion Theory and Simulation, Zhejiang University, Hangzhou 310027, People's Republic of China
[3] WCI Center for Fusion Theory, National Fusion Research Institute, Gwahangno 113, Yusung-gu, Daejeon 305–333, Korea
[4] Department of Modern Physics, USTC, Hefei 230026, People's Republic of China

E-mail: xuyh@swip.ac.cn

Received 1 July 2014, revised 14 September 2014
Accepted for publication 24 September 2014
Published 28 November 2014

Abstract

In recent experiments at the HL-2A tokamak, dynamic features across the low–intermediate–high (L–I–H) confinement transition have been investigated in detail. Experimental evidence shows two types of opposite limit cycles (dubbed type-Y and type-J) between the radial electric field (E_r) and turbulence evolution during the intermediate I-phase. Whereas for type-Y the turbulence grows prior to the change in E_r, for type-J the oscillation in E_r leads turbulence. It has been found that the type-Y usually appears first after an L–I transition, followed by type-J before the transition to the H-mode phase. Possible roles played by zonal flows and the enhanced pressure-gradient-induced flow shear in suppressing turbulence, respectively, in the type-Y and type-J periods have been identified. In addition, during the I-phase of the L–I–H discharges a kink-type MHD mode routinely occurs and crashes rapidly just prior to the I → H transition. The mode crash evokes substantial energy release from the core to plasma boundary and further increases the edge pressure gradient and E_r shear, which eventually results in confinement improvement into the H-mode.

Keywords: L–H transition, zonal flow, tokamak

(Some figures may appear in colour only in the online journal)

1. Introduction

Understanding the dynamic interplay between radially sheared $E_r \times B$ flows and plasma turbulence across the transition to improved confinement is of paramount importance for future fusion reactors such as ITER. In recent years, identification of key plasma parameters that control or determine the low (L) to high (H) confinement transition, in particular, during the low–intermediate–high (L–I–H) confinement transition process, has been intensively carried out in several fusion devices [1–5]. Meanwhile, the limit cycle oscillation (LCO) near the transition threshold has also been studied both theoretically and experimentally in terms of the bifurcation [6] and/or the predator–prey model for zonal flows and turbulence [7]. In recent experiments at the HL-2A tokamak, dynamic features across the L–I–H confinement transition have been investigated in detail. Experimental evidence shows two different types of limit cycles (referred as *type-Y* and *type-J*) in the LCO time ordering between the radial electric field (E_r) and turbulence intensity evolution. In this paper, we report the

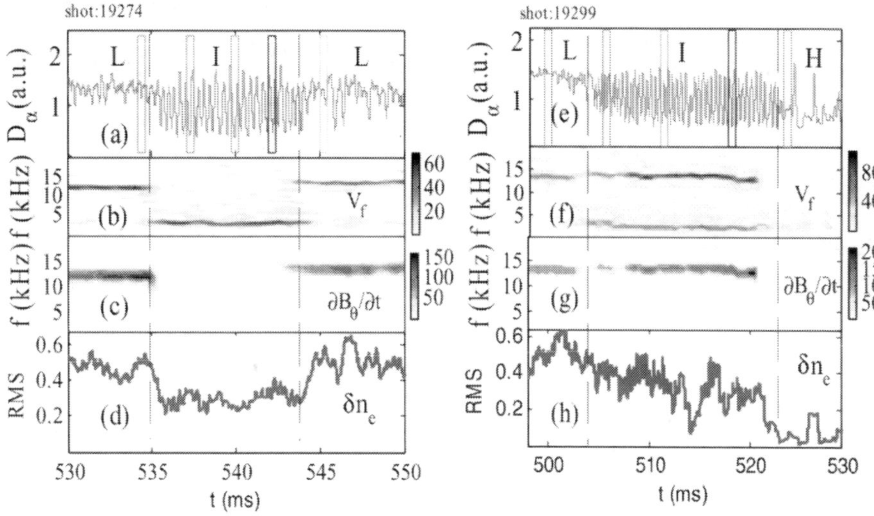

Figure 1. Comparison of time evolutions of plasma parameters in the L–I–L (left column) and L–I–H (right column) discharges. Shown in (*a*, *e*) are D_α emission signals; (*b*, *f*) contour plot of frequency spectrum of floating potential fluctuations; (*c*, *g*) contour plot of frequency spectrum of magnetic fluctuations; (*d*, *h*) RMS value of edge density fluctuations measured at ~5 mm inside the separatrix.

analysis of the temporal dynamics among turbulence, E_r, pressure gradient along with properties of their nonlinear interaction. Moreover, a kink-type MHD mode has been routinely observed in the I-phase of the L–I–H discharges. The crash of the mode evokes substantial energy release from the core to plasma boundary and further increases significantly the edge pressure gradient and $E_r \times B$ sheared flows, which eventually results in confinement improvement into the H-mode.

2. Experimental set-up

The experiments were performed in the NBI heated deuterium plasmas in the HL-2A tokamak with a single-null divertor configuration. Typical discharge parameters are R = 1.65 m, a = 0.4 m, B_T = 1.4 T, I_p = 180–185 kA, P_{NBI} = 1 MW and $< n_{e0} >$ = (2.0–3.0) × 10^{19} m^{-3}. For measuring edge equilibrium density (n_e), electron temperature (T_e) and floating potential (V_f) as well as fluctuating quantities, a four-step Langmuir probe array (connected in a triple probe model) has been employed with a sampling rate of 1 MHz [8]. The radial electric field E_r is deduced from the radial derivative of the plasma potential $V_p = V_f + \alpha T_e$, where $\alpha \approx 2.8$ for deuterium plasmas [9]. The magnetic fluctuations were detected by Mirnov coils at a sampling rate of 1 MHz. In addition, a multi-channel ECE diagnostic is used to measure the electron temperature profile and a soft x-ray (SXR) system, consisting of two pinhole cameras, to measure the soft x-ray emission intensity.

3. Results and discussion

3.1. Dynamic features of edge limit-cycle-oscillation (LCO) during the intermediate confinement phase (I-phase)

In the HL-2A experiments, the intermediate confinement phase (I-phase) has been frequently observed across the low–intermediate–high (L–I–H) confinement transition. Also, this sort of oscillating I-phase appears often in the L–I–L discharges. Figure 1 plots the time evolution of plasma parameters in these two type discharges with a constant NBI heating power. The left column illustrates an L–I–L shot while the right column displays an L–I–H discharge. In figure 1(*a*), the variation in the D_α emission magnitude indicates a transition from L to I confinement and then returns to the L-mode. In figure 1(*e*), the continuous reduction in the D_α radiation level reveals an L → I → H transition process. The I-phase, shown in between two vertical dashed lines, is characterized by a quasi-periodic oscillation in the D_α signal. This oscillation is present as well in edge density and floating potential signals at a frequency about 2.6 kHz, as can be seen in figures 1(*b*) and (*f*) in the potential fluctuations. In addition, a prominent electromagnetic frequency ~13.7 kHz shows up in both potential and magnetic fluctuations during the L-mode phase of L–I–L [see figures 1(*b*) and (*c*)] and the L/I-mode phases of L–I–H [see figures 1(*f*) and (*g*)], respectively. Figures 1(*d*) and (*h*) show time sequences of the RMS value of edge density fluctuations (δn_e) measured by Langmuir probes located at 5 mm inside the separatrix across the L–I–L and L–I–H transition, respectively. In case of L–I–L, the density fluctuation level decreases from the L to I-phase, and rises back from the I to L-mode. The scenario in L–I–H is different. The density fluctuations continuously reduce across the entire L–I–H period, suggesting an improved confinement from the L to I, and then to the H-mode. The rapid reduction in the RMS value of δn_e at time t ≈ 522 ms [see figure 1(*h*)] prior to the I → H transition is related to the crash of the MHD mode, which will be discussed later.

In order to understand the I-phase dynamics and its role in inducing the L–H transition, we explore first the regulative effects of the oscillating E_r on ambient turbulence. During the I-phase, the low-frequency electrostatic oscillation at $f \approx 2.6$ kHz [see figures 1(*b*) and (*f*)] has been associated with large-scale zonal flows, which gain energy from small-scale

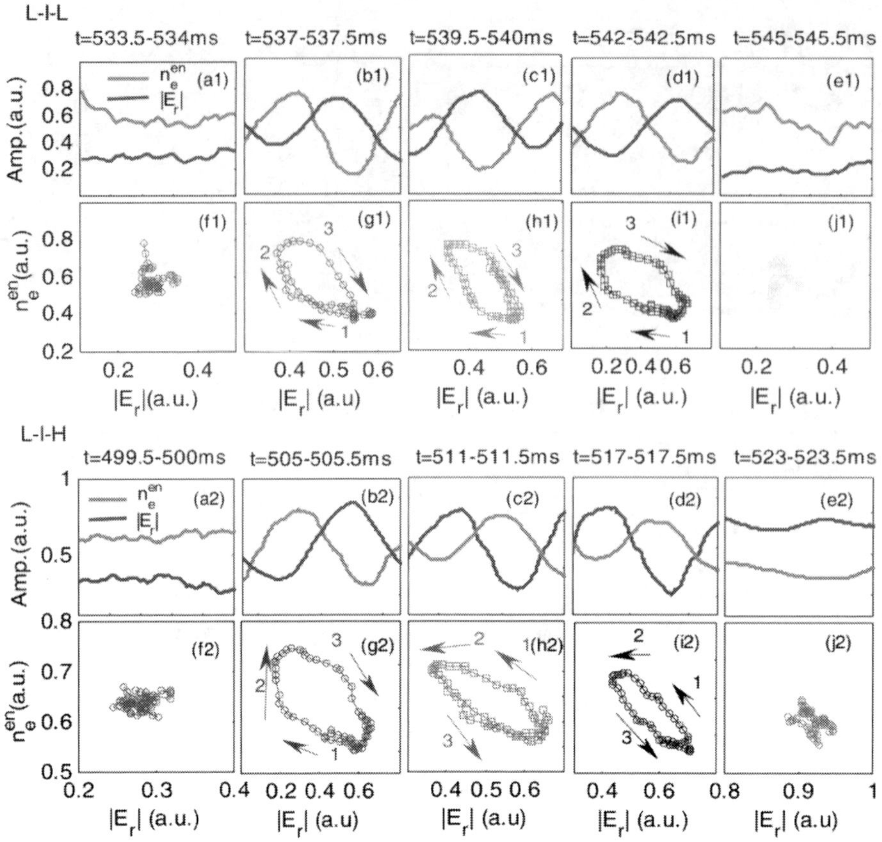

Figure 2. Comparison of time traces between the amplitude of fluctuating E_r (blue curves) and the envelope of density fluctuations (red curves) at five different times marked in figures 1(*a*) and (*e*) across the L–I–L transition (top row) and L–I–H transition (third row), respectively. Drawn in the second and bottom row are the corresponding trajectories in phase space between the E_r amplitude and density fluctuation envelope depicted at the first and third row, respectively.

turbulence via nonlinear coupling, as shown in figure 3(*a*) in the auto-bicoherence spectrum of potential fluctuations. According to theories [7], such a nonlinear interplay between zonal flows and turbulence essentially reflects the predator-prey dynamic process, for which the zonal flow acts as a predator whereas the turbulence is a prey. With a certain time delay, the predator-prey regime exhibits a limit cycle oscillation (LCO) in the time evolution between the zonal flow amplitude and the turbulence level, where the variation in prey normally leads changes in predator. In our study, the $E_r \times B$ zonal flow is estimated by the fluctuating E_r amplitude, $|E_r|$, while the ambient turbulence energy is measured by the envelope of density fluctuations, n_e^{en}, as defined in [10]. The time traces of $|E_r|$ and n_e^{en}, selected in five time windows as marked in figures 1(*a*) and (*e*), are depicted in the first and third row in figure 2 for the L–I–L and L–I–H shot, respectively. In either L-mode or H-mode, the $|E_r|$ and n_e^{en}do not show any oscillation feature, as illustrated in figures 2(*a*1), (*e*1), (*a*2) and (*e*2). During the I-phase of L–I–L, figures 2(*b*1), (*c*1) and (*d*1) show clearly oscillations in zonal flows and turbulence intensity and the variation in n_e^{en}leads the change in $|E_r|$. The corresponding LCOs in phase space [see figures 2(*g*1), (*h*1) and (*i*1)] all rotate in the clockwise direction, in agreement with the prediction of the predator–prey model [7]. This kind of LCO is dubbed type-Y. For the L–I–H transition, the initial stage of I-phase also displays the type-Y LCO, as seen in figures 2(*b*2) and (*g*2). However, in the late period of the I-phase, the time ordering between the $|E_r|$ and $n_e^{en}$is reversed and the LCO changes rotation direction to counter-clockwise [see figures 2(*c*2), (*d*2), (*h*2) and (*i*2)]. Apparently, this type of LCO is in contrast with the conventional predator–prey model and hence dubbed type-J.

To gain more insight into the LCO physics, we have calculated the nonlinear auto-bicoherence [10] of background turbulence for the type-Y and type-J LCOs observed during the I-phase across the L–I–H transition. According to the theory [11], the zonal flows are generated by small-scale fluctuations via three-wave coupling under resonance conditions, which can be identified using bispectral analysis technique for the strength of the coupling [12]. The contour-plots of auto-bispectrum of potential fluctuations are shown in figure 3, where the three bicoherence are computed around the time periods shown in figures 2(*b*2), (*c*2) and (*d*2), respectively. The high level value of bicoherence cascaded from broadband turbulence in figure 3(*a*) verifies that there is strong nonlinear wave-wave coupling in the type-Y LCO phase, suggesting a zonal-flow-type $|E_r|$ oscillation at a frequency of 2.6 kHz, whereas for type-J the $|E_r|$ oscillation

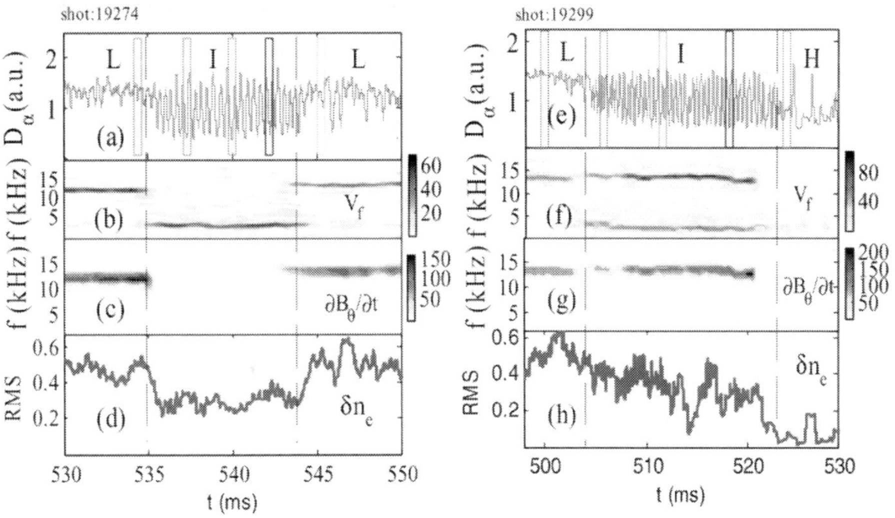

Figure 1. Comparison of time evolutions of plasma parameters in the L–I–L (left column) and L–I–H (right column) discharges. Shown in (*a*, *e*) are D_α emission signals; (*b*, *f*) contour plot of frequency spectrum of floating potential fluctuations; (*c*, *g*) contour plot of frequency spectrum of magnetic fluctuations; (*d*, *h*) RMS value of edge density fluctuations measured at ~5 mm inside the separatrix.

analysis of the temporal dynamics among turbulence, E_r, pressure gradient along with properties of their nonlinear interaction. Moreover, a kink-type MHD mode has been routinely observed in the I-phase of the L–I–H discharges. The crash of the mode evokes substantial energy release from the core to plasma boundary and further increases significantly the edge pressure gradient and $E_r \times B$ sheared flows, which eventually results in confinement improvement into the H-mode.

2. Experimental set-up

The experiments were performed in the NBI heated deuterium plasmas in the HL-2A tokamak with a single-null divertor configuration. Typical discharge parameters are $R = 1.65$ m, $a = 0.4$ m, $B_T = 1.4$ T, $I_p = 180$–185 kA, $P_{NBI} = 1$ MW and $< n_{e0} > = (2.0$–$3.0) \times 10^{19}$ m^{-3}. For measuring edge equilibrium density (n_e), electron temperature (T_e) and floating potential (V_f) as well as fluctuating quantities, a four-step Langmuir probe array (connected in a triple probe model) has been employed with a sampling rate of 1 MHz [8]. The radial electric field E_r is deduced from the radial derivative of the plasma potential $V_p = V_f + \alpha T_e$, where $\alpha \approx 2.8$ for deuterium plasmas [9]. The magnetic fluctuations were detected by Mirnov coils at a sampling rate of 1 MHz. In addition, a multi-channel ECE diagnostic is used to measure the electron temperature profile and a soft x-ray (SXR) system, consisting of two pinhole cameras, to measure the soft x-ray emission intensity.

3. Results and discussion

3.1. Dynamic features of edge limit-cycle-oscillation (LCO) during the intermediate confinement phase (I-phase)

In the HL-2A experiments, the intermediate confinement phase (I-phase) has been frequently observed across the low–intermediate–high (L–I–H) confinement transition. Also, this sort of oscillating I-phase appears often in the L–I–L discharges. Figure 1 plots the time evolution of plasma parameters in these two type discharges with a constant NBI heating power. The left column illustrates an L–I–L shot while the right column displays an L–I–H discharge. In figure 1(*a*), the variation in the D_α emission magnitude indicates a transition from L to I confinement and then returns to the L-mode. In figure 1(*e*), the continuous reduction in the D_α radiation level reveals an L → I → H transition process. The I-phase, shown in between two vertical dashed lines, is characterized by a quasi-periodic oscillation in the D_α signal. This oscillation is present as well in edge density and floating potential signals at a frequency about 2.6 kHz, as can be seen in figures 1(*b*) and (*f*) in the potential fluctuations. In addition, a prominent electromagnetic frequency ~13.7 kHz shows up in both potential and magnetic fluctuations during the L-mode phase of L–I–L [see figures 1(*b*) and (*c*)] and the L/I-mode phases of L–I–H [see figures 1(*f*) and (*g*)], respectively. Figures 1(*d*) and (*h*) show time sequences of the RMS value of edge density fluctuations (δn_e) measured by Langmuir probes located at 5 mm inside the separatrix across the L–I–L and L–I–H transition, respectively. In case of L–I–L, the density fluctuation level decreases from the L to I-phase, and rises back from the I to L-mode. The scenario in L–I–H is different. The density fluctuations continuously reduce across the entire L–I–H period, suggesting an improved confinement from the L to I, and then to the H-mode. The rapid reduction in the RMS value of δn_e at time t ≈ 522 ms [see figure 1(*h*)] prior to the I → H transition is related to the crash of the MHD mode, which will be discussed later.

In order to understand the I-phase dynamics and its role in inducing the L–H transition, we explore first the regulative effects of the oscillating E_r on ambient turbulence. During the I-phase, the low-frequency electrostatic oscillation at $f \approx 2.6$ kHz [see figures 1(*b*) and (*f*)] has been associated with large-scale zonal flows, which gain energy from small-scale

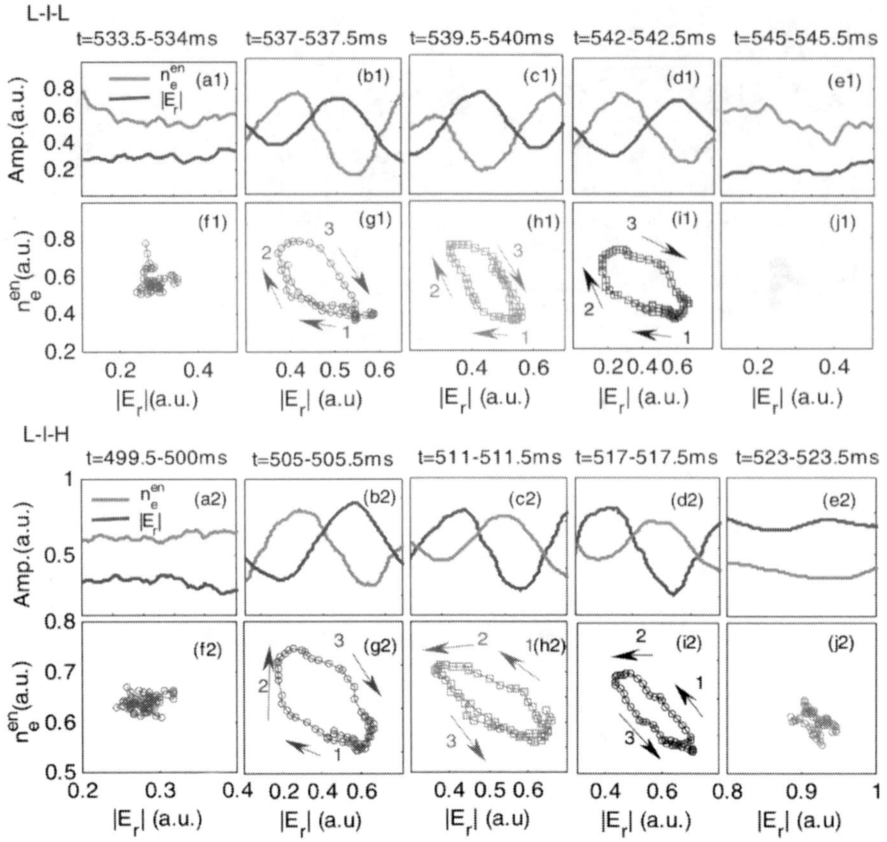

Figure 2. Comparison of time traces between the amplitude of fluctuating E_r (blue curves) and the envelope of density fluctuations (red curves) at five different times marked in figures 1(*a*) and (*e*) across the L–I–L transition (top row) and L–I–H transition (third row), respectively. Drawn in the second and bottom row are the corresponding trajectories in phase space between the E_r amplitude and density fluctuation envelope depicted at the first and third row, respectively.

turbulence via nonlinear coupling, as shown in figure 3(*a*) in the auto-bicoherence spectrum of potential fluctuations. According to theories [7], such a nonlinear interplay between zonal flows and turbulence essentially reflects the predator-prey dynamic process, for which the zonal flow acts as a predator whereas the turbulence is a prey. With a certain time delay, the predator-prey regime exhibits a limit cycle oscillation (LCO) in the time evolution between the zonal flow amplitude and the turbulence level, where the variation in prey normally leads changes in predator. In our study, the $E_r \times B$ zonal flow is estimated by the fluctuating E_r amplitude, $|E_r|$, while the ambient turbulence energy is measured by the envelope of density fluctuations, n_e^{en}, as defined in [10]. The time traces of $|E_r|$ and n_e^{en}, selected in five time windows as marked in figures 1(*a*) and (*e*), are depicted in the first and third row in figure 2 for the L–I–L and L–I–H shot, respectively. In either L-mode or H-mode, the $|E_r|$ and n_e^{en}do not show any oscillation feature, as illustrated in figures 2(*a*1), (*e*1), (*a*2) and (*e*2). During the I-phase of L–I–L, figures 2(*b*1), (*c*1) and (*d*1) show clearly oscillations in zonal flows and turbulence intensity and the variation in n_e^{en}leads the change in $|E_r|$. The corresponding LCOs in phase space [see figures 2(*g*1), (*h*1) and (*i*1)] all rotate in the clockwise direction, in agreement with the prediction of the predator–prey model [7]. This kind of LCO is dubbed type-Y. For the L–I–H transition, the initial stage of I-phase also displays the type-Y LCO, as seen in figures 2(*b*2) and (*g*2). However, in the late period of the I-phase, the time ordering between the $|E_r|$ and $n_e^{en}$is reversed and the LCO changes rotation direction to counter-clockwise [see figures 2(*c*2), (*d*2), (*h*2) and (*i*2)]. Apparently, this type of LCO is in contrast with the conventional predator–prey model and hence dubbed type-J.

To gain more insight into the LCO physics, we have calculated the nonlinear auto-bicoherence [10] of background turbulence for the type-Y and type-J LCOs observed during the I-phase across the L–I–H transition. According to the theory [11], the zonal flows are generated by small-scale fluctuations via three-wave coupling under resonance conditions, which can be identified using bispectral analysis technique for the strength of the coupling [12]. The contour-plots of auto-bispectrum of potential fluctuations are shown in figure 3, where the three bicoherence are computed around the time periods shown in figures 2(*b*2), (*c*2) and (*d*2), respectively. The high level value of bicoherence cascaded from broadband turbulence in figure 3(*a*) verifies that there is strong nonlinear wave-wave coupling in the type-Y LCO phase, suggesting a zonal-flow-type $|E_r|$ oscillation at a frequency of 2.6 kHz, whereas for type-J the $|E_r|$ oscillation

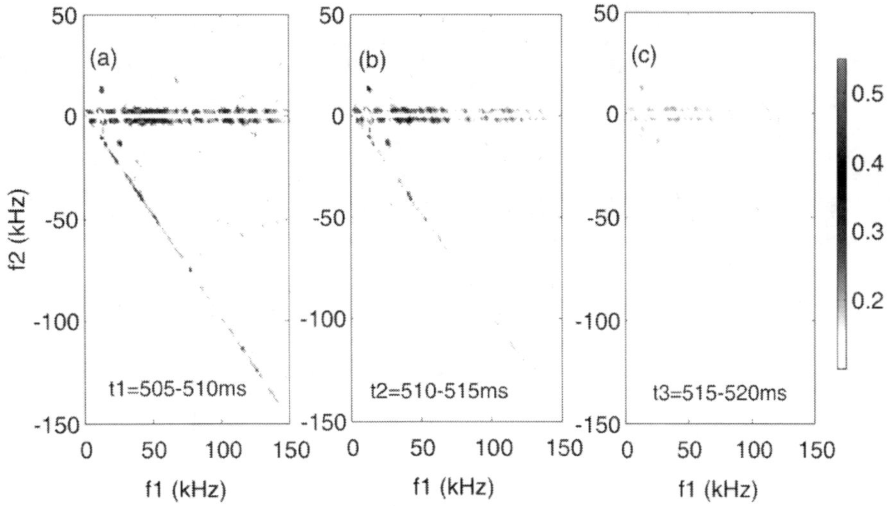

Figure 3. Contour plot of squared auto-bicoherence in potential fluctuations measured at three different times during the I-phase of the L–I–H transition. Figure (*a*) corresponds to type-Y LCO, while (*b*) and (*c*) correspond to type-J.

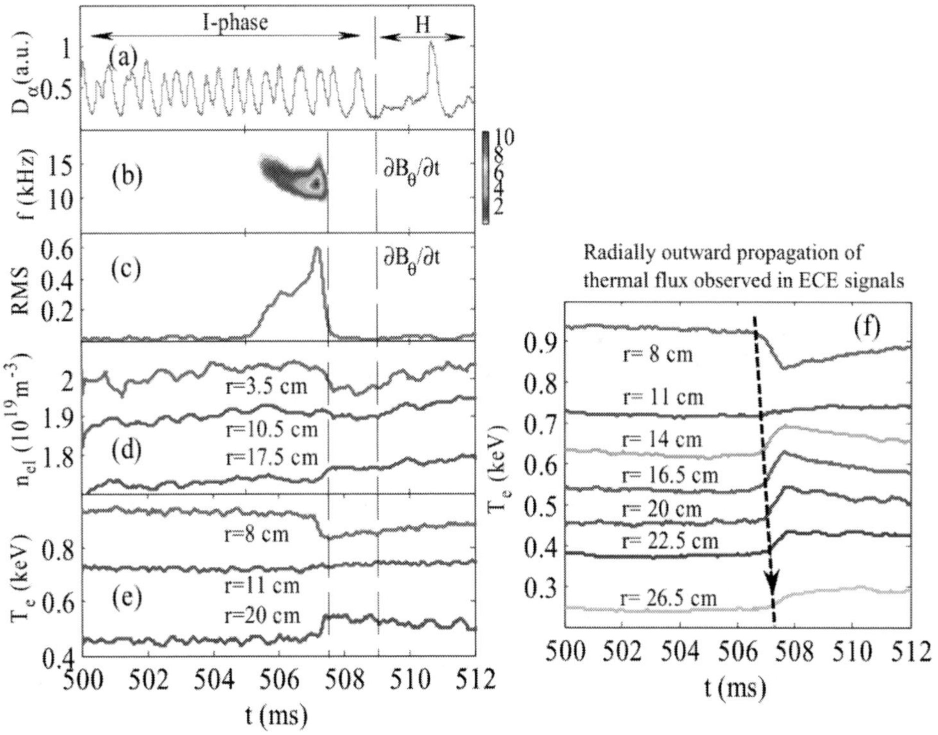

Figure 4. Time histories of plasma parameters across an I → H transition (#19391). (*a*) D_α emission signal; (*b*) contour plot of frequency spectrum of magnetic fluctuations; (*c*) RMS value of magnetic fluctuations; (*d*) line-averaged density detected by HCN chords at three different radial locations; (*e*) local electron temperature measured by the ECE diagnostic at three different radial locations; (*f*) multi-channel ECE signals (smoothed) to show the radially outward propagation of thermal flux.

doesn't belong to zonal flows due to lack of nonlinear three-wave interaction, as exposed in figures 3(*b*) and (*c*). Similar observations of the type-J LCO have also been obtained in other devices [3, 5].

Possible interpretation for the different types of LCOs observed is as follows: during the I-phase, there coexist two sorts of predator-prey regime for the interaction of 'zonal flow—turbulence' and 'turbulence—pressure gradient (∇P)' [13]. In the former case, the zonal flow and turbulence act as the predator and prey, respectively, as we mentioned earlier. For the latter case, turbulence is driven by free energy of the pressure gradient, which is regulated by turbulence-induced transport. A large ∇P leads to high-level turbulence and turbulent transport, resulting in a reduction in ∇P and hence

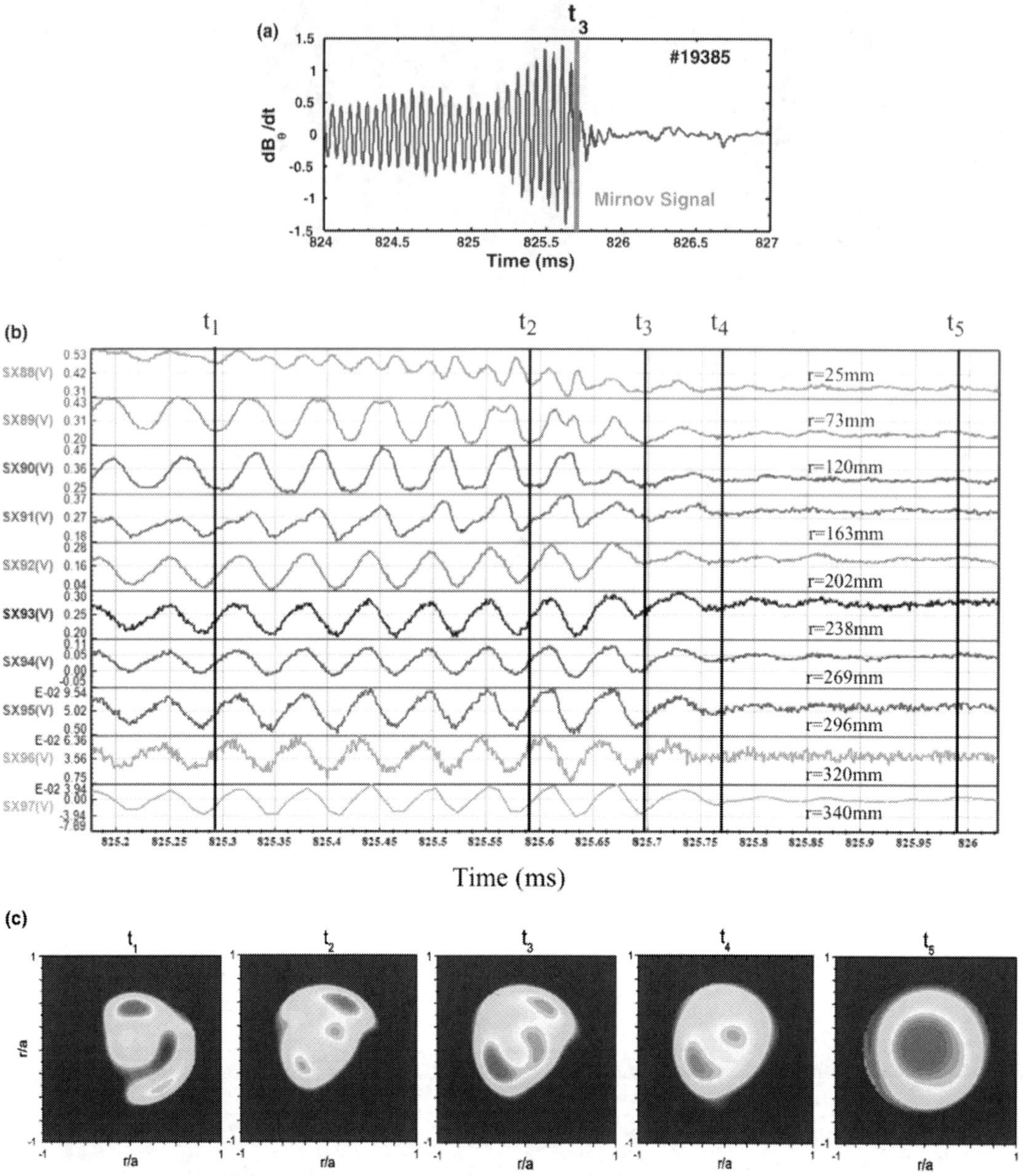

Figure 5. (*a*) Time history of magnetic fluctuations measured by a Mirnov coil to show the collapse of an MHD mode at the time t_3 = 825.7 ms; (*b*) time evolution of the soft x-ray (SXR) signals detected by a pinhole array during the collapse of the MHD mode; (*c*) contour plots of the tomographically reconstructed SXR emission, showing the evolution of a kink-type MHD mode structure at five different times, as marked in the above graph (*b*).

decrease in turbulence (lack of driving). The reduced turbulent transport will establish a large ∇P again and enter the second loop. In such a case, the turbulence and ∇P plays as the predator and prey, respectively. Because the oscillation in ∇P induces an oscillation in the E_r amplitude via the radial force balance equation, finally the change in prey (∇P or E_r) will lead the variation in turbulence intensity. As a consequence, the LCO reverses rotation direction from the 'zonal flow—turbulence' to 'turbulence—∇P' regime. These two regimes are always competitive. In case of L-mode or at the initial stage of the I-phase, the ∇P is relatively weak [8] and thus the 'zonal flow—turbulence' regime prevails. Therefore, the LCO is running in type-Y, as shown in figures 2 (*g*1), (*h*1) and (*i*1) and (*g*2). With the growing of ∇P in the later stage of the I-phase in the L–I–H case (not shown here) for the sake of improved confinement, the 'turbulence—∇P' regime becomes dominant and the LCO runs into type-J, as depicted in figures 2(*h*2) and (*i*2). It appears that the crucial element for the shift of the LCO from type-Y to type-J is the growth of ∇P to a certain threshold value [8, 13], which can only be realized in a good confinement status, such as the I-phase of L–I–H discharges. We have found that during the I-phase of L–I–L shots the ∇P doesn't increase significantly and the LCO usually operates in type-Y, as displayed in figure 2.

3.2. Critical role played by a kink-type MHD mode in triggering the I → H transition

In the L–I–H transition discharges, two branches of magnetic fluctuations coexist in the magnetic signal: one is the MHD mode oscillating at a frequency $f \approx 13.7$ kHz (or higher), while the other is the magnetic fluctuation in LCO at a lower frequency $f \approx 2.6$ kHz, which is also the characteristic frequency of the I-phase oscillation seen in the D_α signal. Because the magnitude of the LCO component ($f \approx 2.6$ kHz) is one/two order smaller than the MHD mode ($f \approx 13.7$ kHz) magnitude [10], the LCO magnetic component ($f \approx 2.6$ kHz) is almost invisible in the contour-plot of frequency spectrum of magnetic fluctuations, as displayed in figure 1(*g*). It has been found that at the LCO frequency ($f \approx 2.6$ kHz) there is large coherence between magnetic fluctuation and pressure gradient signals, suggesting that in type-J LCO the ∇P (and E_r) oscillation is related to magnetic fluctuations via $\nabla P = \mathbf{j} \times \mathbf{B}$ [8, 10].

In this study, we focus on another important feature of the MHD mode at $f \approx 13.7$ kHz present during the I-phase of the L–I–H shots, which is normally absent in the I-phase of the L–I–L discharges, as can be seen in figures 1(*b*), (*c*) and (*f*), (*g*). This mode can be developed at a certain time in the I-phase and it always crashes prior to the I → H transition. Further studies reveal that this mode may play a crucial role in triggering the plasma confinement into the H-mode. To illustrate the underlying physical process, figure 4 plots time histories of the D_α signal, the frequency spectrum and the RMS value in magnetic fluctuations together with multi-chord line-averaged density (n_{el}) and electron temperature (T_e) detected by the HCN and ECE diagnostics at different radial positions. From figures 4(*a*) to (*c*), one can see in the Mirnov signal that a MHD mode is developed and grows rapidly during the I-phase. At about $t = 507.3$ ms, the mode collapses, and concurrently, dramatic changes occur in the density and temperature signals. In the inner region of the core plasma, both n_{el} and T_e quickly reduce [see green curves in figures 4(*d*) and (*e*)] while in the outer area both of them increase (see brown curves). These results clearly show a release of plasma energy from the core to the edge region induced by the MHD mode crash, which can be further verified in the multi-channel ECE signals (smoothed) showing the radially outward propagation of the thermal flux (see figure 4(*f*)). The quick release in plasma density and temperature alters the pressure profile considerably from a peaked to flattened one, resulting in a significant increase in the edge pressure gradient and mean E_r as well. The enhanced mean $E_r \times B$ flow shear will then suppress the turbulence and turbulent transport via flow-shear-decorrelation effects on turbulence eddies [14], as indicated in figure 1(*h*) by a sudden drop in the density fluctuation level at $t \approx 522$ ms after the MHD mode crash. The plasma therefore transits from I-phase into the H-mode.

The properties of the MHD mode structure have also been surveyed using the soft x-ray (SXR) system, consisting of two pinhole arrays at one poloidal cross-section, in the HL-2A tokamak. Figure 5 depicts time evolution of the SXR signals (measure by one array) along with the tomography results for a time duration across the I to H transition. As shown in the Mirnov signal at the top graph (a), the MHD mode collapses at time $t_3 = 825.7$ ms. Figure 5(*b*) plots the time series of the SXR signals detected by a pinhole camera passing through various radial-chord positions. In the outer region ($r = 73$–340 mm), all channels display a mode perturbation at a frequency of $f \approx 17$ kHz, except for the inner one at $r = 25$ mm where the oscillation frequency is twice due to double view in the central chord compared with the side one. For further investigating the mode structure, a tomography reconstruction technique, the Cormack method [15], has been applied to the soft x-ray emission detected by two pinhole arrays in one cross-section. Shown in figure 5(*c*) are contour plots of the tomography results, exhibiting the temporal evolution of the MHD mode structure at five different times marked in the above graph (b). At time t_1, the mode shows an $m = 2$, $n = 1$ structure. From t_2 to t_3, the $m = 1$, $n = 1$ mode is developed in the center and gradually grows to the maximal size prior to the collapse. No phase inversion in the SXR signals has been observed for the $m = 1$ and $m = 2$ modes, suggesting a kink-type nature of the MHD activity. More details on the mode properties will be studied further [16]. After the mode crash at t_3, the $m = 1$ structure disappears, which evokes a rapid energy release from the core towards outside, and hence, flattens the plasma profile, as can be seen at t_4. In the steady H-mode phase, the tomography at t_5 indicates a smooth plasma profile without any MHD mode structure.

4. Conclusion

Experimental investigation on the dynamic features across the low–intermediate–high (L–I–H) confinement transition has been executed in the HL-2A tokamak. During the I-phase, two types of LCOs, i. e., type-Y and type-J, between the radial electric field (E_r) and turbulence evolution have been identified. For type-Y the turbulence grows prior to the change in E_r whereas for type-J the oscillation in E_r leads changes in turbulence. It has been found that the type-Y usually appears (i) in the I-phase of L–I–L discharges; (ii) at the beginning of the I-phase in L–I–H shots, followed by type-J before the transition from I to H-mode. Possible roles played by zonal flows and the enhanced pressure-gradient-induced flow shear in suppressing turbulence and converting the LCO from type-Y to type-J have been analyzed. In addition, during the I-phase of the L–I–H discharges a kink-type MHD mode routinely occurs and crashes rapidly just prior to the I → H transition. The collapse of the mode evokes substantial energy release from the core to plasma boundary and further increases the edge pressure gradient and E_r shear, which eventually leads to confinement improvement into the H-mode.

Acknowledgment

This work is supported by Chinese National Fusion Project for ITER under grant no. 2014GB108000. It is also partially supported by the National Science Foundation of China under grant no. 10990213 and 11375054, the National Magnetic

Confinement Fusion Science Program of China under grant no. 2013GB112008, 2013GB107002 and the China-Korean joint foundation under the grant no. 2012DFG02230 and by KAKENHI (21224014, 23244113) and Asada Science Foundation of Japan.

References

[1] Conway G D *et al* 2011 *Phys. Rev. Lett.* **106** 065001
[2] Estrada T *et al* 2011 *Phys. Rev. Lett.* **107** 245004
[3] Kobayashi T *et al* 2013 *Phys. Rev. Lett.* **111** 035002
[4] Schmitz L *et al* 2014 *Nucl. Fusion* **54** 073012
[5] Muller S H *et al* 2014 *Phys. Plasmas* **21** 042301
[6] Itoh S-I and Itoh K 1988 *Phys. Rev. Lett.* **60** 2276
[7] Kim E J and Diamond P H 2003 *Phys. Rev. Lett.* **90** 185006
[8] Cheng J and Dong J Q *et al* 2013 *Phys. Rev. Lett.* **110** 265002
[9] Stangeby P C and McCracken G M 1990 *Nucl. Fusion* **30** 1225
[10] Cheng J *et al* 2014 Low–intermediate–high confinement transition in HL-2A tokamak plasmas *Nucl. Fusion* at press
[11] Diamond P H *et al* 2005 *Plasma Phys. Control. Fusion* **47** R35
[12] Kim Y and Powers E J 1979 *IEEE Trans. Plasma Sci.* **7** 120
[13] Guo Z B and Diamond P H *et al* 2014 *Phy. Rev.* E **89** 041101(R)
[14] Biglari H *et al* 1990 *Phys. Fluids* B **2** 1
[15] Granetz R S and Camacho J F 1985 *Nucl. Fusion* **25** 727
[16] Liu Y private communication

Plasma Phys. Control. Fusion **57** (2015) 014029 (12pp) doi:10.1088/0741-3335/57/1/014029

Pedestal and edge localized mode characteristics with different first wall materials and nitrogen seeding in ASDEX Upgrade

P A Schneider, L Barrera Orte, A Burckhart, M G Dunne, C Fuchs, A Gude, B Kurzan, W Suttrop, E Wolfrum and the ASDEX Upgrade Team

Max-Planck-Institut für Plasmaphysik, Boltzmannstr. 2, 85748 Garching, Germany

E-mail: philip.schneider@ipp.mpg.de

Received 26 June 2014, revised 16 September 2014
Accepted for publication 30 September 2014
Published 28 November 2014

Abstract

A comparison of ASDEX Upgrade (AUG) discharges performed with carbon and the full tungsten wall shows that the pedestal performance at low triangularity is not altered without gas puffing. The pedestal electron pressure is the same for both wall materials as is the confinement. With the tungsten wall the natural density is higher even without an additional gas puff. In typical operation with gas puffing the density is again higher in tungsten. This results in a higher collisionality with the tungsten wall. Pedestal pressure and plasma confinement, however, are not degraded until very large amounts of deuterium are puffed.

The edge localized mode (ELM) crash in typical AUG discharges is observed to be composed of two independent phases. This is observed for both the carbon and the tungsten wall. The 1st phase of the crash is unaffected by scans of the plasma parameters as long as the pedestal pressure remains constant. The duration of the 2nd phase is strongly anti-correlated with the separatrix density and can be suppressed by the application of nitrogen seeding for divertor cooling. A consistent explanation for the two phases of the ELM crash does not seem possible when considering only the pre-ELM pedestal profiles. The scrape off layer (SOL) plasma provides the necessary free parameter for a consistent explanation, indicating the importance of the SOL in understanding the ELM crash evolution.

Keywords: plasma physics, tokamak, macroinstabilities, edge localised modes, magnetic diagnostics

(Some figures may appear in colour only in the online journal)

1. Introduction

In magnetically confined fusion research the focus of the experiments is shifting more and more towards ITER and corresponding reactor relevant machine requirements and plasma operation. Examples are the change from a carbon to a tungsten wall in ASDEX Upgrade (AUG) or the installation of the tungsten-beryllium wall (ITER like wall—ILW) in JET, as well as mitigation of edge localised modes (ELMs) and radiative cooling. Operational scenarios involving radiative divertor cooling with impurity seeding or mitigation of ELMs with magnetic perturbations applied by external coils are intensively studied. In this contribution we discuss the pedestal and ELM characteristics with and without these additional actuators.

A special focus is put on the comparison of discharges from AUG with mainly carbon (AUG-C) and with full tungsten (AUG-W) first wall. Here, the main interest lies in the energy confinement which in H-mode is dominated by the pedestal and the ELM behaviour. One can distinguish between two categories of consequences a different wall material can have on ELMs and confinement; first, there are direct consequences like the impurity content, e.g. less low-Z and more high-Z impurities with a metal wall. Second, restrictions/modifications of typical plasma operation triggered by the changed boundary conditions. Changes in plasma operation in general should impact the plasma in the same way regardless of the wall material. Therefore, the first category of direct consequences is more relevant to understand the influence of the wall material on ELMs and plasma performance. Typically, it is difficult to clearly separate between both effects because often necessary comparison experiments are missing in the AUG-C database or reference experiments cannot be reproduced easily in AUG-W.

Previous studies at AUG, for which comparison experiments exist—mainly at low triangularities—showed similar confinement factors [1–4]. However, in the broader database of AUG high performance discharges the trend is that in AUG-C the maximal achievable energy confinement was above that in AUG-W [1, 3]. One possible explanation is the necessity to operate with significant gas puffing for ELM tailoring and thereby impurity control [5] to prevent tungsten accumulation in the plasma center which can lead to a radiative collapse of the plasma [1]. High levels of gas puffing were reported to deteriorate confinement both in devices with a carbon and with a metal wall [6–8]. Due to the increased gas puffing, discharges in carbon with very high performance $H_{98(y,2)} \sim 1.5$ at very low plasma collisionality were difficult to reproduce [3]. These were only recently reproduced and, therefore, are not included in the presented study.

It is an active field of research whether changes in the plasma operation can already explain observed differences in energy confinement between carbon and metal machines or additional phenomena need to be considered. It is long known that high-Z wall materials can cause severe energy losses when accumulating in the plasma center, however, in normal plasma operation the high-Z impurity content is kept low enough. The low-Z impurities carbon and nitrogen were reported to show an impact on global confinement when they were artificially seeded into a plasma with very low natural impurity content [9–11]. The variation of impurity content in these seeding experiments is similar to experiments performed in carbon and metal devices, therefore, it was argued that an effect on confinement could have the same origin in both types of comparison experiments [10]. There are studies where the impurity content could indeed provide an explanation for the observed plasma performance, however, other studies with similar differences in the impurity content do not show any change of the energy confinement. Examples for the latter category are low triangularity discharges at constant gas puff levels which showed comparable plasma performance with carbon and metal wall in AUG and JET [2, 8]. The former are, in particular, high triangularity JET-ILW baseline discharges [8, 10].

Most of the cited studies were performed using a database approach which is well suited to identify general trends. This paper tries to identify the different actuators by a careful selection of comparison experiments. Instead of global confinement, the focus is on the pedestal structure and the ELM behavior. In section 2 the pedestal structure in carbon is compared to the one in tungsten without gas puffing at low triangularity. The influence of different deuterium fuelling levels is then presented. Finally, one example of nitrogen seeding in AUG-W is given. In section 3 it will be shown that ELMs in AUG can exhibit two distinct phases. The duration, size and magnetic signature of these phases are compared for both wall materials and the different gas fuelling scenarios. Section 4 summarises the paper with a concluding discussion.

2. Pre-ELM pedestal profiles

This section discusses the pedestal parameters, temperature and density, to document any effect of the different wall materials on the plasma edge. The discussion is first restricted to the pre-ELM pedestal profiles which are assumed to be at the MHD stability limit of ELMs. The influence on the ELMs itself will be presented in section 3. First similar discharges without additional gas puffing are compared and then the influence of different gas fuelling scenarios with deuterium and nitrogen is documented.

2.1. Comparison of carbon and tungsten

Due to the restrictions of operational space as described in the introduction, the number of good comparison experiments is quite small. Suitable candidates were already described in [12 and 2] with emphasis on global confinement and are now used for a detailed comparison in the pedestal region. The discharges are all 1 MA, low triangularity discharges without external gas puffing and heating powers between 6 and 14 MW. The differences between the discharges which could not be avoided are a variation of q_{95} from 4.55 in carbon to 4.85 in tungsten[1], a reduction in the effective charge number $Z_{\rm eff}$ from about 2.0 in carbon to 1.5 in tungsten and a different heating mix with NBI and ICRF in the carbon case and NBI and ECRH in the tungsten case. The ICRF is replaced by ECRH and one additional beam source. This results in a larger core rotation, though, the difference in rotation vanishes towards the edge. Despite the different heating mix, the heat flux profiles—determined with the transport code TRANSP [13]—are found to be the same for the ions on the whole radius and for the electrons on the outer half radius.

To illustrate the effect of the different wall materials two phases with moderate (8 MW) and high (13 MW) heating power are compared. Figure 1 shows ELM synchronised pre-ELM profiles composed of data from electron cyclotron emission (ECE) [14], Li-beam [15] and Thomson scattering (TS) [16] measurements. For the ELM synchronisation profile data in a 2 ms interval before the ELM onset is used, for the presented cases this corresponds to about the last 25% of the ELM cycle. The profiles illustrate the different behaviour of electron temperature and density. In the tungsten case the edge plasma is colder and more dense. The ion temperatures in the plasma core follow the same trend as the electron temperature,

[1] Throughout this paper one focus is on the comparison of plasma operation in AUG with carbon and full tungsten wall. To allow for more fluid reading terms like carbon/tungsten case or discharge will be used. This refers only to the wall material, the discharges were run all with deuterium as main gas.

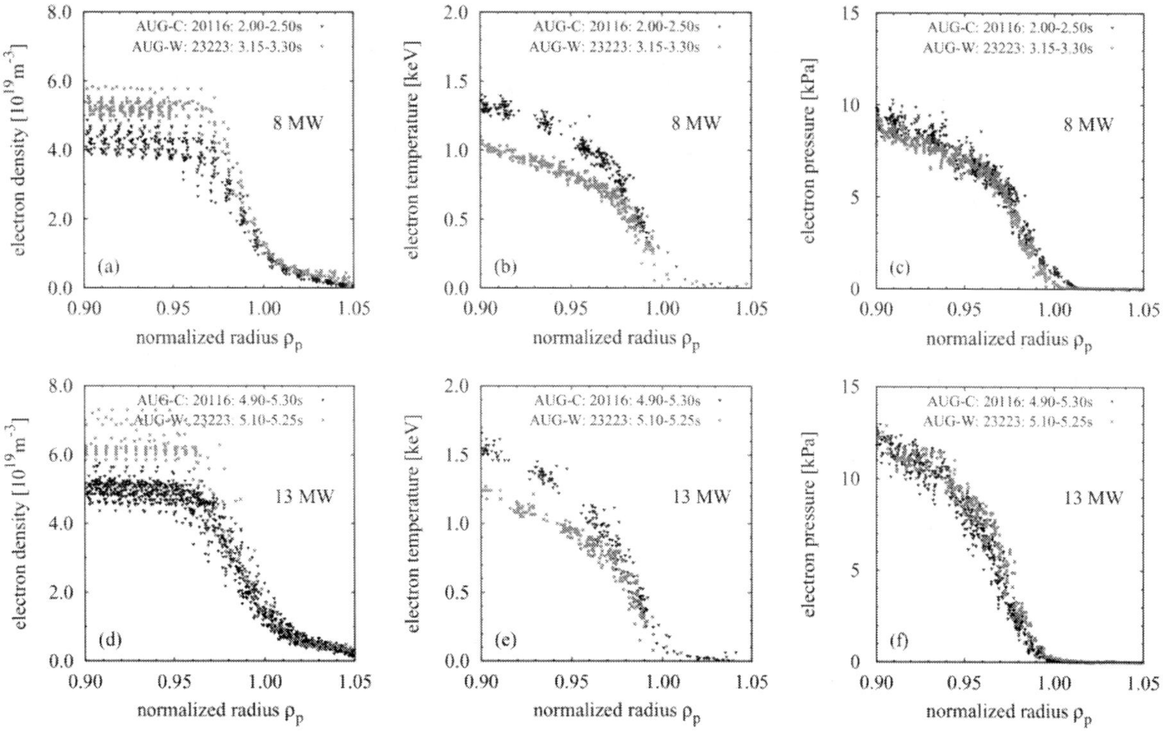

Figure 1. Edge pedestal profiles of electron density (*a*, *d*), temperature (*b*, *e*) and pressure (*c*, *f*) compared for similar discharges in carbon (▾) and tungsten (▪) and different heating power, 8 MW in the top row and 13 MW in the bottom row.

detailed edge measurements of T_i were not available for older discharges with the carbon wall. The electron pressure is the same for 8 MW and also similarly increases with higher heating power. The electron density exhibits several interesting features which are observed for both heating powers in figures 1(*a*) and (*d*). First, the scrape off layer (SOL) density up to the separatrix in carbon is essentially the same as in tungsten. Second, the density gradient in the pedestal is about 25% larger in tungsten than it is in carbon. Since the pedestal width is almost the same for both wall materials, this results in a roughly 25% higher pedestal top density. The density gradient length $L_{ne} = n_e/\nabla n_e$ is reduced for the discharge in tungsten. These are indications that the particle transport in the plasma edge might change. Unfortunately, the particle source profiles cannot be measured and are mainly unknown preventing a definite experimental answer to this question. However, there is no external gas puffing in these discharges so the main particle sources are recycling from the walls and fuelling by the neutral beam injection. Although, the tungsten discharges have one additional neutral beam source to substitute the missing ICRF power, simulations of the beams show fuelling rates of $5 \cdot 10^{20}\ s^{-1}$ to $6 \cdot 10^{20}\ s^{-1}$. This difference is 1–2 orders of magnitude below normal external fuelling levels and therefore too small to account for the density variation. The contribution of the wall recycling to the fuelling might be different which would not be unexpected considering the different wall materials. In principle, carbon is known to have a better pumping capability, i.e. to store more deuterium during the discharge and release it in between discharges [17]. The SOL density, however, is the same in both cases which argues against a strong difference in the wall recycling. Although, it cannot be experimentally verified at this point, an improved particle confinement remains a possible explanation for the higher density in the tungsten case.

Assuming the limiting factor in the pedestal are the ELMs which pose an upper limit to the pressure, as suggested by peeling-ballooning theory [18], the higher density would result in a reduced electron temperature. However, a reversed causality cannot be excluded; if the pedestal temperature is more limited in tungsten and the pressure again clamped by the ELMs the density would be higher.

The plasma response to the increase in heating power from 8 to 13 MW for carbon and tungsten is very similar. If this were not the case, the pedestal stability would be different for carbon compared to the full metal wall. The density increase in the case of higher heating power can be attributed to a SOL effect, because the density rises by about 50% at the separatrix and the neutral pressure measured with pressure gauges in the divertor even doubles. The general picture of the similarity between carbon and tungsten can be drawn with four more discharges of similar type and each with phases of different heating powers. For all these discharges the pre-ELM pedestal profiles are characterised with the two-line method [19] and the results are plotted in figure 2. The density at the pedestal top is larger for all tungsten cases (*a*), although the separatrix density (*d*) is similar for both wall materials at constant heating power. The temperature (*b*) is smaller by the amount needed to have constant $p_{e,ped}(c)$ at a given heating power. When normalising the confinement time to the ITER98 scaling most of the discharge phases show a similar *H*-factor around 1.1. There are

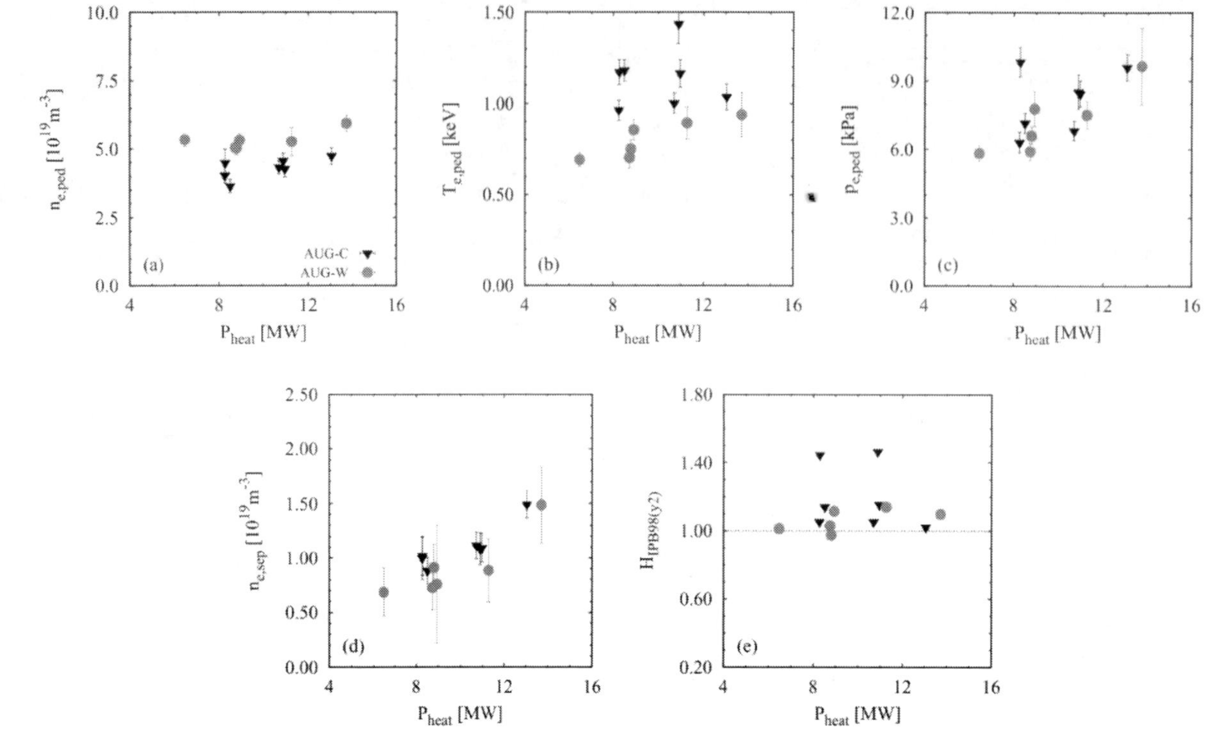

Figure 2. Pedestal top values of the electron density (*a*), temperature (*b*), pressure (*c*), the separatrix electron density (*d*) and the IPB confinement factor (*e*) in carbon (▼) and tungsten (●) plotted against the heating power.

two exceptions with a larger *H*-factor of 1.5 which belong to the same discharge, in [12] the reason for this was attributed to differences in the safety factor profile, namely the shear at mid-radius. The different *q*-profiles influenced the occurrence of instabilities like fishbones and NTMs which can have a significant impact on the core confinement. As indicated in the introduction, in carbon operational windows with very good confinement can be found. Up to now not all of them could be reproduced in the tungsten device. However, in an operational space which overlaps for carbon and tungsten no confinement degradation was found due to the metal wall at low triangularity. In particular, this is true in operation without additional gas puffing. The influence of gas puffing is described in the next section. Observing similar confinement with the same fuelling level in carbon and tungsten is consistent with findings at JET for low triangularity [8]. For high triangularities one has to distinguish between baseline ($q_{95} < 3.5, \beta_N \sim 1.0 - 2.0$) and hybrid ($q_{95} > 3.5, \beta_N \sim 2.0 - 3.0$) discharges [8]—hybrids typically have lower collisionalities than baseline plasmas. While for the hybrids the global confinement is again comparable both in carbon and ILW, the baseline plasmas have strongly reduced confinement in the ILW, even for similar fuelling levels. In both cases the pedestal contribution to the global confinement changes from JET-C to JET-ILW. A consistent explanation for these observations is still missing.

2.2. Influence of external gas puffing

The examples presented in the previous section are fairly special cases and, although, they nicely illustrate the similarities between the C and W discharges without gas puffing, they are not representative for most of the AUG-W operation. Typically, gas is puffed into the machine to increase the ELM frequency and thereby moderate the impurity content in the plasma [5]. The influence of additional gas puffing on the pedestal profiles is documented in the following section.

2.2.1. Deuterium. To avoid a bias by plasma shape or current, discharges with a similar discharge setup as in section 2.1 are selected which have an additional external gas puff. The power scan is achieved only by increasing the NBI power, the ECRH is constant at about 1.2 MW for all cases. The nominal gas puff levels are classified as follows, finite puff rates below $5 \cdot 10^{21}$ electrons s^{-1} are labeled 'low', between 5–$10 \cdot 10^{21}$ electrons s^{-1} 'medium' and larger gas puffs 'high'. The influence of these levels on the plasma density can vary with gas puff location—divertor or midplane—and with the machine condition for a particular discharge. Since all discharges used here are from the same experimental campaign, additional factors like a different first wall structure, i.e. changes in pumping/fuelling channels, do not bias the analysis. However, this can be an issue when comparing discharges from different campaigns. Therefore, the values above do not provide a general classification.

The results of these gas scan experiments are shown in figure 3. By adding additional gas the plasma density increases compared to the unpuffed cases. The electron temperature at the pedestal top decreases while the pedestal electron pressure shows no systematic deviation due to the gas puffing. The separatrix electron density shown in figure 3(*d*)

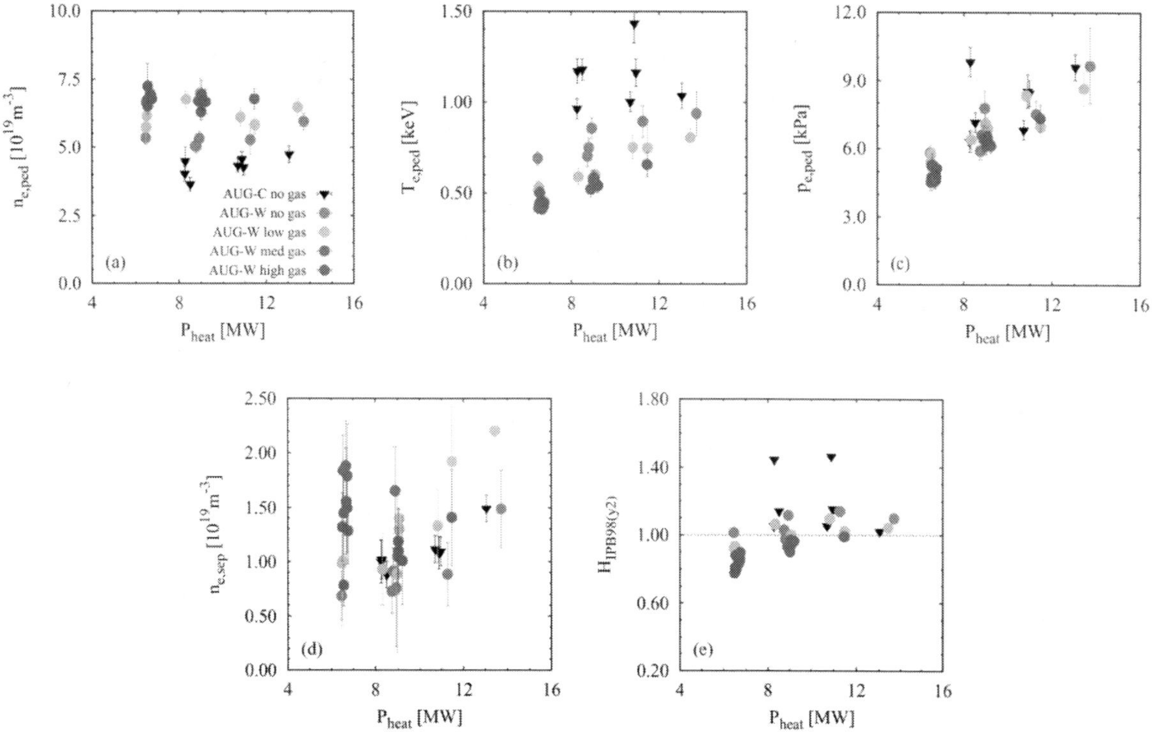

Figure 3. Pedestal top values of the electron density (*a*), temperature (*b*), pressure (*c*), the separatrix electron density (*d*) and the IPB confinement factor (*e*) plotted against the heating power without gas puff in carbon (▾) and tungsten (●) and with low (●), medium (●) and large (●) gas puff in tungsten.

is increasing with the external gas puff for most of the cases. This is also consistent with a larger dataset of AUG discharges presented in [20]. Except for the points with the high gas puff the *H*-factor shown in figure 3(*e*) is fairly constant regardless of the density. The discharges with high gas puffing all deviate from the scaling towards a 20% lower *H*-factor. When carbon and tungsten discharges show different confinement factors, this is in most cases accompanied by a much larger gas puff in tungsten which is necessary for stable operation [8, 21]. Reasons for the lower confinement with very large gas puff might be a changed ELM behaviour as will be discussed in section 3 or the degrading *H*-mode regime at high densities as observed at AUG [7]. The observation of how external particle sources impact the separatrix density, supports the hypothesis that the higher density in the un-puffed tungsten discharges originates in the pedestal.

The trend between heating power and pedestal pressure shown in figure 3(*c*) and its independence of temperature and density allows an estimate for the pedestal pressure, based on engineering variables, to be made. For a larger set of AUG discharges, which includes the same data set as before with additional data at different plasma current and magnetic field, a good estimate for the pedestal pressure in AUG *H*-mode discharges is found with

$$p_{e,ped}[\mathrm{kPa}] = 3.35 \cdot P_{heat}^{0.54} \langle B_p \rangle^{0.87} f_q^{1.65} \qquad (1)$$

where P_{heat} [MW] is the total heating power including the ohmic contribution but without correction for the radiation, $\langle B_p \rangle = \mu_0 I_p / l$ [T] is the flux surface averaged poloidal magnetic field at the separatrix with the separatrix contour length l, $f_q = q_{95}/q_{cyl}$ as discussed in [20]. The comparison of this estimate with experimental data is shown in figure 4 and illustrates a decent match for pedestal pressures between 2 and 10 kPa. The discharge with modified current profile discussed in section 2.1 is the prominent outlier at a pedestal electron pressure of 10 kPa, indicating that these effects are not covered by this estimate. In section 3 it will be shown that for an improved estimate the heating power should be corrected by the ELM power losses and likely by the core radiation. This means the governing parameter for the pedestal top might be

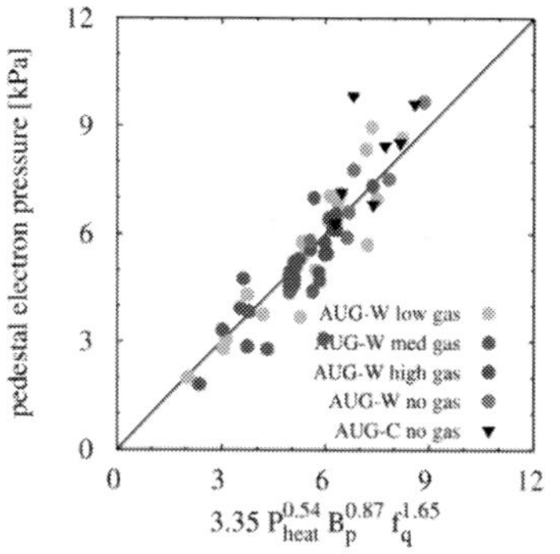

Figure 4. Measured pedestal top electron pressure plotted against an estimate with engineering variables—without gas puff in carbon (▾) and tungsten (●) and with low (●), medium (●) and large (●) gas puff in tungsten.

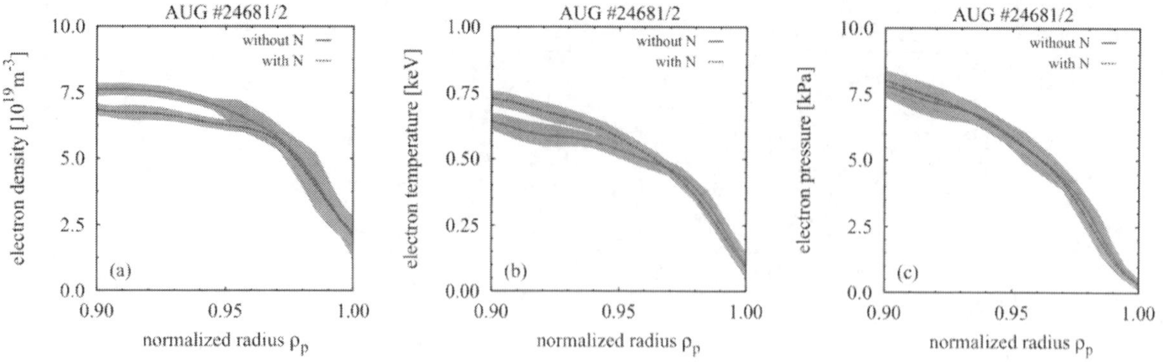

Figure 5. Fits to electron density (*a*), temperature (*b*) and pressure (*c*) for the case with nitrogen (- - -) and without (—) [22].

the power leaving the confined region by steady state transport mechanisms. This is not applied in the estimate in equation (1) because this would include physics parameters in the scaling which are in general not known before the discharge.

2.2.2. Nitrogen seeding. The consequences of nitrogen seeding on the pedestal profiles have been thoroughly discussed e.g. in [3, 10, 22, 23]. Here, a short overview is given in the context of the observations made in the previous sections. In particular, the focus is put on an example with low heating power, 8 MW, and only marginal confinement improvement (<10%). This is done because the pedestal pressure is altered only little while the ELMs already change significantly as will be discussed in section 3. In general, the pedestal response to the nitrogen seeding is more diverse, this can be seen in the references above, a common observation in most cases is a reduction of the ratio n_e/T_e. It has to be noted that when increasing the amount of seeded nitrogen above a certain level divertor detachment is triggered and the pedestal and ELM characteristics discussed in this contribution are not observed any more [24]. The amounts of nitrogen seeded in the discharges presented here are well below that limit.

In the example shown in figure 5 the electron density (*a*) decreases while the electron temperature (*b*) increases resulting in a constant electron pressure. It has been argued that the nitrogen might act as substitute for the missing carbon impurities in the tungsten discharges [10]. The trend in the pedestal top values is similar as observed for the pedestal comparison between tungsten and carbon. The $Z_{\rm eff}$ is also increasing from 2.0 to 2.5 with the nitrogen seeding. However, within the uncertainties no difference in the pedestal gradients is observed for n_e and T_e as was the case in section 2.1. Rather, the density pedestal gets narrower. The additional impact of nitrogen seeding on the divertor plasma and the fuelling efficiency into the main plasma prevents a more definite answer regarding the comparability between carbon and nitrogen as main impurity. Further, there are significant differences in the ELM behaviour which will be discussed in the next sections. A further complication is the necessity of deuterium puffing in nitrogen seeded discharges. Without the deuterium puff too much nitrogen would enter the main plasma and prevent stable operation. Therefore, any comparison experiment with nitrogen seeding includes also a deuterium puff. Depending on the target plasma nitrogen seeding improves the confinement by up to 50% . The exact level of improvement seems to depend on the confinement of the reference discharge, the largest improvement up to now was found in a discharge with high deuterium fuelling and moderate confinement $H_{98,y2} \sim 0.8$ in the reference [10]. A reference discharge with half the amount of deuterium fuelling and a normal confinement of $H_{98,y2} \sim 1.0$ shows a confinement improvement of about 20%, resulting in confinement factors for N seeded discharges between 1.0–1.3 [3].

3. ELM characteristics

This section shows that a typical type-I ELM crash in AUG consists of two distinct phases. The characteristics of the two phases is discussed by means of the comparison experiments presented in the previous sections as well as a large set of AUG standard *H*-modes. Perturbations of the magnetic field due to the ELM crash will be introduced as the main tool for the characterisation of the two ELM crash phases.

3.1. Nitrogen seeding

The existence of two distinct phases in an ELM crash was first reported in [22]. With the application of nitrogen seeding it is possible to suppress the second phase of the ELM crash, while leaving the first phase mainly unchanged. This concept is illustrated with time traces of magnetic pick-up coils, divertor current measurements and edge soft-x ray radiation in figure 6. The two phases of the crash are highlighted by colored shading in green and red along with the other phases of the ELM cycle. The first ELM phase exhibits a high frequency perturbation in the magnetic field. The second phase has a strong low frequent component which is also overlaid with a faster modulation. These characteristics are best analysed in a spectrogram of the pick-up coil measurements which is shown in figure 7(*a*) for an individual ELM. The average over multiple ELMs (*b*) pronounces the main features more clearly: 1st phase of a few 100 μs with strong broadband fluctuation between 0–400 kHz, the 2nd phase of a duration of up to 2 ms and broadband fluctuations up to 250 kHz with a strong low frequent component below 10 kHz. Because it is

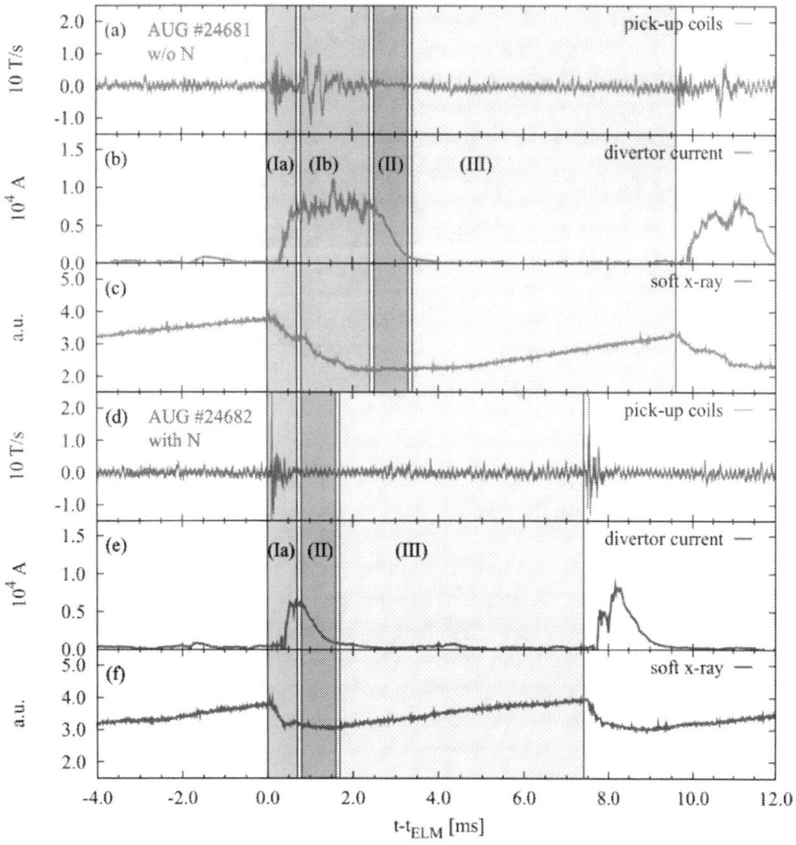

Figure 6. The time traces of pick-up coils (*a*), the divertor current (*b*) and edge soft x-ray emission (*c*) are shown for the reference case without nitrogen gas puff. (*d*)–(*f*) show the same diagnostics in the presence of a nitrogen gas puff. Different phases in the ELM cycle are highlighted—green and red for the ELM crash, blue for the intermediate phase and grey for the recovery phase [22].

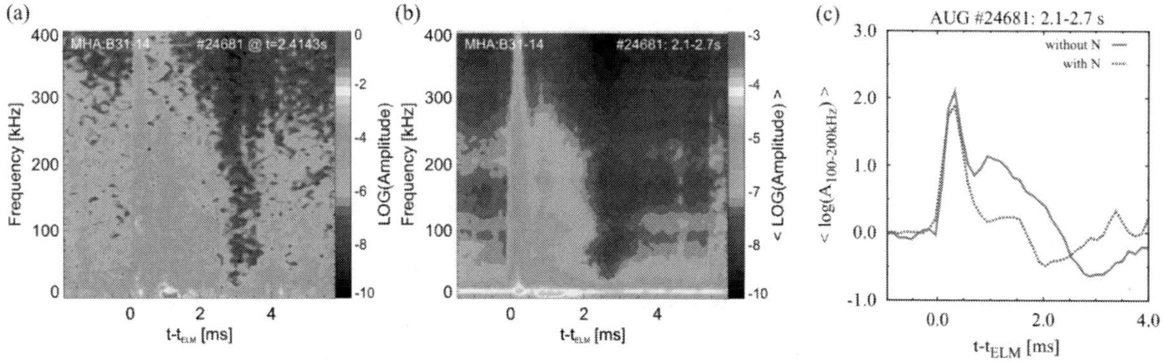

Figure 7. Spectrogram of perturbations in the magnetic field due to an individual ELM with two phases (*a*), a characteristic spectrogram averaged for 52 ELMs and the averaged logarithmic amplitude between 100–200 kHz (*c*) for the case with only the 1st phase (- - -) and with 1st and 2nd phase of the ELM crash (—).

difficult to compare multiple spectrograms an averaged amplitude between 100–200 kHz is normalized to the pre-ELM value and used for comparison of different ELM types. Low frequencies have to be excluded because they often show perturbations of the magnetic field which are not ELM related. In figure 7(*c*) the averaged amplitude is about one order of magnitude larger for the 1st compared to the 2nd phase. The 1st phase has the same averaged magnetic signature both with and without nitrogen which strongly suggests it has the same origin in both cases. This is supported by the detailed comparisons of profile measurements and analysis of soft x-ray radiation [22]. On the one hand it was shown in section 2.2.2 that the pedestal pressure does not vary significantly with nitrogen seeding in the presented case. On the other hand, the divertor/SOL plasma is altered significantly [25]. This suggests an influence of the divertor/SOL plasma on the 2nd phase while

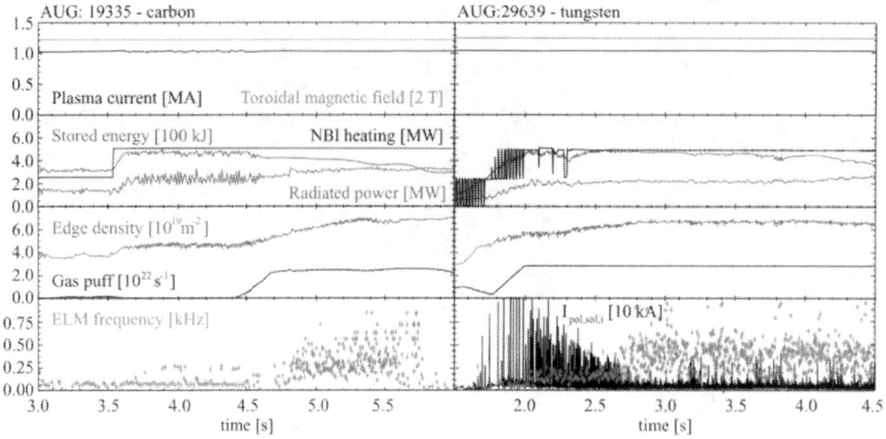

Figure 8. Time traces of a standard *H*-mode with carbon wall (left) and a standard *H*-mode with tungsten wall (right).

the 1st phase is the same at constant pedestal pressure—consistent with peeling-ballooning theory.

This analysis of the magnetic ELM signature is used in the following sections for different reasons besides the capability to easily distinguish between both crash phases. The magnetic measurements are less affected by the plasma density or impurity content than radiation measurements are and the diagnostic was also available for discharges in carbon which is not the case for measurements of the divertor currents.

3.2. Deuterium gas puff

To study the influence of the deuterium gas puff on the ELMs a large number of discharges is necessary which have as little as possible variation in heating power and plasma shape. The AUG 'standard *H*-mode' [26] provides the needed high number of similar discharges. It is a standardised type of discharge which was established in AUG to allow long term comparability of plasma operation. It is run at the beginning of nearly every experimental day and for this section discharges from the last 10 years are utilised which span over the upgrade to the full metal wall. In figure 8 time traces of two standard *H*-mode discharges—one in carbon and one in tungsten—are shown which illustrate the typical behaviour. Plasma current is 1 MA and the toroidal magnetic field −2.5 T. The heating power is ramped up to two neutral beam sources resulting in a stored energy of about 500 kJ which corresponds to a moderate confinement factor of $H_{98,y2} \sim 0.9$. The main difference in the parameters of the two discharges is the timing of the deuterium gas puff. In the carbon case there is an early phase with no gas puff, while in the tungsten case the gas puff is present for almost the entire discharge. This is to increase the ELM frequency in order to flush out impurities thus avoid tungsten accumulation. The gas puff level of $2 \cdot 10^{22}$ electrons s^{-1} is larger than that of the high gas puff cases discussed in section 2.2.1.

In the ELM frequency the impact of the rising density is nicely illustrated, this is observed similarly in carbon and tungsten. In the beginning ELMs with a frequency of below 100 Hz are observed, this frequency increases with rising density up to a point where the ELMs occur very irregularly with inter ELM times from 2–10 ms. At the same time the ELMs are becoming smaller which can be observed in the lower right panel of figure 8, where the peak in the divertor currents serves as indicator for the ELM size. The appearance of smaller ELMs at larger electron density is not a new observation and recent discussions can be found e.g. in [27, 28] for individual discharges. In this section the high number of analysed standard *H*-mode discharges (>200), allows to illustrate the influence of the density on ELMs with a broader database and a particular focus on the two ELM phases. The transition to the small or mitigated ELM regime, with losses below 1% of the total stored energy, will not be discussed further, as these ELMs are too small for the analysis method used here and the reader is refered to [28, 29].

Before the averaging technique can be applied to spectrograms of the magnetic measurements, as described in section 3.1, the ELMs have to be grouped, otherwise the averaging over big and small ELMs would erase most information. For this purpose the peak value of the divertor current during the ELM is utilized. This is only possible for the tungsten standard *H*-modes where divertor current measurements are available. From the time traces in figure 8 no indication for a different ELM behaviour due to the wall material can be inferred. Nevertheless, a detailed comparison of carbon and tungsten will be discussed in section 3.3. For the analysis in the remainder of this section the time between 2.5–3.5 s is divided into 100 ms time-intervals which are each analysed individually to track the density evolution.

The ELMs are sorted in four current-intervals between $I_{pol,sol} = 0, 3, 5, 8, \infty$ kA. The relative frequency of ELMs with these peak divertor currents is illustrated in figure 9 for different densities. The first two current-intervals are combined because they yield similar results. The remaining intervals in $I_{pol,sol}$ divide the ELMs in short figure 9(*a*), medium figure 9(*b*) and long figure 9(*c*) ELMs. The reason for this labeling is discussed in the next paragraph. Figure 9 shows that in principle ELMs of all sizes can be observed to occur simultaneously in tungsten AUG, however, certain conditions favour one or the other type. The histograms show the relative frequency of

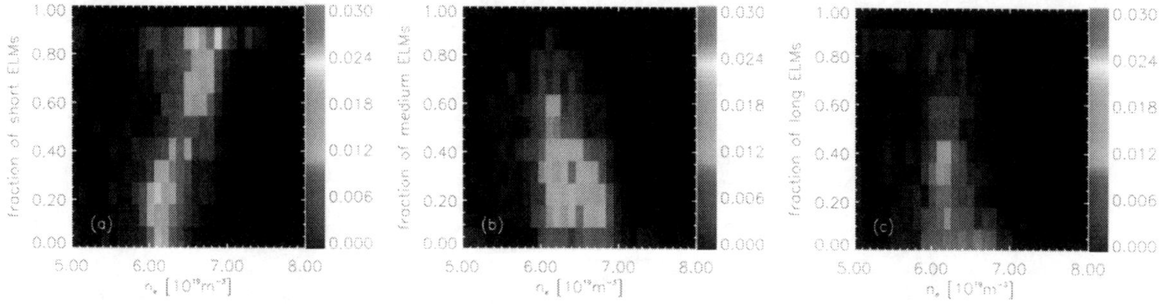

Figure 9. Histograms of the relative frequency of ELM with different peak divertor currents for line averaged edge densities between 5–8 · 10^{19} m^{-3} normalized to the number of analysed time intervals for short (*a*), medium (*b*) and long ELMs (*c*).

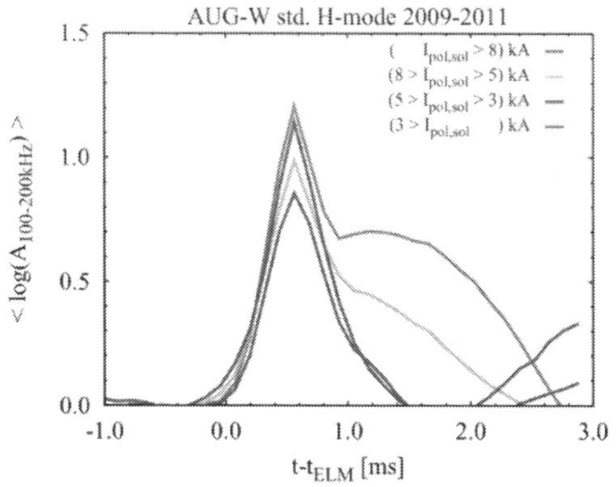

Figure 10. Averaged logarithmic amplitude of perturbations in the magnetic field between 100–200 kHz, synchronised to the ELM onset and averaged over ELMs of different peak divertor current for AUG standard *H*-modes between 2009–2011.

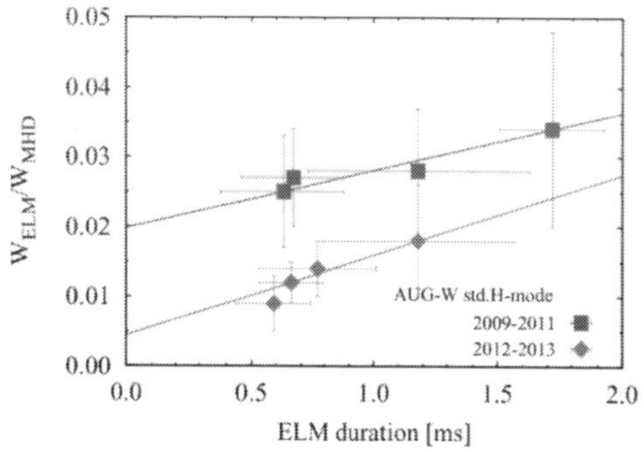

Figure 11. Relative ELM energy loss plotted versus ELM duration for characteristic ELMs in AUG standard *H*-modes between 2009–2011 (■) and 2012–2013 (◆).

each ELM type for different line averaged edge densities [30]. Each bin of the histogram gives the percentage of all analysed time-intervals where a given ELM type has a certain relative frequency. Figure 9(*a*) shows that below 6.5 · 10^{19} m^{-3} the majority of time-intervals consist only of about 20% short ELMs while at higher density 60–90% of all ELMs are short. Similarly, long ELMs only rarely occur above 6.5 · 10^{19} m^{-3}, while the ELMs with a medium $I_{pol,sol,a}$ exist for all densities mostly with a relative frequency below 50% . It should be noted that the density of the confined plasma used here is strongly correlated with the separatrix density as shown in section 2.2.1. The only reason for choosing the edge density is that it is more easily measured.

The naming of the four intervals for the peak divertor current in the previous paragraph was already motivated by the analysis of the magnetic spectograms. The averages between 100–200 kHz for ELMs with certain peak divertor currents is shown in figure 10. For the highest peak $I_{pol,sol}$, therefore the lowest densities, the two phases in the ELM crash are clearly visible, with increasing density and lower $I_{pol,sol}$ the 2nd phase shrinks until only the 1st phase remains for $I_{pol,sol}$ < 5 kA. Therefore, the ELMs in the two intervals with the smallest $I_{pol,sol}$ are named short ELMs. When plotting the FWHM of the averaged spectral amplitude one finds that the ELM energy losses determined via the magnetic equilibrium reconstruction scale linearly with the duration of the 2nd phase in the ELM crash. This is shown in figure 11 for standard *H*-modes between 2009–2011 and 2012–2013 which are both with the full tungsten wall. Although, the ELM energy losses are not the same for the two time periods the increase with ELM duration due to the 2nd phase is comparable as indicated by the linear fits in figure 11. This suggests that the losses which can be attributed to the 2nd phase of the ELM crash are comparable in both cases. Consequently, the 1st phase of the ELM is responsible for the difference in the ELM energy losses between the two time periods. The reason for the variation in ELM losses could not be identified up to now. Different ELM frequencies cannot explain the observations because both sets of discharges show similar ELM frequencies as indicated in figure 12. The main differences between the two time periods are a higher $T_{e,ped}$ at constant $n_{e,ped}$ for the 2012–2013 discharges and a change in gas puff location, from the midplane to the divertor. The divertor gas puffing is correlated in time with an increase of the midplane neutral pressure. Further, the structure of the first wall was successively altered with the installation of magnetic perturbation coils at the low field side. Indications that the wall clearance of the plasma can influence the ELM behaviour were reported before [31]. To explain the variation in ELM losses dedicated experiments will be performed in the future.

With the available data it can be excluded that a diagnostic artifact in the magnetic measurements, which determines the stored energy, is responsible for the change in ELM losses. This is supported by the observation that the electron temperature

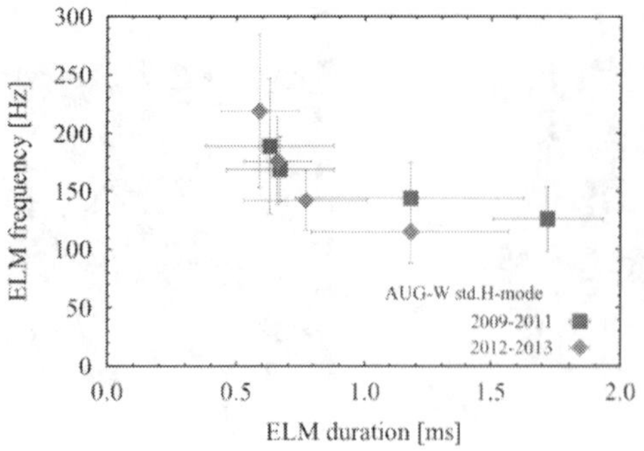

Figure 12. ELM frequency plotted versus ELM duration for characteristic ELMs in AUG standard *H*-modes between 2009–2011 (■) and 2012–2013 (◆).

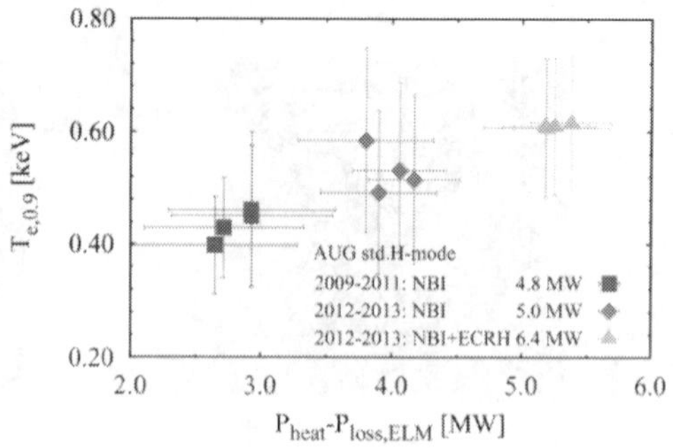

Figure 13. Edge electron temperature plotted against $P_{heat} - P_{loss,ELM}$ for different standard *H*-modes. For each group of standard *H*-modes the 4 points correspond to different durations of the 2nd phase as in figure 11.

follows nicely the trend expected from equation (1) when correcting P_{heat} for the larger ELM power losses $P_{loss,ELM}$. Additional evidence is shown in figure 13. There the electron temperature for $\rho_p \sim 0.9$ is plotted against $P_{heat} - P_{loss,ELM}$ which is a proxy for the power crossing the separatrix by diffusion or convection, if the core radiation is assumed to be constant. Two results can be inferred from this plot; first the duration of the 2nd phase does not contribute to the ELM power loss. This is seen by the clusters of 4 points each, their proximity illustrates the similarity of the ELM power losses despite the different duration of the 2nd phase. With a longer duration of the 2nd phase and higher energy losses per ELM the frequency of the ELMs reduces keeping the power losses constant. Second, a variation in the ELM power loss is equivalent to a change of heating power of the same magnitude regarding their impact on T_e. The ELM power loss can change with the 1st phase of the ELM crash because the ELM frequency is independent of the energy loss in this phase. The difference in energy loss at constant frequency sums up to a power loss which is larger by ~1 MW for the 2009–2011 standard *H*-modes. This is correlated with a reduction of the temperature by 0.1 keV. Additional ECRH heating of ~1 MW in some of the 2012–2013 standard *H*-modes results in a comparable temperature increase of ~0.1 keV.

3.3. Carbon and tungsten

In the previous sections it was shown that the ELM consists of two phases in tungsten and that the duration of the 2nd phase correlates either with $n_{e,ped}$ or with $n_{e,sep}$. This section will show observations of the two crash phases in carbon and bring arguments for the separatrix density being the dominant quantity for the duration of the 2nd phase of the ELM crash.

The ELM characteristics of the carbon-tungsten comparison discharges discussed in section 2.1 is also useful for the ELM comparison. In this comparison $p_{e,ped}$ was constant and $n_{e,ped}$ increased while $n_{e,sep}$ remained constant. In figure 14 it is shown that the ELMs do not change under these conditions despite the different wall materials. In carbon the trace of the magnetic pick-up coils shows the two distinct ELM phases, with the same length as in the tungsten reference. The ELM duration is shown in figure 15 supporting the observation that neither the ELM energy losses nor the duration of the two phases exhibit significant differences due to the wall material. As comparison also the nitrogen seeded discharge with its reference is plotted in green. In the nitrogen seeding experiment $p_{e,ped}$ was also the same for both cases, but the ELM characteristics changed significantly. Z_{eff} and n_e/T_e changed similarly in both comparisons which suggests that these quantities do not play a dominant role in the suppression of the 2nd phase in the ELM crash. The main difference between the carbon/tungsten and N seeding/no seeding is the divertor and SOL plasma. The former showed similar SOL densities, while in the latter the nitrogen seeding is known to have a strong impact on the divertor and SOL plasma.

One can distinguish between a hot, thin and a thick, cold inter ELM SOL/divertor plasma where the latter is achieved with strong deuterium puffing or nitrogen seeding [32]. Nonlinear 2D reduced magnetohydrodynamic simulations suggest a stochastisation of the plasma edge due to ELMs [33, 34] and thereby mixing the pedestal and SOL plasma. Even when starting with the same pedestal pressure, the pedestal-SOL mixture can be quite different and influence the evolution of the ELM crash. This could explain why the ELMs are not affected in the carbon/tungsten comparison with the same pedestal pressure profiles and similar SOL plasmas. At the same time nitrogen seeding changes the ELMs significantly, despite unchanged electron pedestal pressure profiles, because with the seeding the SOL plasma becomes more dense and cools down. Detailed nonlinear MHD simulations are in preparation to test this hypothesis.

4. Conclusions

This paper gave an overview of pedestal profiles of temperature and density and the ELM crash for AUG *H*-mode operation at low triangularity with different boundary conditions, namely different wall materials: dominantly carbon and full tungsten; and different external gas fuelling: without external gas puffing and with gas puffing of deuterium and nitrogen.

It was shown that for all these different boundary conditions the pedestal electron pressure can stay essentially

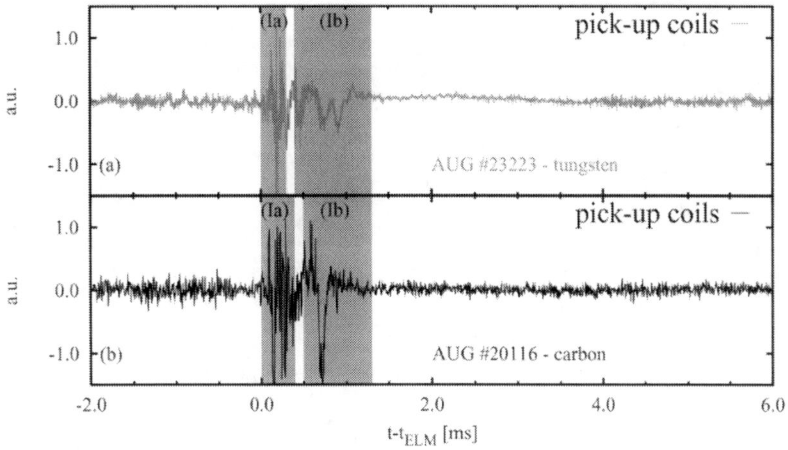

Figure 14. Time traces of the pick-up coil signal for individual ELMs for the comparison discharges between carbon and tungsten wall discussed in section 2.1.

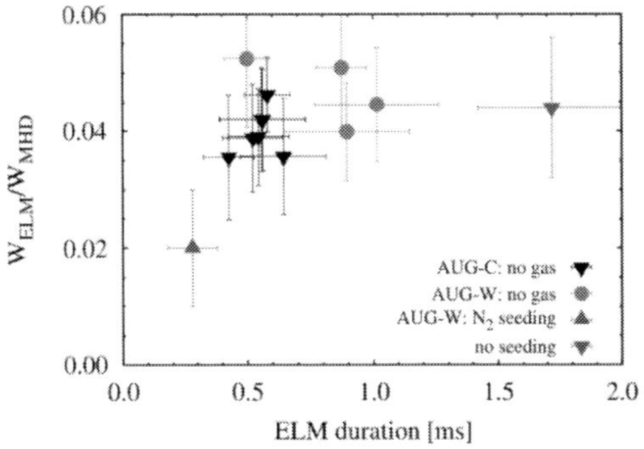

Figure 15. Relative ELM energy loss plotted versus ELM duration for unpuffed discharges in carbon (▼) and tungsten (●). The changes due to nitrogen seeding are added for comparison (▲, ▼).

constant when the heating power, plasma current and plasma shape were not changed. In particular, operation with the full metal and the carbon wall showed no significant differences in plasma performance without gas puffing at low triangularity. The effect of the boundary conditions manifests in a different composition of the pedestal pressure, namely the electron temperature and density. In tungsten the electron density is significantly larger than in a comparison discharge in carbon—both without gas puffing. The higher density was caused by a steeper density gradient in the pedestal. The steeper gradient could be due to improved particle confinement in the pedestal for the tungsten case. Although accurate measurements of the particle source profiles are missing to support this hypothesis experimentally, it remains a possible explanation. This is supported by the observation that external gas sources influence the separatrix density which was not observed in the carbon-tungsten comparison where the density at the separatrix was fairly similar. The higher natural density is one reason why certain high performance regimes observed in the carbon device at low collisionality were only recently reproduced with the tungsten first wall.

The nitrogen seeded case was specially chosen to exhibit constant electron pressure and still show a significant impact on the ELM characteristics. In general, the pressure is not constant, in particular, for higher heating power a larger pedestal pressure is observed with the seeding.

The impact of the impurity content on the pedestal stability might be mitigated/enhanced by additional parameters. One candidate could be the collisionality which changes a lot over the presented discharges and is also the parameter changing most in the JET baseline/hybrid comparison at high triangularity [8]. In the pedestal and the core the hybrid plasmas at low collisionality showed a quite different response to the change from JET-C to JET-ILW compared to the baseline plasmas at higher collisionality.

ELMs in standard AUG discharges exhibit two distinct phases in an ELM crash with different properties. The 1st phase remains the same when the pedestal pressure was not changed. The 2nd phase can be reduced in size by deuterium fuelling and increasing of the separatrix density or it can be suppressed completely by applying nitrogen seeding. Both phases are observed also in carbon discharges with similar size as in the tungsten comparison discharge. There were no indications found that ELMs in carbon and tungsten are fundamentally different. Also their response to additional deuterium puffing and the resulting density increase is similarly observed in standard *H*-modes in carbon and tungsten.

These observations of the pedestal structure and ELM behavior do not suggest that nitrogen simply acts as substitute for carbon as impurity in the pedestal. Nitrogen seeding suppresses the 2nd ELM crash phase while it is unaffected by carbon. In the AUG-C/AUG-W comparison carbon reduces the density gradients while nitrogen seeding in AUG-W is accompanied with smaller pedestal widths at constant gradients. However, the nitrogen seeding also has a strong impact on the SOL plasma, which is not similarly pronounced in the AUG-C/AUG-W comparison.

Both phases of the ELM crash appear to be largely independent of each other. The 2nd phase can have different size and duration while the 1st phase remains unaffected and vice versa. Differences are found in their correlation with ELM power losses. Changes in the duration and size of the 2nd phase anti-correlate with the ELM frequency resulting in the same ELM power loss. In standard *H*-mode plasmas of two consecutive campaigns different sizes were observed for the 1st phase at constant frequency and thereby reducing the ELM power losses. The difference in ELM power loss was found to have a comparable impact on T_e as the same amount of power would have by means of external heating.

The reason causing the reduction in power loss, could not be identified with the available data. Influences on the ELM size like pedestal collisionality or plasma shape can be excluded as reason. The plasma shape did not change in the standard *H*-modes and the collisionality was varied significantly in different comparison experiments without an impact on the 1st phase of the ELM crash.

When only considering the pre-ELM pedestal parameters, a general explanation for the ELM characteristics does not seem possible. Adding the SOL plasma into the description allows to provide a hypothesis: The 1st phase in the ELM crash is set by a stability limit of the pedestal. With the 1st phase the plasma edge becomes stochasticised, thereby mixing the pedestal and SOL plasma—which is supported by reduced magnetohydrodynamic simulations [33, 34]. The 2nd phase is then not governed by the pre-ELM pedestal plasma alone, but the mixture with the SOL.

The observed coupling between the edge temperature and the ELM power loss corrected heating power could influence the quality of extrapolations for the pedestal temperature which are only based on the nominal heating power in future reactor like machines. However, the general trend of these extrapolations should still be correct when the ELM power losses do not vary too much.

The observations about different ELM phases could also impact predictions and extrapolations to future machines. Considering the need for radiative cooling in the SOL and divertor with nitrogen seeding [25] one would expect—based on above hypothesis—the 2nd ELM crash phase to be suppressed by the cold, dense SOL plasma. At the same time no effect on the 1st phase was observed. This would mean extrapolations of ELM sizes might be misleading if they are based on data which does not properly separate both phases of the ELM crash.

Acknowledgments

The authors want to thank M Beurskens, F Ryter and J Schweinzer for fruitful discussions about nitrogen seeding and standard *H*-modes. This work has been carried out within the framework of the EUROfusion Consortium and has received funding from the European Union's Horizon 2020 research and innovation programme under grant agreement number 633053. The views and opinions expressed herein do not necessarily reflect those of the European Commission.

References

[1] Gruber O *et al* 2009 *Nucl. Fusion* **49** 115014
[2] Maggi C *et al* 2010 *Nucl. Fusion* **50** 025023
[3] schweinzer J *et al* 2011 *Nucl. Fusion* **51** 113003
[4] Neu R *et al* 2013 *J. Nucl. Mater.* **438** S34
[5] Pütterich T *et al* 2011 *J. Nucl. Mater.* **415** S334
[6] Saibene G *et al* 2002 *Plasma Phys. Control. Fusion* **44** 1769
[7] Bernert M *et al* 2015 *Plasma Phys. Control. Fusion* **57** 014038
[8] Beurskens M N A *et al* 2014 *Nucl. Fusion* **54** 43001
[9] Rathgeber S K *et al* 2010 *Plasma Phys. Control. Fusion* **52** 095008
[10] Beurskens M N A *et al* 2013 *Plasma Phys. Control. Fusion* **55** 124043
[11] Giroud C *et al* 2013 *Nucl. Fusion* **53** 113025
[12] Stober J *et al* 2007 *Nucl. Fusion* **47** 728
[13] Pankin A *et al* 2004 *Comput. Phys. Commun.* **159** 157
[14] Salmon A N 1994 *Int. J. Infrared Millim. Waves* **15** 53
[15] Fischer R *et al* 2008 *Plasma Phys. Control. Fusion* **50** 085009
[16] Kurzan B *et al* 2011 *Rev. Sci. Instrum.* **82** 103501
[17] Rohde V *et al* 2009 *Nucl. Fusion* **49** 85031
[18] Connor J W *et al* 1998 *Phys. Plasmas* **5** 2687
[19] Schneider P A *et al* 2012 *Plasma Phys. Control. Fusion* **54** 105009
[20] Schneider P A *et al* 2013 *Nucl. Fusion* **53** 73039
[21] Schweinzer J *et al* 2013 *40th EPS Conf. on Plasma Physics (Espoo, Finland, 1–5 July 2013)* P4.113 (http://ocs.ciemat.es/EPS2013PAP/pdf/P4.134.pdf)
[22] Schneider P A *et al* 2014 *Plasma Phys. Control. Fusion* **56** 25011
[23] Dunne M G 2014 Impact of Te and ne on edge current density profiles in ELM mitigated regimes on ASDEX Upgrade *Nucl. Fusion* submitted
[24] Reimold F 2014 Experimental studies and modeling of complete H-mode divertor detachment in ASDEX Upgrade *J. Nucl. Mater.* submitted
[25] Kallenbach A *et al* 2010 *Plasma Phys. Control. Fusion* **52** 55002
[26] Ryter F *et al* 2002 *Plasma Phys. Control. Fusion* **44** A407
[27] Boom J E *et al* 2012 *Nucl. Fusion* **52** 114004
[28] Rathgeber S K *et al* 2014 *Nucl. Fusion* **54** 93011
[29] Barrera L *et al* 2013 *40th EPS Conf. on Plasma Physics (Espoo, Finland, 1–5 July 2013)* P4.113 (http://ocs.ciemat.es/EPS2013PAP/pdf/P4.113.pdf)
[30] Mlynek A *et al* 2010 *Rev. Sci. Instrum.* **81** 033507
[31] Burckhart A *et al* 2010 *Plasma Phys. Control. Fusion* **52** 105010
[32] MüLler H *et al* 2011 *Nucl. Fusion* **51** 073023
[33] Holzl M *et al* 2012 *Phys. Plasmas* **19** 82505
[34] Krebs I *et al* 2013 *Phys. Plasmas* **20** 82506

IOP Publishing Plasma Physics and Controlled Fusion

Plasma Phys. Control. Fusion **57** (2015) 014030 (7pp) doi:10.1088/0741-3335/57/1/014030

Advanced scheme for high-yield laser driven nuclear reactions

D Margarone[1], A Picciotto[2], A Velyhan[1], J Krasa[1], M Kucharik[3], A Mangione[4], A Szydlowsky[5,6], A Malinowska[6], G Bertuccio[7], Y Shi[7], M Crivellari[2], J Ullschmied[8], P Bellutti[2] and G Korn[1]

[1] Institute of Physics ASCR, v.v.i (FZU), ELI-Beamlines project, Prague, Czech Republic
[2] Micro Nano Facility, Fondazione Bruno Kessler, Trento, Italy
[3] Czech Technical University in Prague, FNSPE, Czech Republic
[4] Institute of Advanced Technologies, Trapani, Italy
[5] Institute of Plasma Physics and Laser Microfusion, Warsaw, Poland
[6] National Centre for Nuclear Research, Otwock, Poland
[7] Politecnico di Milano, Department of Electronics Information and Bioengineering, Como, Italy
[8] Institute of Plasma Physics of the ASCR, PALS laboratory, Prague, Czech Republic

E-mail: daniele.margarone@eli-beams.eu

Received 22 July 2014, revised 17 September 2014
Accepted for publication 2 October 2014
Published 28 November 2014

Abstract

The use of a low contrast nanosecond laser pulse with a relatively low intensity ($3 \times 10^{16}\,\mathrm{W\,cm^{-2}}$) allowed the enhancing of the yield of induced nuclear reactions in advanced solid targets. In particular the 'ultraclean' proton–boron fusion reaction, producing energetic alpha particles without neutron generation, was chosen. A spatially well-defined layer of boron dopants in a hydrogen-enriched silicon substrate was used as a target. A combination of the specific target composition and the laser pulse temporal shape allowed the enhancing of the yield of alpha particles up to 10^9 per steradian. This result can be ascribed to the interaction of the long-laser pre-pulse with the target and to the optimal target geometry and composition.

Keywords: laser plasma, nuclear reaction, laser fusion

(Some figures may appear in colour only in the online journal)

1. Introduction

Laser-induced nuclear fusion reactions are nowadays widely investigated as an alternative approach for the production of fusion energy, which could potentially have a high societal impact. Although the most investigated laser fusion reactions are the DD and DT due their high cross section at relatively low temperatures [1], other nuclear reactions have been theoretically and experimentally studied [2–5].

The proton–boron nuclear fusion reaction was first studied in the 1930s [6, 7]. Such a reaction allows the production of energetic alpha particles without neutron generation [3, 8, 9]. The first laser-induced proton–boron reaction was experimentally demonstrated with a picosecond laser pulse (with an intensity of around $2 \times 10^{18}\,\mathrm{W\,cm^{-2}}$) and a boron-enriched polymeric target. The total alpha particle yield produced per pulse was around 10^3 per steradian [4]. The use of a more sophisticated experimental setup and similar laser intensity recently allowed the achievement of a much higher alpha particle yield per pulse (around 10^7 per steradian [5]). This reaction has a main channel with a maximum cross section for protons with energies of 600–700 keV [10–12]. Typically, this channel generates alpha particles with an energy distribution ranging from 2.5 to 5.5 MeV with a maximum of around 4.3 MeV.

The main goal of our experiment was to enhance the alpha particle yield of the proton (p) – boron (^{11}B) nuclear reaction ($^{11}B + p \rightarrow 3\alpha + 8.7\,MeV$) by the optimization of the target composition and by using a nanosecond laser pulse. In fact the enhancement of this nuclear reaction (or similar fusion reaction) products would be beneficial as an *in-situ* diagnostic tool for laser-generated plasmas [13] or for the investigation of ion stopping power in plasmas or high energy density physics [14, 15] by using a very simple experimental setup and the non-stringent laser requirements. Moreover, the aneutronic nature of such a reaction reduces the issues connected to the ionizing radiation protection (shielding and activation of the surrounding materials) during the experiment.

2. Experimental setup

The Prague Asterix Laser System (PALS) in Prague was used for the experiment [16]. The energy delivered on target was around 500 J in 0.3 ns (full-width half-maximum, FWHM) through linearly polarized laser pulses with a wavelength of 1315 nm. The laser focal spot was 80 μm and the nominal intensity around 3×10^{16} W cm^{-2}. The incidence angle between the laser beam axis and the target normal was 30°. Despite the modest nominal laser intensity, in comparison with picosecond and femtosecond laser systems, the maximum proton energies of a few MeV have been shown in recent experiments performed at PALS [17].

Most of the irradiated targets were previously enriched with hydrogen and boron atoms through thermal annealing and 50 keV ion-implantation doping processes performed at the Micro-Nano Facility (MNF) at Fondazione Bruno Kessler in Trento [18, 19]. The well-known chemical–physical processes used in microelectronics allowed the idea to develop of using boron–hydrogen–silicon targets for triggering a specific fusion reaction by precisely controlling the concentration and implantation/diffusion depth of the doping element. Both massive silicon samples (0.5 mm thick) and thin foils (2 μm) were used. The thick samples were implanted with boron atoms (Atomic Mass Unit = 11) located at a depth of 190 nm and having a concentration of about 10^{22} cm^{-3} (the boron layer was approximately 100 nm thick). The concentration of hydrogen in the samples was increased through a thermal process where the hydrogen atoms diffusing into the silicon matrix can link to it by forming Si–H bonds, which neutralize the silicon dangling bonds [20–23], thus reaching a concentration higher than 10^{20} cm^{-3}. Both the thick and thin silicon samples where the boron atoms were homogenously diffused into the substrate via the thermal annealing process (1 h at 1000 °C) were also prepared and used. In the latter case the B-atoms were distributed in a much thicker layer starting from the sample surface up to a certain depth (> 10 μm in the case of massive target) resulting in a lower local boron concentration (~10^{20} cm^{-3}). The silicon 'pure' samples (not artificially hydrogenated and without boron doping) were also irradiated for comparison. The used targets with different compositions are shown in the inset of figure 1(*a*).

The energy distribution of the produced proton and alpha particle beams were measured in various directions with respect to the target normal by a Thomson parabola (TP) spectrometer, a time-of-flight (TOF) silicon carbide (SiC) detector [24–26] and solid-state nuclear track detectors of the PM-355 type [27, 28]. A schematic description of the experimental setup is reported in figure 1(*a*). The PM-355 detectors were placed at different angles (the values reported in figure 1 refer to the angle with respect to the target normal) and at a distance of 50 cm from the target surface; the SiC-TOF detector was placed at a large angle (60 degrees) and at a distance of 124 cm; the TP spectrometer was placed along the target normal direction. An additional PM-355 detector was placed in the forward direction, along the target normal direction, when a thin target was used (in a few cases the TP spectrometer was also placed in the forward direction).

The ns-long-laser pulse's temporal shape is shown in figure 1(*b*). The spatially-resolved pulse time profile was determined on the basis of the infrared streak-camera pictures taken in the near field zone. The pre-pulse contrast was measured separately in a special series of laser shots. The shot-to-shot fluctuations on the intensity of this contrast were below 10%. Although approximately 500 J were contained in the main pulse (0.3 ns at FWHM), the residual laser energy (about 100 J) was stored in the nanosecond pedestal (around 50 J of the pulse energy came before the maximum intensity peak was reached). Thus, three different intensity regions were present: 0-I is the low-intensity (3×10^{10} W cm^{-2}) laser precursor, I-II is the high-intensity (10^{15} W cm^{-2}) pre-pulse and II-III is the high-intensity (nominally 3×10^{16} W cm^{-2}) main pulse.

3. Hydrodynamic simulations

The hydrodynamic simulations were performed by using a 2D PALE (Prague Arbitrary Lagrangian–Eulerian) code [29] in order to predict the action of the nanosecond pre-pulse on the irradiated target. This code employs the compatible staggered Lagrangian numerical scheme [30] and a rezoning/remapping mechanism improving the robustness of the simulations. The thermal conductivity is treated using a mimetic method with the standard Spitzer–Harm heat conductivity coefficient and heat flux limiter. Ray tracing is used for the laser absorption modeling. The simulations shown in this paper were performed in the Lagrangian regime, thus the layers of the targets remained separated throughout the whole simulation. The initial computational mesh was adjusted to the layered targets using a finer mesh resolution at the target surface. The simulations were run from 1 to 1.85 ns (see figure 1(*c*) as a reference). Three different massive target geometries were investigated: (a) A B-implanted layer in a Si substrate with the above described characteristics; (b) a B-diffused layer in a Si substrate with the above described characteristics; (c) a B layer deposited on a Si substrate surface with the same atomic concentration as in (a). Figure 2 shows the results of these hydrodynamic simulations: z is the plasma expansion in the longitudinal direction (parallel to the target normal) and r is the plasma expansion in the transverse direction (parallel to the target surface). The simulation outputs clearly show that the B-plasma is better confined (200 μm from the target surface) and has a higher density (~10^{18} cm^{-3}) in (a) than in (b) and (c). Thus, in (a) the accelerated protons can subsequently interact with higher density plasma and have a higher probability of inducing nuclear reactions than in (b) and (c). Figure 2(*d*) shows the time dependence of the maximum boron plasma density in case (a).

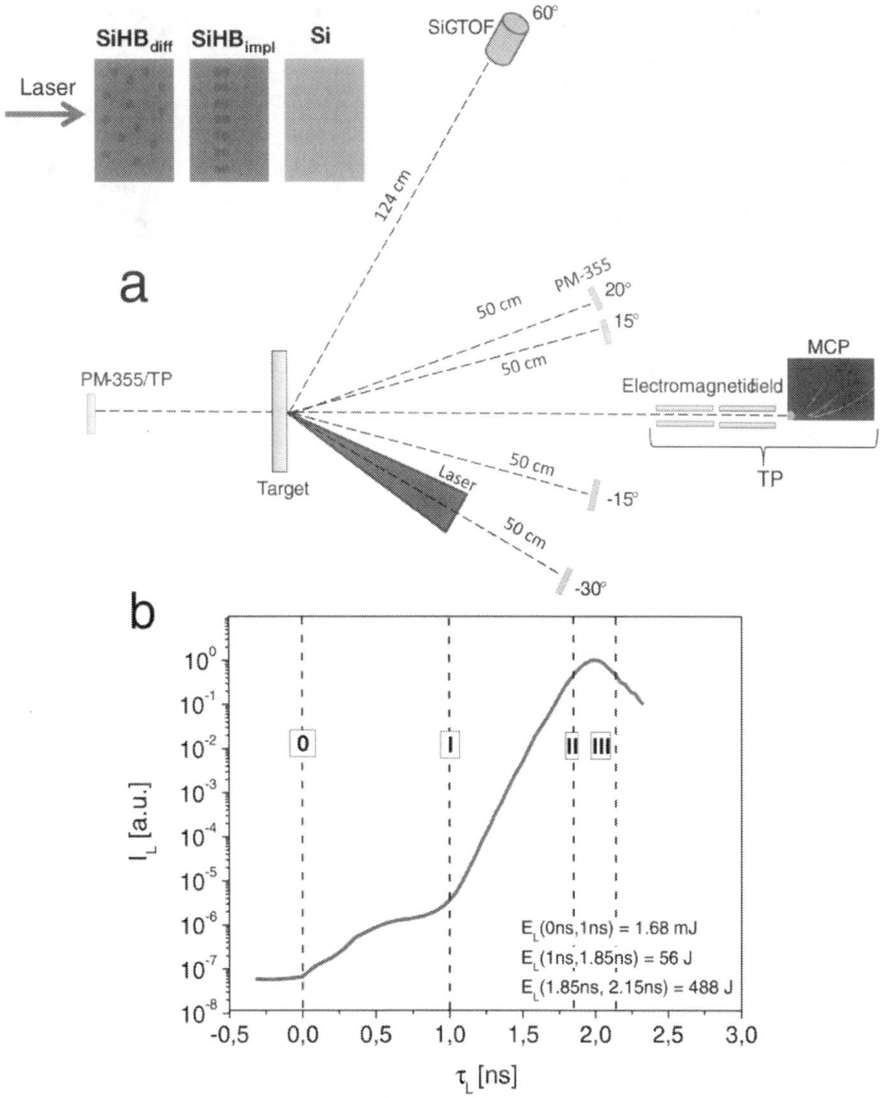

Figure 1. (*a*) The experimental setup used in the experiment with both thick and thin targets (the TP spectrometer was moved in the forward direction in the case of the thin target); the inset shows a simple sketch of the three different target geometries used in the experiment. (*b*) Temporal (τ_L) profile of the laser pulse intensity (I_L).

4. Results and discussion

The PM-355 nuclear track detectors were previously calibrated both for a 1 and 2 h etching time by using mono-energetic ions provided by a standard accelerator, as shown in figure 3(*a*). They were placed at a distance of about 50 cm from the target surface and at different angles (from −30° to 20° with respect to the target normal) and covered by aluminum foils of different thicknesses (6–30 μm) to filter the low-energy plasma ions out and to avoid overlapping between neighboring ion tracks. Figures 3(*b*) and 3(*c*) show the typical PM-355 snapshots (1 and 2 h etching times, respectively) obtained when the massive boron implanted hydrogenated silicon (Si-H-B_{impl}) samples were irradiated. Craters caused by the alpha particles (large dark spots) and by the protons (small light spots) passing through the 15 μm-thick Al foil were clearly distinguished. Figures 3(d) and 3(*e*) show the typical snapshots (1 and 2 h etching times, respectively) obtained when the pure silicon samples (Si) were irradiated for comparison and only proton-generated craters were distinguishable on the PM-355 detectors.

The angular distribution of the alpha particles emitted backward when the thick Si-H-B_{impl} target was irradiated is shown in figure 4. PM-355 detectors placed at different angles with respect to the target normal were used to measure this angular distribution in the same laser shot.

Thin SiHB foils with homogeneously diffused boron atoms were also irradiated, and the corresponding backward and forward PM-355 snapshots are shown in figures 5(*a*) and 5(*b*), respectively. The alpha-particle fluence measured backward (~1 × 10^5 cm^{-2}) is lower than the forward one (~2 × 10^5 cm^{-2}) and is also lower than the fluence measured when using massive targets (~1 × 10^6 cm^{-2}).

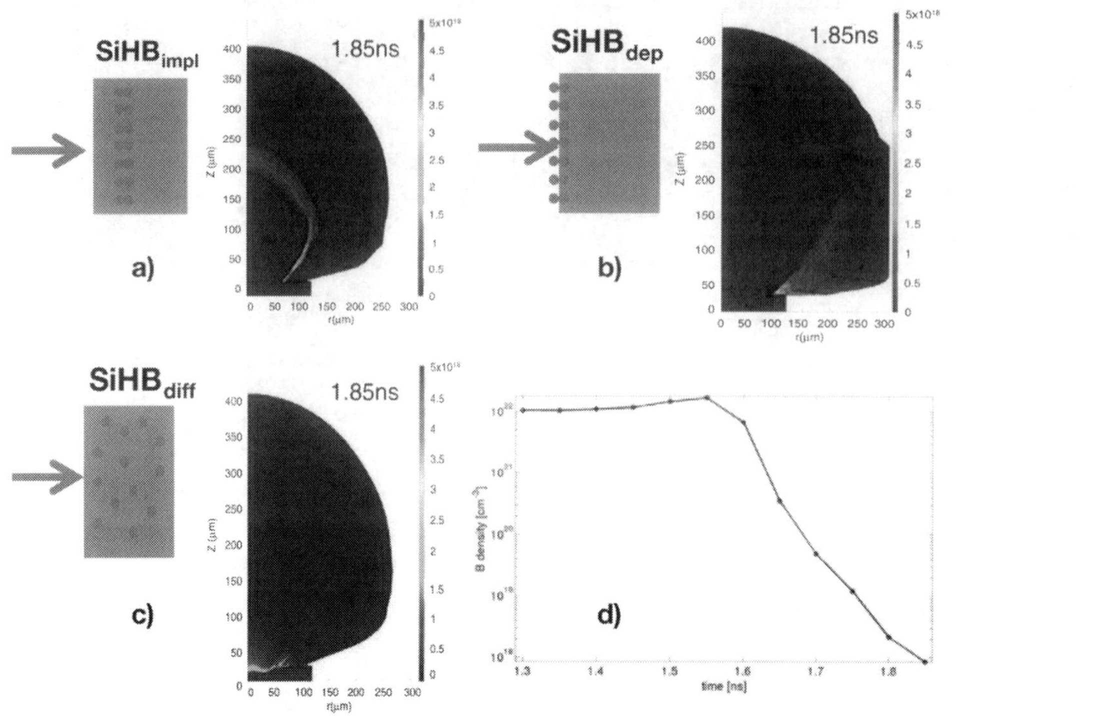

Figure 2. 2D PALE hydrodynamic simulations of the boron plasma expansion after the action of the laser pulse (in the range 1–1.85 ns) for the B layer implanted in the SiH substrate (*a*), the B layer deposited in the SiH substrate (*b*), the B layer diffused in the SiH substrate (*c*) (the gradient scale on the right-hand side shows the B-plasma density). The time dependence of the maximum B-plasma density corresponding to sample (*a*) is plotted in (*d*).

The alpha particles' angular distribution in the case of the massive boron implanted hydrogenated target shows a maximum yield of about 10^9 particles per steradian and a total number of alpha particles produced per laser pulse of about 4×10^8. The total number of alpha particles produced is about 2×10^3 times higher than the previous results in a similar laser irradiation geometry reported in [8] (considering the actual yield suggested in [31]), thus suggesting that our experimental scheme provides an extremely high-efficiency nuclear reaction. An alpha particle yield of about 10^7 per steradian has been recently reported by using a more sophisticated experimental setup with two laser beams (ns and ps) and two different targets [9].

A rough estimation of the theoretical alpha particle yield in the case of a massive target was performed by using the following expression for the number of reactions (ΔN) in a certain time interval (Δt):

$$\Delta N = \rho(B) \times \rho(H) \times <\sigma\nu> \times V \times \Delta t \times 3 \qquad (1)$$

where $\rho(B)$ is the boron density in cm^{-3}, $\rho(H)$ is the hydrogen density in cm^{-3}, $<\sigma\nu>$ is the average reaction rate in $cm^3\,s^{-1}$, V is the interacting volume in cm^{-3} and Δt is the time interval when the nuclear reaction occurs. Our hydrodynamic simulations show that even if the B-plasma is expanding during the laser pre-pulse interaction, it remains very well confined and in a first approximation it is possible to assume that all the protons produced during the interaction and expanding backward (about 10^{11} in the energy range relevant to the nuclear reaction measured in previous experiments [32]) will interact with the B plasma cloud. In this case the previous formula can be reduced to the following one:

$$\Delta N = \rho(B) \times N(H) \times <\sigma\nu> \times \Delta t \times 3 \qquad (2)$$

where $N(H)$ is the total number of protons in the relevant energy range. Assuming a B-plasma density in proximity to the target of about $2 \times 10^{21}\,cm^{-3}$ estimated by hydrodynamic simulations and a Δt of about 1 ns, the estimated total number of alpha particles per laser pulse is approximately 7×10^8 in very good agreement with the value measured experimentally if we take into account the fact that our detectors were placed in front of the target, thus measuring only alpha particles propagating backward (approximately half of the alpha particles propagated forward and were stopped in the massive target).

The energy distribution of the protons was observed far from the irradiated target using a Thomson parabola spectrometer. Two typical snapshots measured in the backward and forward directions are shown in figures 6(*a*) and 6(*b*), respectively. Figure 6(*c*) shows a collection of proton energy distribution calculated from the TP spectra for different laser shots and different target geometries. The proton energy spectra do not show strong fluctuations and the cutoff energy is around 1 MeV.

The alpha particle spectrum was also measured by the silicon carbide detector (in TOF geometry) which provided a full-range energy distribution with a much higher energy

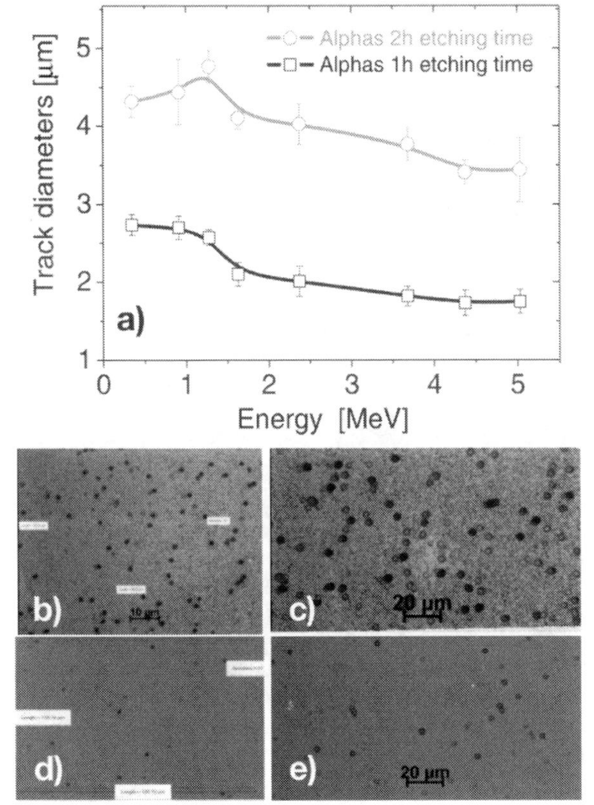

Figure 3. (*a*) The alpha particles' calibration curve of the PM-355 film used in the experiment after 1 and 2 h etching times; (*b*) PM-355 snapshot showing the craters produced by the incident protons (small light spots) and alpha particles (large black spots) for the thick Si-H-B_{impl} target after a 1 h etching time; (*c*) PM-355 snapshot showing the craters produced by the incident protons and alpha particles for the thick Si-H-B_{impl} target after a 2 h etching time; (*d*) PM-355 snapshot showing the craters produced by the incident protons for the thick Si target after a 1 h etching time; (*e*) PM-355 snapshot showing the craters produced by the incident protons for the thick Si target after a 2 h etching time. Al foil 15 μm thick was placed in front of the PM-355 films.

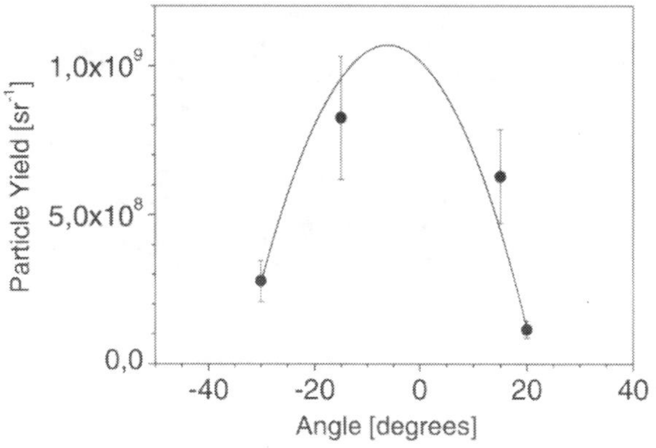

Figure 4. The alpha-particles' angular distribution for the thick Si-H-B_{impl} target measured by the PM-355 detectors placed at different angles with respect to the target normal in the same laser shot.

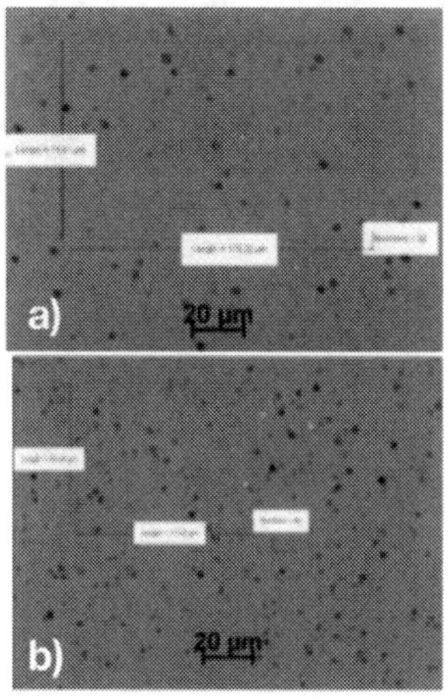

Figure 5. (*a*) PM-355 snapshots showing the craters produced by the incident protons (small light spots) and alpha particles (large black spots) for the thin Si-H-$B_{diff.}$ target in the backward direction; (*b*) PM-355 snapshots showing the craters produced by the incident protons and alpha particles for the thin Si-H-$B_{diff.}$ target in the forward direction.

resolution than the previous technique using PM-355 detectors, which was nevertheless fundamental for estimating the absolute number of produced particles.

A typical signal of the SiC detector screened from low energy plasma ions with an 8 μm-thick aluminum foil is shown in figure 7. This detector was placed at an angle of 60 degrees with respect to the target normal direction, thus the measured plasma ion yield was strongly reduced. The TOF spectrum corresponding to the massive SiHB target with B implanted clearly shows one main ion peak due to the alpha particles coming from the $^{11}B(p,\alpha)2\alpha$ reaction main channel (α_1).This peak is centered at around 95 ns (corresponding to ~3.5 MeV) and contains alpha particles with energies between 2 and 6 MeV. It is also possible to distinguish a small peak, probably attributable to the alpha particles coming from the nuclear reaction's secondary channel (α_2) [5]. This peak is centered at about 55 ns (~10 MeV). The TOF corresponding to the single layer Si target is also shown for comparison. The absence of peaks corresponding to the alpha particles from the nuclear reaction is evident. Just a small plasma ion current (protons and/or silicon ions) was measured.

In the following a tentative explanation of the mechanism which allowed such an enhanced alpha particle yield to be obtained is given. Our proposed mechanism describes a simplified picture of the long-laser pulse interaction with the H-enriched and thick B-doped silicon target. At a time of 2 ns before the maximum intensity, the laser pulse starts to interact with the Si-H layer causing the ablation and ionization of the target material during the 0-I period (the laser fluence is about 30 J cm^{-2}, i.e. well above the ablation threshold for Si [33]), thus sweeping away the Si-H layer on the target surface. The target ablation continues during the I-II period affecting the implanted boron layer, as shown by the hydrodynamic simulations (see figure 2). In fact the ablation depth numerically estimated at the end of this period (corresponding to 1.85 ns in figure 1(*c*)) is around 10 μm,

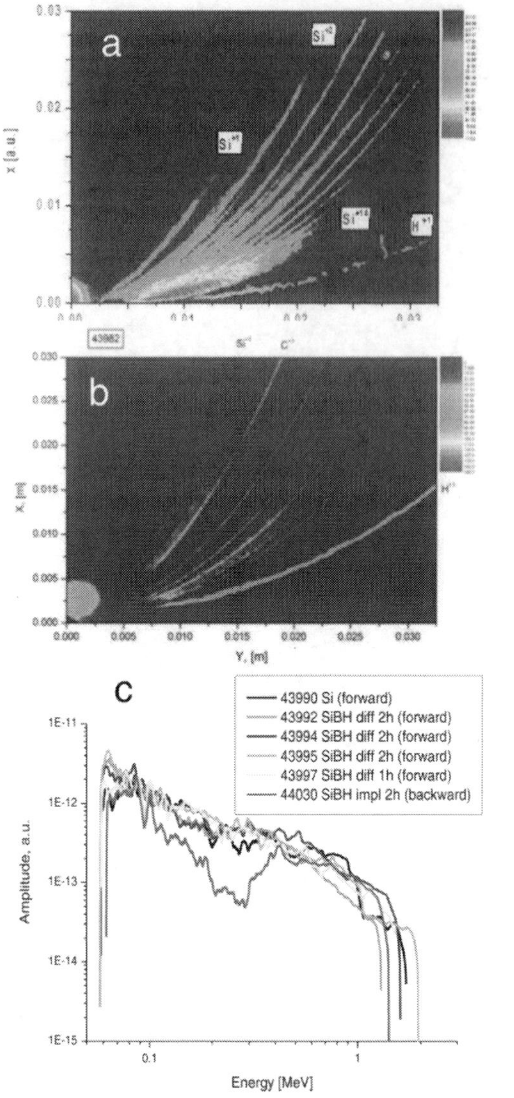

Figure 6. (*a*) and (*b*) Typical TP spectrometer snapshots showing the presence of backward and forward, respectively, accelerated protons and heavier plasma ions; (*c*) proton energy distributions (calculated from TP spectra) for different laser shots.

thus the B-plasma has clearly developed (see figure 2(*a*)) and the protons are already accelerated from the Si-H inner layer. During the II-III period the protons accelerated backward catch the B ions and the 'beam-target' nuclear reaction occurs. In fact the B-ions can be assumed to be at rest since the proton velocity is about $1 \times 10^7 \mathrm{m\,s^{-1}}$ and the boron plasma one is about $2 \times 10^6 \mathrm{m\,s^{-1}}$. Our theoretical model is simplified since the involved processes are dynamic and there is a clear interplay between the B-plasma density (hence the distance of the B-ion cloud from the target surface) and the proton beam energy. In the case of thin (2 μm) B-diffused silicon target the mechanism is similar, with the main difference being that the protons accelerated forward (inside the target) can propagate through a high density B-plasma and produce

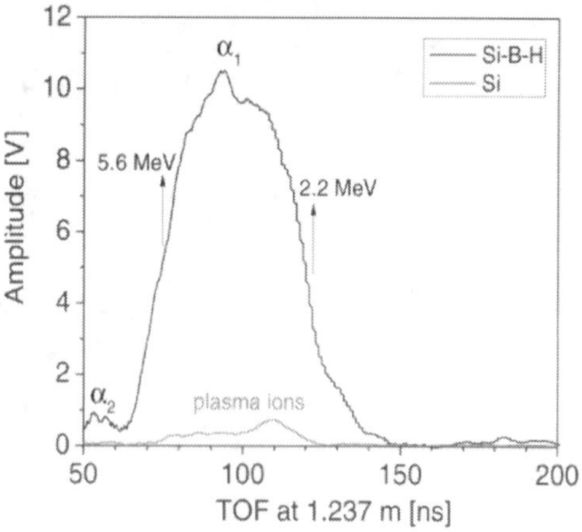

Figure 7. The ion TOF distribution for the thick Si-H-$B_{impl.}$ target (black curve) with the alpha particle signal (α_1 and α_2) and the plasma ions measured by the SiC detector covered by an 8 μm-thick aluminum foil and placed at 60 degrees with respect to the target normal. The ion TOF distribution for the thick Si target (red curve) showing the presence of only plasma ions (no fusion event occurred) is also reported for comparison.

additional alpha particles, which emerge from the target rear side and are detected in the forward direction. In the latter case the total number of alpha particles measured with thin targets was lower than the thick target case due to the lower local boron concentration in the sample (the B atoms were diffused and not implanted in this geometry).

5. Conclusions

The ^{11}B(p,α)2α fusion reaction was enhanced by using modest intensity laser pulses interacting with advanced multi-layer targets. A fusion rate of about 10^9 alpha particles per steradian per pulse was achieved by using low-contrast, long-laser pulses and boron–hydrogen enriched silicon targets with different thicknesses. This result is attributable to the high current proton beam generated by the long (nanosecond) laser pulse, which presented an optimal energy spectrum with a plateau around the maximum of the nuclear reaction cross section. Furthermore, the multilayer target geometry with a high boron concentration at a given depth of the hydrogenated silicon sample allowed the nuclear reaction to occur when the boron ion density was still high (i.e. still close to the target surface), thus ensuring a very high fusion rate.

This (or similar) nuclear fusion reactions can be investigated and further enhanced through the optimization of both laser (nanosecond pulse shape, contrast and energy) and target (overall thickness, B and H concentration, B layer depth) characteristics. Such a scheme could be attractive for the generation of high directionality, high current ion beams with very compact systems to be used '*in-situ*' in laser plasma experiments as a diagnostic tool, e.g. to investigate the so-called 'warm dense matter,' or to determine the ion stopping power in plasmas, which is also of interest for the ignition and burn of an inertial confinement fusion capsule [13–15]. Furthermore, the potential restrictions due to ionizing radiation protection protocols during the experiment can be relaxed thanks to the aneutronic nature of the $p^{11}B$ fusion reaction.

The advantage of our proposed experimental method lies in the use of a moderate laser power (2 TW) and intensity ($3 \times 10^{16}\,W\,cm^{-2}$), thus enabling future possibilities to use compact laser systems and simple irradiation geometries. Moreover, our approach does not imply special laser techniques for compressing the pulse such as the chirped pulse amplification method. Thus, a less sophisticated and cheaper system can be used. In fact, such relatively long pulses do not need vacuum beam transport lines since the laser fluence is below the threshold for laser-induced breakdown sparks in air. Moreover, the used laser intensity was relatively low, thus the spot size on the target was much higher than the diffraction limit, which allowed the generation of a high-yield B-plasma thanks to the volumetric effects increasing, in turn, the number of fusion events.

Furthermore, our hydrodynamic simulations suggest that the relevant physical mechanisms which enable the 'beam-target' fusion reaction occur before the main laser pulse peak. In fact, it was numerically estimated that the boron plasma density value drops below $10^{18}\,cm^{-3}$ after 1.85 ns, hence the number of alpha-particles generated from the fusion reaction decreases drastically before the main laser peak intensity is reached on the target. This implies that only a small part of the laser energy (~50 J) is used for realizing the nuclear reaction; thus, in principle, this experiment can be performed using small and cost-effective laser systems (tens of joules, nanosecond-class and high repetition rate), for instance based on newly established diode-pumped laser technology, which might pave the way for more advanced technological applications, e.g laser fusion propulsion systems, where the efficiency of the process plays an important role [34].

Acknowledgments

The authors thank the technical personnel C Gazzin and S Tomasi of the Micro Nano Facility of Fondazione Bruno Kessler for their support during the ion implantation of the samples and also R Dudzak, M Pfeifer, J Skala and the technical personnel of the PALS laboratory for their support during the experimental work, C Lanzieri, S Masci, L Martinelli and D Puglisi for their collaboration in the SiC detector manufacturing and assembling and S V Bulanov and O Klimo for their scientific discussions and suggestions. Dr M Kucharik was supported in part by the Czech Ministry of Education, Youth and Sports through the COST project LD14089. Moreover, this research was supported by the CHARPAC Work Package of the Laserlab-Europe III (Project No. 284464), by the Ministry of Education, Youth and Sports of the Czech Republic (ELI-Beamlines reg. No. CZ.1.05/1.1.00/02.0061, ECOP3 reg. No. CZ.1.07/2.3.00/20.0279 and PALS project reg. No. LM2010014) and by the Academy of Sciences of the Czech Republic (M100101210).

References

[1] Lindl J 1995 *Phys. Plasmas* **2** 3933
[2] Atzeni S and Meyer-ter-Vehn J 2014 *The Physics of Inertial Fusion: Beam Plasma Interaction, Hydrodynamics, Hot Dense Matter* (Oxford: Oxford Science)
[3] Hora H *et al* 2010 *Energy Environ. Sci.* **3** 479
[4] Belyaev V S *et al* 2005 *Phys. Rev.* E **72** 026406
[5] Labaune C *et al* 2013 *Nat. Commun.* **4** 2506
[6] Oliphant M and Rutherford L 1933 *Proc. R. Soc.* A **141** 259
[7] Dee P and Gilbert C 1936 *Proc. R. Soc.* A **154** 279
[8] Rostoker N, Binderbauer M W, Monkhorst H J 1997 *Science* **278** 1419
[9] Kulcinski G L and Santarius J F 1998 *Nature* **396** 725
[10] Nevins W M and Swain R 2000 *Nucl. Fusion* **40** 865
[11] Moreau D C 1977 *Nucl. Fusion* **17** 1
[12] Stave S *et al* 2011 *Phys. Lett.* B **696** 26
[13] Li C K 2007 *PRL* **99** 055001
[14] Singleton R L 2008 *Phys. Plasmas* **15** 056302
[15] Deutsch C *et al* 2010 *Open Plasma Phys. J.* **3** 88
[16] Jungwirth K *et al* 2001 *Phys. Plasmas* **8** 2495
[17] Margarone D *et al* 2010 *Rev. Sci. Instrum.* **81** 02A506
[18] Picciotto A *et al* 2010 *Europhys. Lett.* **92** 34008
[19] Picciotto A *et al* 2011 *J. Appl. Phys. Express* **4** 126401
[20] Chang K J and Chadi D J 1989 *Phys. Rev.* B **40** 11644
[21] Arrhenius S Z 1889 *Phys. Chem.* **4** 226
[22] Van Wieringen A and Warmoltz N 1956 *Physica* **22** 849
[23] Fick A 1855 *Phil. Mag.* **10** 30
[24] Margarone D *et al* 2011 *J. Appl. Phys.* **109** 103302
[25] Bertuccio G *et al* 2013 *Appl. Surf. Sci.* **272** 128
[26] Bertuccio G 2010 *Proc. SPIE Int. Soc. Opt. Eng.* **7679** 76790T
[27] Szydłowski A *et al* 2001 *W. Radiat. Meas.* **34** 325
[28] Malinowska A *et al* 2012 *Nucl. Instrum. Methods* B **281** 56
[29] Liska R *et al* 2011 ALE method for simulations of laser-produced plasmas *Finite Volumes for Complex Applications VI: Problems and Perspectives* vol 4, ed J Fořt *et al* (Berlin: Springer) pp 857–73
[30] Caramana E J *et al* 1998 *J. Comput. Phys.* **146** 227–62
[31] Kimura S *et al* 2009 *Phys. Rev.* E **79** 038401
[32] Margarone D *et al* 2011 *Nucl. Instrum. Methods Phys. Res. Sec.* A **653** 159
[33] Bulgakov A V *et al* 2004 *Thin Solid Films* **453** 557
[34] Ohlandt C, Cammash T, Powell K G 2003 A design study of p-^{11}B gasdynamic mirror fusion propulsion system *CP654 Space Technology and Applications International Forum, STAIF 2003* ed M S El-Genk (College Park, MD: American Institute of Physics) p 490

IOP Publishing

Plasma Physics and Controlled Fusion

Plasma Phys. Control. Fusion **57** (2015) 014031 (10pp)

doi:10.1088/0741-3335/57/1/014031

Theoretical description of heavy impurity transport and its application to the modelling of tungsten in JET and ASDEX upgrade

F J Casson[1,2], C Angioni[2], E A Belli[3], R Bilato[2], P Mantica[4], T Odstrcil[2], T Pütterich[2], M Valisa[5], L Garzotti[1], C Giroud[1], J Hobirk[2], C F Maggi[2], J Mlynar[6], M L Reinke[7], JET EFDA Contributors[8] and ASDEX-Upgrade Team[9]

JET-EFDA, Culham Science centre, Abingdon, OX14 3DB, UK
[1] CCFE, Culham Science Centre, Abingdon, Oxon, OX14 3DB, UK
[2] Max-Planck-Institut für Plasmaphysik, Garching, Germany
[3] General Atomics, PO Box 85608, San Diego, CA 92186-5608, USA
[4] Istituto di Fisica del Plasma, CNR/ENEA, Milano, Italy
[5] Consorzio RFX-CNR/ENEA, I-35127 Padova, Italy
[6] IPP.CR, Inst. of Plasma Physics AS CR, Prague, Czech Republic
[7] University of York, Department of Physics, Heslington, York, YO10 5DD, UK

Received 26 June 2014, revised 11 September 2014
Accepted for publication 24 September 2014
Published 28 November 2014

Abstract

The effects of poloidal asymmetries and heated minority species are shown to be necessary to accurately describe heavy impurity transport in present experiments in JET and ASDEX Upgrade. Plasma rotation, or any small background electrostatic field in the plasma, such as that generated by anisotropic external heating can generate strong poloidal density variation of heavy impurities. These asymmetries have recently been added to numerical tools describing both neoclassical and turbulent transport and can increase neoclassical tungsten transport by an order of magnitude. Modelling predictions of the steady-state two-dimensional tungsten impurity distribution are compared with tomography from soft x-ray diagnostics. The modelling identifies neoclassical transport enhanced by poloidal asymmetries as the dominant mechanism responsible for tungsten accumulation in the central core of the plasma. Depending on the bulk plasma profiles, turbulent diffusion and neoclassical temperature screening can prevent accumulation. Externally heated minority species can significantly enhance temperature screening in ICRH plasmas.

Keywords: tokamak, impurity, transport, neoclassical, validation, modelling, tungsten

(Some figures may appear in colour only in the online journal)

1. Introduction

Tungsten (W) has good properties as a plasma facing component due to its high heat tolerance, low erosion rate and low hydrogen retention. Tungsten will be used in ITER, is a candidate material for a fusion reactor and is presently used in the ASDEX Upgrade (AUG) tokamak and the recently installed ITER-like wall (ILW) at JET. Since tungsten and other high-Z ions radiate strongly, their concentration in a fusion plasma must be minimised and central accumulation must be avoided to ensure stable operation and good performance. For ITER scenario planning, it is therefore vital to have an understanding of impurity transport underpinned by comprehensive theoretical models [1]. As a prerequisite for reliable predictions, it is important that these models be quantitatively validated against existing experiments.

[8] See appendix of Romanelli F *et al* 2012 *Proc. 24th IAEA Fusion Energy Conf. 2012 (San Diego, CA)*.
[9] See Kallenbach A *et al* 2012 *Proc. 24th IAEA Fusion Energy Conf. 2012 (San Diego, CA)*.

Due to their large mass and charge, heavy impurities such as W experience strong inertial and electrostatic forces, with the result that their densities are not flux functions, but have strong poloidal asymmetries. In a rotating plasma, the centrifugal force (CF) is well known since [2, 3] to cause impurity localisation on the low field side (LFS). The associated increase in neoclassical transport has long been worked out in analytic models, [2, 4–9] but has not usually been included in the numerical tools used for scenario modelling and validation studies [10, 11]. More recently, temperature anisotropies in a minority species heated by Ion Cyclotron Resonance Heating (ICRH) have been observed to create a poloidal electric field leading to high field side (HFS) localisation of heavy impurities [12, 13]. The theory of ICRH induced anisotropy has since been clarified [14] and impurity transport theories have been extended to account for these effects [15–20].

For light impurities, where turbulence dominates neoclassical transport, model validation is progressing well [21–25]. Meanwhile, results from the JET-ILW have renewed interest in heavy impurity transport and now motivated the application [26] of the transport codes GKW [27] and NEO [28, 29][10] which both include comprehensive treatments of poloidal asymmetries [30, 31].

The first validation of the GKW + NEO model for heavy impurities was made in [26], in which the model quantitatively explained the evolution of core W in the JET hybrid H-mode (NBI heating only). There, neoclassical transport enhanced by CF effects was shown to be the primary cause of W accumulation (defined here as strongly peaked W profiles in the *central core*) and the need to include poloidal asymmetries in the impurity transport models was demonstrated.

In this work, GKW + NEO model validation is extended by application to the improved H-mode scenario with current overshoot in AUG (section 4) and the ICRH heated baseline H-mode in JET (section 5). New minority heating effects are included in the model for the JET cases, where central ICRH heating can prevent central W accumulation [32–34] and can reverse the sign of impurity convection [11, 35]. Predicted 2D impurity density distributions are compared with tomography from soft x-ray diagnostics. Section 2 outlines the effects of poloidal asymmetries on neoclassical transport, section 3 describes the modelling setup and new results are presented in sections 4 (AUG) and 5 (JET).

2. Impact of poloidal asymmetries on neoclassical transport

In this section, we summarize the (significant) effects of poloidal asymmetries on neoclassical transport. The asymmetry effects on turbulent transport are also included in our GKW modelling, but their impact on turbulence is less dramatic (see figure 4) and can go in both directions, due to subtle interactions between kinetic profiles and magnetic field shear [15–17, 20].

Neoclassical transport is a flux surface average of local flux vectors which reverse sign from HFS to LFS, so changes in the poloidal density distribution re-weight this average, changing both the sign and magnitude of the net flux [2, 4–9]. We use the model for poloidal asymmetries, presented in [14]; solving the parallel force balance, an anisotropically heated species approximated by a bi-Maxwellian (with $T_\parallel$, $T_\perp$) has poloidally varying equilibrium density

$$n(\theta) = n_{R0}\frac{T_\perp(\theta)}{T_{\perp R0}}\cdot \exp\left(-\frac{eZ\Phi(\theta)}{T_\parallel}+\frac{m\Omega^2(R(\theta)^2-R_0^2)}{2T_\parallel}\right) \tag{1}$$

where θ is poloidal angle, Ω is plasma angular rotation frequency, R is major radius, R_0 represents LFS values and

$$\frac{T_\perp(\theta)}{T_{\perp R0}} = \left[\frac{T_{\perp R0}}{T_\parallel}+\left(1-\frac{T_{\perp R0}}{T_\parallel}\right)\frac{B_{R0}}{B(\theta)}\right]^{-1}. \tag{2}$$

A minority species with $T_\perp > T_\parallel$ is localized on the LFS and creates a poloidally varying potential Φ which pushes high Z impurities towards the HFS (if stronger than the centrifugal force). Equation (1) is also valid for all isotropic species, which have $T_\perp/T_\parallel = T_\perp(\theta)/T_{\perp R0} = 1$. Both GKW and NEO solve for Φ for an arbitrary number of species using a quasi-neutral root-finding algorithm [36].

Neoclassical impurity transport theory has recently been updated to elaborate the case of HFS impurity localisation [18]: When trace impurities are in the deep Pfirsch–Schlüter (PS) regime and Deuterium is in the Banana regime, the neoclassical impurity transport (with a simplified collision model valid at large aspect ratio) can be summarized as [18]

$$R\langle \mathbf{\Gamma}_z^{\rm neo}\cdot\nabla r\rangle \propto n_iT_i\nu_{ii}Z\left[P_A\left(-\frac{R}{L_{n_i}}+\frac{1}{2}\frac{R}{L_{T_i}}+\frac{1}{Z}\frac{R}{L_{n_z}}\right) -0.33P_Bf_c\frac{R}{L_{T_i}}\right] \tag{3}$$

where f_c is the circulating (non-trapped) fraction and P_A, P_B are geometrical factors related to the poloidal asymmetry

$$2P_A\epsilon^2 = \left\langle\frac{n_z}{B^2}\right\rangle\frac{\langle B^2\rangle}{\langle n_z\rangle}-\left[\left\langle\frac{B^2}{n_z}\right\rangle\frac{\langle n_z\rangle}{\langle B^2\rangle}\right]^{-1}, \tag{4}$$

$$2P_B\epsilon^2 = 1-\left[\left\langle\frac{B^2}{n_z}\right\rangle\frac{\langle n_z\rangle}{\langle B^2\rangle}\right]^{-1}. \tag{5}$$

For clarity, we have here re-introduced the diffusive term which is ordered small at large Z (and was dropped in [18]). The usual neoclassical pinch, temperature screening and diffusion (respectively) then appear multiplied by the factor P_A. In addition, a term $\propto P_B$ is present, which reduces the temperature screening, with the coefficient 0.33 applying in the trace limit with D in the Banana regime. For the poloidally symetric case, $P_A = 1$, $P_B = 0$ and standard neoclassical impurity transport is recovered.

In [18], the asymmetry factors P_A, P_B, were calculated for a circular plasma in the limits of weak and strong poloidal

[10] NEO should not be confused with NEOART, another neoclassical code.

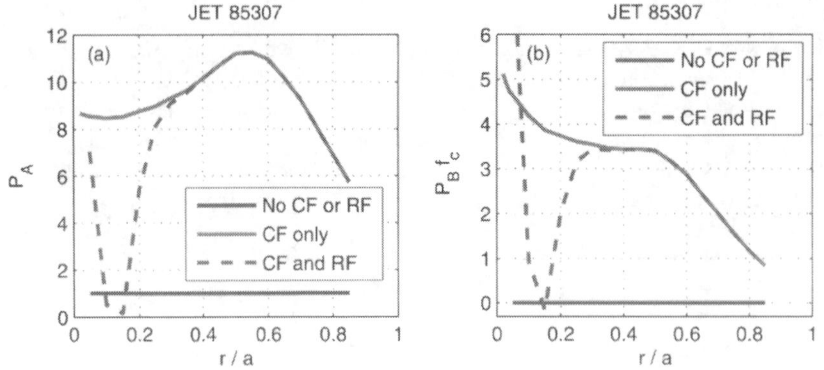

Figure 1. Poloidal asymmetry geometrical factors P_A and $P_B f_c$ for neoclassical transport for the JET case with central ICRH in section 5. Poloidal asymmetries can be generated by rotation (CF) or minority heating (RF).

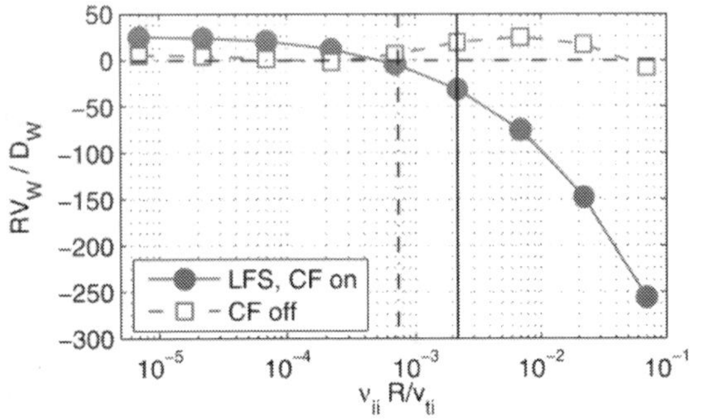

Figure 2. Collisionality scan of the NEO-only peaking factor ($R/L_{nW} = -RV_W/D_W$) at mid-radius for the JET hybrid case presented in [26], both with and without centrifugal effects. The vertical lines indicate the collisionality in hybrid (dashed) and baseline (solid) H-modes.

asymmetries. Here, we present the values in full geometry, with realistic anisotropy calculated by GKW (figure 1) for the JET NBI + ICRH case in section 5. From P_A (figure 1(*a*)), it is evident that CF effects greatly increase the neoclassical pinch and diffusion; from P_B (figure 1(*b*)) it is clear that the neoclassical *V*/*D* ratio can also be changed, since the extra $f_c P_B$ term (largest at small *r*/*a*) reduces the effective temperature screening *relative to the other terms* (figure 2): At high collisionality, with W in the deep PS regime, [18] applies and the effective temperature screening is reduced by CF effects, making the convection more inward. At lower collisionality, as the impurities move out of the PS regime, [18] no longer applies and the numerical NEO results show that the CF effects can reverse sign and reduce the neoclassical $R/L_{nw} = -RV_W/D_W$ (which might be beneficial in a hotter reactor). For JET H-modes, typical collisionalities are marked in figure 1 and indicate that the JET hybrid scenario in [18] is close to a crossover where R/L_{nw} is not significantly affected by the CF effects (although both *V* and *D* are increased by an order of magnitude). For the AUG improved H-mode in section 5, the collisionality is similar to the JET hybrid, but the parameters differ such that the CF effects decrease R/L_{nw}. For the JET baseline H-mode (as in section 5 and [37]), the CF effects (P_B term) reduce temperature screening and increase R/L_{nw}, with a stronger effect at smaller minor radius. Given this collisionality and parameter dependence, it is clear that there is no simple scaling fix for less sophisticated neoclassical models that exclude CF effects and that poloidal asymmetries cannot be neglected in calculations of heavy impurity transport.

3. Modelling methodology

We model steady-state H-mode plasmas using gyrokinetic and neoclassical models including both the rotation-induced and anistropy-induced poloidal asymmetries discussed above. The turbulent transport is computed with the gyrokinetic code GKW [27] including all rotational effects [16, 30, 38, 39], here run in its local, quasilinear (6 modes) and electrostatic limits. The neoclassical transport is computed with the local drift kinetic code NEO [28, 29, 31]. In both codes, ions, electrons and impurities are all modelled kinetically, with W in the trace limit (this limit is valid for the W concentrations $< 10^{-4}$ in these shots [26]). At each radial location, the W impurity is modelled in a single average charge state Z_W between 24 (edge) and 46 (core) of the coronal equilibrium (the charge state range is narrow $\Delta Z < 5$ at the relevant T_e). In GKW, Z_{eff} is used only in the collision operator and other impurities are omitted; in AUG and JET-ILW the plasma is clean ($Z_{eff} < 1.3$), dilution is ~10% and its effects are negligible in the quasilinear ratios [40]. For NEO, an additional species Be (for JET) or B (for AUG) is included to match the measured Z_{eff} profile. For the JET cases, the hydrogen minority is also present in all simulations at concentrations determined from the isotope shift in the edge Balmer-α spectroscopy.

The trace limit allows linearisation of the W transport and is appropriate for most conditions, since W concentrations are usually small ($n_W/n_e < 10^{-4}$ at LFS), except at the end of extreme accumulation phases [26, 32]. The impurity transport is then linearly decomposed into convective and diffusive components

$$R\frac{\Gamma_Z}{n_Z} = D_Z^{\rm GKW}\frac{R}{L_{n_Z,R0}} + D_Z^{\rm NEO}\frac{R}{L_{n_Z,R0}} + RV_Z^{\rm GKW} + RV_Z^{\rm NEO} \quad (6)$$

which are extracted from the two codes using the fluxes of trace species with different gradients. For a poloidally asymmetric distribution, R/L_{n_Z} depends on θ; in equation (6) we use the value defined *at the LFS* (most convenient for the codes). This choice also defines D and V; for transport codes which use flux surface averaged densities, post-processing transformations for D and V are required (defined in [26]). The kinetic profiles and rotation of the bulk plasma (and minority, in section 5) are modelling inputs and the four transport coefficients in equation (6) are outputs. The modelling then combines turbulent and neoclassical transport channels using the anomalous heat diffusivity χ_i^{an} from an interpretive power balance calculation (here using JETTO [41, 42] or ASTRA [43]) to normalize the two transport channels relative to each other [22, 24, 26]. The ratio of combined convection to combined diffusion is a prediction of the steady-state impurity logarithmic density gradient at the low field side

$$\frac{R}{L_{n_Z}} = -\frac{\dfrac{\chi_{\text{i an}}}{\chi_{\text{i NEO}}}\cdot\dfrac{RV_{\text{Z GKW}}}{\chi_{\text{i GKW}}} + \dfrac{RV_{\text{Z NEO}}}{\chi_{\text{i NEO}}}}{\dfrac{\chi_{\text{i an}}}{\chi_{\text{i NEO}}}\cdot\dfrac{D_{\text{Z GKW}}}{\chi_{\text{i GKW}}} + \dfrac{D_{\text{Z NEO}}}{\chi_{\text{i NEO}}}}. \tag{7}$$

The modelling is performed at up to 20 radial locations from $r/a = 0.02$ to $r/a = 0.85$. Given a boundary value, the LFS density gradient is integrated across the profile to predict a LFS impurity profile. Finally, the poloidal variation is integrated using the outputs of the quasi-neutrality solver and equation (1), to produce a 2D prediction of the impurity distribution. For comparison to soft x-ray (SXR) measurements, the SXR emission is forward modelled by a simple multiplication with a T_e-dependent cooling factor and the n_e profile.

To finish this section, we offer some general comments on the modelling sensitivities. An example sensitivity test is shown in figure 7, but we do not have space to present detailed sensitivity studies here. The key sensitivities are to the logarithmic gradient inputs of bulk ion density $n_i \propto n_e$ and temperature T_i, which determine both turbulent stability and neoclassical transport. In the method described above, the usual sensitivity of turbulence to gradients is removed by the power balance normalisation, but in the marginally stable region, changes in the gradients can move the turbulence boundary by $\sim\pm$ 0.1 r/a; in this region, if the micro-instabilities are stable but the power balance transport is anomalous, only neoclassical transport is used, on the assumption that the instability thresholds are more accurate than the power balance calculation. In these plasmas, the dominant micro-instabilities are always ITG modes, so that once unstable, the quasilinear turbulent transport ratios are robust to 10% changes in input gradients. In regions with turbulent diffusion, the W profile is always relatively flat and is not sensitive to the details of the GKW simulations; the details of the neoclassical convection and the boundary between neoclassical and turbulent regions have a greater effect on the profile predictions.

In our experience, the central region of the plasma $r/a < 0.3$ is particularly challenging for quantitative validation for a combination of reasons: In this region, where turbulence is usually absent, the delicate balance between density and temperature gradients ($\sim R/L_n - 0.5R/L_{T_i}$) makes neoclassical convection very sensitive to input profiles. Kinetic measurements in the deep core (vital as inputs for these simulations) are often unavailable or inaccurate and the profile fits are particularly sensitive to the choice of boundary conditions and the location of the magnetic axis in the equilibrium reconstruction. The steady-state required for simple profile prediction cannot be reached in the presence of sawteeth. The validity of the neoclassical model close to the axis (often questioned) is a relatively minor problem by contrast: in the JET cases presented here the size of the potato orbit region is around 1 cm for D and 0.4 cm for W.

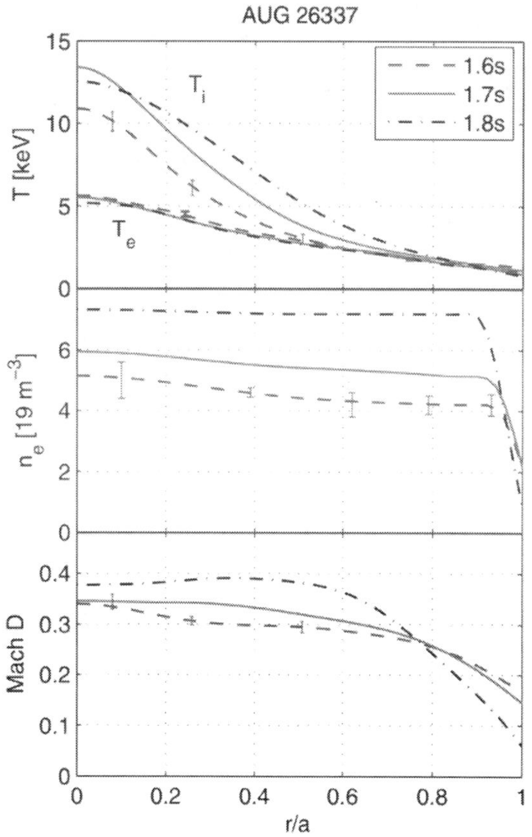

Figure 3. Input profiles for simulated timeslices in AUG 26337, with indicative error bars for selected points. Mach no. and T_i are measured by charge exchange, the density is inverted from interferometry, T_e is measured by electron cyclotron emission.

4. W transport under NBI heating, ASDEX upgrade improved H-mode

In this section we present modelling of the AUG improved H-mode discharge 26337 presented in [44]. In these discharges, the 'current overshoot' ramp-up technique is used to produce a very flat central q-profile $\sim$ 1 and a transient period of improving confinement. Tungsten is not observed to accumulate, suggested in [44] to be due to the enhancement of neoclassical transport due to the rotation. To examine this hypothesis, we model three time slices at the start of the current flattop (ELM-free H-mode), during which the confinement is

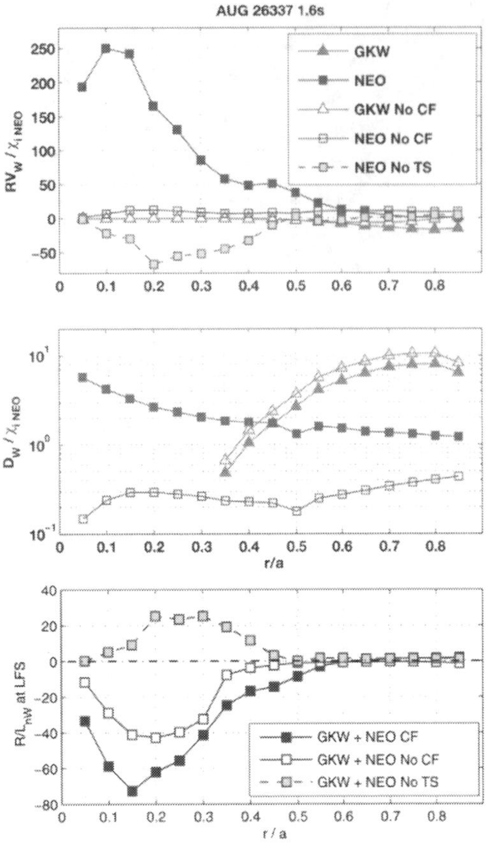

Figure 4. Predicted W transport coefficients (LFS) and R / L_{nW} for AUG 26337 at t = 1.6 s, with additional simulations excluding the centrifugal force (No CF) and excluding neoclassical temperature screening (No TS).

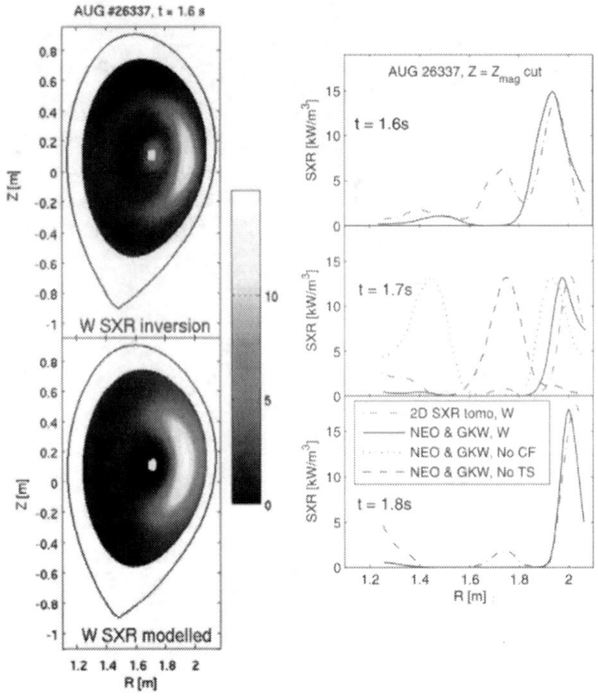

Figure 5. Comparison of tomographic inversions of SXR emission with Brehmstrahlung subtracted (top left) and W SXR emission forward modelled from the predicted W distribution (bottom left) for t = 1.6 s of AUG 26337. The maximum value is used to normalise the modelled result to the tomography. (right) The same data is cut horizontally through the magnetic axis (all three timeslices). For t = 1.7 s, modelling excluding CF effects and temperature screening is also shown.

improving as the NBI power is stepped up (t = 1.6 s: 5 MW, t = 1.7 s: 7.5 MW, t = 1.8 s: 10 MW). The density profile is quite flat but the temperature profile is strongly peaked, with maximum peaking at 1.7 s. (figure 3). The low densities and high NBI power (much larger than the 800 kW central ECRH) result in large plasma rotation, with some of the highest thermal Mach numbers ($M_D = \Omega R / \sqrt{2T_D / m_D}$) for AUG, reaching 0.3–0.4 in the core.

The predicted transport coefficients in figure 4 show that these inputs lead to a strongly outward neoclassical convection over the whole profile, which dominates turbulent convection for $r/a < 0.7$. For the diffusive transport, the turbulence dominates from $r/a > 0.45$.

To validate these predictions, we compare predicted soft x-ray (SXR) emission (forward modelled from the predicted 2D W density) with SXR tomography with Bremsstrahlung radiation subtracted (for the modelled region only), under the assumption that W dominates the remaining emission [45]. Here, high quality SXR tomography is made possible by the high temperatures in this shot (in cooler AUG plasmas W emission falls below the filter cut-off at ~2 keV) and the recent application to AUG of the tomographic method described in [46]. In the LFS outer half of the plasma, the comparison in figure 5 shows agreement well within the uncertainties in both the radial gradients and poloidal structure of the radiation and provides an additional qualitative validation of the model. In the earliest phase, the SXR near the axis is undergoing a fast transient and has not yet reached the predicted steady-state. For the later two phases, following the sensitivity discussion in section 3, uncertainties in the core n_i profile are enough to account for the remaining differences between prediction and tomography near the axis. The disagreements at the HFS are thought to be due to inaccuracies in the rotation measurement causing an overestimate of the predicted asymmetry.

To investigate the components of the model that are required, additional simulations are presented (see figure 4): When CF effects are removed, the neoclassical transport drops by an order of magnitude and no longer dominates the turbulent transport, while the turbulent transport is relatively unaffected. If instead the temperature screening is removed, (and CF effects are kept), the neoclassical transport remains enhanced but reverses sign, which would lead to strong central accumulation. In removing either effect, the comparison to the tomography shows qualitative disagreement (also figure 5), indicating that both components are essential to the model.

To summarize, this case provides a further validation of the GKW + NEO model, in an advanced scenario with strong rotation, strong temperature gradients and weak density gradients. Improved confinement is usually associated

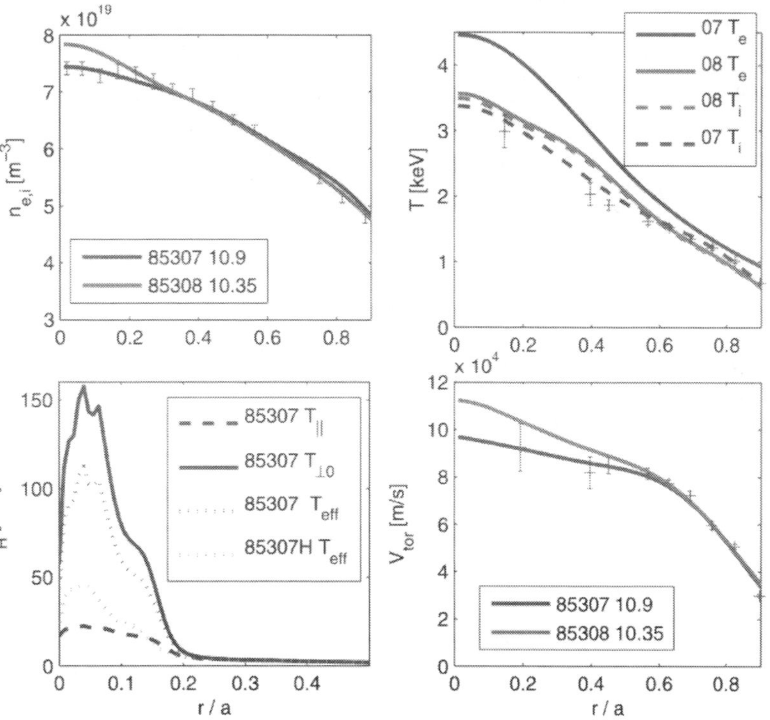

Figure 6. Input profiles for the JET modelling, with indicitive selected data and error bars. n_e and T_e are measured by Thompson scattering (LIDAR), T_i and V_{tor} are measured by charge exchange. (Bottom left) H minority temperatures produced by TORIC-SSFPQL for the ICRH case, inner radii only (the case 85307H is with half ICRH power).

with impurity accumulation [26] but this case provides a transient counter-example in which neoclassical temperature screening alone can trap W in the outer LFS region of the plasma. In this brief phase, neoclassical transport dominates over the entire profile but core W accumulation is avoided.

The present case contrasts with the picture for the stationary standard H-mode in AUG in which central ECRH prevents W accumulation by increasing turbulent impurity diffusion [47–49]. The present work deals with a different scenario and does not invalidate the explanation of [48, 49]. However, future modelling with GKW + NEO (including poloidal asymmetries) should revisit the AUG standard H-mode with and without ECRH to accurately quantify its influence on the components of W transport.

5. W transport under ICRH and NBI heating, JET baseline H-mode

In this section we model W in a pair of JET baseline H-modes in an ICRH power scan. These shots are a follow-up to [11], where it was observed that central ICRH can reverse central impurity convection from inward to outward. The discharges have approximately the same total heating power; 14.7 MW NBI with 4.9 MW central ICRH in 85307 and 19.1 MW NBI in 85308 and both include an H minority at ~9% concentration. In both cases the time selected for modelling was just prior to a sawtooth crash.

The model for poloidal asymmetry of W induced by anisotropic heating of the minority species (section 2) requires inputs of $T_\parallel$ and $T_\perp$ for the minority species. These are not measured directly, but are simulated for 85307 using the the wave code TORIC [50] iteratively coupled [51] to the Fokker–Planck solver SSFPQL [52]. The simulations were performed for a pure plasma using the same kinetic profiles and full geometry as the GKW + NEO simulations, with additional inputs of ICRH power, frequency and antenna phasing. The minority temperature after the collisional slowing down is a nonlinear function of the absorbed power per particle. These simulations do not include the interaction of NBI with ICRH, which may reduce the temperature and the anisotropy of the minority, or finite orbit effects, which may widen the deposition profile and reduce the gradients.

The modelling inputs are shown in figure 6. Discharge 85307 has hotter electrons in the core, since more ICRH power goes to the electrons, but T_i, which determines the W transport, is similar. The higher rotation and more peaked density in 85308 are the key differences which determine the different predictions in figures 7 and 9(*a*) and (*b*). Also shown in figure 6 are the anisotropic H minority temperatures produced from TORIC-SSFPQL.

In the first stage of modelling, the simulations included CF effects only (with $T_H = T_D$), as in the previous section. Both predicted profiles show central W peaking (figure 7), enhanced by CF effects due to the reduction in temperature screening relative to the pinch. The CF effects have a slightly larger impact in 85308 due to the larger rotation (figures 9(*a*) and (*b*)). Without CF effects, the NEO-only R / L_{n_W} closely follows a simple neoclassical estimate $\propto R / L_{n_i} - 0.5R / L_{T_i}$ for the PS regime; already here we see that 85308, without ICRH, shows stronger central peaking for two reasons: First, the lower T_i gradients between $0.2 < r/a < 0.4$ give a boundary of the turbulent region at larger r/a and second, the more peaked density profile increases the inward neoclassical convection. (The reasons for the more peaked density profile in 85308 are not investigated in this work, but are likely due to less central turbulence offsetting the Ware pinch and an increased particle source from NBI [33, 53].)

For 85308, without ICRH, the 2D W SXR prediction shows good qualitative agreement with the interpreted SXR tomography (figure 8) using the tool developed for [32, 54]. For the reasons discussed in section 3 (particularly the presence of large sawteeth), the comparison does not show the same level of quantitative agreement over the full profile as the AUG results above, but nevertheless demonstrates that, for the case without ICRH, the model including CF effects correctly predicts W accumulation.

In contrast, for 85307, with CF effects only, the centrally peaked density profile does not agree with the tomography (figures 11(*a*) versus (*d*)) and indicates a possible missing piece in the modelling, motivating the progressive inclusion of the minority heating effects (figure 9(*c*)):

First, the effective isotropic minority temperature from TORIC-SSFPQL is added to the minority species which is kept isotropic with $T_{\rm eff} = (T_{\parallel} + 2\,T_{\perp R0})/3$. For the GKW simulations, the increased minority temperature gradient shifts the stability boundary slightly inward, but the impact is much larger on the neoclassical transport. The heated minority does not change the neoclassical diffusivity (figure 9), but switches the neoclassical convection to strongly outward in the region of the ICRH absorption ($0.1 < r/a < 0.3$), due to an additional temperature screening from collisions between W and H. Notably, this additional temperature screening becomes negative at $r/a < 0.1$, in exactly the region where $R / L_{T_{\rm eff}} < 0$ for the minority. The ion-impurity friction which drives temperature screening [18] scales as $\propto n_i T_i \nu_{iZ} R / L_{T_i}$. For the H–W and D–W collisions with $Z_W = 46$, these parameters are given in table 1 and demonstrate that the minority H contributes a screening of the same order of magnitude as the bulk D at $r/a = 0.2 - 0.25$, effectively doubling the strength of the screening. We note that at the very high T_H, the minority collisions decouple (in both table 1 and figure 9) and the maximum minority screening effect is not at the ICRH resonance at $r/a = 0.07$, but at the edges of the heated region. For this reason, this additional screening is very sensitive to the exact details of the minority temperature profile from TORIC-SSFPQL.

Second, the minority is made anisotropic using the simulated $T_{\parallel}$, $T_{\perp}$ as inputs to the model of equation (1). The result (figure 9(*c*)) is a strong reduction in neoclassical diffusivity, as expected from section 2, due to the reduction of the P_A factor (figure 1). Additionally, the minority temperature screening effect is strongly enhanced in the regions where $P_A \gg 1$. In these regions, the CF asymmetry dominates, producing LFS W localisation, so both W and H are localised on the LFS, increasing their local collision frequency and amplifying the minority temperature screening effect (the details of this synergy remain to be clarified).

The end result of the additional temperature screening is to significantly flatten the central W profile (figure 10) with the reversal of the minority temperature screening even causing a second, central, peak in qualitative agreement with the tomography (figures 11(*b*) and (*d*)). The effects of the anisotropy (figure 11(*c*)) appear to overly exaggerate the dip in n_W close to the axis. Given the lack of finite orbit effects in TORIC-SSFPQL, both minority effects in our results should be

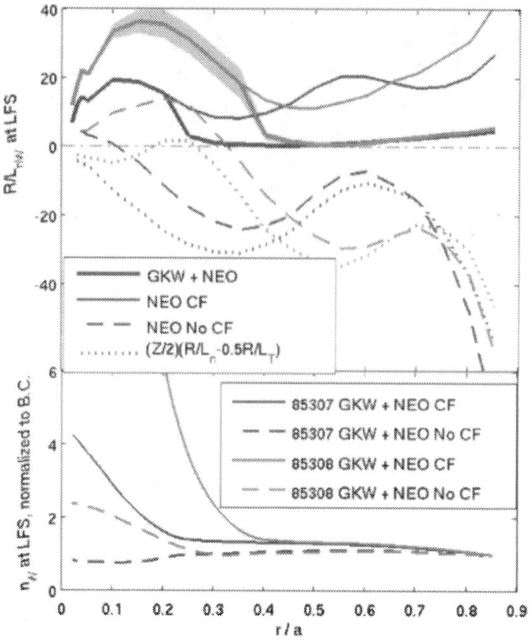

Figure 7. Predicted R / L_{n_W} (top) and integrated n_W profiles (bottom) for JET 85308 w/o ICRH (red) and 85307 with ICRH (blue), with CF effects but no ICRH minority effects. (top) For 85308, the red band indicates sensitivity to ± 10% changes in both R / L_{n_i} and R / L_{T_i} inputs. A simple analytic estimate of neoclassical peaking (dots) closely follows the NEO result w/o CF effects (dashes).

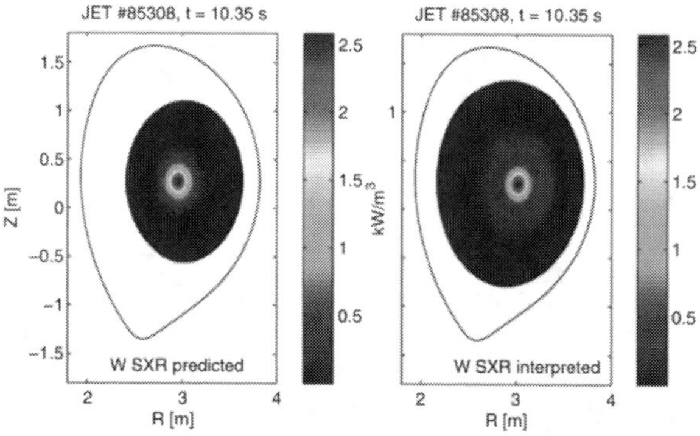

Figure 8. Comparison of predicted and interpreted SXR emission from W for JET 85308 (NBI only). The predicted scale is normalised to the interpreted value at the central maximum.

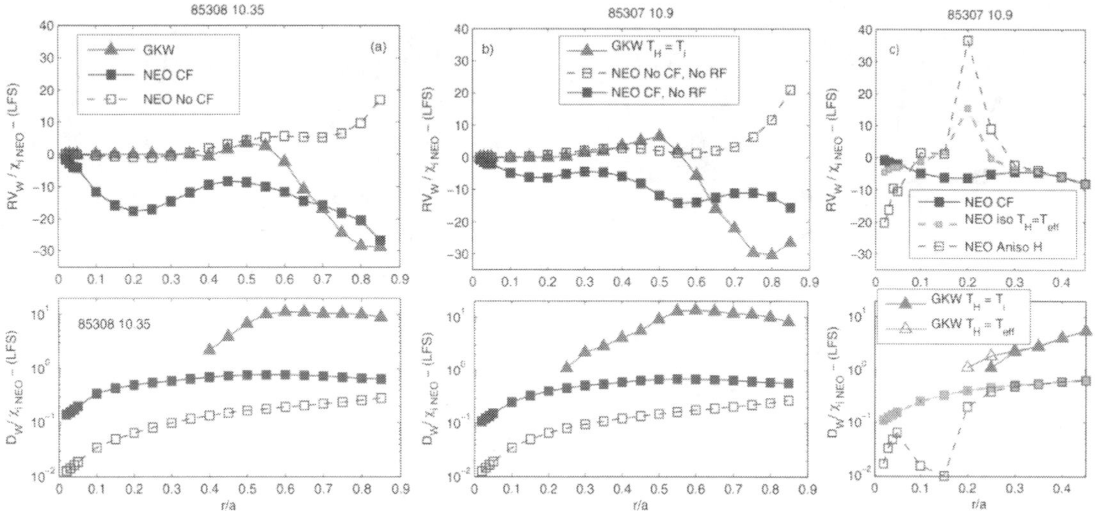

Figure 9. Predicted W transport coefficients for the JET cases. For 85307, results include CF effects only ($T_H = T_i$, middle), CF + heated isotropic minority ($T_H = T_{eff}$, right) and CF + heated anisotropic minority (Aniso H, NEO only, right).

Table 1. Comparision of parameters in ion-W screening for collisions with H and D ions. For readable numbers, $n_W = 10^{19}\ m^{-3}$ (arbitrary) was used for ν_{iW}.

Ion	r/a	$n_i\ [10^{19}\ m^{-3}]$	$T_i\ [keV]$	$\frac{R}{L_{T_i}}$	$\frac{\nu_{iW}}{v_{th,i}/R}$	$n_iT_i\frac{\nu_{iW}}{v_{th,i}/R}\frac{R}{L_{T_i}}$
H	0.10	0.664	63.5	30.7	0.0016	2.1
H	0.15	0.658	45.2	50.3	0.0032	4.7
H	0.20	0.650	7.76	97.8	0.1038	51.2
H	0.25	0.642	3.48	37.5	0.5156	43.3
D	0.10	6.72	3.26	2.16	0.61	29.0
D	0.15	6.65	3.13	2.99	0.66	41.6
D	0.20	6.57	2.97	3.76	0.70	52.1
D	0.25	6.49	2.79	4.21	0.80	61.3

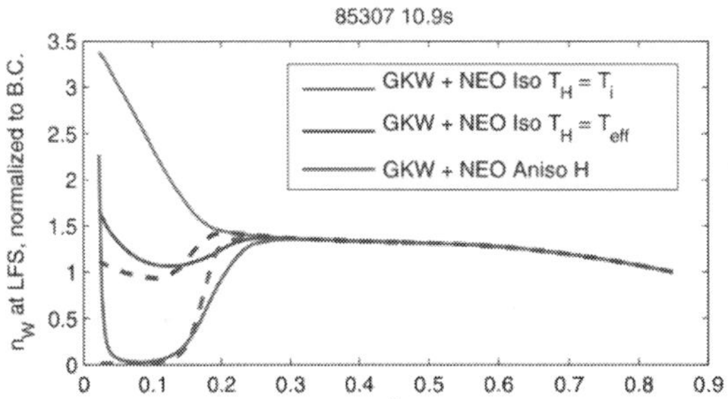

Figure 10. Comparison of LFS predicted profiles for JET 85307 with ICRH minority effects (labels as in figure 9). The dashed curves indicate simulations with half ICRH power using the input labelled 85307H in figure 6.

considered an upper estimate. In sensitivity tests with half ICRH power we observe that the minority effects are qualitatively robust (figure 10).

We note that the ICRH minority effects described here are consistent with the reversal of the convection described in [11]; future work will compare D_{Mo} and V_{Mo} predictions to laser blow off fits and should include these transport coefficients in time evolution of W integrated modelling. The minority screening effect combined with the anisotropy may also explain the strong Mo peaking at $r/a = 0.55$ in [20]; in that case, if P_A is negative due to the HFS impurity localisation, all neoclassical transport including the minority screening would reverse; we leave confirmation for future work. These effects should also be quantified for NBI fast ions.

6. Conclusions

In this work, we have modelled turbulent and neoclassical heavy impurity transport using theory-based numerical tools (GKW and NEO respectively) with comprehensive treatment of poloidal asymmetries, to predict core W distributions in JET and AUG. Our results demonstrate that the impact of poloidal assymetries on neoclassical convection depends strongly on collisionality and plasma gradients, such that models which exclude these asymmetries cannot be used to accurately describe heavy impurity transport.

In the ASDEX-Upgrade improved H-mode with current overshoot, the flat density profiles mean that neoclassical temperature screening is sufficient to prevent accumulation and

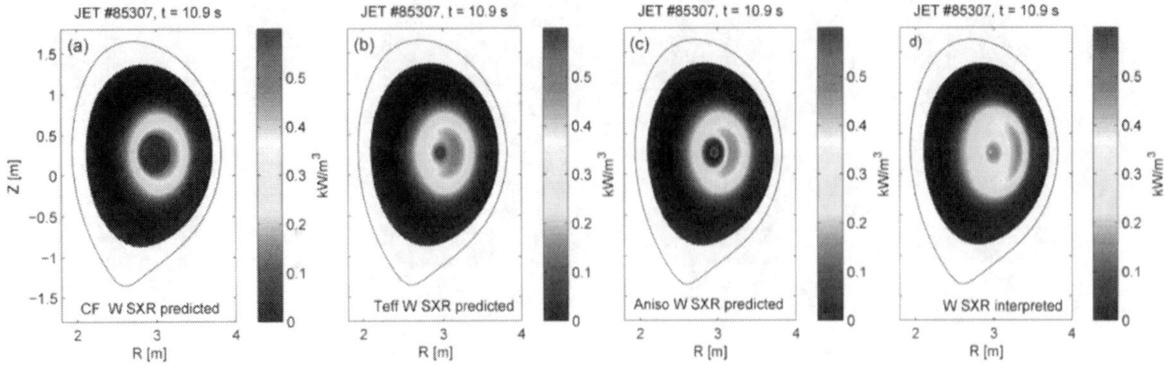

Figure 11. Comparison of predicted and interpreted (*d*) SXR emission from W for JET 85307 (NBI + ICRH). From left to right the predictions include: CF effects only (*a*), CF effects with heated isotropic minority (*b*) and CF effects with heated anisotropic minority (NEO only) (*c*). The predictions are normalised to the interpreted value at $r/a = 0.35$, in the outer LFS maximum.

trap W in the outer half of the plasma. Here, centrifugal effects decrease W peaking and enhance neoclassical transport by an order of magnitude such that it dominates impurity turbulent transport over most of the plasma radius.

In JET baseline H-modes, centrifugal effects increase W peaking, in contrast to the hybrid scenario in which their effect on the overall peaking is smaller [26]. With ICRH, the strong minority heating enhances neoclassical impurity temperature screening and reverses the convection in the region of the ICRH (in agreement with [11]). In addition, the anisotropy-induced poloidal asymmetry reduces neoclassical impurity diffusivity and the minority-impurity temperature screening may be enhanced when both species are localised at the LFS. These effects are complementary to flatter density profiles and a wider region of turbulent diffusion in ICRH plasmas, which all help to prevent central W accumulation.

Comparing our predictions with tomographic inversions from soft x-ray measurements, we have demonstrated further validation of these models over a greater range of plasma conditions. This validation re-emphasizes that poloidal asymmetries are an essential ingredient for accurate modelling of (particularly neoclassical) heavy impurity transport. Additionally, we have shown that the temperature gradients of externally heated species can contribute significantly to impurity temperature screening and should also be included in neoclassical modelling. Experiments with off-axis heating may be able to further probe and isolate these effects.

Acknowledgments

FJC would like to thank Colin Roach for helpful comments and Arthur Peeters for many helpful discussions and for making the GKW code available. This project has received funding from the European Union's Horizon 2020 research and innovation programme under grant agreement number 633053 and by the RCUK Energy Programme [grant number EP/I501045] and the Max Planck Institute. The views and opinions expressed herein do not necessarily reflect those of the European Commission.

References

[1] Shimada M *et al* 2009 *J. Nucl. Mater.* **390–1** 282
[2] Hinton F L and Wong S K 1985 *Phys. Fluids* **28** 3082
[3] Wesson J 1997 *Nucl. Fusion* **37** 577
[4] Chang C S 1983 *Phys. Fluids* **26** 2140
[5] Helander P 1998 *Phys. Plasmas* **5** 1209
[6] Helander P 1998 *Phys. Plasmas* **5** 3999
[7] Romanelli M and Ottaviani M 1998 *Plasma Phys. Control. Fusion* **40** 1767
[8] Fülöp T and Helander P 1999 *Phys. Plasmas* **6** 3066
[9] Fülöp T, Helander P and Catto P J 2002 *Phys. Rev. Lett.* **89** 225003
[10] Parisot T *et al* 2008 *Plasma Phys. Control. Fusion* **50** 055010
[11] Valisa M *et al* 2011 *Nucl. Fusion* **51** 033002
[12] Ingesson L C *et al* 2000 *Plasma Phys. Control. Fusion* **42** 161
[13] Reinke M L *et al* 2012 *Plasma Phys. Control. Fusion* **54** 045004
[14] Bilato R, Maj O and Angioni C 2014 *Nucl. Fusion* **54** 072003
[15] Fülöp T and Moradi S 2011 *Phys. Plasmas* **18** 030703
[16] Angioni C *et al* 2012 *Phys. Plasmas* **19** 122311
[17] Mollén A *et al* 2012 *Phys. Plasmas* **19** 052307
[18] Angioni C and Helander P 2014 *Plasma Phys. Control. Fusion* **56** 124001
[19] Belli E and Candy J 2014 *Plasma Phys. Control. Fusion* **56** 124002
[20] Mollén A *et al* 2014 *Plasma Phys. Control. Fusion* **56** 124005
[21] Guirlet R *et al* 2009 *Nucl. Fusion* **49** 055007
[22] Angioni C *et al* 2011 *Nucl. Fusion* **51** 023006
[23] Howard N *et al* 2012 *Nucl. Fusion* **52** 063002
[24] Casson F J *et al* 2013 *Nucl. Fusion* **53** 063026
[25] Henderson S S *et al* 2014 *Nucl. Fusion* **54** 093013
[26] Angioni C *et al* 2014 *Nucl. Fusion* **54** 083028
[27] Peeters A G *et al* 2009 *Comput. Phys. Commun.* **180** 2650
[28] Belli E A and Candy J 2008 *Plasma Phys. Control. Fusion* **50** 095010
[29] Belli E A and Candy J 2012 *Plasma Phys. Control. Fusion* **54** 015015
[30] Casson F J *et al* 2010 *Phys. Plasmas* **17** 102305
[31] Belli E A and Candy J 2009 *Plasma Phys. Control. Fusion* **51** 075018

[32] Pütterich T *et al* 2013 *Plasma Phys. Control. Fusion* **55** 124036
[33] Giroud C 2014 *Plasma Phys. Control. Fusion* accepted
[34] Goniche M *et al* 2014 *Proc. 41st EPS Conf. Plasma Physics (Berlin, 23–27 June 2014)* O4.129
[35] Puiatti M E *et al* 2006 *Phys. Plasmas* **13** 042501
[36] Casson F J 2011 Turbulent transport in rotating tokamak plasmas *PhD Thesis* University of Warwick (http://wrap.warwick.ac.uk/36765/)
[37] Mantica P *et al* 2014 *Proc. 41st EPS Conf. Plasma Physics (Berlin, 23–27 June 2014)* p P1.017
[38] Peeters A G *et al* 2009 *Phys. Plasmas* **16** 042310
[39] Camenen Y *et al* 2009 *Phys. Plasmas* **16** 012503
[40] Estrada-Mila C, Candy J and Waltz R E 2005 *Phys. Plasmas* **12** 022305
[41] Cenacchi G and Taroni A 1988 *JET Internal Report* JET-IR(88)03
[42] Romanelli M *et al* 2014 *Plasma Fusion Res.* **9** 3403023
[43] Fable E *et al* 2012 *Nucl. Fusion* **52** 063017
[44] Hobirk J *et al* 2012 *IAEA Fusion Energy Conf. (San Diego, USA 8–13 October 2012)* EX/P2–03
[45] Pütterich T *et al* 2008 *Plasma Phys. Control. Fusion* **50** 085016
[46] Odstrcil M *et al* 2012 *Nucl. Instrum. Methods Phys. Res.* A **686** 156
[47] Neu R *et al* 2002 *Plasma Phys. Control. Fusion* **44** 811
[48] Neu R *et al* 2003 *J. Nucl. Mater.* **313–6** 116
[49] Dux R *et al* 2003 *Plasma Phys. Control. Fusion* **45** 1815
[50] Brambilla M 1999 *Plasma Phys. Control. Fusion* **41** 1
[51] Bilato R *et al* 2011 *Nucl. Fusion* **51** 103034
[52] Brambilla M 1994 *Nucl. Fusion* **34** 1121
[53] Angioni C *et al* 2009 *Plasma Phys. Control. Fusion* **51** 124017
[54] Pütterich T *et al* 2012 *IAEA Fusion Energy Conf. (San Diego, USA 8–13 October 2012)* EX/P3–15
[55] Romanelli *et al* 2012 *Proc. 24th IAEA Fusion Energy Conf. (San Diego, USA, 8–13 October 2012)*
[56] Kallenbach *et al* 2012 *Proc. 24th IAEA Fusion Energy Conf. (San Diego, USA, 8–13 October 2012)*

IOP Publishing | Plasma Physics and Controlled Fusion
Plasma Phys. Control. Fusion **57** (2015) 014032 (11pp) | doi:10.1088/0741-3335/57/1/014032

Electromagnetic stabilization of tokamak microturbulence in a high-β regime

J Citrin[1,2], J Garcia[2], T Görler[3], F Jenko[3], P Mantica[4], D Told[3], C Bourdelle[2], D R Hatch[5], G M D Hogeweij[1], T Johnson[6], M J Pueschel[7], M Schneider[2] and JET-EFDA Contributors[8]

JET-EFDA, Culham Science Centre, Abingdon, OX14 3DB, UK
[1] FOM Institute DIFFER—Dutch Institute for Fundamental Energy Research, PO Box 1207, 3430 BE Nieuwegein, The Netherlands
[2] CEA, IRFM, F-13108 Saint Paul Lez Durance, France
[3] Max Planck Institute for Plasma Physics, Boltzmannstr. 2, 85748 Garching, Germany
[4] Istituto di Fisica del Plasma 'P. Caldirola', Milano, Italy
[5] Institute for Fusion Studies, University of Texas at Austin, Austin, Texas 78712, USA
[6] EES, KTH, Osquldas väg 10 100 44 Stockholm Sweden
[7] Department of Physics, University of Wisconsin-Madison, Madison, Wisconsin 53706, USA

Received 27 June 2014, revised 9 September 2014
Accepted for publication 23 September 2014
Published 28 November 2014

Abstract

The impact of electromagnetic stabilization and flow shear stabilization on ITG turbulence is investigated. Analysis of a low-β JET L-mode discharge illustrates the relation between ITG stabilization and proximity to the electromagnetic instability threshold. This threshold is reduced by suprathermal pressure gradients, highlighting the effectiveness of fast ions in ITG stabilization. Extensive linear and nonlinear gyrokinetic simulations are then carried out for the high-β JET hybrid discharge 75225, at two separate locations at inner and outer radii. It is found that at the inner radius, nonlinear electromagnetic stabilization is dominant and is critical for achieving simulated heat fluxes in agreement with the experiment. The enhancement of this effect by suprathermal pressure also remains significant. It is also found that flow shear stabilization is not effective at the inner radii. However, at outer radii the situation is reversed. Electromagnetic stabilization is negligible while the flow shear stabilization is significant. These results constitute the high-β generalization of comparable observations found at low-β at JET. This is encouraging for the extrapolation of electromagnetic ITG stabilization to future devices. An estimation of the impact of this effect on the ITER hybrid scenario leads to a 20% fusion power improvement.

Keywords: tokamak, gyrokinetic, simulation

(Some figures may appear in colour only in the online journal)

1. Introduction

The ion-temperature-gradient (ITG) instability [1, 2] is well established as a primary driver for ion and electron turbulent heat transport in tokamak plasmas [3]. While long studied, open questions remain on aspects of ITG turbulence. Specifically, experimental observations in JET tokamak L-mode discharges of significantly reduced ion temperature profile stiffness [4, 5] within the tokamak plasma half-radius—hitherto unexplained by direct numerical simulation—have recently triggered intense investigation. 'Stiffness' is defined here as the gradient of the gyroBohm normalized ion heat flux with respect to the driving normalized logarithmic ion temperature gradient R/L_{T_i}. While the original experimental hypothesis posed a connection between the reduced stiffness and concomitant low magnetic shear and high $E \times B$ flow shear, this was not borne out by dedicated nonlinear gyrokinetic simulations. Rather, following an extensive numerical study using

[8] See the appendix of Romanelli F *et al* 2012 *Proc. 24th IAEA Fusion Energy Conf. (San Diego, CA)*.

the GENE gyrokinetic code [6], the physical mechanism able to explain the observations is nonlinear electromagnetic stabilization of ITG turbulence [7, 8].

Linear stabilization of ITG modes by electromagnetic (EM) effects (i.e. finite-β) is well known [9, 10]. However, recent *nonlinear* simulations of finite-β ITG turbulence have highlighted that nonlinear EM-stabilization—i.e. the relative reduction of the heat fluxes compared with the electrostatic (ES) case—is significantly greater than the relative reduction of the linear growth rates between the EM and ES cases [11, 12]. This is correlated with an increase in zonal flow drive in the EM case [13].

When including fast ions in the system, the suprathermal pressure gradients further increase the EM coupling, augmenting the stabilization effect. This was first observed in linear calculations [14]. The nonlinear enhancement of the total EM-stabilization over the linear stabilization is maintained in the system including fast ions. In [7, 8] this was key to explain the low T_i-stiffness observations in [4, 5]. There, the low stiffness plasmas were dominated by suprathermal pressure within half-radius. This finding has positive ramifications for extrapolation to ITER and reactor performance, where rotation is expected to be low but fusion α-particles will provide a significant source of suprathermal pressure.

Since the discharges studied in [7, 8] were relatively low-β ($\approx$0.5%), a natural question arises: does this physics change in more reactor-relevant, high-β plasmas? An investigation of this question is the topic of the present paper. A JET hybrid scenario, 75225 (with C-wall) [15], was chosen for detailed linear and nonlinear gyrokinetic study. The importance of fast ions on ITG stabilization in this discharge is already seen in previous linear studies [16]. In general, the enhancement of plasma diamagnetism by the suprathermal pressure gradient is believed to be essential for attaining hybrid regimes [17]. This encouraged the comparison of the ITG stabilization physics in this discharge with the aforementioned L-mode studies. Hybrid scenarios are natural candidates for such analysis due to: a significant suprathermal pressure fraction, arising from relatively low plasma density; low magnetic shear in the inner half-radius, experimentally observed to be correlated with reduced T_i stiffness [18]; and no deleterious MHD activity, which otherwise would interfere with microturbulence transport studies. The nonlinear studies shown here constitute an extension of results shown in [19], which also discusses the importance of fast ions for edge confinement improvement, leading to a beneficial core-edge coupling. The importance of fast ion enhanced linear EM-stabilization in this discharge is also predicted by the TGLF quasilinear transport model [20] when applied in an integrated modeling framework [21].

This paper concentrates on the comparison between well resolved and extensive direct numerical simulation with experimental power balance fluxes and the ramifications thereof on relevant transport questions and extrapolations to future devices. The fundamental theory of the physics mechanisms discussed here is to a large extent still an open question and left for future work.

The rest of the paper is organized as follows. In section 2 the simulation and discharge parameters are reviewed. Section 3 discusses the relevant parametrization of the EM-stabilization. Section 4 reviews the linear analysis of discharge 75225. Section 5 describes the nonlinear simulations of discharge 75225. Conclusions are presented in section 6.

2. Simulation setup and discharge parameters

GENE solves the gyrokinetic Vlasov equation, coupled self-consistently to Maxwell's equations, within a δf formulation. GENE works in field line coordinates, where x is the radial coordinate, z is the coordinate along the field line and y is the binormal coordinate. Both an analytical circular geometry model (derived in [22]) as well as a numerical geometry were used in this work. All simulations carried out were local. Collisions are modeled using a linearized Landau–Boltzmann operator. Both linear and nonlinear initial value simulations were performed. Typical grid parameters were as follows: perpendicular box sizes $[L_x, L_y] = [250, 125]$ in units of ion Larmor radii, perpendicular grid discretisations $[n_x, n_{ky}] = [256, 32]$, 32 point discretisation in the parallel direction, 48 points in the parallel velocity direction and 12 magnetic moments. The high L_x and n_x values were necessary to satisfy the boundary conditions in the low magnetic shear simulations at low radii and were relaxed for simulations with higher magnetic shear. Parallel magnetic fluctuations were included in the simulations of JET discharge 75225. This cannot be neglected, due to the relatively high β and is important for setting the strength of electromagnetic coupling, as seen in dedicated checks.

JET discharge 75225 was analyzed at two separate radial locations, at $\rho = 0.33$ and $\rho = 0.64$, where ρ is the normalized toroidal flux coordinate. The discharge is characterized by a peaked T_i profile in the inner half-radius, motivating this choice of separating the analysis to the two separate regions and examining the relative impact of EM-stabilization and $E \times B$ flow shear stabilization in each region. For preparing the GENE input, interpretative simulations of the discharge with the CRONOS integrated modeling suite [23] were carried out. These simulations included calculations of the NBI heat deposition and fast ion profile using the NEMO/SPOT code [24], power balance analysis, current diffusion and magnetic equilibrium calculations with HELENA [25]. The dimensionless parameters for the discharge in the two regions, included as input for the GENE simulations, are shown in table 1. The fast ions were approximated in GENE as hot isotropic Maxwellians, taking the average T_{fast} from the calculated slowing down distribution.

3. Parametrization of EM-stabilization

Before proceeding to the analysis of JET discharge 75225, we first discuss the parametrization of the strength of the linear and nonlinear EM-stabilization. This is to gain intuition on the relevant parameter regimes of the effect. Since the linear stabilization of the toroidal ITG mode is closely linked with

Table 1. Discharge 75225 dimensionless parameters as input into the GENE simulations. The values are averaged between 6.0–6.5 s, before the onset of a deleterious $n = 3$ neoclassical tearing mode. ν^* is the normalized collisionality: $\nu^* \equiv \nu_{ei}\frac{qR}{\epsilon^{1.5}v_{te}}$, with $\epsilon = a/R$ and $v_{te} \equiv \sqrt{T_e/m_e}$. γ_E is the $E \times B$ shear rate, where $\gamma_E \equiv \frac{r}{q}\frac{\mathrm{d}\Omega}{\mathrm{d}r} / \left(\frac{v_{th}}{R}\right)$ for the purely toroidal rotation assumed here for the NBI driven cases. $v_{th} \equiv \sqrt{T_e/m_i}$. The gradient lengths are defined taking the radial coordinate as the toroidal flux coordinate. We note that the β_j of all ionic species are scaled with β_e, according to the pressure ratios p_j/p_e.

Location	R/L_{Ti}	R/L_{Te}	R/L_{ne}	T_i/T_e	β_e [%]	$\hat{s}$	q	ν^*	$Z_{\rm eff}$	γ_E	$n_{\rm fast}/n_e$	$T_{\rm fast}/T_e$	$R/L_{T fast}$	$R/L_{n fast}$
$\rho = 0.33$	6.2	3.4	2.4	1.2	1.8	0.16	1.1	0.023	1.55	0.2	0.12	7.3	0.33	7.0
$\rho = 0.64$	3.4	5.8	3.05	1.07	0.75	1.44	1.74	0.018	1.90	0.3	0.055	8.65	2.15	8.9

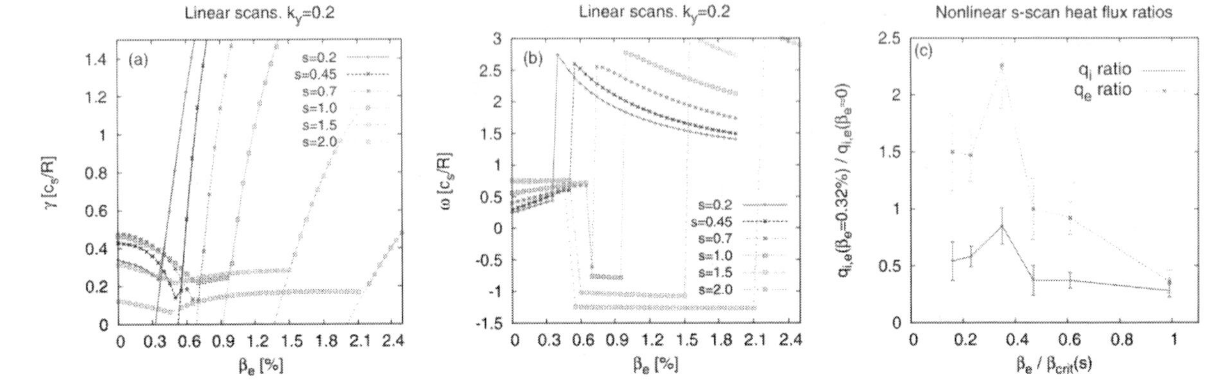

Figure 1. β scans for various $\hat{s}$ values for the low-stiffness JET discharge 66404, at $\rho = 0.33$. The nominal β_e and $\hat{s}$ is 0.32% and 0.7 respectively. The left panel (*a*) shows the linear growth rates and the quadratic fit to the $\beta_{\rm crit}$ KBM threshold (the curves meeting the *x*-axis). The center panel (*b*) shows the corresponding frequencies and the right panel (*c*) shows the ratio between the EM-case and ES-case heat fluxes from the corresponding nonlinear simulations.

the increase of the electromagnetic instability drive [26], we take $\beta/\beta_{\rm crit}$ as a valid parameter of merit. This will be justified in this section.

$\beta_{\rm crit}$ is the kinetic β limit of the EM ion mode, e.g. Kinetic Ballooning Modes (KBM), fast ion driven Alfvénic modes, or more generally a hybrid case. $\beta_{\rm crit}$ is a function of the plasma parameters, including the driving gradient lengths. In the fluid limit the stability threshold is parametrized by $\alpha_{\rm MHD}$, where $\alpha_{\rm MHD} \equiv q^2R\beta' = q^2\sum_j \beta_j\left(\frac{R}{L_{Tj}} + \frac{R}{L_{nj}}\right)$, summing over all species. $\beta_{\rm crit}$ and $\alpha_{\rm crit}$ are thus intrinsically linked, where $\alpha_{\rm crit}$ effectively captures a subset of the parameter dependencies in $\beta_{\rm crit}$. These parametrizations approximately carry over into the kinetic system [11] and thus an increase of the ITG stabilization with pressure gradients is expected (through β'). This is the key factor underlying the mechanism of fast ion stabilization of ITG, since the suprathermal pressure can increase the total pressure gradient (and thus β') and increase the stabilization factor, while not simultaneously adding to the underlying ITG drive. We stress that this effect is independent from the impact of the pressure profile on the magnetic drift frequency (ω_d) and from self-consistent modifications of the magnetic equilibrium. The former can be particularly effective at low magnetic shear [27] and in $\hat{s} - \alpha$ geometry is directly linked to the $\alpha_{\rm MHD}$ dependence of ω_d. In all cases discussed in this section, the analytical unshifted circular geometry formulation was used, with thus no impact of β' on the magnetic equilibrium. While the ∇P in the ω_d calculation did scale with β, test cases with a constant $\nabla P = 0$ in the ω_d calculation were run. Only small differences (<10%) in $\beta_{\rm crit}$ were observed, with the same qualitative behavior throughout.

A study of a subset of the $\beta_{\rm crit}$ dependencies was carried out for the low-stiffness low-β JET discharge 66404 at $\rho = 0.33$, discussed in detail in [7, 8, 28]. Of particular interest is the magnetic shear ($\hat{s}$) dependence, motivated by the observation of increased destiffening at low-$\hat{s}$ [18]. In figure 1(*a*), a β scan is shown for various levels of $\hat{s}$, where $\hat{s} = 0.7$ is the nominal value for the discharge. These scans include both linear calculations and corresponding nonlinear simulations. In all cases here, only the thermal species are included, although nominally 66404 has a significant fast ion population. This is an extension of the $\hat{s}$ -scan for this discharge shown in [7, 8]. The linear runs where carried out at $k_y = 0.2$, the wavenumber corresponding to the most unstable KBM mode. k_y is normalized by $1/\rho_s$

For all values of $\hat{s}$, a transition from ITG modes to KBM is observed, characterized by a sharp upswing of the growth rates for increasing β and a jump in the real frequencies of the dominant mode. Positive ω is defined as the ion diamagnetic direction in GENE. For $\hat{s}$>1, an intermediate TEM zone is apparent, characterized by the negative frequencies. The KBM thresholds for each case were extrapolated by a quadratic fit to the first several KBM growth rates, shown in figure 1(*a*).

An $\hat{s}$ -scan of nonlinear simulations—both EM and ES—was then carried out, with the nominal $\beta_e = 0.32$% corresponding to discharge 66404 at $\rho = 0.33$. The ratio between

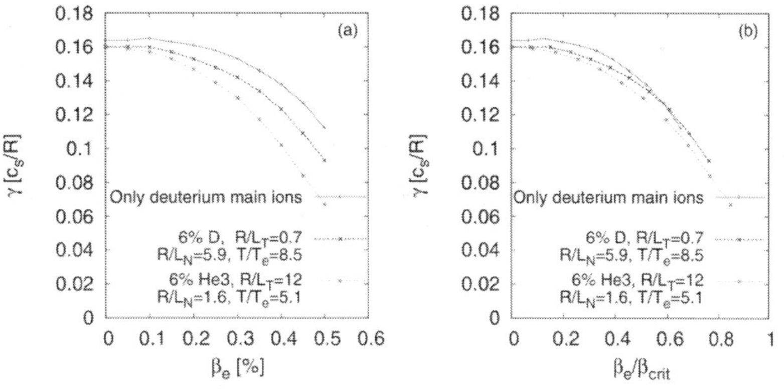

Figure 2. Linear β scans for JET discharge 66404 at $\rho = 0.33$, for thermal species only and when including either NBI or ICRH driven fast ions. The scans are plotted as a function of $\beta_e(a)$ and as a function of $\beta_e/\beta_{crit}(b)$, where β_{crit} was separately calculated for each case.

the EM-case and ES-case heat fluxes at each point in the scan is shown in figure 1(*c*), plotted against the β_e/β_{crit} for each parameter set, where β_{crit} was taken from the fit of the linear scans. Two important points should be emphasized: (a) the β scans and β_{crit} shown here correspond to electron β, but the β_s of all ionic species in the simulations are scaled self-consistently with β_e, according to the pressure ratios p_i/p_e; (b) the $\hat{s}$ -dependent β_{crit} is the parameter which changes within the scan in figure 1(*c*), not β_e. While the intermittency in the simulations leads to significant error bars, a clear separation in this flux ratio is seen between high and low β_e/β_{crit}, with a lower ratio at higher β_e/β_{crit}. This supports the adoption of β/β_{crit} as a valid parameter of merit also for the nonlinear ITG EM-stabilization, suggesting that the EM modification of the linear modes is also linked to the enhanced nonlinear stabilization. For the ion heat flux, the inclusion of EM effects is always stabilizing, consistent with ITG stabilization. However, the situation is more complicated with the electron heat flux. At higher $\hat{s}$ (higher β_{crit} and thus lower β_e/β_{crit}), the inclusion of EM effects leads to an increase of electron heat flux in this specific case. This is due to a combination of magnetic flutter transport and likely an increased relative impact of the trapped electron drive. More detailed analysis is necessary to untangle the various effects. We note however that all these cases remain ion heat flux dominated and even for the EM simulations, the q_i/q_e ratio remains within the range of 2–4.

In figure 2, the relevance of the β/β_{crit} parametrization of EM-stabilization in the presence of fast ions is examined. Figure 2(*a*) displays a β-scan for three separate cases of discharge 66404 at $\rho = 0.33$: with thermal species only, with a fast D species matching the modeled NBI profile for the discharge and with a fast ^{3}He species matching the modeled ICRH profile. In this scan, a separation of the curves are evident, with the EM-stabilization occurring at different rates with increasing β_e. Continuing the calculation of the curves down to the KBM limit (not shown for brevity) we obtain $\beta_{crit} = 0.77\%$, 0.66%, 0.59% respectively for the three cases. When rescaling the *x*-axis of the plot to β_e/β_{crit}, as shown in figure 2(*b*), the overlap between the three curves is much more apparent, illustrating the suitability of β/β_{crit} as a parameter of merit.

To summarize, β/β_{crit} is shown to be a valid parameter of merit for both the linear and nonlinear EM-stabilization. β_{crit} is reduced at lower-$\hat{s}$ and with increased suprathermal pressure gradients. Thus, for a given thermal β, enhanced EM-stabilization of ITG turbulence is expected at both low-$\hat{s}$ and high suprathermal pressure. This statement is relevant for the interpretation of transport in JET discharge 75225, as shown in the following sections.

We conclude this section with a remark on the nature of β_{crit}. We have assumed that $\beta_{crit} = \beta_{KBM}$. However, an additional β-limit is apparent from numerical simulations—β_{NZT} (non-zonal-transition)—leading to a sharp rise in flux due to a reduction in zonal flow activity [29, 30]. For highly driven systems, this occurs for $\beta_{NZT} < \beta_{KBM}$. However, for experimental parameter sets studied thus far, including all those studied in this paper, $\beta_{KBM} < \beta_{NZT}$. This is an encouraging observation since we have seen that access to $\beta \sim \beta_{KBM}$ has positive ramifications for ITG stabilization. If this observation can be shown to be general, this would mean that a prediction to the degree of EM-stabilization in future devices such as ITER could be carried out from linear analysis of the projected profiles and determining β/β_{crit} where β_{crit} is taken as β_{KBM}. This is left for future work.

4. Linear simulations of JET 75225

In this section we describe the linear analysis of the JET high-β hybrid discharge 75225. The analysis is split into 2 subsections concentrating on different radii, $\rho = 0.33$ and $\rho = 0.64$.

4.1. Linear calculations at $\rho = 0.33$

In figure 3 we plot the k_y spectrum of the linear growth rates and frequencies at $\rho = 0.33$. The plot shows 4 separate cases: a case without fast ions, a case including the full fast ion pressure gradient, a case with the fast ion pressure gradient reduced by 30% and a case where β_e was set to 10^{-5}, which is effectively an electrostatic case. The spectrum in all cases is dominated by ITG modes, characterized by the positive frequencies in the diamagnetic drift frequency range. From the comparison of the EM and ES cases, a strong linear EM-stabilization is evident, particularly at the higher k_y values. In all scans shown here, the ∇P taken for the ω_d calculations are set to be consistent with the input β and logarithmic gradients in each run.

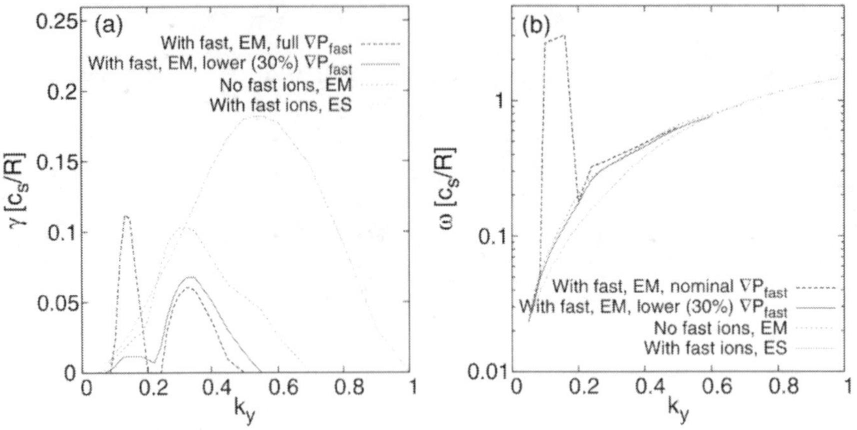

Figure 3. Linear growth rate (a) and frequency (b) spectra as a function of k_y, for JET 75225 at $\rho = 0.33$. The calculations included numerical geometry, collisions, C impurities, EM fluctuations and fast ions, unless otherwise noted.

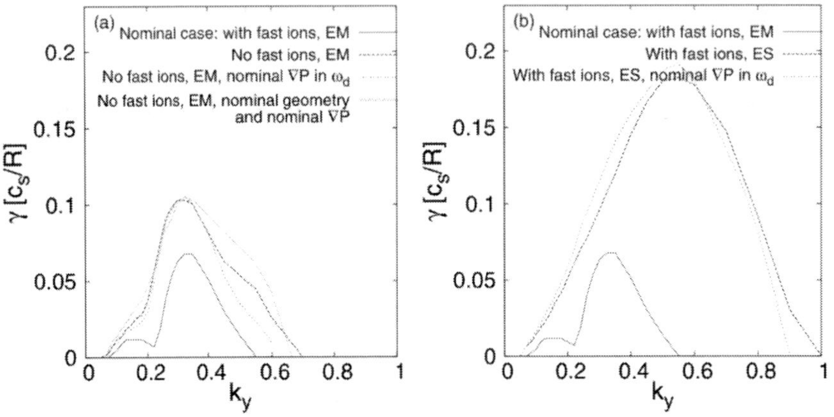

Figure 4. Linear growth rate spectra as a function of k_y, for JET 75225 at $\rho = 0.33$. The impact of EM-stabilization is isolated when comparing cases with and without fast ions (a) and when comparing the EM and ES cases (b).

We also utilize two separate input geometry files, for the cases with and without fast ions respectively. These were calculated from the Grad–Shafranov solver in the CRONOS interpretative runs for each case, at the nominal R/L_{Ti}, where the suprathermal pressure component was removed when calculating the 'no fast ion' magnetic equilibrium.

An additional mode is clearly apparent for the parameters corresponding to including the full suprathermal pressure, at $k_y < 0.2$, at significantly higher frequencies than the ITG range. This mode is stabilized by lowering the fast ion pressure gradient by 30%, illustrating the fast ion drive of this mode. From a Rosenbluth–Hinton test [31] it was determined that the frequency of these modes are within 5% of the GAM frequency for these parameters. This supports the identification of this mode as a Beta induced Alfvén Eigenmode (BAE) [32], which is known to be degenerate with the GAM frequency [33]. However, these modes are likely coupled with the KBM modes which share a similar frequency range. Thus, in lieu of deeper analysis regarding the precise characterization, in the following we refer to this mode as a hybrid BAE/KBM. As will be shown in section 5, the heat fluxes in the simulation of the system with unstable BAE/KBMs is highly inconsistent with the experimental power balance values. Thus, it is assumed that the system self-organizes to a state just below the BAE/KBM stability boundary. We maintain the lowered fast ion pressure for all the nonlinear simulations and consider this the nominal parameter set.

Comparing the ITG growth rates of the cases with and without fast ions shows that the fast ions provide an additional stabilization of ITG beyond the thermal EM-stabilization, as expected from section 3. This stabilization has been isolated to be electromagnetic in nature and not due to the concomitant modifications of ∇P in the ω_d calculation, or the different magnetic equilibrium. This is seen in figure 4(a), where the nominal 'with fast ion' case is compared to various 'no fast ion' cases. The ∇P in ω_d and then the magnetic equilibrium are respectively modified to equate with those of the nominal case. The growth rate differences are minor compared with the EM-stabilization. Interestingly, switching to the higher shifted magnetic equilibrium of the 'with fast ion' case leads to destabilization, i.e. increased growth rates. While increased Shafranov shift typically stabilizes the ITG drive, in this case it also stabilizes the KBM/BAE drive and increases β_{crit}, thus reducing the degree of EM-stabilization of ITG, leading to a net effect of higher growth rates. This impact of geometry on the KBM is linked with the observed improvement

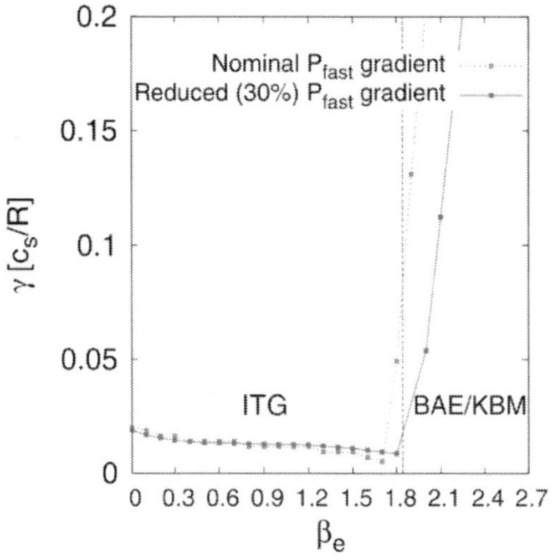

Figure 5. β-scan of linear growth rate at $k_y = 0.1$ for JET 75225 at $\rho = 0.33$. Two separate cases are shown, one with nominal parameters and the second with the fast ion pressure reduced by 30% . The nominal $\beta_e = 1.84\%$ is signified by the vertical black line.

of peeling-ballooning stability in the tokamak edge region in cases with fast ions, which leads to a beneficial core-edge transport coupling [19]. In figure 4(*b*) we also isolate the EM-stabilization effect when comparing the EM and ES cases with fast ions. By replacing the $\nabla P \approx 0$ in the ES case ω_d calculation (due to setting ∇P with respect to the input $\beta \approx 0$ and gradients) with the nominal ∇P from the EM case, we only obtain a minor difference in calculated growth rates. The input magnetic equilibrium was already the same for both cases.

The behavior of the BAE/KBM mode destabilization at $k_y = 0.1$ is shown in figure 5, for the two separate cases with the full and lowered fast ion pressure gradient. For both cases the ITG growth rates are stabilized with higher β_e until the BAE/KBM is eventually destabilized. For the full gradients, this destabilization occurs for a β_e lower than the experimental β_e of 1.84% . For the lowered fast ion pressure gradient, the instability boundary is just above the experimental value. Thus, $\beta/\beta_{\text{crit}} \approx 1$ in this case and according to the results in section 3 we can thus expect significant nonlinear EM-stabilization as well. We reiterate that in the scan, the β_s of all ionic species are also scaled in the scan with β_e, according to the pressure ratios p_j/p_e.

4.2. *Linear calculations at $\rho = 0.64$*

In figure 6 we show the linear calculations at $\rho = 0.64$. Three cases are shown: the nominal case including fast ions, a case without fast ions and an electrostatic case. Three distinct instabilities are apparent. In the EM cases, a micro-tearing-mode (MTM)—identified from its mode parity and structure—is unstable at $k_y = 0.05$. For intermediate k_y ITGs are unstable and for $k_y > 1.1$ trapped electron modes (TEMs) are unstable. As opposed to the $\rho = 0.33$ case, the EM-stabilization is much weaker in the lower, transport driving k_y values. This is reflected by the relatively lower $\beta/\beta_{\text{crit}} \approx 0.3$ values calculated for this case.

We note that these linear calculations were carried out for modified input parameters compared with table 1, with R/L_{Te} and R/L_{ne} reduced by 20% and R/L_{Ti} increased by 20% . This was necessary to reach agreement with the experimental power balance values and q_i/q_e heat flux ratio in the nonlinear simulations. For nominal parameters, a strongly driven TEM regime was predicted at all k_y, leading to $q_e > q_i$, in disagreement with the observations. This modification of the driving gradients is at the limits of the experimental error bars.

5. Nonlinear simulations of JET 75225

In this section we describe the nonlinear simulations of JET discharge 75225, split into 2 subsections concentrating on different radii, $\rho = 0.33$ and $\rho = 0.64$.

5.1. *Nonlinear simulations at $\rho = 0.33$*

In figure 7 the GENE saturated nonlinear transport fluxes are shown for various cases: with fast ions (with 30% reduced suprathermal pressure gradient), without fast ions and with fast ions but electrostatic. In addition, the experimental ion and electron heat fluxes from power balance is shown. The GENE error bars are derived from intermittency and uncertainties in the precise saturated potential level. The power balance ion heat flux uncertainties include statistical errors and the propagation of $T_{i,e}$ errors on the ion-electron heat exchange term.

The most striking observation is that the nonlinear EM-stabilization is absolutely critical for achieving agreement between the GENE and experimental flux values. The nonlinear EM-stabilization is here manifested mostly as a stiffness reduction. The approximate factor 20 reduction in heat flux is much greater than the analogous growth rate reduction, reflecting the enhanced nonlinear nature of the EM-stabilization. This result is thus a reactor-relevant high-β extension of the results reported in [5, 7]. In addition to the stiffness reduction, a slight increase in the value of the Dimits shift [34] is observed between the ES and EM case with fast ions. This is seen by comparing the linear R/L_{Ti} thresholds for these cases, calculated separately and shown in the figure, with the extrapolation to zero-flux of the nonlinear stiffness curves. This Dimits shift extension is consistent with previous findings of nonlinear EM-stabilization [11, 12]. In this case, as opposed to the low-β cases in [7], the majority of the nonlinear EM-stabilization is due to the thermal component. This is consistent with the lower suprathermal pressure fraction in these high-β cases compared with the low-β cases, on the order of 30% here. However, the addition of fast ions does still provide significant additional stabilization, leading to an additional increase of R/L_{Ti} by approximately 10–20%, for the same heat flux level.

The level of agreement between the experimental power balance and GENE predictions is remarkable and provides a highly encouraging validation of nonlinear gyrokinetics in the high-β regime. This is only possible due to the increasing capacity in high performance supercomputing capabilities. For example, the simulations shown in figure 7 necessitated

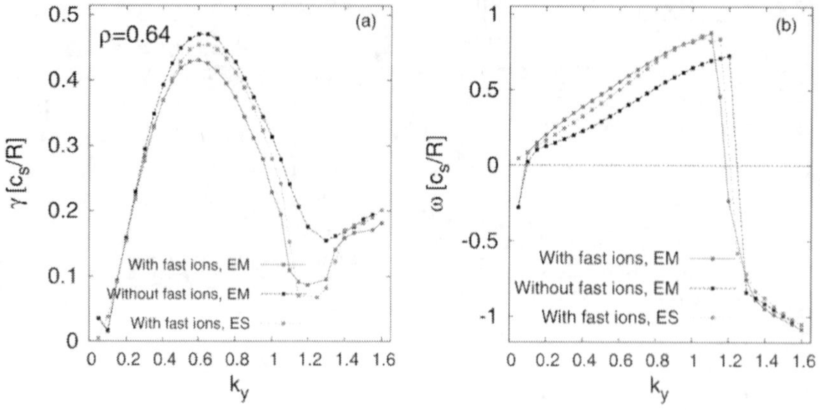

Figure 6. Linear growth rate (*a*) and frequency (*b*) spectra as a function of k_y, for JET 75225 at $\rho = 0.64$. The calculations included numerical geometry, collisions, C impurities, EM fluctuations and fast ions, unless otherwise noted.

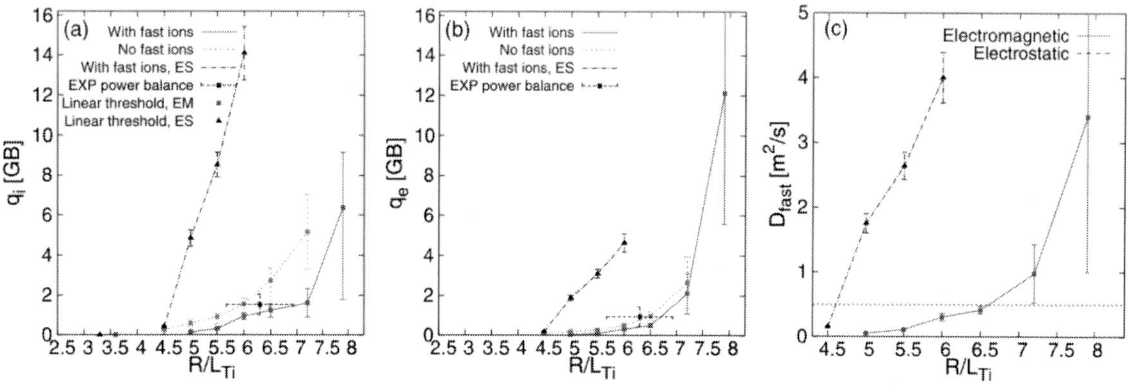

Figure 7. Ion heat fluxes (*a*), electron heat fluxes (*b*) and fast ion particle diffusivity (*c*) from GENE nonlinear simulations of JET 75225 at $\rho = 0.33$.

around 4 million CPU hours of calculation time, not including extensive convergence tests.

The rightmost datapoint in figure 7, with its sharp uptick in heat fluxes at the highest R/L_{Ti}, is a consequence of the destabilization of the BAE/KBM mode, which is also destabilized by the main ion gradients. This is also seen by the strong increase in fast ion diffusivity for that case. The particularly sharp uptick of the electron heat flux is mostly due to increased magnetic flutter. While the decrease in $\beta/\beta_{\mathrm{crit}}$ at increased R/L_{Ti} due to the increase in BAE/KBM drive may help maintain the low-stiffness curve at higher R/L_{Ti}, this high heat flux data point is an indication of the eventual saturation of the EM-stabilization mechanism due to the crossing of the BAE/KBM boundary ($\beta/\beta_{\mathrm{crit}} > 1$).

We recall that these simulations all used a suprathermal pressure gradient reduced by 30% . The anomalous diffusivity prescribed in the SPOT code to provide such a suprathermal pressure reduction in the CRONOS interpretative simulations was 0.5 $m^2\ s^{-1}$. This level is actually consistent with the GENE predicted fast ion diffusivity for the simulation at nominal R/L_{Ti}, which also led to heat flux agreement with power balance. Thus, the self-consistency of this assumption is verified.

The decrease in EM-stabilization when transitioning between the 'with fast ions' to 'no fast ions' curves in figures 7(*a*) and (*b*), at each given R/L_{Ti} point, is correlated with a corresponding decrease in $\beta/\beta_{\mathrm{crit}}$. This is seen in figure 8 for both the ion and electron heat fluxes. β_{crit} was calculated for each case by a dedicated linear β scan at $k_y = 0.1$. This result provides further evidence of the relevance of $\beta/\beta_{\mathrm{crit}}$ as an organizing parameter for the impact of both linear and nonlinear EM-stabilization. However, $\beta/\beta_{\mathrm{crit}}$ does not correlate with $q_{i,e}(\beta_e = 1.84\%)/q_{i,e}(\beta_e \approx 0)$ as β_{crit} changes due to R/L_{Ti} in the R/L_{Ti} scans, particularly for the case including fast ions. This can be understood to be due to the proximity to turbulence threshold and the complication of the increased Dimits shift of the case including fast ions. As R/L_{Ti} is reduced, the case with fast ions drops to the threshold level faster than the electrostatic case, decreasing $q_{i,e}(\beta_e = 1.84\%)/q_{i,e}(\beta_e \approx 0)$ in spite of the lower $\beta/\beta_{\mathrm{crit}}$. Finally, we note that in this case, as opposed to [7], the difference in flux between the 'with fast ions' and 'no fast ions' nonlinear EM simulations is on the same order of the difference in the linear spectra between the 'with fast ions' and 'no fast ions' cases in figure 3. This suggests that while the total degree of EM-stabilization cannot be explained purely by linear theory, the additional boost

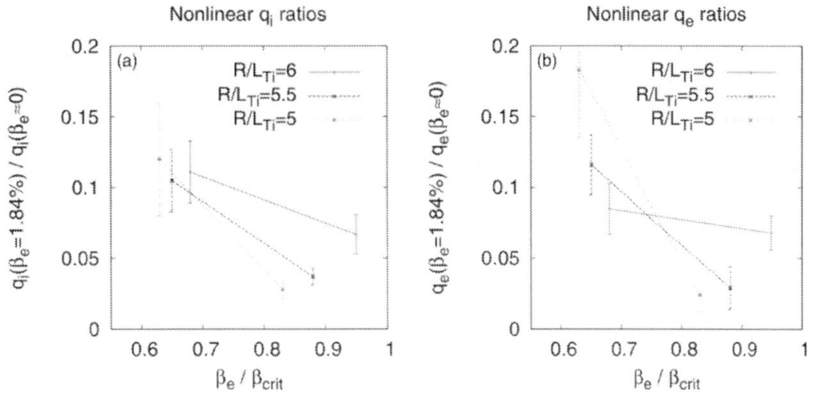

Figure 8. Ratio of ion heat fluxes (*a*) and electron heat fluxes (*b*) between the electromagnetic and electrostatic simulations shown in figure 7, plotted against β/β_{crit}, for 3 separate R/L_{Ti} inputs.

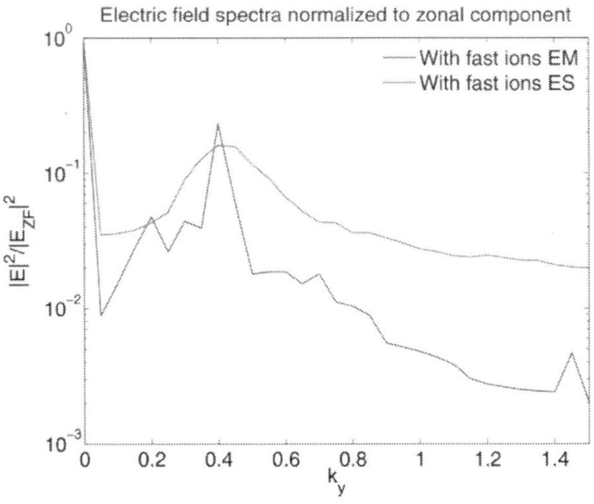

Figure 9. Amplitude of the normalized electric field fluctuations averaged over the saturated nonlinear phase for two separate cases (EM and ES) of JET 75225 at $\rho = 0.33$. The amplitudes at each k_y are averaged for all k_x modes and over the parallel direction.

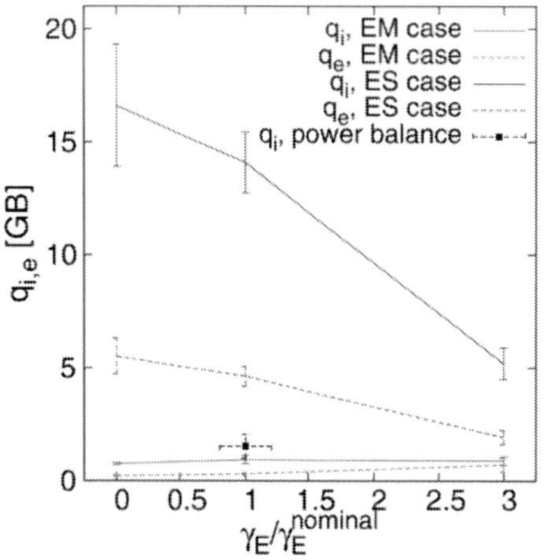

Figure 10. γ_E scans for JET 75225 at $\rho = 0.33$, for both EM (nominal) and assumed ES cases. The nominal γ_E is 0.2. The simulations were carried out for $R/L_{Ti} = 6$ and the ion heat fluxes are compared against the power balance value.

provided by fast ions can be reproduced quasilinearly *for this specific case*, as confirmed in [21].

In ITG turbulence, zonal flows (ZFs) are a well known saturation mechanism [35]. The correlation of the nonlinear EM-stabilization and increased destabilization of ZF secondaries has been reported [13]. This connection is verified in these simulations as well. This is seen by analyzing the ratio of the electric field energy of the ZFs (all $k_y = 0$ modes) to the sum of electric field energies of all drift wave modes. This value, $|E_{ZF}^2|/\sum_{k_y>0}|E_{DW}^2|$, serves a proxy for the relative impact of ZFs in the nonlinear system. This ratio is ≈1.5 for the EM case with fast ions, ≈1.4 for the EM case without fast ions and ≈0.65 for the ES case. This trend correlates well with the trend of the relative heat flux ratios. In figure 9, the normalized electric field energy spectra is shown for the nonlinear simulations of the nominal EM and ES cases. The relative increased energy in the $k_y = 0$ ZF modes in the EM case is clearly evident in the plot.

The relative role of EM-stabilization and flow shear stabilization was investigated by γ_E scans. This is shown in figure 10. The γ_E scans, where $\gamma_E = 0.2$ is the nominal value, were carried out for both EM and ES cases. A number of striking observations come out from this scan. Firstly, for the (nominal) EM case, no flow shear stabilization was evident at all. A slight increase in fluxes is even observed, particularly in the electron channel. This increased q_e is due to increased magnetic flutter, which may suggest increased fast ion mode destabilization with rotation. For the ES case, a more classical trend is observed, with stabilization at increased γ_E. However, for the nominal $\gamma_E = 0.2$, this stabilization remains weak. Even at $\gamma_E = 0.6$, the flux levels are significantly above the power balance levels. This suggests that rotation is not an important stabilization factor in this regime, which is dominated by EM-stabilization, at least for inner radii such as $\rho = 0.33$. The explanation for the complete lack of flow shear stabilization in the EM case remains unknown and is a topic for future study. A conjecture is that this may be related to the much shorter autocorrelation times of the turbulent structures in the EM cases compared with the ES cases.

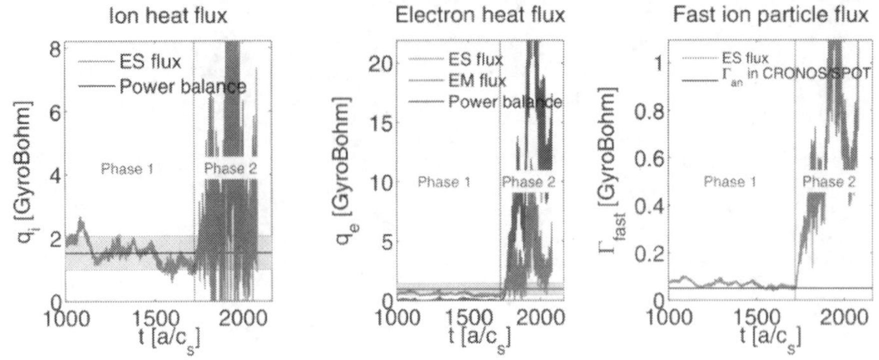

Figure 11. Time traces of GENE simulated ion heat flux (*a*), electron heat flux (*b*) and fast ion diffusivity (*c*) for JET 75225 at $\rho = 0.33$ during a phase with a 30% reduced fast ion pressure gradient (phase 1), followed by a transition back to the full fast ion pressure gradient (phase 2). The time traces are compared with the power balance heat fluxes, whose 1σ confidence region is shown by the grey shaded zone.

We conclude this subsection with a brief discussion of the combined ITG-BAE/KBM turbulent state. This was carried out by carrying out a simulation using the full fast pressure gradient, i.e. the case in which the BAE/KBM mode is seen to be unstable in figure 3. The time dependent nonlinear fluxes are shown in figure 11. In the plot, an initial phase is evident, corresponding to the case with reduced fast pressure gradient. The simulation input is then shifted to the full fast pressure gradient and the simulation is restarted. This then corresponds to phase 2, where all fluxes rise to a degree significantly above the power balance levels. In particular, the EM component (magnetic flutter) of the electron heat flux rises sharply. This EM transport component is correlated with tearing parity fluctuations at $k_y > 0.3$. MTMs are linearly stable at this region and this observation is potentially linked to heightened nonlinear mode-mixing to tearing parity modes [36] when the BAE/KBM is unstable. This is a subject of future study. However, this regime as an operating point is not experimentally relevant due to the high fluxes and this supports the use of 'stiff' Alfvén Eigenmode models in reduced modeling frameworks [37]. Whether the system is maintained constantly below the BAE/KBM mode limit, or has limit cycle dynamics, is an open question and necessitates flux driven simulations. Thus, we cannot fully rule out the experimental relevance of this state. Finally, we note that the phase 1 time trace shown in figure 11 is the tail end of a significantly longer simulation, with sufficient statistics to determine a saturated nonlinear state. However, for phase 2, there are insufficient statistics for such a determination. Nonetheless, we deem the qualitative observation of significantly higher fluxes than the power balance as robust.

5.2. Nonlinear simulations at $\rho = 0.64$

In this subsection we describe the nonlinear analysis at $\rho = 0.64$ and compare the trends observed with the analogous simulations at $\rho = 0.33$. The R/L_{Ti} scans are shown in figure 12. Numerous scans were carried out, isolating the relative roles of EM-stabilization (thermal and with fast ions) and $E \times B$ flow shear stabilization. It is clear that at this location, the EM-stabilization is extremely weak, with the ES and EM

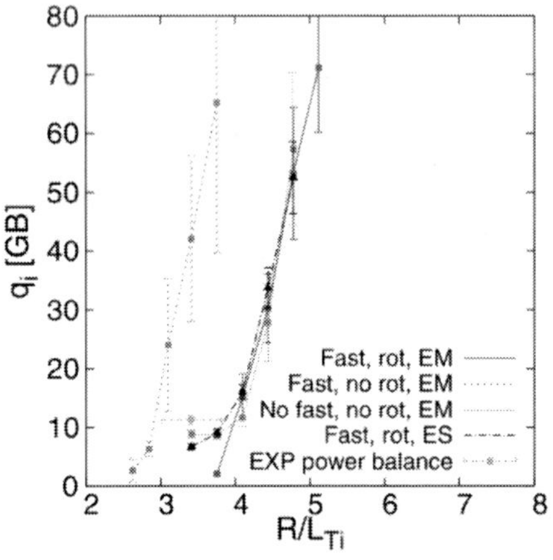

Figure 12. Ion heat fluxes from GENE nonlinear simulations of JET 75225 at $\rho = 0.64$, for various cases including EM and ES simulations and without $E \times B$ flow shear.

curves in close proximity. This is likely due to the much lower $\beta/\beta_{crit} \equiv 0.3$ in this case compared with $\rho = 0.33$.

The main observation is that flow shear stabilization is effective for these parameters and can significantly increase the R/L_{Ti} values which match the power balance fluxes. In general, the stiffness level is much higher for these parameters compared with $\rho = 0.33$, likely due to the higher $\hat{s}$ and q values as well as lower EM-stabilization. The $E \times B$ stabilization provides a threshold shift.

We reiterate that these nonlinear simulations were carried out for modified input parameters compared to the nominal parameters from table 1, with R/L_{Te} and R/L_{ne} reduced by 20% and R/L_{Ti} increased by 20% . This was to avoid a strongly driven TEM regime and match the experimental q_i/q_e and reach the vicinity of the absolute flux values.

5.3. Extrapolation to ITER

We conclude this section with a rough extrapolation of the inner core EM-stabilization effect to ITER, with the aim of judging the scale of potential impact. Theory based extrapolations carried out until now have not included the enhanced

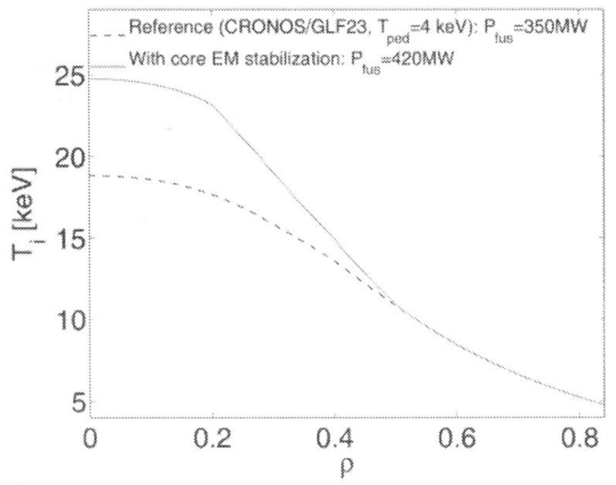

Figure 13. Ion temperature profiles from a reference CRONOS/GLF23 ITER hybrid scenario simulation, compared with an estimate of a peaked T_i profile when including core EM-stabilization effects. The fusion powers of the two cases are compared.

nonlinear EM-stabilization effects. This extrapolation was based on a T_i profile predicted for an ITER hybrid scenario in previous theory based integrated modeling [38]. This modeling consisted of a CRONOS predictive simulation with the GLF23 transport model [39, 40] and a pedestal height of 4 keV. As seen in figure 13, this reference T_i is compared with an estimate—based on the same profile—of T_i peaking in the inner core between $\rho = 0.2 - 0.5$. This peaking was estimated by a simple $\Delta R/L_{Ti} = 2$ upshift, based on the results of JET 75225, which shares a similar β and suprathermal pressure fraction as predicted for this ITER hybrid scenario reference case. The lower cutoff point of $\rho = 0.2$ is based on a conservative assumption that access to the ITG EM-stabilization zone may be hindered by extremely low-$\hat{s}$ in the very inner core which decreases the $\beta_{\rm crit}$ of the BAE/KBMs [41]. Based on the $n_{D,\,T}$ from the reference case, the fusion power corresponding to the two separate T_i profiles can be compared. The reference $P_{\rm fus} = 350$ MW increases by 20% to $P_{\rm fus} = 420$ MW. The impact of core EM-stabilization is thus estimated to be significant.

6. Conclusions

Detailed linear and nonlinear local gyrokinetic calculations of the JET high-β hybrid discharge 75225 at $\rho = 0.33$ and $\rho = 0.64$ were carried out. Good agreement between the simulated and power balance heat fluxes was achieved. It was found that at $\rho = 0.33$, electromagnetic (EM) stabilization of ITG turbulence is critical for reaching agreement with power balance heat fluxes, while $E \times B$ flow shear stabilization was not effective. Nonlinear effects and the inclusion of suprathermal pressure both significantly enhance the EM-stabilization. At $\rho = 0.64$, on the other hand, the EM-stabilization was ineffective while the $E \times B$ flow shear stabilization was important. A likely explanation for the difference in EM-stabilization effectiveness at the two locations is the value of the $\beta/\beta_{\rm crit}$ EM-stabilization parameter of merit, which is ≈ 1 and 0.3 respectively for the two cases. $\beta_{\rm crit}$ is the critical β value for the onset of EM instabilities.

These results provide a high-β extension of previous results of enhanced nonlinear EM-stabilization at low-β. The reactor relevance of this particular discharge is encouraging for extrapolations to future devices, which will have low rotation but a significant suprathermal pressure fraction due to fusion α-particles. An estimate of the impact of this effect on ITER extrapolation was carried out, with a 20% increase of fusion power predicted for an ITER hybrid scenario. This is a strong indication that reduced models employed for transport extrapolation to ITER and reactors must include suprathermal pressure and EM-stabilization effects validated by nonlinear gyrokinetic simulations.

Future work will concentrate on the physical mechanism of the enhanced nonlinear EM-stabilization, which is likely related to either (or both) increased secondary ZF growth rates or reduced tertiary damping of ZFs. However, a precise characterization of the effect is still lacking.

Acknowledgments

This project has received funding from the European Unions Horizon 2020 research and innovation programme under grant agreement number 633053. The views and opinions expressed herein do not necessarily reflect those of the European Commission. This work received financial support from NWO and is supported by NWO-RFBR Centre-of-Excellence on Fusion Physics and Technology (Grant no. 047.018.002). This work is part of the research programme 'Fellowships for Young Energy Scientists' (YES!) of the Foundation for Fundamental Research on Matter (FOM), which is financially supported by the Netherlands Organisation for Scientific Research (NWO). The authors would like to thank E Highcock, S Moradi, G Plunk, A Schekochihin and E Westerhof for stimulating discussions. Resources of the HELIOS supercomputer at IFERC-CSC are acknowledged. This research used computational resources at the National Research Scientific Computing Center, which is supported by the Office of Science of the US Department of Energy under Contract No. DE-AC02-05CH11231. The authors are grateful to D R Mikkelsen for assistance.

References

[1] Romanelli F 1989 *Phys. Fluids* B **1** 1018
[2] Guo S C and Romanelli F 1993 *Phys. Fluids* B **5** 520
[3] Doyle E J *et al* 2007 Progress in the ITER physics basis chapter 2: plasma confinement and transport *Nucl. Fusion* **47** S18
[4] Mantica P *et al* 2009 *Phys. Rev. Lett.* **102** 175002
[5] Mantica P *et al* 2011 *Phys. Rev. Lett.* **107** 135004
[6] Jenko F, Dorland W, Kotschenreuther M and Rogers B N 2000 *Phys. Plasmas* **7** 1904 (http://gene.rzg.mpg.de)
[7] Citrin J *et al* 2013 *Phys. Rev. Lett.* **111** 155001
[8] Citrin J *et al* 2014 *Nucl. Fusion* **54** 023008
[9] Weiland J and Hirose A 1992 *Nucl. Fusion* **32** 151
[10] Hirose A 2000 *Phys. Plasmas* **7** 433

[11] Pueschel M J, Kammerer M and Jenko F 2008 *Phys. Plasmas* **15** 102310
[12] Pueschel M J and Jenko F 2010 *Phys. Plasmas* **17** 062307
[13] Pueschel M J, Görler T, Jenko F, Hatch D R and Cianciara A J 2013 *Phys. Plasmas* **20** 102308
[14] Romanelli M, Zocco A, Crisanti F and JET-EFDA Contributors 2010 *Plasma Phys. Control. Fusion* **52** 045007
[15] Hobirk J *et al* 2012 *Plasma Phys. Control. Fusion* **54** 095001
[16] Garcia J *et al* 2013 *Nucl. Fusion* **53** 043023
[17] Garcia J and Giruzzi G 2010 *Phys. Rev. Lett.* **104** 205003
[18] Mantica P *et al* 2011 *Plasma Phys. Control. Fusion* **53** 124033
[19] Garcia J 2014 *Nucl. Fusion* submitted
[20] Staebler G M *et al* 2005 *Phys. Plasmas* **12** 102508
[21] Baiocchi B *et al* 2014 *Plasma Phys. Control. Fusion* submitted
[22] Lapillonne X *et al* 2009 *Phys. Plasmas* **16** 032308
[23] Artaud J F *et al* 2010 *Nucl. Fusion* **50** 043001
[24] Schneider M *et al* 2011 *Nucl. Fusion* **51** 063019
[25] Huysmans G T A, Goedbloed J P and Kerner W 1991 *Conf. on Computational Physics* (*Amsterdam, The Netherlands, 1 May 1991*) (Singapore: Word Scientific) p 371
[26] Kim J Y, Horton W and Dong J Q 1993 *Phys. Fluids* B **5** 4030
[27] Bourdelle C *et al* 2005 *Nucl. Fusion* **45** 110
[28] Ryter F *et al* 2011 *Nucl. Fusion* **51** 113016
[29] Pueschel M J *et al* 2013 *Phys. Rev. Lett.* **110** 155005
[30] Pueschel M J *et al* 2013 *Phys. Plasmas* **20** 102301
[31] Rosenbluth M N and Hinton F L 1998 *Phys. Rev. Lett.* **80** 724
[32] Zonca F *et al* 1996 *Plasma Phys. Control. Fusion* **38** 2011
[33] Nguyen C, Garbet X and Smolyakov A I 2008 *Phys. Plasmas* **15** 112502
[34] Dimits A M *et al* 2000 *Phys. Plasmas* **7** 969
[35] Diamond P, Itoh S-I, Itoh K and Hahm T S 2005 *Plasma Phys. Control. Fusion* **47** R31
[36] Hatch D R *et al* 2012 *Phys. Rev. Lett.* **108** 235002
[37] Bass E M and Waltz R E 2010 *Phys. Plasmas* **17** 112319
[38] Citrin J *et al* 2010 *Nucl. Fusion* **50** 115007
[39] Waltz R E, Staebler G M, Dorland W and Hammett G W 1997 *Phys. Plasmas* **4** 2482
[40] Kinsey J E, Staebler G M and Waltz R E 2005 *Phys. Plasmas* **47** 052503
[41] Moradi S *et al* 2014 *Nucl. Fusion* accepted
[42] Romanelli F *et al* 2012 *Proc. 24th IAEA Fusion Energy Conf.* (*San Diego, USA, 8–13 October 2012*)

Plasma Phys. Control. Fusion **57** (2015) 014033 (14pp) doi:10.1088/0741-3335/57/1/014033

Sawtooth control in JET with ITER relevant low field side resonance ion cyclotron resonance heating and ITER-like wall

J P Graves[1], M Lennholm[2], I T Chapman[3], E Lerche[4], M Reich[5], B Alper[3], V Bobkov[5], R Dumont[6], J M Faustin[1], P Jacquet[3], F Jaulmes[7], T Johnson[8], D L Keeling[3], Yueqiang Liu[3], T Nicolas[6], S Tholerus[8], T Blackman[3], I S Carvalho[9], R Coelho[9], D Van Eester[4], R Felton[3], M Goniche[6], V Kiptily[3], I Monakhov[3], M F F Nave[9], C Perez von Thun[10], R Sabot[6], C Sozzi[11], M Tsalas[7] and JET EFDA Contributors[12]

JET-EFDA, Culham Science Centre, Abingdon, OX14 3DB, UK
[1] École Polytechnique Fédérale de Lausanne (EPFL), Centre de Recherches en Physique des Plasmas (CRPP), 1015 Lausanne, Switzerland
[2] EFDA-JET CSU, Culham Science Centre, Abingdon, OX14 3DB, UK
[3] CCFE Fusion Association, Culham Science Centre, Abingdon, UK
[4] LPP-ERM/KMS, TEC Partner, Brussels, Belgium
[5] Max-Planck-Institut für Plasma Physik, Boltzmannstr. 2, 85748 Garching, Germany
[6] CEA-Cadarache, F-13108 St. Paul lez Durance, France
[7] FOM Institute DIFFER Dutch Institute For Fundamental Energy Research, FOM, Nieuwegein, The Netherlands
[8] Association VR, KTH, SE-100 44 Stockholm, Sweden
[9] Instituto de Plasmas e Fus ao Nuclear, Instituto Superior Técnico, Universidade de Lisboa, P-1049-001, Lisboa, Portugal
[10] University of Innsbruck, Österreichische Akademie der Wissenshaften (ÖAW), 1010 Wien, Austria
[11] Isituto di Fisica del Plasma CNR, 20125 Milano, Italy

E-mail: jonathan.graves@epfl.ch

Received 27 June 2014, revised 8 September 2014
Accepted for publication 30 September 2014
Published 28 November 2014

Abstract

New experiments at JET with the ITER-like wall show for the first time that ITER-relevant low field side resonance first harmonic ion cyclotron resonance heating (ICRH) can be used to control sawteeth that have been initially lengthened by fast particles. In contrast to previous (Graves *et al* 2012 *Nat. Commun.* **3** 624) high field side resonance sawtooth control experiments undertaken at JET, it is found that the sawteeth of L-mode plasmas can be controlled with less accurate alignment between the resonance layer and the sawtooth inversion radius. This advantage, as well as the discovery that sawteeth can be shortened with various antenna phasings, including dipole, indicates that ICRH is a particularly effective and versatile tool that can be used in future fusion machines for controlling sawteeth. Without sawtooth control, neoclassical tearing modes (NTMs) and locked modes were triggered at very low normalised beta. High power H-mode experiments show the extent to which ICRH can be tuned to control sawteeth and NTMs while simultaneously providing effective electron heating with improved flushing of high Z core impurities. Dedicated ICRH simulations using SELFO, SCENIC and EVE, including wide drift orbit effects, explain why sawtooth control is effective with various antenna phasings and show that the sawtooth control mechanism cannot be explained by enhancement of the magnetic shear. Hybrid kinetic-magnetohydrodynamic

[12] See the appendix of Romanelli F *et al* 2012 *Proc. 24th IAEA Fusion Energy Conf. 2012 (San Diego, CA)*.

stability calculations using MISHKA and HAGIS unravel the optimal sawtooth control regimes in these ITER relevant plasma conditions.

Keywords: sawteeth, ICRH, tungsten, NTMs, tokamaks, MHD

(Some figures may appear in colour only in the online journal)

1. Introduction

The stabilising role of trapped alpha particles are expected to cause long sawtooth periods in ITER. Without adequate sawtooth control techniques, long sawteeth are predicted [1] to trigger fast growing neoclassical tearing modes (NTMs) during standard H-mode operation. Experiments at JET [2, 3] have shown that long period sawteeth can trigger NTMs deeply in L-mode when the normalised beta is only around $\beta_N = 0.8$, a value around half that expected during standard baseline ITER operation. NTMs could be avoided in these experiments by controlling sawteeth with ion cyclotron resonance heating (ICRH) resonance close to the $q = 1$ surface on the high field side, and the control mechanism was isolated and identified as being the kinetic effect associated with finite orbit widths [4].

That NTMs are so easily triggered by sawteeth at low beta in JET is due to the fact that ICRH dominant heating plasmas can produce very long sawteeth and probably because in such conditions the toroidal plasma rotation is weak. Although unproven, it is probable that NTMs couple to the sawtooth precursor and/or harmonics generated at the crash, more easily with weak rotation shear. This is clearly of great concern because ITER will operate with very low toroidal rotation velocity relative to the thermal velocity of the thermal ions, even during operation with full tangential NBI power. With this in mind it is important to explore ITER relevant sawtooth and NTM control experiments in JET. The experiments reported here are undertaken in the 'upgraded' JET with tungsten/beryllium wall, thus operating in conditions which have to contend with potential tungsten accumulation in the core. The experiments employ for the first time in JET a hydrogen minority heating scheme (in majority deuterium) with first harmonic ICRH resonance on the low field side (LFS), close to the $q = 1$ surface on the outer mid-plane. Low field side resonance was employed many years ago in JET [5], but at high minority concentration (10–30%, thus operating in the margins of mode conversion) and more recently second harmonic was used to control sawteeth and avoid NTMs [6]. Such high concentration does not make for efficient ICRH and in addition we have shown that sawtooth control is highly efficient when the concentration is low [3]. Employing LFS resonance is particularly ITER relevant because the ICRH antenna design of ITER does not allow for high field side resonance at full magnetic field. Recent JET experiments reported in [2, 3] employed high field side (HFS) resonance and it was found that very close alignment of the RF resonance position and the inversion radius was required and in addition care needed to be taken with the choice of phasing of the antenna. In contrast, the new experiments reported here investigate to what extent sawteeth can be controlled with (1) less accurate alignment of the ICRH resonance and (2) with various antenna phasings including dipole. The first point here is important if sawtooth control can be achieved routinely without complex control algorithms such as are employed with real time control techniques [7]. The second point is important because sawtooth control on the high field side relies on phased (−90°) ICRH waves, which, relative to dipole phasing, has enhanced plasma-wall interaction via radio frequency-sheath redistribution effects [8]. ITER dedicated simulations [9] of ICRH and internal kink stability with fast ion effects have shown that in contrast to HFS experiments, LFS resonance should provide control with either +90° or −90° phasing. The fast ion populations modelled are first harmonic ICRH minority ions and neutral beam injected ions without second harmonic acceleration. The experiments reported here investigate this lack of sensitivity to the direction of the phasing, which clearly indicate that dipole waves should also provide access to control.

In addition to the avoidance of NTMs at low beta, frequent sawteeth have the advantage that impurities are flushed more regularly from the core of the plasma. While avoiding core impurity build up in an ITER fusion plasma is crucial, it is also very important in JET with the ITER-like wall, where radiative collapse or disruption sometimes occur, in part, when edge localized modes (ELMs) do not redistribute impurities sufficiently regularly and efficiently towards the edge. In addition, large intermittent ELMs are a source of impurities. Probably in line with localised electron cyclotron heating in ASDEX-Upgrade [10], it has recently been shown in JET [11] that centrally deposited ICRH can significantly enhance the plasma beta in the core, improve the general confinement and reduce the central peaking of the tungsten impurities. In the experiments reported here, we show to what extent the core is flushed of impurities when using ICRH on the LFS just inside the $q = 1$ surface. Of particular interest is to see whether impurities that are flushed out of the core by the sawtooth crash are inhibited from returning inside $q = 1$ because of the localised electron temperature gradient produced close to the the ICRH resonance location. Without this effective impurity barrier, impurities can return easily to the core during the sawtooth ramping phase. Clearly, ideally, $q = 1$ localised ICRH would simultaneously shorten sawteeth (and thus avoid NTMs) and prevent impurity build up in the $q < 1$ region.

The paper is organised as follows. In section 2 the LFS ICRH experimental configuration is described and low power L-mode experimental discharges are shown for conditions where short sawteeth are obtained and where long sawteeth trigger NTMs and mode locking. Section 3 examines the mechanism responsible for the control of sawteeth in these experiments. The effect of ICRH antenna phasing on sawtooth

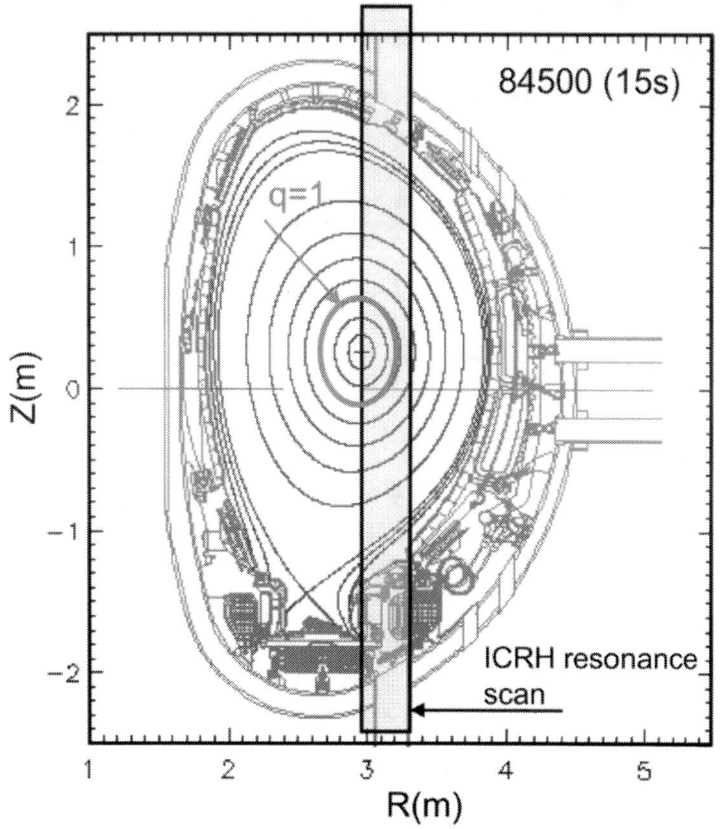

Figure 1. Showing the configuration of JET pulse 84 500 at 15 s (similar for all the pulses shown in this paper). The red curve indicates the q = 1 surface and the shaded box indicates the region over which the ICRH resonance is scanned over the pulse.

period is explored in the experiments and with simulations of the ICRH heating [12–14]. The ICRH distributions are used to estimate the ICRH current drive and corresponding magnetic shear and the effect of fast ions on the internal kink stability. In particular the HAGIS [15] code calculates the kinetic effects described in [2–4], which for this application, are the stabilising effects from NBI and ICRH trapped ions and the potentially destabilising effects of asymmetrically distributed passing ions with finite orbit widths. Section 4 explores high performance discharges in JET employing up to 19 MW of NBI heating in conjunction with up to 5 MW of LFS resonant ICRH with dipole phasing. The control of sawteeth in conjunction with impurity flushing is examined by moving the resonance through the q = 1 surface. The sawtooth impurity flushing is compared with full Lorenz orbit test particle simulations [16] in the evolving electromagnetic fields associated with Kadomtsev [17] reconnection. Conclusions are drawn in section 5.

2. Sawtooth control with resonance on the low field side

The experimental pulses described in this paper follow the approach of [2, 3] by sweeping the ICRH resonance across the q = 1 surface, usually from outside the q = 1 minor radius, to inside it. The difference from [2, 3] is that this is done for the first time on the LFS, with fundamental resonance and with minority hydrogen in majority deuterium (concentration around 3–5%), as opposed to ^{3}He minority with HFS resonance. Fundamental resonance at frequency 37 MHz was employed for all the L-mode pulses of this section. The configuration used in this paper is shown in figure 1, for the specific case of pulse 84 500 at 15 s in figure 2. The resonance position is swept through the q = 1 surface, as indicated in figure 2, by ramping magnetic field linearly from 2.72 T to 2.36 T. The q profile is maintained as constant as possible by ramping the toroidal current proportionally from 2.4 MA to 2.07 MA, with q_{95} = 3.45 and the sawtooth inversion major radius at around $R_{inv} \approx 3.28$ m. For the HFS pulses of [2, 3] it was shown that sawteeth could be controlled (shortened) by using −90° phasing over a narrow variation of the resonance position of about 3 or 4% of the plasma radius (about 3 or 4 centimetres). Figure 2 essentially extends figure 1 of [2] by again using −90° phasing, but sweeping this time the resonance position through q = 1 on the LFS. Once again, this ICRH resonance sweep is undertaken with a pre-existing NBI power of nearly 3 MW, which lengthens the sawtooth period to about 300 ms. It is seen that short sawteeth are obtained over a much wider range of the sweep, in this case around 20% of the plasma radius, which is clearly an advantage over the HFS configuration of [2, 3]. We now examine what happens when sawteeth are not controlled: as the resonance position approaches the magnetic axis, an n = 2 tearing mode is triggered and then locks (hence the vanishing n = 2 magnetics signal in figure 2). The mode causes

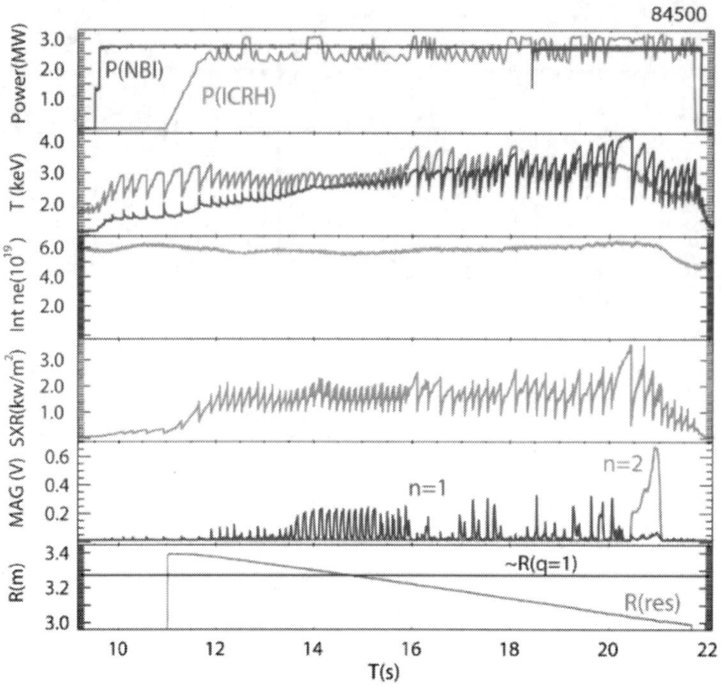

Figure 2. Showing JET pulse 84 500 which employed −90° phasing ICRH of around 3 MW and 2.8 MW of NBI, during a sweep in the toroidal field and toroidal current. Also shown are two (moving) ECE channels giving the temperature (the red T_e trace begins on the magnetic axis and ends off-axis and the blue T_e trace begins off-axis and end on-axis), the line averaged density, a vertical core aligned soft x-ray channel, magnetic signals associated with $n = 1$ and $n = 2$ modes, the ICRH resonance position and an indication of the $q = 1$ position. The toroidal magnetic field was ramped linearly down from 2.72 T to 2.36 T and in proportion the toroidal current was ramped from 2.4 MA to 2.07 MA.

reduction in plasma performance, though the plasma recovers after the auxiliary power is switched off. The mode was triggered at 20.4 s by an uncontrolled long sawtooth of 450 ms and at that moment the normalised beta was only $\beta_N = 0.5$, which is more than three times lower than the standard baseline scenarios envisaged for ITER [1]. The L-mode sawtoothing pulses described in this section and in section 3 employ more ICRH power than NBI power and as such the toroidal rotation velocity is weak. This could explain why NTMs are triggered so easily and this in turn implies that sawtooth control should be taken seriously in ITER plasmas where toroidal rotation will be relatively weak compared to present day tokamak plasmas.

It is perhaps surprising that −90° phasing is effective for sawtooth control on the LFS. The result is contrary to the LFS experiments of Bhatnagar *et al* [5] which showed that the Fisch [18] currents associated with −90° phasing on the LFS should reduce the magnetic shear and hence lengthen the sawtooth period, which indeed was observed. However, while Bhatnagar *et al* [5] operated at very high H concentration, our experiments operate at much lower concentration, where Fisch currents are weaker and the hydrogen minority ions are more energetic. As a result the fast ion orbit width effects described in [2, 4] can be expected to be the dominant mechanism for sawtooth control providing that the fast ion distribution is asymmetric in the parallel velocity. As will be seen in the modelling shown in next section, the fast ion distribution functions and in particular the asymmetry of the distribution functions, are quite similar for +90° and −90° phasing (at this low level of minority concentration). The asymmetries (which can be measured e.g. by the fast ion current density) are caused by the same finite orbit width and detrapping processes for both phasings and point to the possibility that dipole phased ICRH waves may produce similar fast ion distributions. This is indeed verified in the modelling shown in the next section, together with the new result that dipole phasing can also be used to control sawteeth. This last latter result is demonstrated empirically in figure 3 which shows two pulses where dipole ICRH is employed at about 2.5 MW in the absence of NBI. These pulses form the most basic sawtooth control configuration possible on JET. The pulses differ in that the large sweeps in the toroidal field, toroidal current and thus the resonance position, are in exact opposite directions. It is seen that in both cases sawteeth that are shorter than those during the Ohmic phase are obtained when the resonance position is close the inversion radius. It is noted that in pulse 85 932, the line integrated density was not controlled at the same level as 85 930. The enhanced density led to the reduced temperature observable in figure 3. Nevertheless, the characteristics of the sawteeth did not appear to be significantly affected and in particular, any hysteresis observable in figure 3 is weak.

3. Isolation of the mechanism of sawtooth control

We have now seen that sawteeth can be controlled with −90° phasing on the LFS, which is contrary to the results of [5] which employed high concentration hydrogen. It is clearly of interest to compare the effect of antenna phasing on sawtooth control and

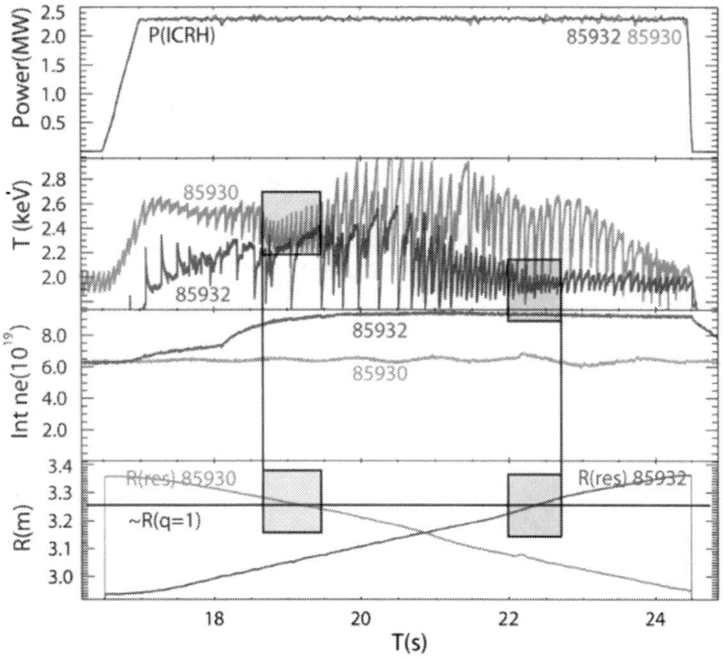

Figure 3. Showing two JET pulses employing dipole phased ICRH and no NBI power. Pulse 85 930 (red) ramped the toroidal magnetic field and toroidal current respectively down from 2.7 T to 2.33 T and from 2.4 MA to 2.1 A. Pulse 85 932 ramped in the exact opposite direction. The plot shows identical ICRH power, an ECE channel measuring the central temperature midway through the pulses (about 21 s), the line integrated density and the resonance positions relative to an approximate position of the measured inversion radius. Small sawteeth and corresponding resonance positions are indicated in shaded boxes.

also to model the equilibrium hydrogen minority fast ion population and the stability of the internal kink mode in order to isolate the mechanism for sawtooth control and thus to understand the empirical results. Figure 4 shows three discharges which differ only in the antenna phasing. One of the pulses, pulse 84 500 is the −90° phased pulse also of figure 2, pulse 84 497 employs +90° phasing and 84 616 employs dipole phasing. The pulses commence with nearly 3MW of NBI which lengthen the sawteeth to about 300 ms. As described earlier for figure 2, the toroidal field and toroidal current are ramped on application of about 3 MW of ICRH, so that the RF resonance position moves from outside to inside $q = 1$. Each of the pulses triggered NTMs by the longest saweeth near the end of the heating phase despite each having normalised beta $\beta_N \approx 0.5$. In pulse 84 497, the $n = 2$ mode locked, as indicated in figure 2, caused early rampdown of the ICRH and a step down in the NBI. The effect of the resonance position sweep on the sawteeth can be compared across the three phasings. It is seen that the effect on the sawteeth is similar regardless of the phasing, though the effect appears to be slightly stronger for +90° phasing: the window of short sawteeth is widest and at around 15 s the sawteeth are shorter than any of the sawteeth in the other two pulses. These observations, together with the surprising finding that the window for short sawteeth is quite wide for all phasings, will be investigated in this section.

In order to model the effect of ICRH on the stability of the internal kink mode, the equilibrium fast ion distribution function f_h of the minority hydrogen ions needs first to be established. This has been undertaken with the codes, SELFO [12], EVE [13] and SCENIC [14], each having their own advantages and deficiencies. Figure 5 compares the energy density $(2\pi)^{-1}\int d\theta\, d^3v\, f_h m_h v^2/2$ obtained for these three codes at 15.2 s in the −90° phasing pulse 84 497. The off-axis peak in each approximately coincides with the position of the resonance mapped on to the normalised minor radius ρ (where $\rho = 1$ marks the plasma edge). Agreement between the codes for this particular velocity moment is reasonably good and it will be seen that the stability properties, which depend on the full phase space properties associated with the distribution functions, also agree quite well across the codes.

The stability associated with the energetic ions can be established via hybrid kinetic-magnetohydrodynamic (MHD). The ideal internal kink mode, in the absence of two-fluid effects and resistivity (but including collisionless fast ion effects) is unstable for

$$\delta\hat{W} < 0, \qquad (1)$$

where the ideal growth rate normalised to the Alfvén frequency $\gamma_I / \omega_A = -\pi\delta\hat{W} / s_1$ with s_1 the magnetic shear at the $q = 1$ surface and $\delta\hat{W} = \delta W / (2\pi^2\xi_r^2 R_0 B_0^2 \epsilon_1^2)$. Here ϵ_1 is the local inverse aspect ratio at the $q = 1$ surface, B_0 and R_0 are the magnetic field and major radius at the magnetic axis and ξ_r is the leading order radial eigenfunction at the magnetic axis for the $m = n = 1$ mode. For certain conditions it has been argued

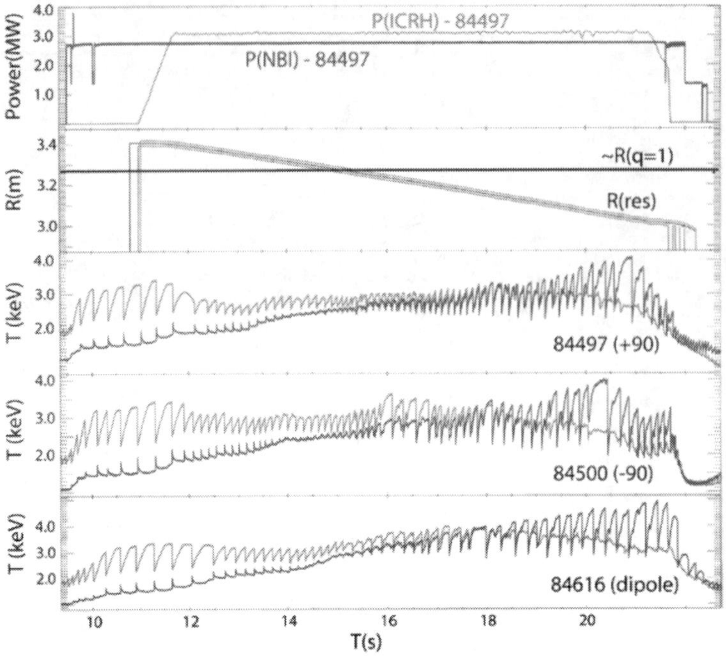

Figure 4. Showing three otherwise similar JET pulses: 84497 employs +90° phasing, 84500 employs −90° phasing and 84616 employs dipole phasing. Each of the pulses commence with almost 3 MW of centrally deposited NBI and then 3.2 MW or ICRH is applied during ramps in the toroidal magnetic field from 2.72 T to 2.36 T and in the toroidal current from 2.4 MA to 2.07 MA.

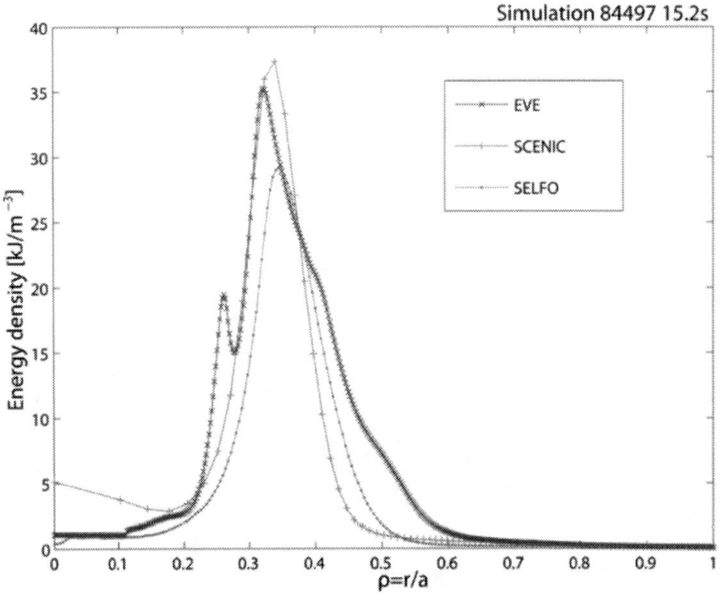

Figure 5. Comparing calculations of the energy density from the three codes EVE, SELFO and SCENIC for the +90° pulse 84497 at 15.2 s.

that resistive two fluid effects provide a stabilising effect, so that the ideal growth rate γ_I needs to be larger for instability and thus δW should be more negative. In this case, the instability condition can be written [19]

$$s_c < s_1 \quad \text{where} \quad s_c = C\delta\hat{W} \tag{2}$$

where C is a positive definite constant, the magnitude of which does not matter for what follows. We are clearly interested in what impact the ICRH ions will have on the internal kink stability criterion. One way in which the ICRH ions can modify equation (2) is through modifying s_1. A particular feature of high energy ICRH ion populations, generated by minority heating, is that they form a fairly strong local current perturbation, which if positioned correctly, can affect s_1 [20]. For ICRH ions to control sawteeth via this 'shear mechanism' we would expect the magnetic shear at

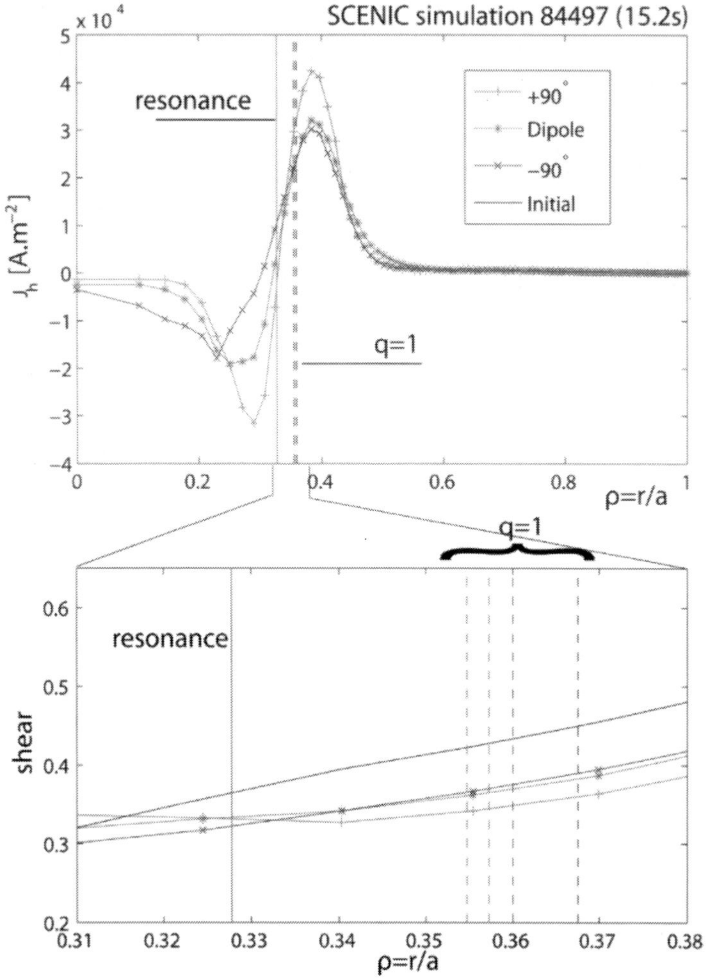

Figure 6. SCENIC calculations of the ICRH contribution to the current density and corresponding magnetic shear for pulses 84 497 (+90°), 84 500 (−90°) and 84 616 (dipole) each at 15.2 s. All take into account the plasma drag current [5, 20]. The shear profile in the absence of the fast ion currents is also shown.

$q = 1$ to be enhanced by the ICRH ions for all three phasings at e.g. 15 s for the pulses in figure 4. Modelling of the current produced by ICRH, including the plasma drag [5, 20], has been undertaken for these pulses at 15.2 s and is shown in figure 6. The dipole structure of these currents is similar for all three phasings, although the current is strongest and most localised for the +90° case. The fast ion induced current introduces changes to the equilibrium q profile, which we model by adding the fast ion current density to the initial current density (common to all three pulses) in the absence of fast ion current and then recalculate the toroidal equilibrium. Crucially, it is seen that these fast ion currents actually reduce the magnetic shear in the vicinity of the $q = 1$ surface, relative to the magnetic shear obtained in the equilibrium reconstruction obtained in the absence of the effect of the fast ions. It is thus seen that the 'shear mechanism' that has previously been used to interpret [5, 20] sawtooth control certainly cannot hold for the minority ion experiments reported here. According to equation (2) the shear effect is actually stabilising to the internal kink mode and thus the present experiments are even more contrary to the 'shear mechanism' of sawtooth control than the HFS experiments using ^{3}He minority in [3]. The latter experiments were designed to have vanishing shear perturbation from the fast ions. The present experiments and the experiments of [2, 3] can instead be understood through the effect of the fast ions on δW. We expect that where sawteeth are controlled, the ICRH ions will produce a strongly negative contribution to $\delta\hat{W}$, which if sufficiently large in amplitude (compared to contributions to δW from the bulk plasma and the NBI ions), would satisfy both the ideal instability criterion of equation (1) and the resistive two-fluid instability criterion of equation (2).

The hot ion contribution to δW is ([2] and references therein):

$$\delta W_h = -\frac{1}{2}\int \mathrm{d}^3x \boldsymbol{\xi}^* \cdot \mathcal{F}_h,$$

where $\boldsymbol{\xi}^*$ is the complex conjugate of the plasma displacement $\boldsymbol{\xi}$. The perturbed force

$$\mathcal{F}_h = -\nabla \cdot \underline{\underline{\delta P_h}}\,.$$

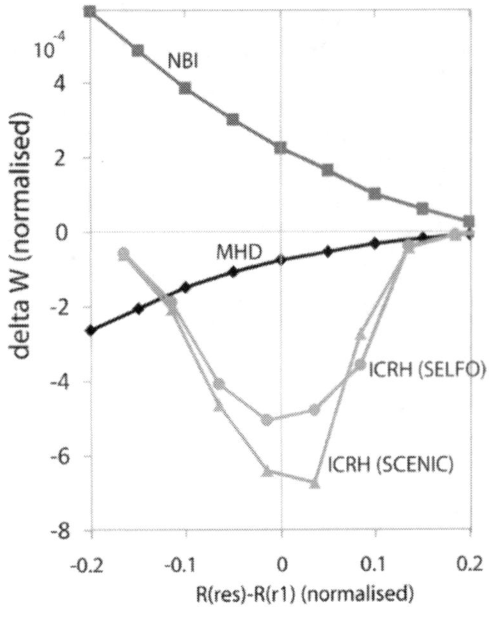

Figure 7. HAGIS simulation of $\delta\hat{W}$ for +90° pulse 84 497, plotted as a function of normalised resonance position relative to $q = 1$ radius $(R\ (res) - R(q = 1))/(R(r = a) - R_0)$. Also shown is the stabilising contribution from NBI ions and destabiling MHD contribution. ICRH markers are taken from both SELFO and SCENIC.

is thus seen to depend on the hot ion contribution to the pressure tensor $\underline{\underline{\delta P_h}}$, which is obtained by taking moments of the solution δf_h of the drift kinetic equation, i.e. the Vlasov equation in reduced phase space. The perturbation of the distribution function about the equilibrium distribution f_h can be written [21] in terms of perturbations of the three quantities that are conserved in the equilibrium state:

$$\delta f_h = -\delta P_\phi \frac{\partial f_h}{\partial P_\phi} - \delta\mu \frac{\partial f_h}{\partial \mu} - \delta\mathcal{E}\frac{\partial f_h}{\partial \mathcal{E}}, \tag{3}$$

where P_ϕ is the toroidal canonical momentum, $\mathcal{E}$ is the kinetic energy of a particle and μ is the magnetic moment. Figure 7 plots the $\delta\hat{W}$ contributions associated with the ICRH distribution function and the NBI distribution function for the +90° pulse 84 497 at 15.2 s. In this plot the MHD contribution to δW is evaluated with the MISHKA code [22] and the fast ion contributions are evaluated with HAGIS [15]. The latter code takes distributions from both SCENIC [14] and SELFO [12] and the eigenfunctions from MISHKA for varying $q = 1$ radius r_1. In HAGIS the mode frequency and growth rate is taken to be zero and the imaginary part of δW is neglected. The plots are shown as a function of the difference between the resonance location and the $q = 1$ radius and thus reflects the evolution of the pulse where the resonance position is swept through $q = 1$ from outside to inside. The δW contributions from SCENIC and SELFO compare well, with both showing a deep minimum in δW when the resonance position is aligned with or just outside $q = 1$. The strong destabilising contribution from ICRH is contrasted with the weak destabilising effect from MHD and the strong stabilising effect from NBI, where the NBI ion distribution has been obtained from the TRANSP [23] code. The NBI and MHD contributions vary in the plot because of their dependence on the $q = 1$ radius, though it is seen that this dependence is weak compared to that of the ICRH distribution.

As discussed in [2–4], the reason for the strong minimum in δW as the RF position passes close to the $q = 1$ surface has to do with the asymmetry in $v_\parallel$ of the equilibrium ICRH fast ion distribution function and in particular its radial variation. Passing ions with large radial drift excursion that intersect the $q = 1$ surface strongly affect stability. A net effect from a population occurs if the number of co and counter passing ions is not balanced, especially when the ratio of co and counter passing ions varies through the $q = 1$ surface. A measure of the latter is the radial gradient of the current associated with the fast ions d $J_h/\mathrm{d}r$ where $J_h = e\, Z_h \int \mathrm{d}^3v\, v_\parallel f_h$ and f_h is the fast ion distribution function associated with ICRH (without the plasma drag current subtracted), evaluated for example by SCENIC. In [4] it was shown that for the ICRH population, with resonance close to the $q = 1$ surface,

$$\delta W_h \propto -\frac{\mathrm{d}}{\mathrm{d}r} J_h\,(\text{passing})\,.$$

We can visualise the radial variation of the distribution function and thus understand the trend in δW exhibited in figure 7 by inspecting figure 8. The parallel current profile generated by passing ions (without the drag current subtracted) is shown for the dipole pulses 84 616 at 15.2 s and three positions are indicated in the profile: one at the minimum in J_h (line (a)), one at the resonance position where the current is small, (line (b)) and one at the peak in the current (line (c)). Also shown are contours of the distribution function from SCENIC labelled (*a*)–(*c*) according to the positions indicated in the current profile, plotted as a function of $v_\parallel$ and $v_\perp$ on the outboard midplane at these three radial positions. Both passing and trapped fractions (the trapped-passing boundary is indicated by the straight white lines) are shown to be asymmetric in $v_\parallel$ at the radial locations of the peak and trough of J_h.

Figure 9 compares the total δW, including the effects of ICRH ions, NBI ions and MHD, with the sawtooth period measured for the three pulses of figure 9. All are plotted with respect to $(R\ (\text{res}) - R(q = 1))/(R(r = a) - R_0) \approx [r\ (\text{res}) - r_1]/a$ with a the plasma minor radius. Note that $R(r = a) - R_0$ is about 88 cm and for the experimental data, the $q = 1$ radius is measured by the inversion radius for each sawtooth event. The ICRH distribution functions for the three phasings were evaluated with the SCENIC code. The general trends of the simulations are in reasonable agreement with the experimental data. For example the strongest destabilisation occurs for +90° phasing and the weakest for dipole phasing. All cases show a region of destabilisation (short sawteeth, or negative δW) which are centred about a value of resonance position R (res) a few centimetres outside the $q = 1$ radius. In addition the region of destabilisation extends to more negative values of R (res) $-R(q = 1)$ for +90° phasing.

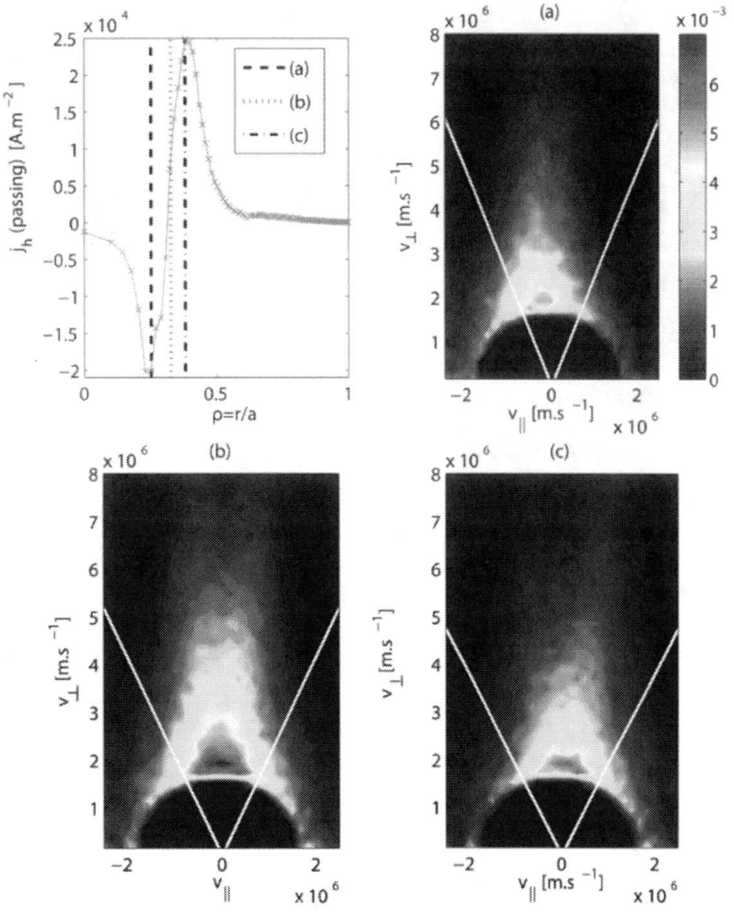

Figure 8. Fast passing ion current J_h (passing) for dipole pulse 84616 at 15.2 s and marked positions for contour plots of the distribution function from the SCENIC code. Distribution function plots undertaken on the outboard midplane show parallel velocity asymmeties are consistent with expectation from the local sign and amplitude of the current.

4. Sawtooth control and impurity flushing during higher performance pulses

The applicability of ICRH control of sawteeth on the low field side is routinely effective during L-mode plasmas at JET. However, H-mode plasmas with the new all-metal wall encompass new challenges that have to be handled at the same time as the sawtooth control. In order to prevent impurities from penetrating the core plasma and to avoid radiative collapse, the ELM frequency has to be sufficiently large soon after H-mode generation and during the rest of the pulse. In the plasmas described here, the ELM frequency was maintained at about 25 Hz by using a novel technique involving realtime dosing of the plasma [24]. The fuelling is increased in realtime if an ELM is not detected quickly enough after the previous ELM, or fueling is reduced if the next ELM comes too soon [24]. Figure 10 shows regular ELMs in pulse 85080 where this active ELM feedback was undertaken. The central SXR signal indicates nevertheless that impurities accumulate steadily in the plasma core within about two seconds after application of the heating. In this pulse the NBI power was held reasonably constant at about 19 MW, while dipole phased ICRH of 5 MW was swept through the $q = 1$ surface from outside to inside. When the resonance position passed a few centimetres inside the measured inversion radius the SXR signal reduced to the extent that a few consecutive sawteeth could barely be seen through variation of the central SXR channel displayed in figure 10. This pulse and all the other H-mode pulses referred to in this section employed dipole phasing using fundamental resonance with frequency 42 MHz. In 85080, the magnetic field was ramped down from 3.1 T to 2.7 T. The q profile is maintained constant by ramping the toroidal current proportionally from 2.8 MA to 2.45 MA, with $q_{95} = 3.45$.

In figure 10 it is seen that the addition of 5 MW of ICRH power to the pre-existing NBI power was offset by an enhancement to the radiative power of almost the same amount. The virtues of ICRH are observed as the resonance passes inside $q = 1$ when the central temperature increases by alomst 20% over one sawtooth cycle. This is particularly evident in figure 11(*f*) where temperature profiles for 85080 are plotted just before the sawtooth crash at 12.04 s (*a*) and just before the consecutive sawtooth crash at 12.43 s (*e*). Figure 11 also shows radiative power tomography [25], just before (*a*) and just after (*b*) the sawtooth crash at 12.04 s, as well as just before the next crash at 12.43 s. The extremely peaked

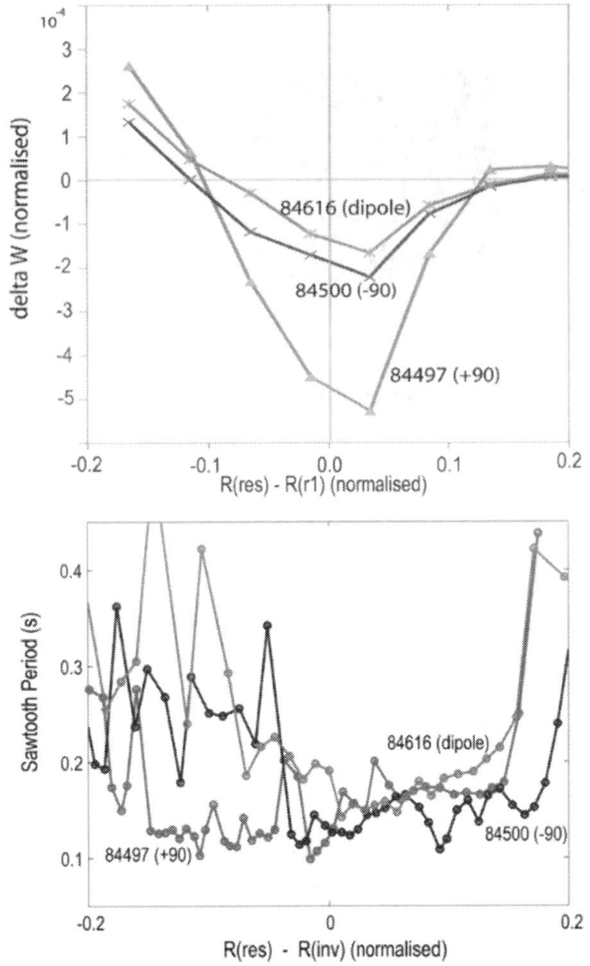

Figure 9. Showing simulations of the total δW as a function of $(R\ (\text{res}) - R(q=1))/(R(r=a) - R_0)$ for the three pulses 84497 (+90°), 84500 (−90°) and 84616 (dipole). These results are compared with experimental deduction of the sawtooth period, plotted as a function of the $(R\ (\text{res}) - R(q=1))/(R(r=a) - R_0)$ where $R(q=1)$ is measured by the inversion radius for each sawtooth event.

impurity population, clearly visible just prior to 12.04 s, is redistributed outside $q = 1$ by the sawtooth event, in agreement with recent 3d sawtooth cycle simulations [26]. The ICRH resonance position being a few centimetres inside $q = 1$ appears to prevent the impurities from re-entering the core; influx during the sawtooth ramp would ordinarily be expected to occur. It is assumed that the strong temperature gradient near $q = 1$ due to the local ICRH resonance introduces a strong inward impurity transport barrier [27]. It thus seen that a combination of sawtooth reconnection and localised thermal gradients is particularly effective, where the former ejects particles almost instantaneously (but intermittently) and the latter reduces the inward transport of impurities over the sawtooth quiescent phase. As seen in figure 10, when the resonance position moves close to the magnetic axis near the end of the pulse, the SXR signal grows again, partly because the slightly increased central temperature naturally enhances the central SXR signal, but also possibly because impurities are once again able to penetrate the $q = 1$ surface and reach the core during the sawtooth ramp phase.

The outward motion of impurity ions during the sawtooth crash has been modelled using the EBdyna_go code [16], which accounts for the full Lorenz motion of test particles in evolving electromagnetic fields. These fields are derived from the Kadomtsev model and it is found that the particles drift outwards during the crash, as shown in figure 11 for the simulations of pulse 85 080 just before (*c*) and after (*d*) the sawtooth crash at 12.04 s. The code does not model the enhanced anomalous transport due to strong temperature gradients [27], nor does it include centrifugal effects on the massive toroidally rotating tungsten ions, which are clearly visible in the experimental data via the enhanced SXR signals on the outboard midplane at all times.

The question still to be answered is whether experiments can be constructed that prevent both density accumulation during the sawtooth ramping phase and control the sawteeth in order to avoid unwanted secondary MHD events like NTMs, which can also further increase impurity fluxes to the core. It appears that this ought to be possible because the L-mode pulses of the previous section showed that the sawteeth can be controlled with the resonance position just inside $q = 1$, which we have seen is good location for preventing impurity build up. This was achieved in for pulse 85082, shown in figure 12, by starting with the resonance closer to $q = 1$ (than in 85 080) and slowing the ramp in the toroidal magnetic field and the toroidal current. As the resonance moves inside $q = 1$ it is seen that the central SXR strongly reduces (though for unknown reasons not as strongly as in 85 080) and two consecutive moderately shorter sawteeth are produced. It appears that in order to trigger short sawteeth in these high power pulses the resonance position has be more accurately placed than for the L-mode pulses described earlier. This is likely to be due to the fact that relative to the NBI power, the ICRH power is much lower in these high performance pulses and also that the effects of impurity accumulation and its dependence on both sawteeth and ICRH resonance position, clouds the ICRH sawtooth control mechanism described in the previous section. Nevertheless a statistical examination of eight consecutive high power pulses shows that sawteeth were controlled over a narrow range of resonance radius *R* (res). Figure 13 indicates that the sawtooth period was reduced by about 30%, with the optimal resonance position being about 6 centimetres inside the inversion radius *R* (inv), a position which appears also to be favourable for keeping impurities outside the $q = 1$ surface and thus reducing core radiation. Efforts also continue to fine tune the resonance position and to enhance the ICRH power, so that the sawteeth can be controlled as effectively as the L-mode pulses described earlier. It should be pointed out, however, that the pulses in figure 13 with the strongest effect of ICRH resonance position on the sawteeth were at the lowest NBI power (11–16 MW) and lowest ICRH power (2–3.5 MW). It has already been shown that sawtooth control works very successfully in low power L-mode, where in fact about 1 MW of ICRH power is capable of controlling the sawteeth. But, sawtooth control with ICRH appears to become much

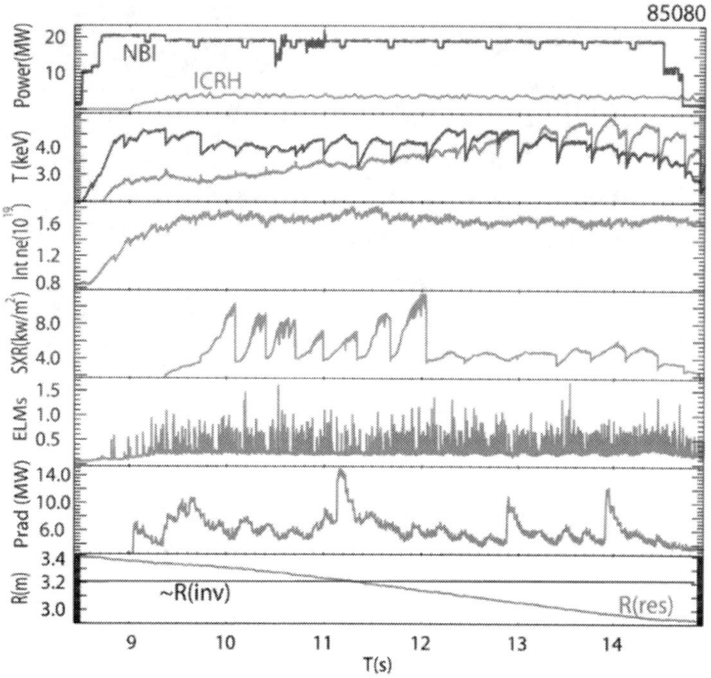

Figure 10. Showing high performance H-mode pulse 85 080, by plotting NBI and ICRH power, central electron temperature, line integrated density, central SXR, ELM fluctuations, radiation power and ICRH resonance position relative to $q = 1$.

more challenging deeply in H-mode and it is not yet clear, especially e.g. from the trends seen in figure 13, whether ICRH power much larger than the maximum in figure 13 would provide, for example, a reduction of the sawteeth by much more than the maximum observed in figure 13, i.e. about 30%.

5. Conclusions

Sawtooth control experiments have been conducted for the first time at JET employing low concentration minority heating with fundamental resonance in the low field side (LFS). This ITER relevant configuration has been shown to be favourable compared to high field side (HFS) sawtooth control experiments in at least two respects. First, in L-mode plasmas, the resonance position does not have to be placed very accurately, which ultimately means that realtime movement of the resonance position is required less than for the HFS experiments reported in [2, 3]. Secondly, it is found that the sawteeth can be controlled with various ICRH antenna phasings, which in practice enables sawtooth control techniques to be adopted in high performance H-mode pulses, where dipole phasing is preferred (due to less plasma wall interaction by RF sheaths [8]) in order to best protect plasma facing components and the all-metal wall. When sawteeth were allowed to be long in these L-mode pulses, NTMs and locked modes were triggered by the long sawteeth at very low normalised beta, $\beta_N \approx 0.5$. This is three times lower than the normalised beta envisaged for baseline scenarios in ITER. Similarly to ITER plasmas, the L-mode plasmas reported here have low toroidal plasma flow shear, prime conditions for mode coupling across neighbouring rational surfaces.

Dedicated internal kink stability simulations of these experiments show that all phasings of the ICRH are effective for sawtooth control, although it is found that +90° phasing is the most efficient phasing, in agreement with experiment. These stability calculations take ICRH marker distributions from the SELFO and SCENIC codes and it is once again [2, 3] found that fast ion orbit widths and parallel velocity asymmetry in the ion distribution function are probably responsible for the sawtooth control. In contrast, calculation of the fast ion driven current, indicates that when the ICRH resonance position is close to the $q = 1$ surface, the shear at $q = 1$ is actually reduced by the ICRH induced current for all phasings. This should be stabilising to the internal kink mode, which thus indicates that the traditional shear mechanism for sawtooth control [5, 20] is outweighed by the fast ion mechanism [2, 4], or perhaps some other mechanism that has not been discovered, nor the physics accounted for in simulations.

Sawteeth were routinely and easily controlled in L-mode plasmas, including plasmas near the L–H threshold, but achieving sawtooth control with ICRH in H-mode remains challenging. In order to produce frequent ELMs that redistribute impurities in the edge, and to avoid intermittent ELMs which sputter tungsten from the wall, H-mode plasmas employ high power NBI, typically more than 15 MW, often in conjunction with realtime ELM control techniques [24]. This five fold increase in NBI power (compared to the L-mode plasmas reported here), dwarves the ICRH power coupled in these ELMy plasmas and as a result the relative amplitude of the sawtooth control actuator is much reduced. Nevertheless

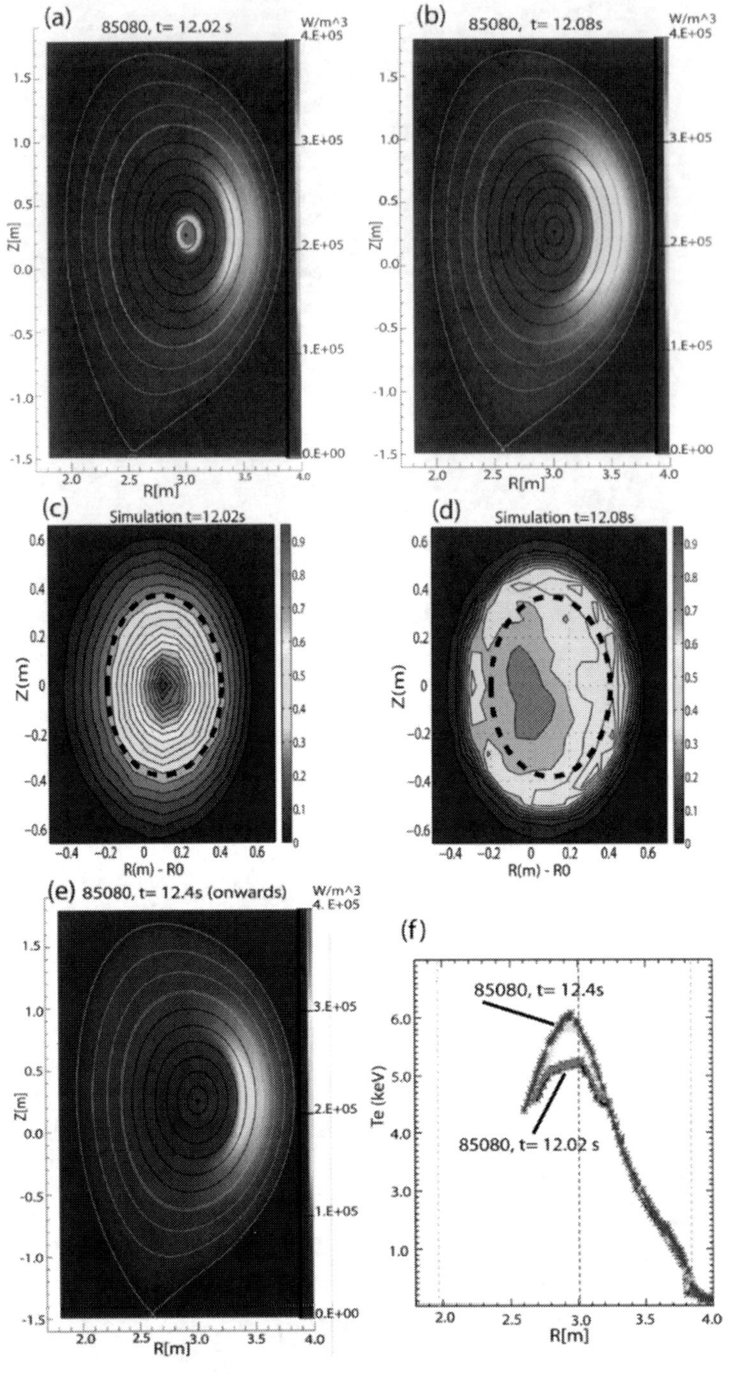

Figure 11. Showing radiative power tomography from 85 080, before (*a*) and after (*b*) the sawtooth crash at 12.04 s. These can be compared with simulations of impurity density at the same times (*c*) and (*d*). Also shown (*e*) is SXR tomography just prior to the next sawtooth crash at 12.43 s. The electron temperature profile is plotted in (*f*) at times just before the consecutive sawtooth crashes at 12.04 s and 12.43 s. The $q = 1$ surface lies just inside the cenfrifugally peaked impurities on the low field side and runs along the nearest black contour in the tomography figures of (*a*), (*b*) and (*e*).

this paper reports pulses where sawtooth control has been achieved in H-mode. It has been found that the resonance position must be closely constrained to be a few centimetres inside the inversion radius and that this ICRH location comes with the crucial advantage that impurities displaced outside $q = 1$ during the sawtooth crash appear to reinvade the core at a much lower extent (between consecutive sawtooth crashes) than when the resonance position is outside $q = 1$. The creation of an inward impurity transport barrier near the $q = 1$ surface is suggested to be consistent with the establishment of a strong electron temperature gradient [27] near the local ICRH resonance. Clearly the barrier must be positioned inside the sawtooth inversion radius, with the optimal position intuitively being just inside the inversion radius. Future experiments

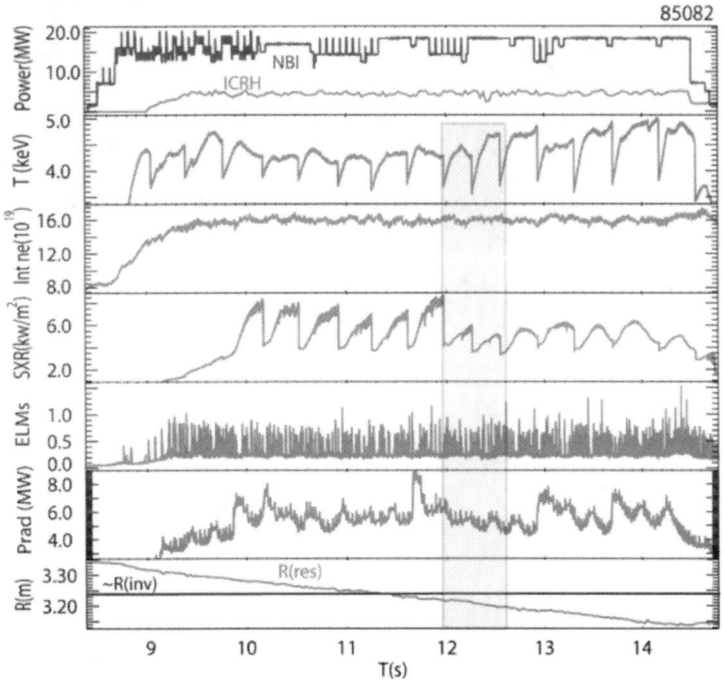

Figure 12. Showing 85 082, a similar pulse to 85 080, figure 10, but with a slower ramp, so that sawteeth were controlled when the resonance passed a few centimetres inside $q = 1$ and impurities were flushed from the core.

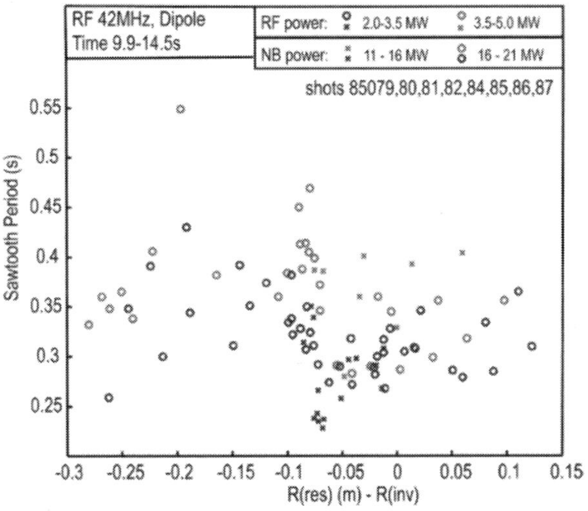

Figure 13. Showing sawtooth control over eight consecutive high power H-mode pulses. The sawtooth period is plotted as a function of R (res) – R (inv) during the decreasing ramps in magnetic field 3.1 T < Bt < 2.7 T and reducing toroidal plasma current 2.8 MA < Ip < 2.45 MA, with $q_{95} = 3.45$.

will concentrate on increasing the relative amplitude of the ICRH power in order to control the sawteeth more efficiently in H-mode, while at the same time assessing the efficiency of the local ICRH resonance for creating an impurity barrier in the core and for enhancing core temperature. In addition realtime sawtooth control techniques [7] will be developed in these LFS experiments. Such techniques will have the greatest application in H-mode pulses where the resonance position must be positioned accurately through the discharge.

Acknowledgments

This work, supported by the Swiss National Science Foundation and by the European Communities under contract of Association between EURATOM and Confédération Suisse, was carried out within the framework of the European Fusion Development Agreement. The views and opinions expressed herein do not necessarily reflect those of the European Commission. SCENIC simulations were performed on BlueGene cluster at EPFL, Lausanne (CH), on ROSA XE6 cluster at CSCS, Manno (CH) and on Helios at CSC, Rokkasho (Japan).

References

[1] Chapman I T *et al* 2010 *Nucl. Fusion* **50** 102001
[2] Graves J P *et al* 2012 *Nat. Commun.* **3** 624
[3] Graves J P *et al* 2010 *Nucl. Fusion* **50** 052002
[4] Graves J P *et al* 2009 *Phys. Rev. Lett.* **102** 065005
[5] Bhatnagar V P *et al* 1994 *Nucl. Fusion* **34** 1579
[6] Sauter O *et al* 2002 *Phys. Rev. Lett.* **88** 105001
[7] Lennholm M *et al* 2011 *Nucl. Fusion* **51** 073032
[8] Czarnecka A *et al* 2011 *Plasma Phys. Control. Fusion* **53** 035009
[9] Chapman I T *et al* 2011 *Plasma Phys. Control. Fusion* **53** 124003
[10] NEU R *et al* 2002 *Plasma Phys. Control. Fusion* **44** 811
[11] Goniche M *et al* 2014 *Proc. 41st EPS Conf. Plasma Phys. (Berlin, Germany, 23rd–27th June 2014)* O4.129 (http://ocs.ciemat.es/EPS2014PAP/pdf/O4.129.pdf)

[12] Hedin J *et al* 2002 *Nucl. Fusion* **42** 527
[13] Dumont R and Zarzoso D 2013 *Nucl. Fusion* **53** 013002
[14] Jucker M *et al* 2011 *Plasma Phys. Control. Fusion* **53** 054010
[15] Pinches S D 1998 *Comput. Phys. Commun.* **111** 133
[16] Jaulmes F *et al* 2014 *Nucl. Fusion* **54** 104013
[17] Kadomtsev B B 1975 *Sov. J. Plasma. Phys.* **1** 389
[18] Fisch N J *et al* 1987 *Rev. Mod. Phys.* **59** 175
[19] Porcelli F *et al* 1996 *Plasma Phys. Control. Fusion* **38** 2163
[20] Eriksson L-G *et al* 2006 *Nucl. Fusion* **46** S951
[21] Helander P *et al* 1997 *Phys. Plasmas* **4** 2181
[22] Chapman I T *et al* 2006 *Phys. Plasmas* **13** 062511
[23] Budny R V 1982 *Nucl. Fusion* **32** 429
[24] Lennhom M *et al* 2014 *Proc. 25th IAEA Fusion Energy Conf. (St Petersburg, Russia, 13–18 October 2014)*
[25] Pütterich T *et al* 2013 *Plasma Phys. Control. Fusion* **55** 124036
[26] Nicolas T *et al* 2014 *Phys. Plasmas* **21** 012507
[27] Valisa M *et al* 2011 *Nuc. Fusion* **51** 033002
[28] Romanelli F *et al* 2012 *Proc. of the 24th IAEA Fusion Energy Conf.* (*San Diego, USA, 8–13 October 2012*)

IOP Publishing Plasma Physics and Controlled Fusion

Plasma Phys. Control. Fusion **57** (2015) 014034 (7pp) doi:10.1088/0741-3335/57/1/014034

Explosive reconnection of the double tearing mode in relativistic plasmas with application to the Crab nebula

J Pétri[1], M Takamoto[2], H Baty[1] and S Zenitani[3]

[1] Observatoire Astronomique de Strasbourg, Université de Strasbourg, CNRS, UMR 7550, 11 rue de l'Université, F-67000 Strasbourg, France
[2] Max-Planck Institute for Nuclear Physics, Saupfercheckweg 1, D-69117 Heidelberg, Germany
[3] National Astronomical Observatory of Japan, 2-21-1 Osawa, Mitaka, Tokyo 181-8588, Japan

E-mail: jerome.petri@astro.unistra.fr

Received 26 June 2014, revised 13 September 2014
Accepted for publication 9 October 2014
Published 28 November 2014

Abstract

The Crab pulsar and its surrounding nebula is a well-known relic of a massive star that exploded in 1054 AD. The Crab nebula was generally believed to be a good standard candle in gamma rays. Recently, this view has been challenged by sudden increases in the gamma-ray flux in a narrow spectral band within a few hundred MeV. These flares are short but powerful; their duration is between a few hours and up to several days with a rising/falling time of a few hours/days. To date it is neither clear what mechanism powers these flares nor where exactly in the nebula they should be located. However, recent models seem to favor emission sites inside the nebula. In the present work, we study the magneto-hydrodynamic tearing instability occurring in a double current sheet configuration with application to the Crab flares. This is investigated by means of resistive relativistic magneto-hydrodynamic simulations. These put some constraints on the maximum Lorentz factor of the striped wind, $\Gamma \lesssim 150$ and on the localization of the emission region, $r \approx 50\, r_L$ where $r_L = c/\Omega$ is the light-cylinder radius, c is the speed of light and Ω is the rotation speed of the pulsar. Sites close to but outside the light-cylinder are favored in our model.

Keywords: plasma, reconnection, relativity, astrophysics, high-energy, gamma-rays

(Some figures may appear in colour only in the online journal)

1. Introduction

The Crab pulsar and its surrounding nebula is a well-known relic of a massive star that exploded in 1054 AD and was already reported by the Chinese at that time. Current wisdom explains the nebula by the formation of a highly magnetized rotating neutron star, a so-called pulsar, located in the center of the nebula and feeding it with a supermagnetosonic wind made of ultra-relativistic electrons and positrons pairs. These leptons, created in the inner part of the pulsar's magnetosphere, are responsible for the bright emission emanating from the Crab pulsar and nebula, seen through the whole electromagnetic spectrum, from radio through optical/x-rays, up to gamma-rays and very high-energies. In addition, several time scales can be associated with the Crab, from the shortest being the pulsar's rotational period of 33 ms up to the secular evolution of the nebula occurring on the year time scale. In between, the recent flares seen in gamma-rays by Fermi/LAT (Abdo *et al* 2011, Buehler *et al* 2012, Striani *et al* 2013) and lasting for about a day came as an astonishing event. Indeed, so far the Crab nebula was believed to be a good standard candle in gamma rays. This view has been challenged by sudden increases in the gamma-ray flux in a narrow spectral band within a few hundred MeV. These flares are short but powerful; their duration is between a few hours and up to several days and with a rising/falling time of a few hours/days. The duration of these flares is on one hand too large to be associated with the rotation of the neutron star, and on the other hand too short for typical nebula dynamical time scales. This short duration points towards a small emitting region (compared to the nebula), therefore associated with an efficient particle accelerator. To date it is neither clear what mechanism powers these flares nor where exactly in the nebula (or maybe in the wind) they should be located. However, recent models seem to favor emission sites inside the nebula.

Relativistic magnetic reconnection is usually believed to be an efficient mechanism to explain these flares (Clausen-Brown and Lyutikov 2012). The fundamental models of magnetic reconnection in the non-relativistic case have been described by Parker (1957), Sweet (1958) and Petschek (1964). It is currently admitted that magnetic reconnection is responsible for non-thermal particle acceleration in Earth's magnetosphere leading to substorms as well as some activity in the solar corona and wind. The presence of current sheets in many astrophysical plasmas drives the process of magnetic reconnection. For instance, magnetic dissipation occurs in gamma-ray bursts (GRB), active galactic nuclei (AGN) jets, neutron star magnetospheres, and winds. In such systems, due to the high magnetization, the plasma becomes relativistic, i.e. its thermal energy exceeds its rest mass energy and the Alfvén velocity approaches light speed. Lyubarsky (2005) extended the Sweet-Parker and Petschek reconnection solutions to the relativistic regime.

The tearing instability of current sheets is the simplest way of triggering magnetic reconnection. The latter instability was first investigated by Lyutikov (2003) in the magnetodynamics approach of a relativistic and highly magnetized plasma. This force-free magnetic field is filled with a transverse guiding field to achieve an equilibrium. This linear study has been confirmed and extended by the non-linear numerical simulations performed by Komissarov *et al* (2007).

In this paper, we extend these results occurring in a single current sheet to a double current sheet configuration, where the guided field is replaced by plasma pressure, removing the force-free assumption and with potential application to the pulsar striped wind. This is investigated by means of resistive relativistic magneto-hydrodynamic simulations (RRMHD). In a preliminary study, Baty *et al* (2013) have shown that after an early linear growth followed by a saturation state of the magnetic islands, a fast and energetic secondary stage of the double tearing mode (abbreviated as DTM in the remainder of the paper) is non-linearly driven. In this paper we explore reconnection regimes more appropriate to the physical conditions in the pulsar striped wind. We also discuss the possible role of such explosive reconnection dynamics to explain the MeV flares observed in the Crab nebula (Buehler *et al* 2012). The observed time scale of the rising and falling part of the flares gives us constraints on the Lorentz factor of the wind and on the location of the emission region.

2. Overview of the Crab

2.1. The pulsar and Its nebula

The Crab pulsar is a neutron star with a radius of around 12 km, rotating at a period of 33 ms and slowly breaking with a spin-down luminosity of $\dot{E} \approx 4 \times 10^{31}$ W. Its surface magnetic field is about 4×10^8 T. It was formed in a supernova explosion already reported by the Chinese in 1054. The observed coherent radio emission is supposed to come from its surface, whereas the high-energy counterpart through curvature, synchrotron, and inverse Compton radiation at higher altitudes in the magnetosphere. Quantum electrodynamical processes such as pair creation/annihilation and quantum synchro-curvature photon emission are the engines of its main activity. Its pair multiplicity factor is unknown but is expected to be as high as $\kappa \approx 10^4$–10^6. Remember that κ is defined by the number of secondary electron-positron pairs produced in a magnetic polar cap cascade by a single primary electron or positron. It is a measure of the efficiency of the leptonic production rate in pulsar magnetospheres. High efficiency implies high pair multiplicity, $\kappa \gg 1$. The pulsar launches an ultra-relativistic magnetized flow via a cold magnetohydrodynamic (MHD) wind feeding its surrounding nebula with fresh relativistic particles indirectly detected through their synchrotron and inverse Compton emission. The cold MHD wind is stopped and heated by crossing the termination shock. It is unclear how the magnetization of this wind could be converted into particle acceleration at the shock and/or before the shock. Even in this case, magnetic reconnection seems to play a central role. In summary, the rotational kinetic energy of the neutron star is converted into relativistic pairs through a highly magnetized wind and eventually into a synchrotron/inverse Compton nebula at large distances from the pulsar.

2.2. Gamma-ray flares

Until the recent discovery of the Crab nebula flares, it was believed to be a steady state emitter. The variability of these flares ranging from a few hours to a few days is on one hand too long to be immediately connected to the pulsar activity with period of 33 ms, and on the other hand too short to be attributed to the secular evolution of the nebula itself. These facts put severe constraints on the possible models.

Several scenarios have been proposed to explain them. On one hand, Lyutikov *et al* (2012) claimed that an oblique shock subject to magnetosonic wave perturbations from the downstream flow leads to a time-dependent mildly relativistic post-shock flow producing synchrotron emission. Lyubarsky (2012) attributed the flares to some kink instability in the polar region of the pulsar wind where dissipation is irrelevant. The flow remains relativistic and magnetized even after crossing the shock. On the other hand, many models rely on relativistic magnetic reconnection, but differ mainly on the location where to trigger it. For instance, Clausen-Brown and Lyutikov (2012) invoked magnetic reconnection in the nebula producing relativistic mini-jets sometimes beaming towards the observer and detected as sporadic flares. Cerutti *et al* (2012) investigated in detail test particle acceleration in a magnetic reconnection layer, including the radiation reaction from synchrotron emission.

In our model (Baty *et al* 2013), magnetic energy release due to reconnection takes place in the unshocked wind, the so-called striped wind, well within the nebula but still outside

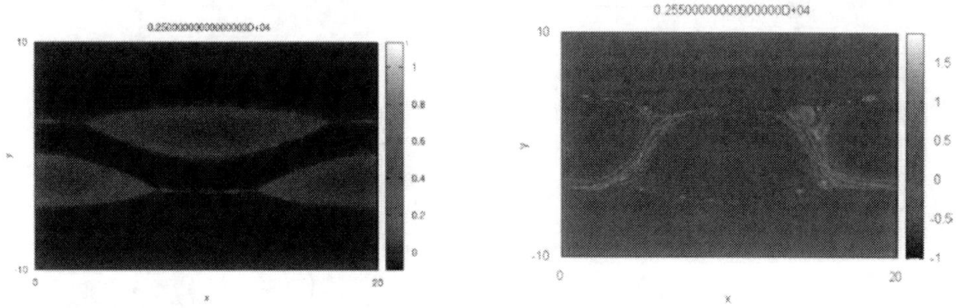

Figure 1. Snapshots of the double tearing mode evolution around the maximum of the instability, at $t = 2500$ on the left and $t = 2550$ on the right. The color scale corresponds to the temperature in the logarithmic scale. The parameters are $\sigma = 12$, $S = 3200$ and $y_0 = 3\ L$.

the pulsar's magnetosphere. The overall implications of the striped pulsar wind can be found in Kirk *et al* (2009). This is a competing model to explain pulsed high-energy emission from pulsars, especially the optical polarization of the striped wind (Pétri and Kirk 2005) and the gamma-ray pulsars (Petri 2012).

3. Explosive reconnection simulations

The electromagnetic structure of the pulsar striped wind shows many magnetic polarity reversals, i.e. underlying current sheets that we hypothesize to be at the origin of the reconnection process. In an MHD model, reconnection requires a dissipation parameter represented by some plasma resistivity generally called η. This resistivity is probably of an anomalous type instead of a collisional origin in the striped wind scenario. It is difficult to put physical constraints on this resistivity as we do not know the particle density number and the effect of the turbulent flow in the stripes. However, our approach constitutes a lowest-order approximation for the collisionless plasma and the starting point of our study. Nevertheless MHD can already give a good insight into the physics of collisionless plasma, as proven by the solar wind for example. We emphasize that we focus on a local description of the dynamics, neglecting the radial dependence (in r^{-1}) of the toroidal magnetic component. We restrict ourselves to the equatorial plane where the distance between the successive current layers is constant.

3.1. Numerical setup

As in our previous work (Baty *et al* 2013), we start with a double current sheet configuration in MHD equilibrium in a two-dimensional cartesian geometry. The magnetic field structure is unidirectional, along the x-axis. It follows the double Harris sheet according to a tangent hyperbolic variation:

$$B_x = B_0\left[1 + \tanh\left(\frac{y - y_0}{L}\right) - \tanh\left(\frac{y + y_0}{L}\right)\right] \qquad (1)$$

where B_0 is the magnitude of the magnetic field, L is the half-thickness of one current sheet, and $2\ y_0$ is the separation between both current sheets. The temperature being uniform in the simulation box, the density profile is separated into two components. One component ensures the equilibrium of the plasma. The other component is a constant background value ρ_b and determines the magnetization $\sigma = B^2/\mu_0\ \rho\ h$, where $h = c^2 + \gamma\ p/(\rho\ (\gamma - 1))$ is the specific enthalpy, γ is the adiabatic index, and p is the pressure. The simulations are started by adding a small appropriate magnetic perturbation.

The simulations are performed with the resistive relativistic MHD code described by Takamoto and Inoue (2011), which can deal with higher magnetization and lower resistivity, and using a rectangular box of fixed size $L_x \times L_y$. The boundary conditions along the longitudinal x direction are periodic and free in the transversal y direction. Most of the simulations use a grid resolution of 2048^2. Based on a classical MHD analysis of the linear double tearing mode by Otto and Birk (1992), the ratio L_x/L is chosen in order to get an unstable mode close to the fastest one. We set $L_x/L = 20$ and $y_0 = 3\ L$. The ratio L_y/L is also equal to 20 in order to have a boundary sufficiently far away from the center to minimize its influence on the dynamics. Our normalizations are $B_0 = \sqrt{2}$, $c = 1$, $L = 1$, $\mu_0 = 1$, $T = 1$.

The mildly relativistic, mildly magnetized, and relatively low Lundquist number, $S = c\ L/\eta$, plasma regime investigated by Baty *et al* (2013) is extended to higher values for Lorentz factors reaching $\Gamma = 11$, high magnetization $\sigma = 120$ and $S = 3200$. These values are believed to be representative of pulsar winds. Note that in our definition of the Lundquist number, we use the speed of light and not the Alfvèn speed, as the latter is close to the former in the ultra-relativistic regime. Moreover we normalize S by using the current sheet width L as a characteristic length. This is the conventional normalization employed in tearing mode analysis (Pucci and Velli 2014).

3.2. Results

The time evolution of a typical run with a high magnetization $\sigma = 12$ is shown in figure 1. The plasma always exhibits four distinct phases; three growing ones, and a last relaxation one.

The first stage corresponds to the early growth of the double tearing mode, which manifests as two islands located at the two current sheets, in our case at $y_0 = \pm\ 3\ L$. The observed mode corresponds to an antisymmetric pattern in agreement with the linear theory. This stage is characterized by a Sweet-Parker-like process developing on a time scale $\sim S^{1/2}$

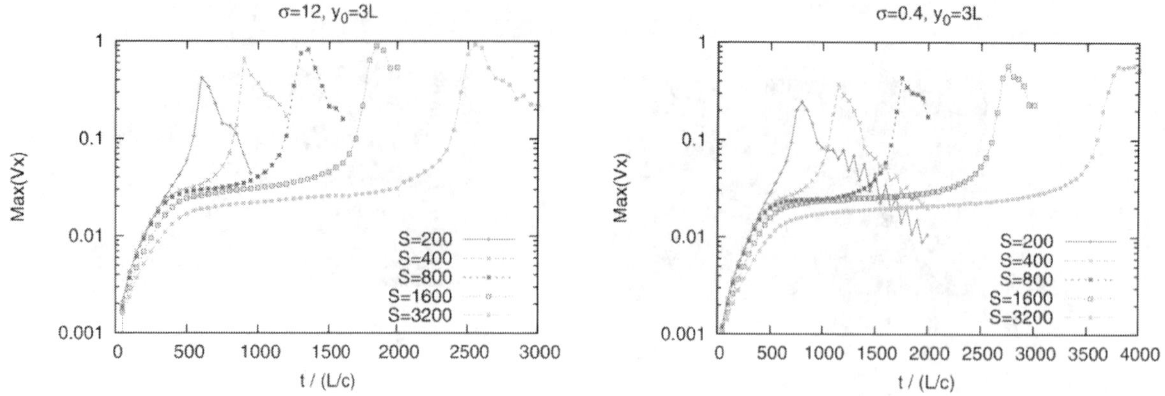

Figure 2. Maximum flow velocity V_x as a function of time for different Lundquist numbers: $S = 200, 400, 800, 1600, 3200$. A mild magnetized case is shown on the right in figure $\sigma = 0.4$ and a high magnetization case on the left in figure $\sigma = 12$.

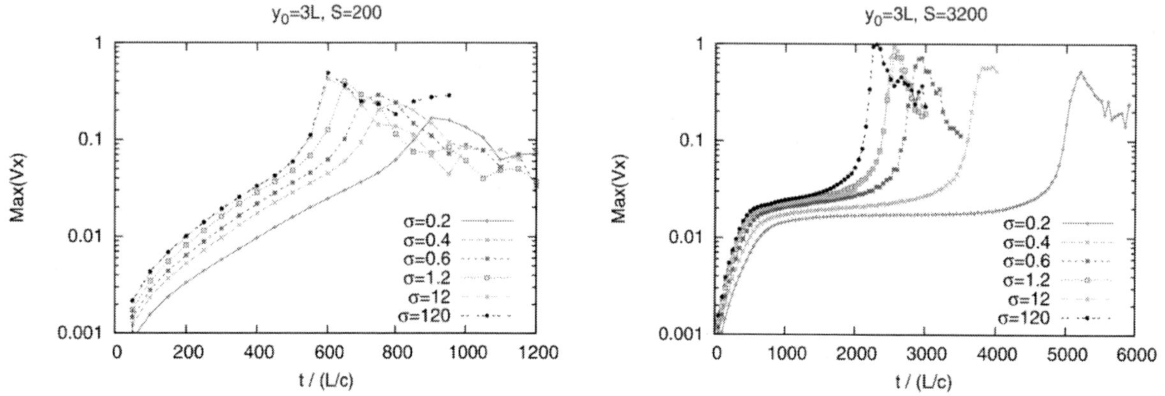

Figure 3. Maximum flow velocity V_x as a function of time for runs using different magnetizations: σ. The Lundquist number is $S = 200$ on the left in figure and $S = 3200$ on the right in figure.

(see also figure 2). In this figure, as a diagnostic, we plot the maximum of the flow velocity V_x as a measure of the magnitude of the instability. Usually, in DTM investigations, the total kinetic energy is reported to account for the diagnostic. Here we use a simplified estimate that should not be confused with the reconnection rate. See below for an investigation of the influence of the Lundquist number for several S.

A second phase of saturation then ensues on a slower time scale, which can be fitted with a scaling close to a purely resistive dependence $\sim S$. This later phase is equivalent to the well-known Rutherford regime of classical MHD (Wang *et al* 2007), in which the magnetic islands continue to grow on a non-linear algebraic resistive timescale (White 1986).

In a third phase, the two islands grow in an explosive way. During this phase, the islands interact strongly and form a triangular deformation (Janvier *et al* 2011b). This structural deformation drives a magnetic reconnection event characterized by a merging between the two islands at a time around the peak of V_x, as shown in figures 2 and 3. This mechanism has been observed in numerical simulations done by Janvier *et al* (2011a) using a non-relativistic simplified MHD model. They found a threshold for triggering this explosive phase. More precisely this threshold is $L_x/y_0 \gtrsim 6$ for fixed boundary conditions. Note that a critical value closer to 4 has also been reported by Yan *et al* (1994), but in the case of periodic boundary conditions. In our simulations we found a value of around 5. The underlying physical mechanism is driven by an imbalance in the total pressure around the islands (Janvier *et al* 2011a).

In the left panel in figure 1, the island originating from the upper layer moves towards the X-point of the second island from the lower layer. Meanwhile this X-point is transformed into a current layer exhibiting many small-scale secondary islands. These islands can be interpreted as plasmoids because the local Lundquist number, defined by $S_l = c\ l/\eta \approx 3 \times 10^4$, where l is the length of the instantaneous current layer (the one giving rise to the abovementioned plasmoids). The length l is estimated to be about 10, obtained from an inspection of the left panel in figure 1, at the time $t = 2500$. Indeed, this local Lundquist number S_l exceeds the critical value of order 10^4, which is necessary to produce the plasmoid instability (Takamoto 2013).

In the fourth and final phase, the magnetic configuration relaxes toward a new stable state. This final state is free from magnetic island and current layer, and the corresponding magnetic field lines become nearly straight and unidirectional.

During the whole process, the magnetic energy is converted into kinetic bulk flow and thermal energy, in roughly equal proportion.

3.3. Scaling with S and σ

We inspect the time scale involved during the explosive growth (phase 3). In the low magnetization regime $\sigma \lesssim 1$, we confirm that it weakly depends on the resistivity with the scaling close to $S^{0.2-0.3}$. Our simulations show a behavior similar to what is expected from non-relativistic investigations (Janvier *et al* 2011a, 2011b, Zhang and Ma 2011). Within the high magnetization limit, the scaling shows a transition to an even weaker S^{α} dependence close to zero $\alpha \approx 0$.

In the same way, we can infer the dependence of the time scales on different magnetizations σ and the given Lundquist number S by inspecting figure 3 for $S = 200$ and for $S = 3200$. The results show a dependence with a scaling according to $\sigma^{0.3}$ for $\sigma \lesssim 1$, in agreement with our previous results. However, for high magnetization $\sigma \gg 1$, it becomes nearly independent of σ.

For low resistivity such as $S \gtrsim 1000$ and strong magnetizations $\sigma \gtrsim 1$, relativistic flows are easily seen. These simulations unambiguously show that the outflow becomes relativistic with $V_x \lesssim c$.

The triggering of the explosive stage depends on the size of the perturbation L_p with respect to the current sheet separation y_0. We found that the elapsed time between the beginning of the computational run and the start of the DTM increases when the ratio L_p/y_0 decreases.

4. Application to the Crab

From the above performed simulations, we have seen that the explosive character of the magnetic reconnection is induced by the double current sheet layer. These structures are believed to exist in the pulsar striped wind. We therefore expect that this mechanism quickly releases a substantial fraction of the energy flux entailed in the Poynting dominated flow, and we remember that it is believed that close to the neutron star the flow is strongly magnetized, $\sigma \gg 1$. This scenario differs from other models because in our case reconnection is activated in the unshocked striped wind before entering the termination shock. The reconnection process creates a radially sheared flow producing internal shocks as in some gamma-ray bursts (GRB) models. A substantial fraction of the pulsar spin-down energy could be released by a clump of matter moving outwards at relativistic speed and supplemented with fresh particles from the inner part of the wind, actually the pulsar magnetosphere. Our suggested location of the flares in the pulsar wind, close to the light-cylinder, makes sense because (Petri 2012) already showed that the synchrotron emission in these stripes can explain the pulsed GeV photons seen by Fermi/LAT, and most importantly they are at the right luminosity level. The estimates about the wind properties and the energetics of the flares made below are fully consistent with such a picture. On a larger scale, in principle, our model could also be applied to the reconnection sites located in the pulsar wind nebulae (PWNe) because our MHD description is scale-free. The essential requirement is that the stripes survive after crossing the termination shock. As reconnection scenarios in PWNe have already been discussed by other groups, and because (Pétri and Lyubarsky 2007) showed that for the Crab pulsar, magnetic dissipation at the termination shock is extremely efficient and destroys the stripes, we do not focus on this possibly in the present study.

Based on Baty *et al* (2013), we follow the same procedure to estimate the orders of magnitude of the rising time of the explosive reconnection phase. The only difference is that this phase now lasts around 300 units of time for the parameters $\sigma = 120$ and $S = 3200$. These parameters are indeed relevant for the pulsar wind, which should be strongly magnetized but weakly resistive. Remember that in the striped wind interpretation, the separation between the two current sheets, $2\ y_0$, is equal to half a wavelength of the wind, corresponding to $\lambda_L \approx 2\,\pi\,r_L$ where $r_L = c\ P/2\ \pi$ is the light-cylinder radius, and P is the period of the pulsar. Note also that we assume that reconnection occurs in the equatorial plane, where the current sheets are equidistant. According to Lorentz-time dilation and the length contraction between the wind frame and the observer frame, the duration of the explosive rise as seen by a distant observer is:

$$\Delta T \approx 50\Gamma^2 P \tag{2}$$

which is essentially the same as in Baty *et al* (2013). However, here we took $y_0 = 3\ L$ from the above simulations, Γ being the Lorentz factor of the wind. Applied to the Crab with a period $P = 33$ ms and a typical rising time of about $\tau_r \approx 10$ h (Striani *et al* 2013), the duration of the explosive time should be less than $\Delta T \lesssim \tau_r$ h. We therefore infer a maximum Lorentz factor of $\Gamma \lesssim 150$, in agreement with our previous results. This Lorentz factor is consistent with the magnitude expected at the base of the wind, close but outside the light-cylinder, where from independent models of pulsed optical polarization we know that $\Gamma \approx 20$–50 (Pétri and Kirk 2005). As a consequence, explosive reconnection occurs outside the light-cylinder, but not necessarily very far away from the pulsar, well within the radius of the termination shock.

The plasmoids move relativistically, dragged by the bulk motion of the wind. Doppler beaming and frequency shifts are therefore important. The peak energy observed at 380 MeV is thus only around 1–2 MeV in the fluid frame and consistent with the synchrotron uppermost limit at about 236 MeV imposed by the radiation reaction limit. Relativistic beaming is favored by the correlation between the peak energy of the flare and its flux as discussed by Buehler *et al* (2012), which is especially true for the April 2011 flare. For a given flare, applying the Doppler boosting, the luminosity in the observer frame is:

$$L = \mathcal{D}^4 L' \tag{3}$$

where L' is the luminosity in the wind frame and $\mathcal{D} = 1/\Gamma(1-\beta\cos\vartheta)$ is the Doppler factor. ϑ is the angle between the velocity $\vec{v}$ of the moving plasma blob and the direction $\vec{n}$ of the line of sight, thus $\beta\cos\vartheta = \vec{\beta}\cdot\vec{n}$. The cutoff energy in the wind frame ε'_c and in the observer frame ε_c are related by $\varepsilon_c = \mathcal{D}\varepsilon_c'$, and thus $L \propto \varepsilon_c^4$. This is consistent with the reported value of $L_{100\mathrm{MeV}} \propto E_c^{3.42\pm0.86}$ by Buehler *et al* (2012).

Concerning the energetics of the flares, the high magnetization flow entails a magnetic energy reservoir that dominates the spin-down luminosity of the pulsar. Going into the numerics, the typical energies of the flares lasting around one day, $t_f \approx 1$ d, are $E_f \approx 10^{34}$ J. The reconnection site dimension is estimated to be a layer of square shape and of side length similar to one wavelength of the striped wind, λ_L. During a typical flare, the energy passing through the reconnection site is approximately $E_f \approx B^2\lambda_L^2 ct_f / \mu_0$. Observational constraints require energies of $E_f \approx 10^{34}$ J, so we deduce the local magnetic field intensity in the flare to be around 2 T. Taking into account the wave nature of the striped wind with mostly a toroidal component, we get $B \approx B_L\, r_L/r$, and for the Crab we have $B_L \approx 100$ T. Applications to flares thus gives an emission region located in the vicinity of $r \approx 50\, r_L$. The associated mean luminosity would be $L_f = E_f/t_f = 10^{29}$ W, which is about 1% of the total Crab spin-down luminosity. As a consequence, flares could emanate from the same site as the one producing the pulsed gamma-ray emission; the location and Lorentz factor are consistent in both cases. At this stage of our model, it is impossible to discuss their respective spectra because this requires a kinetic description of the reconnection site, probably including non-negligible radiation losses, which is beyond the scope of this study. We conclude that although we performed new simulations more appropriate to pulsar winds, the order of magnitude found in Baty *et al* (2013) are fairly robust and not drastically modified.

Furthermore, let us comment on the expected observed properties of flare occurrence. To simplify the discussion, let us assume that all the flares in the comoving frame possess the same characteristics, essentially the same time scales for the explosive phase. Although this is oversimplified, it helps to understand the natural consequences of our model. Indeed, if the emission duration in the rest frame of the wind is given by $\Delta t'$, then the duration of a flare in the observer frame is subject to time dilation and given by equation (2). Because of the motion of the magnetic islands, which is not necessarily relativistic, the duration of the flare Δt_{obs} is Doppler shifted according to $\Delta t_{\mathrm{obs}} = \Delta T\,(1 - \beta\cos\vartheta)$. The blobs pointing towards us, i.e. with $\vartheta \approx 0$, and therefore $\mathcal{D} \approx 2\Gamma$, are the shortest, whereas those making a substantial angle between their velocity and our line of sight will appear longer. Moreover, from the luminosity relation equation (3), the former will appear brightest, whereas the latter will appear fainter. Consequently, we attribute the distinction between waves and flares as a simple consequence of the combination between geometric effects and Doppler beaming. There is no basic physical ground on which they should be distinguished.

To finish this discussion on comparison with observations, let us say that at this stage of our investigations, we are not able to give a detailed account of the statistical properties of the Crab flares. Indeed, we performed a local analysis of the DTM mechanism, which focused on a particular configuration in the plasma comoving frame. Obtaining a precise picture of the frequency of the occurrence of these flares would require a global study of the DTM instability in full 3D according to the anisotropic structure of the striped wind. This is beyond the scope of our simulations presented here.

5. Conclusions

We performed new RRMHD simulations of the DTM of magnetically dominated plasmas shifting to more realistic σ and S parameters. The growth time scale switches from an early $S^{1/2}$ Sweet-Parker-like phase to a slower S^1 Rutherford regime. The third explosive phase is favored when the longitudinal wavelength of the perturbation exceeds a critical value. Our results also indicate a weak dependence of the time scale on the resistivity and the magnetization parameter, $\sim S^{0.2-0.3}$ and $\sim \sigma^{0.3}$ respectively. We focused on the DTM instability using a simple local configuration in the wind. However, pulsar winds exhibit more complex multiple current sheets, thus multiple tearing modes (MTM) should be investigated. Moreover, the spiral structure could also be taken into account by global simulations in the equatorial plane, but this is left for future work. This would also include the wave nature of the field configuration. The hypothesis of equidistant current sheets could also be left for more asymmetric configurations as required when looking above or below the equatorial plane.

DTM and possibly MTM are good candidates to explain the energetics of the Crab flares in the context of magnetic reconnection. Indeed, the explosive growth phase clearly involves a rapid variability accompanied by a relativistic bulk flow that is not reproducible via a simple tearing mode. Moreover, it puts some constraints on the maximum Lorentz factor of the striped wind, $\Gamma \lesssim 150$, and on the localization of the emission region, respectively, $r \approx 50\, r_L$. This provides an unified picture between the pulsed gamma-ray emission and flares both directly related to the pulsar striped wind.

Acknowledgments

JP and HB acknowledge the support by the French National Research Agency (ANR) through the grant No. ANR-13-JS05-0003-01 (project EMPERE). They also benefited from the computational facilities available at Equip@Meso at the Université de Strasbourg. MT acknowledges the support by the Postdoctoral Fellowships for Research Abroad program by the Japan Society for the Promotion of Science No. 20130253.

References

Abdo A A *et al* 2011 *Science* **331** 739

Baty H, Petri J and Zenitani S 2013 Explosive reconnection of double tearing modes in relativistic plasmas: application to the Crab flares *Mon. Not. R. Astron. Soc.* **436** L20–4

Buehler R *et al* 2012 Gamma-ray activity in the Crab nebula: the exceptional flare of 2011 April *Astrophys. J.* **749** 26

Clausen-Brown E and Lyutikov M 2012 Crab nebula gamma-ray flares as relativistic reconnection minijets *Mon. Not. R. Astron. Soc.* **426** 1374–84

Cerutti B, Uzdensky D A and Begelman M C 2012 Extreme particle acceleration in magnetic reconnection layers: application to the Gamma-ray flares in the Crab nebula *Astrophys. J.* **746** 148

Janvier M, Ishizawa A, Li J and Kishimoto Y 2011a Role of the pressure force in the explosive dynamics of magnetic islands in double tearing modes *Phys. Plasmas* **18** 102112

Janvier M, Kishimoto Y and Li J Q 2011b Structure-driven nonlinear instability as the origin of the explosive reconnection dynamics in resistive double tearing modes *Phys. Rev. Lett.* **107** 195001

Komissarov S S, Barkov M and Lyutikov M 2007 Tearing instability in relativistic magnetically dominated plasmas *Mon. Not. R. Astron. Soc.* **374** 415–26

Kirk J G, Lyubarsky Y and Petri J 2009 The theory of pulsar winds and nebulae *Astrophys. Space Sci. Library* **357** 421

Lyutikov M 2003 Explosive reconnection in magnetars *Mon. Not. R. Astron. Soc.* **346** 540–54

Lyutikov M, Balsara D and Matthews C 2012 Crab GeV flares from the corrugated termination shock *Mon. Not. R. Astron. Soc.* **422** 3118–29

Lyubarsky Y E 2005 On the relativistic magnetic reconnection *Mon. Not. R. Astron. Soc.* **358** 113–9

Lyubarsky Y E 2012 Highly magnetized region in pulsar wind nebulae and origin of the Crab gamma-ray flares *Mon. Not. R. Astron. Soc.* **427** 1497–502

Otto A and Birk G T 1992 Resistive instability of periodic current sheets *Phys. Fluids* B **4** 3811–3

Parker E N 1957 Sweet's mechanism for merging magnetic fields in conducting fluids *J. Geophys. Res.* **62** 509–20

Petschek H E 1964 Magnetic field annihilation *NASA Special Publ.* **50** 425

Petri J 2012 High-energy emission from the pulsar striped wind: a synchrotron model for gamma-ray pulsars *Mon. Not. R. Astron. Soc.* **424** 2023–7

Pétri J and Kirk J G 2005 The polarization of high-energy pulsar radiation in the striped wind model *Astrophys. J. Lett.* **627** L37–40

Pétri J and Lyubarsky Y 2007 Magnetic reconnection at the termination shock in a striped pulsar wind *Astron. Astrophys.* **473** 683–700

Pucci F and Velli M 2014 Reconnection of quasi-singular current sheets: the 'Ideal' tearing mode *Astrophys. J. Lett.* **780** L19

Striani E *et al* 2013 Variable gamma-ray emission from the Crab nebula: short flares and long 'waves' *Astrophys. J.* **765** 52

Sweet P A 1958 The neutral point theory of solar flares *Electromagnetic Phenomena in Cosmical Physics IAU Symp.* vol 6, ed B Lehnert (Cambridge: Cambridge University Press) p 123

Takamoto M 2013 Evolution of relativistic plasmoid chains in a poynting-dominated plasma *Astrophys. J.* **775** 50

Takamoto M and Inoue T 2011 A new numerical scheme for resistive relativistic magnetohydrodynamics using method of characteristics *Astrophys. J.* **735** 113

White R B 1986 Resistive reconnection *Rev. Mod. Phys.* **58** 183–207

Wang Z X, Wang X G, Dong J Q, Lei Y A, Long Y X, Mou Z Z and Qu W X 2007 Fast resistive reconnection regime in the nonlinear evolution of double tearing modes *Phys. Rev. Lett.* **99** 185004

Yan M, Otto A, Muzzell D and Lee L C 1994 Tearing mode instability in a multiple current sheet system *J. Geophys. Res.* **99** 8657–69

Zhang C L and Ma Z W 2011 Nonlinear evolution of double tearing mode with guiding magnetic field *Phys. Plasmas* **18** 052303

IOP Publishing

Plasma Physics and Controlled Fusion

Plasma Phys. Control. Fusion **57** (2015) 014035 (9pp)

doi:10.1088/0741-3335/57/1/014035

Impact of magnetic perturbation coils on the edge radial electric field and turbulence in ASDEX Upgrade

G D Conway, S Fietz, H W Müller, T Lunt, P Simon, W Suttrop, M Maraschek, T Happel, E Viezzer and the ASDEX Upgrade Team

Max-Planck Institut für Plasmaphysik, Boltzmannstasse 2, 85748 Garching, Germany

E-mail: garrard.conway@ipp.mpg.de

Received 27 June 2014, revised 24 September 2014
Accepted for publication 10 October 2014
Published 28 November 2014

Abstract

The impact of applied magnetic perturbations (MPs) on tokamak edge parameters has been investigated in ASDEX Upgrade low collisionallity L-mode discharges using a flexible set of in-vessel saddle coils (capable of generating $n = 0$, 1, 2 and 4 toroidal modes) and an extensive set of high resolution edge diagnostics. Doppler reflectometry is used principally to probe the MP field penetration and structure via the radial electric field E_r and density fluctuation δn_e behaviour. Different MP response behaviour are observed for the near scrape-off-layer, SOL, (where E_r flattens) and in the confinement region (where the negative E_r well reverses). The radial structure of E_r and δn_e are particularly sensitive to the degree of MP resonance with the edge rational field-lines. Specifically, the edge turbulence is enhanced for strongly resonant MPs and reduced when non-resonant. The toroidal structure of the MP response has also been probed for various MP configurations by rotating the MP field toroidally and is found to be different for the edge and near SOL regions.

Keywords: magnetic perturbations, radial electric field, ASDEX upgrade, turbulence

(Some figures may appear in colour only in the online journal)

1. Introduction

The use of magnetic perturbation (MP) control coils for modifying the ELM and MHD behaviour in tokamak H-mode plasmas has become again an intense topic of investigation in recent years with many machines operating a variety of in- and ex-vessel saddle coils [1–5]. Although the theory and application of small perturbing magnetic fields is not new, see TEXT and JFT-2M results from the 1990s [6–8], the non-linear response of the plasma to the MPs is rather complex and there remain many open questions. For example, a recent aspect is the role of the MP on the edge turbulence and zonal flows and their impact on the ELM cycle [9]. Nevertheless, to investigate the fundamental plasma response to MPs it is generally more convenient to operate in regimes where the plasma response is strongest and clearest to observe—namely low density L-mode plasmas, see [10–13]. ASDEX Upgrade (AUG) is particularly well suited to this task with its flexible set of in-vessel saddle coils (detailed in section 2) which are capable of generating MPs with (predominant) toroidal mode numbers $n = 0$, 1, 2 and 4; and its extensive set of high resolution edge diagnostics.

Langmuir probes have been widely used to measure plasma potential and density variations (which are seen to be particularly sensitive indicators of the MP effect) in the plasma scrape-off-layer (SOL) and separatrix regions of tokamaks [12, 14–16], while deeper probing has been achieved with reflectometry [17], CXRS [11, 18], BES [9] and HIBP [7] for example. In this work, Doppler reflectometry (described in section 3) is used on AUG to directly measure the radial electric field E_r, its fluctuations, as well as the density turbulence δn_e behaviour with high spatial resolution from the far SOL to inside the edge gradient region of low collisionality L-mode discharges.

Here, new results are presented showing different MP plasma response behaviour for the open field-line SOL—where

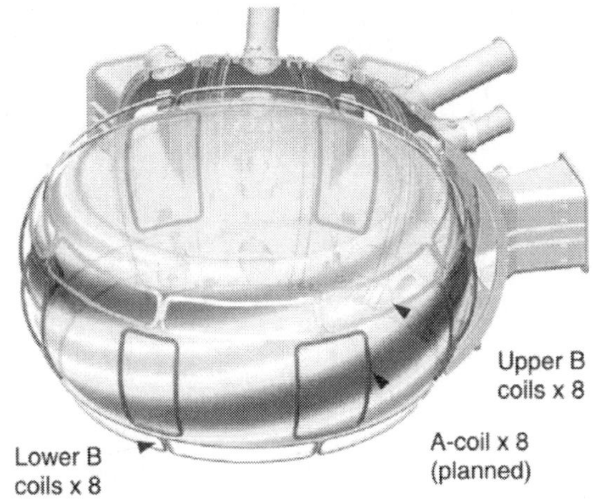

Figure 1. A 3D view of the AUG MP coils and plasma with representative perturbation field together with cut-away of the vacuum vessel.

flattening of the E_r and n_e profiles is seen in the near SOL—and for the stochastic confinement region where the negative E_r well reverses (sections 4 and 5). The extent and radial structure of the E_r and δn_e are found to be particularly sensitive to the MP configuration, as well as the MP parity/phasing and the degree of MP field resonance with the edge rational field-lines. Specifically, the turbulence is locally enhanced when the MP configuration is significantly resonant (section 6). The 3D structure of the MP has also been investigated for various MP configurations by rotating the MP field toroidally. The toroidal variation of the E_r and δn_e radial profiles are found to be different for the edge and near SOL regions (section 7). Although the focus here is on $n = 2$ MPs, the overall behaviour is similar for $n = 1$ and $n = 4$ MPs.

2. Magnetic perturbation (MP) coils

AUG currently has installed sixteen in-vessel MP saddle coils in the form of two toroidal sets of eight coils (termed B-coils) above and below the tokamak low-field-side (LFS) mid-plane—spanning the full toroidal circumference, as shown in figure 1. A further eight 'picture-frame' (A-coils) are also planned [19] for future installation. The coils are mounted on the upper and lower passive stabilizing loops (PSL) close to the plasma boundary, as shown in the poloidal cross-sectional view of the vessel in figure 2. Each coil consists of an encased, single 5-turn water cooled Cu conductor with separate coaxial feeders, which can be connected in (anti-)series to create MPs with toroidal mode numbers of $n = 0$, 1, 2 and 4. The MP poloidal mode number m is defined by the coil's poloidal dimensions and their spacing etc. Generally, there is no single corresponding m but a broad spectrum of modes and harmonics. The coils are powered by two independent supplies capable of generating 5 kAt, i.e. a maximum perturbation field b_r of around 3 mT at the plasma boundary in-front of an upper coil, which corresponds to a normalized perturbation field of $b_r/B \sim 10^{-3}$ for typical −2.5 T operation.

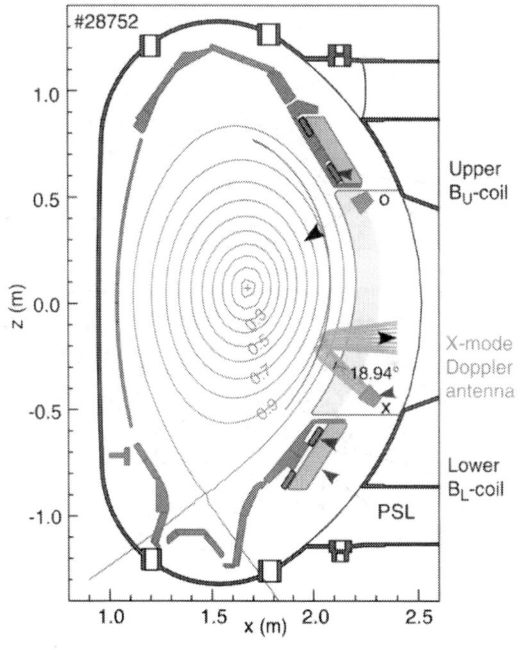

Figure 2. Poloidal cross section of the AUG vessel together with an axisymmetric flux surface plot. Marked are the locations of the upper and lower B-coils mounted on the passive stabilizing loops (PSL) together with the Doppler reflectometer antenna and a 64 GHz probing beam trajectory for shot #28752.

For the static current results reported here, the toroidal phase Φ of the MP pattern relative to the tokamak can be varied in steps of 45° for $n = 1$, 90° for $n = 2$ and 180° for $n = 4$ by changing the current polarity of the individual coils via a switch-board panel. A half Φ step can also be achieved by powering only alternate coils. With programmed dynamic coil current waveforms it is also possible to simulate finer toroidal steps—but this was not used here. The relative phase $\Delta\Phi$ between the upper and lower coil set (i.e. coil parity) can also be varied to change the alignment of the MP field with respect to the tokamak field lines and hence the degree of MP resonance with rational q surfaces.

Figure 3(*a*) shows a contour plot of the *vacuum*, i.e. external, MP magnetic field b_n normal to the flux surface (Biot-Savart equation; red is positive outward pointing b_n) [20] as a function of machine toroidal angle ϕ and poloidal straight field line angle θ^* calculated at the $q = -3$ surface for an $n = 2$ MP configuration with $\Delta\Phi = 90°$ and $\Phi = 180°$ (here $\Phi = 0$ is defined as a positive going b_n with a null between coils Bu8 and Bu1). The line indicates the field line inclination, which connects here opposite polarity coils in the upper and lower rows. The arrow indicates the Doppler reflectometer position. Below, in figure 3(*b*) is the corresponding 2D Fourier (m, n) mode spectrum. The resonant spectral condition is marked by the inclined line (for the left-hand field line helicity $q < 0$) which in this case does not intersect the main spectral peaks and thus indicates a non-resonant or non-field-aligned situation. Due to the low magnetic shear at the outboard mid-plane the MP will be simultaneously (non-)resonant for many edge rational surfaces [20], as shown by the poloidal mode m spectra for the $n = 2$ component as a function of poloidal

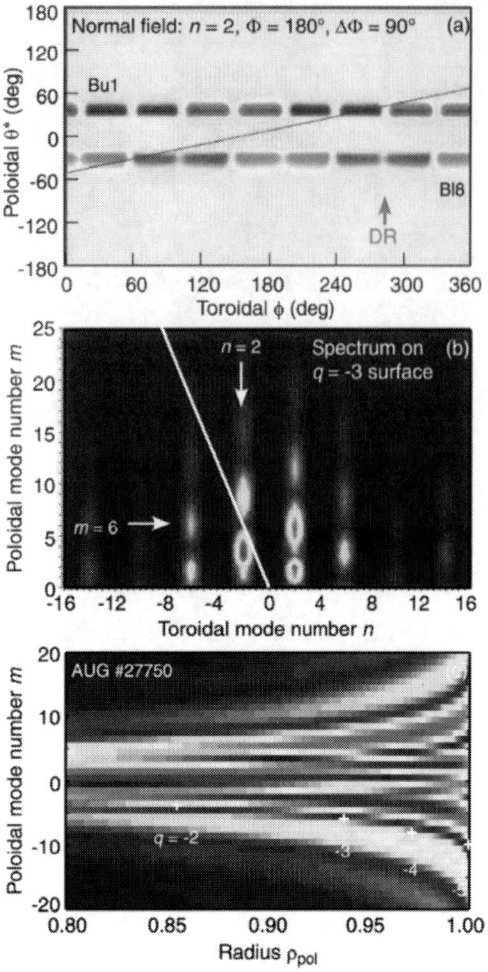

Figure 3. (*a*) MP vacuum field b_n for non-resonant $n = 2$ configuration on $q = -3$ surface (blue = negative or inward pointing b_n, red = positive) in straight field-line coordinates, (*b*) corresponding 2D Fourier (m, n) spectrum and (*c*) poloidal spectrum of $n = 2$ component versus radius. White crosses mark rational q surfaces.

flux radius ρ_{pol} in figure 3(*c*). The crosses mark rational surface positions—which all lie in a trough (blue is low) in the spectral power. One side-effect of the plasma response to the MP field may be a modification of the resonance condition. However, in the absence of reliable modelling of the plasma screening for these discharges the resonance is discussed in terms of the vacuum component alone.

3. Doppler reflectometer diagnostic

An X-mode stepped frequency (50–75 GHz) Doppler reflectometer [21] is used to measure the *local* radial electric field profile $E_r \simeq -u_\perp B$ from the Doppler frequency shift ω_D in the backscattered microwave beam (where $u_\perp = \omega_D/k_\perp = v_{E\times B} + v_{ph}$ and the usual condition of the turbulence phase velocity $v_{ph} \ll v_{E\times B}$ & $k_\parallel \ll k_\perp$) [22]. In addition, the density fluctuation level $|\delta n_e(k_\perp)|^2$ at the probed perpendicular wavenumber $k_\perp$ can be obtained from the integrated

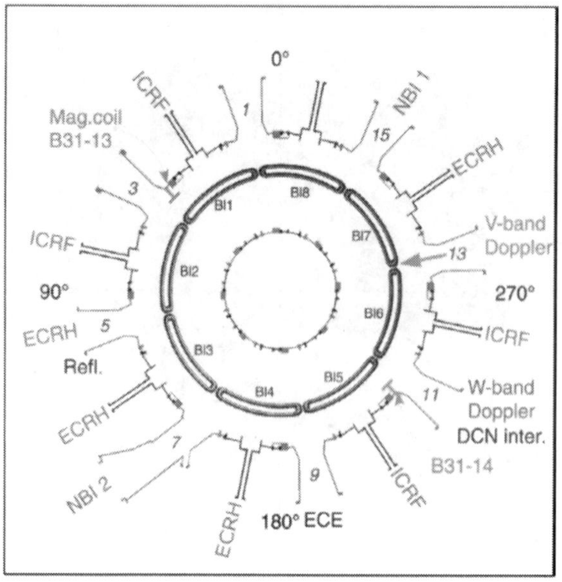

Figure 4. Schematic top view of the AUG vessel showing the toroidal locations of the B-coils with selected diagnostics and heating systems.

spectral peak $S(k_\perp)$. In the experiment the measurement location (beam turning point) and the probed $k_\perp = -2N_\perp k_o$ are obtained using a beam tracing code e.g. TORBEAM [23], with smooth density profiles fitted to experimental data and *axisymmetric* equilibria (i.e. without accounting for any 3D impact of the MP fields on the equilibrium). For typical L-mode conditions the diagnostic generally has a radial coverage from the far SOL to inside the edge gradient region, but at very low n_e the diagnostic can reach to around $\rho_{pol} \sim 0.6$ in radius.

The reflectometer bistatic antenna pair are located just above the lower B-coil row with an upward line-of-sight, as shown in the vessel cross-section in figure 2. Overlaid are flux surfaces and a 64 GHz X-mode beam trace for a $n_{eo} = 2 \times 10^{19}$ m^{-3}, $B_t = -1.9$ T shot, showing the Doppler reflectometer measures the LFS E_r poloidally somewhat below the magnetic axis. Toroidally the reflectometer probes between coils 6 and 7 (sector 13), as shown in the tokamak plan view of figure 4. Thus, by rotating the MP coil phasing relative to the fixed diagnostic, the MP toroidal response may be investigated.

4. Edge E_r response to MP

Figure 5 shows the response of various edge plasma profiles to the application of an $n = 2$ odd-parity ($\Delta\Phi = 180°$) edge 'mostly resonant' MP for a typical $n_{eo} = 2 \times 10^{19}$ m^{-3}, $B_t = -2.5$ T, $I_p = 1.0$ MA, $q_{95} = -4.2$, L-mode discharge. The schematic shows the MP configuration (b_n directions). With a (close to maximum) B-coil current of $I_B = 0.9$ kA the MP impact is very pronounced in E_r, figure 5(*a*) (red with and blue without MP) across both the edge gradient region (3) where the usual negative E_r well is inverted and in the near SOL region (2) where there is a distinct reduction of the E_r peak just outside the separatrix. In the far SOL (1) there is

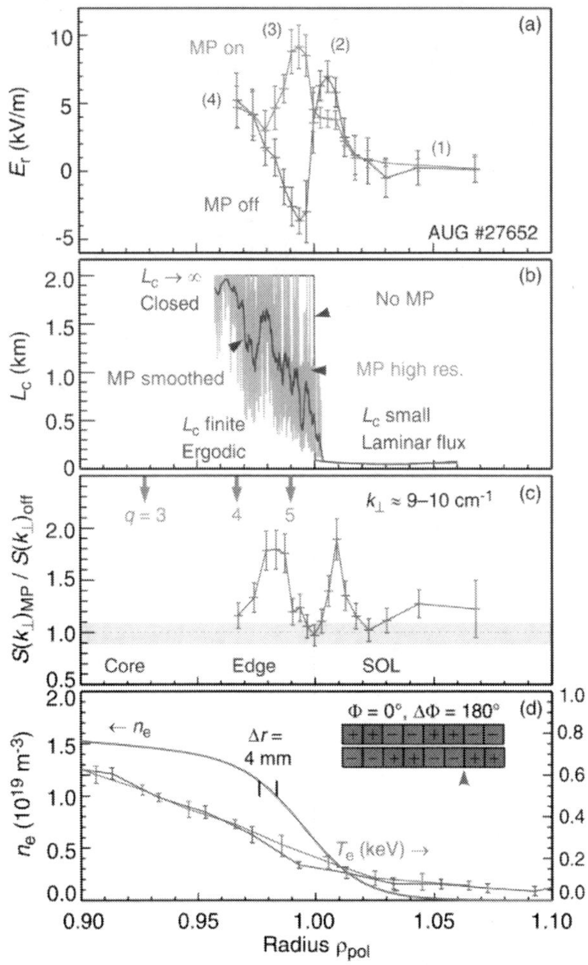

Figure 5. (*a*) L-mode E_r, (*b*) vacuum connection length L_c, (*c*) turbulence ratio and (*d*) n_e and T_e radial profiles with (red) and without (blue) $n = 2$, 'resonant'($I_B = 0.9$ kA, $\Phi = 0°$, $\Delta\Phi = 180°$) odd-parity MPs (−2.5 T, 1.0 MA, $q_{95} = -4.2$, $P_{ech} = 0.7$ MW, $T_{eo} = 5$ keV, $n_{eo} = 2 \times 10^{19}$ m^{-3}).

little or no observable impact; and again inside of the edge gradient (4) where the E_r profiles re-converge. The E_r well reversal is consistent with TEXT results (HIBP) [7], Tore Supra (Scattering) [13], TEXTOR (reflectometer) [17] and (probes) [14] and MAST (probes) [12]—indicating that the Doppler reflectometer measurements are robust.

The usual explanation for the E_r well inversion is the ambipolar plasma response to the opening and/or mixing (stochastization) of the edge magnetic field lines resulting from the MP field penetration [7]. The creation of a stochastic field region in the edge is also indicated by Poincoré plots of the computed vacuum field [24] and by the field line connection length L_c (from inner to outer divertor), as shown in figure 5(*b*). Without the MPs L_c is small in the SOL and infinity for the closed field line region (blue curve). With MPs the field lines inside the separatrix begin to break and L_c drops to around 1–2 km (red curves). The radial structure in L_c is on the mm scale—comparable to typical island widths. Interpreting the width of the E_r inversion region (note the outer inversion radii is close to the computed last closed flux surface LCFS from the axisymmetric equilibrium) as an indication of the MP field penetration gives an estimated depth of ~1.5 cm—which is somewhat narrower than the apparent vacuum field line mixing region from the L_c calculation and is suggestive of non-negligible plasma screening effects.

Generally the L-mode edge density turbulence is spectrally broad-band—up to several hundred kHz and several cm^{-1} [25]. Its response to the MP is shown in figure 5(*c*) in the form of a turbulence enhancement factor calculated from the integrated spectral Doppler intensity $S(k_\perp) \propto |\delta n_e|^2$ (at the probed $k_\perp \sim 9$ cm^{-1}—which is representative of the overall fluctuation behaviour) with MPs, normalized to the non-MP $S(k_\perp)$ (from the same discharge). In the two regions where the E_r profile is impacted the turbulence is also enhanced by a factor of 2. In both the near SOL and in the edge gradient the peak turbulence enhancement aligns with regions where the negative E_r shear has either been reduced (at the SOL peak) or reversed (in the gradient region). Elsewhere, in the far SOL and around the separatrix, the turbulence level remains unaffected; which is generally consistent with far SOL probe measurements [16].

The corresponding kinetic profiles, electron temperature T_e from ECE and density n_e—in this case a modified hyperbolic tangent fitted to Lithium beam, Thomson Scattering and FMCW reflectometry data (which describes well the edge n_e profile shape) are shown in figure 5(*d*). There is little change in the kinetic profiles with the MPs (within measurement errors) and no evidence of any significant edge density pump-out associated with the MPs at these low densities in L-mode. However, the far SOL does show a slight n_e enhancement together with divertor strike-point splitting [15]. T_i measurements are unfortunately not available in these experiments.

5. MP thresholds

It has long been understood that the plasma attempts to resist the penetration of the externally applied MP field via the formation of shielding currents [26]. However, the plasma screening is difficult to calculate and is a topic of much theoretical study [27–33]. For the low density, low-rotation, L-mode conditions of interest here calculations of the plasma screening are currently not available. But, it is known that screening leads to a threshold or critical field for the MP penetration. This has been investigated for different collisionality regimes, but generally scales linear with n_e and inverse with B_t [34]. A critical field of $(b_n / B)_{crit} \sim \mathcal{O}(10^{-4})$ is typically expected for usual tokamak conditions. Experimentally, clear threshold effects have been observed in MAST [12], but are less clear in other machines [7, 11, 14, 35]. Density and plasma rotation dependences have also been reported [13, 27].

Figure 6 shows such a critical threshold in the E_r for an $n = 2$ edge 'resonant' MP. The MP coil current is stepped from 0 to 0.9 kA at constant density and electron temperature (shown in figure 11(*e*)) and energy confinement. Both the near SOL E_r peak and the edge gradient E_r well begin to respond slowly to the increasing I_B, but it is only at the highest current (red curve) that the E_r well rises sharply and the SOL peak

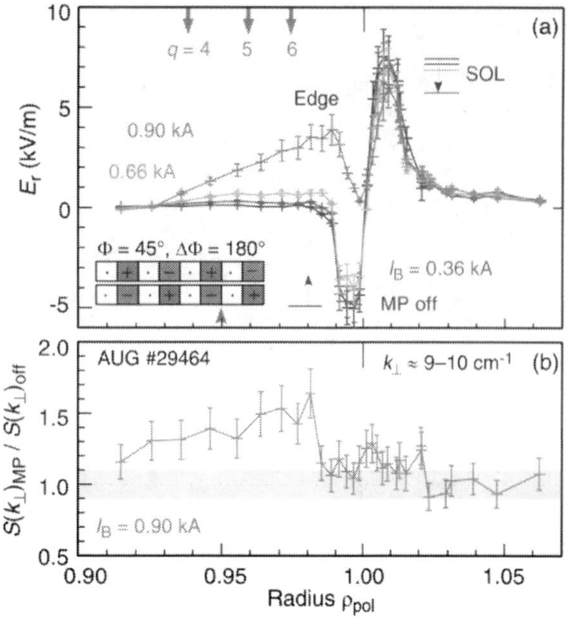

Figure 6. (*a*) Edge E_r profiles at fixed L-mode plasma parameters (−2.5 T, 0.8 MA, $q_{95} = -5.2$, $P_{ech} = 0.55$ MW, $T_{eo} = 4.5$ keV, $n_{eo} \approx 1.5 \times 10^{19}$ m^{-3}) with $n = 2$, 'resonant' (Φ = 45°, ΔΦ = 180°), odd-parity, MP I_B current steps; (*b*) turbulence enhancement factor for $I_B = 0.9$ kA MP step.

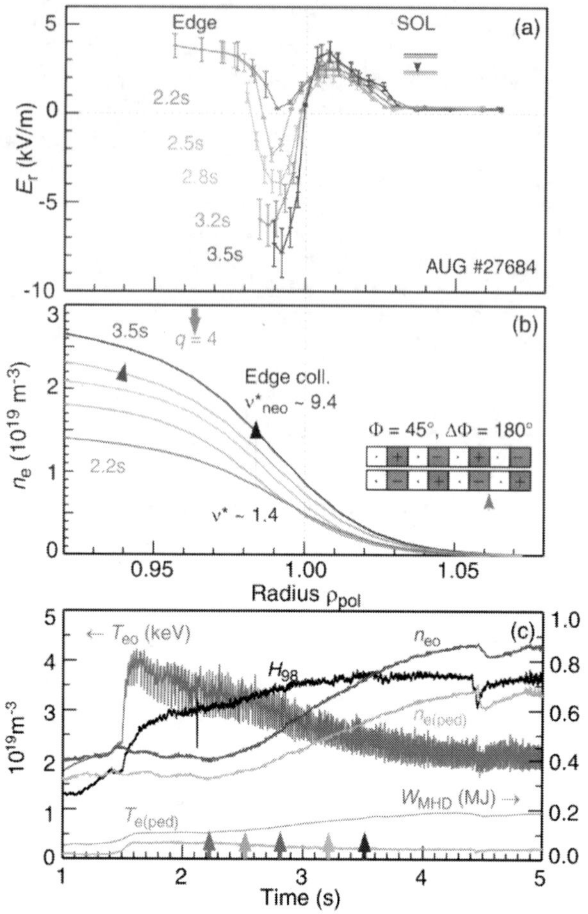

Figure 7. (*a*) Edge E_r, (*b*) n_e profiles and (*c*) time traces of n_e, T_e etc. during an L-mode density ramp with $n = 2$ 'resonant'($I_B = 0.9$ kA, Φ = 45°, ΔΦ = 180°) odd-parity MPs (−2.5 T, 1.0 MA, $q_{95} = -4.3$, $P_{ech} = 0.5$ MW.).

drops (outside of error bars). The turbulence is again enhanced in regions of reduced negative E_r shear—particularly inside of the E_r well where the shear becomes positive.

The expected density dependence of the MP threshold is displayed in the edge E_r profile evolution in figure 7(*a*) during an L-mode density ramp of $n_{eo} = 2 \rightarrow 5 \times 10^{19}$ m^{-3} with a fixed $I_B = 0.9$ kA and an $n = 2$ edge 'resonant' MP configuration. As n_e rises, T_e falls, figures 7(*b*) and (*c*), (with a corresponding increase in collisionality $\nu^*_{neo} = q_{95} R_o \epsilon^{-3/2} \lambda_e^{-1}$) the MP impact disappears, both in the near SOL where the E_r peak flips between reduced and non-reduced states when n_e exceeds a certain value (compare curves for 2.8 and 3.5 s); and in the E_r well, which conversely gradually reverts to the no-MP depth with increasing n_e. Note that without MPs the variation in E_r with density is less than 1 kVm^{-1}. The different density responses for the SOL (threshold) and E_r well regions (gradual transition) suggests distinct MP mechanisms for the laminar SOL and stochastic edge regions. The absence of an MP impact at high density on T_i and the E_r well has also been observed using CXRS [18].

6. MP resonance

The degree of MP resonance is defined as the alignment of the perturbation field with the equilibrium rational surface field lines. This can be varied via the plasma safety factor q profile (i.e. via B_t or I_p), or by changing the toroidal and poloidal spectra of the MP field. For a selected n configuration this can be achieved by shifting the relative ΔΦ phase between the upper and lower B-coil sets [19].

Figure 8 compares the edge E_r (top row, blue without and red with MP) and turbulence behaviour (middle row) from very similar $n_{eo} \approx 1.3 \times 10^{19}$ m^{-3}, L-mode discharges for three possible differential phases: (*a*) ΔΦ = 0° (even parity), (*b*) 90° (mid parity) and (*c*) 180° (odd parity) for an $n = 2$ MP with Φ = 0°. The bottom row of figure 8 shows the corresponding MP poloidal vacuum spectra for each configuration and shows how changing the differential phase changes the MP spectral components ($n = 2$, $m = 8$ & 10) from moderate, to weak, to strong at the $q = -4$ ($\rho_{pol} \approx 0.98$) and $q = -5$ ($\rho_{pol} \approx 0.99$) surfaces respectively. All three parities display some degree of E_r impact, the weakest being for the 'non-resonant' case and the strongest for the 'mostly resonant' case. The turbulence enhancement only being significantly enhanced in the most resonant configuration around the E_r reversal, again associated with the region of negative E_r shear loss.

Figure 9 shows the effect of changing the q profile. The edge E_r profiles are for a pair of matched −2.5 T L-mode discharges at $I_p = 1.0$ (red/solid) and 0.6 MA (blue/dashed) with $n_{eo} \approx 1.5 \times 10^{19}$ m^{-3}) and $n = 2$ edge 'resonant'($I_B = 1.0$ kA, Φ= 45°, ΔΦ = 180°) MP. For the 1.0 MA case the E_r well is strongly impacted, but is unaffected for the 0.6 MA case.

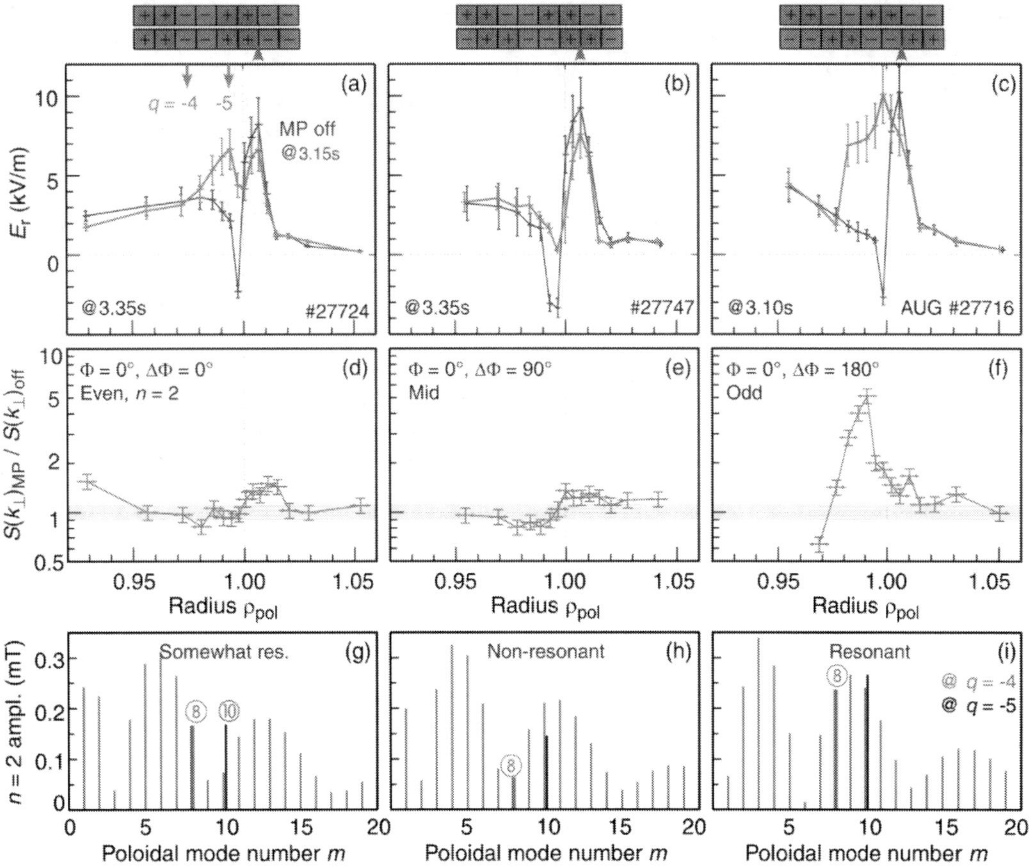

Figure 8. Top: E_r profiles, middle: turbulence ratio, bottom: MP m-spectra from matched L-mode shots (−2.3 T, 1.0 MA, $q_{95} \approx -4$, $P_{ech} = 0.6$ MW, $T_{eo} = 4.5$ keV, $n_{eo} \approx 1.3 \times 10^{19}$ m^{-3}) with $n = 2$ ($I_B = 0.9$ kA, $\Phi = 0°$) MPs for (a) $\Delta\Phi = 0°$ (even), (b) $\Delta\Phi = 90°$ (mid) and (c) $\Delta\Phi = 180°$ (odd).

Below in figure 9(b) are the corresponding q profiles from two equilibrium reconstruction codes; FPP and CLISTE. For $I_p = 1.0$ MA the $q = -4$ and −5 rational surfaces are close to the edge and accessible to the MP field. The MP vacuum poloidal spectrum in figure 9(c) confirms that the ($n = 2$, $m = 10$) spectral component is significant at the $q = -5$ surface. Lowering the I_p, raises the edge q profile, such that the low order rational surfaces are now much deeper and the edge is filled with high order surfaces. The ($m = 10$ @ $q = -5$) component is much reduced (hardly resonant) and the higher ($m = 14$ @ $q = -7$ and $m = 16$ @ $q = -8$) components are negligible.

7. MP toroidal phase

The MP field is helical in form with both toroidal and poloidal structure. This structure can be sampled by shifting the MP coil phase, i.e. a rigid toroidal rotation of the MP field, relative to the fixed diagnostic measurement point. The poloidal structure, however, is less easy to access with only two poloidal coil rows and one measurement location.

The toroidal variation is illustrated in figure 10 with a fixed $n = 2$, $\Delta\Phi = 90°$ edge 'non-resonant' configuration for four $\Phi = 0$, 90, 180, 270° static MP phases during a single discharge with constant edge n_e and T_e etc. Both the near SOL E_r and the edge E_r well clearly display individual reactions to the MP orientation—which is particularly evident in the $\Phi = 180°$ and 270° orientations. This suggests separate, but not necessarily unrelated, toroidal 3D structures for the SOL (laminar) and edge (stochastic) regions.

As in previous examples the MP affected E_r regions are also accompanied by changes in the density turbulence level, figure 10 (bottom row). However, in this 'non-resonant' MP configuration only the near SOL shows significant (factor 2) enhancements in the turbulence level when the SOL E_r peak is suppressed. In the edge gradient region the turbulence level is actually decreased. Contrasting this behaviour with figures 5(c), 6(b) and 8(f): all resonant, with figure 8(e): non-resonant, suggests that the edge region δn_e may be increased (by a factor of 10 for strongly resonant MPs) or decreased depending on the MP configuration.

In principle, rotating the MP phase Φ should not affect the MP spectra or degree of resonance. However, observations with $n = 1$ MPs [36, 37] as well as recent measurements with $n = 2$ indicate that certain Φ can either promote or hinder mode locking, implying toroidal asymmetries or subtle interactions between the MP and intrinsic static fields, such as the presence of an $n = 2$ error field. Note that lock-mode cases are explicitly excluded in this paper.

3D vacuum field-line tracing including MPs predict a toroidally sinusoidal radial displacement of the low-field-side separatrix position [38]. The displacement amplitude depends

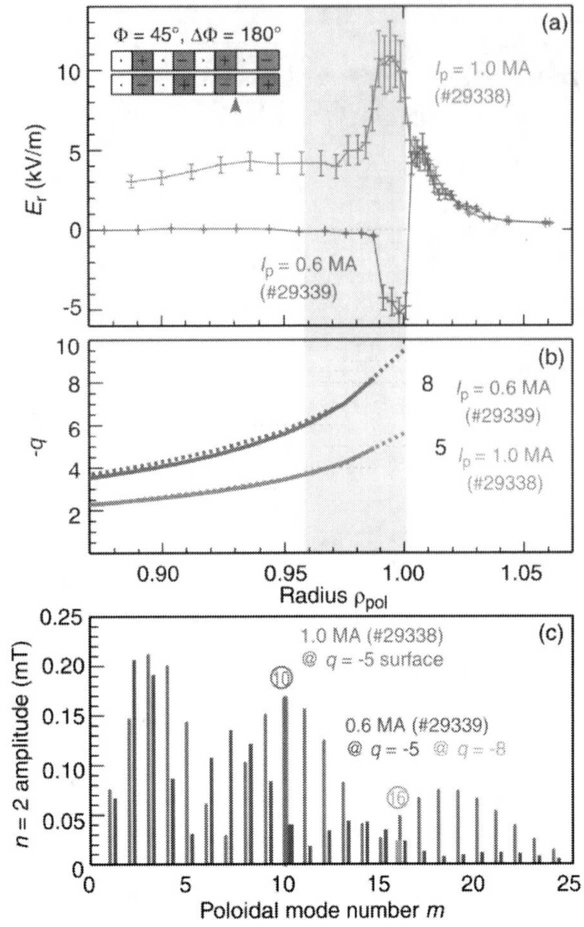

Figure 9. (*a*) E_r, (*b*) q-profiles and (*c*) poloidal mode spectra for matched L-mode (red) $I_p = 1.0$ MA and (blue) 0.6 MA shots (−2.5 T, $P_{ech} = 0.6$ MW, $T_{eo} \approx 5$ keV, $n_{eo} \approx 1.5 \times 10^{19}$ m^{-3}) with odd-parity $n = 2$, $I_B = 1.0$ kA, $\Phi = 45°$, $\Delta\Phi = 180°$ MPs.

strongly on the MP configuration but, can be several mm in worst cases—leading to profile mapping issues [39, 40]. Compression of the flux surfaces is not critical for the Doppler reflectometer measurement position since it follows the correspondingly displaced density profile, which is seen to remain roughly constant in the flux surface coordinate ρ_{pol} (ignoring small changes in $B(R)$ for X-mode probing). A $\Delta r \sim 4$ mm bar (maximum separatrix displacement at the diagnostic location) is shown for comparison in figure 5(*d*)—which is comparable in magnitude to the Doppler beam scattering width, i.e. measurement error. This allows, with care, the use of the unperturbed (non-MP) density profile and an axisymmetric equilibria in the beam tracing procedure for the Doppler reflectometer analysis. Plasma screening may of course introduce localized non-uniformities in the edge equilibrium, which are not accounted for here. Flux surface distortion will, however, affect the potential gradient, e.g. E_r, to a small degree.

8. Edge flow perturbation

Geodesic Acoustic Modes (GAMs) are universally observed in the edge confinement region of AUG low collisionality L-mode discharges, see [41]. Figures 11(*a*) and (*b*) shows the effect of applying an edge 'resonant' $n = 2$ MP on the $E_r \times B$ flow and n_e fluctuation spectra at the GAM maximal location ($\rho_{pol} \approx 0.99$) for the discharge shown in figure 6. For the last I_B step with the impacted E_r well, the coherent flow spectral peak is up-shifted in frequency from 14.6 kHz to 15.9 kHz and weakened in amplitude. Overall, δE_r falls while the δn_e spectra is broadly enhanced. The non-MP GAM peak generally extends from the separatrix across the negative E_r shear region, increasing in frequency with temperature in a staircase fashion. Each frequency stair corresponding to a GAM zonal layer [41]. The GAM peaks in magnitude close to the stair inner edge. With MPs the mode extends somewhat deeper, but peaks further out, closer to the E_r minimum.

Accompanying the up-shifted flow perturbation is a corresponding magnetic perturbation, as illustrated in figures 11(*c*) and (*d*) with the spectrogram of a LFS midplane Mirnov coil δb_r signal and the 12–20 kHz band-limited spectral power. A weak magnetic signature appears in the penultimate I_B step at the GAM frequency and then smoothly transitions in frequency and amplitude with the full MP I_B step. Note there is no corresponding rise in the edge T_e to account for a classic GAM frequency jump. T_i was not measured here, but would need to jump by >30% to account for the frequency step. GAMs in AUG generally display no δb_r signature, but often a δb_{pol} component [42] with a not too dissimilar to the predicted [43] and measured $m = \pm 2$ structure in TCV [44].

There is extensive theory on the penetration of resonant MP components leading to island formation at resonant rational surfaces and seeding neoclassical tearing modes (NTMs), as well as inducing toroidal torques etc. see [32, 34]. In the case of MHD modes reacting to MPs in AUG they are seen to slow down and eventually lock [45]. However, in the $n = 2$ cases discussed here the mode spins-up in frequency with the application of sufficient I_B. Although there is a minor burp in frequency (vertical line) in figure 11(*c*) coinciding with a T_e dip is hard to distinguish cause or effect. In addition, signatures remain even when clear mode-locking has taken place. On MAST, similar frequency up-shifting of flow perturbations were observed using probes [46]. As there was no magnetic signature (evidence against tearing modes) they were identified as GAMs with the frequency shift attributed to an increase in the edge plasma Mach number.

The AUG data suggests that the GAM is structurally modified by the MP. The δb_r signatures have an ($n = 2$, $m > 7$) structure, but the toroidal behaviour is complex with a strong dependence on the MP configuration and phase Φ. This suggests the intrinsic ($n = 0$) GAM b-field interacts with the applied MP b_n field creating a non-axisymmetric ($n \neq 0$) GAM and the observed δb_r signature. The GAM is weakened and radially eroded by the progressive field-line stochatization of the edge (not inconsistent with TEXTOR long-range correlation results [47]). However, further measurements and modeling are required to establish the extent of stochasticity needed to destroy the GAM.

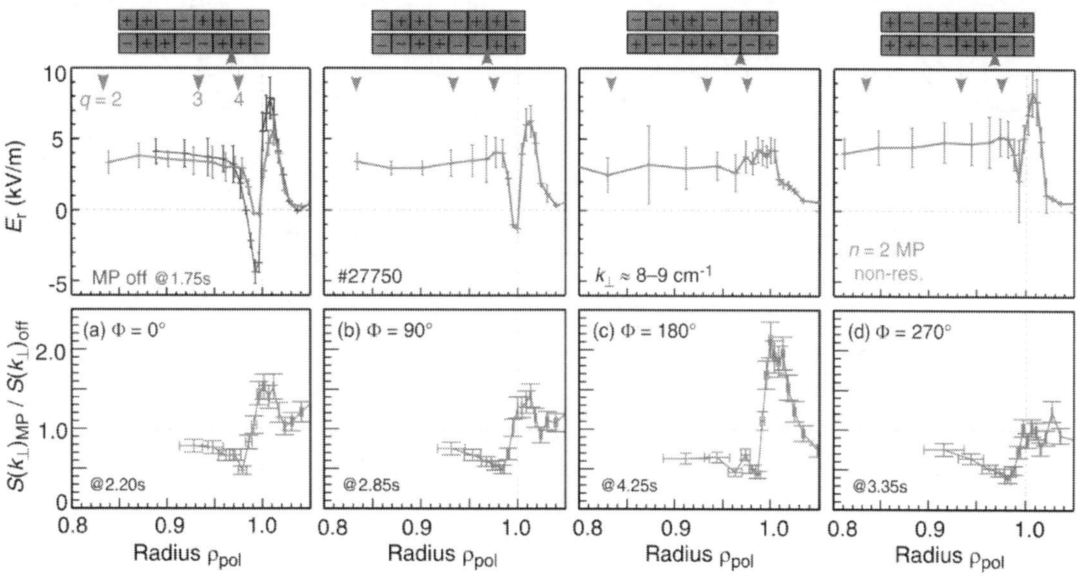

Figure 10. (Top row) E_r profiles and (bottom row) turbulence enhancement factor with (red) and without (blue) $n = 2$, 'non-resonant'($I_B = 0.9$ kA, $\Delta\Phi = 90°$) MPs for $\Phi = 0°$, 90°, 180° and 270° toroidal phases. (−2.5 T, 1.0 MA, $q_{95} \approx -4$, $P_{ech} = 0.65$ MW, with constant $T_{eo} = 4.5$ keV and $n_{eo} \approx 1.5 \times 10^{19}$ m^{-3}).

9. Summary and discussion

The high resolution AUG Doppler reflectometer results presented here both support previous observations as well as reveal new features. Specifically the reduction and reversal of the E_r well and the dependence on n_e and b_n thresholds linked to the MP penetration are identified. Different threshold dependencies for the SOL and confinement regions are also noted. Such differences may be attributed to the laminar versus stochastic field line nature of the two regions.

$n = 2$ MP penetration into the confinement region appears to be effective only at relatively low plasma density and in general extends only a few cm beyond the separatrix. However, at very low density and/or for $n = 1$ MPs the penetration can be significant and the promotion of mode locking for certain MP phases is pronounced. Between mode locking and total field exclusion is a density range where much detailed information on the MP field structure and profile impact can be obtained. In this operational range, despite the evident impact on the edge E_r and turbulence, the MPs only mildly affect the kinetic profiles and energy confinement.

For the confinement region, the MP spectral content and degree of resonance with the edge rational surfaces plays a significant role. A high degree of MP resonance has a stronger impact with larger and wider E_r well reversals. The radial limit of the MP field penetration is particularly well illustrated when raising the q profile and filling the edge with high order rational surfaces. To properly quantify the MP penetration and the extent of the plasma response in modifying the resonance condition, requires modelling of the AUG geometry and L-mode conditions. This is now beginning. Likewise, the effect of plasma rotation on screening is also a topic of further investigation.

The 3D structure of the MP field has also been probed by rotating the MP coil phase toroidally. Significant variations, as well as discrepancies, are revealed in the toroidal pattern in

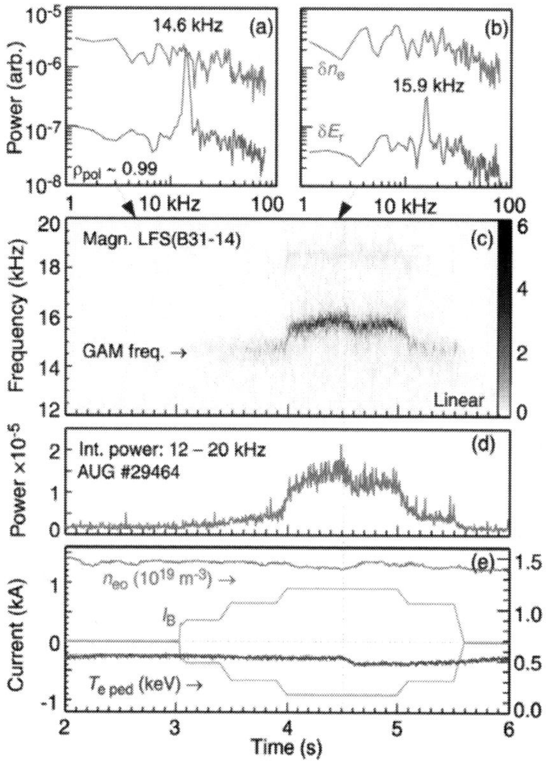

Figure 11. E_r and n_e spectra at $\rho_{pol} \approx 0.99$ without (*a*) and with (*b*) MP, (*c*) LFS Mirnov (B31-14) spectrogram with linear power scale, (*d*) 12–20 kHz integrated spectral power and (*e*) core n_{eo}, edge T_{edge} and MP current I_B time traces for L-mode discharge #29464. Parameters as per figure 6.

both the near SOL and the confinement regions—which again highlights the different nature of the two regions. The spatial sampling of the toroidal structure was limited here (four Φ for $n = 2$), however, finer toroidal resolution measurements using dynamic coil currents have recently been achieved (to be presented shortly).

The turbulence behaviour is more complicated. δn_e is generally enhanced in the near SOL for all MP configurations, but can be enhanced or reduced across the edge depending on the degree of MP resonance and/or configuration. Nevertheless, there is a clear linkage between the δn_e behaviour and the negative E_r shear. Coincident with the E_r impact, the edge flow (GAM) properties are also modified. A coherent flow perturbation continues to exist across the edge region with a frequency which chirps up from the preceding GAM. The flow perturbation amplitude decreases, but is accompanied by an enhanced magnetic signature, consistent with an non-axisymmetric GAM. Finally, although only results for $n = 2$ MPs were shown, similar overall behaviour is also observed for $n = 1$ and $n = 4$ MPs, which will be presented in future work.

Acknowledgments

This work has been carried out within the framework of the EUROfusion Consortium and has received funding from the European Unions Horizon 2020 research and innovation programme under grant agreement number 633053. The views and opinions expressed herein do not necessarily reflect those of the European Commission. This work has also been performed in the framework of the Helmholtz Virtual Institute on plasma dynamical processes and turbulence using advanced microwave diagnostics.

References

[1] Evans T E *et al* 2005 *Nucl. Fusion* **45** 225004
[2] Kirk A *et al* 2010 *Nucl. Fusion* **50** 034008
[3] Canik J M *et al* 2010 *Phys. Rev. Lett.* **104** 045001
[4] Suttrop W *et al* 2011 *Phys. Rev. Lett.* **106** 225004
[5] Jeon Y M *et al* 2012 *Phys. Rev. Lett.* **109** 035004
[6] Wootton A J 1990 *J. Nucl. Mater.* **176–177** 77
[7] Yan X Z *et al* 1991 *Phys. Fluids* B **3** 3448
[8] Shoji T *et al* 1992 *J. Nucl. Mater.* **196–8** 296
[9] McKee G R *et al* 2013 *Nucl. Fusion* **53** 113011
[10] Xu Y *et al* 2006 *Phys. Rev. Lett.* **97** 165003
[11] Unterberg B *et al* 2007 *J. Nucl. Mater.* **363–5** 698
[12] Tamain P *et al* 2010 *Plasma Phys. Control. Fusion* **52** 075017
[13] Payan J *et al* 1995 *Nucl. Fusion* **35** 1357
[14] Xu Y *et al* 2007 *Nucl. Fusion* **47** 1696
[15] Müller H W *et al* 2013 *J. Nucl. Mater.* **438** 564
[16] Müller H W *et al* 2014 *Contrib. Plasma Phys.* **54** 261
[17] Krämer-Flecken A *et al* 2006 *Nucl. Fusion* **46** S730
[18] Viezzer E *et al* 2013 *Nucl. Fusion* **53** 053005
[19] Suttrop W *et al* 2009 *Fusion Eng. Des.* **84** 290
[20] Suttrop W *et al* 2012 *Proc. 39th EPS Conf.* (*Stockholm, Sweden*, 2–6 July 2012) vol 36F, P 2.092
[21] Conway G D *et al* 2004 *Plasma Phys. Control. Fusion* **46** 951
[22] Conway G D *et al* 2010 *Plasma Fusion Res.* **5** S2005
[23] Poli E *et al* 2001 *Comput. Phys. Commun.* **136** 90
[24] Lunt T *et al* 2012 *Nucl. Fusion* **52** 054013
[25] Conway G D *et al* 2008 *Proc. 22nd IAEA Fusion Energy Conf.* (*Geneva, Switzerland, 13–18 October 2008*) IAEA-CN-165/EX/P5-38
[26] Fitzpatrick R 1993 *Nucl. Fusion* **33** 1049
[27] De Bock M F M *et al* 2008 *Nucl. Fusion* **48** 015007
[28] Waelbroeck F L 2009 *Nucl. Fusion* **49** 104025
[29] Yu Q *et al* 2009 *Phys. Plasma* **16** 042301
[30] Park G *et al* 2010 *Phys. Plasma* **17** 102503
[31] Liu Y *et al* 2010 *Phys. Plasma* **17** 122502
[32] Callen J D 2011 *Nucl. Fusion* **51** 094026
[33] Kaveeva E and Rozhansky V 2012 *Nucl. Fusion* **52** 054011
[34] Fitzpatrick R 2012 *Plasma Phys. Control. Fusion* **54** 094002
[35] Coenen J W *et al* 2011 *Nucl. Fusion* **51** 063030
[36] Conway G D *et al* 2013 *Proc. 40th EPS Conf.* (*Helsinki, 1–5 July 2013*) vol 37D, P5.175
[37] Maraschek M *et al* 2013 *Proc. 40th EPS Conf.* (*Helsinki, 1–5 July 2013*) vol 37D, P4.127
[38] Fuchs J C *et al* 2011 *Proc. 40th EPS Conf.* (*Helsinki, 1–5 July 2013*) vol 35G, P1.090
[39] Fuchs J C *et al* 2013 *Proc. 40th EPS Conf.* (*Helsinki, 1–5 July 2013*) vol 37D, P4.126
[40] Fischer R *et al* 2012 *Plasma Phys. Control. Fusion* **54** 115008
[41] Conway G D *et al* 2008 *Plasma Phys. Control. Fusion* **50** 055009
[42] Simon P 2014 Private communication
[43] Wahlberg C 2009 *Plasma Phys. Control. Fusion* **51** 085006
[44] Huang Z *et al* 2013 *Proc. 40th EPS Conf.* (*Helsinki , 1–5 July 2013*) vol 37D, P2.175
[45] Fietz S *et al* 2013 *Proc. 40th EPS Conf.* (*Helsinki, 1–5 July 2013*) vol 37D, P5.158
[46] Robinson J R *et al* 2012 *Plasma Phys. Control. Fusion* **54** 105007
[47] Xu Y *et al* 2010 *Proc. 23rd IAEA Fusion Energy Conf.* (*Daejon, 11–16 October 2010*) EXC/9-3

IOP Publishing

Plasma Physics and Controlled Fusion

Plasma Phys. Control. Fusion **57** (2015) 014036 (9pp)

doi:10.1088/0741-3335/57/1/014036

Topology bifurcation of a magnetic flux surface in toroidal plasmas

K Ida[1], T Kobayashi[1], S Inagaki[2], Y Suzuki[1], S Sakakibara[1], K Itoh[1], H Tsuchiya[1], C Suzuki[1], M Yoshinuma[1], Y Narushima[1], M Yokoyama[1], S-I Itoh[2] and the LHD Experiment Group[1,2]

[1] National Institute for Fusion Science, Toki, Gifu 509-5292, Japan

[2] Research Institute for Applied Mechanics, Kyushu University, Kasuga 816-8560, Japan

E-mail: ida@nifs.ac.jp

Received 26 June 2014, revised 4 October 2014
Accepted for publication 13 October 2014
Published 28 November 2014

Abstract

The bifurcation of magnetic topology is identified by a significant change in the heat pulse propagation properties in magnetized plasmas. Clear evidence of stochastization of the magnetic surfaces near a rational surface is observed in the core plasma with weak magnetic shear in the Large Helical Device by slowly decreasing the magnetic shear and measured by applying heat pulses driven by modulated electron cyclotron heating (MECH). Three topologies of the magnetic flux surfaces (a nested magnetic island and partial and full stochastization) are identified by the patterns of heat pulse propagation observed in the flat temperature region in the plasma. Slow heat pulse propagation exhibiting a non-monotonically increasing delay time is evidence of a magnetic island, while the fast heat pulse propagation observed in the plasma with medium magnetic shear is evidence of the stochastization of the magnetic surfaces. The region with the fast heat pulse propagation varies with a slight change of magnetic shear. There are two types of stochastization of the magnetic surfaces. In one, the stochastization region is localized near the rational surface (partial stochastization) and in the other, the region is extended to the magnetic axis (full stochastization). The appearance of a stochastic magnetic field is not caused by MHD instability. The significant increase of the ratio of electron thermal diffusivity to ion thermal diffusivity is consistent with that expected by stochastization of the magnetic field.

Keywords: plasma, magnetic topology, magnetic island, stochastization

(Some figures may appear in colour only in the online journal)

1. Introduction

A global stochastization of the magnetic surfaces is expected to be induced when the magnetic islands are overlapped and their width exceeds a threshold in toroidal plasmas and this typically occurs in the core plasma in a reversed-field pinch (RFP) [1] and in the ergodic region near the plasma periphery of tokamaks with perturbation fields [2]. Stochastization of the magnetic surfaces has been considered to be important [3], because the partial stochastization of the magnetic surface due to the perturbation fields can contribute to the mitigation of Edge-Localized-Modes (ELMs). Moreover, a spontaneous transition from the stochastic state to a helical equilibrium in an RFP [4] and a bifurcation phenomenon of a magnetic island at a rational surface in the Large Helical Device (LHD) [5] suggest that the stochastization of the magnetic surfaces as well as the formation of a magnetic island is a self-organized transition phenomenon.

In spite of the importance of the experimental confirmation of the stochastization of the magnetic surface, it is difficult to measure the stochastization of the magnetic surface directly and it is inferred from the magnetic field measured outside the plasma. Since the heat conductivity parallel to the magnetic field is much larger than that perpendicular to the magnetic field, the radial heat conductivity (heat diffusivity) should be extremely large. One piece of evidence of

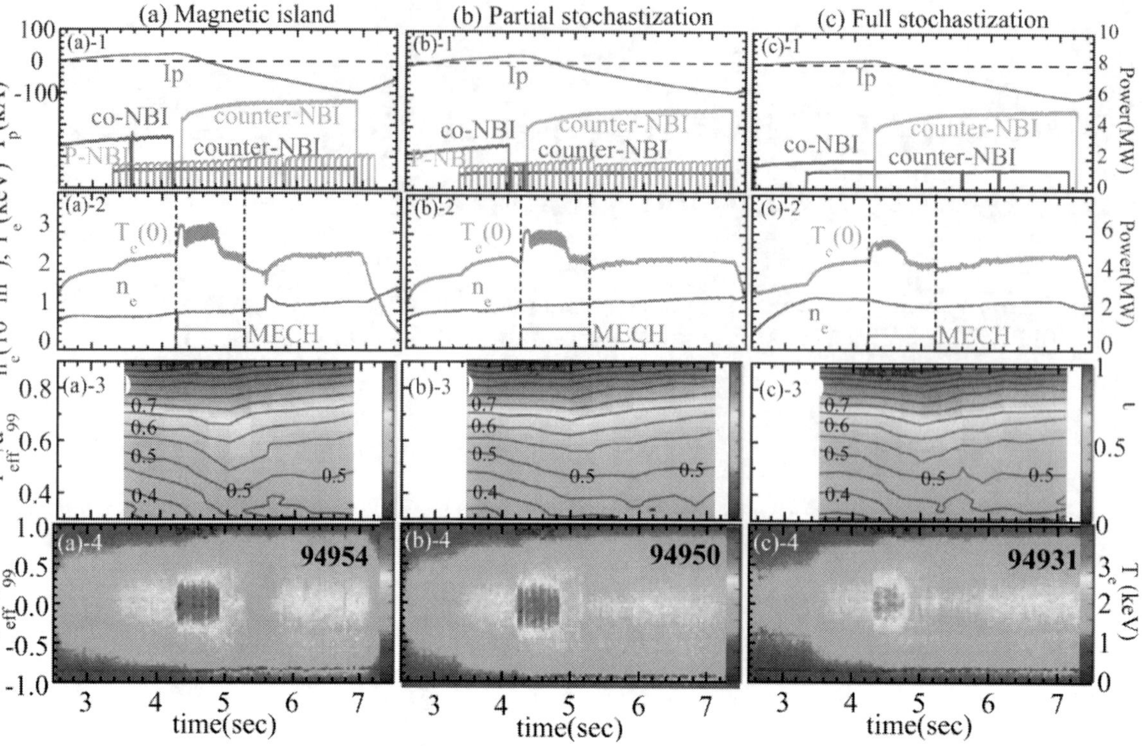

Figure 1. Time evolution of (1) the plasma current, power of neutral beam injection in the co-direction, counter-direction and perpendicular-direction (co-NBI, counter-NBI and P-NBI, respectively), where the counter-NBI with 5MW is to drive the plasma current to negative (in the counter-direction) and to decrease the magnetic shear, while the counter-NBI with a power of 1.5 MW is for the MSE measurements, (2) central electron temperature, line averaged density and contour of (3) rotational transform (ι) and (4) electron temperature in the discharge with (*a*) a magnetic island, (*b*) partial stochastization and (*c*) full stochastization. The ι values are indicated in the contour plot of rotational transform.

stochastization of the magnetic surface is the flattening of the T_e profile. However, flattening can occur due to the lack of heat flux crossing the magnetic flux as seen in the magnetic island, even if the magnetic flux surface is nested inside.

In RFP plasmas, fine structures of the internal magnetic field and current density have been measured with the Faraday effect [6, 7] and direct measurement of the internal magnetic field structure associated with a 3D helical equilibrium generated spontaneously in the core of an axisymmetric toroidal plasma clearly demonstrated the bifurcation from a stochastic magnetic field to a helical flux surface [8]. However, direct measurements of the stochastization of the magnetic field surfaces (non-integrability of the magnetic field, B) require very precise measurements of all components of B in a 3D geometry, which is not possible with current B-field measurement techniques.

In the LHD, the change of the magnetic topology, for example, the change from nested magnetic flux surfaces to a magnetic island, can be identified by heat pulse propagation [9, 10]. These experiments suggest that the stochastization of the magnetic flux surfaces [11] can also be identified by the characteristics of heat pulse propagation [12]. In this paper, we describe the topology bifurcation of magnetic topology (ex. magnetic flux with normal helical equilibrium, magnetic island and stochastic magnetic field region) in the core plasma in the LHD identified by a significant change in the heat pulse propagation properties.

2. Magnetic shear control by NBCD

The LHD is a heliotron-type device which has three tangential neutral beams, two of which are used to drive the plasma current in a direction parallel (coinjection) or antiparallel (counterinjection) to the equivalent plasma current. The radial profiles of the rotational transform, ι [$\approx \langle RB_\theta/(rB_\phi)\rangle$], are derived from the polarization angle measured with the Motional Stark Effect (MSE) spectroscopy using an equilibrium code [13]. The electron temperature (T_e) is measured with a YAG Thomson scattering system and an electron cyclotron emission (ECE) radiometer. The ion temperature (T_i) is measured with charge exchange spectroscopy [14].

Figure 1 shows the time evolution of the plasma current, power of the neutral beam injection in co-direction, counter-direction and perpendicular-direction, the central electron temperature, the line averaged density and contour of rotational transform (ι) and the electron temperature in the discharge with (*a*) the magnetic island, (*b*) partial stochastization and (*c*) full stochastization. In this experiment, the magnetic field, B, is 2.705 T at the magnetic axis in the vacuum field, R_{ax}, is 3.6 m The direction of NBI is changed from co-injection to counter-injection during the discharge. The counter-NBI with 5 MW is to drive the plasma current to negative levels (in the counter-direction) and to decrease the magnetic shear, while the counter-NBI with the power of 1.5 MW is for the MSE measurements. In discharges (*a*) and (*b*) the perpendicular NBI (P-NBI) is injected for 3.3–7.3 s with 10 Hz modulation in order to measure the ion temperature with charge exchange spectroscopy. The 10 Hz variations in the electron temperature, T_e in cases (*a*) and (*b*) are due to the modulation of P-NBI. The MECH heating power in the discharge (*c*) is lower than that in discharges (*a*) and (*b*).

The total plasma current driven by the neutral beam is in the range of 100 kA (counterdirection) to 50 kA (codirection), which is only 3–6% of the equivalent plasma current (1.8 MA) produced by the external helical coils. The line averaged electron density is $\sim 1 \times 10^{19}$ m^{-3} and the central electron temperature is 2 ~ 3 keV for these three discharges. The central electron temperature increases after the onset of modulated electron cyclotron heating (MECH) ($t = 4.2$ s) and an abrupt drop of electron temperature is observed at the $t \sim 4.8$ section. Although the change near the plasma edge ($\iota > 0.8$) is due to the neutral beam current drive and is consistent with that of the total plasma current, the change in rotational transform at the plasma core region ($\iota < 0.6$) is the opposite to that at the plasma edge.

As seen in the contour of rotational transform (ι), after the switch of the NBI from co-direction to counter-direction, the core ι increases while the edge ι decreases and the gradient of ι (magnetic shear) near the rational surface ($r_{\rm eff}/a_{99} \sim 0.5$, $\iota = 0.5$) becomes smaller. The increase of the core ι is due to the inductive current compensating the toroidal current driven by the neutral beam. The increase of core ι is larger (magnetic shear is smaller) in the plasma with the magnetic island (case a) than in the plasma with stochastization (cases b and c). The decay of the magnetic shear is slightly faster in the discharge with the magnetic island than in the discharges with stochastization. This slight difference of rotational transform profile causes a transition to magnetic island or stochastization. The change in the core region is due to the inductive current compensating the toroidal current driven by the neutral beam. The contour of electron temperature shows sudden flattening at $t = 4.8$ s (0.5 s after the switch of NBI direction) for these three discharges. The differences in the magnetic topology (magnetic island, partial/full stochastization) are identified by the characteristics of the heat pulse propagation, which will be discussed below.

Figure 2 shows the relation between the electron temperature gradient measured with a YAG Thomson scattering system and magnetic shear measured with MSE at the rational surface of $\iota = 0.5$ in the discharge with the magnetic island, partial stochastization and full stochastization. After the direction of the injected neutral beam switches from parallel (codirection) to antiparallel (counterdirection) to the equivalent plasma current or vice versa at $t = 4.3$ s, the magnetic shear at $\iota = 0.5$ ($r_{\rm eff}/a_{99} \sim 0.4$) gradually decreases from 0.9–1.0 to 0.2–0.3. When the magnetic shear reaches the critical values (0.45 for the magnetic island formation and 0.55 for the stochastization), temperature flattening suddenly takes place and temperature gradients drop to zero. It is an important fact that the critical magnetic shear for the transition to stochastization as indicated by the sudden drop of temperature gradient is higher than that for the transition to the magnetic island.

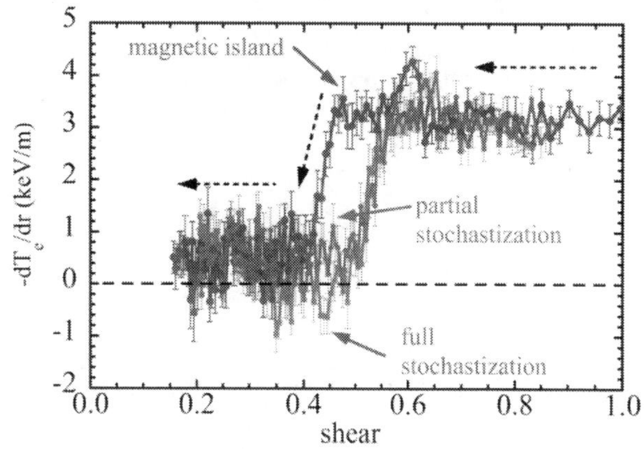

Figure 2. The relation between the electron temperature gradient and magnetic shear at the rotational surface of $\iota = 0.5$ in the discharge with a magnetic island, partial stochastization and full stochastization.

3. Characteristics of magnetic field perturbation

Because the appearance of a magnetic island and stochastization cause temperature flattening at a rational surface, it is important to investigate the magnetic field perturbation to distinguish the magnetic topology. Figure 3 shows the time evolution of the $m/n = 2/1$ and $m/n = 1/1$ component of the perturbation flux of the radial magnetic field estimated from saddle loop measurements for the discharges with (*a*) a magnetic island, (*b*) partial stochastization and (*c*) full stochastization. The poloidal distribution of the perturbation flux and the curve fitted with $m = 1$ and $m = 2$ Fourier components are also plotted. Before the temperature flattening ($t = 4.5$ s) the perturbation flux of the radial magnetic field is small for all poloidal angles. At a sudden drop of the temperature gradient, a perturbation field of $m/n = 2/1$ or 1/1 starts to increase ($t = 4.8$ in (*a*) and ($t = 4.6$ s in (*b*) and (*c*)). A clear difference in the perturbation magnetic field with the lowest mode number of $m/n = 2/1$ is observed. In the discharge (*a*), a perturbation of the $m/n = 2/1$ mode starts to increase at the $t = 4.8$ section. This fast growth of the $m/n = 2/1$ perturbation magnetic field measured with the saddle loop in a discharge with a fast magnetic shear drop indicates the growth of the 2/1 magnetic island. The number of the saddle coil signal is only 5 in the toroidal direction. Therefore, the saddle coil is only sensitive to $n = 1$ and cannot detect the perturbation of the magnetic field with $n > 1$ mode.

In contrast, in the discharges (*b*) and (*c*) with slightly slower magnetic shear drop, there is no significant increase of perturbation of the $m/n = 2/1$ mode, and the increase of the perturbation of the $m/n = 1/1$ mode at $t = 4.6$ s is more pronounced in the discharges (*b*) and (*c*) with a slightly slow magnetic shear drop. The increase of the perturbation of the $m/n = 1/1$ mode is due to the growth of the $m/n = 1/1$ magnetic island near the edge at $r_{\rm eff}/a_{99} \sim 0.9$ and $\iota = 1$. Although the size of the $m/n = 1/1$ magnetic island is much smaller than that of $m/n = 2/1$, the observed perturbation magnetic field of the $m/n = 1/1$ magnetic island becomes comparable to that of the $m/n = 2/1$ magnetic island because of the difference in the distance between the flux loop

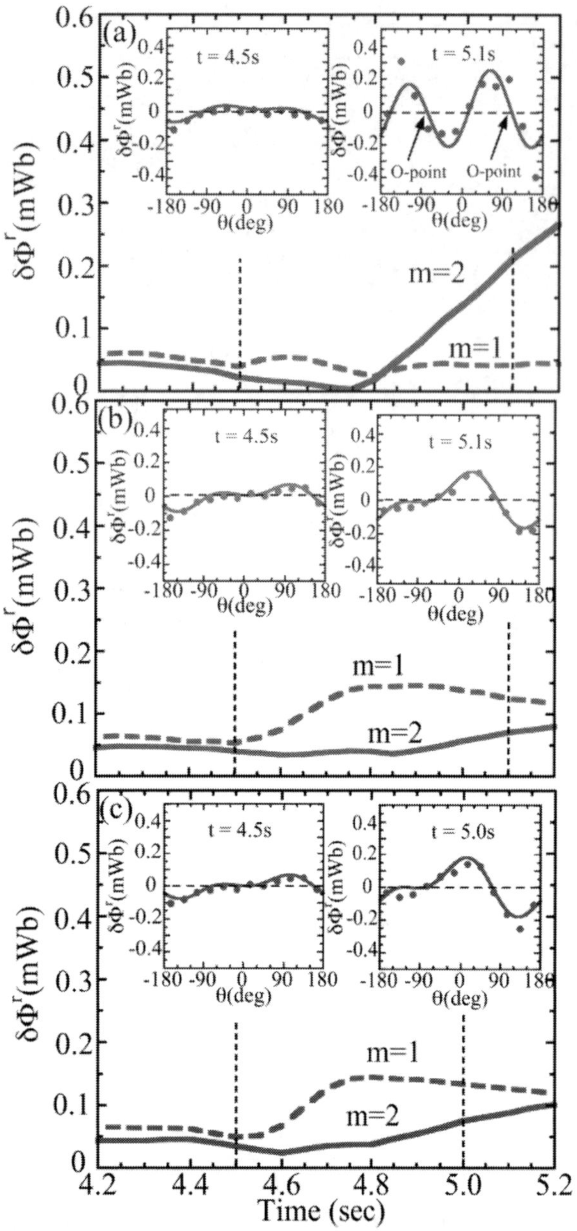

Figure 3. Time evolution of the $m/n = 2/1$ and $m/n = 1/1$ components of the perturbation flux of the radial magnetic field estimated from saddle loop measurements for the discharges with (*a*) magnetic island, (*b*) partial stochastization and (*c*) full stochastization. The poloidal distribution of the perturbation flux and the fitted curve with $m = 1$ and $m = 2$ Fourier components are also plotted.

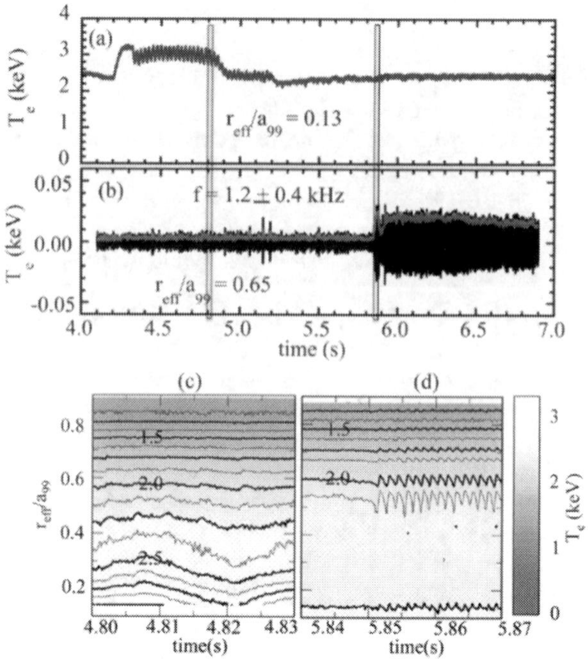

Figure 4. Time evolution of (*a*) electron temperature near the magnetic axis ($r_{eff}/a_{99} = 0.13$), (*b*) temperature fluctuation outside the temperature flattening region ($r_{eff}/a_{99} = 0.65$) and its envelop in the frequency range of 0.6–3.0 kHz and the temperature contour, (*c*) before temperature flattening ($t = 4.80$–4.83 s) and (*d*) at the onset of the interchange mode ($t = 5.84$–5.87 s) for the discharge with partial stochastization. The T_e values are indicated in the contour plot (*c*) and (*d*).

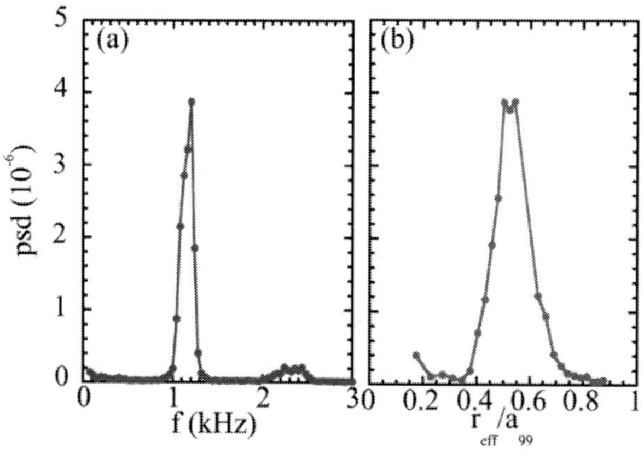

Figure 5. Square of fluctuation amplitude as a function of (*a*) frequency and (*b*) normalized minor radius at $t = 6.0$–6.4 s for the discharge with partial stochastization

outside of the plasma and the locations of the magnetic islands ($r_{eff}/a_{99} \sim 0.5$ for the $m/n = 2/1$ magnetic island and $r_{eff}/a_{99} \sim 0.9$ for the $m/n = 1/1$ magnetic island). There is no detectable difference in the saddle loop signal between the cases of (*b*) partial stochastization and (*c*) full stochastization. This is because the transition from partial to full stochastization is due to the radial extension of higher harmonic modes to the magnetic axis. The saddle loops are insensitive to the change of the magnetic field near the plasma centre. Figure 3(*a*) shows that a magnetic island with the lowest mode number of $m/n = 2/1$ develops, while figures 3(*b*) and (*c*) indicate that the stochastic magnetic field due to the overlapping of the magnetic islands with higher order at the rational surface of $\iota = 0.5$ becomes dominant. Please note that there are only 12 saddle loops in the poloidal direction, which may not be enough to identify the poloidal mode number of the higher harmonic modes.

It is interesting whether or not there is an MHD activity before the flattening of the temperature. There is no significant

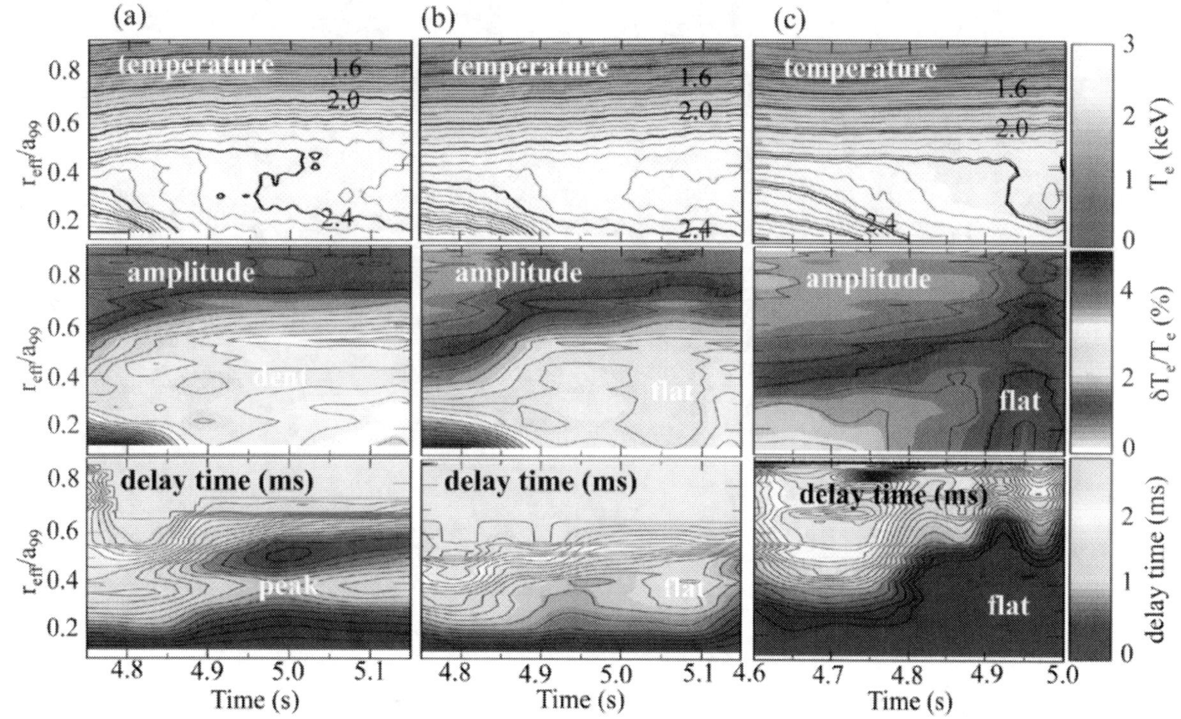

Figure 6. Contour of electron temperature, temperature modulation amplitude ($\delta T_e/T_e$) and delay time of a fundamental component (39 Hz) of a heat pulse in space and time for the discharges with (*a*) magnetic island, (*b*) partial stochastization and (*c*) full stochastization. The T_e values are indicated in the contour plot of temperature.

MHD activity in the frequency range of 0.6–3.0 kHz just before the drop of the central temperature (t = 4.80–4.83 s) as seen in figure 4. The gradual change in temperature is due to the modulated ECH with a frequency of 39 Hz. The displacement of the magnetic flux in this mode is ~3 cm ($\delta r_{eff}/a_{99} \sim 0.05$) and is in phase for all chords. This is in contrast to that of the tearing mode where the displacement of the magnetic flux is out-of-phase between the inner and outer side of the rational surface. Therefore, the in-phase displacement of the magnetic flux is considered to be an interchange mode [15]. It is important to investigate the characteristics of MHD instability, such as frequency spectra and radial profiles of amplitude. As seen in figure 5, this interchange mode shows a very sharp peak at a frequency of 1.2 kHz and is localized at $r_{eff}/a_{99} = 0.6 \pm 0.1$ near the rational surface of ι of 0.5, which is also consistent with the characteristics of the interchange mode. It is also interesting that the appearance of this mode is abrupt within one oscillation. The delay in the appearance of interchange modes suggests that the flattening of the electron temperature is due to the formation of a magnetic island or to their overlapping rather than the growth of the interchange modes.

4. Heat pulse propagation driven by modulated ECH

As discussed in the introduction, the heat pulse propagation experiment is a useful tool to identify the magnetic topology in further detail. Figure 6 shows a contour of electron temperature, temperature modulation amplitude ($\delta T_e/T_e$) and delay time of the heat pulse in space and time for the discharges with (*a*) a magnetic island, (*b*) partial stochastization and (*c*) full stochastization. The delay time is derived from the phase delay of the fundamental modulation frequency relative to the innermost channel, where the MECH power deposits. There is a slight difference in temperature flattening. The temperature flattening starts at $r_{eff}/a_{99} = 0.35$ and the flat region expands in time in the discharge with a magnetic island, while it starts at $r_{eff}/a_{99} = 0.4$ and the flat region extends mainly inwards in the discharges with stochastization.

When the magnetic island exists in the plasma, the heat pulse propagation becomes complicated and does not show the monotonic change both in the temperature modulation amplitude ($\delta T_e/T_e$) and delay time. The heat pulse propagates radially faster through the X-point of the magnetic island than through the O-point of the magnetic island. This is because the magnetic field lines of the inner and outer boundaries of the magnetic island cross each other at the X-point. Therefore, the heat pulse arrives almost simultaneously at two points separated in the radius (inner and outer boundary of the magnetic island). Then the heat pulse propagates slowly from the boundary of the magnetic island to the O-point of the magnetic island, which can be detected by the bi-directional heat pulse propagation at the mid-plane where the O-point of the magnetic island is located. Since the heat pulse propagates through the X-point, the temperature modulation amplitude ($\delta T_e/T_e$) should be reduced, which will be detected as a dent of the modulation amplitude in the radial profile of $\delta T_e/T_e$. In fact, the small dent of the modulation amplitude appears at $r_{eff}/a_{99} = 0.4$ (two peaks at $r_{eff}/a_{99} = 0.26$ and 0.54) after

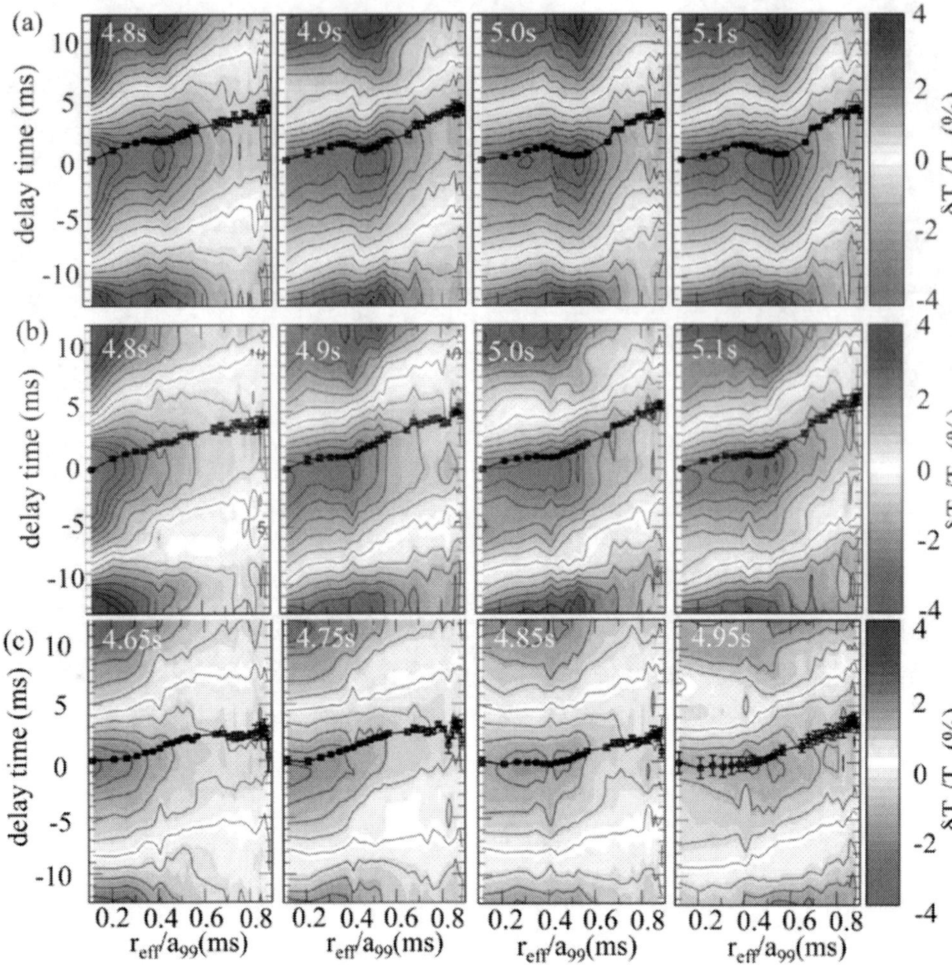

Figure 7. Contour of the relative modulation amplitude of electron temperature in space and time with a conditioning average of four modulations (~100 ms) for the discharges with (*a*) magnetic island, (*b*) partial stochastization and (*c*) full stochastization. The radial profile of the delay time of a fundamental component (39 Hz) of the heat pulse is also plotted.

t = 5.0 as seen in figure 6(*a*). The delay time shows a peak at r_{eff}/a_{99} = 0.4 and this peak becomes bigger as the magnetic island grows after t = 5.0. This observation is consistent with the growth of the m/n = 2/1 perturbation magnetic field measured with the saddle loop in figure 3. The flattening of the modulation amplitude and the delay time are observed in the discharges with stochastization. The flattening region of the delay time is located partially in the plasma of r_{eff}/a_{99} = 0.3–0.5 in the discharge with partial stochastization, while it becomes wider and then extends to the magnetic axis at r_{eff}/a_{99} < 0.6.

Figure 7 shows a contour of the relative modulation amplitude of electron temperature in space and time with a conditioning average of four modulations (~100 ms) for the discharges with (*a*) a magnetic island, (*b*) partial stochastization and (*c*) full stochastization. The velocity of the heat pulse propagation outside the stochastic region is an order of the transport time scale, while the velocity in the stochastic region is much faster. The radial profiles of the delay time with a fundamental frequency of 39 Hz are also plotted. Before the flattening of the temperature at t = 4.5 s, the radial propagation of the modulation amplitude of the electron temperature shows characteristics of diffusive transport and the delay time increases towards the plasma edge, which indicates normal nested magnetic flux surfaces. During the temperature flattening, the radial propagation of the modulation amplitude becomes complicated. The inflection of the curve of the delay time between 0.4 < r_{eff}/a_{99} < 0.6 in figure 7(*a*) is a clear signature of the presence of a m/n = 2/1 magnetic island. The inflection of the curve of the delay time becomes much weaker in the plasma with partial stochastization in figure 7(*b*). There is no inflection observed in the plasma with full stochastization as seen in figure 7(*c*). It should be noted that although the radial propagations of the relative modulation amplitude are almost identical for these three discharges, they show quite different characteristics during and after the temperature flattening.

The radial profile of electron temperature measured with YAG Thomson scattering and the delay time measured with ECE after the temperature flattening for these three discharges are plotted in figure 8. The location of the rational surfaces of ι = 0.5 and 1 are also indicated. The flattening of electron temperature is observed at the rational surface of ι = 1 in the

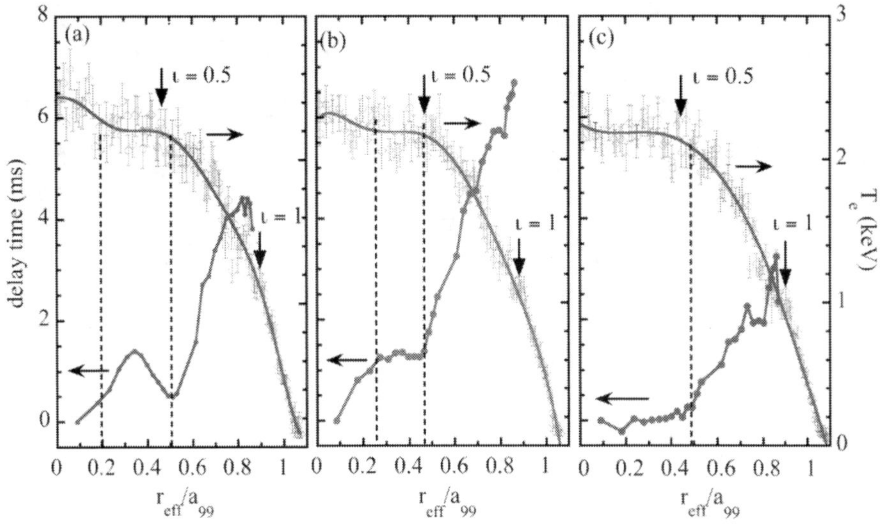

Figure 8. Radial profile of electron temperature and delay time for the discharges with (*a*) magnetic island, (*b*) partial stochastization and (*c*) full stochastization.

discharges with stochastization, which is consistent with the growth of the $m/n = 1/1$ perturbation magnetic field measured with the saddle loop in figure 2. The temperature profiles in the discharge (*a*) show a finite temperature gradient near the magnetic axis, which indicates the normal nested magnetic flux near the magnetic axis in the plasma with a magnetic island. However, this temperature gradient becomes small in the plasma with partial stochastization and disappears in the plasma with full stochastization, which suggests that the stochastization extends to the magnetic axis. These results are consistent with the finite temperature gradient in the temperature contour measured with ECE for the discharge with a magnetic island and partial stochastization, as shown in figure 5. In the case (*a*), the radial profile of the delay time shows a peak of 1.5 ms at $r_{eff}/a_{99} = 0.35$, which is clear evidence of the presence of a magnetic island. The time delay near the magnetic axis $r_{eff}/a_{99} < 0.2$ decreases after the formation of the magnetic island (see figure 6(*a*)), which implies the change in transport near the magnetic axis. This is in contrast to that of the radial profile of the delay time which shows a flattening of 1.5 ms at $0.27 < r_{eff}/a_{99} < 0.48$ in the plasma with partial stochastization (case (*b*)). In the plasma with partial stochastization, the time delay near the magnetic axis $r_{eff}/a_{99} \sim 0.2$ after the stochastization is similar to that before the stochastization (see figure 6(*b*)). Then the delay time of the flat region is comparable to the peak in the plasma with a magnetic island. In the case of (*c*), the delay time is close to zero in the wide region of the plasma $r_{eff}/a_{99} < 0.5$, which proves that the stochastization extends to the magnetic axis where the ECH power is deposited.

The magnitude of the radial magnetic field can be evaluated from the perturbation field measured with a saddle loop of the width of the magnetic island or the stochastization region. Figure 9(*a*) shows the time evolution of the ratio of the perturbed radial magnetic field to the toroidal magnetic field evaluated from the $m/n = 2/1$ component of the perturbation flux measured with saddle loop and from the width of the magnetic island or the partial stochastization region measured with heat pulse propagation. The magnitude of the perturbed radial magnetic field normalized by toroidal magnetic field reaches the level of 10^{-3} in the discharge with a magnetic island. In the plasma with a magnetic island, the perturbed radial magnetic fields evaluated from a saddle loop and heat pulse propagation agree with each other. However, in the plasma with partial stochastization, the perturbed radial magnetic field evaluated from a saddle loop is smaller than that evaluated from heat pulse propagation. This is because the saddle loops are insensitive to higher order modes of perturbation in the radial magnetic field. Figure 9(*b*) shows the relation between the island and partial stochastization width normalized by minor radius and a peak of the delay time of the heat pulse. Here the stochastization width is derived from the width of the flattening region of the electron temperature profile. In the plasma with a magnetic island the peak of the delay time near $r_{eff}/a_{99} \sim 0.4$ increases as the width of the magnetic island is increased and the delay time reaches 1–1.5 ms when the width of the magnetic island is 40% of the minor radius. However, in the plasma with partial stochastization, the delay time is less than 0.1 ms, which is smaller than that in the plasma with a magnetic island by one order of magnitude. This is because the heat pulse propagating parallel to the magnetic field line (the stochastic case) is much faster than the cross-field thermal diffusion that occurs within the magnetic island.

This is a distinct difference between the magnetic island and the stochastic domain in the region $\iota \sim 0.5$. The delay time in the plasma with full stochastization is not included in figure 9 because the ECH power deposits inside the stochastic region in this case and the modulation amplitude becomes too small to evaluate the delay time accurately. There are two traces with different delay times (0.9 ms and 1.5 ms at $W/a = 0.3$) in the plasma with a magnetic island, which indicates the different heat propagation speed inside the magnetic island. This data suggests that there are multi

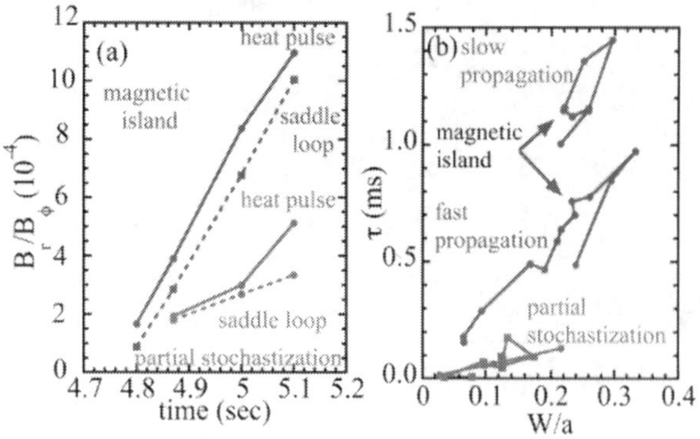

Figure 9. (*a*)Time evolution of the ratio of the perturbed radial magnetic field to the toroidal magnetic field evaluated from the $m/n = 2/1$ component of the perturbation flux measured with a saddle loop and from the width of the island or partial stochastization region measured with heat pulse propagation. (*b*) Relation between the island and partial stochastization width normalized by minor radius and the increase of the delay time of the heat pulse inside the temperature flat region.

states of completeness of the nesting of the magnetic field inside the magnetic island. Slower propagation of the heat pulse implies that the magnetic field inside the magnetic island is more completely nested.

5. The effect of stochastization on ion transport

It is important to compare the effect of stochastization on electron and ion transport. Figure 10 shows the flux gradient relation at $r_{eff}/a_{99} = 0.35$ in (*a*) electron and (*b*) ion heat transport during the flattening for the discharge with full stochastization. The radial profiles of the electron temperature and ion temperature before ($t = 5.41$ s in T_e, 5.84 s in T_i) and after ($t = 6.51$ s in T_e, 6.64 s in T_i) and the stochastization of the magnetic field are also plotted. In this discharge, co-NBI is injected at $t = 3.3$–5.3 s, then the two ctr-NBIs are injected at the $t = 5.3$–7.3 section The increase of electron radial heat flux from 10 to 20 keVm s^{-1} at the beginning is due to this increase in heating power. Both the electron and ion temperature gradients decrease in time even with the constant radial heat flux associated with the stochastization. It is an important finding that there is a different magnitude of flattening between the electron and ion temperature profiles. The electron temperature gradient decreases almost to zero, while the finite temperature gradient is observed in the ion temperature profile.

The thermal diffusivity which can be defined by the ratio of normalized heat flux to the temperature gradient is evaluated for ion and electron transport. The electron thermal diffusivity (χ_e) at $r_{eff}/a_{99} = 0.35$ is 4 m^2 s^{-1} (blue points) before the stochastization. It increases by more than one order of magnitude and exceeds 10^2 m^2 s^{-1} (red points) after the stochastization while the ion thermal diffusivity (χ_i) increases only by a factor of 1.6 ($\chi_i = 4$ m^2 s^{-1} → 6.5 m^2 s^{-1}). Theoretically, the electron and ion thermal diffusivity in the stochastic region can be evaluated as $\chi^{st}_{i,e} = D_M v_{e,i}$, where v_e and v_i are the thermal velocities of electrons and ions, respectively and D_M is the diffusion

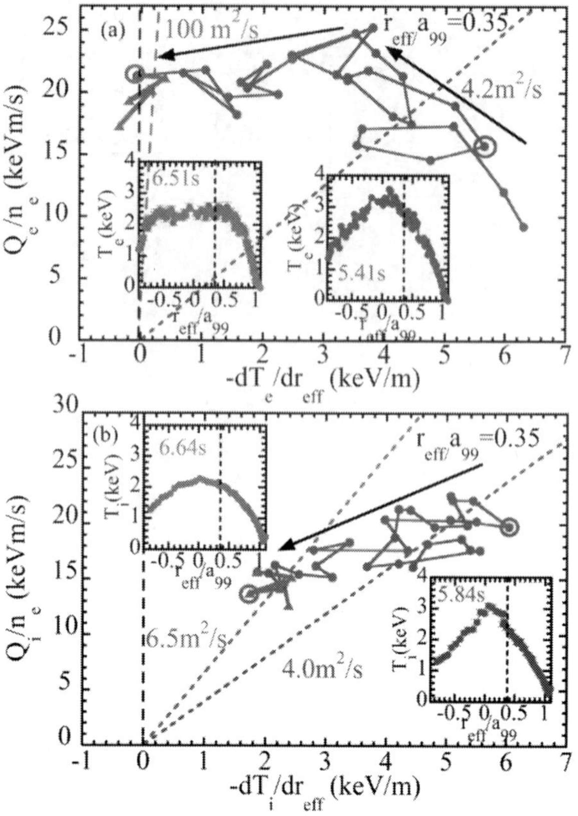

Figure 10. Flux gradient relation in (*a*) electron and (*b*) ion heat transport during the flattening for the discharge with full stochastization. The blue, green and red points refer to the data before during and after the stochastization. Profiles for circled points are also plotted.

of the field line defined by $2\pi (r_s^2 / R) B_n^2$, r_s is the radius of the resonant surface and B_n [$=(RB_r)/(r_s B_\phi)$] is the normalized perturbation field [3]. Then χ_e in the stochastic region is expected to be much larger than that of the ions by $(m_i/m_e)^{1/2}$ because of its larger thermal velocity [16]. The thermal diffusivity in the stochastic region χ^{st} evaluated from similar discharges [17] is 2–3 × 10^2 m^2 s^{-1} for electrons and 5–7 × m^2 s^{-1} for ions, which is consistent with this experimental observation.

6. Conclusion and discussion

Clear evidence of partial and full stochastization of the magnetic surfaces near a rational surface is observed in the core plasma with weak magnetic shear in the LHD by applying heat pulses driven by MECH. There are significant differences (slow and fast) in the heat pulse propagation in the electron temperature flat region in the LHD between the plasma with medium magnetic shear and weak magnetic shear. A slow heat pulse propagation observed in the weak shear is evidence of a nested magnetic island, while the fast heat pulse propagation observed in the plasma with medium magnetic shear is evidence of stochastization of the magnetic surfaces. The present experiment implies that there are three states of the magnetic flux surface (a nested magnetic island and partial and full stochastization) besides normal nested flux surfaces, as predicted by a theoretical model [18].

The effective electron thermal diffusivity (χ_e) and ion thermal diffusivity (χ_e) are comparable before the stochastization of the magnetic field ($\chi_e \sim \chi_i \sim 4$ m^2 s^{-1}). After the stochastization of the magnetic field, the increase of χ_e is much larger than that of the ions ($\chi_e/\chi_i > 10$–15) because of the difference in thermal velocity, which is in qualitative agreement with the Rochester model (~40) [19].

Acknowledgments

The authors would like to thank the technical staff in LHD for supporting these experiments. This work is partly supported by a Grant-in-Aid for Specially-Promoted Research (No. 21224014) and a Grant-in-aid for Scientific Research (No. 23246164) of JSPS Japan. This work is also partly supported by the National Institute for Fusion Science grant administrative budget, NIFS10ULHH021.

References

[1] Angelo F D' and Paccagnella R 1996 *Phys. Plasmas* **3** 2353
[2] Liang Y *et al* 2007 *Phys. Rev. Lett.* **98** 265004
[3] Lichtenberg A J *et al* 1992 *Nucl. Fusion* **32** 495
[4] Lorenzini R *et al* 2009 *Nat. Phys.* **5** 570
[5] Ida K *et al* 2008 *Phys. Rev. Lett.* **100** 045003
[6] Ding W X *et al* 2004 *Phys. Rev. Lett.* **93** 045002
[7] Ding W X *et al* 2009 *Phys. Rev. Lett.* **103** 025001
[8] Bergerson W F *et al* 2011 *Phys. Rev. Lett.* **107** 255001
[9] Inagaki S *et al* 2004 *Phys. Rev. Lett.* **92** 055002
[10] Yakovlev M *et al* 2005 *Phys. Plasmas* **12** 092506
[11] Suzuki Y *et al* 2013 *Plasma Phys. Control. Fusion* **55** 124042
[12] Kobayashi T *et al* 2011 *Plasma Phys. Control. Fusion* **53** 095012
[13] Ida K *et al* 2010 *Fusion Sci. Technol.* **58** 383
[14] Yoshinuma Y *et al* 2010 *Fusion Sci. Technol.* **58** 375
[15] Isayama A *et al* 2006 *Plasma Phys. Control. Fusion* **48** L45
[16] Itoh S-I *et al* 1996 *Plasma Phys. Control. Fusion* **38** 527
[17] Ida K *et al* 2013 *New J. Phys.* **15** 013061
[18] Itoh K *et al* 1992 *Nucl. Fusion* **32** 1851
[19] Rechester A B and Rosenbluth M N 1978 *Phys. Rev. Lett.* **40** 38

OPEN ACCESS
IOP Publishing
Plasma Physics and Controlled Fusion
Plasma Phys. Control. Fusion 57 (2015) 014037 (11pp)
doi:10.1088/0741-3335/57/1/014037

Characterising dust in JET with the new ITER-like wall

J C Flanagan[1,10,11], M Sertoli[2], M Bacharis[3], G F Matthews[1], P C de Vries[4], A Widdowson[1], I H Coffey[5], G Arnoux[1], B Sieglin[2], S Brezinsek[6], J W Coenen[6], S Marsen[2], T Craciunescu[7], A Murari[8], D Harting[1], A Cackett[1], E Hodille[9] and JET-EFDA Contributors[12]

JET-EFDA, Culham Science Centre, Abingdon, OX14 3DB, UK
[1] CCFE, Culham Science Centre, Abingdon, Oxon, OX14 3DB, UK
[2] Max-Planck-Institut für Plasmaphysik, EURATOM Association, 85748 Garching, Germany
[3] Imperial College of Science and Technology, London, SW7 2AZ, UK
[4] ITER Organization, Route de Vinon-sur-Verdon, CS 90 046, 13067 St. Paul Lez Durance Cedex, France
[5] Queen's University, Belfast, BT7 1NN, UK
[6] Forschungszentrum Juelich GmbH, EURATOM Association, 52425 Juelich, Germany
[7] EURATOM-MEdC Assoc., Inst. for Laser, Plasma and Radiation Physics, Romania
[8] Consorzio RFX-Associazione EURATOM ENEA per la Fusione, I-35127 Padova, Italy
[9] IRFM, CEA Cadarache, 13108 Saint Paul Lez Durance, France

E-mail: joanne.flanagan@ccfe.ac.uk

Received 24 June 2014, revised 12 September 2014
Accepted for publication 25 October 2014
Published 28 November 2014

Abstract

Recent studies dedicated to the characterisation of in-vessel dust in JET with the new ITER-like wall (ILW) show that dust levels are orders of magnitude lower compared with the latter stages of the carbon-wall (CW) period and are decreasing with operational time. Less than 1 g of dust was recovered in a recent inspection, compared with more than 200 g of material recovered at the end of the JET-CW life. Recent inspection of the ILW shows low rates of re-deposition with only small areas of damage of a type likely to create particulate matter. Quantifiers from laser scattering techniques also indicate an order of magnitude reduction in dust relative to the JET-CW and show that the amount of dust mobilized after a disruption is proportional to the dynamic vessel forces. It is not possible to infer what fraction of dust (if any) might be created by disruptions. However, disruption mitigation is found to reduce the amount of dust seen after moderate disruptions by a factor of 4. Analysis of the transient impurity events (TIEs) associated with dust show that tungsten dominates. A significant contribution to TIEs is also seen from iron, nickel and chromium (probably from steel and Inconel components). The incidence of severe negative effects on operations from TIEs is found to be relatively rare, with <1% of ILW disruptions linked to TIEs. The evolution of the TIE rate closely follows changes in the laser scattering dust quantifiers; both trend downwards in time but peak during periods of higher disruption rate (thought to be primarily driven by the mobilization of existing dust).

Keywords: ITER-like wall (ILW), dust, transient impurity events, JET, tokamak, UFO

(Some figures may appear in colour only in the online journal)

[10] Invited speaker.
[11] Author to whom any correspondence should be addressed.
[12] See the appendix of Romanelli *et al* 2012 *Proc. of the 24th IAEA Fusion Energy Conf.* (*San Diego, USA*).

1. Introduction

1.1. Dust

In this paper we focus on results from studies of dust in JET with the new ITER-like wall. However, in-vessel dust in magnetic fusion devices has been the topic of a large number of

Figure 1. Photograph of the JET ITER-Like Wall (ILW). Coloured shading indicates material composition.

studies and for a review of this area in general we direct the reader to (Krasheninnikov 2011, Ratynskaia 2011). In-vessel dust is loosely defined as particulate matter up to ~1 mm in size and is produced via plasma interaction with plasma-facing components and also via physical trauma during in-vessel maintenance / enhancement works (Rosanvallon 2009). As such, dust primarily consists of the materials that makeup the in-vessel environment. In JET's current ITER-like wall (ILW) configuration, the plasma facing components (PFCs) principally consist of beryllium (Be) and tungsten (W), with bulk Be tiles and W-coated carbon fibre composite (CFC) tiles and bulk W tiles used in the higher heat flux areas. Nickel, iron and chromium are also present via steel and Inconel components (i.e. supporting structures). Figure 1 shows JET's ITER-like wall, with the different materials shaded according to their composition. For a more detailed description we refer the reader to (Matthews 2011) and references therein.

In-vessel dust has important implications for plasma operations and safety in the tokamak environment, particularly in future long-pulse devices such as ITER (Rosanvallon 2009). Transient impurity events (TIEs), thought to be caused by small particles or dust contaminating the core plasma, are typified by a sharp increase in radiated power (Coenen 2013, Sertoli 2014). TIEs can harm plasma performance and have the potential to lead to instability events and can abruptly terminate the plasma discharge (Ekedahl 2009, de Vries 2014, Sertoli 2014). Large volumes of dust also present safety issues in terms of tritium retention and the potential for volatile interaction in the event of air or water ingress into a hot vessel (Bell 2002, Roth 2009). The current ITER strategy in response to these concerns is to measure and minimise the dust inventory (Rosanvallon 2009). Building an understanding of dust characteristics and their impact on plasma operations within current devices is thus essential to this goal.

1.2. Disruptions

During a disruption, the growth of plasma instabilities impairs confinement and causes a rapid and large loss of thermal energy (referred to as the thermal quench). This typically leads to the loss of plasma current (the current quench) and is often accompanied by a loss of vertical stability, called a vertical displacement event (de Vries 2014). The fast release of magnetic energy often results in large forces transferred to the mechanical structure of the vessel, causing it to shake. Disruption forces of up to 4 MN have been recorded at JET (in both CW and ILW operational periods), with disruption forces of >1 MN commonplace. Measurement of the vessel forces on JET are made by strain gauges placed on each of the vessel's 32 support legs (Marchese 1996, Lescure 2009). During the thermal quench, high heat-loads can also be transferred onto the PFCs. The relatively delicate nature of JET's ILW (Be melts at ~1300 °C) has necessitated the development of disruption mitigation methods based on a massive gas injection (MGI), delivered just prior to plasma termination (Lehnen 2010). Mitigation by MGI has been shown to significantly decrease the sideways forces during vertical displacement events. The heat loads during the thermal quench can also be reduced by the enhancement in energy radiated from the plasma that is seen with MGI (Lehnen 2010, de Vries 2014). Timing parameters relating to JET disruptions are defined here as: the disruption time t_d, which is equal to the onset of the thermal quench; and the time of maximum vessel displacement t_m, which marks the point at which one may expect to see disruption related mobilization occurring. Both of these times are given relative to the start of the plasma at $t = 0$ s.

1.3. Overview

In this work, we aim to provide answers to some key questions on the topic of in-vessel dust for JET's new ILW; e.g. how much dust is there and what is it made from? How does this compare to previous phases of operations with the CFC wall? What are the sources of dust and what impact does the current dust inventory have on plasma operations? Are there particular events that we should strive to avoid in terms of dust production / mobilization? How is the amount of dust changing with time? At what point does the dust inventory become a real problem and is active mitigation necessary? What does this mean for future devices, such as ITER? Answering these questions is only possible by studying the properties of the in-vessel dust at frequent intervals, including dust related events occurring during plasma operations. Consequently, remote methods of dust and TIE analysis are essential and are considered in addition to the assessment of information and material gathered during in-vessel studies (for which manned access is, briefly, necessary). The methods considered here are:

(a) Images of events during operations from JET's numerous camera-based monitoring systems
(b) VUV spectroscopy of TIEs during plasma discharges
(c) Analysis of photo surveys and dust collection performed during a recent in-vessel inspection
(d) Laser scattering from dust mobilized by disruptions

The experimental methods of observing dust and related events are described in detail in section 2, together with

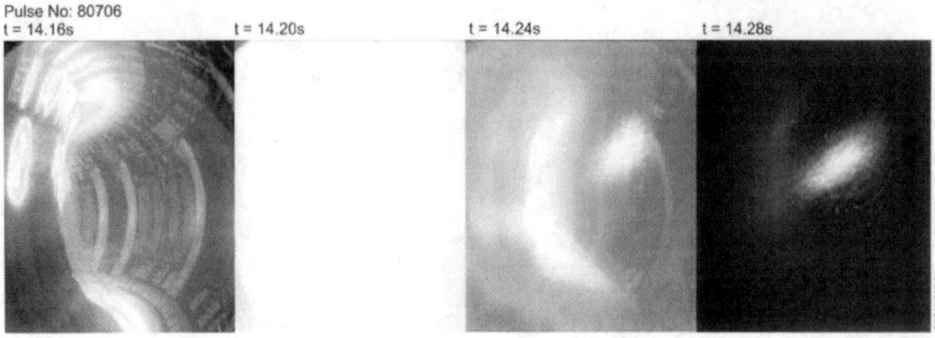

Figure 2. Frames from images from JET's COHU colour CCD camera (KL1) from octant 4 for JET pulse 80706. The plasma is terminated by a disruption with t_d = 14.17 s and t_m = 14.25 s. Induced vessel forces are recorded as 2.52 MN. A cloud of debris is seen at t = 14.24 s and in subsequent frames.

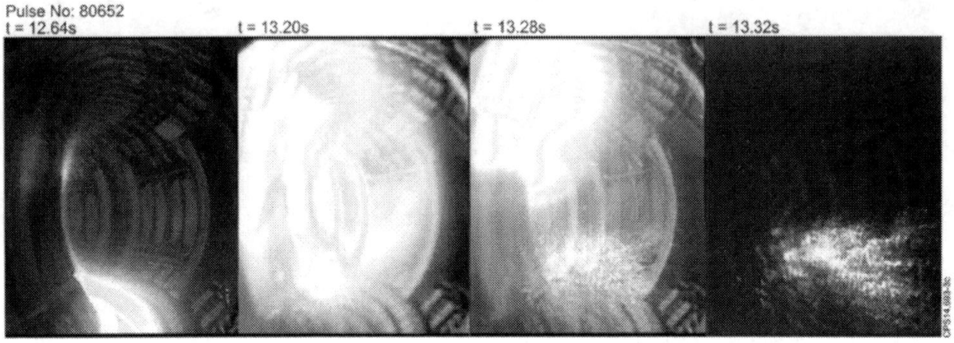

Figure 3. Images from JET's COHU colour CCD camera (KL1) from octant 4 for JET pulse 80652. The plasma is terminated by a disruption with t_d = 13.14 s and t_m = 13.18 s. Induced vessel forces are recorded as 0.43 MN. A cloud of debris is seen at t = 13.20 s and in subsequent frames. Note that the images in frames after $t = t_d$ show distortion due to camera shaking.

results for JET's recent operational period with the ILW. Where possible, comparisons to results from JET's carbon-wall (CW) period are made. Theoretical modelling of the interaction of plasma and dust is another valuable source of information; one that has the potential to bridge the gaps in knowledge obtained through remote observations and can also give a deeper understanding of the physical mechanisms responsible for impurity radiation events. Results from numerical modelling work on the dust-plasma interaction in JET presented in (Sertoli 2014) are highlighted for comparison within section 2 where relevant. Discussion and conclusions are presented in section 3.

2. Experimental methods and results

A summary of results from four methods of monitoring in-vessel dust are presented here; a) camera imaging, b) VUV spectroscopy, c) in-vessel inspection (including physical dust collection) and d) laser scattering after disruptions.

2.1. Camera systems

JET houses numerous visible and infrared camera systems that make real-time observations of the plasma chamber; to monitor the plasma and the plasma facing components. A systematic analysis of this type of data over the many thousands of JET pulses is a formidable task, and not one that we consider here. Instead, we look to this source of information for a few events of particular interest; to assist in the understanding and the benchmarking of data from other sources and to highlight events for further study. However, we note that the automatic analysis of defects (such as particulate matter) on the horizontal surfaces of JET's divertor tiles has been demonstrated (Craciunescu 2014a 2014b).

JET's camera systems are often particularly illuminating in the case of disruptions and often capture particles shaken loose by the vessel displacement, as seen in figures 2 and 3, which show sequential video frames from the KL1 camera for JET pulse 80706 and 80652 respectively. In both instances a cloud of glowing debris is visible just after the disruption. When sequential frames are viewed, the dust seen in figure 2 appears to be rotating, centred around the middle of the vessel, while the dust in figure 3 moves from the vessel centre to the outer divertor region and does not rise above the mid-plane of the vessel. The fact that dust does not uniformly fill the vessel in these events is important when interpreting data from diagnostic systems with narrow lines of sight. Disruptions can also lead to high heat-loads onto the PFC, which are closely monitored by JET's infrared protection camera systems (Arnoux 2010). In extreme heat-load events, melting of parts of the Beryllium PFCs as the plasma terminates has been observed. In JET pulse 85943, an experiment studying runaway electrons disrupted after the runaways doubled past the intended

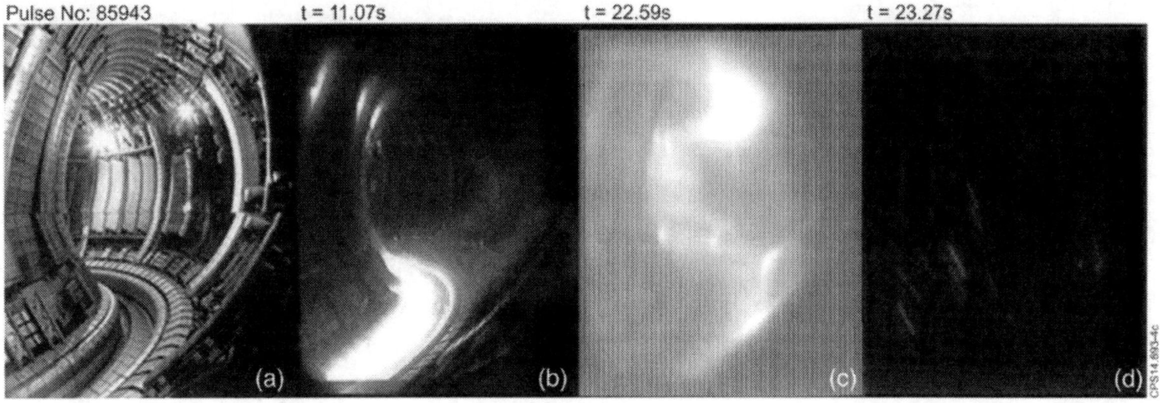

Figure 4. (*a*) View from octant 5, as seen by the high resolution DSLR camera (KL14) with the internal vessel lights on. (*b*)–(*d*) Frames from KL14 for JET pulse 85943. The plasma is terminated by a disruption with $t_d = 22.53$ s and $t_m = 22.59$ s. Induced vessel forces are recorded as 0.90 MN. (*d*) Sometime after the initial disruption (at $t = 23.10$ s onwards), particles can be seen falling from an upper part of the vessel into the divertor region.

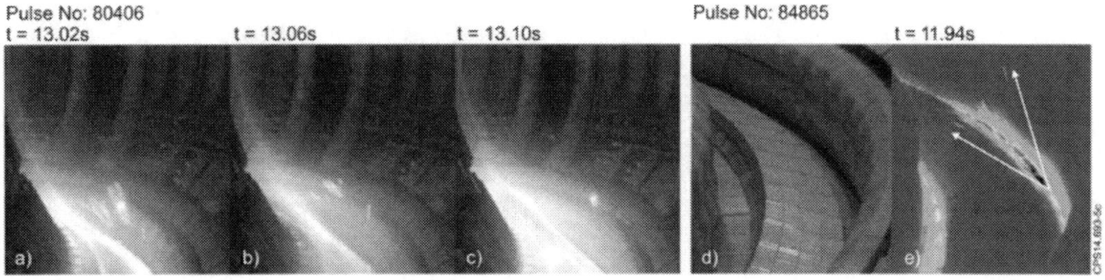

Figure 5. Images showing influx of particulate matter into the JET plasma. (*a*)–(*c*) JET's COHU colour CCD camera (KL1) from octant 4 for JET pulse 80406, (*d*) Divertor view from octant 1 as seen by the AVT-PIKE camera (KL11) (Huber 2012), with the internal vessel lights on (*e*) frame from KL11 for JET pulse 84865.

number. The resultant high heat-load on the upper dump plates caused partial melting. Droplets of beryllium were seen to rain down into the divertor region, as shown by the video frames from KL14, a high resolution DSLR camera, in figure 4. This is JET's most sensitive visible camera; the slight distortion in the image is due to the complex optics of this system, which includes in-vessel mirrors (Clever 2013).

The influx of particles into the JET plasma is also sometimes captured by camera systems; see examples in figure 5. JET pulse 80406 disrupted soon after the particle influx shown in figures 5(*a*)–(*c*), while the particles seen entering the plasma in JET pulse 84645 (figure 5(*e*)) resulted in a large TIE and significantly lowered the pedestal temperature. In both of these cases, VUV analysis identified tungsten as the responsible impurity (see section 2.2 for more details). Images of this type allow the particle trajectory and velocity at input to be approximated; parameters which will serve as useful inputs to any modelling effort (as described in section 2.3).

2.2. VUV spectroscopy of impurity events

Transient impurity events (TIEs) are also known as unidentified flying objects (UFOs) as they often appear as bright flying objects and are thought to result from small particles or dust contaminating the core plasma. TIEs are characterised by large, often short-lived, spikes in radiated power (Coenen 2013, Sertoli 2014). An example of these impurity driven radiation events is shown in figures 6(*a*)–(*c*). Please refer to the figure caption for more details. These radiation events have a typical rise time of order ms and can, in some cases, lead to a long-lasting growth in radiated power. In severe cases, this can add up to ~1 MW to the steady-state bulk radiated power—a significant contribution. In the most extreme cases, these radiation events can lead to plasma instabilities and even termination of the plasma discharge (an example of the latter is pulse 81781 from figure 5) (Ekedahl 2009, de Vries 2014, Sertoli 2014).

Data from two of JETs VUV survey SPRED spectrometers have been used to investigate the elemental composition of the impurities related to TIEs (Sertoli 2014). These systems monitor the impurity content of the JET plasma across the spectral ranges 100–1000 Å and 140–443 Å, along lines-of-sight at octant 6 (shown in the poloidal cross-section in figure 6(*a*)) (Fonck 1982, Wolf 1995, Coffey 2004). By integrating the VUV signal over known spectral emission lines, events originating from Ni, Fe, Cr, Cu, Al and W impurities can be distinguished. Example data for three sequential TIEs in the same pulse resulting from W impurities is shown in figure 6(*c*).

The occurrence of TIEs has been analysed for JET pulse 80128–85699, which starts with the very first campaign of JET's new ILW (Sertoli 2014). A total of 3388 events have been detected in 4144 plasma discharges (~23 h of plasma), with an average of 1 event every 25 s of plasma. This study shows that the major contributor to TIEs is W, accounting for 51% of all impurity events, with W usually occurring in isolation.

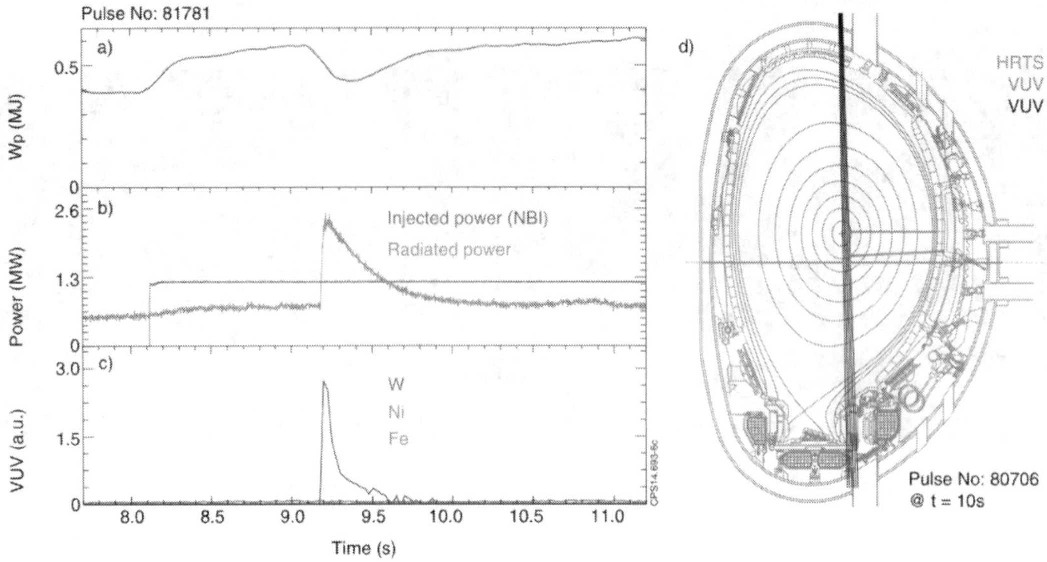

Figure 6. (*a*)–(*c*) Example of a TIE occurring in JET pulse 81781 (*a*) plasma energy W_p, (*b*) power from neutral beam injection (NBI) and radiated power, (*c*) VUV data for W, Ni and Fe lines. (*d*) A cross-sectional view of JET, showing the diagnostic lines-of-sight for HRTS (octant 5) and two VUV diagnostics (both octant 6). Magnetic flux lines for JET pulse 80706 at t = 10 s are shown for illustrative purposes.

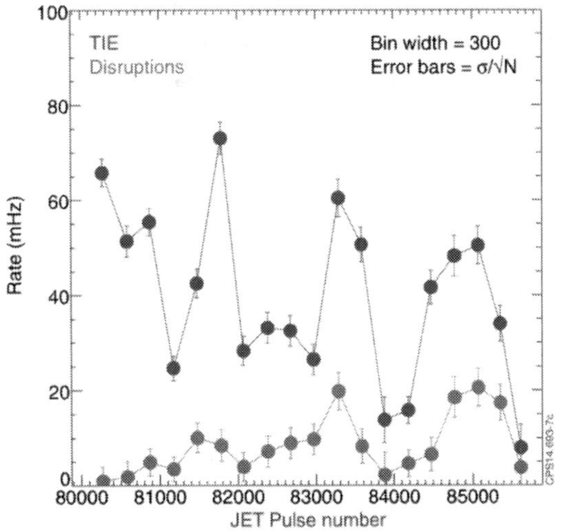

Figure 7. The TIE and disruption rate (normalised to plasma operational time) as a function of JET pulse number (JPN).

26% of all events are correlated with the simultaneous presence of Ni, Fe and Cr, which almost certainly originate from Inconel or steel in-vessel components. The remaining percentage of events corresponds to either Al or Cu, or to events with no clear VUV signature in the spectral range considered (these may result from lighter elements, such as carbon and beryllium). Over the initial ILW operation period, 22 impurity events are induced by material flaking from the reciprocating probe (RCP) diagnostic, used to measure the electron temperature at the upper edge of the JET plasma at octant 5 (Davies 1996). These events are dominated by Ni, Fe and Cr, which is consistent with the material of the probe shaft (Inconel / steel).

A database containing various parameters for every plasma discharge in the TIE analysis pulse range has also been collated in order to explore parametric dependencies (Sertoli 2014). A clear correlation with disruption events is found; TIEs are most likely to occur in the few discharges following a disruption, with the TIE rate 50% higher in the 4 discharges following a disruption. Figure 7 shows the TIE and disruption rate (defined as the number of events per bin, normalised to the total plasma operational time for the discharges in that bin), with a bin width of 300 JET pulses. A strong correlation between the TIE and disruption rate is apparent, with three discreet peaks seen in each data set. The peaks in disruption rate correspond to intentional disruptions associated with experimental sessions studying disruptions, disruption mitigation techniques and runaway electron events. The majority of disruption events in these periods are intentionally induced with typically high displacement forces. The TIE rate follows a very similar trend, with an additional overall downward drift; one that is particularly noticeable for JET pulse < 81200, where the TIE rate decreases by a factor of three in spite of the rise that occurs in the disruption rate over this time. This initial drop is also mirrored in the HRTS dust signal (see section 2.4 for more details). Taken together, these results suggest that an initial inventory of dust, present from the very start of ILW operations, is being gradually removed from circulation. The rises associated with periods of frequent disruptions may indicate that disruptions redistribute dust into more accessible areas. It is not possible to infer from this if disruptions create new dust material. The study in (Sertoli 2014) has also shown that TIEs are far more likely in diverted plasma configurations with the probability of a TIE occurring significantly reduced if the strike point is located on the vertical surface of a divertor tile, as opposed to a horizontal surface. This seems logical, given that dust tends to collect on the horizontal surfaces of the divertor region. Overall, the probability of a severe outcome from a TIE is low; only 25 of the 3388 TIEs studied result in a significant

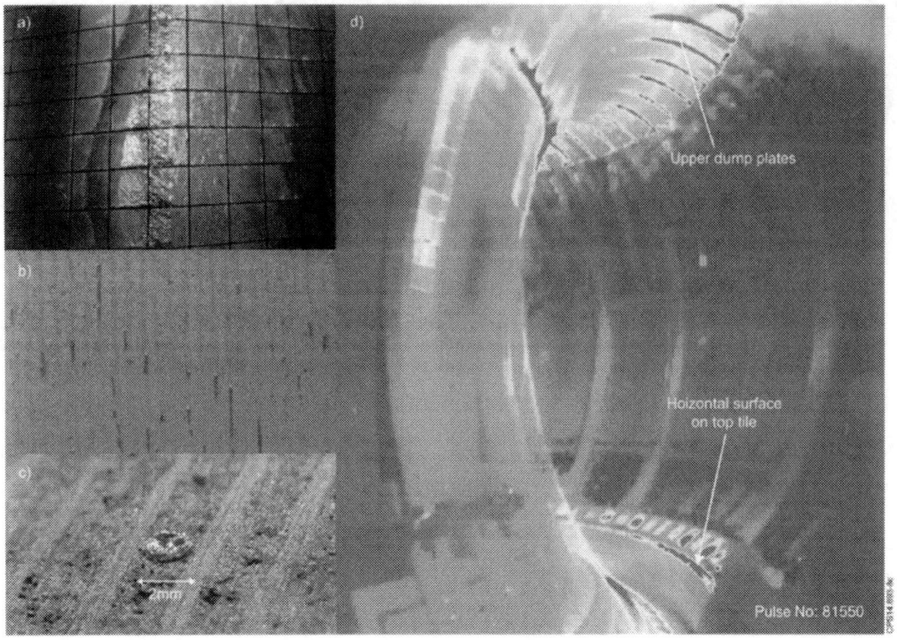

Figure 8. (*a*) Upper dump plate (bulk-Be) showing melting in the centre and the results of arcing on the right hand side (Arnoux 2014). (*b*) and (*c*) show images from the horizontal surface on top tile 8 (W-coated CFC). In (*b*), delamination of the W-coating with small beads of W are visible. (*c*) Shows a small Be droplet. (*d*) Infrared image from the KL7 camera showing the heat loads on the JET ILW during a density limit disruption pulse; JET pulse 81550.

(>60%) drop in plasma energy. And of these, less than 10 result in a disruption in the few seconds after the TIE (cf with ~700 disruptions over the ILW period).

2.3. *In-vessel inspection*

There are many erosion-deposition diagnostic tools installed into JET's ILW, including beryllium and tungsten marker tiles and wall probes in the main chamber and divertor (Rubel 2013). During a recent intervention period with in-vessel access (just after JET pulse 83934), these diagnostic tools were recovered for analysis. Detailed in-vessel photographic surveys and dust collection was also performed (Widdowson 2014). Here, we present a brief summary of some of the initial dust related findings from these studies. The post-mortem analysis from these studies will continue for some time.

A selection of images from the detailed photographic survey are shown in figures 8(*a*)–(*c*) together with an infrared image from the KL7 camera in (*d*) showing the heat loads on the JET ILW during a density limit disruption pulse (JET pulse 81550). This is shown to illustrate some of the areas of JETs PFCs most at risk. The image in (*a*) shows damage due to melting and arcing on one of the bulk Be upper dump plate tiles. The images in figures 8(*b*) and (*c*) correspond to the upper horizontal surface of the divertor tile 8. In (*b*) delamination of the W coating is seen, with small bead of W visible at the ends of some of the damaged strips. These regions of delamination lie along the CFC fibres and are not seen on all tiles: some are noticeably worse than others. This probably results from the fact that the CFC proved difficult to coat with W (Matthews 2011). In (*c*), a small droplet of Be can be seen, possibly originating from the upper dump plate above.

Dust collection from a portion of the divertor region yielded ~1 g of material; significantly less than the 200 g recovered from the same area during the CW period (Widdowson 2014). Compositional analysis on the dust extracted from the divertor is in progress. Studies of deposition diagnostics reveal that the quantity of deposits in the divertor region is an order of magnitude less than for the CW, with deposits of only a few 100 nm thick observed so far (Widdowson 2014). This reduction in migrated material may be linked to the drop in dust collected to date since there will be a significantly increased time for layers to reach mechanical stability limits. The overall picture from this analysis is that the ILW is a much cleaner environment than the CW, but that mechanisms do exist for the production of metal particulates.

From inspection of the TIE data, we see that there are 1697 events that can be attributed to W across the operational period considered for the ILW. If we assume that the W concentration spikes to approximately 10^{-4} (an upper estimate) in each instance, and assume a volume average electron density of order $10^{20}\,m^{-3}$, we can estimate that there are somewhere in the region of 10^{18} W atoms per TIE (JET plasma volume ~$100\,m^3$). A maximum estimate of the W particle mass required for all observed W TIEs is then ~0.3 g. The 1 g of dust recovered from a portion of divertor thus demonstrates that there is certainly sufficient material (assuming the dust has a representative proportion of W as seen in the TIEs).

However, many parameters influence how impurity particles interact with the plasma (i.e. particle size, injection velocity, plasma parameters) and screening may prevent

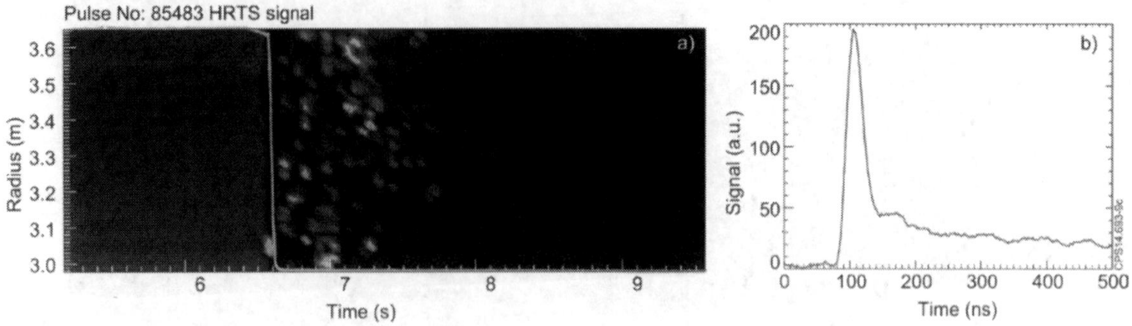

Figure 9. HRTS (*a*) Colour levels show the HRTS signal for JET pulse 85483 as a function of time. The plasma is terminated by a disruption with $t_d = 6.56$ s and $t_m = 6.65$ s. Induced vessel forces are recorded as 0.75 MN. At the time of the disruption, the plasma current (superimposed schematically in green) plummets to zero. At this time, the HRTS signal changes from Thomson scattered light, visible on all spectrometers, to an intermittent dust signal with a characteristic temporal signature, shown in the example in (*b*).

them from entering the core at all. A deeper understanding of this interaction can only be answered with a more sophisticated analysis; such as the numerical modelling described in (Sertoli 2014), which looks at the dust-plasma interaction for the same JET ILW period. In this study, the transport and ablation of tungsten dust particles in the JET plasma has been modelled using the 2D code DTOKS (Bacharis 2010). This work considers W particles injected from the outer strike point with an initial velocity of $10\,\mathrm{m\,s^{-1}}$ (correlating well with observed behaviour, as shown in figure 5). The study considers particle sizes from 10–100μm and predicts that a maximum of ~50% of a particle's mass is ablated inside the plasma. The 1D STRAHL code has also been used to simulate the impurity transport occurring after the dust ablation from this DTOKS study (Dux 2006, Sertoli 2014). Estimates from this approach indicate that somewhere in the order of 2×10^{17} ions are needed to generate the observed increase in radiated power, which equates to a W particle of ~90μm in radius. This is a minimum estimate since it does not include screening effects (i.e. ablation with in the scrape-off-layer en route to the plasma core). Allowing for a screening of 50%, this study finds that 0.14 g of W dust would account for the 1697 detected W events; in good agreement with the previous crude estimates. Note also that this estimate of particle size is of the order of the droplets seen during the bulk-W melt experiments (Coenen 2014).

2.4. Laser scattering

JETs high resolution Thomson scattering (HRTS) diagnostic is a conventional Thomson scattering system (Pasqualotto 2004), which has been routinely producing electron density and temperature profiles of the JET plasma since 2007. This system uses a 5 J pulsed Nd:YAG laser fired into the JET vessel through a vacuum window located roughly mid-plane at octant 5 (see figure 6(*d*)). The HRTS diagnostic provides 35 s of data from the start of a JET discharge with a spatial resolution of approx. 1.5 cm. During a plasma discharge, Thomson scattered light from the laser chord is analysed by a bank of spectrometers. In the few seconds immediately following a disruption, large spikes are often seen in the HRTS signal, as shown in figure 9(*a*). These signal spikes originate from the interaction between the laser beam and dust mobilized by the disruption (Giovannozzi 2010). The long decay time of the dust signal detected by the HRTS spectrometers, shown in figure 9(*b*), is typical of light emitted from a source that has been heated or ablated by a laser pulse. Typically, dust signals are detected for a period of approx. 1–3 s following a disruption.

The HRTS system thus provides an automatic measure of the quantity of dust mobilized by a disruption event. However, limitations of this technique should be appreciated; the volume of the JET vessel sampled by the HRTS laser chord is small and signal is only detected when dust crosses the chord. Cases are noted where dust / debris visible on JETs camera systems does not intersect the HRTS chord (an example was shown in figure 3). In addition, it is not simple to infer particle size from the event magnitude, since different materials are expected to have different intrinsic brightness when interacting with the pulsed laser. Surface quality and any surface coatings present may also play a role in determining the brightness of any given dust-laser interaction event. Furthermore, it is not possible to distinguish between many particles in close proximity and a single large event, and a particle moving along the laser chord will be recorded as multiple events. When added to the fact that this measure of the dust inventory is only made infrequently (i.e. with disruptions), one must be cautious about interpreting the resulting data. However, by analysing the HRTS signal across many disruption events, we find that it is possible to extract trends and explore relationships.

The parameters that are extracted from the HRTS dust signal for analysis are:

(a) The total signal (T_{sig}): this is the integrated signal level of all dust events detected in a single disruption; i.e. the total 'brightness'.
(b) The total number of events (N_{dust}): this is the total number of signal spikes detected in a single disruption; i.e. the number of dust particles (to first order approximation).

These parameters can be generated automatically using a variant of the code used to provide the electron temperature and density data from this diagnostic. HRTS dust data is available for 363 disruptions in the very latter part of the CW period (for JET pulse 73337–79853) and for 695 disruptions

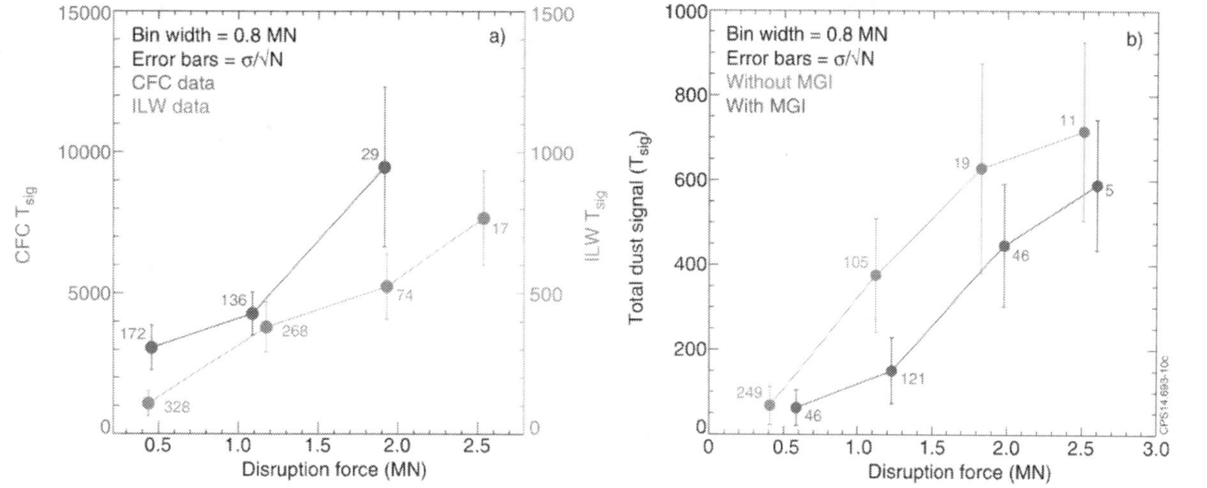

Figure 10. (*a*) HRTS total dust signal T_{sig} for both the CW (blue) and ILW (red) as a function of disruption force. Note that the CW data is plotted on a vertical scale 10 times larger than for the ILW data. Each data point corresponds to the average value in a bin of width 0.8 MN with error bars of $\sigma/\sqrt{N}$, where N is the number of disruption events in that bin and σ is their standard deviation. The value of N is shown on the figure, next to the corresponding data point. (*b*) A similar plot as (*a*), but for ILW data only, for disruptions with (blue) and without (red) use of disruption mitigation via massive gas injection (MGI). Data in (*b*) only includes pulses for which the disruption mitigation system was fully active: JET pulse 81560–85977.

in the period of operation to date with the ILW (for JET pulse 80128–85978). The most striking result from this analysis is that the amount of dust seen is significantly smaller for the ILW than for the CW period, by a factor of 20 (on average) for the total dust signal and a factor of 30 (on average) for the number of dust events. This is in rough agreement with the differences in the mass of dust collected from the vessel during CW and ILW intervention periods (see section 2.3 for more detail).

Further analysis shows that there is a clear positive correlation between the disruption force and the HRTS dust parameters, as shown in figure 10(*a*) which shows T_{sig} for both the CW (blue) and ILW (red) as a function of disruption force. Please see the figure caption for further details. A very similar relationship is seen for N_{dust} (not shown here for brevity). This relationship confirms a somewhat intuitive expectation; that the amount of dust shaken free is proportional to the vessel displacement forces. However, as for the case of the TIE analysis, this provides no information on what fraction of dust might be created by a disruption as opposed to merely mobilized.

Information on the effect of disruption mitigation via massive gas injection (MGI), known to reduce forces and heat loads, can also be gained from the HRTS dust data. In figure 10(*b*), the ILW T_{sig} data from (*a*) is re-plotted, with all events split into two data sets: accounting for whether or not MGI was used to mitigate the disruptions. Both data sets show the same increasing trend with disruption force, with a difference in magnitude. The amount of data here is somewhat limited, with error bars producing overlapping data sets for much of the force range. However, for the second force bin, located at ~1.25 MN, it can be seen that there is a clear drop with the average T_{sig} reduced by a factor of 1.6 (on average) in the case of MGI (N_{dust} is similarly reduced, by a factor of 1.9). Note that this observed reduction in dust is for disruptions of the *same* force. When the force reduction of the MGI is taken into account, this demonstrates that the use of MGI can reduce the dust mobilized by a disruption by approximately a factor of 4.

The evolution of the HRTS dust signals over the first five campaigns of the ILW period (covering JET pulse 80128–85977) is shown in figure 11. Light blue and red points with vertical markers show data for individual disruptions. Dark blue and red points show average values for a bin width of 300 pulses. The individual disruptions with values far higher than the global average are explained in figure 12(*a*) and surrounding text. In the initial period (up to JET pulse 81400), there is a rapid decline, with the drop in N_{dust} closely mirroring the fall in the rate of TIEs seen over this period. As previously mentioned, this suggests that material present prior to the start of the ILW plasma operations has been gradually removed from circulation. Small particles of debris visible on images of divertor tiles taken before ILW operations began add weight to this theory. Images of the same tiles taken just after JET pulse 82929, when both the TIE and HRTS signal levels are significantly reduced, are clear of debris. Inset, in Figures 11(*b*) and 12, the evolution of T_{sig} past the initial period of steep decline is shown (for JET pulse > 81400). Here, several peaks in T_{sig} can be seen, superimposed on an overall downward trend, which match the fluctuations in the TIE rate over the same period (plotted in blue). This correlation adds further weight to the theory that TIEs are caused by dust particles.

While T_{sig} and N_{dust} show a strong correlation (on average) and both display similar trends with disruption force, their ratio shows a large degree of variation. In an attempt to further understand this, we find it useful to plot T_{sig} as a function of N_{dust} for each disruption, as shown in figure 12. Displaying the data in this way allows the most extreme events to be clearly

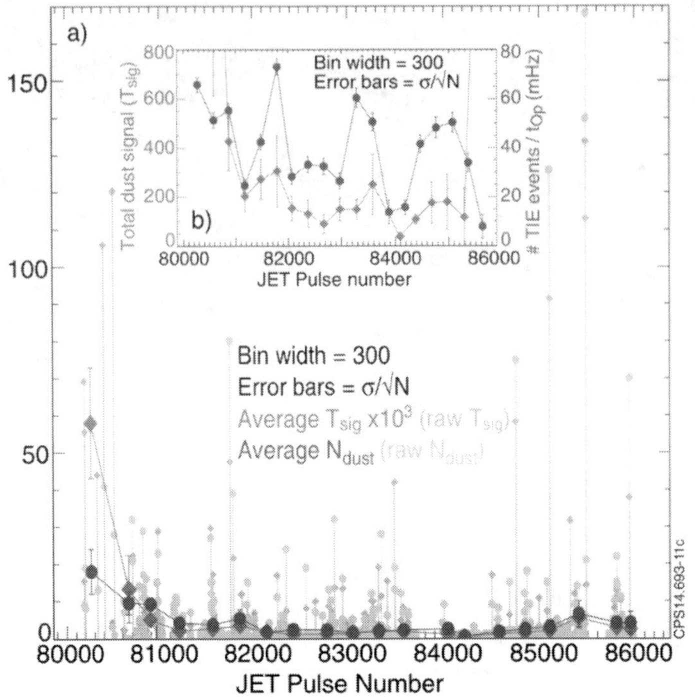

Figure 11. (*a*) HRTS T_{sig} (diamonds) and N_{dust} (circles) as a function of JET pulse number (JPN) 80128–85977, corresponding to the first 5 campaigns of the ILW period. Light blue and red points with vertical markers show raw data for individual disruptions. Dark blue and red points show average values over 300 pulses. Error bars are defined in the caption to figure 10. (*b*) inset, shows T_{sig} on a smaller scale, together with TIE rate data from figure 7.

identified. By exploring the details of these events, we see patterns emerge; the events in figure 12 with high values of both T_{sig} and N_{dust} (circled) correspond to discharges in which the reciprocating probe (RCP) diagnostic was malfunctioning, introducing material into the vessel. As mentioned in section 2.2, VUV analysis has shown that impurities associated with these events are consistent with the materials of the probe shaft (Sertoli 2014).

The disruptions with very high values of T_{sig} but lower values of N_{dust}, are all found to lie in the very early part of the ILW operational period—within the first 10 disruptions. On closer inspection of the HRTS data we see that, in each case, the very high signal level can often be attributed in the major part to just a few dust events. This suggests that the high values of T_{sig} after disruptions at this time period may have resulted from a few large dust particles that were later fragmented / eroded. Other high magnitude events are also found to be associated with an influx of material, within that same pulse, from a range of sources. These are indicated in the legend on figure 12(*a*) and include tungsten influx from the divertor region (as shown in figure 5), beryllium droplets from melt damage (as described in figures 4 and 8(*a*) and surrounding text), in addition to copper influx (possibly from the neutral beam injection system).

3. Discussion and conclusions

The results presented here have shown that the amount of dust for the ILW is low and, at present, is decreasing with operational time. These results also suggest that disruptions primarily act to mobilize the existing quota of dust, moving a subset of this material into areas more accessible to the plasma. These re-positioned dust particles are then more likely to appear as transient impurity events (TIEs) in subsequent plasma discharges, where they are partially consumed and later re-deposited as thin films onto the plasma facing components. It should be noted that the methods used here are very poor indicators of how much of the dust seen after a disruption might be created in that event, as opposed to merely mobilized. The fact that dust levels are decreasing with time despite relative frequent disruptions (many of them deliberately induced for studies of disruption physics etc) suggests that little, if any, dust is created during a disruption. However, for moderate disruption forces, the amount of dust seen for mitigated disruptions is notably reduced. Since mitigated disruptions lower heat loads to the plasma facing components, this may indicate that unmitigated disruptions do indeed create some material.

The high TIE rate and (relatively) large amounts of dust seen after disruptions during the early period of operation with the ILW points to material left in the vessel after the installation of the ILW was completed. Over time, these particles appear to have been gradually removed from circulation (via processes of erosion / fragmentation and possible re-positioning into in-accessible areas, such as the spaces behind the first wall tiles). Some events during plasma operations have also been seen to add to the quota of particulate matter, but, at present, the rate at which this material is removed from

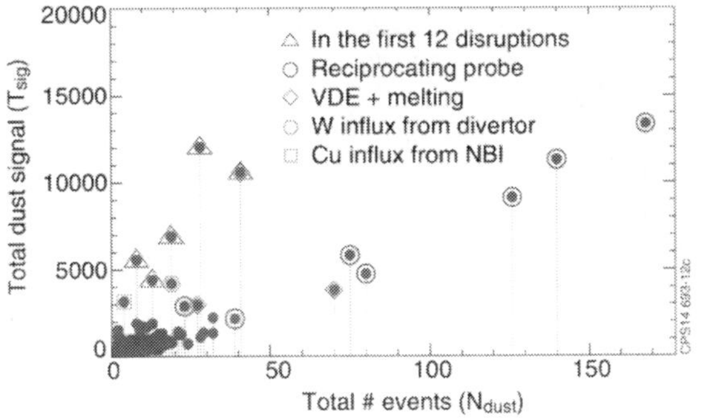

Figure 12. HRTS dust data: Total signal versus total number of events. Each data point represents a single disruption event. Additional details for extreme events are noted in the legend.

circulation is greater than any accumulation. As re-deposited layers build up and reach mechanical stability limits, this may change. However, the rates of material migration are orders of magnitude lower than for the carbon-wall and this may not happen for some time.

The effect of dust particles on plasma operations seems to be tolerable, with large losses to plasma energy or confinement relative rare. However, a few (of the handful) of impurity induced disruptions account for some of the most extreme disruptions in the ILW history to date; the highest disruption forces for an unintentional disruption, of 3.21 MN, occurred after a large W-influx from the divertor region. However, it should be noted that in this case, the disruption was unmitigated. The, now more routine, use of a massive gas injection in response to disruption precursors will obviously lessen the impact of such events.

Numerical modelling of the transport and ablation of W dust particles in the JET plasma confirm that W particles of a feasible size are capable of generating radiation events of the magnitude seen in JET plasmas (Sertoli 2014). This modelling work also helps to confirm that there is sufficient material generated from the (relatively small amounts) of damage seen to the tungsten parts of the plasma facing components to account for the number of TIEs seen during the operation period of the ILW to date. Further studies of particle influx in images from JET camera systems and analysis of the collected dust in terms of size and composition should allow for further refinement of input parameters to these models.

In conclusion, although impurities from in-vessel dust can, in extreme cases, have undesirable effects on plasma performance, taken together, the results shown here are encouraging for future devices such as ITER. The recent operational period of JET with the ILW has demonstrated that despite an initial inventory of material, and with repeated additions to this inventory, there exist efficient mechanisms that act to remove dust from circulation during normal plasma operations. Whether these mechanisms will succeed in balancing dust production as JET continues operations or in larger devices is, of course, another question and one that requires significantly more study to understand. However, since dust particle sizes are unlikely to scale with the vessel dimensions, the relative impact of individual impurities events should reduce for larger devices, such as ITER.

Acknowledgments

This work was supported by EURATOM and carried out within the framework of the European Fusion Development Agreement and the RCUK Energy Programme [grant number EP/I501045]. The views and opinions expressed herein do not necessarily reflect those of the European Commission or ITER.

References

Arnoux G 2010 Heat load measurements on the JET first wall during disruptions *Proc. 19th Int. Conf. on Plasma Surface Interactions (San Diego, CA)*

Arnoux G 2014 Power handling of the JET ITER-like wall *Phys. Scr.* **2014** 014009

Bacharis M 2010 Dust in tokamaks: an overview of the physical model of the dust in tokamaks code *Phys. Plasmas* **17** 042505

Bell A 2002 Tritium inventory control—the experience with DT tokamaks and its relevance for future machines *Proc. SOFT 2002 (Helsinki, Finland)* p 913

Clever M 2013 A wide angle view imaging diagnostic with all reflective, in-vessel optics at JET *Fusion Eng. Des.* **88** 1342–6

Coenen J 2013 Long-term evolution of the impurity composition and impurity events with the ITER-like wall at JET *Nucl. Fusion* **53** 073043

Coenen J 2014 ELM induced tungsten melting and its impact on tokamak operation *Proc. 21st Int. Conf. on Plasma Surface Interactions (PSI) (Kanazawa, Japan)*

Coffey I 2004 First tritium operation of ITER-prototype VUV spectroscopy on JET *Rev. Sci. Instrum.* **75** 3737

Craciunescu T 2014a An original method for spot detection and analysis for large surveys of videos in JET *IEEE Trans. Plasma Sci.* **42** 1358–66

Craciunescu T 2014b Overview of image processing tools to extract physical information from JET videos *Proc. 8th Workshop on Fusion Data Processing, Validation and Analysis (4–6 November 2013)*

Davies S 1996 The JET reciprocating probe systems—performance and data analysis *Contrib. Plasma Phys.* **36** 117–23

de Vries P C 2014 The influence of an ITER-like wall on disruptions at JET *Phys. Plasmas* **21** 056101

Dux R 2006 STRAHL User Manual, IPP Garching: Laborberiicht 10/30

Ekedahl A 2009 Analysis of radiative disruptions in RF-heated Tore Supra plasmas using infrared imaging *J. Nucl. Mater.* **390–391** 806–9

Fonck R 1982 Multichannel grazing incidence spectrometer for plasma impurity diagnosis: SPRED *Appl. Opt.* **21** 2115

Giovannozzi E 2010 Detection of dust on JET with the high resolution Thomson scattering system *Rev. Sci. Instrum.* **81** 10E131

Huber A 2012 Development of a mirror-based endoscope for divertor spectroscopy on JET with the new ITER-like wall *Rev Sci. Instrum.* **83** 10D511

Krasheninnikov S I 2011 Dust in magnetic fusion devices *Plasma Phys. Control. Fusion* **53** 083001

Lehnen M 2010 Disruption mitigation by massive gas injection in JET *Proc. 23rd IAEA Fusion Energy Conf. (Daejon, Republic of Korea)*

Lescure C 2009 Measurement of disruption forces in JET using fiber-optic sensors *Proc. 36th Int. Conf. on Plasma Science and 23rd Symp. on Fusion Engineering Conf. ICOPS-SOFE 2009 (San Diego, CA, 31 May 2009–5 June 2009)*

Marchese V 1996 Enhancement of JET machine instrumentation and coil protection *Proc. 19th SOFT 1996 (Lisbon, Portugal)* p 747

Matthews G F 2011 JET ITER-like wall—overview and experimental programme *Phys. Scr.* **2011** 014001

Pasqualotto R 2004 High resolution Thomson scattering for Joint European Torus (JET) *Rev. Sci. Instrum.* **75** 3891

Ratynskaia S 2011 Diagnostics of mobile dust in scrape-off layer plasmas *Plasma Phys. Control. Fusion* **53** 074009

Rosanvallon S 2009 Dust limit management strategy in tokamaks *J. Nucl. Mater.* **390–391** 57–60

Roth J 2009 Recent analysis of key plasma wall interactions issues for ITER *J. Nucl. Mater.* **390–391** 1–9

Rubel M 2013 Overview of erosion-deposition diagnostic tools for the ITER-like wall in the JET tokamak *J. Nucl. Mater.* **438** S1204–7

Sertoli M 2014 *Impact of W Events and Dust on JET-ILW Operation* (Japan: Ongaku-do, Kanazawa Ishikawa)

Widdowson A 2014 Material migration patterns and overview of first surface analysis of the JET ITER-like wall *Proc. 14th Int. Conf. of Plasma-Facing Materials and Components for Fusion Applications (Julich, Germany)*

Wolf R 1995 A vacuum ultra-violet spectrometer (Double_SPRED) for the observations of the JET divertor plasma *Rev. Sci. Instrum.* submitted

IOP Publishing
Plasma Physics and Controlled Fusion
Plasma Phys. Control. Fusion **57** (2015) 014038 (12pp)
doi:10.1088/0741-3335/57/1/014038

The H-mode density limit in the full tungsten ASDEX Upgrade tokamak

M Bernert[1], T Eich[1], A Kallenbach[1], D Carralero[1], A Huber[2], P T Lang[1], S Potzel[1], F Reimold[1], J Schweinzer[1], E Viezzer[1], H Zohm[1] and the ASDEX Upgrade team

[1] Max Planck Institute for Plasma Physics, Boltzmannstraße 2, 85748 Garching, Germany
[2] Forschungszentrum Juelich GmbH, Wilhelm-Johnen-Straße, 52425 Juelich, Germany

E-mail: matthias.bernert@ipp.mpg.de

Received 27 June 2014, revised 17 October 2014
Accepted for publication 24 October 2014
Published 28 November 2014

Abstract

The high confinement mode (H-mode) is the operational scenario foreseen for ITER, DEMO and future fusion power plants. At high densities, which are favorable in order to maximize the fusion power, a back transition from the H-mode to the low confinement mode (L-mode) is observed. In present tokamaks, this H-mode density limit (HDL) occurs at densities on the order of, but below, the Greenwald density.

In gas ramp discharges at the fully tungsten covered ASDEX Upgrade tokamak (AUG), four distinct operational phases are identified in the approach towards the HDL. These phases are a stable H-mode, a degrading H-mode, the breakdown of the H-mode and an L-mode. They are reproducible, quasi-stable plasma regimes and provide a framework in which the HDL can be further analyzed. During the evolution, energy losses are increased and a fueling limit is encountered. The latter is correlated to a plateau of electron density in the scrape-off layer (SOL). The well-known extension of the good confinement at high density with high triangularity is reflected in this scheme by extending the first phase to higher densities.

In this work, two mechanisms are proposed, which can explain the experimental observations. The fueling limit is most likely correlated to an outward shift of the ionization profile. The additional energy loss channel is presumably linked to a regime of increased radial filament transport in the SOL. The SOL and divertor plasmas play a key role for both mechanisms, in line with the previous hypothesis that the HDL is edge-determined.

The four phases are also observed in carbon covered AUG, although the HDL density exhibits a different dependency on the heating power and plasma current. This can be attributed to a changed energy loss channel in the presented scheme.

Keywords: density limit, H-mode, tokamak, ASDEX Upgrade

(Some figures may appear in colour only in the online journal)

1. Introduction

For a sufficient energy production and fusion performance, DEMO and future fusion power plants are foreseen to operate at highest densities in the high confinement mode (H-mode) [37]. The H-mode operation is accessed by applying sufficient heating power to the plasma [23]. However, the operation of a tokamak at high densities is limited by the so-called Greenwald limit [13, 12]

$$n_{\mathrm{GW}} = \frac{I_{\mathrm{p}}}{\pi a^2},$$

where n_{GW} is the line averaged density in units of 10^{20} m^{-3}, I_{p} the plasma current in MA and a the minor radius in m. At Greenwald fractions ($f_{\mathrm{GW}} = \frac{\bar{n}_e}{n_{\mathrm{GW}}}$) of about 0.8–1.2, the plasma is observed to end in a disruption. Therefore, the Greenwald limit represents a hard limit.

At high densities also a back transition from the H-mode to the low confinement mode (L-mode) is observed. This H-mode density limit (HDL) typically takes place at Greenwald fractions of 0.8–1 [12]. The HDL represents a soft limit, as the plasma operation can be continued at the lower confinement and a renewed transition into H-mode is in principle possible. In typical gas ramp discharges, the density increases further after the HDL until the Greenwald limit is reached.

Both the Greenwald limit and the HDL can be exceeded by centrally elevated density profiles, as they are achieved in todays experiments by pellet fueling [20].

This paper focuses on the characterization of the H-mode density limit by a classification of the different sequences of temporal evolution of the plasma edge parameters and gives an initial interpretation of the observed effects.

The HDL has to be discriminated against an H-L transition, which is caused by the reduction of the heating power below the threshold, since it is reported to be independent of the heating power [15, 25].

Studies at ASDEX Upgrade (AUG) and JET, both with a fully carbon covered wall, attribute the HDL to the full detachment of the divertor [5, 7]. In these studies, the H-L back transition is linked to a radiation instability close to or inside the confined region of the plasma, which is triggered by the detachment.

At AUG and JET, the first wall material was changed to tungsten and tungsten and beryllium, respectively, which has a significant impact on the radiation characteristics of the plasma. Therefore, the HDL is investigated in the presented work at the fully tungsten covered ASDEX Upgrade (full-W AUG).

This paper is organized as follows. The performed experiments and their gas fueling schemes are described in section 2. The four phases, which are identified in the approach towards the HDL, are introduced and characterized in section 3. Several mechanisms, which were previously proposed to cause the HDL, are examined in full-W AUG in section 4. A regression analysis of the maximum achieved densities in H-mode is given in section 5. A new interpretation of the HDL, based on the observation of the four phases and related mechanisms are proposed in section 6. The impact of centrally elevated density profiles investigated in section 7 and the presented work is summarized in section 8.

2. Experiments

2.1. Discharge setup

Between 2011 and 2014 over 30 dedicated gas ramp discharges were conducted at AUG in order to study the HDL in a full-W tokamak. The standard scenario for these studies is a gas ramp discharge with a plasma current of $I_p = 0.8$ MA, a toroidal magnetic field on-axis of $B_t = -2.5$ T and hence a safety factor q_{95} of about 6. The averaged triangularity is in the low to medium range of $\delta_{av} \approx 0.23$. The discharges

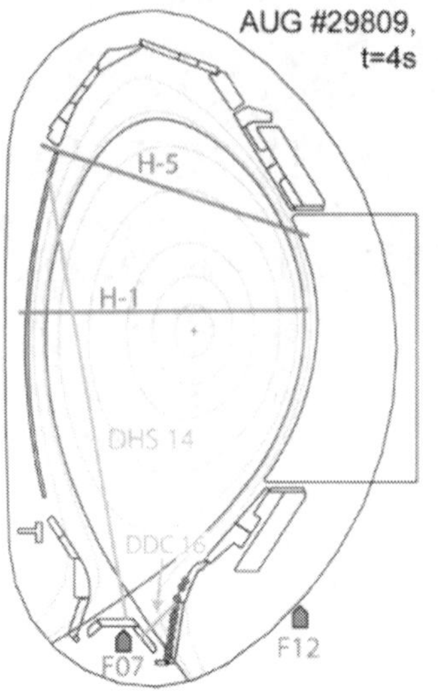

Figure 1. Magnetic equilibrium (blue, cyan) of an HDL discharge and the location of the main measurements used in the presented work: edge (H-5) and core (H-1) interferometry (red), ionisation gauges (green), Langmuir probes in the outer divertor (purple) and two lines of sight of the AXUV diode system (orange).

are heated by neutral beam injection (NBI) with a standard input power of $P_{heat} = 7.5$ MW, which is well above the L-H power threshold of about 3.3 MW (at a separatrix surface of 43.6 m^2).

The magnetic equilibrium of the standard scenario and the location of the main measurements can be seen in figure 1. The electron density measured by interferometry is given throughout this publication in line integrated rather than line averaged measurements. This is necessary due to a significant electron density in the scrape-off layer (see figures 9(*a*) and 10), which would otherwise be erroneously interpreted as density within the confined region. For an estimation of the line averaged densities, a chord length of the central chord (H-1) of 1.02 ± 0.03 m and of the edge chord (H-5) of 0.79 ± 0.02 m can be used, which are (within the error bars) the same for all presented discharges of full-W AUG.

The dataset of 30 discharges contains separate scans of safety factor, plasma current, external heating power and triangularity, which are discussed in section 5.

2.2. High density operation and fueling

The HDL discharges are only fueled externally by valves in the private flux region of the lower divertor, fueling by the NBI is negligible. A neutral pressure in the divertor of up to 8 Pa is necessary to achieve the plasma densities relevant for the HDL. AUG has to be operated solely with the turbo pump system and without the cryo pump system, reducing the pumping capacity by a factor of 8, in order to achieve these high pressures.

Figure 2 shows the gas fueling scheme for a typical HDL discharge. Due to the reduced pumping, the constant gas

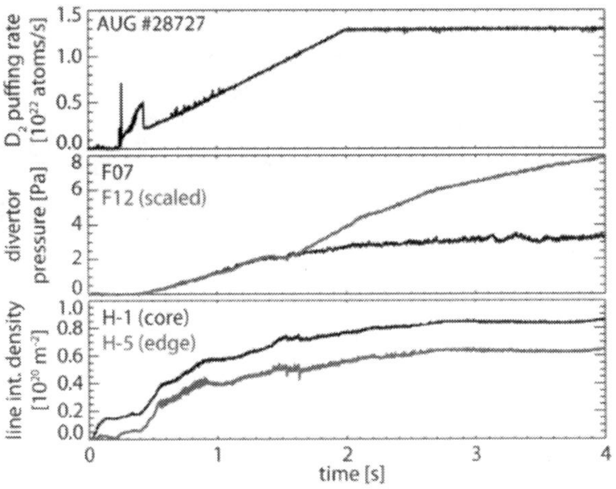

Figure 2. Typical time traces of neutral gas fueling rate, neutral gas pressure in the divertor measured by ionization gauges and line integrated electron density of an HDL discharge. The ionization gauge below the roof baffle (F07) saturates at around 1.6 s, the neutral gas pressure can be estimated out of the measurements of a gauge situated in the pump chamber (F12), which is scaled to match the unsaturated measurement below the roof baffle.

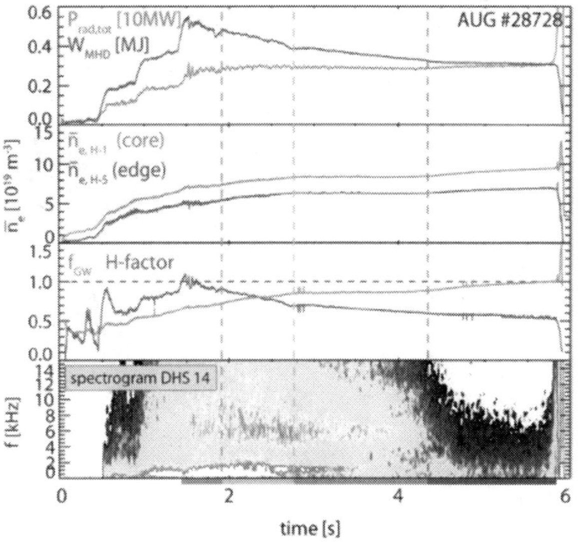

Figure 4. Time traces of radiated power and stored energy, edge and central line integrated electron density, Greenwald fraction and H-factor and a spectrogram of an AXUV diode signal observing the X-point of HDL discharge AUG #28728. The colored bars on bottom indicate the four phases discussed in section 3.

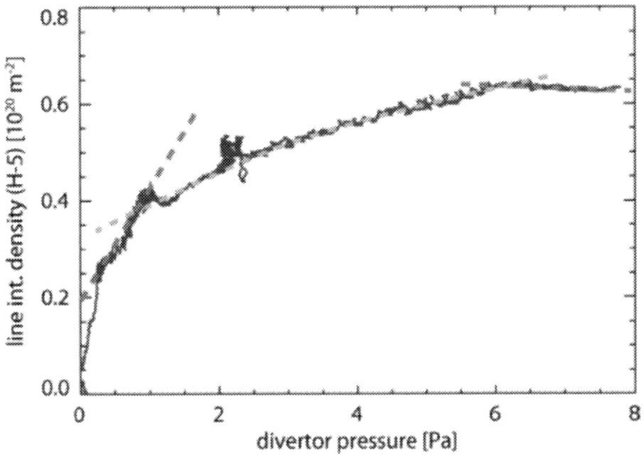

Figure 3. Line integrated edge density plotted versus the neutral gas pressure in the divertor as shown in figure 2. The plasma density saturates and reaches a fueling limit (red line).

puff leads to a monotonic increase of the divertor neutral gas pressure.

A neutral gas pressure of several Pa is necessary for the HDL discharges since a fueling limit is reached. The fueling limit corresponds to a saturation of the plasma density relative to the neutral gas pressure in the divertor. In figure 3 the electron density is plotted versus the neutral pressure. At low densities, the two measurements are almost proportional to each other (typical operational range, indicated by the green line). At medium to high densities, this relation becomes non-linear and a much higher divertor pressure is needed to achieve the same increase of plasma density (orange line). When the fueling limit is finally reached, the plasma density does not increase despite an increase of the neutral gas pressure by 30% (red line).

2.3. Evolution of a typical HDL gas ramp discharge

Figure 4 shows several time traces of a typical HDL gas ramp discharge. The ramp-up of plasma shape, current and heating power is finished at 1.5 s. From this point on, all external parameters are kept constant except the neutral gas pressure, which is increased by the constant gas puff (similar to figure 2).

The plasma stored energy decreases constantly, indicating a reduction of the energy confinement time. This can also be observed in the reduction of the H-factor, which gives the energy confinement time with respect to the ITERH98P(y,2) scaling [27]. The total radiated power, determined out of bolometry measurements [4, 11], is constant or even slightly reducing during the discharge up to the disruption. Due to the high gas puff is the tungsten concentration in general below $5 \cdot 10^{-5}$ (in most cases below $1 \cdot 10^{-5}$) for the HDL discharges and has no significant effect on the plasma radiation.

The transition from type-I to type-III edge localized modes (ELMs), which is typically observed at high densities [34, 38], takes place in the presented discharge between 1.5 and 1.7 s. The spectrogram of an AXUV diode channel [4] observing the X-point shows a broad frequency band around 6 kHz. This is characteristic for the X-point fluctuations, which are not specific to the H-mode and are most likely correlated with the regime of divertor detachment [29]. The full detachment of the outer divertor, which is for full-C AUG reported to trigger the HDL [7], takes place immediately before the disruption, here at around 5.8 s. The full detachment is correlated with the development of a MARFE [21] at the X-point, which subsequently moves upwards. Along with this movement a tearing

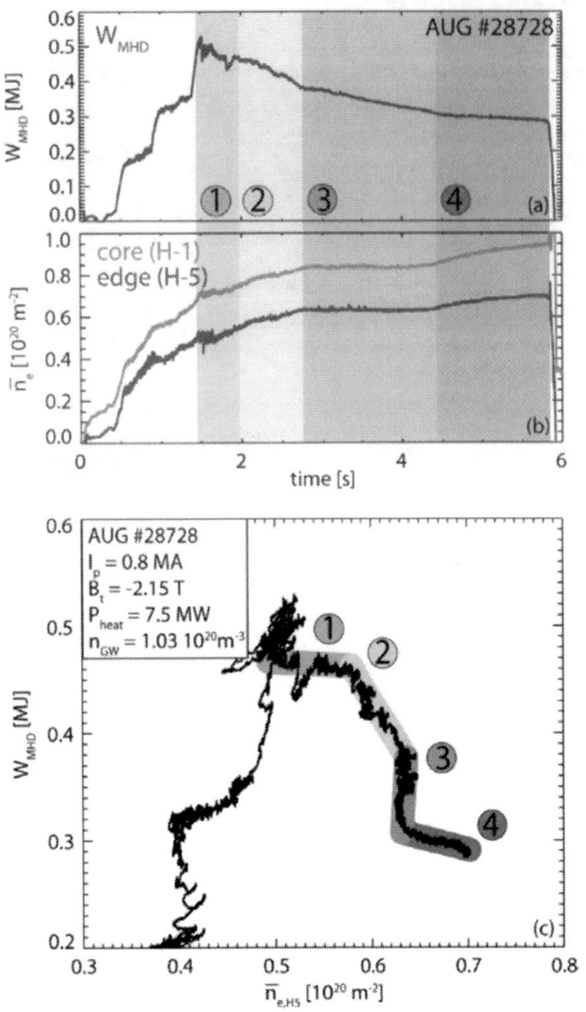

Figure 5. (a, b) Time traces of stored energy and line integrated density of HDL discharge AUG #28728. The background colors indicate the time duration of the four phases identified in figure (c). (c) Stored energy plotted versus edge line integrated density.

mode (typically 2-1) is triggered, which finally causes the disruption, resembling the 'Greenwald-like' L-mode density limit.

3. The four phases of the H-mode density limit

In the gas ramp discharges at AUG, four distinct phases are identified in the approach towards the HDL. These phases are distinguished by their impact on the plasma stored energy (W_{MHD}) and the edge line integrated electron density ($\bar{n}_{e,H-5}$). In figure 5 these two parameters are plotted against each other, highlighting the different phases. The four phases are identified as:

(1) A stable H-mode (green),
(2) a degrading H-mode (orange),
(3) the breakdown of H-mode (red) and
(4) an L-mode (blue).

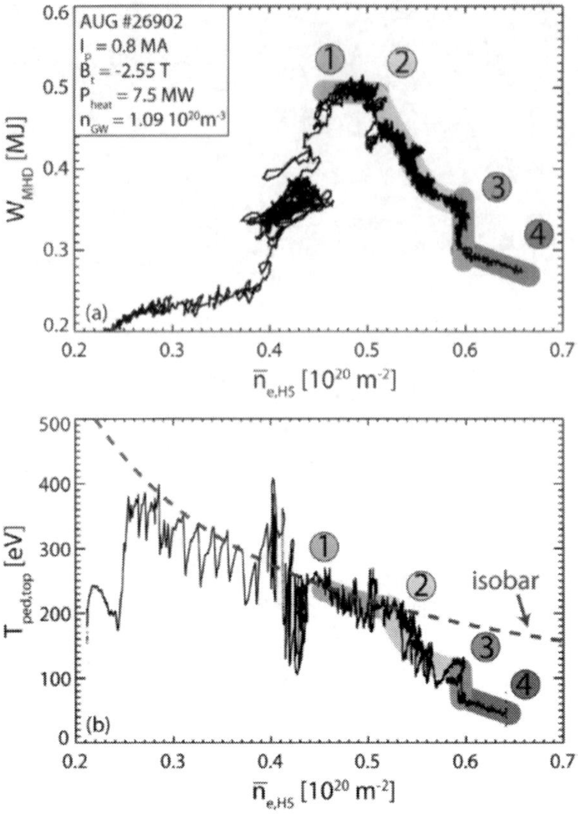

Figure 6. (a) $W_{\mathrm{MHD}} - \bar{n}_{e,H-5}$ plot for HDL discharge AUG #26902. (b) Estimate of the pedestal top temperature plotted versus the edge density. The four phases can also be distinguished, the first phase evolves along an isobaric line. The strong oscillations are caused by heating steps and ELMs.

The phases can also be distinguished by their impact on the pedestal top temperature, as shown in figure 6(b). The first phase evolves along an isobaric line, while in the second and third phases the pressure reduces and thus the confinement degrades.

The four phases are observed in all discharges dedicated to the HDL studies at full-W AUG, independent of the discharge parameters and fueling approaches. However, the timing and duration of the phases change significantly. Figure 7 shows the $W_{\mathrm{MHD}} - \bar{n}_{e,H-5}$ plots for various discharge parameters in full-W AUG, all exhibiting similar characteristics. The same observation can be made for the HDL discharges in full-C AUG, as seen in figure 8.

The instantaneous change of the plasma behaviour (specifically, the change of the stored energy relative to the density increase, as seen in figures 5–8) at the transition from one phase to another indicates that the individual phases are discrete plasma states. This instantaneous change can also be seen in the change of gradients of the time traces shown in figures 4 and 5(a) and (b).

Every phase can be accessed directly (e.g. by an increase of heating power, seen in figures 7(c) and (d)) and is stable against ELMs or short trips of heating power. This indicates, that the four phases are reproducible, discrete and

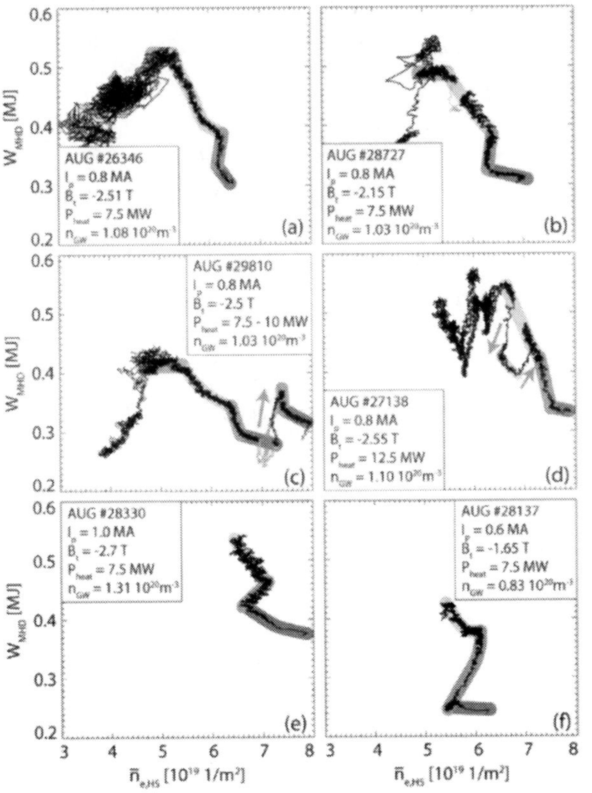

Figure 7. $W_{MHD} - \bar{n}_{e,H-5}$ plots of various HDL discharges in full-W AUG. Trips of heating power are shaded in grey, NBI heating steps ($\Delta P_{heat} = 2.5$ MW) indicated by cyan arrows. The four phases are always reproduced, quasi-stable and can directly be accessed e.g. by heating steps as seen in figures (*c*) and (*d*).

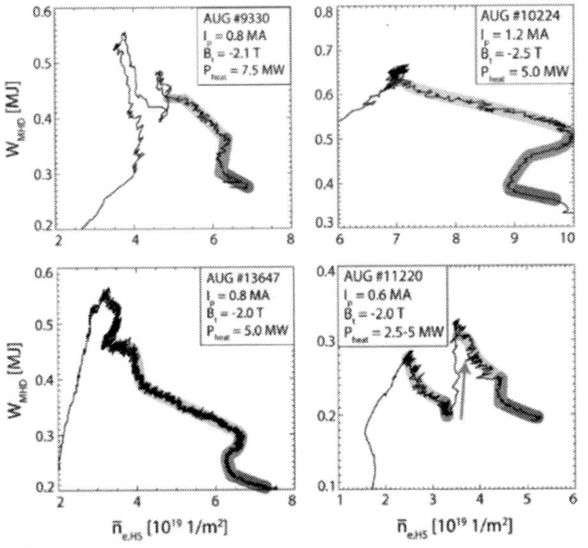

Figure 8. $W_{MHD} - \bar{n}_{e,H-5}$ plots of various HDL discharges in full-C AUG. The four phases are also observed with the different first wall material.

quasi-stable plasma regimes, which are present independent of the first wall material and discharge parameters and, thus, characteristic for the high density operation at AUG.

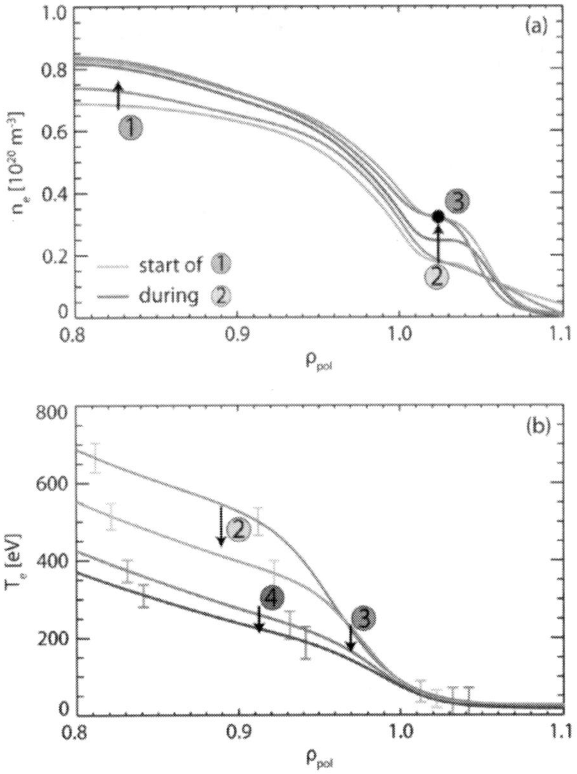

Figure 9. Radial profiles of electron density ((*a*), AUG #28726, fitted to data from the interferometry, Thomson scattering and Lithium beam diagnostics, details shown in figure 10) and electron temperature ((*b*), AUG #29809, fitted to data from the ECE and Thomson scattering diagnostics) at the end of the corresponding phases. The arrows and numbers indicate the radial position and the phase in which the predominant changes in the profiles occur.

They provide a framework in which the HDL can be further analyzed.

The characteristics of the phases and their impact on the electron density and temperature profiles (see figure 9) are listed in the following:

(1) The **stable H-mode** is the standard operational H-mode regime: the plasma density increases at a stable confinement. The pressure is constant and the plasma evolves along an isobaric line (see figure 6). The increase in electron density takes place in the plasma core and edge and, since the pressure is constant, the electron temperature decreases accordingly.
The transition from type-I to type-III ELMs takes place typically, but not exclusively, in this phase.

(2) In the second phase, the **degrading H-mode** the confinement reduces, corresponding to a loss of pressure. The electron density in the confined region stays almost constant while a plateau of electron density evolves in the scrape-off layer (SOL) up to about 40–50% of the pedestal top density. Meanwhile, the width of the temperature pedestal reduces, leading to a lower pedestal top and central electron temperature (see figure 9).

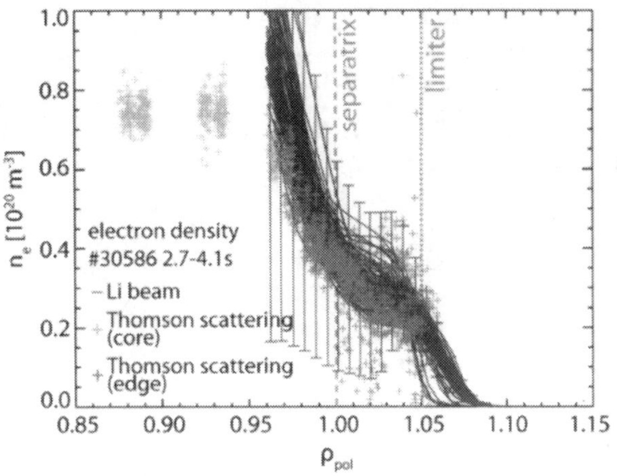

Figure 10. Electron density measurements during the breakdown phase of HDL discharge AUG #30586. The profile does not change during the whole phase.

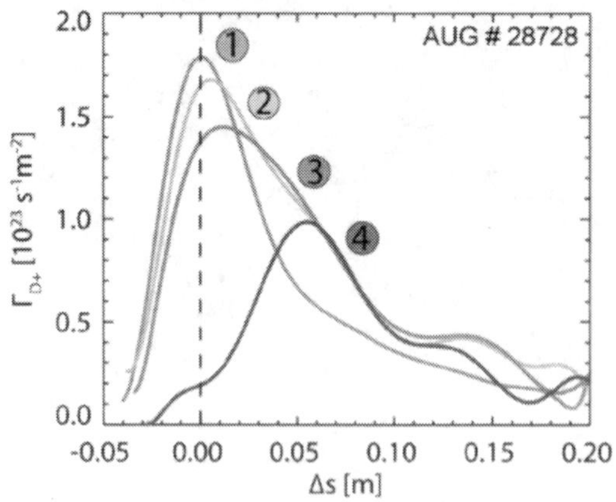

Figure 11. Averaged profiles of ion flux density to the divertor target plates for the different phases.

(3) The third phase, the **breakdown of the H-mode** is characterized by an overall fixed radial electron density profile. Measurements of the electron density by the Li-beam and Thomson Scattering diagnostics during this phase are compared in figure 10.
The confinement keeps decreasing at constant density, which leads to the step-like function in the $W_{\mathrm{MHD}} - \bar{n}_{e,H-5}$ plot. The gradient of the electron temperature in the pedestal is also constantly decreasing during this phase, which leads to a reduced temperature and pressure in the confined region and, thus, to the reduction of confinement. Along with the electron temperature gradient, also the radial electric field (E_r) well depth at the pedestal is continuously decreasing up to about -15 kV m^{-1} (see section 4). The whole phase is still defined as an H-mode, since ELMs exist and E_r stays below the reported threshold of -15 kV m^{-1} [32] up to the end of this phase.
The duration of breakdown phase varies between 0.1 s and 4 s. Therefore, it also resembles a quasi-stable plasma regime. This phase is here defined as the actual H-mode density limit, since it resembles the H-mode with the highest edge density and used for the regression analysis in section 5. However, all four phases have to be regarded in order to find an explanation for the HDL.

(4) The last phase resembles an **L-mode** with a typical H-factor of about 0.5. The confinement stays almost constant in this phase and the plasma density increases again. During this phase, the outer divertor fully detaches and a MARFE evolves at the X-point. With the increasing density the L-mode density limit is reached, the MARFE moves upwards and a tearing mode is triggered, which finally leads to the disruption.

4. Previously proposed mechanisms for the HDL

In previous studies several mechanisms were proposed to cause the HDL. These mechanisms are revised with the new database of full-W AUG, using the improved diagnostic capabilities.

4.1. Detachment

Full detachment of the outer divertor and related radiation losses were identified to trigger the HDL in full-C AUG [6]. In full-W AUG, the roll-over of the ion saturation current measurements at the outer strikeline starts during the breakdown of the H-mode and the outer divertor is in the L-mode phase in a partially detached regime. Full detachment is only achieved shortly before the disruption. The evolution of the ion flux density profiles for the different phases is shown in figure 11. Note, that the high frequency ELMs during the second and third phase also affect the divertor condition.

In addition, stable H-modes are established in full-W AUG, which are fully detached during the inter-ELM phases [30]. Therefore, the full detachment can be refuted as a cause for the HDL in full-W AUG. However, it might play a significant role in the HDL of full-C AUG.

4.2. Radiation and MARFEs

Radiation in the confined plasma can increase the energy losses, reduce the power flux over the separatrix and lead to a back transition into L-mode. Also MARFEs, which evolve in full-W AUG typically in the X-point region, lead to such an increased radiation. In plasmas with a high impurity content, MARFEs are observed to trigger an H-L back transition by radiative cooling of the pedestal. However, in the dedicated HDL discharges, which have a low impurity concentration due to the high deuterium gas puff, the MARFE does not start evolving before the L-mode phase and, thus, is not related to the HDL.

Furthermore, the poloidal radiation distribution (see figure 12) and the total radiated power (see figures 4 and 17) stay constant (or even decreases) during the first three phases of the HDL discharges at full-W AUG. Therefore, in full-W AUG, no additional energy losses from the confined region are created by radiation.

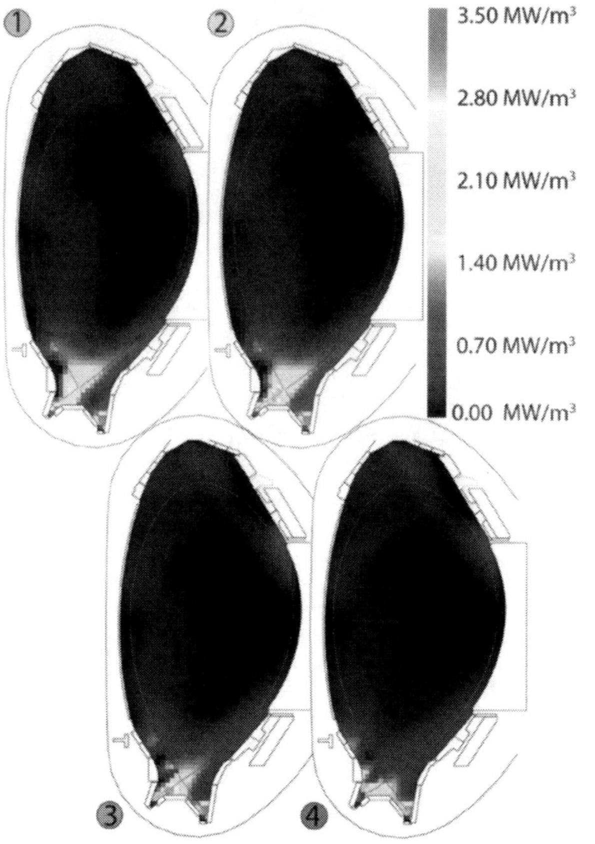

Figure 12. Deconvolution of the poloidal radiation distribution measured by bolometry for the four phases in AUG #29809. The radiation stays constant (or even decreases) throughout the discharge.

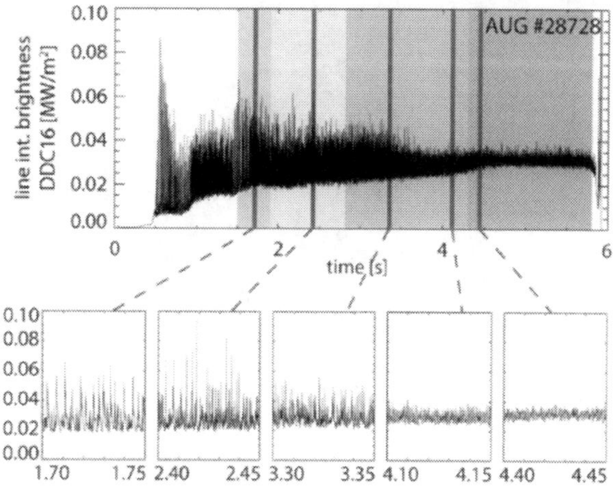

Figure 13. Measurement of an AXUV channel observing the outer divertor. The ELMs vanish during the third phase (third and fourth bottom plots), only small high frequent fluctuations remain, which are also present in the L-mode phase.

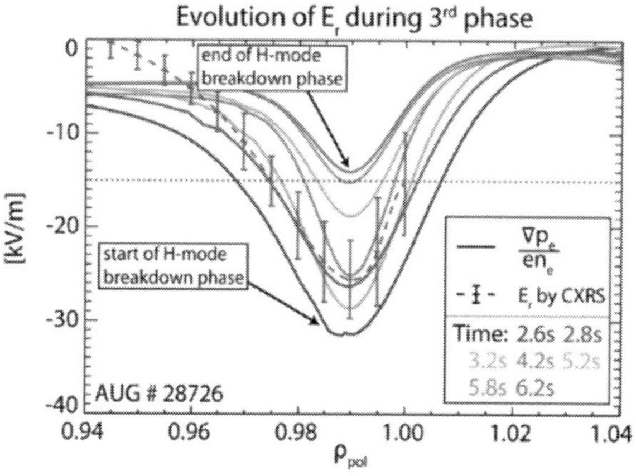

Figure 14. Measurement of the radial electric field (dashed line) and evolution of an estimate for the radial electric field well during the breakdown phase of the H-mode in AUG #28726. The reported threshold [32] is indicated by the dotted line.

4.3. ELMs

ELMs are discussed in references [7, 12] to trigger the HDL: The pedestal might be more eroded by high frequency ELMs than can be recovered in the short inter-ELM phase.

The ELM frequency increases in full-W AUG strongly when approaching the HDL, but also the ELM size (relative loss of stored energy per ELM, also seen in the ELM-induced radiation, figure 13) reduces. While the degradation of the confinement and, thus, the energy losses are ongoing the ELMs are occasionally observed to vanish during the breakdown of the H-mode (see figure 13 and in the low frequency range of the bottom plot of figure 4). Therefore, there must be another effect leading to the energy loss.

Furthermore, the transition from type-I to type-III ELMs (not shown here) is not related to any of the HDL phases and is observed to take place in the first or in the third phase.

4.4. Thresholds of heating power and radial electric field

H-L back transitions take place if the heating power is reduced below the L-H power threshold (taking also into account the L-H-L hysteresis).

For the HDL discharges in full-W AUG, the divertor loss power is significantly above the L-H power threshold scaling ($P_{L-H} \approx 3.3$ MW, [23]) during the first three phases. The divertor loss power is the sum of the power radiated in the divertor volume, measured by bolometry and the power deposited on the target plates, measured by thermography and must be lower or equal to the power flowing over the separatrix. Therefore, additional energy losses from the confined region do not reduce the power flux over the separatrix below the reported threshold. However, the threshold might be important for the final transition from the breakdown phase of the H-mode to the L-mode, where the divertor loss power is close to the threshold.

In reference [32] a threshold for the depth of the radial electric field (E_r) well in the pedestal of -15 kV m^{-1} is reported for the L- to H-mode transition at similar discharge parameters. Figure 14 shows the evolution of the diamagnetic term of the radial force balance, which is an estimate of E_r ($E_r \approx \frac{\nabla p_i}{en_i} \approx \frac{\nabla p_e}{en_e}$, [35]), during the breakdown phase of the

H-mode. At the onset of this phase, the E_r well is significantly below the threshold, as also seen by direct E_r measurements using charge exchange recombination spectroscopy (CXRS) [36]. Only at the end of this phase, the E_r well is close to the threshold. Therefore, the E_r is not related to the onset of the HDL, but might play a significant role at the final transition to L-mode.

It is seen that the threshold for the E_r well depth and for the heating power behave similar at the HDL of full-W AUG. Both might be determined by the same effect. The E_r threshold might also be interpreted as the local parameter which is described by the global L-H power threshold.

5. Regression analysis

The onset of the different phases of the HDL could yet not be assigned to a fixed value of a specific parameter, such as temperature, plasma density or neutral gas pressure. In order to identify the parameter dependencies of the HDL, the 30 conducted discharges include separate scans of

- heating power (6 MW $\leqslant P_{\text{heat}} \leqslant$ 12.5 MW),
- safety factor ($3.5 \leqslant q_{95} \leqslant 6$),
- plasma current (0.6 MA $\leqslant I_p \leqslant$ 1.2 MA),
- toroidal magnetic field (1.45 T $\leqslant B_t \leqslant$ 2.7 T) and
- triangularity ($0.23 \leqslant \delta \leqslant 0.37$).

A regression analysis of the achieved line integrated edge density ($\bar{n}_{e,H-5}$ in 10^{20} m^{-2}) at the onset third phase is performed. This represents the highest achieved pedestal top density in H-mode and, thus, a physical limit. The onset of the degrading H-mode phase, which might be more important as operational point of future reactors, is not regressed here and subject to future studies.

The log-linear regression results in:

$$\bar{n}_{e,H5,\text{regr.}} = 0.506 \pm 0.192 \frac{P_{\text{heat}}^{0.396\pm0.13}\, I_{\text{p}}^{0.265\pm0.14}}{q_{95}^{0.323\pm0.14}}, \qquad (1)$$

where the regression values are given with their 95% confidence intervals. The standard deviation is $\sigma = 0.0311$, regressed versus measured densities are shown in figure 15.

The regression of the line integrated edge density ($\bar{n}_{e,H-5}$) leads to a more accurate result than the regression of the line averaged central density ($\langle n_{e,\ H-1}\rangle$, $\sigma_{H-1} = 0.058$). The scaling factor between the two values is around $\frac{\bar{n}_{e,H-5}}{\langle n_{e,H-1}\rangle} = 0.78 \pm 0.05$ m for these flat density profiles. Note, however, that the density plateau in the SOL does lead to a wrong interpretation of the line averaged density (see section 2.1).

Any combination of two parameters out of I_p, B_t and q_{95} can be used in the regression and leads to a similar fit quality due to the collinearity of the three parameters. The normalized pressure β_N does not lead to a better regression than the heating power P_{heat}. The triangularity does not have an effect on the density of the third phase and thus is omitted in the regression.

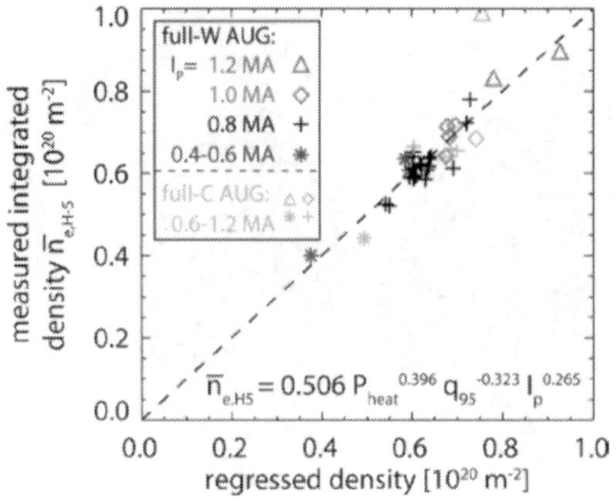

Figure 15. Regressed versus measured density at the onset of the third HDL phase, also including discharges of full-C AUG (orange), which were not included in the regression analysis.

All dependencies in equation (1) were reproduced in separate scans of the given parameters. Notably, the pronounced heating power dependence, which is for full-C AUG and JET reported to be not existent or insignificant [6, 12, 15, 24], is seen within a discharge: In discharge AUG #29810, shown in figure 7(*c*), the heating power was increased in the fourth phase, which lead to a transition back into the breakdown phase of the H-mode at a higher density. This demonstrates directly to power dependence of the HDL.

The regression yields a significantly lower dependence on the plasma current than the Greenwald limit [13]. The empirical scaling of full-C AUG and JET [6] deviates from the above given regression in the dependence from the heating power and the plasma current as well as the triangularity. A more detailed comparison of these scalings is given in reference [3].

A variation of the triangularity reveals, that the density of the breakdown of the H-mode is not depending on the triangularity. However, figure 16 shows, that the onset of the degrading H-mode phase is at a higher density for the higher triangularity, i.e. the stable H-mode phase extends to higher densities. This is in line with the observations in references [31, 33], which report no influence of the triangularity on the confinement for low densities, but a significant increase with triangularity at high densities. There is a contradiction with observations at the DIII-D tokamak [28]. There, a reduction of the pressure is seen with increasing density, similar to the degrading H-mode phase. However, the density, at which this reduction sets in, is at DIII-D independent of the triangularity (figure 4 of [28]).

6. Interpretation of the four phases

All mechanisms discussed in section 4 can be refuted as a cause for the HDL in full-W AUG. Instead, the observation of the four phases, described in section 3, provides a framework for a new interpretation of the underlying mechanisms of the HDL.

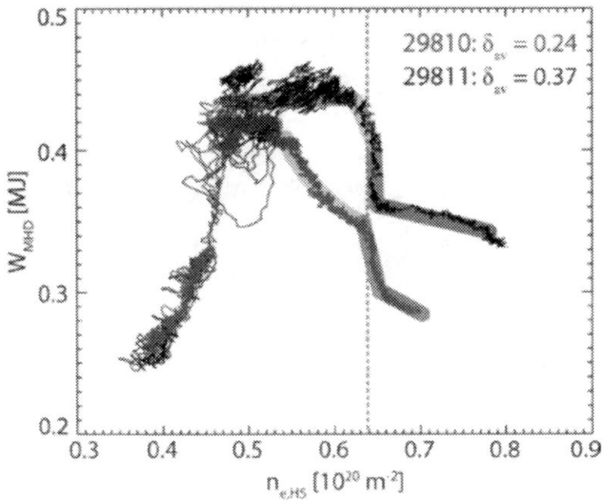

Figure 16. $W_{\mathrm{MHD}} - \bar{n}_{e,H-5}$ plots for two discharges at different triangularities. The breakdown of the H-mode takes place at the same density, while the onset of the degradation is at a higher density for higher triangularity.

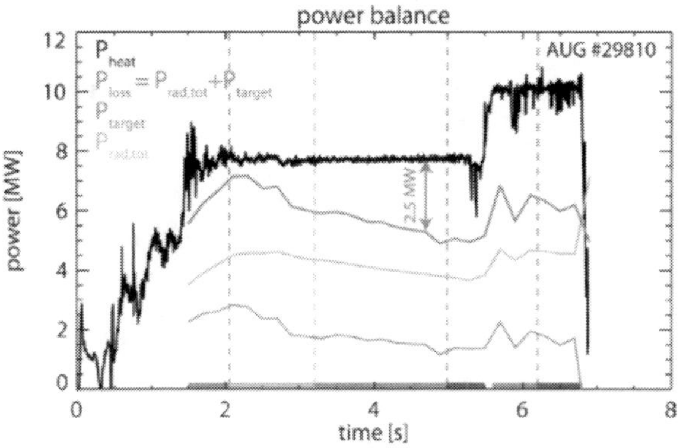

Figure 17. Time traces of the global power balance during HDL discharge AUG #29810: Total heating power (black), total loss power (red), power deposited on the divertor target plates (green) and total radiated power (orange). The four phases are indicated by the colored bars on the bottom.

The influence of the phases on the plasma density and the stored energy has to be considered:

Phase	Plasma energy	Plasma density
1	constant	increases
2	decreases	constant inside sep.
3	decreases	constant
4	constant	increases

The first and last phases are comparable to standard operation in H- or L-mode, respectively. The second and third phase exhibit an increased energy loss and a non-linear density evolution. The observation of the four phases is caused by a coupling of these two effects.

The increased energy losses are evident in the global power balance of the plasma, shown in figure 17. The power balance compares the total heating power (NBI and ohmic) with the sum of the power deposited on the divertor target plates and the total radiated power. In the first phase the heating power is matched to about 90%, the expected value for standard operation in H-mode [14]. During the second to fourth phase, the power balance shows a significant mismatch of up to 3 MW, corresponding to almost 40% of the heating power. This is significantly higher than the neutral beam losses ouside the confined plasma, which are estimated to be below 1 MW. The missing power is neither radiated nor deposited on the divertor target plates. Therefore, a significant part of this power must be transported to the main chamber walls. To prove the latter, a direct measurement of the heat deposition on the first wall is in preparation.

The mismatch of the power balance indicates that an additional energy loss channel opens up. Such a high power deposition on the first wall might be caused by a regime of increased radial filament velocity in the SOL [9, 26], which increases the radial SOL transport significantly. This regime, which is observed for high densities at AUG [8], JET [9] and Alcator C-Mod [18], is depending on the collisionality [26] along the magnetic field lines and is most likely triggered by the partial detachment of the outer divertor [9]. Although this is in line with the observations at the HDL, this transport mechanism describes only the SOL plasma and does not explain how additional energy is ejected from the confined plasma.

At high densities, the fueling of the plasma changes, which leads to the non-linear density evolution. In the present work, this is considered to be an H-mode fueling limit, as introduced in section 2.2. First, the density of the confined plasma saturates. The further density increase takes place only in the SOL, where an electron density plateau evolves between the separatrix and the limiter. In the third phase, the density does also not increase any more in the SOL.

The density plateau in the SOL might influence the fueling significantly. Measurements by the fast reciprocating probe in the SOL indicate that the measured density (see figure 10) consists of a background density and an averaged contribution by filaments. First simulations of the radial neutral density and ionization profiles with the KN1D code [17] indicate an outward shift of the ionization profile. This is in line with the observed fueling limit, i.e. the saturated density of the confined plasma and evolution of the background density in the SOL. With the increased SOL density, the ionization profile is shifted even further outside and neutral particles might not penetrate the confined plasma any more.

Although we introduced two mechanisms, which potentially explain each of the two effects of the HDL, these mechanisms have to be coupled to each other. Otherwise, it is unlikely that both effects always start and end simultaneously, which leads to the sharp bends in the graphs of figures 7 and 8. The coupling of the two mechanisms will be subject to future studies.

The four phases are also observed in the full-C AUG (see figure 8). This indicates, that also with carbon as a wall material, a combination of an energy loss channel and a fueling limit leads to the HDL.

However, for full-C AUG it is reported [5, 6], that the full detachment causes the HDL. It is likely, that the change of the wall material lead to a change or modification of the

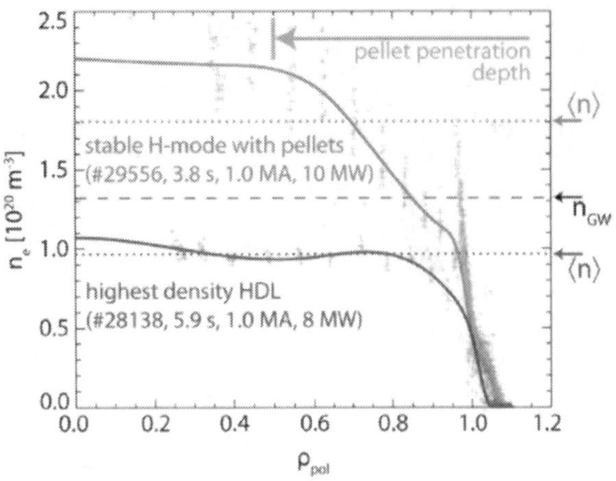

Figure 18. Density profiles of the HDL discharge with the highest density and a pellet fueled H-mode discharge. The central line averaged densities (H-1, dotted) and Greenwald density (dashed) are indicated.

energy loss mechanism, while keeping the according combination of fueling limit and energy loss channel. The full detachment of the divertor might occur at full-C AUG at lower densities than the energy loss channel observed for full-W AUG. The observed change in the parameter dependencies of the HDL for the different wall materials can also be assigned to a changed energy loss mechanism. A replacement of the energy loss channel might be attributed to the changed radiation characteristics of a carbon device, which also impacts the detachment behaviour. According to this, it might be possible to reproduce the behaviour of full-C AUG at full-W AUG with nitrogen seeding. However, this could not be tested as nitrogen seeding is not directly applicable in the HDL discharges at full-W AUG due to the requirements of the non-standard operation at the highest densities (see section 2.2).

7. Exceeding the H-mode density limit with centrally elevated density profiles

In present tokamaks, at high densities, the density profiles of gas puffed discharges are flat due to the high collisionality [2]. For flat profiles, the edge, core and line averaged density are inherently related to each other. Thus, the density limits (Greenwald limit and HDL), which are most likely determined by edge parameters, can be correlated with the line averaged density. With centrally elevated density profiles, it is possible to disentangle these densities.

At AUG, such centrally elevated density profiles can be achieved by pellet fueling [19]. In figure 18 the density profile of a gas puffed HDL discharge and a pellet fueled H-mode are compared. The gas puffed discharge exhibits a flat density profile and the line averaged density stays below the Greenwald density. With the pellet fueling, a centrally elevated density profile is achieved and the line averaged density exceeds the Greenwald density significantly, while the edge density stays well below the Greenwald density. The density gradient in the confined region ($\rho_{pol} < 0.95$) seems to be correlated to the radial range of the pellet particle source.

The energy confinement time τ_E stays constant for the pellet fueled discharge with and without pellets. The centrally increased density does not have an effect on the confinement, but the density dependence of the confinement scalings (e.g. ITERH98P(y,2) [27]) cannot be reproduced at these high densities [20].

Several experiments show that the Greenwald limit is determined to plasma edge parameters (e.g. JT-60 [16], DIII-D [22], Tore Supra [10]). Also, H-modes with centrally elevated density profiles achieved Greenwald fractions above 1 (figure 18 or [20, 22]), while keeping a moderate confinement. This suggests that both density limits, the Greenwald limit and the H-mode density limit, are correlated to the plasma parameters at and outside of the pedestal top ($\rho_{pol} \gtrsim 0.95$).

Future tokamak fusion devices will operate with lower collisionality and, therefore, exhibit intrinsically peaked density profiles [1]. Consequently, it is possible that these tokamaks operate in a stable H-mode regime at Greenwald fractions above 1.

8. Conclusion

In gas ramp discharges performed at the fully tungsten covered ASDEX Upgrade tokamak, four discrete, quasi-stable plasma regimes are identified in the approach towards the H-mode density limit. These phases are distinguished by their impact on the plasma density and stored energy. They are identified as

- a **stable H-mode** where the plasma density increases at a stable confinement,
- a **degrading H-mode** where the confinement decreases while the density in the confined region stays constant and a plateau of electron density evolves in the SOL,
- the **breakdown of the H-mode** where the overall radial density profile stays fixed while the confinement is decreasing and, finally,
- an **L-mode** where the density increases again at almost constant, but low, confinement.

Several previously discussed mechanisms, such as the full detachment, increased radiation, MARFEs and ELMs are refuted as a cause for the HDL in full-W AUG. Measurements indicate that a threshold of the depth of the radial electric field or of the power flux into the divertor, which aim to empirically describe the onset of the H-mode (either locally or globally), determines the final transition back to L-mode. However, both quantities are significantly above their thresholds at the onset of the HDL.

The observation of the four phases reveals, that an additional or increased energy loss channel must be present, which is seen in the decrease of stored energy in the second and third phase. This loss channel erodes the electron temperature gradient and the E_r well. Thus, the pedestal is deteriorated until the L-mode regime is reached. Along with this energy degradation the fueling of the plasma changes and a plateau of electron density evolves in the SOL. In particular, the overall density increase is interrupted at the third phase of the HDL until the pedestal is fully eroded. Both effects, the additional loss channel and the change of the plasma fueling, appear to be coupled to each other and their combination creates the four quasi-stable plasma regimes.

A promising candidate for the energy loss channel is a regime of increased radial filament transport in the SOL at high densities [9]. However, this model does not directly explain the increased energy losses inside the separatrix. The density evolution and the related H-mode fueling limit might be connected to an outward shift of the ionization profile. Both effects are subject to further studies as the triggering and coupling mechanisms have to be identified.

With centrally elevated density profiles, e.g. by pellet fueling, it is possible to exceed the HDL and Greenwald limit scalings at a reasonable H-mode confinement [20]. This is strong evidence that both limits are determined by the plasma parameters at the pedestal top and further outside, as it is also indicated by the two proposed mechanisms of the HDL.

The latter finding might allow for future fusion power plants to operate in a stable H-mode regime at central densities above the Greenwald limit and, thus, increase the fusion power production. An identification of the mechanisms, which lead to the energy losses and the fueling limit and their scaling to large scale devices is essential to find the optimal operational point for these machines with the highest attainable density and confinement.

Acknowledgments

This project has received funding from the European Union's Horizon 2020 research and innovation program under grant agreement number 633053. The views and opinions expressed herein do not necessarily reflect those of the European Commission.

References

[1] Angioni C, Fable E, Greenwald M, Maslov M, Peeters A G, Takenaga H and Weisen H 2009 Particle transport in tokamak plasmas, theory and experiment *Plasma Phys. Control. Fusion* **51** 124017

[2] Angioni C *et al* 2011 Intrinsic toroidal rotation, density peaking and turbulence regimes in the core of tokamak plasmas *Phys. Rev. Lett.* **107** 215003

[3] Bernert M 2013 Analysis of the H-mode density limit in the ASDEX Upgrade tokamak using bolometry *PhD Thesis* LMU München (http://edoc.ub.uni-muenchen.de/16262/)

[4] Bernert M, Eich T, Burckhart A, Fuchs J C, Giannone L, Kallenbach A, McDermott R M, Sieglin B and the ASDEX Upgrade Team 2014 Application of AXUV diode detectors at ASDEX Upgrade *Rev. Sci. Instrum.* **85** 033503

[5] Borrass K, Lingertat J and Schneider R 1998 A scrape-off layer based density limit for JET ELMy H-modes *Contrib. Plasma Phys.* **38** 130–5

[6] Borrass K *et al* 2004 Recent H-mode density limit studies at JET *Nucl. Fusion* **44** 752

[7] Borrass K, Schneider R and Farengo R 1997 A scrape-off layer based model for Hugill–Greenwald type density limits *Nucl. Fusion* **37** 523

[8] Carralero D *et al* An experimental investigation on the high density transition of the scrape-off-layer transport in ASDEX Upgrade *Nucl. Fusion* accepted

[9] Carralero D *et al* Implications of high density operation on SOL transport: a multimachine investigation *J. Nucl. Mater.* at press

[10] Commaux N, Graud A, Pgouri B, Clairet F, Gil C, Gunn J, Gros G, Joffrin E and Hertout P 2006 Pellet fuelling experiments above the Greenwald density in Tore Supra *Europhysics Conf. Abstracts Proc. of the 33rd EPS Conf. on Controlled Fusion and Plasma Physics (Rome, 19–23 June 2006)* vol 30I, P 5.105 (http://epsppd.epfl.ch/Roma/pdf/P5_105.pdf)

[11] Giannone L, Queen D, Hellman F and Fuchs J C 2005 Prototype of a radiation hard resistive bolometer for ITER *Plasma Phys. Control. Fusion* **47** 2123

[12] Greenwald M 2002 Density limits in toroidal plasmas *Plasma Phys. Control. Fusion* **44** R27

[13] Greenwald M, Terry J, Wolfe S, Ejima S, Bell M, Kaye S and Neilson G 1988 A new look at density limits in tokamaks *Nucl. Fusion* **28** 2199

[14] Herrmann A, Eich T, Rohde V, Fuchs J C, Neuhauser J and the ASDEX Upgrade Team 2004 Power deposition outside the divertor in ASDEX Upgrade *Plasma Phys. Control. Fusion* **46** 971

[15] Huber A *et al* 2013 Impact of the ITER-like wall on divertor detachment and on the density limit in the JET tokamak *J. Nucl. Mater.* **438** S139–47 (*Proc. of the 20th Int. Conf. on Plasma-Surface Interactions in Controlled Fusion Devices (Aachen, Germany, 21–25 May 2012)*)

[16] Kamada Y, Hosogane N, Yoshino R, Hirayama T and Tsunematsu T 1991 Study of the density limit with pellet fuelling in JT-60 *Nucl. Fusion* **31** 1827

[17] LaBombard B 2001 KN1D: A 1-D space, 2-D velocity, kinetic transport algorithm for atomic and molecular hydrogen in an ionizing plasma *Technical Report PSFC/RR-01-3 PSFC Report*, MIT, Boston, USA (www.psfc.mit.edu/library1/catalog/reports/2000/01rr/01rr003/01003_full.pdf)

[18] LaBombard B, Boivin R L, Greenwald M, Hughes J, Lipschultz B, Mossessian D, Pitcher C S, Terry J L, Zweben S J and the Alcator Group 2001 Particle transport in the scrape-off layer and Its relationship to discharge density limit in Alcator C-Mod *Phys. Plasmas* **8** 2107–17

[19] Lang P *et al* 2014 ELM pacing and high-density operation using pellet injection in the ASDEX Upgrade all-metal-wall tokamak *Nucl. Fusion* **54** 083009

[20] Lang P *et al* 2012 High-density H-mode operation by pellet injection and ELM mitigation with the new active in-vessel saddle coils in ASDEX Upgrade *Nucl. Fusion* **52** 023017

[21] Lipschultz B, LaBombard B, Marmar E, Pickrell M, Terry J, Watterson R and Wolfe S 1984 Marfe: an edge plasma phenomenon *Nucl. Fusion* **24** 977

[22] Mahdavi M *et al* 2002 High performance H mode plasmas at densities above the Greenwald limit *Nucl. Fusion* **42** 52

[23] Martin Y R, Takizuka T and the ITPA CDBM H-mode Threshold Database Working Group 2008 Power requirement for accessing the H-mode in ITER *J. Phys.: Conf. Ser.* **123** 012033

[24] Mertens V, Borrass K, Gafert J, Laux M, Schweinzer J and the ASDEX Upgrade Team 2000 Operational limits of ASDEX Upgrade H-mode discharges in the new closed divertor-II configuration *Nucl. Fusion* **40** 1839

[25] Mertens V *et al* 1997 High density operation close to Greenwald limit and H-mode limit in ASDEX Upgrade *Nucl. Fusion* **37** 1607

[26] Myra J R, Russell D A and D'Ippolito D A 2006 Collisionality and magnetic geometry effects on tokamak edge turbulent

transport I A two-region model with application to blobs *Phys. Plasmas* **13** 112502

[27] ITER Physics Expert Group on Confinement, Transport , ITER Physics Expert Group on Confinement Modelling, Database and and ITER Physics Basis Editors 1999 Chapter 2: plasma confinement and transport *Nucl. Fusion* **39** 2175

[28] Osborne T H *et al* 2000 The effect of plasma shape on H-mode pedestal characteristics on DIII-D *Plasma Phys. Control. Fusion* **42** A175

[29] Potzel S, Wischmeier M, Bernert M, Dux R, Müller H and Scarabosio A 2013 Characterization of the fluctuating detachment state in ASDEX Upgrade *J. Nucl. Mater.* **438** S285–90 (*Proc. of the 20th Int. Conf. on Plasma-Surface Interactions in Controlled Fusion Devices* (*Aachen, Germany, 21–25 May 2012*))

[30] Reimold F *et al*Divertor studies in nitrogen induced, completely detached H-modes in full-W ASDEX Upgrade *Nucl. Fusion* submitted

[31] Saibene G *et al* 1999 The influence of isotope mass, edge magnetic shear and input power on high density ELMy H modes in JET *Nucl. Fusion* **39** 1133

[32] Sauter P *et al* 2012 L- to H-mode transitions at low density in ASDEX Upgrade *Nucl. Fusion* **52** 012001

[33] Stober J, Gruber O, Mertens V, Ryter F, Suttrop W, Treutterer W and the ASDEX Upgrade Team 1999 Dependence of confinement and transport on triangularity in ASDEX Upgrade *Europhysics Conf. Abstracts* **23J** 1401–4 (*Proc. of the 26th EPS Conf. on Controlled Fusion and Plasma Physics* (*Maastricht, 14–18 June 1999*) (http://epsppd.epfl.ch/Maas/web/pdf/p4009.pdf)

[34] Suttrop W, Mertens V, Murmann H, Neuhauser J, Schweinzer J and the ASDEX Upgrade Team 1999 Operational limits for high edge density H-mode tokamak operation *J. Nucl. Mater.* **266–9** 118–23

[35] Viezzer E *et al* 2013 High-accuracy characterization of the edge radial electric field at ASDEX Upgrade *Nucl. Fusion* **53** 053005

[36] Viezzer E, Pütterich T, Dux R, McDermott R M and the ASDEX Upgrade Team 2012 High-resolution charge exchange measurements at ASDEX Upgrade *Rev. Sci. Instrum.* **83** 103501

[37] Wagner F *et al* 1982 Regime of improved confinement and high beta in neutral-beam-heated divertor discharges of the ASDEX tokamak *Phys. Rev. Lett.* **49** 1408–12

[38] Zohm H 1996 Edge localized modes (ELMs) *Plasma Phys. Control. Fusion* **38** 105